Multifunctional Pharmaceutical Nanocarriers

FUNDAMENTAL BIOMEDICAL TECHNOLOGIES

Series Editor:
Mauro Ferrari, Ph.D. Houston, TX.

Vladimir Torchilin

Editor

Multifunctional Pharmaceutical Nanocarriers

 Springer

Editor
Vladimir Torchilin
Northeastern University
Boston, MA
USA
v.torchilin@neu.edu

ISBN: 978-0-387-76551-8   e-ISBN: 978-0-387-76554-9
DOI: 10.1007/978-0-387-76554-9

Library of Congress Control Number: 2008920068

Printed on acid-free paper

9 8 7 6 5 4 3 2 1

springer.com

Preface

The use of various pharmaceutical carriers to enhance the in vivo efficiency of many drugs and drug administration protocols has been well established during the last decade in both pharmaceutical research and clinical setting. Surface modification of pharmaceutical nanocarriers, such as liposome, micelles, nanocapsules, polymeric nanoparticles, solid lipid particles, and niosomes, is normally used to control their biological properties in a desirable fashion and to simultaneously make them perform various therapeutically or diagnostically important functions. The most important results of such modification include an increased stability and half-life of drug carriers in the circulation, required biodistribution, passive or active targeting into the required pathological zone, responsiveness to local physiological stimuli, and ability to serve as contrast agents for various imaging modalities (gamma-scintigraphy, magnetic resonance imaging, computed tomography, ultra-sonography). Frequent surface modifiers (used separately or simultaneously) include soluble synthetic polymers (to achieve carrier longevity); specific ligands, such as antibodies, peptides, folate, transferrin, and sugar moieties (to achieve targeting effect); pH- or temperature-sensitive lipids or polymers (to impart stimuli sensitivity); chelating compounds, such as EDTA, DTPA, and deferoxamine (to add a heavy metal-based diagnostic/contrast moiety onto a drug carrier).

Certainly, new or modified pharmaceutical carriers (nanocarriers) as well as their use for the delivery of various drugs and genes are still described in many publications. However, looking into the future of the whole field of drug delivery, we have to think about the development of the next generation of pharmaceutical nanocarriers, combining variety of properties and allowing for the simultaneous performance of multiple functions. The current level of engineering pharmaceutical carriers in some cases allows for drug delivery systems, demonstrating a combination of several desired properties. Long-circulating immunoliposomes represent a good example of this approach as they combine the ability to remain in the circulation for a long time with the ability to specifically accumulate in target areas. One may add pH-sensitive long-circulating liposomes and micelles, or nanocarriers simultaneously loaded with a drug and an imaging agent, to the list. Such nanocarriers belong to the new, "smart" generation of drug delivery systems. In principle, we can imagine drug delivery systems, which, depending on the immediate requirements, can simultaneously or sequentially demonstrate the following properties: (1) circulate long in the blood or, more generally, stay long in the body; (2) specifically target the site of the disease (accumulate there) via both nonspecific and/or specific mechanisms, such as enhanced permeability and retention (EPR) effect and ligand-mediated recognition; (3) respond local stimuli characteristic of the pathological site, such as intrinsic abnormal pH values or temperature or externally applied heat, magnetic field, and ultrasound, by, for example, releasing an entrapped drug or deleting a protective coating and facilitating the contact between drug-loaded nano-carriers and target pathological cells; (4) provide an enhanced intracellular delivery of an entrapped drug in case the drug is expected to exert its action inside the cell (gene delivery to the nuclei or delivery of proapoptotic drugs to the mitochondria surface are good examples); (5) supply real-time information about the carrier (and drug) biodistribution and target accumulation as well as the outcome of the therapy due to the presence within the structure of the carrier of a certain reporter/contrast group. Some other less significant and more exotic functions can also be "attached." Strictly speaking, the term "multifunctionality" may also be applicable to pharmaceutical carriers simultaneously loaded with more than one drug type. To meet the requirements listed above, drug carrier should simultaneously carry various moieties capable of functioning in a certain orchestrated and coordinated fashion. Thus, for example, if a system that can provide the combination of longevity (allowing for the target accumulation via the EPR effect) and specific cell binding (allowing for its internalization by target cells) has to be constructed, two requirements have to be met. First, the half-life of the carrier in the circulation should be long to fit EPR effect requirements. Second, the internalization of the carrier within the target cells should proceed fast to avoid carrier degradation and drug loss in the interstitial space. We have to agree that systems like this still represent a challenge, although a certain work in this direction has already been done and certain examples of multifunctional matrices for oral and tumoral delivery already exist.

This book attempts to cover an emerging area of multifunctional pharmaceutical carriers. It includes 15 chapters describing different aspects of this approach, from stimuli-responsive long-circulating micelles to magnetically sensitive drug carriers, which can be simultaneously used as imaging agent. Certainly, a single book cannot include all the currently available information, and the potential reader may discover that certain areas of interest are absent in this volume. Still, I feel that it is a good beginning.

I am deeply grateful to all my friends and colleagues who have contributed to this book. As an editor, I am open to comments and advices from our readers and I believe that they will find this book useful.

Boston, MA   Vladimir Torchilin

Contents

Contributors

Christine Allen
Departments of Pharmaceutical Sciences, Chemistry, and Chemical Engineering
and Applied Chemistry, University of Toronto, ON, Canada, cj.allen@utoronto.ca

Mansoor M. Amiji
Department of Pharmaceutical Sciences, School of Pharmacy, Northeastern
University, 110 Mugar Life Sciences Building, Boston, MA 02115, USA,
m.amiji@neu.edu

You Han Bae
Department of Pharmaceutics and Pharmaceutical Chemistry, The University of
Utah, 421 Wakara Way, Suite 318, Salt Lake City, UT 84108, USA,
you.bae@utah.edu

Ande Bao
Department of Radiology, The University of Texas Health Science Center at
San Antonio, 7703 Floyd Curl Dr., San Antonio, TX 78284-7800, USA,
Bao@uthscsa.edu

Keith L. Black
Department of Neurosurgery, Sinai Medical Center, 8631 W. Third Street, Suite
800E, Los Angeles, CA 90048, USA, blackk@cshs.org

Sarathi Boddapati
Department of Pharmaceutical Sciences, Northeastern University, 360 Huntington
Ave, 312 Mugar Building, Boston, MA 02115, USA

Myrra G. Carstens
Department of Pharmaceutics, Utrecht Institute for Pharmaceutical Sciences,
Utrecht, University, P.O. Box 80082, 3508 TB Utrecht, The Netherlands

Hari Krishna Devalapally
Department of Pharmaceutical Sciences, The Nanomedicine Education and
Research Consortium, Northeastern University, Boston, MA 02115, USA

Gerard D'Souza
Department of Pharmaceutical Sciences, Northeastern University, 360 Huntington Ave, 312 Mugar Building, Boston, MA 02115, USA

O.B. Garbuzenko
Department of Pharmaceutics, Ernest Mario School of Pharmacy, Rutgers, The State University of New Jersey, Piscataway, NJ 08854, USA

Hamidreza Ghandehari
Center for Nanomedicine and Cellular Delivery, University of Maryland, Baltimore, MD 21201, USA
Department of Pharmaceutics & Pharmaceutical Chemistry and Bioengineering, University of Utah, Salt Lake City, UT 84108, USA, hamid.ghandehari@pharm.utah.edu

Beth Goins
Department of Radiology, The University of Texas Health Science Center at San Antonio, 7703 Floyd Curl Dr., San Antonio, TX 78284-7800, USA, Goins@uthscsa.edu

Wim. E. Hennink
Department of Pharmaceutics, Utrecht Institute for Pharmaceutical Sciences, Utrecht, University, P.O. Box 80082, 3508 TB Utrecht, The Netherlands, w.e.hennink@uu.nl

Allan S. Hoffman
Department of Bioengineering, University of Washington, Seattle WA 98195, USA

Eggehard Holler
Institut für Biophysik und Physikalische Biochemie der Universität Regensburg, 93040 Regensburg, Germany, eggehard.holler@biologie.uni-regensburg.de

Richard W. Horobin
Division of Neurosciences and Biomedical Systems, IBLS University of Glasgow, Glasgow, Scotland, UK

David A. Jaffray
Departments of Medical Biophysics and Radiation Oncology, University of Toronto, Toronto, ON, Canada
Department of Radiation Physics, Radiation Medicine Program, Princess Margaret Hospital, Toronto, ON, Canada

Alexander V. Kabanov
Department of Pharmaceutical Sciences, Center for Drug Delivery and Nanomedicine, College of Pharmacy, University of Nebraska Medical Center, 985830 Nebraska Medical Center, Omaha, NE 68198-5830, USA

Han Chang Kang
Department of Pharmaceutics and Pharmaceutical Chemistry, The University of
Utah, 421 Wakara Way, Suite 318, Salt Lake City, UT 84108, USA

J.J. Khandare
Department of Pharmaceutics, Ernest Mario School of Pharmacy, Rutgers,
The State University of New Jersey, Piscataway, NJ 08854, USA

Rohit Kolhatkar
Center for Nanomedicine and Cellular Delivery, University of Maryland,
Baltimore, MD 21201-1075, USA
Department of Pharmaceutical Sciences, University of Maryland, Baltimore, MD
21201-1075, USA

Jindřich Kopeček
Departments of Pharmaceutics and Pharmaceutical Chemistry, and of
Bioengineering, University of Utah, Salt Lake City, UT 84112, USA,
Jindrich.Kopecek@utah.edu

Eun Seong Lee
Department of Pharmaceutics and Pharmaceutical Chemistry, The University of
Utah, 421 Wakara Way, Suite 318, Salt Lake City, UT 84108, USA

Claus-Michael Lehr
Biopharmaceutics and Pharmaceutical Technology, Saarland University, 66041
Saarbrücken, Germany

Alexander V. Ljubimov
Ophthalmology Research Laboratories, Cedars-Sinai Medical Center, 8700
Beverly Boulevard, D-2025, Los Angeles, CA 90048, USA
David Geffen School of Medicine, University of California Los Angeles,
Los Angeles, CA 90095, USA, ljubimov@cshs.org

Julia Y. Ljubimova
Department of Neurosurgery, Sinai Medical Center, 8631 W. Third Street, Suite
800E, Los Angeles, CA 90048, USA, ljubimovaj@cshs.org

Padmaja Magadala
Department of Pharmaceutical Sciences, School of Pharmacy, Northeastern
University, 110 Mugar Life Sciences Building, Boston, MA 02115, USA

Tamara Minko
Department of Pharmaceutics, Ernest Mario School of Pharmacy, Rutgers,
The State University of New Jersey, 160 Frelinghuysen Road, Piscataway, NJ
08854-8020, USA, minko@rci.rutgers.edu

Kun Na
Division of Biotechnology, The Catholic University of Korea, 43-1 Yeokgok
2-dong, Wonmi-gu, Bucheon-si, Gyeonggi-do 420-743, Republic of Korea

Noha Nafee
Biopharmaceutics and Pharmaceutical Technology, Saarland University, 66041
Saarbrücken, Germany

Dattatri Nagesha
Department of Physics, The Nanomedicine Education and Research Consortium,
Northeastern University, Boston, MA 02115, USA

Cornelus. F. van Nostrum
Department of Pharmaceutics, Utrecht Institute for Pharmaceutical Sciences,
Utrecht, University, P.O. Box 80082, 3508 TB Utrecht, The Netherlands

Huaizhong Pan
Department of Pharmaceutics and Pharmaceutical Chemistry, Department of
Bioengineering, University of Utah, Salt Lake City, UT 84112, USA,
Huaizhong.Pan@utah.edu

V.P. Pozharov
Department of Pharmaceutics, Ernest Mario School of Pharmacy, Rutgers,
The State University of New Jersey, Piscataway, NJ 08854, USA

William Phillips
Department of Radiology, The University of Texas Health Science Center at
San Antonio, 7703 Floyd Curl Dr., San Antonio, TX 78284-7800, USA,
Phillips@uthscsa.edu

Cristianne J.F. Rijcken
Department of Pharmaceutics, Utrecht Institute for Pharmaceutical Sciences,
Utrecht, University, P.O. Box 80082, 3508 TB Utrecht, The Netherlands

M. Saad
Department of Pharmaceutics, Ernest Mario School of Pharmacy, Rutgers,
The State University of New Jersey, Piscataway, NJ 08854, USA

Marc Schneider
Pharmaceutical Nanotechnology, Institute of Biopharmaceutics, Saarland
University, 66041 Saarbrücken, Germany

Aliasgar Shahiwala
Department of Pharmaceutical Sciences, School of Pharmacy, Northeastern
University, 110 Mugar Life Sciences Building, Boston, MA 02115, USA

V.A. Soldatenkov
Department of Radiation Medicine, Lombardi Comprehensive Cancer Center,
Georgetown University Medical Center, Washington, DC 2007, USA

Srinivas Sridhar
Department of Physics, The Nanomedicine Education and Research Consortium,
Northeastern University, Boston, MA 02115, USA

Patrick S. Stayton
Department of Bioengineering, University of Washington, Box 355061, Seattle
WA 98195, USA, stayton@u.washington.edu

Deborah Sweet
Center for Nanomedicine and Cellular Delivery, University of Maryland,
Baltimore, MD 21201, USA
Fischell Department of Bioengineering, University of Maryland, College Park,
MD 20742, USA

Vladimir P. Torchilin
Department of Pharmaceutical Sciences and Center for Pharmaceutical
Biotechnology and Nanomedicine, Northeastern University, Boston, MA 02115,
USA

A.A. Vetcher
National Center for Biodefense and Infectious Diseases, George Mason
University, Manassas, VA 20110, USA

Serguei V. Vinogradov
Department of Pharmaceutical Sciences, Center for Drug Delivery and
Nanomedicine and College of Pharmacy, University of Nebraska Medical
Center, 985830 Nebraska Medical Center, Omaha, NE 68198-5830, USA

Lilian E. van Vlerken
Department of Pharmaceutical Sciences, School of Pharmacy, Northeastern
University, 110 Mugar Life Sciences Building, Boston, MA 02115, USA

Volkmar Weissig
Department of Pharmaceutical Sciences, Northeastern University, 360 Huntington
Ave, 312 Mugar Building, Boston, MA 02115, USA
vweissig@hotmail.com

Jinzi Zheng
Department of Medical Biophysics, University of Toronto, Toronto, ON, Canada
Department of Radiation Physics, Radiation Medicine Program, Princess
Margaret Hospital, Toronto, ON, Canada

Multifunctional Pharmaceutical Nanocarriers: Development of the Concept

Vladimir P. Torchilin

1 Multifunctionality of Pharmaceutical Carriers: What to Expect?

The use of nanoparticulate pharmaceutical carriers to enhance the in vivo efficiency of many drugs and drug administration protocols well established itself over the past decade both in pharmaceutical research and in clinical setting. Certainly, new or modified nanocarriers as well as their combinations with various drugs and genes are still described in multiple publications. However, looking into the future of the field of drug delivery, we have to think about the development of the next generation of pharmaceutical nanocarriers combining variety of properties and allowing for the simultaneous performance of multiple functions. Considering the task within our current level of understanding on what is good in drug delivery systems (DDSs), we can imagine a DDS, which, depending on the immediate requirements, can simultaneously or sequentially demonstrate the following properties: (1) circulate long in the blood or, more generally, stay long in the body; (2) specifically target the site of the disease (accumulate there) via both non-specific and specific mechanisms, such as enhanced permeability and retention (EPR) effect and ligand-mediated recognition; (3) respond to the local stimuli characteristic of the pathological site, such as intrinsic abnormal pH values or temperature or externally applied heat, magnetic field or ultrasound, by, for example, releasing an entrapped drug or deleting a protective coating and thus facilitating the contact between drug-loaded nanocarriers and cancer cells; (4) provide an enhanced intracellular delivery of an entrapped drug in case the drug is expected to exert its action inside the cell (gene delivery to the nuclei or delivery of proapoptotic drugs to the mitochondria surface); (5) supply a real-time information about the carrier (and drug) biodistribution and target accumulation as well as about the outcome of the therapy due to the presence of a certain reporter or contrast moiety within the structure of the carrier. Some other, less significant and more exotic functions can also be "attached." Strictly speaking, the term *multifunctionality* may also be applicable to pharmaceutical carriers simultaneously loaded with more than one drug type, but such systems have not be discussed here. Any way, to be able to meet the requirement listed earlier, a drug carrier should simultaneously carry on its surface various moieties capable of

V. Torchilin (ed.), *Multifunctional Pharmaceutical Nanocarriers*,
© Springer Science+Business Media, LLC 2008

functioning in a certain orchestrated order. The schematic structure of such pharmaceutical carriers is shown in Fig. 1. We have to agree that systems like this still represent a challenge, although a certain work in this direction is already done and certain examples of multifunctional matrices for oral and tumoral delivery have even been already reviewed (Bernkop-Schnurch and Walker, 2001; Torchilin, 2006a; van Vlerken and Amiji, 2006).

Various pharmaceutical nanocarriers, such as nanospheres, nanocapsules, liposomes, micelles, cell ghosts and lipoproteins, are widely used for experimental (and already clinical) delivery of therapeutic and diagnostic agents (Domb et al., 2007; Thassu et al., 2007; Torchilin, 2006b). Surface modification of these carriers is often used to control their properties in a desirable fashion and make them to

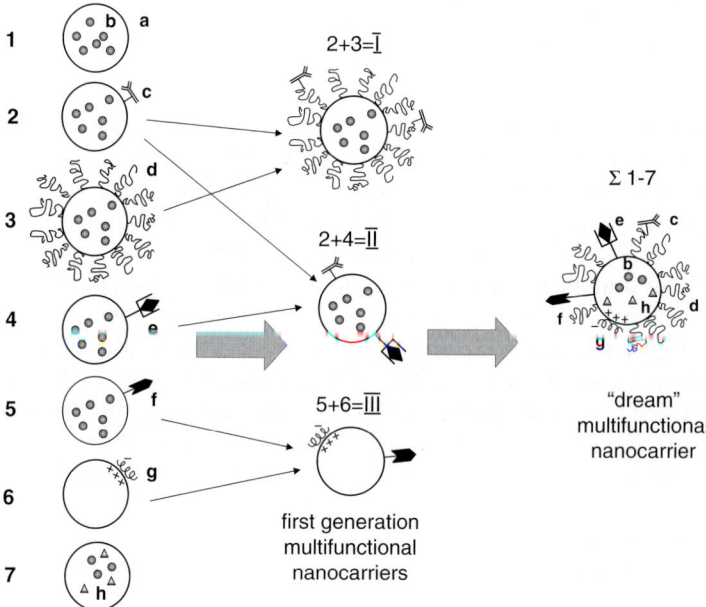

Fig. 1 Typical representatives of monofunctional pharmaceutical nanocarriers: (1) traditional "plain" nanocarrier (a, nanocarrier; b, drug loaded into the carrier); (2) targeted nanocarrier or immunocarrier (c, specific targeting ligand, usually a monoclonal antibody, attached to the carrier surface); (3) long-circulating nanocarrier [d, surface-attached protecting polymer (usually PEG) allowing for prolonged circulation of the nanocarrier in the blood]; (4) contrast nanocarier for imaging purposes (e, heavy metal atom – [111]In, ^{99m}Tc, Gd, Mn – loaded onto the nanocarrier via the carrier-incorporated chelating moiety for γ- or MR imaging application); (5) cell-penetrating nanocarrier (f, cell-penetrating peptide, CPP, attached to the carrier surface and allowing for the carrier enhanced uptake by the cells); (6) DNA-carrying nanocarrier such as lipoplex or polyplex (g, DNA complexed by the carrier via the carrier surface positive charge); (7) magnetic nanocarrier (h, magnetic particles loaded into the carrier together with the drug and allowing for the carrier sensitivity towards the external magnetic field and its use as a contrast agent for magnetic resonance imaging). First-generation multifunctional pharmaceutical nanocarriers may include different combinations of individual functions – see examples I, II and III. Hypothetical "dream" multifunctional pharmaceutical nanocarrier combines the properties of all monofunctional carriers 1–7

simultaneously perform several different functions. The most important results of such modification(s) include increased longevity and stability of the carrier (and carrier-incorporated drug) in the circulation, favorably changed biodistribution, targeting effect, stimuli (pH or temperature)-sensitivity and contrast properties. Frequent surface modifiers (used separately or simulatenously) include soluble synthetic polymers (to achieve carrier longevity); specific ligands, such as antibodies, peptides, folate, transferrin and sugar moieties (to achieve targeting effect); pH- or temperature-sensitive lipids or polymers (to impart stimuli-sensitivity); chelating compounds, such as EDTA, diethylene triamine penta acetic acid (DTPA) or deferoxamine (to add a heavy-metal-based diagnostic or contrast moiety onto a drug carrier). Evidently, different modifiers can present on the surface of the same nanoparticular drug carrier in different combinations, providing it with a set of useful properties required in each particular case (e.g. longevity and targetability, targetability and stimuli-sensitivity, or longevity, targetability and contrast properties). See the general scheme in Fig. 1.

Chemical or physical conjugation of proteins, peptides, polymers and other molecules to the carrier surface required to produce multifunctional pharmaceutical nanocarriers with controlled properties can proceed covalently or non-covalently, via the hydrophobic adsorption of certain intrinsic or specially inserted hydrophobic groups in the ligands. Thus, amphiphilic polymers or hydrophobically modified proteins can adsorb on the hydrophobic surface of polystyrene nanoparticles (Yuan et al., 1995) or incorporate into the phospholipid membrane of liposomes (Torchilin, 1998) or hydrophobic core of micelles (Torchilin, 2001). The attachment can also be performed chemically, via the reaction of reactive groups generated on the carrier surface and certain groups in the molecule to be attached. In many cases, the conjugation methodology is based on three efficient and selective reactions: reaction between activated carboxyl groups and amino groups yielding an amide bond, reaction between pyridyldithiols and thiols yielding disulfide bonds and reaction between maleimide derivatives and thiols yielding thioether bonds (Torchilin and Klibanov, 1993). Some other approaches also exist, for example yielding the carbamate bond via the reaction of the p-nitropheylcarbonyl groups introduced onto the surface of nanocarriers with amino group of various ligands (Torchilin et al., 2001b). The detailed review of numerous coupling procedures and protocols used for attaching the whole variety of surface modifiers to drug carriers can be found in Klibanov et al. (2003) and Torchilin et al. (2003c).

It was shown, for example, that carboxylic groups of immunoglobulins can be activated by water-soluble carbodiimide; activated protein then can be bound to free amino-group-containing surfaces, such as phosphatidyl ethanolamine (PE)-containing liposomes (Dunnick et al., 1975). For further ligand attachment, corresponding reactive groups on the surface of nanocarriers can be pre-modified with the aid of heterobifunctional cross-linking reagents, such as popular N-succinimidyl-3 (2-pyridyldithio)propionate (SPDP) reagent, which was used for the synthesis of a PE derivative further used for the coupling to SH-containing proteins (Leserman et al., 1980). Another possibility is to rely on the reaction of the thiol groups on a ligand (protein) with the maleimide-group-carrying surfaces. This approach (Martin and Papahadjopoulos, 1982) is now one of the most widely used in research and practical applications. For example, various high and low molecular weight compounds have been attached to liposomes, the most popular drug carrier, by using pyridyldithiopropionyl-PE or maleimide reagents (Klibanov et al., 2003; Torchilin et al., 2003c). The most interesting is the application of free thiol groups on immuno globulin Fab fragments. The main advantages of this procedure are its simplicity and the possibility of controlling the progress of the reaction.

Some ligands carry carbohydrate residues, which can be easily oxidized to yield aldehyde groups that can react with surface aminogroups, for example with liposomal aminophospholipids (e.g. PE), with the formation of the Schiff bases (Heath et al., 1980). Liposomes containing carboxyl-bearing derivatives of PE were used for the attachment of different ligands (Kung and Redemann, 1986) after the activation with water-soluble carbodiimide directly prior to ligand addition. Same chemical reactions can be used to attach non-modified proteins and peptides to various nanocarriers, including pre-formed liposomes, containing membrane-incorporated reactive lipid derivatives, such as N-glutaryl-PE or glutaryl-cardiolipin (Bogdanov et al., 1988; Weissig and Gregoriadis, 1992; Weissig et al., 1990). The use of a four-tailed hydrophobic cardiolipin derivative instead of a two-tailed PE derivative allows for a decrease in the number of amino groups involved in the conjugation reaction at the same degree of hydrophobicity. This results in better activity preservation by the hydrophobized and liposome-attached protein (Niedermann et al., 1991; Weissig et al., 1986). Some methods for attaching various ligands to nanocarriers are reviewed in Nobs et al. (2004).

2 Longevity of Pharmaceutical Nanocarriers in the Blood

For the body defence system, "plain" pharmaceutical nanocarriers usually represent foreign particles. As a result, they become easily opsonised and eliminated from the circulation long before the completion of their function. Thus, the longevity function of pharmaceutical nanocarriers becomes prerequisite, and long-circulating pharmaceuticals and pharmaceutical carriers represent currently an important and still growing area of biomedical research (Cohen and Bernstein, 1996; Lasic and Martin, 1995; Moghimi and Szebeni, 2003; Torchilin, 1996, 1998; Torchilin and Trubetskoy, 1995b). The longevity of drug carriers allows maintaining a required level of a pharmaceutical agent in the blood for extended time intervals. In addition, long-circulating drug-containing microparticulates or large macromolecular aggregates can slowly accumulate [EPR effect, also termed as *passive* targeting or accumulation via an impaired filtration mechanism; see Maeda (2001) and Maeda et al. (2000)] in pathological sites with compromised and leaky vasculature (such as tumors, inflammations and infarcted areas), and facilitate drug delivery in those areas (Gabizon, 1995; Maeda, 2001; Maeda et al., 2000). In addition, the prolonged circulation can help to achieve a better targeting effect for targeted (specific ligand-modified) drugs and drug carriers, allowing more time for their interaction with the target (Torchilin, 1996).

The most frequent way to impart the in vivo longevity to drug carriers is their chemical modification with certain synthetic polymers, such as poly(ethylene glycol) (PEG), as was first suggested for liposomes (Allen et al., 1991; Klibanov et al., 1990; Maruyama et al., 1991; Papahadjopoulos et al., 1991; Senior et al., 1991). Hydrophilic polymers have been shown to protect individual molecules and solid particulates from interaction with different solutes providing what is named *steric stabilization* (Napper, 1983). Coating nanoparticles with PEG sterically hinders interactions of blood components with their surface and reduces the binding of plasma proteins with nanoparticles as was demonstrated for liposomes (Allen, 1994; Chonn et al., 1991, 1992; Lasic et al., 1991; Senior et al., 1991; Woodle, 1993), thus preventing drug carrier interaction with opsonins and their fast capture by RES (Senior, 1987) due to the formation of the polymeric layer over the particle surface, which is impermeable for other solutes even at relatively low polymer concentrations (Gabizon and Papahadjopoulos, 1992; Torchilin et al., 1994). Currently, there exist many chemical approaches to synthesize activated derivatives of PEG and to couple these derivatives with a variety of drugs and drug carriers [see reviews in Torchilin (2002), Veronese (2001) and Zalipsky (1995)]. Thus, for example, to make PEG capable of incorporating into the liposomal membrane, the reactive derivative of hydrophilic PEG is single terminus modified with hydrophobic moiety (usually, the residue of PE or long-chain fatty acid is attached to PEG-hydroxysuccinimide ester) (Klibanov et al., 1990, 1991). In majority of protocols, PEG-PE is used, which must be added to the lipid mixture prior to liposome formation. Alternatively, it was suggested to synthesize single end-reactive derivatives of PEG able to be coupled with certain reactive groups (such as maleimide) on the surface of already prepared liposomes, referred to as the post-coating method (Maruyama et al., 1995). Spontaneous incorporation of PEG-lipid conjugates into the liposome membrane from PEG-lipid micelles was also shown to be very effective and did not disturb the vesicles (Sou et al., 2000).

Although PEG is the golden standard in making long-circulating drugs and drug carriers, quite a few other biocompatible, soluble, and hydrophilic polymers have also been suggested as alternative steric protectors for nanoparticular drug carriers (Torchilin and Trubetskoy, 1995b; Torchilin et al., 1995b), such as single terminus lipid-modified poly(acryl amide) and poly(vinyl pyrrolidone) (Chonn et al., 1992; Lasic et al., 1991), poly(acryloyl morpholine) (Monfardini et al., 1995; Ranucci et al., 1994; Sartore et al., 1994), phospholipid (PE)-modified poly(2-methyl-2-oxazoline) or poly(2-ethyl-2-oxazoline) (Woodle et al., 1994), phosphatidyl polyglycerols (Maruyama et al., 1994) and polyvinyl alcohol (Takeuchi et al., 1999).

Surface modification of hydrophobic polymeric nanoparticles can be performed by physical adsorption of a protecting polymer on a particle surface, or by chemical grafting of polymer chains onto a particle. Possible examples of the first case include the absorption of series of polyethylene oxide and polypropylene oxide copolymers (Pluronic/Tetronic™ or Poloxamer/Poloxamine™ surfactants) on the surface of polystyrene latex particles via the hydrophobic interaction mechanism, and resulting polymer-coated nanoparticles also become protected from the uptake by reticulo-endothelial system upon intravenous injection (Illum and Davis, 1983). The absorption of the above copolymers leads not only to the decrease of particle uptake by resident macrophages in liver but, after coating with some specific copolymers, can redirect the injected nanoparticles to other organs (Porter et al., 1992). For example, the coating of 60-nm polystyrene latex with Poloxamer 407 results in increased particle accumulation in bone marrow. The same group has demonstrated that analogous procedure also helps substantially to alter the biodistribution of subcutaneously injected nanospheres. Coating of 60-nm diameter polystyrene nanospheres with certain Poloxamer/Poloxamine copolymers results in their increased accumulation in regional lymph nodes. The optimal length of the copolymer polyoxyethylene block for this particular purpose has been found to be 5–15 oxyethylene units. Non-coated particles normally stay at the injection site while particles coated with longer polyoxyethylene-containing copolymers are not retaining in the nodes and eventually appearing in systemic circulation (Moghimi et al., 1994). Surface modification of polystyrene latexes with PEG was also successfully applied to make long-circulating particles and study their penetration into tumors (Hobbs et al., 1998; Monsky et al., 1999; Yuan et al., 1995).

Another important type of polymeric nanoparticles is based on the block-copolymer of PEG and polylactide-glycolide (PEG-PLAGA) (Gref et al., 1994, 1995; Krause et al., 1985). Using PLAGA-PEG copolymer, one can prepare long-circulating particles with insoluble (solid) PLAGA core and water-soluble PEG shell covalently linked to the core (Gref et al., 1994, 1995). Similar effects on longevity and biodistribution of microparticular drug carriers might be achieved by direct chemical attachment of protective polyethylene oxide chains onto the surface of preformed particles (Harper et al., 1991). Similarly, coating polycyanoacrylate particles with PEG resulted in their increased longevity in the circulation, allowing even for their diffusion into the brain tissue (Calvo et al., 2001; Peracchia et al., 1999). Fluorouracil-containing dendrimer nanoparticles modified with PEG demonstrated better drug retention and less hemolytic activity (Bhadra et al., 2003). Grafting PEG onto the surface of gold particles via mercaptosilanes expectedly resulted in decreased protein adsorption onto modified particles and less platelet adhesion (Zhang et al., 2001).

Thus, the most significant biological consequence of nanocarrier modification with protecting polymers is a sharp increase in its circulation time and decrease in their RES (liver) accumulation (Klibanov et al., 1990; Torchilin, 1998; Torchilin et al., 1994). This fact is very important clinically, since various long-circulating nanocarriers have been shown to effectively accumulate in many tumors via the EPR effect (Gabizon and Papahadjopoulos, 1988; Gabizon, 1995; Maeda, 2001; Maeda et al., 2000). Long-circulating liposomes were prepared containing various anticancer agents, such as doxorubicine, arabinofuranosylcytosine, adriamycin, and vincristin (Allen et al., 1992; Boman et al., 1994; Gabizon et al., 1994; Huang et al., 1994). PEG-liposome-incorporated doxorubicine (Doxil®) has already demonstrated very good clinical results (Ewer et al., 2004; Gabizon, 1995; Rose, 2005). From a pharmacokinetic point of view, the presence of protective polymer on the carrier surface further improves the parameters favorably influenced by drug association with nanocarriers, such as delayed drug absorption, restricted drug biodistribution, decreased volume of drug biodistribution, delayed drug clearance and retarded drug metabolism (Allen et al., 1995; Hwang, 1987; Senior et al., 1991; Torchilin, 1998).

3 Long-Circulating Targeted Pharmaceutical Nanocarriers

The most evident approach to imparting more than one function to a pharmaceutical nanocarrier is to add the property of the specific target recognition to the carrier's ability to circulate for long. Targeting of drug carriers with the aid of ligands selective to cell-surface receptors allows for the selective drug delivery to those cells. There are, however, certain considerations to be taken into account in the design of ligand-coated long-circulating drug carriers: (1) the ligand (antibody, another protein, peptide or carbohydrate) attached to the carrier surface may increase the rate of its uptake in the liver and spleen, despite the presence of a PEG brush or another "long-circulating" molecule on the carrier surface; see, for example, Klibanov (1998); (2) ligand-bearing long-circulating nanocarriers could facilitate the development of an unwanted immune response in the body against the ligand or other carrier components, as was shown with the raise of anti-liposome antibodies, the extent of which depends on the character of the ligand (small peptide or Fv fragment is less immunogenic than a complete foreign IgG molecule) and the liposome composition (Benhar et al., 1994; Harding et al., 1997; Park et al., 2001); (3) the amount of ligand attached to the carrier may be critical to ensure successful binding with the target while maintaining the extended circulation of the carrier. Thus, the use of drug carriers with lower surface density of the ligand may allow to extend the carrier circulation time and to improve the overall in vivo targeting efficacy to smaller targets with limited blood flow. Such carriers, however, may not be the best to bind to the same target in vitro, when compared with the ones fully coated with ligand.

To obtain targeted nanocarriers, a variety of methods have been developed to attach corresponding vectors (antibodies, peptides, sugar moieties, folate and other ligands) to the carrier surface. Thus, for example, numerous methods for antibody coupling to liposomes have been reviewed long ago (Torchilin, 1984, 1985). Modification with specific antibodies was also successfully applied to non-liposomal nanocarriers. Thus, nanoparticles made of gelatin and human serum albumin, were modified with HER2 receptor-specific antibody trastuzumab via avidin–biotin system (Wartlick et al., 2004). These surface-modified nanoparticles were effectively endocytosed by HER2-overexpressing cells. Anti-CD3 antibodies are attached to gelatin particles via the same avidin–biotin system in order to enhance the interaction of these particles with lymphocytes (Balthasar et al., 2005). Antibodies specific for CD14 and prostate-specific membrane antigen were used to modify the surface of dendrimer nanoparticles (Thomas et al., 2004), which acquired the ability to specifically bind to the cells overexpressing corresponding antigens.

To achieve better selective targeting by long-circulating PEGylated nanoparticulates, targeting ligands were attached to nanocarriers via the PEG spacer arm, so that the ligand is extended outside of the dense PEG brush excluding steric hindrances for its binding to the target receptors. With this in mind, potential ligands were attached to the activated far (distal) ends of some liposome-grafted polymeric chain (Blume et al. 1993; Torchilin et al., 2001a). For this purpose several types of end-group functionalized lipopolymers of general formula X-PEG-PE (Zalipsky, 1995; Zalipsky et al., 1998), where X represents a reactive functional-group-containing moiety and PEG-PE represents the conjugate of PE and PEG, were introduced. Most of the end-group functionalized PEG-lipids were synthesized from heterobifunctional PEG derivatives containing hydroxyl and carboxyl or amino groups. To further simplify the coupling procedure and to make it applicable for single-step binding of a large variety of amino-group-containing ligands (including antibodies, proteins and small molecules) to the distal end of nanocarrier-attached polymeric chains, amphiphilic PEG derivative, *p*-nitrophenylcarbonyl-PEG-PE (*p*NP-PEG-PE), was introduced (Torchilin et al., 2000, 2001a, 2003b). *p*NP-PEG-PE readily adsorbs on hydrophobic nanoparticles or incorporates into liposomes and micelles via its phospholipid residue, and easily binds any amino-group-containing compound via its water-exposed *p*NP group forming stable and non-toxic urethane (carbamate) bond.

Several strategies have been suggested to engineer specific ligand-bearing long-circulating PEGylated nanocarriers, first of all, liposomes (Zalipsky et al., 1997, 1998). The first approach involves the modification of preformed nanocarriers, including liposomes, containing a certain number of reactive groups exposed into the aqueous surroundings. According to the second approach, pure ligand-PEG-lipid conjugate is mixed with other liposomal matrix-forming components, for example lecithin and cholesterol, and then made into unilamellar vesicles (DeFrees et al., 1996; Gabizon et al., 1999; Wong et al., 1997; Zalipsky et al., 1997). According to the third approach, *m*PEG-DSPE or any ligand modified with the reactive PEG-PE is post-inserted into preformed liposomes (Yoshioka, 1991; Zalipsky et al., 1997).

The majority of research in this area relates to cancer targeting, which utilizes a variety of monoclonal antibodies. Internalizing antibodies are required to achieve a really improved therapeutic efficacy of antibody-targeted liposomal drugs as was shown using B-lymphoma cells and internalizable epitopes (CD19) as an example (Sapra and Allen, 2002). An interesting concept was developed to target HER2-overexpressing tumors using anti-HER2 liposomes (Park et al., 2001). Antibody CC52 against rat colon adenocarcinoma CC531 attached to PEGylated liposomes provided specific accumulation of liposomes in rat model of metastatic CC531 (Kamps et al., 2000).

A nucleosome-specific monoclonal antibody (mAb 2C5) capable of recognition of various tumor cells via the tumor-cell surface-bound nucleosomes significantly improved Doxil® targeting to tumor cells and increased its cytotoxicity (Lukyanov et al., 2004a) both in vitro and in vivo in different test systems, including intracranial human brain U-87 tumor xenograft in nude mice (Gupta and Torchilin, 2007).

The same antibody was also used to effectively target long-circulating PEG-liposomes loaded with an agent for tumor photo-dynamic therapy (PDT) both to multiple cancer cells in vitro and to experimental tumors in vivo and provide a significantly enhanced tumor-cell killing under the conditions of PDT (Roby et al., 2007); see Fig. 2.

GD2-targeted immunoliposomes with novel antitumoral drug, fenretinide, inducing apoptosis in neuroblastoma and melanoma cell lines, demonstrated strong anti-neuroblastoma activity both in vitro and in vivo in mice (Raffaghello et al., 2003). Combination of immunoliposome and endosome-disruptive peptide improves the cytosolic delivery of the liposomal drug, increases cytotoxicity, and opens new approach to constructing targeted liposomal systems as shown with diphteria toxin A chain incorporated together with pH-dependent fusogenic peptide diINF-7 into liposomes specific towards ovarian carcinoma (Mastrobattista et al., 2002).

Surface modification with antibodies was also applied to make other pharmaceutical nanocarriers targeted, in particular cancer-targeted; see Brannon-Peppas and Blanchette (2004) for review. Nanoparticles made of poly(lactic acid) were surface-modified with PEG and with anti-transferrin receptor monoclonal antibody to produce PEGylated immunoparticles with the size of about 120 nm and containing ca. 65 bound antibody molecules per single particle (Olivier et al., 2002). Mammalian cells (NIH3T3, 32D, Ba/F3, hybridoma 9E10) were surface-modified with distal terminus-activated oleyl-PEG and various proteins (streptavidin, EGFP and antibody) were successfully attached to the activated PEG termini (Kato et al., 2004), producing potentially interesting multifunctional (long-circulating and targeted) drug delivery system.

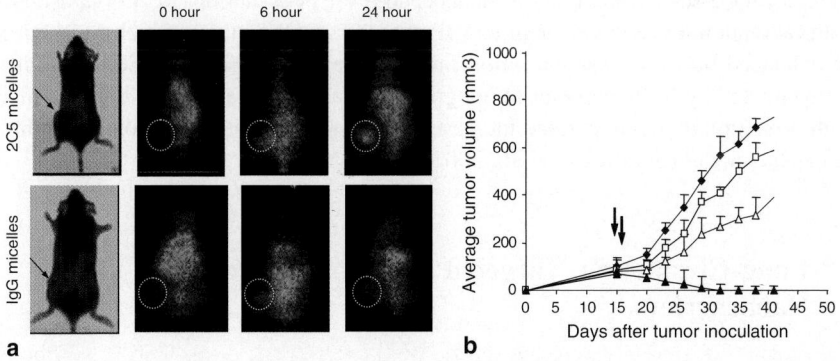

Fig. 2 **a** Tumor accumulation of micellar preparations in LLC tumor-bearing mice C57BL/6 at 0, 6 and 24 h post-injection via tail vein with ^{111}In-labelled TPP-loaded PEG-PE micelles by γ-scintigraphy: *upper row* – mAb 2C5-modified micelles; *bottom row* – micelles modified with non-specific IgG. **b** In vivo therapeutic effect of various TPP preparations in LLC tumor-bearing mice after i.v. injection of 1 mg/kg of TPP 14 days post-tumor inoculation (the data expressed as tumor volumes): *filled diamond*, control untreated animals; *open square*, animals treated with free TPP; *open triangle*, animals treated with TPP in PEG-PE micelles; *filled triangle*, animals treated with TPP in mAb 2C5-PEG-PE-immunomicelles. Arrows indicate TPP injection (*left arrow*) and irradiation (*right arrow*) time points

Similar combination of longevity and targetability can also be achieved by using some other specific ligands attached to long-circulating preparations. Thus, since transferrin (Tf) receptor (TfR) is overexpressed on the surface of many tumor cells, antibodies against TfR as well as Tf itself are among popular ligands for targeting various nanoparticular drug carriers, including liposomes to tumors and inside tumor cells (Hatakeyama et al., 2004). Recent studies involve the coupling of Tf to PEG on PEGylated liposomes in order to combine longevity and targetability for drug delivery into solid tumors (Ishida et al., 2001). Immunoliposomes with OX26 monoclonal antibody to the rat TfR were found to concentrate on brain microvascular endothelium (Huwyler et al., 1996).

Targeting tumors with folate-modified nanocarriers also represents a popular approach, since folate receptor (FR) expression is frequently overexpressed in many tumor cells (Gabizon et al., 2004; Leamon and Low, 1991; Lee and Low, 1994; Lu and Low, 2002). Folate-targeted liposomes have been suggested as delivery vehicles for boron neutron capture therapy (Stephenson et al., 2003). Folate was also attached to the surface of cyanoacrylate-based nanoparticles via activated PEG blocks (Stella et al., 2000). Similarly, PEG-polycaprolactone-based particles were surface-modified with folate and, after being loaded with paclitaxel, demonstrated increased cytotoxicity (Park et al., 2005). Superparamagnetic magnetite nanoparticles were modified with folate (with or without PEG spacer) and demonstrated better uptake by cancer cells, which can be used for both diagnostic (magnetic resonance imaging agents) and therapeutic purposes (Choi et al., 2004; Zhang et al., 2002).

Other specific ligands attached to long-circulating nanocarriers have also been described. Thus, hyaluronan-modified long-circulating liposomes loaded with mitomycin C are active against tumors overexpressing hyaluronan receptors (Peer and Margalit, 2004). Vasoactive intestinal peptide (VIP) was attached to PEG-liposomes with radionuclides to target them to VIP-receptors of the tumor, which resulted in an enhanced breast cancer inhibition in rats (Dagar et al., 2003). PEG-liposomes were targeted by RGD peptides to integrins of tumor vasculature and, being loaded with doxorubicin, demonstrated increased efficiency against C26 colon carcinoma in murine model (Schiffelers et al., 2003).

4 Long-Circulating, Targeted and Stimuli-Sensitive Nanocarriers

An additional function can be added to long-circulating PEGylated pharmaceutical carriers, which allows for the detachment of protecting polymer (PEG) chains under the action of certain local stimuli characteristic of pathological areas, such as decreased pH value or increased temperature usually noted for inflamed and neoplastic areas. In fact, the stability of PEGylated nanocarriers may not always be favorable for drug delivery. If drug-containing nanocarriers accumulate inside the tumor, they may be unable to easily release the drug to kill the tumor cells. Likewise, if the carrier has to be taken up by a cell via an endocytic pathway, the presence of the PEG coat on its surface may preclude the contents from escaping the endosome and being delivered in the cytoplasm. To solve these problems, for example, in the case of long-circulating liposomes, the chemistry was developed to detach PEG from the lipid anchor in the desired conditions. Labile linkage that would degrade only in the acidic conditions characteristic of the endocytic vacuole or the acidotic tumor mass can be based on the diorto esters (Guo and Szoka, 2001), vinyl esters (Boomer and Thompson, 1999), cystein-cleavable lipopolymers (Zalipsky et al., 1999), double esters and hydra-zones that are quite stable at pH around 7.5 but hydrolyzed relatively fast at pH values of 6 and below (Guo and Szoka, 2001; Kratz et al., 1999; Zhang et al., 2004). When the PEG brush is cleaved (e.g. from the liposome surface), the membrane destabilization should occur, and the liposome contents would be delivered to its target (e.g. by escaping from the primary endosome into the cell cytoplasm). Polymeric components with pH-sensitive (pH-cleavable) bonds are used to produce stimuli-responsive drug delivery systems that are stable in the circulation or in normal tissues, however, acquire the ability to degrade and release the entrapped drugs in body areas or cell compartments with lowered pH, such as tumors, infarcts, inflammation zones or cell cytoplasm or endosomes (Roux et al., 2002a, 2004;Simoes et al., 2004). A variety of liposomes (Leroux et al., 2001; Roux et al., 2002b) and polymeric micelles (Lee et al., 2003a,b;; Sudimack et al., 2002) have been described that include the components with acid-labile bonds as well as variety of drug conjugates capable of releasing such drugs as adriamycin (Jones et al., 2003), paclitaxel (Suzawa et al., 2002), doxorubicin (Potineni et al., 2003), and DNA (Cheung et al., 2001; Venugopalan et al., 2002; Yoo et al., 2002) in acidic cell compartments (endosomes) and pathological body areas under acidosis. Serum-stable, long-circulating PEGylated pH-sensitive liposomes were also prepared using the combination of PEG and pH-sensitive terminally alkylated copolymer of N isopropylacrylamide and methacrylic (Roux et al., 2004) on the same liposome, since the attachment of the pH-sensitive polymer to the surface of liposomes might facilitate liposome destabilization and drug release in compartments with decreased pH values. Combination of liposome pH-sensitivity and specific ligand targeting for cytosolic drug delivery utilizing decreased endosomal pH values was described for folate- and Tf-targeted liposomes (Kakudo et al., 2004; Shi et al., 2002; Turk et al., 2002).

Dendrimeric systems derived from diaminobutane poly(propylene imine) with surface-attached PEG and loaded with various drugs demonstrated acid-sensitivity and were capable of releasing incorporated drugs when titrated with acids followed by the addition of sodium chloride solution (Paleos et al., 2004).

The stimuli-sensitivity of PEG coats can also allow for the preparation of multi-functional drug delivery systems with temporarily "hidden" functions, which under normal circumstances, are "shielded" by the protective PEG coat, which however become exposed after PEG detaches (see Intracellular Drug Delivery). Such systems require that multiple functions attached to the surface of the nanocarrier should function in a certain orchestrated and coordinated way. For the above system the following requirements have to be met: (1) the life of the carrier in the circulation should be long enough to fit EPR effect or targeted delivery requirements (i.e. PEG coat mediating the longevity function or specific ligand mediating the targeting function should not be lost by the nanocarrier when in circulation) and

(2) the internalization of the carrier within the target cells should proceed fast, not allowing for the carrier degradation and drug loss in the interstitial space (i.e. local stimuli-dependent removal of the protective function and the exposure of the temporarily hidden second function should proceed fast).

In yet another approach, drug carriers, such as microcapsules, can be loaded not only with the drug alone, but also with magnetic nanoparticles allowing for manipulation of such capsules in magnetic field or with metallic nanoparticles, which can respond to external electromagnetic field and control the rate of drug release by oscillating or heating the carrier (Sukhorukov et al., 2007).

5 Intracellular Drug Delivery by Multifunctional Nanocarriers

Intracellular transport of different biologically active molecules, including various large molecules (proteins, enzymes, antibodies) and even drug-loaded pharmaceutical nanocarriers, is one of the key problems in drug delivery in general. Many pharmaceutical agents need to be delivered intracellularly to exert their therapeutic action inside cytoplasm or onto nucleus or other specific organelles, such as lysosomes, mitochondria or endoplasmic reticulum. This group includes preparations for gene and antisense therapy, which have to reach cell nuclei; pro-apoptotic drugs, which target mitochondria; lysosomal enzymes, which have to reach lysosomal compartment; and some others. For example, intracytoplasmic drug delivery in cancer treatment may overcome such an important obstacle in anti-cancer chemotherapy as multidrug resistance. However, the lipophilic nature of the biological membranes restricts the direct intracellular delivery of such compounds. The cell membrane prevents various soluble small molecules as well as big molecules such as peptides, proteins and DNA from spontaneously entering cells unless there is an active transport mechanism as in the case of some short peptides. Under certain circumstances, these molecules or even small particles can be taken from the extracellular space into cells by the receptor-mediated endocytosis (Varga et al., 2000). However, molecules or particles entering cell via the endocytic pathway become entrapped into endosomes and eventually end in lysosomes, where active degradation processes proceed under the action of the lysosomal enzymes. As a result, only a small fraction of unaffected substance appears in the cell cytoplasm. Another problem is that even after being safely delivered into the cell cytoplasm drugs should still find their way to specific organells where they are expected to utilize their therapeutic potential. This is especially important in case of gene delivery. Viral vectors for DNA delivery suffer from non-specificity and inherent risks of virus-induced complications. Non-viral delivery systems, first of all, cationic lipids or liposomes (Farhood et al., 1995) also have certain drawbacks, such as non-specificity and cytotoxic reactions (Filion and Phillips, 1997; Scheule et al., 1997), though new cationic lipid derivatives with decreased toxicity are currently under development (Tang and Hughes, 1999). Still, the traditional routes of internalization of DNA carriers by endocytosis or pinocytosis with subsequent degradation of the delivered DNA by lysosomal nucleases strongly limit the efficacy of transfection (Xu and Szoka, 1996). From this point of view, the development of a new method that can deliver genetic constructs directly into the cytoplasm of the target cells would be highly desirable.

Evidently, a certain function providing an efficient intracellular penetration and, ideally, even subcellular targeting, is the highly desirable one for multifunctional nanocarriers. The addition of the positive charge to the nanocarrier can significantly enhance its uptake by cells, and the use of cationic lipids and cationic polymers as transfection vectors for efficient intracellular delivery of DNA was suggested about 20 years ago (Wu and Wu, 1987; Xu and Szoka, 1996). Currently this is a vast and well-developed field (see one of the recent reviews in Elouahabi and Ruysschaert (2005). Complexes between cationic lipids (such as Lipofectin®, an equimolar mixture of N-(1-(2,3-dioleyloxy)propyl)-N,N,N-trimethylammonium chloride – DOTMA and dioleoyl phosphatidyl ethanolamine – DOPE) and DNA (lipoplexes) and complexes between cationic polymers, such as polyethyleneimine (PEI) (Kunath et al., 2003), and DNA (polyplexes) are formed because of strong electrostatic interactions between the positively charged carrier and negatively charged DNA. A slight net positive charge of already formed lipoplexes and polyplexes is believed to facilitate their interaction with negatively charged cells and improve transfection efficiency (Sakurai et al., 2000). Endocytosis (including the receptor-mediated endocytosis) was repeatedly confirmed as the main mechanism of lipoplex or polyplex internalization by cells (Ogris et al., 2001). Of special importance is the fact that despite endocytosis-mediated uptake of lipolexes and polyplexes, DNA does not end in lysosomes but gets released into the cytoplasm because of the destabilization of the endosomal membrane provoked by lipid or polymeric component of the complexes.

Early studies in this area of the carriers, which can destabilize the endosomal membrane have been performed with the liposomal drug carriers, when different methods of liposomal content delivery into the cytoplasm have been elaborated by adding the pH-sensitivity function to liposomal preparations, which can already bear some other functions, such as longevity and targetability (Straubinger et al., 1985; Torchilin, 1991). It was believed that such pH-sensitive carriers would destabilize the endosomal membrane when inside endosomes liberating the entrapped drug into the cytoplasm. For example, according to one of these methods, the liposome is made of pH-sensitive components, and after being endocytosed in the intact form, it fuses with the endovacuolar membrane under the action of lowered pH inside the endosome and destabilizes it, releasing its content into the cytoplasm (Torchilin et al., 1993). Thus, endosomes become the gates, enabling transport from the outside into the cell cytoplasm (Sheff, 2004). Cellular drug delivery mediated by pH-sensitive liposomes is not a simple intracellular leakage from the lipid vesicle since the drug has to cross the endosomal membrane also (Asokan and Cho, 2003). The presence of fusogenic lipids in the liposome composition, such as unsaturated DOPE, is usually required to render pH-sensitivity to liposomes (Shalaev and Steponkus, 1999). Multifunctional long-circulating PEGylated DOPE-containing pH-sensitive liposomes, although have a decreased pH-sensitivity, still effectively deliver their contents into cytoplasm [recent review in Simoes et al. (2004)]. Antisense oligonucleotides were delivered into the cells by anionic pH-sensitive PE-containing liposomes, which are stable in the blood, however, undergo phase transition at acidic endosomal pH and facilitate oligo release into cell cytoplasm (recent review in Fattal et al. (2004)). Serum stable, long-circulating PEGylated pH-sensitive liposomes were also prepared using, on the same liposome, the combination of PEG and pH-sensitive terminally alkylated copolymer of *N*-isopropylacrylamide and methacrylic (Roux et al., 2004). Combination of liposome pH-sensitivity and specific ligand targeting for cytosolic drug delivery utilizing decreased endosomal pH values was described for both folate and Tf-targeted liposomes (Xu et al., 2002). Additional modification of pH-sensitive liposomes with an antibody results in pH-sensitive immunoliposomes. A successful application of pH-sensitive immunoliposomes has been demonstrated for the delivery of a variety of molecules including fluorescent dyes, antitumor drugs, proteins and DNA (Torchilin, 2006c). In addition to membrane-destabilizing lipid components, there exists a large family of membrane-destabilizing anionic polymers that can also enhance the endosomal escape of various drugs and biomacromolecules (Yessine and Leroux, 2004). This family includes various carboxylated polymers, copolymers of acrylic and methacrylic acids, copolymers of maleic acid, polymers and copolymers of *N*-isopropylacrylamide, which demonstrate lower critical solution (solubility–insolubility switch) at physiological temperatures, and when precipitated, destabilize biomembranes they are interacting with (Yessine et al., 2003). Such polymers can be attached to the surface of drug or DNA-loaded nanocarriers, allowing for endosome destabilization and cytoplasmic escape.

In case of polyplexes, which cannot directly destabilize the endosomal membrane, the mechanism of DNA escape from endosomes is associated with the ability of polymers, such as PEI, strongly protonate under the acidic pH inside endosome and create a charge gradient eventually provoking a water influx and endosomal swelling and disintegration (Boussif et al., 1995). In both cases, however, DNA-containing complexes when released into the cytosol dissociate, allowing for nuclear entry of free DNA. Nuclear translocation of the plasmid DNA is relatively inefficient because of the barrier function of the nuclear membrane and small size of nuclear pores (ca. 25 nm), in addition DNA degrades rather fast under the action of cytoplasmic nucleases (Pollard et al., 2001), and only 0.1% of palsmids undergo nuclear translocation from the cytosol (Pollard et al., 1998). The attachment of nuclear localization sequences to plasmid DNA may enhance its nuclear translocation and transfection efficiency (Branden et al., 1999). New approaches in using multifunctional carriers for DNA delivery include the application of bimetallic nanorods that can simultaneously bind compacted DNA plasmid and targeting ligands in a spatially defined manner (Salem et al., 2003).

Polymeric micelles can also demonstrate pH-sensitivity and ability to escape from endosomes. Thus, micelles prepared from PEG-poly(aspartate hydrazone adriamycin) easily release an active drug at lowered pH values typical for endosomes and facilitate its cytoplasmic delivery and toxicity against cancer cells (Bae et al., 2005). Alternatively, micelles for intracellular delivery of antisense oligonucleotides (ODN) were prepared from ODN-PEG conjugates complexed with a cationic fusogenic peptide, KALA, and provided much higher intracellular delivery of the ODN, which could be achieved with free ODN (Jeong et al., 2003). One could also enhance an intracellular delivery of drug-loaded micelles by adding to their composition lipid components used in membrane-destabilizing Lipofectin®. The compensation of the negative charge of PEG-lipid micelles (Lukyanov et al., 2004b) by the addition of positively charged lipids to PEG-PE micelles could improve the uptake by cancer cells of drug-loaded mixed PEG-PE or positively charged lipid micelles. After the enhanced endocytosis, such micelles could escape from the endosomes and enter the cytoplasm of cancer cells. This approach was used to increase an intracellular delivery and, thus, the anticancer activity of the micellar paclitaxel by preparing paclitaxel-containing micelles from the mixture of PEG-PE and positively charged lipids (Wang et al., 2005). Multifunctional polymeric micelles capable of pH-dependent dissociation and drug release when loaded with doxorubicin and supplemented with biotin as cancer cell-interacting ligand were also described in Lee et al. (2005).

Another, more recent approach in intracellular drug delivery is based on the modification of drugs and drug carriers (including the multifunctional ones) with certain proteins and peptides demonstrating a unique ability to penetrate into cells (*transduction* phenomenon). This function can be added on top of the longevity and targetability of the pharmaceutical drug-loaded nanocarriers. Thus, the trans-activating transcriptional activator (TAT) protein from HIV-1 enters various cells when added to the surrounding media (Frankel and Pabo, 1988). The same is true about several other cell-penetrating proteins and peptides (CPPs). All CPPs are divided into two classes: the first class consists of amphipathic helical peptides, such as transportan and model amphipathic peptide (MAP), where lysine (Lys) is the main contributor to the positive charge, while the second class includes arginine (Arg)-rich peptides, such as TAT (47–57) and Antp or penetratin (Hallbrink et al., 2001). TAT peptide (TATp) includes a cluster of basic amino acids 47–57 (11-mer; Tyr-Gly-Arg-Lys-Lys-Arg-Arg-Gln-Arg-Arg-Arg), which represents the minimal protein transduction domain (PTD) (Loret et al., 1991; Schwarze et al., 2000). The minimal PTD of Antp, called penetratin, is the 16-mer peptide (43–58 residues) (Derossi et al., 1994). Other CPPs that can be used for the modification of nanocarriers include VP22; transportan, a 27-amino-acid-long chimeric CPP (Pooga et al., 1998); 18-mer amphipathic model peptide with the sequence KLALKLALKALKAALKLA (Oehlke et al., 1998). In terms of the cellular uptake and cargo delivery kinetics, MAP has the fastest uptake, followed by transportan, TATp (48–60), and penetratin. Similarly, MAP has the highest cargo delivery efficiency, followed by transportan, TATp (48–60), and penetratin. The membrane disturbing potential of these peptides is proportional to the hydrophobic moment of the peptide (Hallbrink et al., 2001). The available data assume more than one mechanism for CPPs and CPP-mediated intracellular delivery of various molecules and particles. TAT-mediated intracellular delivery of large molecules and nanoparticles was proved to proceed via the energy-dependent macropinocytosis with subsequent enhanced escape from endosome into the cell cytoplasm (Wadia et al., 2004), while individual CPPs or CPP-conjugated small molecules penetrate cells via electrostatic interactions and hydrogen bonding and do not seem to depend on the energy (Rothbard et al., 2004).

It was shown that CPPs could internalize nanosized particles into the cells (Josephson et al., 1999). Superparamagnetic iron oxide nanoparticles (SPIONs) conjugated with TATp and fluorescein isothiocyanate were taken up quickly by T cells, B cells and macrophages followed by migration of the conjugate primarily to the cytoplasm, which could be tracked readily by MRI (Kaufman et al., 2003). A biocompatible dextran-coated SPION derivatized with TATp were internalized into lymphocytes by over 100-fold more efficiently than were non-modified particles. The characterization on the number of TATp molecules required for an efficient delivery of magnetic nanoparticles revealed that higher numbers of TATp molecules (above 10 per single SPIO particle) enhanced the intracellular accumulation of such particles with the 100-fold increase in cell labeling efficiency (Zhao et al., 2002). The combination of longevity, magnetic properties and ability to penetrate inside cells results in pharmaceutical nanopreparations with new and unique properties, including contrast properties, allowing for MR visualization of cells taking up such particles.

It was also demonstrated that even relatively large particles, such as liposomes, could be delivered into various cells by multiple TATp or other CPP molecules attached to their surface (Gorodetsky et al., 2004; Torchilin et al., 2000; Tseng et al., 2002). The translocation of TATp–liposomes (both plain and PEGylated) into cells required the direct interaction of the liposome-attached TATp with the cell surface (Levchenko et al., 2003; Torchilin et al., 2000). Complexes of TATp–liposomes with a plasmid (plasmid pEGFP-N1 encoding for the green fluorescence protein, GFP) were used for successful in vitro transfection of various tumor and normal cells as well as for in vivo transfection of tumor cells in mice bearing Lewis lung carcinoma (Torchilin et al., 2003b) (the combination of positive charge for DNA complexation and cell-penetrating functions). Antp and TATp coupled to small unilamellar liposomes were accumulated in higher proportions within tumor cells and dendritic cells than when coupled to unmodified control liposomes (Marty et al., 2004). The uptake was time- and concentration-dependent and at least 100 PTD molecules per small unilamellar liposomes were required for efficient uptake inside cells. The uptake of the modified liposomes was inhibited by the preincubation of liposomes with heparin, confirming the role of heparan sulfate proteoglycans in CPP-mediated uptake. Coupling of TATp to the outer surface of liposomes was also described that resulted in an enhanced binding and endocytosis of the liposomes in ovarian carcinoma cells (Fretz et al., 2004). Antp-liposomes have also been considered as a carrier system for an enhanced cell-specific delivery of liposome-entrapped molecules (Marty et al., 2004).

Talking about the multifunctionality, one would like a nanoparticular DDS to be able to (1) specifically accumulate in the required organ or tissue and then (2) penetrate inside target cells delivering its load (drug or DNA) intracellularly. Organ or tissue (tumor, infarct) accumulation could be achieved by the passive targeting via the EPR effect (Maeda et al., 2000; Palmer et al., 1984) assisted by prolonged circulation of such nanocarrier (e.g. as a result of its coating with protecting polymer such as PEG); or by the antibody-mediated active targeting (Jaracz et al., 2005;

Torchilin, 2004), while the intracellular delivery could be mediated by certain internalizable ligands (folate, transferrin) (Gabizon et al., 2004; Widera et al., 2003) or by CPPs (Gupta et al., 2005; Lochmann et al., 2004). Evidently, such a DDS should simultaneously carry on its surface various active moieties, i.e. be multifunctional and possess the ability to "switch on" certain functions (such as intracellular penetration) only when necessary, for example under the action of local stimuli characteristic of the target pathological zone (first of all, increased temperature or lowered pH values characteristic of inflamed, ischemic and neoplastic tissues). These "smart" DDSs should be built in such a way that during the first phase of delivery, a non-specific cell-penetrating function is shielded by the function providing organ- or tissue-specific delivery (sterically-protecting polymer or antibody). Upon accumulation in the target, protecting polymer or antibody attached to the surface of the DDS via the stimuli-sensitive bond should detach under the action of local pathological conditions (abnormal pH or temperature) and expose the previously hidden second function, allowing for the subsequent delivery of the carrier and its cargo inside cells (see the general scheme in Fig. 3). This is especially important for

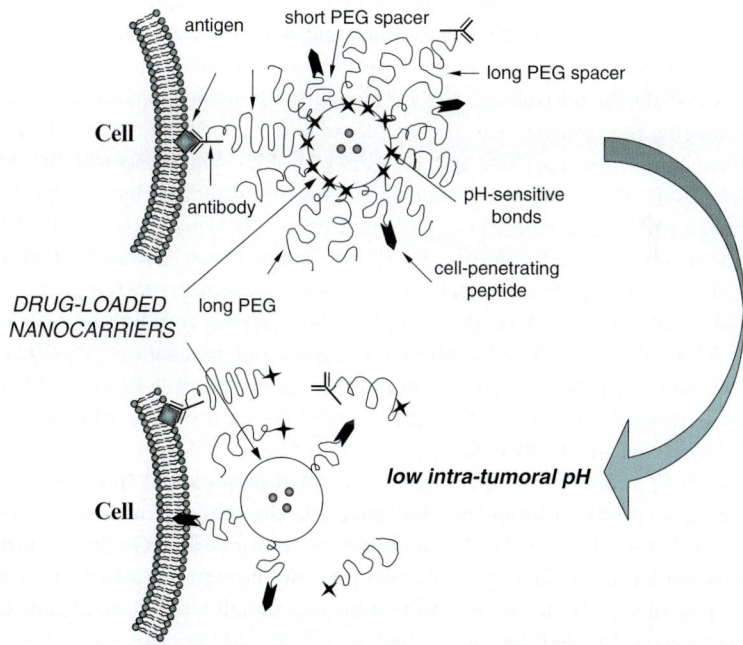

Fig. 3 Schematic representation of a "double-targeted" pharmaceutical carrier with temporarily "hidden" function, for example cell-penetrating peptide; and "shielding" polymeric coat (with or without targeting antibody attached to it) providing longevity in the blood and specific target (tumor) accumulation and preventing the hidden function from the premature interaction with target cells. Polymeric chains are attached to the carrier surface via low pH-degradable bonds. After the accumulation in the tumor due to PEG (longevity) or antibody (specific targeting) or both, pH-dependent de-shielding of the temporarily hidden cell-penetrating function allow for carrier penetration inside tumor cells

CPP-bearing nanocarriers, since all CPPs are highly non-selective and can lead their cargo to any cells, including many non-target ones.

Following this route, "smart" nanoparticular drug delivery system can be prepared capable to accumulate in the required organ or tissue, and then penetrate inside target cells delivering there its load (drug or DNA). The initial target (tumor, infarct) accumulation could be achieved by the passive targeting via the EPR effect or by the specific ligand (antibody)-mediated active targeting, whereas the subsequent intracellular delivery could be mediated by certain internalizable ligands (folate, transferrin) or by CPPs. When in the blood, the cell-penetrating function is temporarily inactivated (sterically shielded) to prevent a non-specific drug delivery into non-target cells; however when inside the target, the nanocarrier loses its protective coat, exposes the cell-penetrating function and provides intracellular drug delivery. We have recently suggested and prepared targeted long-circulating PEGylated liposomes and PEG-phosphatidylethanolamine (PEG-PE)-based micelles possessing several functionalities (Kale and Torchilin, 2007; Sawant et al., 2006). First, such systems were capable of targeting a specific cell or organ by attaching the monoclonal antibody (infarct-specific antimyosin antibody 2G4 or cancer-specific anti-nucleosome antibody 2C5) to their surface via reactive pNP-PEG-PE moieties. Second, these liposomes and micelles were additionally modified with TATp moieties attached to the surface of the nanocarrier by using TATp-short PEG-PE derivatives. PEG-PE used for liposome surface modification or for micelle preparation was made degradable by inserting the pH-sensitive hydrazone bond between PEG and PE (PEG-Hz-PE). Under normal pH values, TATp functions on the surface of nanocarriers were "shielded" by long protecting PEG chains (pH-degradable PEG_{2000}-PE or PEG_{5000}-PE) or by long pNP-PEG-PE moieties used to attach antibodies to the nanocarrier (non-pH-degradable PEG_{3400}-PE or PEG_{5000}-PE). At pH 7.5–8.0, both liposomes and micelles demonstrated high specific binding with antibody substrates, but very limited internalization by NIH/3T3 or U-87 cells. However, upon brief incubation (15–30 min) at lower pH values (pH 5.0–6.0) nanocarriers lost their protective PEG shell because of acidic hydrolysis of PEG-Hz-PE and acquired the ability to be effectively internalized by cells via TATp moieties.

In vivo, TATp-modified pGFP-loaded liposomal preparations have been administered intratumorarly in tumor-bearing mice and the efficacy of tumor-cell transfection was followed after 72 h. The administration of pGFP-TATp-liposomes with non–pH-sensitive PEG coating has resulted in only minimal transfection of tumor cells because of steric hindrances for the liposome-to-cell interaction created by the PEG coat, which shielded the surface-attached TATp. At the same time, the administration of pGFP-TATp-liposomes with the low pH-detachable PEG resulted in the highly efficient transfection since the removal of PEG under the action of the decreased intratumoral pH leads to the exposure of the liposome-attached TATp residues, enhanced penetration of the liposomes inside tumor cells and more effective intracellular delivery of the pGFP (Kale and Torchilin, 2007).

Interesting multifunctional envelope-type devices have been recently described for the cytoplasmic delivery of proteins, DNA and oligonucleotides (Suzuki et al.,

2007). Nanoparticles have been formed by the condensation of the substances to be delivered inside cells with lipid derivatives of CPPs, such as polyarginine, and were efficiently internalized by the cells and released their cargo into the cytosol.

6 Multifunctional Nanocarriers for Image-Guided Drug Delivery and Therapy

In all clinically relevant imaging modalities, to achieve a sufficient attenuation, especially, in case of small lesions, contrast agents are used to absorb certain types of signal (irradiation) much stronger than do the surrounding tissues. The contrast agents are specific for each imaging modality, and as a result of their accumulation in certain sites of interest, those sites may be easily visualized when the appropriate imaging modality is applied (Torchilin, 1995). To still further increase a local spatial concentration of a contrast agent for better imaging, it was a natural progression to use nanoparticulate carriers to carry multiple contrast moieties for an efficient delivery of contrast agents to areas of interest and enhancing a signal from these areas (Morawski et al., 2005; Sullivan and Ferrari, 2004).

To use pharmaceutical nanocarriers for diagnostic or imaging purposes simultaneously with their therapeutic use and to allow for their real-time biodistribution and target accumulation, the contrast reporter moieties can be added to multifunctionalized nanocarriers. Among nanocarriers for contrast agents, liposomes, micelles, and later dendrimers draw a special attention because of their easily controlled properties and good pharmacological characteristics. Liposomes, for example, may incorporate contrast agents in both internal aqueous compartment and membrane. Two general approaches are used to prepare liposomes for γ- and MR-imaging, when heavy metal atoms are used as contrast moieties. The reporter metal could be chelated into a soluble chelator (such as DTPA) and then incorporated into the interior of a liposome (Tilcock et al., 1989). Alternatively, DTPA or a similar chelating compound could be chemically modified with a hydrophobic group, which can anchor the chelating moiety onto the liposome surface during or after liposome preparation (Kabalka et al., 1991a). Different chelators and different hydrophobic anchors were tried for the preparation of ^{111}In, ^{99m}Tc, Mn-, and Gd-loaded liposomes (Glogard et al., 2002; Grant et al., 1989; Kabalka et al., 1991b; Phillips and Goins, 1995; Schwendener et al., 1990; Tilcock, 1993; Torchilin, 1997b; Torchilin and Trubetskoy, 1995a). In the case of MR imaging, for a better MR signal, all reporter atoms should be freely exposed for interaction with water as in the case of membranotropic chelating agents – such as DTPA-stearylamine (DTPA-SA) (Schwendener et al., 1990) or DTPA-phosphatidyl ethanolamine (DTPA-PE) (Kabalka et al., 1991a), which results in better relaxivity of the final preparation when compared with liposome-encapsulated paramagnetic ions (Barsky et al., 1992; Putz et al., 1994; Schwendener, 1994; Unger et al., 1990). The amphiphilic chelating probes (paramagnetic Gd-DTPA-PE and radioactive ^{111}In-DTPA-SA) can also be incorporated into PEG(5 kDa)-PE micelles and used for in vivo MR and scintigraphy imaging. In micelles, the lipid part of the molecule can be anchored in the micelle's hydrophobic core while a more hydrophilic chelate is localized on the hydrophilic shell of the micelle.

To still further increase liposome load with diagnostic moieties, amphiphilic polychelating polymers (PAPs) were synthesized consisting of the main chain with multiple side chelating groups capable of firm binding many reporter metal atoms and hydrophobic terminal group, allowing for polymer adsorption onto hydrophobic nanoparticles or incorporation into hydrophobic domains of liposomes or micelles (Torchilin, 2000). Such surface modification of nanocarriers allows for sharp increase in the number of bound reporter metal atoms per particle and image signal intensity. In case of MR, metal atoms chelated into polymer side groups are directly exposed to the water environment that enhances the relaxivity of the paramagnetic ions and leads to the corresponding enhancement of the vesicle contrast properties (Torchilin, 1997b, 1999; Trubetskoy and Torchilin, 1994).

An intersting example of the application of PAP-nanoparticles for actual in vivo imaging is the MRI of lymphatic system components with Gd-loaded nanocarriers. Liposomes and micelles have been studied as delivery vehicles to the lymphatic (Torchilin et al., 1995c; Trubetskoy and Torchilin, 1996). It has been shown that radioactively labeled small negatively charged liposomes are the most efficient in targeting rat regional lymph nodes after the subcutaneous administration (Patel et al., 1984). The optimal diameter of liposomes that localize in the lymph nodes after the peritoneal administration in rats is approximately 200 nm (Hirano and Hunt, 1985). Liposomes loaded with chelated paramagnetic ions (mostly Gd, Dy, Mn and Fe) have been demonstrated to be useful as MRI contrast agents mostly for the visualization of the macrophage-rich tissues such as organs of the reticuloendothelial system (Unger et al., 1989). The overall performance of Gd-PAP-liposomes or -micelles could be further improved in case of the co-incorporation of amphiphilic PEG onto the liposome membrane or micelle surface, which can be explained by increased relaxivity of PEG-Gd-liposomes because of the presence of increased amount of PEG-associated water protons in the close vicinity of chelated Gd ions located on the liposomal membrane (Torchilin et al., 1995a; Trubetskoy et al., 1995). Multifunctional approach certainly assists here, since in addition to the enhanced relaxivity, the coating of liposome surface with PEG polymer can help in avoiding the contrast agent uptake in the site of injection by resident phagocytic cells. This circumstance might increase penetration of the vesicles into initial lymphatic capillaries and further down the chain of lymph nodes. Increased circulation time (the result of PEGylation) also allows for better accumulation of long-circulating contrast liposomes in areas of leaky vasculature, such as tumors via the EPR effect. In case of multifunctional nanocarriers additionally loaded with a drug, the presence of a contrast moiety allows for the real-time control of drug accumulation in the target. This is also true for other nanoparticulate carriers, including polymeric micelles (Torchilin, 1997a; Trubetskoy et al., 1996). Both PAP-bearing liposomes and micelles additionally containing PEG on their surface can also serve as long-circulating contrast agents for the blood pool γ- or MR-imaging.

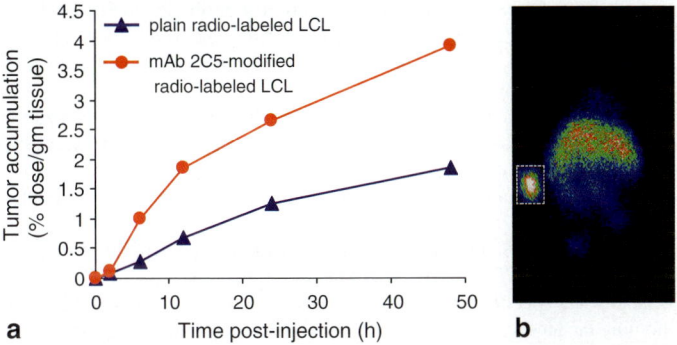

Fig. 4 a In vivo tumor accumulation of ^{111}In-labeled long-circulating liposomes (LCL) in murine Lewis lung carcimona (LLC) model. **b** Whole body γ-imaging of murine breast 4T1 tumor-bearing mice, 6 h after the injection with 2C5-modified ^{111}In-labeled long-circulating liposomes. Circle indicates tumor location. Antibody-modified targeted preparation provides better tumor accumulation and clearly enhanced tumor visualization

The combination of longevity, drug loading, targetability, and contrast properties results in multifunctional nanopharmaceuticals of new generation as was shown, for example, for long-circulating PEGylated liposomes loaded with doxorubicin and additionally decorated with a tumor-specific antibody and contrast moieties (Elbayoumi et al., 2007; Elbayoumi and Torchilin, 2006; Erdogan et al., 2006). The resulting preparations demonstrated an increased therapeutic activity in vivo, and their target accumulation coud be easily followed by γ-scintigraphy (see Fig. 4) or MRI. Multifunctional nanocarriers for image-guided drug delivery, which combine therapeutic and imaging agents merged in one preparation, have also been described in Koning and Krijger (2007) and Rapoport et al. (2007), the last one of these studies combining the ultrasonic tumor imaging with targeted therapy by doxorubicin.

References

Allen, T.M., 1994, The use of glycolipids and hydrophilic polymers in avoiding rapid uptake of liposomes by the mononuclear phagocyte system, *Adv Drug Deliv Rev*, **13**(3):285–309.

Allen, T.M., Hansen, C., Martin, F., Redemann, C. and Yau-Young, A., 1991, Liposomes containing synthetic lipid derivatives of poly(ethylene glycol) show prolonged circulation half-lives in vivo, *Biochim Biophys Acta*, **1066**(1):29–36.

Allen, T.M., Mehra, T., Hansen, C. and Chin, Y.C., 1992, Stealth liposomes: an improved sustained release system for 1-β-D-arabinofuranosylcytosine, *Cancer Res*, **52**(9):2431–9.

Allen, T.M., Hansen, C.B. and de Menenez, D.E.L., 1995, Pharmacokinetics of long-circulating liposomes, *Adv Drug Deliv Rev*, **16**:267–84.

Asokan, A. and Cho, M.J., 2003, Cytosolic delivery of macromolecules. II. Mechanistic studies with pH-sensitive morpholine lipids, *Biochim Biophys Acta*, **1611**(1–2):151–60.

Bae, Y. et al., 2005, Preparation and biological characterization of polymeric micelle drug carriers with intracellular pH-triggered drug release property: tumor permeability, controlled subcellular drug distribution, and enhanced in vivo antitumor efficacy, *Bioconjug Chem*, **16**(1):122–30.

Balthasar, S. et al., 2005, Preparation and characterisation of antibody modified gelatin nanoparticles as drug carrier system for uptake in lymphocytes, *Biomaterials*, **26**(15):2723–32.

Barsky, D., Putz, B., Schulten, K. and Magin, R.L., 1992, Theory of paramagnetic contrast agents in liposome systems, *Magn Reson Med*, **24**(1):1–13.

Benhar, I., Padlan, E.A., Jung, S.H., Lee, B. and Pastan, I., 1994, Rapid humanization of the Fv of monoclonal antibody B3 by using framework exchange of the recombinant immunotoxin B3(Fv)-PE38, *Proc Natl Acad Sci U S A*, **91**(25):12051–5.

Bernkop-Schnurch, A. and Walker, G., 2001, Multifunctional matrices for oral peptide delivery, *Crit Rev Ther Drug Carrier Syst*, **18**(5):459–501.

Bhadra, D., Bhadra, S., Jain, S. and Jain, N.K., 2003, A PEGylated dendritic nanoparticulate carrier of fluorouracil, *Int J Pharm*, **257**(1–2):111–24.

Blume, G. et al., 1993, Specific targeting with poly(ethylene glycol)-modified liposomes: coupling of homing devices to the ends of the polymeric chains combines effective target binding with long circulation times, *Biochim Biophys Acta*, **1149**(1):180–4.

Bogdanov, A.A., Jr., Klibanov, A.L. and Torchilin, V.P., 1988, Protein immobilization on the surface of liposomes via carbodiimide activation in the presence of *N*-hydroxysulfosuccinimide, *FEBS Lett*, **231**(2):381–4.

Boman, N.L., Masin, D., Mayer, L.D., Cullis, P.R. and Bally, M.B., 1994, Liposomal vincristine which exhibits increased drug retention and increased circulation longevity cures mice bearing P388 tumors, *Cancer Res*, **54**(11):2830–3.

Boomer, J.A. and Thompson, D.H., 1999, Synthesis of acid-labile diplasmenyl lipids for drug and gene delivery applications, *Chem Phys Lipids*, **99**(2):145–53.

Boussif, O. et al., 1995, A versatile vector for gene and oligonucleotide transfer into cells in culture and in vivo: polyethylenimine, *Proc Natl Acad Sci U S A*, **92**(16):7297–301.

Branden, L.J., Mohamed, A.J. and Smith, C.I., 1999, A peptide nucleic acid-nuclear localization signal fusion that mediates nuclear transport of DNA, *Nat Biotechnol*, **17**(8):784–7.

Brannon-Peppas, L. and Blanchette, J.O., 2004, Nanoparticle and targeted systems for cancer therapy, *Adv Drug Deliv Rev*, **56**(11):1649–59.

Calvo, P. et al., 2001, Long-circulating PEGylated polycyanoacrylate nanoparticles as new drug carrier for brain delivery, *Pharm Res*, **18**(8):1157–66.

Cheung, C.Y., Murthy, N., Stayton, P.S. and Hoffman, A.S., 2001, A pH-sensitive polymer that enhances cationic lipid-mediated gene transfer, *Bioconjug Chem*, **12**(6):906–10.

Choi, H., Choi, S.R., Zhou, R., Kung, H.F. and Chen, I.W., 2004, Iron oxide nanoparticles as magnetic resonance contrast agent for tumor imaging via folate receptor-targeted delivery, *Acad Radiol*, **11**(9):996–1004.

Chonn, A., Semple, S.C. and Cullis, P.R., 1991, Separation of large unilamellar liposomes from blood components by a spin column procedure: towards identifying plasma proteins which mediate liposome clearance in vivo, *Biochim Biophys Acta*, **1070**(1):215–22.

Chonn, A., Semple, S.C. and Cullis, P.R., 1992, Association of blood proteins with large unilamellar liposomes in vivo. Relation to circulation lifetimes, *J Biol Chem*, **267**(26):18759–65.

Cohen, S. and Bernstein, H. (Editors), 1996, Microparticulate systems for the delivery of proteins and vaccines. Drugs and the Pharmaceutical Sciences, v. 77. Marcel Dekker, New York, ix, 525 pp.

Dagar, S., Krishnadas, A., Rubinstein, I., Blend, M.J. and Onyuksel, H., 2003, VIP grafted sterically stabilized liposomes for targeted imaging of breast cancer: in vivo studies, *J Control Release*, **91**(1–2):123–33.

DeFrees, S.A., Phillips, L., Guo, L. and Zalipsky, S., 1996, Sialyl Lewis x liposomes as a multivalent ligand and inhibitor of E-selectinmediated cellular adhesion, *J Am Chem Soc*, **118**:6101–4.

Derossi, D., Joliot, A.H., Chassaing, G. and Prochiantz, A., 1994, The third helix of the Antennapedia homeodomain translocates through biological membranes, *J Biol Chem*, **269**(14):10444–50.

Domb, A.J., Tabata, Y., Ravi Kumar, M.N.V. and Farber, S. (Editors), 2007, Nanoparticles for Pharmaceutical Applications. American Scientific Publishers, Stevenson Ranch, CA.

Dunnick, J.K., McDougall, I.R., Aragon, S., Goris, M.L. and Kriss, J.P., 1975, Vesicle interactions with polyamino acids and antibody: in vitro and in vivo studies, *J Nucl Med*, **16**(6):483–7.

Elbayoumi, T.A. and Torchilin, V.P., 2006, Enhanced accumulation of long-circulating liposomes modified with the nucleosome-specific monoclonal antibody 2C5 in various tumours in mice: γ-imaging studies, *Eur J Nucl Med Mol Imaging*, **33**(10):1196–205.

Elbayoumi, T.A., Pabba, S., Roby, A. and Torchilin, V.P., 2007, Antinucleosome antibody-modified liposomes and lipid-core micelles for tumor-targeted delivery of therapeutic and diagnostic agents, *J Liposome Res*, **17**(1):1–14.

Elouahabi, A. and Ruysschaert, J.M., 2005, Formation and intracellular trafficking of lipoplexes and polyplexes, *Mol Ther*, **11**(3):336–47.

Erdogan, S., Roby, A. and Torchilin, V.P., 2006, Enhanced tumor visualization by γ-scintigraphy with 111In-labeled polychelating-polymer-containing immunoliposomes, *Mol Pharm*, **3**(5):525–30.

Ewer, M.S. et al., 2004, Cardiac safety of liposomal anthracyclines, *Semin Oncol*, **31**(6 Suppl 13):161–81.

Farhood, H., Serbina, N. and Huang, L., 1995, The role of dioleoyl phosphatidyl ethanolamine in cationic liposome mediated gene transfer, *Biochim Biophys Acta*, **1235**(2):289–95.

Fattal, E., Couvreur, P. and Dubernet, C., 2004, "Smart" delivery of antisense oligonucleotides by anionic pH-sensitive liposomes, *Adv Drug Deliv Rev*, **56**(7):931–46.

Filion, M.C. and Phillips, N.C., 1997, Toxicity and immunomodulatory activity of liposomal vectors formulated with cationic lipids toward immune effector cells, *Biochim Biophys Acta*, **1329**(2):345–56.

Frankel, A.D. and Pabo, C.O., 1988, Cellular uptake of the TAT protein from human immunodeficiency virus, *Cell*, **55**(6):1189–93.

Fretz, M.M., Koning, G.A., Mastrobattista, E., Jiskoot, W. and Storm, G., 2004, OVCAR-3 cells internalize TAT-peptide modified liposomes by endocytosis, *Biochim Biophys Acta*, **1665**(1–2):48–56.

Gabizon, A.A., 1995, Liposome circulation time and tumor targeting: implications for cancer chemotherapy, *Adv Drug Deliv Rev*, **16**:285–94.

Gabizon, A. and Papahadjopoulos, D., 1988, Liposome formulations with prolonged circulation time in blood and enhanced uptake by tumors, *Proc Natl Acad Sci U S A*, **85**(18):6949–53.

Gabizon, A. and Papahadjopoulos, D., 1992, The role of surface charge and hydrophilic groups on liposome clearance in vivo, *Biochim Biophys Acta*, **1103**(1):94–100.

Gabizon, A. et al., 1994, Prolonged circulation time and enhanced accumulation in malignant exudates of doxorubicin encapsulated in polyethylene-glycol coated liposomes, *Cancer Res*, **54**(4):987–92.

Gabizon, A. et al., 1999, Targeting folate receptor with folate linked to extremities of poly(ethylene glycol)-grafted liposomes: in vitro studies, *Bioconjug Chem*, **10**(2):289–98.

Gabizon, A., Shmeeda, H., Horowitz, A.T. and Zalipsky, S., 2004, Tumor cell targeting of liposome-entrapped drugs with phospholipid-anchored folic acid-PEG conjugates, *Adv Drug Deliv Rev*, **56**(8):1177–92.

Glogard, C., Stensrud, G., Hovland, R., Fossheim, S.L. and Klaveness, J., 2002, Liposomes as carriers of amphiphilic gadolinium chelates: the effect of membrane composition on incorporation efficacy and in vitro relaxivity, *Int J Pharm*, **233**(1–2):131–40.

Gorodetsky, R. et al., 2004, Liposome transduction into cells enhanced by haptotactic peptides (Haptides) homologous to fibrinogen C-termini, *J Control Release*, **95**(3):477–88.

Grant, C.W., Karlik, S. and Florio, E., 1989, A liposomal MRI contrast agent: phosphatidylethanolamine-DTPA, *Magn Reson Med*, **11**(2):236–43.

Gref, R. et al., 1994, Biodegradable long-circulating polymeric nanospheres, *Science*, **263**(5153):1600–3.

Gref, R. et al., 1995, The controlled intravenous delivery of drugs using PEG-coated sterically stabilized nanospheres, *Advanced Drug Delivery Reviews*, **16**(2–3):215–233.

Guo, X. and Szoka, F.C., Jr., 2001, Steric stabilization of fusogenic liposomes by a low-pH sensitive PEG-diortho ester–lipid conjugate, *Bioconjugate Chem*, **12**(2):291–300.

Gupta, B. and Torchilin, V.P., 2007, Monoclonal antibody 2C5-modified doxorubicin-loaded liposomes with significantly enhanced therapeutic activity against intracranial human brain U-87 MG tumor xenografts in nude mice, *Cancer Immunol Immunother*, **56**(8):1215–23.

Gupta, B., Levchenko, T.S. and Torchilin, V.P., 2005, Intracellular delivery of large molecules and small particles by cell-penetrating proteins and peptides, *Adv Drug Deliv Rev*, **57**(4):637–51.

Hallbrink, M. et al., 2001, Cargo delivery kinetics of cell-penetrating peptides, *Biochim Biophys Acta*, **1515**(2):101–9.

Harding, J.A., Engbers, C.M., Newman, M.S., Goldstein, N.I. and Zalipsky, S., 1997, Immunogenicity and pharmacokinetic attributes of poly(ethylene glycol)-grafted immunoliposomes, *Biochim Biophys Acta*, **1327**(2):181–92.

Harper, G.R. et al., 1991, Steric stabilization of microspheres with grafted polyethylene oxide reduces phagocytosis by rat Kupffer cells in vitro, *Biomaterials*, **12**(7):695–700.

Hatakeyama, H., Akita, H., Maruyama, K., Suhara, T. and Harashima, H., 2004, Factors governing the in vivo tissue uptake of transferrin-coupled polyethylene glycol liposomes in vivo, *Int J Pharm*, **281**(1–2):25–33.

Heath, T.D., Robertson, D., Birbeck, M.S. and Davies, A.J., 1980, Covalent attachment of horseradish peroxidase to the outer surface of liposomes, *Biochim Biophys Acta*, **599**(1):42–62.

Hirano, K. and Hunt, C.A., 1985, Lymphatic transport of liposome-encapsulated agents: effects of liposome size following intraperitoneal administration, *J Pharm Sci*, **74**(9):915–21.

Hobbs, S.K. et al., 1998, Regulation of transport pathways in tumor vessels: role of tumor type and microenvironment, *Proc Natl Acad Sci U S A*, **95**(8):4607–12.

Huang, S.K. et al., 1994, Liposomes and hyperthermia in mice: increased tumor uptake and therapeutic efficacy of doxorubicin in sterically stabilized liposomes, *Cancer Res*, **54**(8):2186–91.

Huwyler, J., Wu, D. and Pardridge, W.M., 1996, Brain drug delivery of small molecules using immunoliposomes, *Proc Natl Acad Sci U S A*, **93**(24):14164–9.

Hwang, K.J., 1987, Liposome pharmacokinetics. In: M.J. Ostro (Editor), Liposomes: From Biophysics to Therapeutics. Dekker, New York, pp. 109–56.

Illum, S.L. and Davis, S.S., 1983, Effect of the nonionic surfactant poloxamer 338 on the fate and deposition of polystyrene microspheres following intravenous administration, *J Pharm Sci*, **72**(9):1086–9.

Ishida, O. et al., 2001, Liposomes bearing polyethyleneglycol-coupled transferrin with intracellular targeting property to the solid tumors in vivo, *Pharm Res*, **18**(7):1042–8.

Jaracz, S., Chen, J., Kuznetsova, L.V. and Ojima, I., 2005, Recent advances in tumor-targeting anticancer drug conjugates, *Bioorg Med Chem*, **13**(17):5043–54.

Jeong, J.H., Kim, S.W. and Park, T.G., 2003, Novel intracellular delivery system of antisense oligonucleotide by self-assembled hybrid micelles composed of DNA/PEG conjugate and cationic fusogenic peptide, *Bioconjug Chem*, **14**(2):473–9.

Jones, M.C., Ranger, M. and Leroux, J.C., 2003, pH-sensitive unimolecular polymeric micelles: synthesis of a novel drug carrier, *Bioconjug Chem*, **14**(4):774–81.

Josephson, L., Tung, C.H., Moore, A. and Weissleder, R., 1999, High-efficiency intracellular magnetic labeling with novel superparamagnetic-Tat peptide conjugates, *Bioconjug Chem*, **10**(2):186–91.

Kabalka, G.W., Davis, M.A., Holmberg, E., Maruyama, K. and Huang, L., 1991a, Gadolinium-labeled liposomes containing amphiphilic Gd-DTPA derivatives of varying chain length: targeted MRI contrast enhancement agents for the liver, *Magn Reson Imaging*, **9**(3):373–7.

Kabalka, G.W. et al., 1991b, Gadolinium-labeled liposomes containing various amphiphilic Gd-DTPA derivatives: targeted MRI contrast enhancement agents for the liver, *Magn Reson Med*, **19**(2):406–15.

Kakudo, T. et al., 2004, Transferrin-modified liposomes equipped with a pH-sensitive fusogenic peptide: an artificial viral-like delivery system, *Biochemistry*, **43**(19):5618–28.

Kale, A.A. and Torchilin, V.P., 2007, Design, synthesis, and characterization of pH-sensitive PEG-PE conjugates for stimuli-sensitive pharmaceutical nanocarriers: the effect of substitutes at the hydrazone linkage on the ph stability of PEG-PE conjugates, *Bioconjug Chem*, **18**(2):363–70.

Kale, A. and Torchilin, V., 2007, Enhanced transfection of tumor cells in vivo using "Smart" pH-sensitive TAT-modified PEGylated liposomes, *J Drug Target*, **15**(7–8):538–45.

Kamps, J.A. et al., 2000, Uptake of long-circulating immunoliposomes, directed against colon adenocarcinoma cells, by liver metastases of colon cancer, *J Drug Target*, **8**(4):235–45.

Kato, K., Itoh, C., Yasukouchi, T. and Nagamune, T., 2004, Rapid protein anchoring into the membranes of mammalian cells using oleyl chain and poly(ethylene glycol) derivatives, *Biotechnol Prog*, **20**(3):897–904.

Kaufman, C.L. et al., 2003, Superparamagnetic iron oxide particles transactivator protein-fluorescein isothiocyanate particle labeling for in vivo magnetic resonance imaging detection of cell migration: uptake and durability, *Transplantation*, **76**(7):1043–6.

Klibanov, A.L., 1998, Antibody-mediated targeting of PEG-coated liposomes. In: M.C. Woodle and G. Storm (Editors), Long Circulating Liposomes: Old Drugs, New Therapeutics. Biotechnology Intelligence Unit, Springer, Berlin, pp. 269.

Klibanov, A.L., Maruyama, K., Torchilin, V.P. and Huang, L., 1990, Amphipathic polyethyleneglycols effectively prolong the circulation time of liposomes, *FEBS Lett*, **268**(1):235–7.

Klibanov, A.L., Maruyama, K., Beckerleg, A.M., Torchilin, V.P. and Huang, L., 1991, Activity of amphipathic poly(ethylene glycol) 5000 to prolong the circulation time of liposomes depends on the liposome size and is unfavorable for immunoliposome binding to target, *Biochim Biophys Acta*, **1062**(2):142–8.

Klibanov, A.L., Torchilin, V.P. and Zalipsky, S., 2003, Long-circulating sterically protected liposomes. In: V.P. Torchilin and V. Weissig (Editors), Liposomes: A Practical Approach. Practical Approach Series. Oxford University Press, Oxford, New York, pp. 231–265.

Koning, G.A. and Krijger, G.C., 2007, Targeted multifunctional lipid-based nanocarriers for image-guided drug delivery, *Anticancer Agents Med Chem*, **7**(4):425–40.

Kratz, F., Beyer, U. and Schutte, M.T., 1999, Drug–polymer conjugates containing acid-cleavable bonds, *Crit Rev Ther Drug Carrier Syst*, **16**(3):245–88.

Krause, H.J., Schwartz, A. and Rohdewald, P., 1985, Polylactic acid nanoparticles, a colloidal drug delivery system for lipophilic drugs, *Int. J. Pharm.*, **27**(2–3):145–55.

Kunath, K. et al., 2003, Low-molecular-weight polyethylenimine as a non-viral vector for DNA delivery: comparison of physicochemical properties, transfection efficiency and in vivo distribution with high-molecular-weight polyethylenimine, *J Control Release*, **89**(1):113–25.

Kung, V.T. and Redemann, C.T., 1986, Synthesis of carboxyacyl derivatives of phosphatidylethanolamine and use as an efficient method for conjugation of protein to liposomes, *Biochim Biophys Acta*, **862**(2):435–9.

Lasic, D.D. and Martin, F.J. (Editors), 1995, Stealth Liposomes. CRC Press, Boca Raton, 320 pp.

Lasic, D.D., Martin, F.J., Gabizon, A., Huang, S.K. and Papahadjopoulos, D., 1991, Sterically stabilized liposomes: a hypothesis on the molecular origin of the extended circulation times, *Biochim Biophys Acta*, **1070**(1):187–92.

Leamon, C.P. and Low, P.S., 1991, Delivery of macromolecules into living cells: a method that exploits folate receptor endocytosis, *Proc Natl Acad Sci U S A*, **88**(13):5572–6.

Lee, R.J. and Low, P.S., 1994, Delivery of liposomes into cultured KB cells via folate receptor-mediated endocytosis, *J Biol Chem*, **269**(5):3198–204.

Lee, E.S., Na, K. and Bae, Y.H., 2003a, Polymeric micelle for tumor pH and folate-mediated targeting, *J Control Release*, **91**(1–2):103–13.

Lee, E.S., Shin, H.J., Na, K. and Bae, Y.H., 2003b, Poly(L-histidine)-PEG block copolymer micelles and pH-induced destabilization, *J Control Release*, **90**(3):363–74.

Lee, E.S., Na, K. and Bae, Y.H., 2005, Super pH-sensitive multifunctional polymeric micelle, *Nano Lett*, **5**(2):325–9.

Leroux, J., Roux, E., Le Garrec, D., Hong, K. and Drummond, D.C., 2001, N-isopropylacrylamide copolymers for the preparation of pH-sensitive liposomes and polymeric micelles, *J Control Release*, **72**(1–3):71–84.

Leserman, L.D., Barbet, J., Kourilsky, F. and Weinstein, J.N., 1980, Targeting to cells of fluorescent liposomes covalently coupled with monoclonal antibody or protein A, *Nature*, **288**(5791):602–4.

Levchenko, T.S., Rammohan, R., Volodina, N. and Torchilin, V.P., 2003, Tat peptide-mediated intracellular delivery of liposomes, *Methods Enzymol*, **372**:339–49.

Lochmann, D., Jauk, E. and Zimmer, A., 2004, Drug delivery of oligonucleotides by peptides, *Eur J Pharm Biopharm*, **58**(2):237–51.

Loret, E.P. et al., 1991, Activating region of HIV-1 Tat protein: vacuum UV circular dichroism and energy minimization, *Biochemistry*, **30**(24):6013–23.

Lu, Y. and Low, P.S., 2002, Folate-mediated delivery of macromolecular anticancer therapeutic agents, *Adv Drug Deliv Rev*, **54**(5):675–93.

Lukyanov, A.N., Elbayoumi, T.A., Chakilam, A.R. and Torchilin, V.P., 2004a, Tumor-targeted liposomes: doxorubicin-loaded long-circulating liposomes modified with anti-cancer antibody, *J Control Release*, **100**(1):135–44.

Lukyanov, A.N., Hartner, W.C. and Torchilin, V.P., 2004b, Increased accumulation of PEG-PE micelles in the area of experimental myocardial infarction in rabbits, *J Control Release*, **94**(1):187–93.

Maeda, H., 2001, The enhanced permeability and retention (EPR) effect in tumor vasculature: the key role of tumor-selective macromolecular drug targeting, *Adv Enzyme Regul*, **41**:189–207.

Maeda, H., Wu, J., Sawa, T., Matsumura, Y. and Hori, K., 2000, Tumor vascular permeability and the EPR effect in macromolecular therapeutics: a review, *J Control Release*, **65**(1–2): 271–84.

Martin, F.J. and Papahadjopoulos, D., 1982, Irreversible coupling of immunoglobulin fragments to preformed vesicles. An improved method for liposome targeting, *J Biol Chem*, **257**(1):286–8.

Marty, C., Meylan, C., Schott, H., Ballmer-Hofer, K. and Schwendener, R.A., 2004, Enhanced heparan sulfate proteoglycan-mediated uptake of cell-penetrating peptide-modified liposomes, *Cell Mol Life Sci*, **61**(14):1785–94.

Maruyama, K. et al., 1991, Effect of molecular weight in amphipathic polyethyleneglycol on prolonging the circulation time of large unilamellar liposomes, *Chem Pharm Bull (Tokyo)*, **39**(6):1620–2.

Maruyama, K. et al., 1994, Phosphatidyl polyglycerols prolong liposome circulation in vivo, *Int J Pharm*, **111**(1):103–7.

Maruyama, K. et al., 1995, Targetability of novel immunoliposomes modified with amphipathic poly(ethylene glycol)s conjugated at their distal terminals to monoclonal antibodies, *Biochim Biophys Acta*, **1234**(1):74–80.

Mastrobattista, E. et al., 2002, Functional characterization of an endosome-disruptive peptide and its application in cytosolic delivery of immunoliposome-entrapped proteins, *J Biol Chem*, **277**(30):27135–43.

Moghimi, S.M. and Szebeni, J., 2003, Stealth liposomes and long circulating nanoparticles: critical issues in pharmacokinetics, opsonization and protein-binding properties, *Prog Lipid Res*, **42**(6):463–78.

Moghimi, S.M. et al., 1994, Surface engineered nanospheres with enhanced drainage into lymphatics and uptake by macrophages of the regional lymph nodes, *FEBS Lett*, **344**(1):25–30.

Monfardini, C. et al., 1995, A branched monomethoxypoly(ethylene glycol) for protein modification, *Bioconjug Chem*, **6**(1):62–9.

Monsky, W.L. et al., 1999, Augmentation of transvascular transport of macromolecules and nanoparticles in tumors using vascular endothelial growth factor, *Cancer Res*, **59**(16):4129–35.

Morawski, A.M., Lanza, G.A. and Wickline, S.A., 2005, Targeted contrast agents for magnetic resonance imaging and ultrasound, *Curr Opin Biotechnol*, **16**(1):89–92.

Napper, D.H., 1983, Polymeric Stabilization of Colloidal Dispersions. Academic Press, London, New York, xvi, 428 pp.

Niedermann, G., Weissig, V., Sternberg, B. and Lasch, J., 1991, Carboxyacyl derivatives of cardiolipin as four-tailed hydrophobic anchors for the covalent coupling of hydrophilic proteins to liposomes, *Biochim Biophys Acta*, **1070**(2):401–8.

Nobs, L., Buchegger, F., Gurny, R. and Allemann, E., 2004, Current methods for attaching targeting ligands to liposomes and nanoparticles, *J Pharm Sci*, **93**(8):1980–92.

Oehlke, J. et al., 1998, Cellular uptake of an α-helical amphipathic model peptide with the potential to deliver polar compounds into the cell interior non-endocytically, *Biochim Biophys Acta*, **1414**(1–2):127–39.

Ogris, M., Steinlein, P., Carotta, S., Brunner, S. and Wagner, E., 2001, DNA/polyethylenimine transfection particles: influence of ligands, polymer size, and PEGylation on internalization and gene expression, *AAPS Pharm Sci*, **3**(3):E21.

Olivier, J.C., Huertas, R., Lee, H.J., Calon, F. and Pardridge, W.M., 2002, Synthesis of pegylated immunonanoparticles, *Pharm Res*, **19**(8):1137–43.

Paleos, C.M., Tsiourvas, D., Sideratou, Z. and Tziveleka, L., 2004, Acid- and salt-triggered multifunctional poly(propylene imine) dendrimer as a prospective drug delivery system, *Biomacromolecules*, **5**(2):524–9.

Palmer, T.N., Caride, V.J., Caldecourt, M.A., Twickler, J. and Abdullah, V., 1984, The mechanism of liposome accumulation in infarction, *Biochim Biophys Acta*, **797**(3):363–8.

Papahadjopoulos, D. et al., 1991, Sterically stabilized liposomes: improvements in pharmacokinetics and antitumor therapeutic efficacy, *Proc Natl Acad Sci U S A*, **88**(24):11460–4.

Park, J.W. et al., 2001, Tumor targeting using anti-her2 immunoliposomes, *J Control Release*, **74**(1–3):95–113.

Park, E.K., Lee, S.B. and Lee, Y.M., 2005, Preparation and characterization of methoxy poly(ethylene glycol)/poly(epsilon-caprolactone) amphiphilic block copolymeric nanospheres for tumor-specific folate-mediated targeting of anticancer drugs, *Biomaterials*, **26**(9): 1053–61.

Patel, H.M., Boodle, K.M. and Vaughan-Jones, R., 1984, Assessment of the potential uses of liposomes for lymphoscintigraphy and lymphatic drug delivery. Failure of 99m-technetium marker to represent intact liposomes in lymph nodes, *Biochim Biophys Acta*, **801**(1):76–86.

Peer, D. and Margalit, R., 2004, Loading mitomycin C inside long circulating hyaluronan targeted nano-liposomes increases its antitumor activity in three mice tumor models, *Int J Cancer*, **108**(5):780–9.

Peracchia, M.T. et al., 1999, Stealth PEGylated polycyanoacrylate nanoparticles for intravenous administration and splenic targeting, *J Control Release*, **60**(1):121–8.

Phillips, W.T. and Goins, B., 1995, Targeted delivery of imaging agents by liposomes. In: V.P. Torchilin (Editor), Handbook of Targeted Delivery of Imaging Agents. CRC Press, Boca Raton, pp. 149–173.

Pollard, H. et al., 1998, Polyethylenimine but not cationic lipids promotes transgene delivery to the nucleus in mammalian cells, *J Biol Chem*, **273**(13):7507–11.

Pollard, H. et al., 2001, Ca2+-sensitive cytosolic nucleases prevent efficient delivery to the nucleus of injected plasmids, *J Gene Med*, **3**(2):153–64.

Pooga, M., Hallbrink, M., Zorko, M. and Langel, U., 1998, Cell penetration by transportan, *Faseb J*, **12**(1):67–77.

Porter, C.J., Moghimi, S.M., Illum, L. and Davis, S.S., 1992, The polyoxyethylene/polyoxypropylene block co-polymer poloxamer-407 selectively redirects intravenously injected microspheres to sinusoidal endothelial cells of rabbit bone marrow, *FEBS Lett*, **305**(1):62–6.

Potineni, A., Lynn, D.M., Langer, R. and Amiji, M.M., 2003, Poly(ethylene oxide)-modified poly(β-amino ester) nanoparticles as a pH-sensitive biodegradable system for paclitaxel delivery, *J Control Release*, **86**(2–3):223–34.

Putz, B., Barsky, D. and Schulten, K., 1994, Mechanisms of liposomal contrast agents in magnetic resonance imaging, *J. Liposome Res.*, **4**(2):771–808.

Raffaghello, L. et al., 2003, Immunoliposomal fenretinide: a novel antitumoral drug for human neuroblastoma, *Cancer Lett*, **197**(1–2):151–5.

Ranucci, E., Spagnoli, G., Sartore, L. and Ferutti, P., 1994, Synthesis and molecular weight characterization of low molecular weight end-functionalized poly(4-acryloymorpholine), *Macromol Chem Phys*, **195**:3469–79.

Rapoport, N., Gao, Z. and Kennedy, A., 2007, Multifunctional nanoparticles for combining ultrasonic tumor imaging and targeted chemotherapy, *J Natl Cancer Inst*, **99**(14):1095–106.

Roby, A., Erdogan, S. and Torchilin, V.P., 2007, Enhanced in vivo antitumor efficacy of poorly soluble PDT agent, meso-tetraphenylporphine, in PEG-PE-based tumor-targeted immunomicelles, *Cancer Biol Ther*, **6**(7):1136–42.

Rose, P.G., 2005, Pegylated liposomal doxorubicin: optimizing the dosing schedule in ovarian cancer, *Oncologist*, **10**(3):205–14.

Rothbard, J.B., Jessop, T.C., Lewis, R.S., Murray, B.A. and Wender, P.A., 2004, Role of membrane potential and hydrogen bonding in the mechanism of translocation of guanidinium-rich peptides into cells, *J Am Chem Soc*, **126**(31):9506–7.

Roux, E., Francis, M., Winnik, F.M. and Leroux, J.C., 2002a, Polymer based pH-sensitive carriers as a means to improve the cytoplasmic delivery of drugs, *Int J Pharm*, **242**(1–2):25–36.

Roux, E. et al., 2002b, Steric stabilization of liposomes by pH-responsive *N*-isopropylacrylamide copolymer, *J Pharm Sci*, **91**(8):1795–802.

Roux, E., Passirani, C., Scheffold, S., Benoit, J.P. and Leroux, J.C., 2004, Serum-stable and long-circulating, PEGylated, pH-sensitive liposomes, *J Control Release*, **94**(2–3):447–51.

Sakurai, F. et al., 2000, Effect of DNA/liposome mixing ratio on the physicochemical characteristics, cellular uptake and intracellular trafficking of plasmid DNA/cationic liposome complexes and subsequent gene expression, *J Control Release*, **66**(2–3):255–69.

Salem, A.K., Searson, P.C. and Leong, K.W., 2003, Multifunctional nanorods for gene delivery, *Nat Mater*, **2**(10):668–71.

Sapra, P. and Allen, T.M., 2002, Internalizing antibodies are necessary for improved therapeutic efficacy of antibody-targeted liposomal drugs, *Cancer Res*, **62**(24):7190–4.

Sartore, L. et al., 1994, Low molecular weight end-functionalized poly(*N*-vinylpyrrolidone) for the modifications of polypeptide aminogroups, *J Bioact Compact Polym*, **9**:411–27.

Sawant, R.M. et al., 2006, "SMART" drug delivery systems: double-targeted pH-responsive pharmaceutical nanocarriers, *Bioconjug Chem*, **17**(4):943–9.

Scheule, R.K. et al., 1997, Basis of pulmonary toxicity associated with cationic lipid-mediated gene transfer to the mammalian lung, *Hum Gene Ther*, **8**(6):689–707.

Schiffelers, R.M. et al., 2003, Anti-tumor efficacy of tumor vasculature-targeted liposomal doxorubicin, *J Control Release*, **91**(1–2):115–22.

Schwarze, S.R., Hruska, K.A. and Dowdy, S.F., 2000, Protein transduction: unrestricted delivery into all cells?, *Trends Cell Biol*, **10**(7):290–5.

Schwendener, R.A., 1994, Liposomes as carriers for paramagnetic gadolinium chelates as organ specific contrast agents for magnetic resonance imaging (MRI), *J Liposome Res*, **4**(2):837–55.

Schwendener, R.A., Wuthrich, R., Duewell, S., Wehrli, E. and von Schulthess, G.K., 1990, A pharmacokinetic and MRI study of unilamellar gadolinium-, manganese-, and iron-DTPA-stearate liposomes as organ-specific contrast agents, *Invest Radiol*, **25**(8):922–32.

Senior, J.H., 1987, Fate and behavior of liposomes in vivo: a review of controlling factors, *Crit Rev Ther Drug Carrier Syst*, **3**(2):123–93.

Senior, J., Delgado, C., Fisher, D., Tilcock, C. and Gregoriadis, G., 1991, Influence of surface hydrophilicity of liposomes on their interaction with plasma protein and clearance from the circulation: studies with poly(ethylene glycol)-coated vesicles, *Biochim Biophys Acta*, **1062**(1):77–82.

Shalaev, E.Y. and Steponkus, P.L., 1999, Phase diagram of 1,2-dioleoyl phosphatidyl ethanolamine (DOPE): water system at subzero temperatures and at low water contents, *Biochim Biophys Acta*, **1419**(2):229–47.

Sheff, D., 2004, Endosomes as a route for drug delivery in the real world, *Adv Drug Deliv Rev*, **56**(7):927–30.

Shi, G., Guo, W., Stephenson, S.M. and Lee, R.J., 2002, Efficient intracellular drug and gene delivery using folate receptor-targeted pH-sensitive liposomes composed of cationic/anionic lipid combinations, *J Control Release*, **80**(1–3):309–19.

Simoes, S., Moreira, J.N., Fonseca, C., Duzgunes, N. and de Lima, M.C., 2004, On the formulation of pH-sensitive liposomes with long circulation times, *Adv Drug Deliv Rev*, **56**(7): 947–65.

Sou, K., Endo, T., Takeoka, S. and Tsuchida, E., 2000, Poly(ethylene glycol)-modification of the phospholipid vesicles by using the spontaneous incorporation of poly(ethylene glycol)-lipid into the vesicles, *Bioconjug Chem*, **11**(3):372–9.

Stella, B. et al., 2000, Design of folic acid-conjugated nanoparticles for drug targeting, *J Pharm Sci*, **89**(11):1452–64.

Stephenson, S.M. et al., 2003, Folate receptor-targeted liposomes as possible delivery vehicles for boron neutron capture therapy, *Anticancer Res*, **23**(4):3341–5.

Straubinger, R.M., Duzgunes, N. and Papahadjopoulos, D., 1985, pH-sensitive liposomes mediate cytoplasmic delivery of encapsulated macromolecules, *FEBS Lett*, **179**(1):148–54.

Sudimack, J.J., Guo, W., Tjarks, W. and Lee, R.J., 2002, A novel pH-sensitive liposome formulation containing oleyl alcohol, *Biochim Biophys Acta*, **1564**(1):31–7.

Sukhorukov, G.B. et al., 2007, Multifunctionalized polymer microcapsules: novel tools for biological and pharmacological applications, *Small*, **3**(6):944–55.

Sullivan, D.C. and Ferrari, M., 2004, Nanotechnology and tumor imaging: seizing an opportunity, *Mol Imaging*, **3**(4):364–9.

Suzawa, T. et al., 2002, Enhanced tumor cell selectivity of adriamycin-monoclonal antibody conjugate via a poly(ethylene glycol)-based cleavable linker, *J Control Release*, **79**(1–3): 229–42.

Suzuki, R., Yamada, Y. and Harashima, H., 2007, Efficient cytoplasmic protein delivery by means of a multifunctional envelope-type nano device, *Biol Pharm Bull*, **30**(4):758–62.

Takeuchi, H. et al., 1999, Prolonged circulation time of doxorubicin-loaded liposomes coated with a modified polyvinyl alcohol after intravenous injection in rats, *Eur J Pharm Biopharm*, **48**(2):123–9.

Tang, F. and Hughes, J.A., 1999, Use of dithiodiglycolic acid as a tether for cationic lipids decreases the cytotoxicity and increases transgene expression of plasmid DNA in vitro, *Bioconjug Chem*, **10**(5):791–6.

Thassu, D., Deleers, M. and Pathak, Y. (Editors), 2007, Nanoparticulate Drug Delivery Systems. Informa Healthcare USA, New York, NY.

Thomas, T.P. et al., 2004, In vitro targeting of synthesized antibody-conjugated dendrimer nanoparticles, *Biomacromolecules*, **5**(6):2269–74.

Tilcock, C., 1993, Liposomal paramagnetic magnetic resonance contrast agents. In: G. Gregoriadis (Editor), Liposome Technology. CRC Press, Boca Raton, FL, pp. 65–87.

Tilcock, C., Unger, E., Cullis, P. and MacDougall, P., 1989, Liposomal Gd-DTPA: preparation and characterization of relaxivity, *Radiology*, **171**(1):77–80.

Torchilin, V.P., 1984, Immobilization of specific proteins on liposome surface: Systems for drug targeting. In: G. Gregoriadis (Editor), Liposome Technology. CRC Press, Boca Raton, FL, pp. 75–94.

Torchilin, V.P., 1985, Liposomes as targetable drug carriers, *Crit Rev Ther Drug Carrier Syst*, **2**(1):65–115.

Torchilin, V.P. (Editor), 1991, Immobilized enzymes in medicine. Progress in Clinical Biochemistry and Medicine, v. 11. Springer-Verlag, Berlin, New York, viii, 206 pp.

Torchilin, V.P., 1995, Handbook of Targeted Delivery of Imaging Agents. CRC Press, Boca Raton, 732 pp.

Torchilin, V.P., 1996, How do polymers prolong circulation times of liposomes, *J Liposome Res*, **9**:99–116.

Torchilin, V.P., 1997a, Pharmacokinetic considerations in the development of labeled liposomes and micelles for diagnostic imaging, *Q J Nucl Med*, **41**(2):141–53.

Torchilin, V.P., 1997b, Surface-modified liposomes in γ- and MR-imaging, *Adv Drug Deliv Rev*, **24**(2–3):301–13.

Torchilin, V.P., 1998, Polymer-coated long-circulating microparticulate pharmaceuticals, *J Microencapsul*, **15**(1):1–19.

Torchilin, V.P., 1999, Novel polymers in microparticulate diagnostic agents, *Chemtech*, **29**(11):27–34.

Torchilin, V.P., 2000, Polymeric contrast agents for medical imaging, *Curr Pharm Biotechnol*, **1**(2):183–215.

Torchilin, V.P., 2001, Structure and design of polymeric surfactant-based drug delivery systems, *J Control Release*, **73**(2–3):137–72.

Torchilin, V.P., 2002, Strategies and means for drug targeting: an overview. In: V. Muzykantov and V.P. Torchilin (Editors), Biomedical Aspects of Drug Targeting. Kluwer, Boston, pp. 3–26.

Torchilin, V.P., 2004, Targeted polymeric micelles for delivery of poorly soluble drugs, *Cell Mol Life Sci*, **61**(19–20):2549–59.

Torchilin, V.P., 2006a, Multifunctional nanocarriers, *Adv Drug Deliv Rev*, **58**(14):1532–55.

Torchilin, V.P. (Editor), 2006b. Nanoparticualtes as Pharmaceutical Carriers. Imperial College Press, London, UK, 756 pp.

Torchilin, V.P., 2006c, Recent approaches to intracellular delivery of drugs and DNA and organelle targeting, *Annu Rev Biomed Eng*, **8**:343–75.

Torchilin, V. and Klibanov, A., 1993, Coupling and labeling of phospholipids. In: G. Cevc (Editor), Phospholipid Handbook. Marcel Dekker, New York, pp. 293–322.

Torchilin, V.P. and Trubetskoy, V.S., 1995a, In vivo visualizing of organs and tissues with liposomes, *J Liposome Res*, **5**(4):795–812.

Torchilin, V.P. and Trubetskoy, V.S., 1995b, Which polymers can make nanoparticulate drug carriers long-circulating?, *Adv Drug Deliv Rev*, **16**(2):141–55.

Torchilin, V.P., Zhou, F. and Huang, L., 1993, pH-sensitive liposomes, *J Liposome Res*, **3**(2):201–255.

Torchilin, V.P. et al., 1994, Poly(ethylene glycol) on the liposome surface: on the mechanism of polymer-coated liposome longevity, *Biochim Biophys Acta*, **1195**(1):11–20.

Torchilin, V.P., Trubetskoy, V.S., Narula, J. and Khaw, B.A., 1995a, PEG-modified liposomes for γ- and magentic resonance imaging. In: D.D. Lasic and F.J. Martin (Editors), Stealth Liposomes. CRC Press, Boca Raton, pp. 225–31.

Torchilin, V.P. et al., 1995b, New synthetic amphiphilic polymers for steric protection of liposomes in vivo, *J Pharm Sci*, **84**(9):1049–53.

Torchilin, V.P., Trubetskoy, V.S. and Wolf, G.L., 1995c, Magnetic resonance imaging of lymph nodes with GD-containing liposomes. In: V.P. Torchilin (Editor), Handbook of Targeted Delivery of Imaging Agents. CRC Press, Boca Raton, pp. 403–13.

Torchilin, V.P. et al., 2000, PEG-Immunoliposomes: attachment of monoclonal antibody to distal ends of PEG chains via *p*-Nitrophenylcarbonyl groups, 27th International Symposium on Controlled Release of Bioactive Materials. Controlled Release Society, Inc., Paris, pp. 217–18.

Torchilin, V.P. et al., 2001a, *p*-Nitrophenylcarbonyl-PEG-PE-liposomes: fast and simple attachment of specific ligands, including monoclonal antibodies, to distal ends of PEG chains via *p*-nitrophenylcarbonyl groups, *Biochim Biophys Acta*, **1511**(2):397–411.

Torchilin, V.P. et al., 2001b, Amphiphilic poly-*N*-vinylpyrrolidones: synthesis, properties and liposome surface modification, *Biomaterials*, **22**(22):3035–44.

Torchilin, V.P. et al., 2003a, Cell transfection in vitro and in vivo with nontoxic TAT peptide-liposome-DNA complexes, *Proc Natl Acad Sci U S A*, **100**(4):1972–7.

Torchilin, V.P., Lukyanov, A.N., Gao, Z. and Papahadjopoulos-Sternberg, B., 2003b, Immunomicelles: targeted pharmaceutical carriers for poorly soluble drugs, *Proc Natl Acad Sci U S A*, **100**(10):6039–44.

Torchilin, V.P., Weissig, V., Martin, F.J., Heath, T.D. and New, R.R.C., 2003c, Surface modification of liposomes. In: V.P. Torchilin, Weissig, V. (Editor), Liposomes—A Practical Approach. Oxford University press, Oxford, pp. 193–230.

Trubetskoy, V.S. and Torchilin, V.P., 1994, New approaches in the chemical design of Gd-containing liposomes for use in magnetic resonance imaging of lymph nodes, J. Liposome Res., 4:961–980.

Trubetskoy, V.S., Torchilin, V.P., 1995, Use of polyoxyethylene-lipid conjugates as long-circulating carriers for delivery of therapeutic and diagnostoc agents, Adv Drug Deliv Rev, 16:311–20.

Trubetskoy, V.S., Torchilin, V.P., 1996, Polyethyleneglycol based micelles as carriers of therapeutic and diagnostic agents, S.T.P. Pharma Sciences, 6:79–86.

Trubetskoy, V.S., Cannillo, J.A., Milshtein, A., Wolf, G.L. and Torchilin, V.P., 1995, Controlled delivery of Gd-containing liposomes to lymph nodes: surface modification may enhance MRI contrast properties, Magn Reson Imaging, 13(1):31–7.

Trubetskoy, V.S., Frank-Kamenetsky, M.D., Whiteman, K.R., Wolf, G.L. and Torchilin, V.P., 1996, Stable polymeric micelles: lymphangiographic contrast media for γ-scintigraphy and magnetic resonance imaging, Acad Radiol, 3(3):232–8.

Tseng, Y.L., Liu, J.J. and Hong, R.L., 2002, Translocation of liposomes into cancer cells by cell-penetrating peptides penetratin and tat: a kinetic and efficacy study, Mol Pharmacol, 62(4):864–72.

Turk, M.J., Reddy, J.A., Chmielewski, J.A. and Low, P.S., 2002, Characterization of a novel pH-sensitive peptide that enhances drug release from folate-targeted liposomes at endosomal pHs, Biochim Biophys Acta, 1559(1):56–68.

Unger, E.C. et al., 1989, Hepatic metastases: liposomal Gd-DTPA-enhanced MR imaging, Radiology, 171(1):81–5.

Unger, E., Cardenas, D., Zerella, A., Fajardo, L.L. and Tilcock, C., 1990, Biodistribution and clearance of liposomal gadolinium-DTPA, Invest Radiol, 25(6):638–44.

van Vlerken, L.E. and Amiji, M.M., 2006, Multi-functional polymeric nanoparticles for tumour-targeted drug delivery, Expert Opin Drug Deliv, 3(2):205–16.

Varga, C.M., Wickham, T.J. and Lauffenburger, D.A., 2000, Receptor-mediated targeting of gene delivery vectors: insights from molecular mechanisms for improved vehicle design, Biotechnol Bioeng, 70(6):593–605.

Venugopalan, P. et al., 2002, pH-sensitive liposomes: mechanism of triggered release to drug and gene delivery prospects, Pharmazie, 57(10):659–71.

Veronese, F.M., 2001, Peptide and protein PEGylation: a review of problems and solutions, Biomaterials, 22(5):405–17.

Wadia, J.S., Stan, R.V. and Dowdy, S.F., 2004, Transducible TAT-HA fusogenic peptide enhances escape of TAT-fusion proteins after lipid raft macropinocytosis, Nat Med, 10(3):310–15.

Wang, J., Mongayt, D. and Torchilin, V.P., 2005, Polymeric micelles for delivery of poorly soluble drugs: preparation and anticancer activity in vitro of paclitaxel incorporated into mixed micelles based on poly(ethylene glycol)-lipid conjugate and positively charged lipids, J Drug Target, 13(1):73–80.

Wartlick, H. et al., 2004, Highly specific HER2-mediated cellular uptake of antibody-modified nanoparticles in tumour cells, J Drug Target, 12(7):461–71.

Weissig, V. and Gregoriadis, G., 1992, Coupling of aminogroup bearing ligands to liposomes. In: G. Gregoriadis (Editor), Liposome Technology. CRC Press, Boca Raton, pp. 231–48.

Weissig, V., Lasch, J., Klibanov, A.L. and Torchilin, V.P., 1986, A new hydrophobic anchor for the attachment of proteins to liposomal membranes, FEBS Lett, 202(1):86–90.

Weissig, V., Lasch, J. and Gregoriadis, G., 1990, Covalent binding of peptides at liposome surfaces, Die Pharmazie, 45(11):849–50.

Widera, A., Norouziyan, F. and Shen, W.C., 2003, Mechanisms of TfR-mediated transcytosis and sorting in epithelial cells and applications toward drug delivery, Adv Drug Deliv Rev, 55(11):1439–66.

Wong, J.Y., Kuhl, T.L., Israelachvili, J.N., Mullah, N. and Zalipsky, S., 1997, Direct measurement of a tethered ligand–receptor interaction potential, *Science*, **275**(5301):820–2.

Woodle, M.C., 1993, Surface-modified liposomes: assessment and characterization for increased stability and prolonged blood circulation, *Chem Phys Lipids*, **64**(1–3):249–62.

Woodle, M.C., Engbers, C.M. and Zalipsky, S., 1994, New amphipatic polymer–lipid conjugates forming long-circulating reticuloendothelial system-evading liposomes, *Bioconjug Chem*, **5**(6):493–6.

Wu, G.Y. and Wu, C.H., 1987, Receptor-mediated in vitro gene transformation by a soluble DNA carrier system, *J Biol Chem*, **262**(10):4429–32.

Xu, Y. and Szoka, F.C., Jr., 1996, Mechanism of DNA release from cationic liposome/DNA complexes used in cell transfection, *Biochemistry*, **35**(18):5616–23.

Xu, L. et al., 2002, Systemic tumor-targeted gene delivery by anti-transferrin receptor scFv-immunoliposomes, *Mol Cancer Ther*, **1**(5):337–46.

Yessine, M.A. and Leroux, J.C., 2004, Membrane-destabilizing polyanions: interaction with lipid bilayers and endosomal escape of biomacromolecules, *Adv Drug Deliv Rev*, **56**(7):999–1021.

Yessine, M.A., Lafleur, M., Meier, C., Petereit, H.U. and Leroux, J.C., 2003, Characterization of the membrane-destabilizing properties of different pH-sensitive methacrylic acid copolymers, *Biochim Biophys Acta*, **1613**(1–2):28–38.

Yoo, H.S., Lee, E.A. and Park, T.G., 2002, Doxorubicin-conjugated biodegradable polymeric micelles having acid-cleavable linkages, *J Control Release*, **82**(1):17–27.

Yoshioka, H., 1991, Surface modification of haemoglobin-containing liposomes with polyethylene glycol prevents liposome aggregation in blood plasma, *Biomaterials*, **12**(9):861–4.

Yuan, F. et al., 1995, Vascular permeability in a human tumor xenograft: molecular size dependence and cutoff size, *Cancer Res*, **55**(17):3752–6.

Zalipsky, S., 1995, Chemistry of polyethylene glycol conjugates with biologically active molecules, *Adv Drug Deliv Rev*, **16**:157–82.

Zalipsky, S. et al., 1997, Poly(ethylene glycol)-grafted liposomes with oligopeptide or oligosaccharide ligands appended to the termini of the polymer chains, *Bioconjug Chem*, **8**(2):111–18.

Zalipsky, S., Gittelman, J., Mullah, N., Qazen, M.M. and Harding, J.A., 1998, Biologically active ligand-bearing polymer-grafted liposomes. In: G. Gregoriadis (Editor), Targeting of Drugs 6: Strategies for Stealth Therapeutic Systems. NATO ASI Series, Series A, Life sciences, v. 300. Plenum Press, New York, pp. 131–9.

Zalipsky, S. et al., 1999, New detachable poly(ethylene glycol) conjugates: cysteine-cleavable lipopolymers regenerating natural phospholipid, diacyl phosphatidylethanolamine, *Bioconjug Chem*, **10**(5):703–7.

Zhang, F., Kang, E.T., Neoh, K.G. and Huang, W., 2001, Modification of gold surface by grafting of poly(ethylene glycol) for reduction in protein adsorption and platelet adhesion, *J Biomater Sci Polym Ed*, **12**(5):515–31.

Zhang, Y., Kohler, N. and Zhang, M., 2002, Surface modification of superparamagnetic magnetite nanoparticles and their intracellular uptake, *Biomaterials*, **23**(7):1553–61.

Zhang, J.X., Zalipsky, S., Mullah, N., Pechar, M. and Allen, T.M., 2004, Pharmaco attributes of dioleoyl phosphatidyl ethanolamine/cholesteryl hemisuccinate liposomes containing different types of cleavable lipopolymers, *Pharmacol Res*, **49**(2):185–98.

Zhao, M., Kircher, M.F., Josephson, L. and Weissleder, R., 2002, Differential conjugation of TAT peptide to superparamagnetic nanoparticles and its effect on cellular uptake, *Bioconjug Chem*, **13**(4):840–4.

Multifunctional Polymeric Nanosystems for Tumor-Targeted Delivery

Padmaja Magadala, Lilian E. van Vlerken, Aliasgar Shahiwala, and Mansoor M. Amiji

1 Introduction

Cancer is the second leading cause of morbidity and mortality in the United States, with occurrences portraying an upward trend for the future. In 2007, approximately 10 million cases of cancer will occur globally, with a total of around 1.5 million new cancer cases and over 560,000 deaths expected in the United States (U.S. National Institute of Health, 2006). Strikingly, remarkable advances in diagnosis and therapy of cancer have been made over the past few decades resulting from significant advances in fundamental cancer biology. What lacks in this case is clinical translation of these advances into effective therapies. A major hurdle in cancer diagnosis and therapy is the targeted and efficacious delivery of agents to the tumor site, while avoiding adverse damage resulting from systemic administration. While systemic drug delivery already hinges largely on physicochemical properties of the drug, such as size, diffusivity, and plasma protein binding affinity, tumors possess a dense, heterogeneous vasculature and an outward net convective flow that act as hurdles to efficient drug deposition at the target site (Jang et al., 2003). Nanocarrier-mediated delivery has emerged as a successful strategy to enhance delivery of therapeutics and imaging agents to tumors, thereby increasing the potential for diagnosis at an earlier stage or for therapeutic success (or both). Based on the initial observation by Maeda and Matsumura that tumors possess a fenestrated vasculature, with pores on average ranging between 200 and 800 nm, and a lack of lymphatic drainage, together termed the enhanced permeability and retention (EPR) effect, it was found that colloidal carriers in the nanometer size range could target tumors passively, by specific extravasation through these fenestrations, and are retained at the site for prolonged time because of lack of lymphatic drainage (Matsumura and Meada, 1986). This physiological advantage has been used successfully to enhance delivery of diagnostic and therapeutic agents, leading to the U.S. Food and Drug Administration (FDA) approval of nanoparticle formulations such as Feridex® for diagnostic applications and Doxil® and Abraxane® for cancer therapy (U.S. Food and Drug Administration, 2006).

The most basic and simple nanoparticle platform for tumor drug delivery is generally lipid- or polymer-based (Fig. 1). Liposomes are the simplest form of a

V. Torchilin (ed.), *Multifunctional Pharmaceutical Nanocarriers*,
© Springer Science+Business Media, LLC 2008

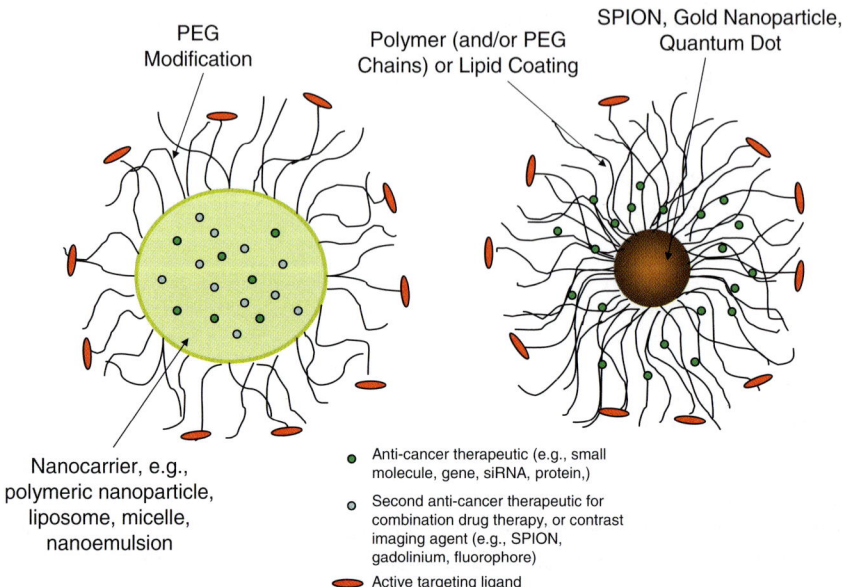

Fig. 1 Typical multifunctional nanoparticle platforms for tumor-targeted therapy

nanoparticle, and became the first system to receive FDA approval for tumor-targeted applications. Constructed from phospholipids as spherical vesicles, they take on the form of aqueous capsules bound by a lipid bilayer, mimicking the plasma membrane of mammalian cells in composition, thereby allowing for great biocompatibility and versatility. Doxil®, a liposomal form of doxorubicin, received FDA approval for the treatment of Kaposi's sarcoma over a decade ago, and is now additionally used against breast cancer and advanced ovarian cancer (U.S. Food and Drug Administration, 2007). Similarly, DaunoXome®, a formulation of daunorubicin, followed suit for the treatment of Kaposi's sarcoma, and a myriad of other liposomal formulations are undergoing preclincal and clinical evaluations as tumor-targeted drug delivery vehicles. Alternatively, micelles have broken through as potential nanocarriers for oncologic applications as well. Micelles are colloidal carriers that spontaneously form through thermodynamically favored aggregation of amphiphiles at or above the critical micellar concentration (CMC). Often such amphiphiles are lipids (lysophopholipids), but amphiphilic polymers and even lipid–polymer hybrids are also frequently used. Micelles are attractive nanocarriers for tumor targeting, due to their small (10–100 nm) size and spontaneous assembly, even though stability of micelles in vivo has been a questionable parameter given their spontaneous disintegration at concentrations below the CMC. Although no micellar formulations have thus far been approved for delivery of anticancer drugs, several are in clinical trials in Asia, and many others are proving quite promising as drug delivery vehicles in early preclinical development. As an example, encapsulation of paclitaxel into a block-copolymer micelle, composed of monomethoxy poly(ethylene oxide)-*block*-poly(lactide), not only increased the maximum tolerated dose (MTD) threefold, but mice bearing subcutaneous MX-1 breast tumors experienced complete tumor regression by day 24 after treatment initiation, while treatment with Taxol®, a clinically used paclitaxel formulation in Cremophore EL®:ethanol mixture, resulted merely in a partial tumor regression followed by complete regrowth by day 24 (Kim et al., 2001). Similar results were seen when the treatment was repeated on mice bearing subcutaneous SKOV3 ovarian tumors (Kim et al., 2001).

On the other spectrum, nanoparticles constructed of natural or synthetic polymers are another group of nanoscale drug delivery systems widely employed in cancer treatment, whose successes to date also include an FDA approved formulation, Abraxane® – paclitaxel bound into albumin nanoparticles (U.S. Food and Drug Administration, 2007). Polymeric nanoparticles offer a particular advantage as drug delivery vehicles since a myriad of different polymers exist or can be developed for the formulation of nanoparticles (Table 1). Over the past several years, our group has developed a variety of polymeric nanoparticles for tumor drug delivery, all leading to an enhanced in vivo therapeutic efficacy. Some examples of these

Table 1 Illustrative examples of multifunctional nanoparticle systems used in cancer therapy

Active ingredients	Nanoparticle platform	Malignancy	Reference
Combination Drug Therapy			
Doxorubicin and combretastatin-A4	Poly(D,L-lactic-*co*-glycolic acid) nanoparticle core in liposome	Lewis lung carcinoma and B16/ F10 melanoma	Sengupta et al. (2005)
Doxorubicin and cyclosporine-A	Poly(alkylcyanoacrylate) nanoparticles	P388 leukemia	Soma et al. (2000)
Doxorubicin and elacridar	Polymer-modified lipid nanoparticles	MDA-MB-435 breast carcinoma	Wong et al. (2006)
Paclitaxel and C$_6$-ceramide	PEO-modified poly(epsilon-caprolactone) nanoparticles	SKOV3 ovarian carcinoma	van Vlerken et al. (2007)
Combination Hyperthermia and Drug Therapy			
TNF-α	PEG-modified gold nanoparticles	MC38 colon carcinoma	Paciotti et al. (2005)
Bleomycin	Microgels	Small intestine	Blanchette and Peppas (2005)
Doxorubicin	PEG-modified liposomes	Hepatic carcinoma	Goldberg et al. (2002)
Combination Imaging and Drug Therapy			
Doxorubicin	Iron oxide nanoparticles inside PEG-poly (L-lactide) micelles	SLK tumor endothelium	Nasongkla et al. (2006)
Doxorubicin	Dermatan sulfate-modified iron oxide nanoparticles	AT1 prostate carcinoma and MX1 breast carcinoma	Ranney et al. (2005)

(continued)

Table 1 (continued)

Active ingredients	Nanoparticle platform	Malignancy	Reference
Methotrexate	Iron oxide nanoparticles	MCF7 breast carcinoma and HeLa cervical carcinoma	Kohler et al. (2005)
Daunorubicin	3-Mercaptopropionic acid – modified gold nanoparticles	K562 leukemia	Li et al. (2007)
Sialyl-Tn and Lewis-y antigens	Carbohydrate-coated gold nanoparticles	n/a	Ojeda et al. (2007)
TNF-α	PEG-modified gold nanoparticles	MC38 colon carcinoma	Paciotti et al. (2004)
Combination Ultrasound and Drug Therapy			
5-Flurouracil	Perfluorocarbon	C32 melanoma	Larina et al. (2005)

PEG poly(ethylene glycol); *PEO* poly(ethylene oxide); *TNF-α* tumor necrosis factor-α

include the delivery of tamoxifen in poly(ethylene oxide)-modified poly(caprolactone) (PEO-PCL) nanoparticles to MDA-MB-231 breast cancer (Shenoy and Amiji, 2005), the delivery of paclitaxel in PEO-modified PCL and PEO-modified poly(β-amino ester) nanoparticles to SKOV3 ovarian cancer (Devalapally et al., 2006), and even the delivery of a gene therapeutic encoding for sFlt-1 or VEGF-R1 to MDA-MB-435 breast cancer from gelatin nanoparticles (Kommareddy and Amiji, 2007). This versatility of polymer platforms allows for fine tuning of the drug delivery formulation to meet specific advantages. For example, the composition of polymeric matrix can be chosen to match the chemical properties of the encapsulated drug(s) to match loading efficiency and release behavior. Or the composition can be tuned to provide precise drug capture or release in response to environmental triggers. Alternatively, the composition can even be optimized to allow for inclusion of multifunctional properties, such as a combination of therapeutics, targeting, and/or imaging modalities, all within one nanoparticle platform.

A versatile function that is applicable to nearly all nanocarrier platforms is the inclusion of active targeting ligands. While the nanoparticle platform enhances targeting of therapeutics or imaging agents through passive means of the EPR effect, active targeting of these nanoparticles to tumor tissue and cellular surface components uniquely present on target cells can aide in the nanoparticle's ability to locate the target cell type in the tumor mass, or can even enhance internalization of these nanoparticles into their target cells. A wide variety of tumor targeting ligands have been successfully used for active targeting of nanoparticles. Depending on the tumor or cell type, surface proteins overexpress or uniquely express, such as the HER2 receptor, prostate-specific membrane antigen, the folate receptor, the thiamine transporter, integrins, and a myriad of other surface factors that can serve as specific targets to active targeting approaches through inclusion of small molecule ligands such as folate (Kim et al., 2005; Sun et al., 2006) and thiamine (Oyewumi et al., 2003), sugar residues such as galactose (Jeon et al., 2005), peptides such as argenine-glycine-aspartic acid (RGD) (Schiffelers et al., 2004), proteins such as transferrin (Bellocq et al., 2003) lectins (Gao et al., 2006), as well as antibodies and antibody fragments (Hayes et al., 2006; Elbayoumi and Torchilin, 2006). However, more recent high-throughput construction and validation have led to the use of aptamers (Farokhzad et al., 2006a) and sequences identified by phage display (Nielsen et al., 2002; Simberg et al., 2007) as alternative active targeting ligands, thereby greatly widening the pool of targeting constructs to direct nanoparticles more specifically. Regardless of the targeting moiety, the principle outcome is essentially the same, mainly improved tumor-cell recognition, improved intracellular penetration, and reduced recognition at nonspecific sites.

Nanoparticle platforms are of great use in tumor targeting for enhanced delivery of anticancer therapeutics, spanning the range from small molecule drugs through biotherapeutics such as genes and peptides or proteins. However, the same principle has been widely applied to cancer detection, where passive or active tumor targeting of fluorescent probes or contrast imaging agents can help increase sensitivity of tumor detection or even metastatic behavior to advance diagnostics to improve patient prognosis from the other spectrum. Current nanoparticle research and development is moving towards multifunctionalization of these nanoparticle platforms for cancer treatment, whereby all the applicable uses of nanoparticles are essentially merged together. These advances lead to therapeutic systems that, from a single dose, administers combination drug therapies, combination therapies of chemotherapeutic drugs with physical stressors (such as thermal therapies, radiation, and photodynamic therapies), or even combines therapeutics with imaging agents for envisioning a "real-time" therapeutic approach. Not only does nanotechnology make these advances possible, but many such successful multifunctional nanoparticle strategies are already in circulation. This chapter describes the most recent approaches in use that employ multifunctional nanoparticle strategies to enhance overall cancer therapy (Fig. 2).

2 Multifunctional Nanocarriers to Overcome Biological Barriers

2.1 Nanocarriers for Oral Absorption

The oral route is one of the most attractive methods of drug administration in the body because of opportunities for self-administration and associated high patient compliance. The oral route is also amenable for administration of different formulations, including solid, semi-solid, and liquid dosage forms. For certain drugs (including majority of anticancer therapeutics), their oral route bioavailability is relatively much lower to provide meaningful therapeutic outcomes. This is partly due to the

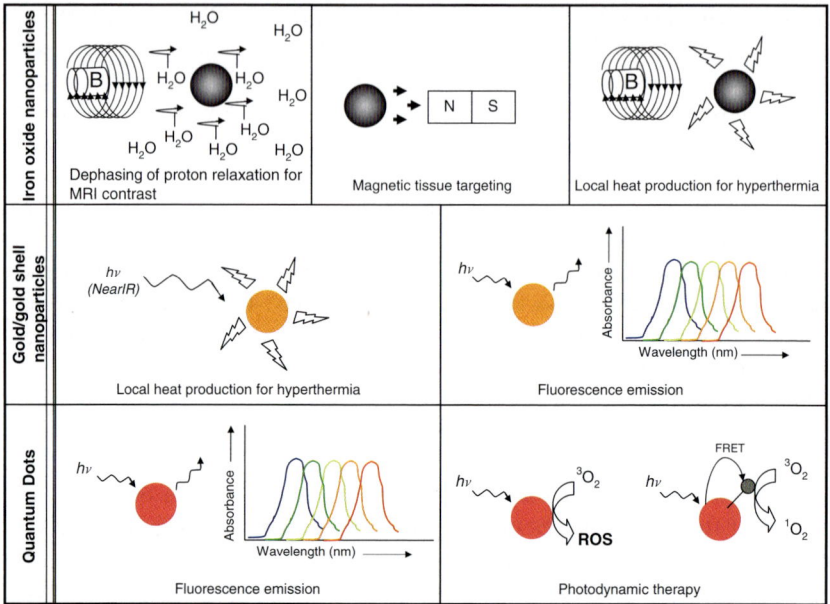

Fig. 2 Opportunity for multiple applications of metallic and semi-conductor nanoparticles in cancer imaging and therapy

presence of large number of multidrug transporters, such as multidrug resistance protein (MRP), p-glycoprotein (p-gp), and the multispecific organic anion transporter (MOAT) on the enterocyte membrane of the gastrointestinal (GI) tract (Taylor, 2002; Thomas and Coley, 2003). These transporters recognize the therapeutic agent as a substrate and actively effluxes the molecule out of the cells. Different types of strategies have been used to enhance oral bioavailability of drugs, including co-administration of p-gp transporter inhibitors and formulation in different nanocarrier delivery systems. Co-administration of a p-gp inhibitor with the active therapeutic agent can decrease the efflux of the agent by preferential binding with the p-gp pump on the cell membrane (Sadeque et al., 2000; Savolainen et al., 2002). However, this strategy has generally shown higher toxicity in vivo, mostly from the high doses of the p-gp inhibitor that are needed, and the additional undesirable pharmacokinetic interactions between the therapeutic of interest and p-gp inhibitor.

Another challenge in oral administration is the presence of high concentrations of metabolizing enzymes in the GI lumen. Besides proteases and nucleases, which can degrade protein and nucleic acid therapeutics, respectively, the GI lumen also expresses cytochrome P-450 metabolizing enzyme systems. Premature drug metabolism at the GI lumen before the active molecule can be absorbed into the systemic circulation significantly limits the bioavailability at the active site. Prodrugs have been designed to improve the stability of therapeutic agents in the GI tract by promoting the conversion to active moiety after absorption in the systemic circulation or, more preferably, at the disease target (Somogyi et al., 1998). In cancer therapy, the prodrug approach can have significant benefit in limiting the toxicity of the agent, if the drug can be selectively activated at the tumor site.

Spray-dried poly(D,L-lactide-*co*-glycolide) (PLGA) nanoparticles have been investigated for oral delivery of amifostine (Pamujula et al., 2004), an organic thiophosphate prodrug that is metabolized by tissue alkaline phosphatase into active thiol metabolite. When administered orally to mice, the amifostine-encapsulated PLGA nanoparticles promoted absorption and the drug was present in blood and other highly-perfused tissues within 30 min of administration. Other polymeric nanoparticles, especially after surface modification to enhance muco- or bioadhesion can be used to enhance the residence time in the GI tract. For instance, tocopheryl poly(ethylene glycol) (PEG) 1,000 succinate (vitamin E-TPGS) modified biodegradable PLGA nanoparticles were proposed for oral administration of paclitaxel. In vitro studies in Caco-2 cells showed a 1.4-fold higher cellular uptake of the TPGS-modified PLGA nanoparticles relative to aqueous solution control.

2.2   Enhancement of Transport Across Other Biological Barriers

Another limitation for drug delivery, especially for systemic brain tumor therapy, is the poor transport across the blood–brain barrier (BBB). The BBB selectively restricts drug transport into the brain because of very tight endothelial cell junctions in the capillary as well as expression of efflux transporters (e.g., *p*-gp) and drug-metabolizing enzymes (Koziara et al., 2004; Ningaraj, 2006). Several studies have shown that poly(alkylcyanoacrylate) nanoparticles promote the delivery of several chemotherapeutic drugs, including doxorubicin, across the BBB when the nanoparticle surfaces are coated with polysorbate (e.g., Tween® 20, 40, 60, and 80) and certain types of poloxamers (e.g., Pluronic® F68). In one example, the therapeutic benefit of doxorubicin administration in sterically stabilized poly(alkylcyanoacrylate) nanoparticulate system was examined in rats bearing intracranial glioblastoma. The investigators proposed that the enhancement in cerebral delivery could probably be due to preferential nanoparticle endocytosis by the low-density lipoprotein receptors on brain capillary endothelial cells after systemic administration. Following cellular internalization, the drug would be able to diffuse out of the nanoparticle matrix and be transported into the brain tissue by transcytosis. The therapeutic potential of this formulation in vivo was studied in rat model with established intracranial 101/8 glioblastoma. Systemic administration of doxorubicin in the polysorbate-modified poly(alkylcyanoacrylate) nanoparticles enabled significantly greater fraction of the animals to survive than did the administration of doxorubicin in solution. Additional opportunities for brain delivery of polymeric nanoparticles can be realized with delivery of combination chemotherapeutic agent and *p*-gp efflux transporter inhibitor. Co-encapsulation of these agents can provide an opportunity to enhance brain delivery of chemotherapeutic agent.

2.3   Enhancement of Drug Availability and Residence at the Tumor Site

For systemic therapy, passive and active targeting strategies are used. Passive targeting relies on the properties of the delivery system and the disease pathology in order to preferentially accumulate the drug at the site of interest and avoid nonspecific distribution. For instance, PEG- or PEO-modified nanocarrier systems can preferentially accumulate in the vicinity of the tumor mass upon intravenous administration based on the hyperpermeability of the newly-formed blood vessels by a process known as EPR effect. Maeda et al. (Maeda, 2001; Jun-Fang et al., 2006) first described the EPR effect in murine solid tumor models and this phenomenon has been confirmed by others. When polymer–drug conjugates are administered, 10–100 fold higher concentrations can be achieved in the tumor (due to EPR effect) than when free drug is administered. Some investigators have also suggested that the EPR effect is present in inflammatory areas and in myocardial infarction. Other approaches for passive targeting involve use of specific stimuli-sensitive delivery system that can release the encapsulated payload only when such a stimuli is present. For instance, the pH around tumor and other hypoxic disease tissues in the body tend to be more acidic (i.e., ~5.5–6.5), relative to physiological pH (i.e., 7.4). We found significant enhancement in drug delivery and accumulation in the tumor mass when pH-sensitive PEO-modified PbAE nanoparticles were used; in contrast drug delivery using non-pH sensitive PEO-PCL nanoparticles in aqueous solution was not as effective (Shenoy et al., 2005). Other approaches for passive targeting involve optimization of nanocarrier size and surface charge modulation. Nanoparticles of <200 nm in diameter and those with positive surface charge are known to preferentially accumulate and reside in the tumor mass for longer duration than do either neutral or negatively charged nanoparticles (van Vlerken and Amiji, 2006). Besides PEG or PEO, other hydrophilic polymers including poly(vinyl alcohol), poly(acryl amide), poly(N-vinylpyrrolidone), poly-(N-(2-hydroxypropyl)methacrylamide), polysorbate 80 (Tween® 80), and block co-polymers such as poloxomer and poloxamine are also being used to modify the physico-chemical properties of the colloidal carriers (Torchilin, 1996; Oupicky et al., 2000; Fenske et al., 2001).

Active targeting to the disease site relies, in addition to PEG modification of nanocarriers to enhance circulation time and achieve passive targeting, coupling of a specific ligand on the surface that will be recognized by the cells present at the disease site. Using solid tumor as an example again, there are several strategies that can be adopted for surface modification of nanocarrier systems for effective targeted delivery to the tumor cells or to endothelial cells of the tumor blood vessels. Since tumor cells are rapidly proliferating, they over-express certain receptors for enhanced uptake of nutrients, including folic acid, vitamins, and sugars. When the surface of nanocarriers is modified with folic acid, they can be targeted to the tumor cells that over-express folate receptors. Tumor and capillary endothelial cells also express specific integrin receptors, such as $\alpha_v\beta_5$ or $\alpha_v\beta_3$ that can bind to RGD tripeptide sequence. RGD-modification, therefore, has been used to direct nanocarriers to tumor cells and capillary endothelial cells of the angiogenic blood vessels (Cegnar et al., 2005; Chiellini et al., 2006; Gabizon et al., 2006). The phage display method has been used to identify specific peptide sequences that can be used for targeting to tumors and other disease areas in the body. Development of monoclonal antibodies against specific epitopes present only on tumor cells allows for other targeting strategies. Using a monoclonal antibody 2C5 that specifically recognizes antinuclear histones, Torchilin's group (Torchilin, 1994; Torchilin et al., 2001, 2003; Lukyanov et al., 2004; Gupta et al., 2005a) has developed various strategies for active targeted delivery of drugs to the tumor mass using liposomes and micellar delivery systems. Other groups have used transferrin, an iron-binding protein, for surface modification of nanocarriers for delivery to tumors. Recently, Farokhzad et al. (2004, 2006a,b) have elegantly described the use of aptamers, nucleic acid constructs that specifically recognize prostate membrane antigen on prostate cancer cells. The aptamer technology provides an additional strategy for active targeting to tumor cells in the body.

2.4   Enhancement of Intracellular Uptake

Once the nanocarriers are delivered to the specific diseased organ or tissue, they may need to enter the cells of interest and ferry the payload to subcellular organelles. In this case, nonspecific or specific cell penetrating strategies need to be adopted. Nonspecific cell uptake of nanocarriers occurs by endocytotic process, where the membrane envelops the nanocarriers to form a vesicle in the cell called an endosome (Panyam and Labhasetwar, 2003). The endosome then shuttles the content in the cell and can fuse with lyososomes, which are highly acidic organelles rich in degrading enzymes. Endocytosed nanocarriers usually travel in a specific direction and converge at the nuclear membrane. Weissig's group (Weissig, 2005; Weissig et al., 2006) has attempted to direct various nanosized delivery systems to mitochondria using delocalized cationic amphiphiles and other mitochondriotropic vector systems. Specific cellular uptake can occur through receptor-mediated endocytosis, where upon binding of the ligand-modified nanocarrier with the cell-surface receptor leads to internalization of the entire nanocarrier–receptor complex and vesicular transport through the endosomes (Panyam and Labhasetwar, 2003). Following dissociation of the nanocarrier–receptor complex, the receptor can be re-cycled back to the cell membrane. Recently, to enhance cellular uptake, a surge of research effort has been directed towards development of argenine-rich cell penetrating peptides (CPPs) (Torchilin, 2002; Torchilin and Levchenko, 2003; Gupta et al., 2005b; Emerich and Thanos, 2006; Gupta and Torchilin, 2006). Based on the initial work of Dowdy's group (Schwarze and Dowdy, 2000; Becker-Hapak et al., 2001) HIV-1 Tat peptide was identified to promote nonspecific intracellular localization of various molecules upon systemic delivery. This observation has been supported by other groups and a number of cationic peptides have been identified, including penetratin, to enhance intracellular delivery. The exact mechanism of how Tat and other CPPs enhance cell permeation is still a subject of controversy, but recent data show that it may be through endocytosis as well. Following cellular internalization, stability of the payload in the cytosol and uptake by specific organelle, such as the nucleus, is also essential for nucleic-acid-based therapeutics. For efficient systemic gene therapy using nonviral vectors, nuclear import of plasmid DNA in nondividing cells is considered to be the major limiting factor.

3 Multifunctional Nanocarriers for Combination Drug Therapy

The versatility of nanocarrier platforms opens up possibilities to incorporate combination therapies into a single drug delivery system. Combination drug therapy for cancer was first proposed in a legendary move by Drs. Frei, Freireich, and Holland who revolutionized cancer therapy by implementing combination chemotherapy to treat acute lymphoblastic leukemia (ALL), a malignancy that prior to 1950 was largely incurable (Frei et al., 1958). In this case, it was hypothesized that concurrent use of multiple drugs with differing mechanisms of action would circumvent the development of drug resistance, the likely cause for prior therapeutic failure in ALL. The success of this strategy caused the approach to quickly gain widespread acknowledgement to become a common consideration in current cancer therapy. Furthermore, the idea has been extended beyond combination chemotherapy to combine drugs with entirely distinct pharmacological targets – e.g., combinations of chemotherapeutic agents with angiogenesis inhibitors, protease inhibitors, immunotherapeutics, hormone therapeutics, and modulators of multidrug resistance (MDR) – therapies largely stemming from advances in cancer molecular and cell biology leading to identification of alternate therapeutic targets.

3.1 Combination Antiangiogenic and Cytotoxic Chemotherapy

Angiogenesis is the process of new blood vessel formation (neovascularization), and has been established as the key factor for tumor growth and development past the primary stage (Folkman, 1972). Anti-angiogenic therapy quickly became a popular alternative in cancer therapeutic development versus conventional chemotherapy (Folkman, 1972), leading to FDA approval of the first anti-angiogenesis drug for cancer therapy in 2004. However, since angiogenesis is only implicated in tumor growth and survival beyond the initial avascular tumor core, to date it is standard practice to combine this treatment option with conventional chemotherapy. Such clinically approved regimens include the combination of the angiogenesis inhibitor bevacizumab with standard chemotherapy (ironotecan, 5-fluorouracil, and leucovorin) for metastatic colorectal cancer and with carboplatin and paclitaxel against non–small-cell lung cancer (Fayette et al., 2005), although an additional variety of such combination therapies also persists in clinical use or clinical trials.

Given the success of nanoparticles in chemotherapeutic drug delivery to tumors, it followed suit that antiangiogenic therapies were delivered in nanoparticles as well, for similar enhancement of tumor-targeting, leading to enhanced therapeutic efficacy, particularly aiding delivery of labile gene therapeutics that have recently found a trend in angiogenesis inhibition. For example, treatment of mice bearing MDA-MD-435 breast tumors with an antiangiogenic gene therapeutic, namely sFlt-1, delivered within long-circulating thiolated gelatin nanoparticles resulted in a nearly sixfold higher transfection efficiency of the gene therapeutic at the tumor-site, a corresponding fourfold decrease in microvessel density in the tumor mass, and a complete tumor growth delay over the course of 25 days (Kommareddy and Amiji, 2007). Similarly, an siRNA therapeutic directed against VEGF-R2 encapsulated within cationic polyplexes bearing an RGD-active targeting moiety also caused a significant inhibition of tumor growth, due to the significant decrease in tumor vascuality (Schiffelers et al., 2004). However, thus far clinical antiangiogenesis therapies are co-administered with chemotherapeutics, leading to an interest in the development of multifunctional nanoparticle formulations for co-administration of the therapeutics. Furthermore, research has alluded to the fact that simultaneous administration of angiogenesis inhibitors and chemotherapeutics may actually cause detrimental effects, where a breakdown of vascularity not only prevents the chemotherapy from accumulating throughout the tumor site, but that it can also lead to tumor hypoxia, which may promote drug resistance and metastasis (Tran et al., 2002). Given this dilemma, it was thought that this form of combination therapy may actually benefit from temporal controlled release, a feat that can well be mediated by using nanoparticles as drug delivery vehicles. On this premise, Sengupta et al. developed a novel multifunctional nanoparticle formulation that, upon localization in the tumor mass, first releases the antiangiogenic drug combretastatin-A4 to shut down tumor vasculature, followed by the sustained release of the cytotoxic agent doxorubicin, already localized within the tumor mass, thereby avoiding the aforementioned problems associated with chemotherapeutic delivery after vascular shutdown (Sengupta et al., 2005). By this mechanism, survival and tumor growth delay of mice bearing either Lewis Lung carcinoma or B16/F10 melanoma models drastically improved when compared with simultaneous nanoparticle administration of the combination therapy lacking temporally controlled release.

3.2 Combination Therapy to Overcome Tumor Drug Resistance

Another treatment target in cancer that greatly benefits from a therapeutic approach that utilizes drug combinations is the treatment of tumors that present with drug resistance, a phenotype whereby the cancer is largely resistant to chemotherapeutic treatment alone. Combination chemotherapy has been extensively used in the clinic to treat cancers that develop resistance, and it is of interest to note that the original use of combination chemotherapy derived by Frei, Freireich, and Holland was intended to circumvent the establishment of drug resistance in ALL. However, treatment with multiple cytotoxic chemotherapeutic agents lacks in benefit, since these potently toxic drugs can provoke detrimental adverse effects in patients, not to mention the fact that the occurrence of MDR, a cross-resistance to structurally and functionally unrelated classes of anticancer drugs, rules out hope for much of combination chemotherapy (Harris and Hochhauser, 1992). Decades of research into the cellular mechanisms that cause drug resistance to develop have opened up a new avenue of therapeutic targets for combination therapy against drug resistance, most notably aimed at inhibiting drug efflux pumps of the ATP-binding cassette (ABC) family of transporters (most notably p-gp/MDR-1), inhibiting drug detoxification mechanisms, and restoring or lowering the apoptotic threshold of MDR cancer cells (Bradley et al., 1988; Harris and Hochhauser, 1992). Initial and some ongoing clinical strategies against MDR used inhibitors of p-gp to revert resistance in combination with chemotherapeutic drugs (Gottesman et al., 2002). This principle was quickly combined with the benefits of nanoparticle drug delivery, as demonstrated by Soma et al. who used poly(alkylcyanoacrylate) nanoparticles for co-administration of doxorubicin with the p-gp inhibitor cyclosporin A to successfully reverse MDR in monocytic leukemia cell line (p388) (Soma et al., 2000). Similarly, Wong et al. used polymer–lipid nanoparticles to co-administer doxorubicin with GG918 (Elacridar – a third generation p-gp inhibitor that has been undergoing testing in clinical trials for the treatment of MDR), to also observe a significantly improved chemosensitization in MDR MDA-MB-435 breast cancer cells (Wong et al., 2006). In the most basic form, intracellular uptake of nanoparticles by endocytotic mechanisms has been explored as a mechanism for chemotherapeutic drugs to bypass drug efflux pumps from the ABC family. Although shown to be a successful approach to chemosensitize MDR cancer types, this benefit of nanoparticles on a cellular level can be used to still deliver a combination therapy against alternate mechanisms of MDR to further improve therapeutic success. We have recently explored this strategy by using PEO-modified PCL nanoparticles to administer a combination therapy of paclitaxel with ceramide, an apoptotic modulator aimed to restore apoptotic signaling in the MDR phenotype. While it was found that the combination therapy significantly improved chemosensitivity in an MDR ovarian cancer model through a restoration of apoptotic activity in response to paclitaxel poisoning, encapsulation of this combination therapy into nanoparticles further enhanced the MDR modulation efficacy on a cellular level, as shown by the multifunctional strategy of simultaneously evading p-gp drug efflux as well (van Vlerken et al., 2007).

4   Multifunctional Nanocarriers for Combination Hyperthermia and Drug Therapy

The National Cancer Institute defines hyperthermia as a form of cancer treatment wherein high temperatures of up to 45 °C are applied to the tumor tissue (National Cancer Institute Fact Sheet, 2005). As opposed to thermal ablation, where significantly higher temperatures (up to 70–80 °C) are used to completely coagulate the tissue for a brief period of time, hyperthermia extends for up to an hour and causes damages to the cellular proteins and organelles, eventually leading to cell death as evident by the tumor shrinkage (National Cancer Institute Fact Sheet, 2005). For general hyperthermia, hot-water baths are commonly used, but local temperature at the tumor site can be raised by 5–8 °C over the physiological temperature using various other thermal techniques such as high radio frequency, ultrasound, infrared, and microwave radiation. High temperatures (43 °C–45 °C) over a fixed period of time (30–60 min) are also used to sensitize tumors to chemotherapy and radiation (National Cancer Institute Fact Sheet, 2005). As such, combination of heat and chemo- or radiotherapy can be used very effectively to augment the therapeutic benefit in cancer leading to better clinical outcomes.

In the context of cancer therapy, hyperthermia has been studied mostly for its increased drug uptake and therapeutic activity enhancement properties. For a long time, hyperthermia has been speculated to preferentially cause changes in tumor metabolism and tumor vasculature by increasing cellular permeability (National Cancer Institute Fact Sheet, 2005). This has been the applied principle in thermal medicine with combination therapy. However, until recently, hyperthermic therapy was not a widely accepted treatment modality due to problems associated in maintaining homogenous temperatures in the target tumor mass and prevention of heat-induced injury to neighboring normal tissue (National Cancer Institute Fact Sheet, 2005). This is particularly challenging in thermotherapy of deep-seeded solid tumors, such as those in the liver, pancreas, prostate, and lung as heat-inducing probes are applied from the exterior, thus making it an invasive and complicated procedure (National Cancer Institute Fact Sheet, 2005). Even to date, clinical hyperthermia is yet to achieve the significance as an adjuvant therapeutic modality. Nevertheless, this treatment option combined with various delivery applications in chemotherapy and radiotherapy can have significant implication in the future.

4.1 Rationale for Combination Thermal Medicine and Drug Therapy

In any cell, hyperthermia triggers the synthesis of heat shock proteins (HSPs), which mediate various cellular defenses, including dynamic protein folding and chaperoning functions throughout the cell, thus inducing thermo-tolerance. Hyperthermic damage to tumor cells is greater when compared to normal cells (van der Zee, 2002), due to various tumor micro-environmental factors, including hypoxia, low pH, and susceptible vasculature, which makes this quite an attractive treatment modality (Ciocca and Calderwood, 2005).

At the molecular level, heat shock factor 1 (HSF-1), in association with heat shock elements (HSEs), mediates the heat shock gene expression (Brade et al., 2000). In the cancerous cell, excessive heat induces production of HSPs leading to cell repair. Cell signaling, apoptosis, and nuclear function involving HSPs have been a prime focus in recent research because of their potential as therapeutic targets.

For example, therapeutic genes that transcribe for cytokines, such as interleukin-2, interleukin-12, and tumor necrosis factor-α (TNF-α), have been successfully targeted in tumor models by using adjuvant hyperthermia (Siddiqui et al., 2007; Visaria et al., 2006). Wild-type HSP-70b promoter was used to control the expression of β-galactosidase reporter gene carried by an adenoviral vector (Brade et al., 2000).

The idea of adjuvant hyperthermia provides some hope as several investigators have recently attempted to address the problems of achieving homogenous therapeutic thermal dose within the tumor interstititum over the necessary period of time. The increased thermal and radio-sensitization brought about by small molecules, which function as radiosensitizers and lower HSF-1 activation, also has shown to cause loss of mitochondrial membrane potential, thus leading to mitochondrial damage (Sekhar et al., 2007).

Systemic chemotherapy has been the most successful mode of cancer therapy for a long time. However, infusing therapeutic doses of cytotoxic drugs into the blood stream and achieving the desired concentration in the tumor without producing toxic effects in the healthy body tissues has been the biggest challenge in cancer chemotherapy. Similarly, gene delivery systems encountered the problem of insufficient uptake, cytotoxicity, and undesirable immunogenic side effects due to the lack of safe tissue- or cell-specific vectors. This problem is being largely addressed by the advent of numerous surface-modified nano-sized drug delivery systems that can escape the reticulo-endothelial system and reach the target tissue with the aid of various target-specific ligands upon systemic administration.

4.2   Select Examples of Nanocarriers for Combination Thermal Medicine and Drug Delivery

Over the last decade, liposomes have been the most studied group of drug delivery systems. Ponce et al. (2007) observed increased uptake of liposomes, loaded with chemotherapeutic drugs when administered in combination with local hyperthermia induced via catheter inserted in the tumor. Further, the drug delivery pattern was observed by magnetic resonance imaging (MRI) to study the antitumor effect of drug loaded liposomes, and reported an increase in the tumor accumulation when administered with hyperthermia (Ponce et al., 2007).

Polymeric nanoparticles are not far behind liposomes or micelles in competing for candidacy of efficient drug delivery systems. Researchers have successfully exploited certain stimuli-responsive polymeric nanocarriers, which undergo thermodynamically reversible lower critical solution temperature (LCST) phase transition, also known as inverted phase transition (Meyer et al., 2001). This means that the polymeric nanocarriers become soluble upon injection in vivo, and then become insoluble only to accumulate at the tumor site due to induction of local hyperthermic state. Poly(N-isopropylacrylamide) and certain elastin-like peptides are ideally suited for such thermally-targeted drug delivery in cancer (Meyer et al., 2001). This strategy employs the high loading capacity of the polymeric carriers and the synergistic effect of macromolecular extravasation by hyperthermia to localize the delivery system at the tumor site (Meyer et al., 2001). Local precipitation of the delivery system in the tumor vasculature due to increase in temperature can also lead to site-specific micro-embolization to prevent oxygen and nutrients diffusing into the tumor mass (Meyer et al., 2001). This is quite an elegant strategy that combines drug delivery with antivascular therapeutic approach to synergistically inhibit tumor growth.

Colloidal gold and microgels are other emerging examples of drug delivery systems that are being used in combination with thermotherapy. A recent study reported that the antitumor efficacy of TNF-α increased significantly upon encapsulation in PEG-coated colloidal gold particles and when administered in combination with thermotherapy (Visaria et al., 2006). In a similar gene-based approach, adjuvant hyperthermia was shown to enhance the anti-angiogenic efficacy of interleukin-12 upon administration with the aid of adenoviral vectors (Siddiqui et al., 2007). Microgels, which are microscopic particles of hydrogels, have gained substantial attention in controlled drug release. Upon cross-linking into mesh systems, various polymers can be used to formulate hydrogels holding large water content (Vinogradov, 2006). Microgels made with polymeric materials that undergo marked volume transitions upon exposure to external stimuli, such as temperature, can be quite useful in chemo-embolization in combination with hyperthermia, especially for liver cancer (Vinogradov, 2006). Certain anticancer drugs such as bleomycin can be encapsulated in the microgels, and upon oral administration, preferential release of the drug at higher pH (of the small intestine) was observed (Blanchette and Peppas, 2005). Further, release of certain encapsulated biomolecules, such as pDNA, can be targeted and controlled by hydrogen or hydrophobic bonding (Vinogradov, 2006).

Radio-frequency (RF) ablation is an image-guided, percutaneous ablative procedure, which applies the principle of tumor necrosis mediated by targeted heat delivery to the tumor mass (Chang, 2003). Electric probes are introduced into the center of the tumor through which high frequency alternating current (up to 550 kHz) is passed to generate heat by agitation of conductive ions, leading to irreversible cellular damage and tumor coagulation (Chang, 2003). Current clinical applications of RF ablation have found importance in the treatment of large lung and liver tumor masses (Chang, 2003).

Recently, RF ablation has become an increasingly popular mode of treatment in malignancies, although tumors larger than 3 cm have shown discouraging outcome (Hines-Peralta et al., 2006). This has prompted several researchers to investigate RFA in conjunction with chemotherapy. Goldberg's group has observed significant increases in tumor accumulation of doxorubicin and antitumor efficacy of the drug encapsulated in PEG-modified liposomes (Doxil®) upon administration in combination with RF ablation. Preliminary results of clinical studies involving patients with hepatic tumors show that high tumor necrosis levels were achieved when RF ablation was used in combination with targeted chemotherapy (Goldberg et al., 2002). Furthermore, favorable results have been reported when the liposomal chemotherapeutic agent was modified to achieve greater tumor coagulation levels.

However, the success of RFA-chemotherapy combination is also influenced by formulation characteristics such as nanoparticle size, the nature, and circulation time of cytotoxic drug delivered there in (Ahmed, 2005). Interestingly, arsenic trioxide (As_2O_3) has been used to enhance RF ablation in solid tumors through its apoptotic activation, antivascular, and thermo-sensitizing properties (Hines-Peralta et al., 2006).

It is critical to evaluate and optimize the current and emerging image-guidance tools applied in tumor ablation with patient-specific temperature maps, also known as isotherms (Wood et al., 2007). Technetium-99 m radiolabeled chemotherapeutic liposomes could be used to monitor drug release, which helps in the calculation of the desired intensity of hyperthermic intervention (Kleiter et al., 2006). X-ray computed tomography may also be used to study the release kinetics of the adjuvant chemotherapeutic agent released in response to thermal trigger at the tumor site and localization in the tumor mass. This is important in the context of physiological changes brought about by the severe tissue damage by RF ablation treatment. Despite the large scope of RF ablation in the treatment of cancer, the size and location of the tumor pose great challenges for successful clinical outcomes as deep-seated solid tumors are often poor targets for RF ablation due to poor visualization, higher probability of incomplete ablation, and potential for cancer relapse (Wood et al., 2007). In deep-seated tumor mass, intense damage to nontarget tissues is also highly possible because of the invasive nature of the treatment modality.

Deep-tissue solid tumors require innovative strategies, which employ minimally invasive, targeted implantation of "thermoseeds" to sensitize the tumor mass to radiotherapy or chemotherapy (or both) (Johannsen et al., 2005). Recently, liposomes or polymeric nanoparticles encapsulating magnetic iron oxide particles have been recruited to induce hyperthermia in tumors and also to serve as MRI contrast agents. In one clinical study, magnetic nanoparticles were evaluated for interstitial thermotherapy where-in iron oxide nanoparticles were suspended in water and administered to 22 patients with pelvic, thoracic, and head-and-neck tumors and exposed to alternating magnetic field to generate local heat (Wust et al., 2006). The results showed good tolerance of magnetic heating of iron oxide nanoparticles by these patients. In a separate study, the same group of investigators evaluated the efficacy of aminosilane-coated iron oxide nanoparticles for thermotherapy of recurrent glioblastoma multiforme and reported that magnetic nanoparticles were safe and efficacious in achieving hyperthermia-mediated tumor control (Maier-Hauff et al., 2007).

5   Multifunctional Nanocarriers for Imaging and Drug Therapy

Perhaps the most common form of nanocarrier multifunctionalization finds itself in the combination of imaging modalities and drug therapy into a single nanoparticle platform. Since the improvement in survival outcome of cancer patients over the last few decades can be largely attributed to improvements in both therapy as well as diagnostics, the combination of both modalities seems obvious, particularly since the tumor targeting properties of nanoparticles would benefit both therapy and imaging. A concept that is readily attainable through nanoparticles, and would be greatly beneficial to cancer patients, is the idea of "real-time" therapy, a situation whereby a clinician can visually track where in the body the administered dose disperses and how much accumulates at the tumor site, and as a result, can either predict therapeutic outcome, or even go as far as to visually monitor tumor shrinkage over time. Multifunctionalization of nanoparticles through the co-inclusion of therapeutics and imaging contrast agents will allow for such major advances.

Superparamagnetic iron oxide nanoparticles are colloidal suspensions of magnetite (Fe_3O_4) that were approved over a decade ago by the FDA for parental use as a contrast agent in MRI. Originally approved for liver imaging, the superparamagnetic nature of iron oxide nanoparticles enhances contrast of their area of accumulation on a T_2 weighted MRI image, a feat that is advantageous in the tumor detection as well. While MRI in itself is a very useful technique for detection of solid tumors, by providing clear anatomical detail and soft tissue contrast, in the past MRI has been quite insensitive for smaller events in cancer imaging, such as the detection of lymph node metastasis and therapeutic efficacy of cancer treatment. Iron oxide nanoparticles were successful in the detection of 90.5% lymph node metastasis in patients with prostate cancer as opposed to 35.4% detection using conventional MRI, a 2.5-fold greater increase in diagnostic sensitivity (Harisinghani et al., 2003). In a more advanced use of contrast imaging, iron oxide nanoparticles have been shown to image cellular events in vivo. Zhao et al. (2001) targeted iron oxide nanoparticles to anionic phospholipids present on the surface of apoptotic cells by incorporating the C2-domain of synaptotagmin I onto the surface of the nanoparticles, allowing for a real-time visualization of apoptotic activity as an indicator of chemotherapeutic efficacy. Magnetite nanoparticles formulated with PLGA have been successful in combining delivery of chemotherapeutic drugs to the tumor, while retaining enough magnetic strength for imaging contrast enhancement, a potential use for real-time tracking of therapeutic efficacy. This potential has also been demonstrated by Reichardt et al. (2005) who used iron oxide nanoparticles as a tumor contrast enhancement in MRI to visualize the tumor therapeutic response of MV522 colon carcinoma xenografts to a VEGF receptor tyrosine kinase inhibitor over time. From this study, they were able to show a statistically significant decrease in relative vascular volume fraction in real-time over the duration of treatment, as measured by sequential MRI of the tumors using these iron oxide nanoparticles as a tumor-imaging enhancer. Similarly, Nasongkla et al. (2006) developed multifunctional polymeric micelles loaded with doxorubicin and superparamagnetic nanoparticles in the core, and surface modified by inclusion of cyclic RGD for active tumor targeting. Self-assembling dermatan sulfate based nanoparticles formulated as a superparamagnetic nanoparticle with inclusion of the chemotherapeutic drug doxorubicin, is another example of a multifunctional nanoparticle for tumor imaging and treatment (Ranney et al., 2005). Not only have these nanoparticles been shown successful in imaging AT1 tumors in vivo by MRI, surprisingly, therapeutic efficacy against MX-1 breast tumor xenografts increased significantly when doxorubicin was delivered encapsulated in these nanoparticles, versus treatment with free doxorubicin, as indicated by the drastic tumor growth delay in 60% of mice and complete tumor regression in 40% of mice treated with the nanoparticle formulation, as opposed to the lack of tumor regression and shorter tumor growth delay in mice treated with doxorubicin alone (Ranney et al., 2005). An alternative approach to a similar multifunctional nanoparticle by Kohler et al. (2005) multifunctionalized iron oxide nanoparticles by binding methotrexate to the surface to produce a targeting construct to folate receptors; however, once internalized by the cancer cell, lysosomal pH cleaved methotrexate from the surface, allowing it to further serve as a chemotherapeutic for cancer eradication, thereby producing a multifunctional system that allows for simultaneous tumor therapy and real-time imaging of drug delivery.

Another MRI contrast agent applicable in nanotechnology is gadolinium. Gadolinium-157 is a stable (nonradioactive) nuclide that is frequently used as a contrast agent in MRI diagnostics, to enhance contrast in T1 weighted images (Aime et al., 2004), for example, in MRI in vivo models of lymph node metastasis (Kobayashi et al., 2006). However, an additional benefit of gadolinium nanoparticles is that upon irradiation with thermal neutrons gadolinium-157 produces cytotoxic γ-ray radiation (Barth and Soloway, 1994), enabling gadolinium for the additional use in neutron capture therapy (NCT) of cancer. Thus, the combined therapeutic and imaging properties of gadolinium make it an excellent candidate for multifunctional cancer treatment. Using gadolinium nanoparticles as such as therapeutic modality, tumor growth was significantly suppressed and survival time increased through NCT in mice bearing a radio-resistant melanoma (Tokomitsu et al., 2000). Delivery of gadolinium through gadopentatic acid (Gd-DTPA) allows for association of gadolinium into polymeric nanoparticles, a principle proven by Tokomitsu et al. (Tokumitsu et al., 2000; Shikata et al., 2002) who utilized this concept to associate gadolinium into chitosan nanoparticles for NCT. Although, prior use of Gd-DTPA as an MRI contrast agent and use of chitosan nanoparticles in delivery of chemotherapeutics such as paclitaxel and doxorubicin to tumors (Nsereko and Amiji, 2002) seem evident, the dual use of these gadolinium-containing chitosan nanoparticles in imaging and therapy is yet to be investigated. Thus far, multifunctionalization of gadolinium nanoparticles has been improved through conjugation of folic acid or thiamine to the surface of gadolinium-containing nanoparticles (through distearoylphosphatidylethanolamine (DSPE) and a PEG spacer), greatly enhancing cell uptake of gadolinium to cancer cells expressing receptors for folate and thiamine respectively in vitro and in vivo, thereby potentially improving localization and tumor eradication by NCT (Oyewumi and Mumper, 2002; Oyewumi et al., 2003, 2004). Already, gadolinium nanoparticles present multifunctional properties in their ability to image and ablate the tumor in one system. However, further multifunctionalization of these vectors by conjugation with tumor-specific targeting ligands and incorporation of a drug load remains to be examined.

As another imaging modality, gold nanoparticles and gold nanoshells (silica core nanoparticles surrounded by a layer of gold coating) are favorable to be used as contrast agents in optical coherence tomography (OCT), since variations in their size and shape allows for precise tuning of their resonance wavelength between near-ultraviolet and mid-infrared (Oldenburg et al., 1999). For example, a gold nanoshell with a 20-nm shell on a 60-nm silica core will resonate at around 700–750 nm, while a nanoshell with a 5-nm shell on the same 60-nm core will resonate at around 1,000–1,050 nm (Loo et al., 2004). In this manner, multifunctionalized gold nanoparticles have been used for tumor imaging and drug delivery. For example, delivery of daunorubicin from gold nanoparticles was shown to cause 15–20% greater inhibition of cell growth of multidrug resistant K562 leukemia cells and over administration of free drug, while retaining imaging capabilities of these cells through fluorescence detection (Li et al., 2007). Similarly, gold nanoparticles have been conjugated to a carbohydrate coating to incorporate glycogen antigens to develop a multifunctional anticancer vaccine (Ojeda et al., 2007). The disperse range permissible to these nanoparticles spans the near-infrared (NIR), and since NIR light experiences maximal tissue penetration with minimal en route absorption, they become beneficial for use in thermal ablation, a property that, like gadolinium nanoparticles, gives these platforms an inherent multifunctional capability in cancer imaging and therapy, aside from the additional inclusion of anticancer drugs. Such combined imaging and therapeutic use of these gold nanoshells has been proven in several cancer models, both in vitro and in vivo (Hirsch et al., 2003; Loo et al., 2004, 2005). It has been shown that thiolated PEG easily assembles onto the nanoshell surface providing a linker for surface incorporation of active tumor targeting moieties or even biotherapeutic agents. Paciotti et al. (2004) have used colloidal gold particles in this manner to successfully deliver TNF-α as an anticancer therapeutic to an MC-38 colon carcinoma in vivo. Mukherjee et al. (2005) reported the inhibition of angiogenesis by gold nanoparticles, through direct binding of the particles to heparin-binding growth factors (VPF or VEGF and FGF specifically), a property that is very useful in halting tumor proliferation. Hainfeld et al. (2004) have shown that gold nanoparticles can help to localize radiotherapy to prolong one-year survival rates of mice bearing EMT-6 mammary carcinomas (86% survival with gold nanoparticles versus 20% survival with x-rays alone). Although the latter three examples used colloidal gold nanoparticles rather than the silicon–gold nanoshells, which bear the combined use of imaging and thermal ablation, future research may allow for the development of a gold nanoshell or a particle that ties together all these uses.

Finally, a more recent nanoparticle platform that emerged for cancer diagnostics, and has further allowed for the multifunctional modality of imaging and therapy is the semiconductor nanocrystal, otherwise known as the quantum dot. Quantum dots are semiconductor-based nanoparticles that function as fluorescent probes for imaging purposes (Gao et al., 2005). Similar to gold nanoshells, quantum dots are favorable imaging agents, that is their absorption properties can be tuned from visible to infrared wavelengths, they emit highly intense signals, and they are chemically, photochemically, and thermally stable (Chan et al., 2002). Quantum dots have the unique property that, from a single excitation wavelength, emission photons can span any wavelength between blue and infrared depending on the nanocrystal size and composition (Voura et al., 2004). Therefore a number of quantum dots, each actively targeted to a different tumor marker, can be visualized simultaneously, a useful property in real-time cancer imaging. This function has been particularly useful in the tracking of metastatic tumors (Voura et al., 2004). Quantum dots, miniscule in size (2–8 nm in diameter), are easily bioconjugated with peptides, antibodies, and small-molecule drugs through polymer linkers without loss of their fluorescence or tumor localization properties (Gao et al., 2005). Typically, high quality quantum dots are prepared in the organic solvent mixture tri-n-octyl phosphine/tri-n-octylphosphine oxide (TOP/TOPO) at high temperatures, which caps the quantum dots with a monolayer of the nonpolar solvent. This capping allows for surface adhesion of amphiphilic polymers (such as PEG and poly(ethylene oxide)-containing block copolymers), which not only facilitate solubility and bioavailability of the nanoparticles, but provide a linker for bioconjuation of peptides, antibodies, oligonucleotides, or small molecule drugs, thereby multifunctionalizing the quantum dot for tumor targeting, tumor imaging, and potential drug delivery. A few examples of such incorporation to quantum dots in this manner include antibodies against HER2 (Wu et al., 2003), prostate specific membrane antigen (Gao et al., 2004), HSPs (Medintz et al., 2005), and p-gp (Sukhanova et al., 2004). Although from this step forward it seems inherent that drugs can be loaded into the bulk of the polymer coating or grafted onto the surface (successful multifunctionalization of this degree), while retaining the imaging, biocompatibility, and bioavailability properties remains to be proven. Nevertheless, through a recent discovery it appears that quantum dots, such as gold and gadolinium nanoparticles, may possess an inherent therapeutic capability, thereby maintaining the combined tumor imaging and therapy functions that makes these nanoparticles multifunctional. It appears that quantum dots can act as photosensitizers in photodynamic therapy (PDT) (Bakalova et al., 2004). PDT utilizes light, oxygen, and a photosensitizer to selectively destroy target tissue by generating reactive oxygen species, which promotes apoptosis of the target cells. In this, Samia (2006) have shown that cadmium selenide quantum dots can generate the singlet oxygen species that take part in PDT, although generation is at a lower rate than conventional photosensitizers. However, with this nanoparticle tumor therapy system the promise exists for a multifunctional imaging and therapeutic approach, with great benefit to cancer treatment.

6   Other Examples of Multifunctional Nanosystems

6.1   Combination Drug Delivery and Ultrasound

Perfluorocarbon emulsion nanoparticles are under investigation as ultrasound contrast agents and ultrasonically enhanced drug delivery vehicles. With a mean diameter on the order of hundreds of nanometers, approximately 10-fold smaller than commercially available microbubble contrast agents, targeting of, and extravasation through tumor endothelium may be superior to microbubbles (Kong et al., 2000, 2001). Compared with microbubbles, liquid-filled nanodroplets are more stable under pressure and mechanical stress and are capable of carrying a larger drug payload, although they are also less echogenic. Both microbubbles and liquid-filled nanoparticles can be encapsulated by a molecularly targeted lipid shell.

Lanza and coworkers (Lanza et al., 2002; Lanza and Wickline, 2003; Wickline and Lanza, 2003) have described the use of perfluorocarbon emulsion nanoparticles as ultrasound contrast agents and have developed theoretical models for estimating acoustic reflectivity of different perfluorocarbon nanoparticle formulations (Marsh et al., 1998, 2002a&b; Hall et al., 2000, 2001). Nanoparticles have low acoustic reflectivity in solution; however, their echogenicity increases when they are deposited in a layer, resulting in a targeted contrast agent that is detectable only when adherent at the target site (Lanza and Wickline, 2003). Perfluorocarbon nanoparticles can also serve as MRI contrast agents when gadolinium is incorporated into their lipid shell useful for multimodality imaging studies (Anderson et al., 2000; Winter et al., 2003a,b; Lanza et al., 2004; Morawski et al., 2004; Cyrus et al., 2005; Schmieder et al., 2005). In addition to their application as ultrasound contrast agents, perfluorocarbon nanoparticles have also been used as therapeutic delivery vehicles for doxorubicin, paclitaxel, and other therapeutic agents (Lanza and Wickline, 2001; Wickline and Lanza, 2003; Larina et al., 2005). Crowder et al. (2005) have shown that ultrasound enhances trans-membrane delivery of fluorescent dye from nanoparticles to C32 melanoma cells. Ultrasonic molecular imaging is unique, that is, the optimal application of these agents depends not only on the surface chemistry but also on the applied ultrasound field, which can increase receptor–ligand binding and membrane fusion (Dayton et al., 1999; Zhao et al., 2004; Rychak et al., 2005). Dayton et al. (1999) and Rychak et al. (2005) have previously demonstrated that acoustic radiation force produced by ultrasound can enhance the efficiency of targeted imaging with microbubble-based agents by deflecting targeted particles to the endothelium and facilitating bond formation. Lum et al. (2006) and Shortencarier et al. (2004) have demonstrated that physically localizing drug delivery vehicles with acoustic radiation force can enhance localized drug delivery. Recently, Crowder et al. (2005) have observed acoustically enhanced dye delivery from perfluorocarbon nanoparticles and postulated that acoustic radiation force is partially responsible for this effect.

Using microbubbles as a carrier particle and attaching nanoparticles containing a higher payload of drug allows the biodistribution of such a carrier particle to be controlled by insonation, using ultrasound pulse schemes that are designed to deflect the vehicle to a target vessel wall and then to rupture the larger lipid carrier. When a traveling ultrasonic wave is absorbed by a particle, the momentum associated with the wave produces a net primary ultrasound radiation force (USRF), whereby the radiating sound wave is transferred to the particle. While incompressible objects do experience USRF, compressible objects such as gas bubbles experience far larger forces and are displaced by low-amplitude ultrasound waves (Aaron et al., 2006). Avidinated neutravidin-coated fluorescent nanobeads bound to the biotinylated lipid shells of preformed microbubbles that specifically targets using USRF and biotin–avidin interactions is demonstrated (Aaron et al., 2006). Targeting of nanobeads was molecularly specific and dependent on, in order of importance, vehicle concentration, wall shear stress, nanobead size, and insonation time. This method of delivery is shown to enable targeted deposition of nanoparticles in shear flow and can be modified to carry therapeutic agents for controlled release in targeted delivery applications.

6.2   Combination Drug Delivery and PDT

Photodynamic therapy (PDT), the activation of a tumor-localized photosensitizer by light, is generally applied as a single modality for the treatment of a variety of solid tumors. Its dominant mechanism of action is the local generation of cytotoxic singlet oxygen, which causes the destruction of tumor cells and damage of the tumor microvasculature (Henderson and Gollnick, 2003). It has been applied to both treatment of superficial tumors (such as cutaneous basal-cell carcinoma and head and neck tumors) and to deeper tumors accessible by endoscopies (including esophageal and lung cancers) (Hopper, 2000). PDT with photofrin has been approved by FDA for the treatment of Barrett's esophagus and endobronchial and esophageal carcinomas, and perhaps the most successful approval of PDT is with verteporfin for injection (Visudyne®) to treat age-related macular degeneration (AMD) (Dolmans et al., 2003).

In general, a photosensitizer is confined within the tumor vasculature initially after injection and PDT that employs a short drug-light interval largely damages tumor vasculature (Veenhuizen et al., 1997). This mechanism is mainly responsible for some of the more successful clinical implementations of PDT today, including AMD treatment with verterpofin (Brown and Mellish, 2001) and prostate cancer treatment with Pd-bacteriopheophorbide TOOKAD (Chen et al., 2002a; Koudinova et al., 2003).

While many studies have explored ways to maximize the therapeutic effect of PDT (Gudgin Dickson et al., 2002), recent efforts are more focused on utilizing targeting strategies that are directed at the tumor vasculature. However, it should be realized that neither vascular targeting nor cellular targeting PDT regime alone is perfect for tumor cell killing. Solely vascular targeting may be a good approach for purely vascular diseases such as AMD (Schmidt-Erfurth et al., 1994), yet it may not be enough for tumors because peripheral tumor vessels are shown to be somewhat resistant to both vascular-targeting agents (Pedley et al., 2001) and PDT-induced vascular effects (Uehara et al., 1998; Chen et al., 2002b; Koudinova et al., 2003).

Despite the extensive central tumor necrosis induced by vascular targeting PDT, tumor vessels or cells can re-grow from the peripheral rim after treatment. The major problem for cellular-targeting PDT is that it suffers from complex issues such as heterogeneity of tumor microenvironment and inhomogeneous photosensitizer distribution. Additionally, tissue hypoxia has been identified as a major obstacle to direct targeting tumor cells by PDT (Dougherty et al., 1998).

Inadequate photosensitizer delivery due to heterogeneous tumor perfusion, vascular permeability, and tumor interstitial pressure can also affect the effectiveness of cellular targeting PDT. Combination of tumor vascular and cellular targeting approaches can be a way to overcome the problem associated with each individual targeting strategy and to achieve maximal opportunity of tumor eradication (Wachsberger et al., 2003). Also, most photosensitizers are hydrophobic and difficult to prepare in an injectable form.

Nanocarriers can provide solution to all the above problems by not only providing a stable dispersion of these drugs into aqueous systems, but also upon systemic administration, these carriers are preferentially taken up by tumor tissues by virtue of the "enhanced permeability and retention effect", which is the property of such tissues to engulf and retain circulating macromolecules and particles owing to their "leaky" vasculature. The carriers include oil dispersions (micelles), liposomes, low-density lipoproteins, polymeric micelles, and hydrophilic drug–polymer complexes. In one study (Qing et al., 2006), Profrin II nanoparticles-PDT results in inhibition of Lovo colon carcinoma growth in post-PDT earlier period in vivo, and were shown to prolong the survival time of nude mice bearing xenografts significantly, whereas Profrin II-PDT could not inhibit the growth of colon tumor completely. In another study (Reddy et al., 2006), multifunctional polymeric nanoparticle consisting of a surface-localized tumor vasculature targeting F3 peptide and encapsulated PDT and imaging agents were shown to specifically bound to the surface of MDA-435 cells in vitro and were internalized conferring photosensitivity to the cells. Treatment of glioma-bearing rats with targeted nanoparticles followed by PDT showed a significant improvement in survival rate when compared with animals who received PDT after administration of nontargeted nanoparticles or systemic photofrin.

Zinc(II) phthalocyanine (ZnPc), a second generation loaded PLGA nanoparticle, was shown to maintain its photo-physical behavior after encapsulation (Ricci-Junior and Marchetti, 2006). Photosensitizer release from nanoparticles was sustained with a moderate burst effect of 15% for 3 days. The photocytotoxicity of ZnPc loaded PLGA Np was evaluated on P388-D1 cells that were incubated with ZnPc loaded Np ($5 \propto M$) by 6h and exposed to red light (675 nm) for 120s, and light dose of $30 J/cm^2$. After 24h of incubation, the cellular viability was determined, obtaining 61% of cellular death. From the physical–chemical, photophysical, and photobiological measurements performed it was concluded that ZnPc loaded PLGA nanoparticles is a promising drug delivery system for PDT. In another study, Prasad's group (Cinteza et al., 2006) described the ceramic-based nanoparticles capable of selectively delivering photosensitizers to tumor cells and damaging them in vitro. These studies establish the role of nanocarriers in PDT.

7   Conclusions

Over the last decade, a wide range of nanocarrier systems, such as liposomes, polymeric nanoparticles, nanoemulsions, micelles, and hydrogels have shown tremendous progress in pharmaceutical applications. These engineered multifunctional nanocarrier systems have successfully evolved to possess some very useful properties such as prolonged circulation in blood, target specificity, and increased cell penetration of the therapeutic drugs and molecules. Prompted by the clinical success of some nanocarriers, most drug delivery research has focused on integrating the various beneficial properties of the nanovectors to make the treatment strategies more direct, specific, stable, less invasive, and in some cases to tackle the problem of MDR.

Current research is also focused on understanding and taking advantage of the features of tumor microenvironment such as pH and temperature changes. Developing nanocarriers that employ various beneficial properties require the assembly of a number of chemical moieties on a single nanoparticle. However, immediate challenges in the formulation of such nanovector system include characteristics such as size, surface charge, cytotoxicity, immunogenicity, cell membrane, and organelle barriers to name a few.

Nanovectors, in their simplest form, could enable deliver a combination of drugs or genes (or both) to take advantage of synergistic or bystander properties of the biomolecules. Currently, various nanocarrier systems undergo surface modification, by synthetic polymers such as PEG and targeting ligands such as peptides, antibodies, or sugar moieties, in order to escape the physiological attack by the reticulo-endothelial sytem in the body and target the disease site. In certain cases, the surface modification could enable the nanovector to pass through the blood–brain barrier. The protective PEG coat of the nanocarriers may inhibit the release of the encapsulated drug, thus encouraging the development of drug delivery systems that could be pH- and temperature-responsive, especially in conditions such as inflammation, infarction, and cancer.

Hyperthermia has been emerging as an important adjuvant mode alongside of chemotherapy, radiotherapy, and surgery. Increased local or whole body temperatures brought about by radiofrequency ablation, ultrasonic waves, or by the administration of magnetic nanoparticles that act in alternating magnetic fields are some of the tested strategies in the clinic. Nanocarriers that enable contrasting agents to transmit a signal drug accumulated site provide for diagnostic and imaging techniques. Gold nanoparticles, quantum dots, liposomes, and micelles are among the successful nanovectors. When formulated in combination, these drug delivery systems could enable imaging and controlled release of drugs or therapeutic molecules in a spatiotemporal pattern. Such multifaceted, versatile nanocarriers and drug delivery systems promise a substantial increase in the efficacy of diagnostic and therapeutic applications in pharmaceutical sciences.

References

Aaron, F. H. L., A. B. Mark, A. D. Paul, et al. (2006). "Ultrasound radiation force enables targeted deposition of model drug carriers loaded on microbubbles." *J Contr Release*. 111: 128–34.

Ahmed, M., A. N. Lukyanov, V. Torchilin, et al. (2005). Combined radiofrequency ablation and adjuvant liposomal chemotherapy: effect of chemotherapeutic agent, nanoparticle size, and circulation time. *Journal of vascular and interventional radiology*. 16(10): 1365–71.

Aime, S., A. Barge, C. Cabella, et al. (2004). "Targeting cells with MR imaging probes based on paramagnetic Gd(III) chelates." *Curr Pharmaceut Biotechnol.* 5: 509–18.

Alyautdin, R. N., E. B. Tezikov, P. Ramge, et al. (1998). "Significant entry of tubocurarine into the brain of rats by adsorption to polysorbate 80-coated polybutylcyanoacrylate nanoparticles: an in situ brain perfusion study." *J Microencapsul.* 15(1): 67–74.

Alyaudtin, R. N., A. Reichel, R. Lobenberg, et al. (2001). "Interaction of poly(butylcyanoacrylate) nanoparticles with the blood–brain barrier in vivo and in vitro." *J Drug Target.* 9(3): 209–21.

Amiji, M. (2006). "Polymeric delivery - Engineered nanosystems for targeted delivery of drugs and genes". *Future Drug Delivery.* http://www.touchbriefings.com/pdf/1859/amiji.pdf (Accessed August 09, 2006).

Anderson, S. A., R. K. Rader, W. F. Westlin, et al. (2000). "Magnetic resonance contrast enhancement of neovasculature with $\alpha_v\beta_3$-targeted nanoparticles." *Magn Reson Med.* 44(3): 433–9.

Bakalova, R., H. Ohba, Z. Zhelev, et al. (2004). "Quantum dots as photosensitizers?" *Nat Biotech.* 22(11): 1360–1.

Bargoni, A., R. Cavalli, G. P. Zara, et al. (2001). "Transmucosal transport of tobramycin incorporated in solid lipid nanoparticles (SLN) after duodenal administration to rats. Part II—tissue distribution." *Pharmacol Res.* 43(5): 497–502.

Barth, R. F. and A. H. Soloway (1994). "Boron neutron capture therapy of primary and metastatic brain tumors." *Mol Chem Neuropathol.* 21: 139–54.

Becker-Hapak, M., S. S. McAllister and S. F. Dowdy (2001). "TAT-mediated protein transduction into mammalian cells." *Methods.* 24(3): 247–56.

Bellocq, N. C., S. H. Pun, G. S. Jensen, et al. (2003). "Transferrin-containing, cyclodextrin polymer-based particles for tumor-targeted gene delivery." *Bioconjugate Chem.* 14(6): 1122–32.

Bidwell, G. L., III, I. Fokt, W. Priebe, et al. (2007). "Development of elastin-like polypeptide for thermally targeted delivery of doxorubicin." *Biochem Pharmacol.* 73(5): 620–31.

Blanchette, J and N. A. Peppas (2005). Oral chemotherapeutic delivery: design and cellular response. *Ann Biomed Eng.*, 33(2):142–9.

Brade, A. M., D. Ngo, P. Szmitco, et al. (2000). Heat-directed gene targeting of adenoviral vectors to tumor cells. *Cancer Gene Ther.* 7(12):1566–74.

Bradley, G., P. F. Juranka and V. Ling (1988). Mechanism of multidrug resitance. *Biochem Biophys Acta.* 948: 87–128.

Brown, S. B. and K. J. Mellish (2001). "Verteporfin: a milestone in opthalmology and photodynamic therapy." *Expert Opin Pharmacother.* 2(2): 351–61.

Calvo, P., B. Gouritin, H. Chacun, et al. (2001a). "Long-circulating PEGylated polycyanoacrylate nanoparticles as new drug carrier for brain delivery." *Pharm Res.* 18(8): 1157–66.

Calvo, P., B. Gouritin, I. Brigger, et al. (2001b). "PEGylated polycyanoacrylate nanoparticles as vector for drug delivery in prion diseases." *J Neurosci Methods.* 111(2): 151–5.

Calvo, P., B. Gouritin, H. Villarroya, et al. (2002). "Quantification and localization of PEGylated polycyanoacrylate nanoparticles in brain and spinal cord during experimental allergic encephalomyelitis in the rat." *Eur J Neurosci.* 15(8): 1317–26.

Cegnar, M., J. Kristl and J. Kos (2005). Nanoscale polymer carriers to deliver chemotherapeutic agents to tumours. *Expert Opinion Biologicy and Therapeutics.* 5(12): 1557–69.

Chan, W. C. W., D. J. Maxwell, X. Gao, et al. (2002). "Luminescent quantum dots for multiplexed biological detection and imaging." *Curr Opin Biotechnol.* 13(1): 40–6.

Chang, I. (2003). Finite element analysis of hepatic radiofrequency ablation probes using temperature-dependent electrical conductivity. *BioMedical Engineering Online,* 2: 12

Chen, B., Y. Xu, T. Roskams, et al. (2001). "Efficacy of antitumoral photodynamic therapy with hypericin: relationship between biodistribution and photodynamic effects in the RIF-1 mouse tumor model." *Int J Cancer.* 93(2): 275–82.

Chen, Q., Z. Huang, D. Luck, et al. (2002a). "Preclinical studies in normal canine prostate of a novel palladium-bacteriopheophorbide (WST09) photosensitizer for photodynamic therapy of prostate cancers." *Photochem Photobiol.* 76(4): 438–45.

Chen, B., T. Roskams and P. A. de Witte (2002b). "Enhancing the antitumoral effect of hypericin-mediated photodynamic therapy by hyperthermia." *Lasers Surg Med.* 31(3): 158–63.

Chen, J., F. Saeki, B. J. Wiley, et al. (2005). "Gold nanocages: bioconjugation and their potential use as optical imaging contrast agents." *Nano Lett.* 5(3): 473–7.

Chiellini, E. E., F. Chiellini and R. Solaro (2006). Bioerodible polymeric nanoparticles for targeted delivery of proteic drugs. *Journal of Nanoscience and Nanotechnology.* 6(9–10): 3040–7.

Cinteza, L. O., T. Y. Ohulchanskyy, Y. Sahoo, et al. (2006). "Diacyllipid micelle-based nanocarrier for magnetically guided delivery of drugs in photodynamic therapy." *Mol Pharm.* 3(4): 415–23.

Ciocca, D. R. and Calderwood, S. K. (2005). Heat shock proteins in cancer: diagnostic, prognostic, predictive, and treatment implications. *Cell Stress & Chaperones.*, Summer, 10(2): 86–103.

Cole, S. P., G. Bhardwaj, J. H. Gerlach, et al. (1992). "Overexpression of a transporter gene in a multidrug-resistant human lung cancer cell line." *Science.* 258(5088): 1650–4.

Crowder, K. C., M. S. Hughes, J. N. Marsh, et al. (2005). "Sonic activation of molecularly-targeted nanoparticles accelerates transmembrane lipid delivery to cancer cells through contact-mediated mechanisms: implications for enhanced local drug delivery." *Ultrasound Med Biol.* 31(12): 1693–700.

Cyrus, T., P. M. Winter, S. D. Caruthers, et al. (2005). "Magnetic resonance nanoparticles for cardiovascular molecular imaging and therapy." *Expert Rev Cardiovasc Ther.* 3(4): 705–15.

Dayton, P., A. Klibanov, G. Brandenburger, et al. (1999). "Acoustic radiation force in vivo: a mechanism to assist targeting of microbubbles." *Ultrasound Med Biol.* 25(8): 1195–201.

Devalapally, H., D. Shenoy, S. Little, et al. (2007). Poly(ethylene oxide)-modified poly(beta-amino ester) nanoparticles as a pH-sensitive system for tumor-targeted delivery of hydrophobic drugs: part 3. Therapeutic efficacy and safety studies in ovarian cancer xenograft model. *Cancer chemotherapy and pharmacology.* 59(4): 477–84.

Dolmans, D. E., D. Fukumura and R. K. Jain (2003). "Photodynamic therapy for cancer." *Nat Rev Cancer.* 3(5): 380–7.

Dougherty, T. J., C. J. Gomer, B. W. Henderson, et al. (1998). "Photodynamic therapy." *J Natl Cancer Inst,* 90(12): 889–905.

Elbayoumi T. A. and V. P. Torchilin (2006). "Enhanced accumulation of long-circulating liposomes modified with the nucleosome-specific monoclonal antibody 2C5 in various tumours in mice: γ-imaging studies." *Eur J Nucl Med Mol Imag.* 33(10): 1196–1205.

Emerich, D. F. and C. G. Thanos (2006). "The pinpoint promise of nanoparticle-based drug delivery and molecular diagnosis." *Biomol Eng.* 23(4): 171–84.

Fang, J., T. Sawa, H. Maeda (2003). Factors and mechanism of "EPR" effect and the enhanced antitumor effects of macromolecular drugs including SMANCS. *Advances in experimental medicine and biology.* 519: 29–49.

Farokhzad, O. C., S. Jon, A. Khademhosseini, et al. (2004). Nanoparticle-aptamer bioconjugates: a new approach for targeting prostate cancer cells. *Cancer Research.* 64(21): 7668–72.

Farokhzad, O. C., J. Cheng, B. A. Teply, et al. (2006a). "Targeted nanoparticle–aptamer bioconjugates for cancer chemotherapy in vivo." *PNAS.* 103(16): 6315–20.

Farokhzad O. C., J. M. Karp and R. Langer (2006b). Nanoparticle-aptamer bioconjugates for cancer targeting. *Expert opinion on drug delivery.* 3(3): 311–24.

Fayette, J., J.-C. Soria and J.-P. Armand (2005). "Use of angiogenesis inhibitors in tumour treatment." *Eur J Canc.* 41(8): 1109–16.

Fellner, S., B. Bauer, D. S. Miller, et al. (2002). "Transport of paclitaxel (Taxol) across the blood–brain barrier in vitro and in vivo." *J Clin Invest.* 110(9): 1309–18.

Fenart, L., A. Casanova, B. Dehouck, et al. (1999). "Evaluation of effect of charge and lipid coating on ability of 60-nm nanoparticles to cross an in vitro model of the blood–brain barrier." *J Pharmacol Exp Ther.* 291(3): 1017–22.

Fenske, D. B., I. MacLachlan and P. R. Cullis (2001). "Long-circulating vectors for the systemic delivery of genes." *Curr Opin Mol Ther.* 3(2): 153–8.

Folkman, J. (1972). "Anti-angiogenesis: new concept for therapy of solid tumors." *Ann Surg.* 175(3): 409–16.

Frei, E., III, J. F. Holland, M. A. Schneiderman, et al. (1958). "A comparative study of two regimens of combination chemotherapy in acute leukemia." *Blood*. 13(12): 1126–48.

Gabizon, A. A., H. Shmeeda, S. Zalipsky (2006). Pros and cons of the liposome platform in cancer drug targeting. *Journal of liposome research*. 16(3): 175–83.

Gao, X., Y. Cui, R. M. Levenson, et al. (2004). "In vivo cancer targeting and imaging with semiconductor quantum dots." *Nat Biotechnol*. 22(8): 969–76.

Gao, X., L. Yang, J. A. Petros, et al. (2005). "In vivo molecular and cellular imaging with quantum dots." *Curr Opin Biotechnol*. 16(1): 63–72.

Gao, X., W. Tao, W. Lu, et al. (2006). "Lectin-conjugated PEG-PLA nanoparticles: preparation and brain delivery after intranasal administration." *Biomaterials*. 27(18): 3482–90.

Genentech Biotechnology. Avastin®, Bevacizumab. Product Information Guide. http://www.avastin.com/avastin/index.jsp?hl=en&lr=&q=Avastin (Accessed September 28, 2006).

Genentech Biotechnology. Herceptin®, Transtuzumab. Product Information Guide. http://www.herceptin.com/herceptin/patient/index.jsp (Accessed September 28, 2006).

Gomez-Lopera, S. A., R. C. Plaza and A. V. Delgado (2001). "Synthesis and characterization of spherical magnetite/biodegradable polymer composite particles." *J Colloid Interface Sci*. 240(1): 40–7.

Gottesman, M. M., Fojo, T. and Bates, S. E. (2002). "Multidrug resistance in cancer: role of ATP-dependent transporters." *Nat Rev Cancer*. 2: 48–58.

Gudgin Dickson, E. F., R. L. Goyan and R. H. Pottier (2002). "New directions in photodynamic therapy." *Cell Mol Biol (Noisy-le-grand)*. 48(8): 939–54.

Gulyaev, A. E., S. E. Gelperina, I. N. Skidan, et al. (1999). "Significant transport of doxorubicin into the brain with polysorbate 80-coated nanoparticles." *Pharm Res*. 16(10): 1564–9.

Gupta, B. and V. P. Torchilin (2006). "Transactivating transcriptional activator-mediated drug delivery." *Expert Opin Drug Deliv*. 3(2): 177–90.

Gupta, B., T. S. Levchenko and V. P. Torchilin (2005). "Intracellular delivery of large molecules and small particles by cell-penetrating proteins and peptides." *Adv Drug Deliv Rev*. 57(4): 637–51.

Hainfeld, J. F., D. N. Slatkin and H. M. Smilowitz (2004). "The use of gold nanoparticles to enhance radiotherapy in mice." *Phys Med Biol*. 49(18): N309–N315.

Hall, C. S., J. N. Marsh, M. J. Scott, et al. (2000). "Time evolution of enhanced ultrasonic reflection using a fibrin-targeted nanoparticulate contrast agent." *J Acoust Soc Am*. 108(6): 3049–57.

Hall, C. S., J. N. Marsh, M. J. Scott, et al. (2001). "Temperature dependence of ultrasonic enhancement with a site-targeted contrast agent." *J Acoust Soc Am*. 110(3, Pt 1): 1677–84.

Harisinghani, M. G., J. Barentsz, P. F. Hahn, et al. (2003). "Noninvasive detection of clinically occult lymph-node metastases in prostate cancer." *N Engl J Med*. 348(25): 2491–9.

Harris, A. L. and D. Hochhauser (1992). Mechanisms of multidrug resistance in cancer treatment. *Acta Oncol*. 31(2): 205–13.

Hayes, M. E., D. C. Drummond, K. Hong, et al. (2006). "Increased target specificity of anti-HER2 genospheres by modification of surface charge and degree of PEGylation." *Mol Pharm*. 3(6): 726–36.

Helm, C. W., C. R. Toler, R. S. Martin, III, et al. (2007). "Cytoreduction and intraperitoneal heated chemotherapy for the treatment of endometrial carcinoma recurrent within the peritoneal cavity." *Int J Gynecol Cancer*. 17(1): 204–9.

Henderson, B. W. and S. O. Gollnick (2003). Mechanistic principles of photodynamic therapy. Boca Raton, CRC Press.

Hildebrandt, B., P. Wust, O. Ahlers, et al. (2002). "The cellular and molecular basis of hyperthermia." *Crit Rev Oncol Hematol*. 43(1): 33–56.

Hines-Peralta, A., V. Sukhatme, M. Regan, et al. (2006). "Improved tumor destruction with arsenic trioxide and radiofrequency ablation in three animal models." *Radiology*. 240(1): 82–9.

Hirsch, L. R., R. J. Stafford, J. A. Bankson, et al. (2003). "Nanoshell-mediated near-infrared thermal therapy of tumors under magnetic resonance guidance." *PNAS*. 100(23): 13549–54.

Hopper, C. (2000). "Photodynamic therapy: a clinical reality in the treatment of cancer." *Lancet Oncol.* 1: 212–19.

Huwyler, J. and W. M. Pardridge (1998). "Examination of blood–brain barrier transferrin receptor by confocal fluorescent microscopy of unfixed isolated rat brain capillaries." *J Neurochem.* 70(2): 883–6.

Huwyler, J., A. Cerletti, G. Fricker, et al. (2002). "By-passing of *p*-glycoprotein using immunoliposomes." *J Drug Target.* 10(1): 73–9.

Iinuma, S., K. T. Schomacker, G. Wagnieres, et al. (1999). "In vivo fluence rate and fractionation effects on tumor response and photobleaching: photodynamic therapy with two photosensitizers in an orthotopic rat tumor model." *Cancer Res.* 59(24): 6164–70.

Illum, L., L. O. Jacobsen, R. H. Muller, et al. (1987). "Surface characteristics and the interaction of colloidal particles with mouse peritoneal macrophages." *Biomaterials.* 8(2): 113–17.

Isomoto, H., A. Ohtsuru, V. Braiden, et al. (2006). "Heat-directed suicide gene therapy mediated by heat shock protein promoter for gastric cancer." *Oncol Rep.* 15(3): 629–35.

Jain, S., V. Mishra, P. Singh, et al. (2003). "RGD-anchored magnetic liposomes for monocytes/neutrophils-mediated brain targeting." *Int J Pharm.* 261(1–2): 43–55.

Jang, S. H., M. G. Wientjes, D. Lu, et al. (2003). "Drug delivery and transport to solid tumors." *Pharm Res.* 20(9): 1337–50.

Jeon, S. I., J. H. L. Andrade and P. G. de Gennes (1991). "Protein-surface interactions in the presence of polyethylene oxide: Simplified theory." *J Colloid Interface Sci.* 142: 149–58.

Jeong, Y. I., S. J. Seo, I. K. Park, et al. (2005). "Cellular recognition of paclitaxel-loaded polymeric nanoparticles composed of poly(γ-benzyl L-glutamate) and poly(ethylene glycol) diblock copolymer endcapped with galactose moiety." *Int J Pharm.* 296(1–2): 151–61.

Jiang, C., N. Koyabu, Y. Yonemitsu, et al. (2003). "In vivo delivery of glial cell-derived neurotrophic factor across the blood–brain barrier by gene transfer into brain capillary endothelial cells." *Hum Gene Ther.* 14(12): 1181–91.

Johannsen, M., U. Gneveckow, L. Eckelt, et al. (2005). "Clinical hyperthermia of prostate cancer using magnetic nanoparticles: presentation of a new interstitial technique." *Int J Hyperthermia.* 21(7): 637–47.

Kakinuma, K., R. Tanaka, H. Takahashi, et al. (1996). "Drug delivery to the brain using thermosensitive liposome and local hyperthermia." *Int J Hyperthermia.* 12(1): 157–65.

Kawashita, M., K. Sadaoka, T. Kokubo, et al. (2006). "Enzymatic preparation of hollow magnetite microspheres for hyperthermic treatment of cancer." *J Mater Sci Mater Med.* 17(7): 605–10.

Kim, S. H., D. W. Kim, Y. H. Shim, et al. (2001). "In vivo evaluation of polymeric micellar paclitaxel formulation: toxicity and efficacy." *J Contr Release.* 72(1–3): 191–202.

Kim, S. H., J. H. Jeong, K. W. Chun, et al. (2005). "Target-specific cellular uptake of PLGA nanoparticles coated with poly(L-lysine)–poly(ethylene glycol)-folate conjugate." *Langmuir.* 21(19): 8852–7.

Kleiter, M. M., D. Yu, L. A. Mohammadian, et al. (2006). A tracer dose of technetium-99m-labeled liposomes can estimate the effect of hyperthermia on intratumoral doxil extravasation. *Clinical cancer research.* 12(22): 6800–7.

Kobayashi, H., S. Kawamoto, M. Bernardo, et al. (2006). "Delivery of gadolinium-labeled nanoparticles to the sentinel lymph node: Comparison of the sentinel node visualization and estimations of intra-nodal gadolinium concentration by the magnetic resonance imaging." *J Contr Release.* 111(3): 343–51.

Kohler, N., C. Sun, J. Wang, et al. (2005). "Methotrexate-modified superparamagnetic nanoparticles and their intracellular uptake into human cancer cells." *Langmuir.* 21: 8858–64.

Kommareddy, S. and M. Amiji (2007). "Antiangiogenic gene therapy with systemically administered sFlt-1 plasmid DNA in engineered gelatin-based nanovectors." *Cancer Gene Ther.* 14(5): 488–98.

Kong, G., R. D. Braun and M. W. Dewhirst (2000). "Hyperthermia enables tumor-specific nanoparticle delivery: effect of particle size." *Cancer Res.* 60(16): 4440–5.

Kong, G., R. D. Braun and M. W. Dewhirst (2001). "Characterization of the effect of hyperthermia on nanoparticle extravasation from tumor vasculature." *Cancer Res.* 61(7): 3027–32.

Koudinova, N. V., J. H. Pinthus, A. Brandis, et al. (2003). "Photodynamic therapy with Pd-Bacteriopheophorbide (TOOKAD): successful in vivo treatment of human prostatic small cell carcinoma xenografts." *Int J Cancer*. 104(6): 782–9.

Koziara, J. M., P. R. Lockman, D. D. Allen, et al. (2004). "Paclitaxel nanoparticles for the potential treatment of brain tumors." *J Contr Release*. 99(2): 259–69.

Kreuter, J. (1994). "Drug targeting with nanoparticles." *Eur J Drug Metab Pharmacokinet*. 19(3): 253–6.

Kreuter, J. (2001). "Nanoparticulate systems for brain delivery of drugs." *Adv Drug Deliv Rev*. 47(1): 65–81.

Kreuter, J. (2004). "Influence of the surface properties on nanoparticle-mediated transport of drugs to the brain." *J Nanosci Nanotechnol*. 4(5): 484–8.

Kreuter, J., R. N. Alyautdin, D. A. Kharkevich, et al. (1995). "Passage of peptides through the blood–brain barrier with colloidal polymer particles (nanoparticles)." *Brain Res*. 674(1): 171–4.

Kreuter, J., P. Ramge, V. Petrov, et al. (2003). "Direct evidence that polysorbate-80-coated poly(butylcyanoacrylate) nanoparticles deliver drugs to the CNS via specific mechanisms requiring prior binding of drug to the nanoparticles." *Pharm Res*. 20(3): 409–16.

Lammers, T., P. Peschke, R. Kuhnlein, et al. (2007). "Effect of radiotherapy and hyperthermia on the tumor accumulation of HPMA copolymer-based drug delivery systems." *J Contr Release*. 117(3): 333–41.

Lanza, G. M. and S. A. Wickline (2001). "Targeted ultrasonic contrast agents for molecular imaging and therapy." *Prog Cardiovasc Dis*. 44(1): 13–31.

Lanza, G. M. and S. A. Wickline (2003). "Targeted ultrasonic contrast agents for molecular imaging and therapy." *Curr Probl Cardiol*. 28(12): 625–53.

Lanza, G. M., D. R. Abendschein, X. Yu, et al. (2002). "Molecular imaging and targeted drug delivery with a novel, ligand-directed paramagnetic nanoparticle technology." *Acad Radiol*. 9 Suppl 2: S330–1.

Lanza, G. M., P. M. Winter, S. D. Caruthers, et al. (2004). "Magnetic resonance molecular imaging with nanoparticles." *J Nucl Cardiol*. 11(6): 733–43.

Larina, I. V., B. M. Evers, T. V. Ashitkov, et al. (2005). "Enhancement of drug delivery in tumors by using interaction of nanoparticles with ultrasound radiation." *Technol Cancer Res Treat*. 4(2): 217–26.

Li, J., X. Wang, C. Wang, et al. (2007). "The enhancement effect of gold nanoparticles in drug delivery and as biomarkers of drug-resistant cancer cells." *ChemMedChem*. 2(3): 374–8.

Liu, W., M. R. Dreher, D. Y. Furgeson, et al. (2006). "Tumor accumulation, degradation and pharmacokinetics of elastin-like polypeptides in nude mice." *J Contr Release*. 116(2): 170–8.

Lockman, P. R., M. O. Oyewumi, J. M. Koziara, et al. (2003). "Brain uptake of thiamine-coated nanoparticles." *J Contr Release*. 93(3): 271–82.

Loo, C., A. Lin, L. Hirsch, et al. (2004). "Nanoshell-enabled photonics-based imaging and therapy of cancer." *Technol Cancer Res Treat*. 3(1): 33–40.

Loo, C., A. Lowery, N. Halas, et al. (2005). "Immunotargeted nanoshells for integrated cancer imaging and therapy." *Nano Lett*. 5(4): 709–11.

Lukyanov, A. N., T. A. Elbayoumi, A. R. Chakilam and V. P. Torchilin (2004). Tumor-targeted liposomes: doxorubicin-loaded long-circulating liposomes modified with anti-cancer antibody. *J Control Release*. 100(1): 135–44.

Lum, A. F., M. A. Borden, P. A. Dayton, et al. (2006). "Ultrasound radiation force enables targeted deposition of model drug carriers loaded on microbubbles." *J Contr Release*. 111(1–2): 128–34.

Maeda, H. (2001). "The enhanced permeability and retention (EPR) effect in tumor vasculature: the key role of tumor-selective macromolecular drug targeting." *Adv Enzyme Regul*. 41: 189–207.

Maier-Hauff, K., R. Rothe, R. Scholz, et al. (2007). "Intracranial thermotherapy using magnetic nanoparticles combined with external beam radiotherapy: results of a feasibility study on patients with glioblastoma multiforme." *J Neurooncol*. 81(1): 53–60.

Marsh, J. N., M. S. Hughes, C. S. Hall, et al. (1998). "Frequency and concentration dependence of the backscatter coefficient of the ultrasound contrast agent Albunex (R)." *J Acoust Soc Am.* 104: 1654–66.

Marsh, J. N., C. S. Hall, M. J. Scott, et al. (2002a). "Improvements in the ultrasonic contrast of targeted perfluorocarbon nanoparticles using an acoustic transmission line model." *IEEE Trans Ultrason Ferroelectr Freq Contr.* 49(1): 29–38.

Marsh, J. N., C. S. Hall, S. A. Wickline, et al. (2002b). "Temperature dependence of acoustic impedance for specific fluorocarbon liquids." *J Acoust Soc Am.* 112(6): 2858–62.

Matsumura, Y. and H. Maeda (1986). "A new concept for macromolecular therapeutics in cancer chemotherapy: Mechanism of tumoritropic accumulation of proteins and the antitumor agent SMANCS." *Canc Res.* 46: 6387–92.

Matsuo, H., T. Okamura, J. Chen, et al. (2000). "Efficient introduction of macromolecules and oligonucleotides into brain capillary endothelial cells using HVJ-liposomes." *J Drug Target.* 8(4): 207–16.

Medintz, I. L., H. T. Uyeda, E. R. Goldman, et al. (2005). "Quantum dot bioconjugates for imaging, labelling and sensing." *Nat Mater.* 4(6): 435–46.

Meyer, D. E., B. C. Shin, G. A. Kong, et al. (2001). Drug targeting using thermally responsive polymers and local hyperthermia. *Journal of controlled release.* 74(1-3): 213–24

Morawski, A. M., P. M. Winter, K. C. Crowder, et al. (2004). "Targeted nanoparticles for quantitative imaging of sparse molecular epitopes with MRI." *Magn Reson Med.* 51(3): 480–6.

Morel, S., E. Terreno, E. Ugazio, et al. (1998). "NMR relaxometric investigations of solid lipid nanoparticles (SLN) containing gadolinium(III) complexes." *Eur J Pharm Biopharm.* 45(2): 157–63.

Mukherjee, P., R. Bhattacharya, P. Wang, et al. (2005). "Antiangiogenic properties of gold nanoparticles." *Clin Cancer Res.* 11(9): 3530–4.

Murray, C. B., D. J. Norris and M. G. Bawendi (1993). "Synthesis and characterization of nearly monodisperse CdE (E = sulfur, selenium, tellurium) semiconductor nanocrystallites." *J Am Chem Soc.* 115(19): 8706–15.

Nasongkla, N., E. Bey, J. Ren, et al. (2006). "Multifunctional polymeric micelles as cancer-targeted, MRI-ultrasensitive drug delivery systems." *Nano Lett.* 6(11): 2427–30.

National Cancer Institute (2004). Hyperthermia in cancer treatment: questions and answers (FS 7.3). Accessed on February 27, 2007 from http://www.cancer.gov/cancertopics/factsheet/Therapy/hyperthermia

National Cancer Institute (2005). "Hyperthermia in cancer treatment: questions and answers (FS 7.3)." http://www.cancer.gov/PDF/FactSheet/fs7_3.pdf (Accessed April 3, 2007).

Nielsen, U. B., D. B. Kirpotin, E. M. Pickering, et al. (2002). "Therapeutic efficacy of anti-ErbB2 immunoliposomes targeted by a phage antibody selected for cellular endocytosis." *Biochem Biophys Acta.* 1591(1–3): 109–18.

Ningaraj, N. S. (2006). Drug delivery to brain tumours: challenges and progress. *Expert opinion on drug delivery.* 3(4): 499–509.

Nsereko, S. and M. Amiji (2002). "Localized delivery of paclitaxel in solid tumors from biodegradable chitin microparticle formulations." *Biomaterials.* 23(13): 2723–31.

Ojeda, R., J. L. de Paz, A. G. Barrientos, et al. (2007). "Preparation of multifunctional glyconanoparticles as a platform for potential carbohydrate-based anticancer vaccines." *Carbohydr Res.* 342(3–4): 448–59.

Olbrich, C., A. Gessner, O. Kayser, et al. (2002). "Lipid–drug-conjugate (LDC) nanoparticles as novel carrier system for the hydrophilic antitrypanosomal drug diminazenediaceturate." *J Drug Target.* 10(5): 387–96.

Oldenburg, S. J., J. B. Jackson, S. L. Westcott, et al. (1999). "Infrared extinction properties of gold nanoshells." *Appl Phys Lett.* 75(19): 2897–9.

Olivier, J. C., L. Fenart, R. Chauvet, et al. (1999). "Indirect evidence that drug brain targeting using polysorbate 80-coated polybutylcyanoacrylate nanoparticles is related to toxicity." *Pharm Res.* 16(12): 1836–42.

Oupicky, D., K. A. Howard, C. Konak, et al. (2000). "Steric stabilization of poly-L-lysine/DNA complexes by the covalent attachment of semitelechelic poly(*N*-(2-hydroxypropyl)methacryla mide)." *Bioconjugate Chem.* 11(4): 492–501.

Oyewumi, M. O. and R. J. Mumper (2002). "Engineering tumor-targeted gadolinium hexanedione nanoparticles for potential application in neutron capture therapy." *Bioconjugate Chem.* 13(6): 1328–35.

Oyewumi, M. O., S. Liu, J. A. Moscow, et al. (2003). "Specific association of thiamine-coated gadolinium nanoparticles with human breast cancer cells expressing thiamine transporters." *Bioconjugate Chem.* 14(2): 404–11.

Oyewumi, M. O., R. A. Yokel, M. Jay, et al. (2004). "Comparison of cell uptake, biodistribution and tumor retention of folate-coated and PEG-coated gadolinium nanoparticles in tumor-bearing mice." *J Contr Release.* 95(3): 613–26.

Paciotti, G. F., L. Myer, D. Weinreich, et al. (2004). "Colloidal gold: a novel nanoparticle vector for tumor directed drug delivery." *Drug Deliv.* 11(3): 169–83.

Pamujula, S., R. A. Graves, T. Freeman, et al. (2004). Oral delivery of spray dried PLGA/amifostine nanoparticles. *The Journal of Pharmacy and Pharmacology.* 56(9): 1119–25.

Panyam, J. and V. Labhasetwar (2003). "Biodegradable nanoparticles for drug and gene delivery to cells and tissue." *Adv Drug Deliv Rev.* 55(3): 329–47.

Pedley, R. B., S. A. Hill, G. M. Boxer, et al. (2001). "Eradication of colorectal xenografts by combined radioimmunotherapy and combretastatin a-4 3-O-phosphate." *Cancer Res.* 61(12): 4716–22.

Peira, E., P. Marzola, V. Podio, et al. (2003). "In vitro and in vivo study of solid lipid nanoparticles loaded with superparamagnetic iron oxide." *J Drug Target.* 11(1): 19–24.

Pelz, J. O., J. Doerfer, W. Hohenberger, et al. (2005). "A new survival model for hyperthermic intraperitoneal chemotherapy (HIPEC) in tumor-bearing rats in the treatment of peritoneal carcinomatosis." *BMC Canc.* 5(1): 56.

Ponce, A. M., B. L. Viglianti, D. Yu, et al. (2007). Magnetic resonance imaging of temperature-sensitive liposome release: drug dose painting and antitumor effects. *Journal of the National Cancer Institute.* 99(1): 53–63.

Qing, S. H., L. Y. Li, X. H. Sheng, et al. (2006). "Photosensitizer nanoparticles photodynamic therapy on LOVO human colon cancer xenografts in athymic mice." *Zhonghua Wei Chang Wai Ke Za Zhi.* 9(6): 530–3.

Ramge, P., J. Kreuter and B. Lemmer (1999). "Circadian phase-dependent antinociceptive reaction in mice determined by the hot-plate test and the tail-flick test after intravenous injection of dalargin-loaded nanoparticles." *Chronobiol Int.* 16(6): 767–77.

Ranney, D., P. Antich, E. Dadey, et al. (2005). "Dermatan carriers for neovascular transport targeting, deep tumor penetration and improved therapy." *J Contr Release.* 109(1–3): 222–35.

Reddy, G. R., M. S. Bhojani, P. McConville, et al. (2006). "Vascular targeted nanoparticles for imaging and treatment of brain tumors." *Clin Cancer Res.* 12(22): 6677–86.

Reichardt, W., D. Hu-Lowe, D. Torres, et al. (2005). "Imaging of VEGF receptor kinase inhibitorinduced antiangiogenic effects in Drug-Resistant Human Adenocarcinoma Model." *Neoplasia.* 7: 847–53.

Ricci-Junior, E. and J. M. Marchetti (2006). "Preparation, characterization, photocytotoxicity assay of PLGA nanoparticles containing zinc (II) phthalocyanine for photodynamic therapy use." *J Microencapsul.* 23(5): 523–38.

Rychak, J. J., A. L. Klibanov and J. A. Hossack (2005). "Acoustic radiation force enhances targeted delivery of ultrasound contrast microbubbles: in vitro verification." *IEEE Trans Ultrason Ferroelectr Freq Contr.* 52(3): 421–33.

Sadeque, A. J., C. Wandel, H. He, et al. (2000). "Increased drug delivery to the brain by *p*-glycoprotein inhibition." *Clin Pharmacol Ther.* 68(3): 231–7.

Samia, A. C. S., S. Dayal and C. Burda (2006). Quantum Dot-based Energy Transfer: Perspectives and Potential for Applications in Photodynamic Therapy. *Photochemistry and Photobiology.* 82(3): 617–625.

Sauer, I., I. R. Dunay, K. Weisgraber, et al. (2005). "An apolipoprotein E-derived peptide mediates uptake of sterically stabilized liposomes into brain capillary endothelial cells." *Biochemistry*. 44(6): 2021–9.

Savolainen, J., J. E. Edwards, M. E. Morgan, et al. (2002). "Effects of a *p*-glycoprotein inhibitor on brain and plasma concentrations of anti-human immunodeficiency virus drugs administered in combination in rats." *Drug Metabol Dispos*. 30(5): 479–82.

Schiffelers, R. M., A. Ansari, J. Xu, et al. (2004). "Cancer siRNA therapy by tumor selective delivery with ligand-targeted sterically stabilized nanoparticle." *Nucleic Acids Res*. 32(19): e149.

Schmidt-Erfurth, U., T. Hasan, E. Gragoudas, et al. (1994). "Vascular targeting in photodynamic occlusion of subretinal vessels." *Ophthalmology*. 101(12): 1953–61.

Schmieder, A. H., P. M. Winter, S. D. Caruthers, et al. (2005). "Molecular MR imaging of melanoma angiogenesis with $\alpha_v\beta_3$-targeted paramagnetic nanoparticles." *Magn Reson Med*. 53(3): 621–7.

Schroeder, U. and B. A. Sabel (1996). "Nanoparticles, a drug carrier system to pass the blood–brain barrier, permit central analgesic effects of i.v. dalargin injections." *Brain Res*. 710(1–2): 121–4.

Schroeder, U., P. Sommerfeld and B. A. Sabel (1998). "Efficacy of oral dalargin-loaded nanoparticle delivery across the blood–brain barrier." *Peptides*. 19(4): 777–80.

Schwarze, S. R. and S. F. Dowdy (2000). "In vivo protein transduction: intracellular delivery of biologically active proteins, compounds and DNA." *Trends Pharmacol Sci*. 21(2): 45–8.

Sekhar, K. R., V. N. Sonar, V. Muthusamy, et al. (2007). Novel chemical enhancers of heat shock increase thermal radiosensitization through a mitotic catastrophe pathway. *Cancer research*. 67(2): 695–701.

Sengupta, S., D. Eavarone, I. Capila, et al. (2005). "Temporal targeting of tumour cells and neo-vasculature with a nanoscale delivery system." *Nature*. 436(7050): 568–72.

Shenoy, D. B. and M. M. Amiji (2005). Poly(ethylene oxide)-modified poly(epsilon-caprolactone) nanoparticles for targeted delivery of tamox ifen in breast cancer. *International journal of pharmaceutics*. 293(1-2):261-70.

Shenoy, D., S. Little, R. Langer and M. Amiji (2005). Poly(ethylene oxide)-modified poly(beta-amino ester) nanoparticles as a pH-sensitive system for tumor-targeted delivery of hydrophobic drugs: part 2. In vivo distribution and tumor localization studies. *Pharmaceutical research*. 22(12): 2107–14.

Shi, N., Y. Zhang, C. Zhu, et al. (2001). "Brain-specific expression of an exogenous gene after i.v. administration." *Proc Natl Acad Sci U S A*. 98(22): 12754–9.

Shikata, F., H. Tokumitsu, H. Ichikawa, et al. (2002). "In vitro cellular accumulation of gadolinium incorporated into chitosan nanoparticles designed for neutron-capture therapy of cancer." *Eur J Pharm Biopharm*. 53(1): 57–63.

Shortencarier, M. J., P. A. Dayton, S. H. Bloch, et al. (2004). "A method for radiation-force localized drug delivery using gas-filled liposheres." *IEEE Trans Ultrason Ferroelectr Freq Contr*. 51(7): 822–31.

Siddiqui, F., C. Y. Li, S. M. Larue, et al. (2007). "A phase I trial of hyperthermia-induced interleukin-12 gene therapy in spontaneously arising feline soft tissue sarcomas." *Mol Cancer Ther*. 6(1): 380–9.

Simberg, D., T. Duza, J. H. Park, et al. (2007). "Biomimetic amplification of nanoparticle homing to tumors." *PNAS*. 104(3): 932–6.

Soma, E. C., C. Dubernet, D. Bentolila, et al. (2000). "Reversion of multidrug resistance by co-encapsulation of doxorubicin and cyclosporin A in polyalkylcyanoacrylate nanoparticles." *Biomaterials*. 21(1): 1–7.

Somogyi, G., L. Prokai and N. Bodor (1998). "Targeted drug delivery to the brain via phosphonate derivatives II. Anionic chemical delivery system for zidovudine (AZT)." *Int J Pharm*. 166: 27–35.

Stewart, F., P. Baas and W. Star (1998). "What does photodynamic therapy have to offer radiation oncologists (or their cancer patients)?" *Radiother Oncol*. 48(3): 233–48.

Storm G., T. Daemen and D. D. Lasic (1995). "Surface modification of nanoparticles to oppose uptake by the mononuclear phagocyte system." *Adv Drug Deliv Rev*. 17: 31–48.

Sukhanova, A., J. Devy, L. Venteo, et al. (2004). "Biocompatible fluorescent nanocrystals for immunolabeling of membrane proteins and cells." *Anal Biochem.* 324(1): 60–7.

Sun, C., R. Sze and M. Zhang (2006). "Folic acid-PEG conjugated superparamagnetic nanoparticles for targeted cellular uptake and detection by MRI." *J Biomed Mater Res* 78(3): 550–7.

Szymanski-Exner, A., N. T. Stowe, R. S. Lazebnik, et al. (2002). "Noninvasive monitoring of local drug release in a rabbit radiofrequency (RF) ablation model using X-ray computed tomography." *J Contr Release.* 83(3): 415–25.

Taylor, E. M. (2002). "The impact of efflux transporters in the brain on the development of drugs for CNS disorders." *Clin Pharmacokinet.* 41(2): 81–92.

Thomas, H. and H. M. Coley (2003). Overcoming multidrug resistance in cancer: an update on the clinical strategy of inhibiting p-glycoprotein. *Cancer control.* 10(2): 159–65.

Tokes, Z. A., A. K. St Peteri and J. A. Todd (1980). "Availability of liposome content to the nervous system. Liposomes and the blood–brain barrier." *Brain Res.* 188(1): 282–6.

Tokumitsu, H., J. Hiratsuka, Y. Sakurai, T. Kobayashi, H. Ichikawa and Y. Fukumori (2000). "Gadolinium neutron-capture therapy using novel gadopentetic acid-chitosan complex nanoparticles: in vivo growth suppression of experimental melanoma solid tumor." *Canc Lett.* 150(2): 177–82.

Torchilin, V. P. (1996). "How do polymers prolong circulation time of liposomes?" *J Liposome Res.* 6: 99–116.

Torchilin, V. P., T. S. Levchenko, A. N. Lukyanov, et al. (2001). p-Nitrophenylcarbonyl-PEG-PE-liposomes: fast and simple attachment of specific ligands, including monoclonal antibodies, to distal ends of PEG chains via p-nitrophenylcarbonyl groups. *Biochim Biophys Acta.* 1511(2): 397–411.

Torchilin, V. P. (2002). "TAT peptide-modified liposomes for intracellular delivery of drugs and DNA." *Cell Mol Biol Lett.* 7(2): 265–7.

Torchilin, V. P. and T. S. Levchenko (2003). "TAT-liposomes: a novel intracellular drug carrier." *Curr Protein Pept Sci.* 4(2): 133–40.

Tran, J., Z. Master, J. L. Yu, et al. (2002). "A role for survivin in chemoresistance of endothelial cells mediated by VEGF." *PNAS.* 99: 4349–54.

Tusji, A., Ed. (2000). The blood–brain barrier and drug delivery to the CNS. New York, Marcel Dekker.

Uehara, M., T. Inokuchi, K. Sano, et al. (1998). "The anti-tumor effect of photodynamic therapy evaluated by bromodeoxyuridine immunohistochemistry." *Int J Oral Maxillofac Surg.* 27(3): 204–8.

U.S. Food and Drug Administration, "Center for Drug Evaluation and Research." http://www.accessdata.fda.gov/scripts/cder/drugsatfda/index.cfm (Accessed April 25, 2006).

U.S. National Institute of Health, "Cancer Statistics" http://www.cancer.gov/statistics/ (Accessed September 15, 2006).

van der Zee, J. (2002). Heating the patient: a promising approach? *Ann Oncol.* 13:1173–84.

van Vlerken, L. E. and M. M. Amiji (2006). Multi-functional polymeric nanoparticles for tumour-targeted drug delivery. *Expert opinion on drug delivery.* 3(2): 205–16.

van Vlerken, L. E., Z. Duan, M. V. Seiden, et al. (2007). "Modulation of intracellular ceramide using polymeric nanoparticles to overcome multidrug resistance in cancer." *Canc Res.* 67(10): 4843–50.

Veenhuizen, R., H. Oppelaar, M. Ruevekamp, et al. (1997). "Does tumour uptake of Foscan determine PDT efficacy?" *Int J Cancer.* 73(2): 236–9.

Vinogradov, S. V. (2006). Colloidal microgels in drug delivery applications. *Current Pharmaceutical Design.* 2006;12(36): 4703–12.

Visaria, R. K., R. J. Griffin, B. W. Williams (2006). Enhancement of tumor thermal therapy using gold nanoparticle-assisted tumor necrosis factor-alpha delivery. *Molecular cancer therapeutics.* 5(4): 1014–20.

Voura, E. B., J. K. Jaiswal, H. Mattoussi, et al. (2004). "Tracking metastatic tumor cell extravasation with quantum dot nanocrystals and fluorescence emission-scanning microscopy." *Nat Med.* 10(9): 993–8.

Wachsberger, P., R. Burd and A. P. Dicker (2003). "Tumor response to ionizing radiation combined with antiangiogenesis or vascular targeting agents: exploring mechanisms of interaction." *Clin Cancer Res.* 9(6): 1957–71.

Wani, M. C., H. L. Taylor, M. E. Wall, et al. (1971). "Plant antitumor agents VI. The isolation and structure of Taxol, a novel antitumor and anitleukemic agent from *Taxus brevifolia.*" *J Am Chem Soc.* 18(3): 242–60.

Weinberg, B. D., E. Blanco, S. F. Lempka, et al. (2007). "Combined radiofrequency ablation and doxorubicin-eluting polymer implants for liver cancer treatment." *J Biomed Mater Res A.* 81(1): 205–13.

Weissig, V. 2005. Targeted drug delivery to mammalian mitochondria in living cells. *Expert Opin Drug Deliv.* 2(1): 89–102.

Weissig, V., S. V. Boddapati, S. M. Cheng and G. G. D'Souza (2006). Liposomes and liposome-like vesicles for drug and DNA delivery to mitochondria. *J Liposome Res.* 16(3): 249–64.

Wickline, S. A. and G. M. Lanza (2003). "Nanotechnology for molecular imaging and targeted therapy." *Circulation.* 107(8): 1092–5.

Winter, P. M., S. D. Caruthers, A. Kassner, et al. (2003a). "Molecular imaging of angiogenesis in nascent Vx-2 rabbit tumors using a novel $\alpha_v\beta_3$-targeted nanoparticle and 1.5 tesla magnetic resonance imaging." *Canc Res.* 63(18): 5838–43.

Winter, P. M., A. M. Morawski, S. D. Caruthers, et al. (2003b). "Molecular imaging of angiogenesis in early-stage atherosclerosis with $\alpha_v\beta_3$-integrin-targeted nanoparticles." *Circulation.* 108(18): 2270–4.

Wong, H. L., R. Bendayan, A. M. Rauth, et al. (2006). "Simultaneous delivery of doxorubicin and GG918 (Elacridar) by new polymer-lipid hybrid nanoparticles (PLN) for enhanced treatment of multidrug-resistant breast cancer." *J Contr Release.* 116(3): 275–84.

Wood, B. J., J. K. Locklin, A. Viswanathan, et al. (2007). Technologies for guidance of radiofrequency ablation in the multimodality interventional suite of the future. *Journal of vascular and interventional radiology.* 18(1 Pt 1): 9–24.

Wu, X., H. Liu, J, Liu, et al. (2003). "Immunofluorescent labeling of cancer marker Her2 and other cellular targets with semiconductor quantum dots." *Nat Biotechnol.* 21(1): 41–6.

Wust, P., B. Hildebrandt, G. Sreenivasa, et al. (2002). "Hyperthermia in combined treatment of cancer." *Lancet Oncol.* 3(8): 487–97.

Wust, P., U. Gneveckow, M. Johannsen, et al. (2006). "Magnetic nanoparticles for interstitial thermotherapy—feasibility, tolerance and achieved temperatures." *Int J Hyperther.* 22(8): 673–85.

Yang, S. C., L. F. Lu, Y. Cai, et al. (1999a). "Body distribution in mice of intravenously injected camptothecin solid lipid nanoparticles and targeting effect on brain." *J Contr Release.* 59(3): 299–307.

Yang, S., J. Zhu, Y. Lu, et al. (1999b). "Body distribution of camptothecin solid lipid nanoparticles after oral administration." *Pharmaceut Res.* 16(5): 751–7.

Zara, G. P., R. Cavalli, A. Bargoni, et al. (2002a). "Intravenous administration to rabbits of non-stealth and stealth doxorubicin-loaded solid lipid nanoparticles at increasing concentrations of stealth agent: pharmacokinetics and distribution of doxorubicin in brain and other tissues." *J Drug Target.* 10(4): 327–35.

Zara, G. P., A. Bargoni, R. Cavalli, et al. (2002b). "Pharmacokinetics and tissue distribution of idarubicin-loaded solid lipid nanoparticles after duodenal administration to rats." *J Pharmaceut Sci.* 91(5): 1324–33.

Zhang, X., J. Xie, S. Li, et al. (2003). "The study on brain targeting of the amphotericin B liposomes." *J Drug Target.* 11(2): 117–22.

Zhao, M., D. A. Beauregard, L. Loizou, et al. (2001). "Non-invasive detection of apoptosis using magnetic resonance imaging and a targeted contrast agent." *Natl Med.* 1: 1241–1244.

Zhao, S., M. Borden, S. H. Bloch, et al. (2004). "Radiation-force assisted targeting facilitates ultrasonic molecular imaging." *Mol Imag.* 3(3): 135–48.

Nanogels as Pharmaceutical Carriers

Alexander V. Kabanov and Serguei V. Vinogradov

1   Introduction

Nanogels are nanosized networks of chemically or physically cross-linked polymers that swell in a good solvent. The term "nanogel" (NanoGel™) was first introduced by us to define cross-linked bifunctional networks of a polyion and a nonionic polymer for delivery of polynucleotides (cross-linked polyethyleneimine (PEI) and poly(ethylene glycol) (PEG) or PEG-*cl*-PEI) (Lemieux et al., 2000; Vinogradov et al., 1999). However, some other studies also described nanoparticles of polymeric hydrogels. For example, work by Akiyoshi and Sunamoto proposed nanosized swollen aggregates of cholesterol-modified polysaccharide (pullulan) for delivery of insulin (Akiyoshi et al., 1998). Altogether, nanogels represent a novel family of nanoscale materials for delivery drugs, genes, and imaging agents. Publications using nanogels in pharmaceutics and nanomedicine have greatly increased after 2002 (Fig. 1), when the first review on this subject was published (Vinogradov et al., 2002). This demonstrates an increasing interest in nanogels by biomaterial and pharmaceutical scientists.

Nanogels are very promising in drug delivery applications due to their high loading capacity that is unique for pharmaceutical nanocarriers. Unloaded nanogels in a swollen state contain considerable amount of water. Loading of biological agents is often achieved through self-assembly mechanisms involving electrostatic, Van der Waals, and/or hydrophobic interactions between the agent and the polymer matrix. As a result, nanogels collapse forming stable nanoparticles, in which biological agent is entrapped. To stabilize nanogels in dispersion, water-soluble nonionic polymers, such as PEG, can be introduced in nanogel structure. Such polymers form protective soluble layer around a collapsed core, similar to a shell of polymeric micelles that prevents phase separation. Novel methods have been proposed for self-assembly and cross-linking of double hydrophilic block copolymers allowing strict control over special distribution of nanogel chains between their internal and external layers. Nanogel surface groups have been modified with ligands to enable receptor-mediated delivery to target cells. Various nanogels have been shown to deliver their payload inside cells and cross biological barriers. Due to elevated stability inside cells, nanogels demonstrate good potential for enhancing oral and brain bioavailability of low molecular drugs and biomacromolecules.

V. Torchilin (ed.), *Multifunctional Pharmaceutical Nanocarriers*,   67
© Springer Science+Business Media, LLC 2008

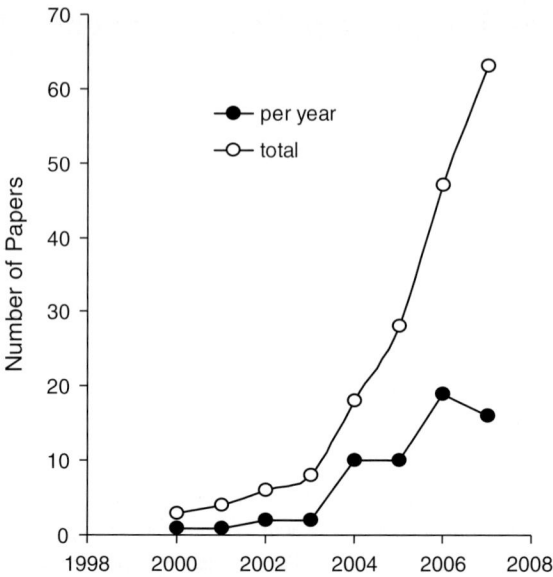

Fig. 1 Recent growth of nanogel publications based on Medline data (data as of July 2007)

2 Synthesis of Nanogels

Current approaches used for preparation of nanogels can be divided into (1) chemical synthesis by polymerization (copolymerization), (2) chemical cross-linking of polymeric chains, and (3) physical self-assembly of polymers. Polymerization or cross-linking is usually carried out in colloidal dispersions to ensure formation of nanoscale-sized species. In this case, cross-linking agents are often introduced to form nanogels. For example, Peppas et al. have been able to produce hydrogels composed of PEG grafted on poly(methacrylic acid) (PMAA), as a nanosphere suspension using a UV-initiated solution/precipitation polymerization method in water (Donini et al., 2002). DeSimone et al. have synthesized cationic nanogels by inverse microemulsion polymerization of 2-acryloxyethyltrimethylammonium chloride (AETMAC), 2-hydroxyethylacrylate (HEA), and poly(ethylene glycol)diacrylate (PEGdiA) in heptane (PAETMAC nanogel) (McAllister et al., 2002). This technique was in essence similar to inverse microemulsion copolymerization of acrylamide adopted for immobilization of enzymes in nanoparticles (Khmelnitsky et al., 1992). Labile bonds are also frequently introduced into hydrogels to make them (bio)degradable to facilitate drug delivery (Park et al., 2007). Frechet and coworkers reported inverse emulsion polymerization utilizing free radical polymerization to prepare degradable acrylamide-based nanogels containing acid-liable acetal cross-linkers for protein, antigen, and DNA delivery (Goh et al., 2004; Kwon et al., 2005b; Murthy et al., 2003). Matyjaszewski et al. used atom transfer radical polymerization (ATRP) in inverse miniemulsion for synthesis of stable cross-linked nanogels of water-soluble polymers (Oh et al., 2006). A disulfide-functionalized cross-linker was used in this work to synthesize cross-linked biodegradable nanogels. Matyjaszewski and Kataoka further extended this approach to synthesize biodegradable, cross-linked poly(oligo(ethylene oxide) monomethylether methacrylate) (POEOMA) nanogels (Oh et al., 2007a). Another study fabricated hyaluronic acid (HA) nanogels containing biodegradable disulfide linkages by inverse water-in-oil emulsion method (Lee et al., 2007).

Chemical cross-linking is a highly versatile method of creating polymeric hydrogels with large pore sizes (Hennink and van Nostrum, 2002). The cross-linking method was used to synthesize the PEG-*cl*-PEI nanogel for polynucleotide delivery (Vinogradov et al., 1999). In this case, a bis-activated PEG was conjugated to a branched PEI (25 kDa) in an oil-in-water emulsion, followed by evaporation of the solvent in vacuo and maturation of the nanogel in aqueous solution. Subsequently, PEI was cross-linked to bis-activated pluronic block copolymers in aqueous micellar solutions (Vinogradov et al., 2006). Reactions were carried out in the presence of pluronic micelles, which, presumably, resulted in the formation of nanogels with pluronic micelle core surrounded by PEI shell. Strict control of spatial distribution of polymer chains in nanogel was achieved by Bronich and coworkers (Bronich et al., 2005). This work developed a procedure, in which block ionomer complexes were initially prepared by self-assembly of ionic blocks of double hydrophilic block copolymers with an oppositely charged condensing agent, followed by chemical cross-linking of ionic blocks in the core and removal of the condensing agent. Nanogels prepared using this approach from PEG-*b*-PMAA diblock copolymers contained a hydrophilic PEG shell and a cross-linked hydrophilic PMA ionic core, which swelled in water and incorporated hydrophilic drugs (Bontha et al., 2006). Similar technique was also used to prepare core–shell nanogels by condensation and cross-linking of PEG-grafted poly(acrylic acid) (PEG-*g*-PAA) (Bronich et al., 2006).

Physical self-assembly of polymers to produce nanogels was used by several groups. These methods usually involve controlled aggregation of hydrophilic polymers containing side groups capable of hydrophobic or electrostatic interactions with each other and/or hydrogen bonding sometimes followed by covalent (e.g., disulfide) bond formation. The sizes of the resulting nanogels can be controlled by proper selection of the concentration of the polymers and environmental parameters of the system, such as pH, ionic strength, and temperature. For example, Yu et al. prepared protein nanogels by temperature-induced gelation of oppositely charged proteins, such as ovalbumin and lysozyme or ovotransferrin (Yu et al., 2006b). Similarly, nanogels were obtained by pH- and temperature-induced gelation of chitosan and ovalbumin (Yu et al., 2006a). A hybrid nanogel has also been developed based on interpenetrating networks of thermosensitive poly-(*N*-isopropylacrylamide) (PNIPAAm) gels and tailored nanoporous silica (Shin et al., 2001). Akiyoshi and Sunamoto prepared hydrogels by hydrophobic association of cholesterol-modified pullulan (Akiyoshi et al., 1998). Gref and colleagues described self-assembly of nanogels of various sizes by association of a lauryl-modified dextran and β-cyclodextrin polymer in aqueous media (Daoud-Mahammed et al., 2007).

3   Swelling of Nanogel Particles

In water, swelling of nanogels is controlled by several factors: the cross-linker concentration, charge concentration (for polyelectrolyte gels), and environmental parameters (such as pH, ionic strength, temperature). It is well recognized that a balance between the osmotic pressure and the polymer elasticity sets the physical dimensions of a hydrogel particle (Ricka and Tanaka, 1984). For poly-electrolyte hydrogels, osmotic pressure results from the net difference in concentration of mobile ions between the interior of the gel particle and exterior solution. The fixed charged groups attract hydrated counterions, which tend to expand the gel, while the conformational entropy elasticity of the cross-linked polymer chains opposes this expansion. Neutralizing the polyelectrolyte gel reduces the net ion concentration difference (osmotic swelling pressure). This reduction in the number of counterions results in dehydration and compression of the gel to an extent when the excluded volume of the polymer chains limits further compression. Hydrophobic interactions and hydrogen bond formation between hydrogel chains and solutes can further contribute to gel condensation and collapse.

For weak polyelectrolytes, variation of pH alters the number of charged groups and results in nanogel particle size change. For example, core–shell nanogels of cross-linked PEG-*b*-PMA swelled as pH increased from 5 to 9 due to ionization of carboxylic groups of PMA (Bontha et al., 2006). Conversely, the size of PEG-*cl*-PEI nanogel decreased as pH increased from ca. 8.5 to 10 due to deprotonation of PEI amino groups (Bronich et al., 2001). Despite relatively high concentration of fixed charged groups, these PEG-*cl*-PEI nanogels exhibited relatively little size variation upon changes in pH. This behavior was attributed to high degree of cross-linking of the PEG-*cl*-PEI network that limits the extent of its swelling. As a general rule, the swelling ratio decreases as the number of cross-links increases (Bontha et al., 2006; McAllister et al., 2002). Swelling of polyelectrolyte hydrogels also depends on ionic strength. For example, at high ionic strengths, the swelling of cationic PAETMAC nanogels was governed by the cross-linker concentration, while at low ionic strengths the swelling was influenced by both the cross-linker and charge concentration (McAllister et al., 2002). Temperature, in selected cases, can affect the quality of the solvent, and, therefore, the nanogel particle size. Temperature sensitivity of swelling was observed, for example, for nanogels of *N*-isopropylacrylamide (NIPAm) copolymers (Shin et al., 2002; Varga et al., 2006). Altogether, one of the advantages of highly dispersed hydrogels is that they usually respond very rapidly to changing environmental conditions (Eichenbaum et al., 1998), which facilitates incorporation and release of biological agents in pharmaceutical applications. In contrast, the swelling equilibrium for the macrogels requires periods in the order of days.

4 Loading with Biological Agents

Biological agents can be incorporated in nanogels by (1) covalent conjugation, (2) physical entrapment, or (3) self-assembly. Covalent conjugation of biological agents can be achieved using preformed nanogels or during nanogel synthesis. For example, enzymes modified with acrylic groups were copolymerized with acrylamide either in inverse microemulsion (Khmelnitsky et al., 1992) or dilute aqueous solutions (Yan et al., 2006, 2007) to obtain nanosized hydrogels. Physical entrapment was employed for incorporation of proteins in cholesterol-modified pullulan nanogels (Akiyoshi et al., 1998) and siRNA in HA nanogels (Lee et al., 2007). In addition, hydrophobic molecules can incorporate into nonpolar domains formed by hydrophobic chains present in selected nanogels. For example, prostaglandin E_2 was solubilized in nanogels of cholesterol-modified pullulan (Kato et al., 2007). In another study, N-hexylcarbamoyl-5-fluorouracil (HCFU) was noncovalently incorporated in cross-linked nanogels of N-isopropylacrylamide (NIPAAm) and N-vinylpyrrolidone (VP) copolymers (PNIPAAm/VP) (Soni et al., 2006). Doxorubicin was also loaded in amphiphilic cross-linked nanogels based on pluronic F127 (Missirlis et al., 2005) and POEOMA (Oh et al., 2007a). In most cases, loading achieved due to hydrophobic interaction of the drug molecules with the nanogel result in relatively low degrees of loading (not more than ca. 10%).

A different approach based on interaction of polyelectrolyte-based nanogels with oppositely charged solutes can produce nanoscale materials, with high content of biological agents. Like linear and block polyelectrolytes, polyelectrolyte-based nanogels were shown to interact with oppositely charged surfactants, synthetic polyions, polynucleotides, and proteins (Bronich et al., 2001; Ogawa et al., 2005, 2007; Vinogradov et al., 1999). These processes are characterized by high binding cooperativity and efficiency. For example, binding of an anionic surfactant, such as sodium tetradecyl sulfate, with cationic PEG-cl-PEI nanogels has an onset at a "critical association concentration" (cac), which is two orders of magnitude lower than critical micelle concentration (cmc) of this surfactant alone (Bronich et al., 2001). Decrease in cac compared to cmc is explained by cooperative stabilization of surfactant aggregates as a result of the surfactant headgroup electrostatic interaction with PEI chains of the nanogel. Charged and amphiphilic biologically active molecules such as sodium oleate (OA), indomethacin, and retinoic acid (RA) were also incorporated into PEG-cl-PEI nanogels (Bronich et al., 2001). For example, RA-loaded nanogels formed nanosized dispersions stable at physiological pH and ionic strength, which could be lyophilized, stored, and then redispersed. This suggested that a useful pharmaceutical formulation of RA can be prepared by its immobilization in nanogel system. In another study, cisplatin was loaded into a cross-linked polyion core of PEG-*b*-PMA nanogels through a reversible covalent attachment of cisplatin to PMA carboxylic groups (Bronich et al., 2005). Furthermore, hydrophobic regions of PEG-*cl*-PEI/surfactant complexes can serve as nonaqueous reservoirs for solubilizing water-insoluble molecules (Bronich et al., 2001). Hence, polyelectrolyte nanogels are a versatile platform technology for incorporation of various low molecular mass compounds via combinations of electrostatic and hydrophobic interactions as well as hydrogen bond formation.

One of the most important features of weakly cross-linked polyelectrolyte nanogels is the ability to incorporate biomacromolecules of opposite charge. Accommodation of biomacromolecules in hydrogels is usually hindered by effects of excluded volume. However, if biomacromolecules and polymer network have opposite charges, they effectively react with each other forming a polyelectrolyte complex. In the cases when the polyelectrolyte chains can penetrate nanogels, the process develops as a frontal reaction between oppositely charged polyions and spreads from the exterior of the gel to its core (Kabanov et al., 2004). As a result, efficient loading with biomacromolecules can be achieved even in the case of a bulk polyelectrolyte network (Fig. 2) (Kabanov et al., 2004; Oh et al., 2007b). This principle has been exploited to immobilize polynucleotides in cationic nanogels. For example, cationic nanogels or PEG-*cl*-PEI and PAETMAC were used for incorporation of single strand antisense oligonucleotides and DNA (McAllister et al., 2002; Vinogradov et al., 1999). Addition of phosphorothioate oligonucleotide to PEG-*cl*-PEI nanogel dispersion at physiological pH resulted in immediate formation of polyelectrolyte complexes between the oligonucleotide and PEI chains (Vinogradov et al., 1999). The nanogel loading capacity with respect to the oligonucleotide was 15–30% by weight. The oligonucleotide binding with nanogel was almost complete, which is important to minimize the loss of agents available in small quantities. Furthermore, anionic

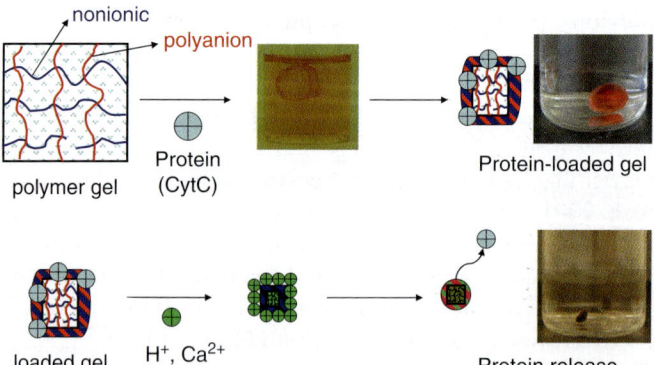

Fig. 2 Loading and release of cationic protein (cytochrome C) in cross-linked PEG and PAA hydrogels. Spontaneous loading is achieved due to polyion complex formation between protein and PAA chains. Acidification or addition of Ca^{2+} ions results in the protein release (Oh et al., 2007b)

compounds bearing only few charges, such as nucleoside analog 5'-triphosphates (NTPs) (5'-triphosphate derivatives of fludarabine, zidovudine, ribavirin, and cytarabine) were also efficiently loaded into nanogels of cross-linked PEI and PEG or PEI and Pluronic block copolymers (Kohli et al., 2007; Vinogradov et al., 2005). In these studies, the loading capacity of nanogels with respect to NTPs was reported to be over 30%. Altogether, approaches based on binding of polyelectrolyte nanogels with oppositely charged solutes are very convenient and efficient and can be used with many biological agents and preformed well-characterized nanogel carriers.

5 Release of Biological Agents

The biological agents can be released from the nanogels as a result of (1) diffusion, (2) nanogel degradation, or (3) displacement by ions present in the environment. Examples include diffusional release of doxorubicin from stable hydrogel nanoparticles based on pluronic block copolymer (Missirlis et al., 2006). This release mechanism is simple and has been successfully employed in various nanomedicines, such as polymeric micelles that have already reached a clinical stage (Kabanov and Alakhov, 2002). At the same time, there is an increased interest in developing nanogels that can release biological agents in response to environmental cues at the targeted site of action. For example, disulfide cross-linked POEOMA nanogels biodegraded into water-soluble polymers in the presence of a glutathione tripeptide, which is commonly found in cells (Oh et al., 2007a). The degradation of these nanogels was shown to trigger the release of encapsulated molecules including rhodamine 6G, a fluorescent dye, and Doxorubicin, an anticancer drug, as well as facilitate the removal of empty vehicles. In another study, a release of siRNA from disulfide cross-linked HA nanogels was facilitated by adding glutathione (GSH), an intracellular reductive agent that induced nanogel dissolution (Lee et al., 2007). Dispersed hydrogels based on acrylamide copolymers with acid-liable acetal cross-linker were shown to be stable at extracellular pH 7.4 but degraded releasing entrapped protein at acidic pH 5.0 (Murthy et al., 2003).

Polyelectrolyte hydrogels that incorporate biological agents via electrostatic bonds allow for release of biological agents in response to environmental changes. For instance, hydrogels of cross-linked PEG and PAA were shown to release an oppositely charged protein upon 1) addition of calcium ions that reacted with carboxylate groups of PAA and displaced the protein or 2) acidification of the media by decreasing pH from 7.4 to 5.5 (Fig. 2) (Oh et al., 2007b). A similar mechanism was proposed for release of oligonucleotides from PEG-cl-PEI nanogels (Vinogradov et al., 1999). In this case, electrostatically bound oligonucleotides are believed to be displaced by negatively charged cellular components. Cell membrane-triggered release of negatively charged drugs from complexes with cationic nanogels was also proposed to explain cellular accumulation of an NTPs drug delivered with nanogels (Vinogradov, 2006).

6   Chemical Modification of Nanogels for Targeted Delivery

Nanocarriers can be delivered to disease-affected sites after injection in the blood stream. Major impediments to this delivery strategy include (1) interaction of nano-carriers with serum proteins resulting in opsonization or agglutination, (2) clearance of nanocarriers by reticuloendothelial system (RES), and (3) clearance of nanocarri-ers through kidney glomerules. To reduce interaction with serum proteins, extend circulation time and decrease renal clearance nanocarrier surface is often modified with hydrophilic inert polymers, such as PEG (Francis et al., 1996). For example, PEG-*cl*-PEI nanogels can be additionally modified by PEG chains grafted to its sur-face (Lemieux et al., 2000; Vinogradov et al., 1999). Similarly, PEG chains can be tethered to polymetacrylate nanogels during the emulsion polymerization procedure (Hayashi et al., 2005). Nanogels with cross-linked polyion cores and PEG corona were also prepared by condensation of block ionomers into polyion complex micelles and subsequent cross-linking of the core (Bronich et al., 2005).

Targeting ligands can be also attached to nanogels to enhance their site-specific delivery in the body. For example, we described biotinylated PEG-*cl*-PEI nanogels that were vectorized via (strepta)avidin by biotinylated ligands (transferrin or insulin) (Vinogradov et al., 2004). Biotin groups were also attached to OH-functionalized POEOMA nanogels (Oh et al., 2007a). However, biotin-(strepta)avidin conjugation of targeting groups is not practical for in vivo delivery because of biological activity of biotin. Hence direct conjugation of nanogels with targeting groups is being devel-oped. For example, 1–5% of primary amino groups in PEG-*cl*-PEI nanogels were modified with folic acid using 1-ethyl-3-(3-dimethylaminopropyl)carbodiimide (EDC) in aqueous media (Vinogradov et al., 2005). Polymetacrylate microgels modified with folate demonstrated increased and selective cellular uptake in cancer cell lines overexpressing folate receptors (Nayak et al., 2004). To reduce problems associated with conjugated folate availability for corresponding cellular receptors, several authors recommended insertion of a polymer linker, e.g., PEG between the folate moiety and drug carrier (Shiokawa et al., 2005). Nanogels were also conju-gated with human transferrin (hTf), a tumor-specific receptor target protein (Vinogradov, 2006). In this method, amino groups in hTf were first reacted with a bifunctional reagent, sulfosuccinimidyl 4-(*N*-meleimidomethyl)-cyclohexane-1-car-boxylate (SMCC), to obtain a maleimide derivative of hTf. Second, thiol groups were introduced into PEG-*cl*-PEI nanogels by reaction with 2-iminothiolane (Trout's reagent). Finally, reaction between maleimide-hTf and thiol-nanogels led to the for-mation of hTf-nanogels with 4–12 hTf molecules per particle. Because of the protein size, most conjugated hTf molecules were located at nanogel surface, which facili-tated easy access to cellular transferrin receptors. Additionally, proteins and peptide ligands were also attached to nanogels through a disulfide bridge using a PEG linker (Vinogradov, 2006). First, a mono-*N*-acetylcystamine-PEG linker was prepared, activated by 1,1'-carbonyldiimidazole, and conjugated to a PEG-*cl*-PEI nanogel. Next, amino groups of proteins or peptides were reacted with *N*-succinimidyl 3-(2-pyridyldithio)-propionate (SPDP), and 2,2'-dipyridyldisulfide to form thiol-specific derivatives. Finally, nanogels were treated with dithiothreitol (DTT), followed by the thiol-specific derivatives to yield the modified nanogels. Several short homing peptides and monoclonal antibodies to prostate-specific membrane antigen (PSMA) were also conjugated to nanogels using this synthetic approach.

7 Delivery of Small Biological Agents Using Nanogels

Significant progress was made in the application of nanogels as for delivery of small biologically active molecules. We encapsulated RA into PEG-*cl*-PEI nanogels (Bronich et al., 2001). This formulation is of interest for application in complex drug therapies since RA is known to provide an anticancer effect (Soprano et al., 2004). More recently, a similar formulation of valproic acid (VA) in PEG-*cl*-PEI nanogels was prepared and studied in a cellular model of the blood–brain barrier (BBB), bovine brain microvessel endothelial cells (BBMEC) monolayers (Vinogradov et al., 2004). At least 70% increase in the transcellular transport of VA in nanogel across BBMEC monolayers was observed compared to a free drug, suggesting that this nanogel formulation may be useful for drug delivery to the brain. In another study, HCFU, a prodrug of 5-fluorouracil (5-FU), has been encapsulated into PNIPAAm/VP nanogels (Soni et al., 2006). After coating with polysorbate 80, these nanogels were shown to target brain tissue across BBB in rabbits. An antileishmaniasis drug, arjunglucoside I, was also incorporated into PNIPAAm/VP nanogels (Tyagi et al., 2005). This formulation showed enhanced therapeutic efficacy against parasites compared to free drug and a similar activity compared to poly(lactic acid) (PLA) nanoparticles. Both nanogel and PLA nanoparticles were effective in reducing hepatotoxicity and nephrotoxicity of the drug.

Another application of nanogels involves delivery of nucleoside analogs as anticancer agents. These agents usually undergo complex biochemical transformations in a cell including (1) phosphorylation into nucleoside 5'-phosphates by intracellular nucleoside kinases (Hatse et al., 1999), (2) formation of nucleoside 5'-diphosphates, (3) conversion of ribonucleotides into deoxyribonucleotides by nucleoside reductases, and (4) synthesis of nucleoside analog NTPs. The latter are actual active molecules, which arrest DNA replication and transcription in cancer cells (Galmarini et al., 2002). Previously, many prospective nucleoside analogs were discarded in earlier preclinical studies, or withdrawn later from clinical studies, because their intracellular conversion into NTPs was insufficient. However, using PEG-*cl*-PEI nanogels as carriers, it became possible to directly deliver NTPs into cancer cells (Kohli et al., 2007; Vinogradov et al., 2005, 2006). For example, fludarabine in folate-conjugated nanogels demonstrate greatly increased uptake and cytotoxicity in several cancer cell lines as well as increased transport across intestinal cells, Caco-2, compared to the free drug (Vinogradov et al., 2005). Furthermore, antiviral NTPs, such as 5'-triphosphorylated ribavirin were evaluated in cationic nanogel formulations and showed increased activity in MDCK cells infected with influenza A virus (Kohli et al., 2007).

8 Delivery of Biomacromolecules Using Nanogels

There are a few examples of in vitro or in vivo delivery of biomacromolecules using nanogels. For instance, nanogels loaded with oligonucleotides were shown to cross cellular barriers. One work evaluated permeability of a phosphorothioate oligonucleotide in polarized Caco-2 cell monolayers as in vitro model of gastrointestinal epithelium (Vinogradov et al., 1999). Incorporation of oligonucleotide into PEG-*cl*-PEI nanogel resulted in drastic increase in cellular permeability compared to the free oligonucleotide. Permeability of ^{3}H-mannitol, a paracellular marker, was not affected suggesting that permeability enhancement was due to transcellular delivery of oligonucleotide/nanogel complexes rather than increased leakiness of the cell monolayers. Furthermore, the oligonucleotide immobilized in nanogel particles crossed cells practically without degradation, while the free oligonucleotide was essentially degraded. Thus, PEG-*cl*-PEI nanogel appears to be a promising carrier for oral delivery of oligonucleotides.

Another study evaluated transport of a phosphorothioate oligonucleotide across polarized BBMEC monolayers, an in vitro model of BBB (Vinogradov et al., 2004). It was found that oligonucleotide in PEG-*cl*-PEI nanogel was effectively transported across the cell monolayers. Permeability was further increased when the surface of the nanogel was modified with hTf or insulin. The oligonucleotides were transported through a transcellular pathway. After release at the abluminal side of the cells, they remained nondegraded and incorporated in the nanogel. Biodistribution studies further demonstrated that brain accumulation of oligonucleotide 1 h after intravenous injection in a mouse was increased by over 15-fold for nanogel-incorporated oligonucleotide compared to the free oligonucleotide. At the same time, liver and spleen accumulation of oligonucleotide was decreased by twofold. Overall, this study suggested that cationic nanogels have potential for delivery of oligonucleotides to the brain.

Frechet and colleagues used acid-degradable cationic nanogels for antigen presentation in vitro and vaccination in vivo (Goh et al., 2004; Kwon et al., 2005a–c; Murthy et al., 2003; Standley et al., 2007). For example, incubation of these nanogels loaded with ovalbumin, as a model antigen, with dendritic cells derived from bone marrow (BMDCs) resulted in enhanced presentation of ovalbumin-derived peptides. Cationic nature of the particles contributed to phagocytosis, while their acid sensitivity served to release ovalbumin in acidic lysosomes and to enable ovalbumin peptides processing and presentation via the major histocompatibility complex (MHC) class I pathway. It was also shown that adjuvant molecules such as unmethylated CpG oligonucleotides and anti-interleukin-10 oligonucleotides could be codelivered with the protein antigen for maximized cellular immune response (Standley et al., 2007).

9 Conclusions

In conclusion, nanogels are promising novel pharmaceutical carriers for small biologically active agents and biomacromolecules. The advantages of these systems include simplicity of formulation with the drugs, high loading capacity, and stability of the resulting formulation in dispersion. These systems allow immobilization of biologically active compounds of diverse structure including charged drugs, low molecular mass hydrophobes, and biopolymers. Furthermore, nanogels can be chemically modified to incorporate various ligands for targeted drug delivery. The in vitro and in vitro studies suggest that nanogels can be used for efficient delivery of biopharmaceuticals in cells as well as for increasing drug delivery across cellular barriers.

Acknowledgments The authors are grateful for the support by National Institutes of Health grants NS36229, NS051335, CA89225, and CA116591 (to AVK) and NS050660 and CA102791 (to SV), Department of Defense grant USAMRMC 06108004 (to AVK) and National Science Foundation grant DMR 0513699 (to AVK).

References

Akiyoshi, K., Kobayashi, S., Shichibe, S., Mix, D., Baudys, M., Kim, S. W., and Sunamoto, J., 1998, Self-assembled hydrogel nanoparticle of cholesterol-bearing pullulan as a carrier of protein drugs: complexation and stabilization of insulin, *J Control Release* **54**:313–320.

Bontha, S., Kabanov, A. V., and Bronich, T. K., 2006, Polymer micelles with cross-linked ionic cores for delivery of anticancer drugs, *J Control Release* **114**:163–174.

Bronich, T., Vinogradov, S., and Kabanov, A. V., 2001, Interaction of nanosized copolymer networks with oppositely charged amphiphilic molecules, *Nano Lett* **1**:535–540.

Bronich, T. K., Keifer, P. A., Shlyakhtenko, L. S., and Kabanov, A. V., 2005, Polymer micelle with cross-linked ionic core, *J Am Chem Soc* **127**:8236–8237.

Bronich, T. K., Bontha, S., Shlyakhtenko, L. S., Bromberg, L., Hatton, T. A., and Kabanov, A. V., 2006, Template-assisted synthesis of nanogels from pluronic-modified poly(acrylic acid), *J Drug Target* **14**:357–366.

Donini, C., Robinson, D. N., Colombo, P., Giordano, F., and Peppas, N. A., 2002, Preparation of poly(methacrylic acid-g-poly(ethylene glycol)) nanospheres from methacrylic monomers for pharmaceutical applications, *Int J Pharm* **245**:83–91.

Daoud-Mahammed, S., Couvreur, P., and Gref, R., 2007, Novel self-assembling nanogels: stability and lyophilisation studies, *Int J Pharm* **332**:185–191.

Eichenbaum, G. M., Kiser, P. F., Simon, S. A., and Needham, D., 1998, pH and ion-triggered volume response of anionic hydrogel microspheres, *Macromolecules* **31**:5084–5093.

Francis, G. E., Delgado, C., Fisher, D., Malik, F., and Agrawal, A. K., 1996, Polyethylene glycol modification: relevance of improved methodology to tumour targeting, *J Drug Target* **3**:321–340.

Galmarini, C. M., Mackey, J. R., and Dumontet, C., 2002, Nucleoside analogues and nucleobases in cancer treatment, *Lancet Oncol* **3**:415–424.

Goh, S. L., Murthy, N., Xu, M., and Frechet, J. M., 2004, Cross-linked microparticles as carriers for the delivery of plasmid DNA for vaccine development, *Bioconjug Chem* **15**:467–474.

Hatse, S., De Clercq, E., and Balzarini, J., 1999, Role of antimetabolites of purine and pyrimidine nucleotide metabolism in tumor cell differentiation, *Biochem Pharmacol* **58**:539–555.

Hayashi, H., Iijima, M., Kataoka, K., and Nagasaki, Y., 2005, pH-Sensitive nanogel possessing reactive PEG tethered chains on the surface, *Macromolecules* **37**:5389–5396.

Hennink, W. E. and van Nostrum, C. F., 2002, Novel crosslinking methods to design hydrogels, *Adv Drug Deliv Rev* **54**:13–36.

Kabanov, A. V. and Alakhov, V. Y., 2002, Pluronic block copolymers in drug delivery: from micellar nanocontainers to biological response modifiers, *Crit Rev Ther Drug Carrier Syst* **19**:1–72.

Kabanov, V. A., Skobeleva, V. B., Rogacheva, V. B., and Zezin, A. B., 2004, Sorption of proteins by slightly cross-linked polyelectrolyte hydrogels: kinetics and mechanism, *J Phys Chem B* **108**:1485–1490.

Kato, N., Hasegawa, U., Morimoto, N., Saita, Y., Nakashima, K., Ezura, Y., Kurosawa, H., Akiyoshi, K., and Noda, M., 2007, Nanogel-based delivery system enhances PGE(2) effects on bone formation, *J Cell Biochem* (in press).

Khmelnitsky, Y. L., Neverova, I. N., Gedrovich, A. V., Polyakov, V. A., Levashov, A. V., and Martinek, K., 1992, Catalysis by alpha-chymotrypsin entrapped into surface-modified polymeric nanogranules in organic solvent, *Eur J Biochem* **210**:751–757.

Kohli, E., Han, H. Y., Zeman, A. D., and Vinogradov, S. V., 2007, Formulations of biodegradable Nanogel carriers with 5'-triphosphates of nucleoside analogs that display a reduced cytotoxicity and enhanced drug activity, *J Control Release*.

Kwon, Y. J., James, E., Shastri, N., and Frechet, J. M., 2005a, In vivo targeting of dendritic cells for activation of cellular immunity using vaccine carriers based on pH-responsive microparticles, *Proc Natl Acad Sci USA* **102**:18264–18268.

Kwon, Y. J., Standley, S. M., Goh, S. L., and Frechet, J. M., 2005b, Enhanced antigen presentation and immunostimulation of dendritic cells using acid-degradable cationic nanoparticles, *J Control Release* **105**:199–212.

Kwon, Y. J., Standley, S. M., Goodwin, A. P., Gillies, E. R., and Frechet, J. M., 2005c, Directed antigen presentation using polymeric microparticulate carriers degradable at lysosomal pH for controlled immune responses, *Mol Pharm* **2**:83–91.

Lee, H., Mok, H., Lee, S., Oh, Y. K., and Park, T. G., 2007, Target-specific intracellular delivery of siRNA using degradable hyaluronic acid nanogels, *J Control Release* **119**:245–252.

Lemieux, P., Vinogradov, S. V., Gebhart, C. L., Guerin, N., Paradis, G., Nguyen, H. K., Ochietti, B., Suzdaltseva, Y. G., Bartakova, E. V., Bronich, T. K., St-Pierre, Y., Alakhov, V. Y., and Kabanov, A. V., 2000, Block and graft copolymers and NanoGel copolymer networks for DNA delivery into cell, *J Drug Target* **8**:91–105.

McAllister, K., Sazani, P., Adam, M., Cho, M. J., Rubinstein, M., Samulski, R. J., and DeSimone, J. M., 2002, Polymeric nanogels produced via inverse microemulsion polymerization as potential gene and antisense delivery agents, *J Am Chem Soc* **124**:15198–15207.

Missirlis, D., Tirelli, N., and Hubbell, J. A., 2005, Amphiphilic hydrogel nanoparticles. Preparation, characterization, and preliminary assessment as new colloidal drug carriers, *Langmuir* **21**:2605–2613.

Missirlis, D., Kawamura, R., Tirelli, N., and Hubbell, J. A., 2006, Doxorubicin encapsulation and diffusional release from stable, polymeric, hydrogel nanoparticles, *Eur J Pharm Sci* **29**:120–129.

Murthy, N., Xu, M., Schuck, S., Kunisawa, J., Shastri, N., and Frechet, J. M., 2003, A macromolecular delivery vehicle for protein-based vaccines: acid-degradable protein-loaded microgels, *Proc Natl Acad Sci USA* **100**:4995–5000.

Nayak, S., Lee, H., Chmielewski, J., and Lyon, L. A., 2004, Folate-mediated cell targeting and cytotoxicity using thermoresponsive microgels, *J Am Chem Soc* **126**:10258–10259.

Ogawa, K., Sato, S., and Kokufuta, E., 2005, Formation of intra- and interparticle polyelectrolyte complexes between cationic nanogel and strong polyanion, *Langmuir* **21**:4830–4836.

Ogawa, K., Sato, S., and Kokufuta, E., 2007, On an intraparticle complex of cationic nanogel with a stoichiometric amount of bound polyanions, *Langmuir* **23**:2095–2102.

Oh, J. K., Tang, C., Gao, H., Tsarevsky, N. V., and Matyjaszewski, K., 2006, Inverse miniemulsion ATRP: a new method for synthesis and functionalization of well-defined water-soluble/cross-linked polymeric particles, *J Am Chem Soc* **128**:5578–5584.

Oh, J. K., Siegwart, D. J., Lee, H. I., Sherwood, G., Peteanu, L., Hollinger, J. O., Kataoka, K., and Matyjaszewski, K., 2007a, Biodegradable nanogels prepared by atom transfer radical polymerization as potential drug delivery carriers: synthesis, biodegradation, in vitro release, and bioconjugation, *J Am Chem Soc* **129**:5939–5945.

Oh, K. T., Bronich, T. K., Kabanov, V. A., and Kabanov, A. V., 2007b, Block polyelectrolyte networks from poly(acrylic acid) and poly(ethylene oxide): sorption and release of cytochrome C, *Biomacromolecules* **8**:490–497.

Park, H., Temenoff, J. S., Tabata, Y., Caplan, A. I., and Mikos, A. G., 2007, Injectable biodegradable hydrogel composites for rabbit marrow mesenchymal stem cell and growth factor delivery for cartilage tissue engineering, *Biomaterials* **28**:3217–3227.

Ricka, J. and Tanaka, T., 1984, Swelling of ionic gels: quantitative performance of the Donnan theory, *Macromolecules* **17**:2916–2921.

Shin, Y., Chang, J. H., Liu, J., Williford, R., Shin, Y., and Exarhos, G. J., 2001, Hybrid nanogels for sustainable positive thermosensitive drug release, *J Control Release* **73**:1–6.

Shin, Y., Liu, J., Chang, J. H., and Exarhos, G. J., 2002, Sustained drug release on temperature-responsive poly(*N*-isopropylacrylamide)-integrated hydroxyapatite, *Chem Commun (Camb)* **16**:1718–1719.

Shiokawa, T., Hattori, Y., Kawano, K., Ohguchi, Y., Kawakami, H., Toma, K., and Maitani, Y., 2005, Effect of polyethylene glycol linker chain length of folate-linked microemulsions loading aclacinomycin A on targeting ability and antitumor effect in vitro and in vivo, *Clin Cancer Res* **11**:2018–2025.

Soni, S., Babbar, A. K., Sharma, R. K., and Maitra, A., 2006, Delivery of hydrophobised 5-fluorouracil derivative to brain tissue through intravenous route using surface modified nanogels, *J Drug Target* **14**:87–95.

Soprano, D. R., Qin, P., and Soprano, K. J., 2004, Retinoic acid receptors and cancers, *Annu Rev Nutr* **24**:201–221.

Standley, S. M., Mende, I., Goh, S. L., Kwon, Y. J., Beaudette, T. T., Engleman, E. G., and Frechet, J. M., 2007, Incorporation of CpG oligonucleotide ligand into protein-loaded particle vaccines promotes antigen-specific CD8 T-cell immunity, *Bioconjug Chem* **18**:77–83.

Tyagi, R., Lala, S., Verma, A. K., Nandy, A. K., Mahato, S. B., Maitra, A., and Basu, M. K., 2005, Targeted delivery of arjunglucoside I using surface hydrophilic and hydrophobic nanocarriers to combat experimental leishmaniasis, *J Drug Target* **13**:161–171.

Varga, I., Szalai, I., Meszaros, R., and Gilanyi, T., 2006, Pulsating pH-responsive nanogels, *J Phys Chem B Condens Matter Mater Surf Interfaces Biophys* **110**:20297–20301.

Vinogradov, S. V., 2006, Colloidal microgels in drug delivery applications, *Curr Pharm Des* **12**:4703–4712.

Vinogradov, S. V., Batrakova, E. V., and Kabanov, A. V., 1999, Poly(ethylene glycol)-polyethyleneimine NanoGel particles: novel drug delivery systems for antisense oligonucleotides, *Colloids Surf B Biointerfaces* **16**:291–304.

Vinogradov, S. V., Bronich, T. K., and Kabanov, A. V., 2002, Nanosized cationic hydrogels for drug delivery: preparation, properties and interactions with cells, *Adv Drug Deliv Rev* **54**:135–147.

Vinogradov, S. V., Batrakova, E. V., and Kabanov, A. V., 2004, Nanogels for oligonucleotide delivery to the brain, *Bioconjug Chem* **15**:50–60.

Vinogradov, S. V., Zeman, A. D., Batrakova, E. V., and Kabanov, A. V., 2005, Polyplex Nanogel formulations for drug delivery of cytotoxic nucleoside analogs, *J Control Release* **107**:143–157.

Vinogradov, S. V., Kohli, E., and Zeman, A. D., 2006, Comparison of nanogel drug carriers and their formulations with nucleoside 5'-triphosphates, *Pharm Res* **23**:920–930.

Yan, M., Ge, J., Liu, Z., and Ouyang, P., 2006, Encapsulation of single enzyme in nanogel with enhanced biocatalytic activity and stability, *J Am Chem Soc* **128**:11008–11009.

Yan, M., Liu, Z., Lu, D., and Liu, Z., 2007, Fabrication of single carbonic anhydrase nanogel against denaturation and aggregation at high temperature, *Biomacromolecules* **8**:560–565.

Yu, S., Hu, J., Pan, X., Yao, P., and Jiang, M., 2006a, Stable and pH-sensitive nanogels prepared by self-assembly of chitosan and ovalbumin, *Langmuir* **22**:2754–2759.

Yu, S., Yao, P., Jiang, M., and Zhang, G., 2006b, Nanogels prepared by self-assembly of oppositely charged globular proteins, *Biopolymers* **83**:148–158.

Multifunctional Water-Soluble Polymers for Drug Delivery

Huaizhong Pan and Jindřich Kopeček

1 Introduction

Water-soluble polymer–drug conjugates are multifunctional nanomedicines at the interface of polymer chemistry and biomedical sciences. Advances in chemistry and applied biology have provided scientists with powerful and flexible tools to tailor the features of synthetic polymers and design functions according to their ultimate usage. The techniques used to synthesize polymer conjugates (copolymerization of polymerizable bioactive compounds and polymeranalogous reactions) afford a vast variety of designs to match their ultimate applications. Incorporation of hydrophilic groups confers polymers with water solubility and improved biocompatibility. Charged groups or hydrophobic groups can be introduced into polymers to endow them with special interactions, or environmental response abilities. Targeting moieties bestow biorecognizability; attachment of drug(s) provides specific pharmaceutical properties. Reporter (labeling) groups are frequently incorporated into the structure to permit the evaluation of the fate of the conjugate. The possibility to insert multiple functions into one macromolecule gives the scientists the opportunity to mimic natural functional macromolecules (Torchilin, 2006a). Multifunctional polymer–drug conjugates have abilities to store inactive drugs as prodrugs or pro-enzymes, protect drugs that do not reach the target place, direct drugs to the proper site by passive or active targeting, activate the drugs at a suitable site, have impact on cellular signaling pathways, block or prompt reactions, etc.

Polymer–drug conjugates (macromolecular therapeutics) emerged half a century ago. Jatzkewitz (1955) first attempted to attach mescaline to polyvinylpyrrolidone through a glycyl-L-leucyl spacer as a drug depot formulation to improve the drug efficiency. In the sixties and seventies, numerous polymer–drug conjugates with the drug covalently bound to the water-soluble polymer carrier have been evaluated (Panarin and Ushakov, 1968). Finally, Ringsdorf (1975) presented the first clear concept of the use of polymers as targetable drug carriers. The model is based on the understanding of internalization and subcellular trafficking of macromolecules in cells. It consists of a polymeric carrier, drug attached via a hydrolytically or enzymatically cleavable spacer, and a targeting group complementary to a receptor/antigen at the target cell (Fig.1).

V. Torchilin (ed.), *Multifunctional Pharmaceutical Nanocarriers*,
© Springer Science+Business Media, LLC 2008

The unique structural, physicochemical, and biological properties of macromolecular therapeutics result in advantageous properties when compared with low molecular weight drugs: (1) improved water solubility of hydrophobic low molecular weight drugs with concomitant improvement of bioavailability; (2) protection of unstable drugs from deterioration; (3) long-lasting circulation in the bloodstream; (4) decreased nonspecific toxicity of the conjugated drug; (5) increased *active* accumulation of the drug at the tumor site by targeting and/or increased *passive* accumulation of the drug at the tumor site by the enhanced permeability and retention (EPR) effect;

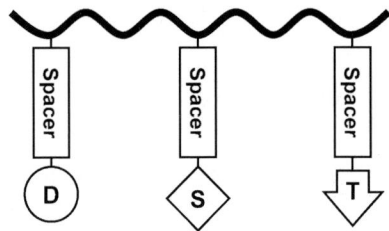

Fig. 1 Polymer–drug conjugate. D: drug; S: solubilizer; T: targeting moiety

Table 1 Polymer–antitumor drug conjugates that entered clinical trials

Conjugate	Name	Status	References
HPMA copolymer–Dox	PK1, FCE28069	Phase II	Vasey et al. (1999)
HPMA copolymer–Dox–galactosamine	PK2, FCE28069	Phase I/II	Seymour et al. (2002)
HPMA copolymer–paclitaxel	PNU166945	Phase I	Meerum Terwogt et al. (2001)
HPMA copolymer–camptothecin	MAG-CPT, PCNU166148	Phase I	Bissett et al. (2004), Sarapa et al. (2003), and Wachters et al. (2004)
HPMA copolymer–platinate	AP5280	Phase I/II	Rademaker-Lakhai et al. (2004)
HPMA copolymer–DACH–platinate	AP5346	Phase I/II	Campone et al. (2007)
HPMA copolymer–Dox	ZencoDox	Phase I	
PEG–camptothecin	Prothecan	Phase II	Rowinsky et al. (2003)
PEG-aspartic acid–Dox	NK911	Phase I	Nakanishi et al. (2001) and Tsukioka et al. (2002)
Polyglutamic acid–paclitaxel	CT-2103, Xyotax	Phase III	Langer (2004a,b) and Sabbatini et al. (2004)
Polyglutamic acid–camptothecin	CT-2106	Phase I	Bhatt et al. (2003)
Dextran–Dox	AD-70, DOX-OXD	Phase I	Danhauser-Riedl et al. (1993)
Carboxymethyldextran–polyalcohol–camptothecin	DE-310	Phase I	Kumazawa and Ochi (2004)

HPMA: *N*-(2-hydroxypropyl)methacrylamide; Dox: doxorubicin; DACH: diaminocyclohexane; PEG: poly(ethylene glycol)

Table 2 Polymer–protein conjugates used in the clinics

Conjugate	Name	Indication	Market	References
PEG–adenosine deaminase	Adagen	SCID syndrome	1990	Levy et al. (1988)
SMANCS	Zinostatin, Stimalmer	Hepatocellular carcinoma	1993	Matsumura and Maeda (1986)
PEG-L-asparaginase	Oncaspar	Acute lymphoblastic leukemia	1994	Graham (2003)
PEG-IFNα 2b	Pegintron	Hepatitis C	2000	Wang et al. (2002)
PEG-IFNα 2a	Pegasys	Hepatitis C	2002	Rajender Reddy et al. (2002)
PEG-HGF	Pegvisomant	Acromegaly	2002	Mukherjee et al. (2003)
PEG-G-CSF	PEG-filgrastim, Neulasta	Chemotherapy associated neutropenia	2002	Molineux (2004)

SMANCS: styrene-maleic anhydride copolymer-neocarzinostatin; INFα: interferon-α; HGF: human growth hormone; G-CSF: granulocyte-colony stimulating factor

(6) ability to deliver two or more drugs with different (complementary) properties to the same target site, enhancing the drug efficiency by cooperative effects.

The systematic research in this area and improved design of conjugates based on a sound biological rationale resulted in the evaluation of numerous conjugates in clinical trials (Table 1) and translation into clinical use (Table 2).

2   Design and Synthesis of the Polymer–Drug Conjugates

Polymer–drug conjugates are composed of distinct parts; the choice and variation of functional parts permit the design of multifunctional conjugates with desired properties. The structure and conformation of the water-soluble main chain has a major impact on water solubility and biocompatibility, the choice of spacers linking the drugs to the polymer backbone controls the drug release kinetics, and the use of targeting moieties may direct the drug to a particular cell or subcellular compartment.

2.1   Water-Soluble Polymeric Carriers

The strength and flexibility of synthetic chemistry provide a tool for scientists to design polymers with different composition, desired architecture (Qiu and Bae, 2006), and tailor-made properties. A water-soluble polymer with good biocompatibility can provide an ideal platform for the design of a variety of polymer–drug conjugates. Although thousands of natural and synthetic polymers exist, only a small subset of polymers has been widely used as drug carriers. These polymers include poly(ethylene glycol), N-(2-hydroxypropyl)methacrylamide copolymers, polyvinylpyrrolidone, polyethyleneimine, polyamidoamine, poly(amino acid)s, dextran, dextrin, and chitosan.

Poly(ethylene glycol) (PEG). PEG is synthesized by the anionic ring-opening polymerization of ethylene oxide. It is water soluble, but can also dissolve in many organic solvents. Its good water solubility and biocompatibility have shown it to be a versatile carrier in polymer–drug conjugates (Garnett, 2001; Khandare and Minko, 2006).

With the development of recombinant technologies, many peptides and proteins have been developed as therapeutic agent, but their intrinsic instability, degradability, and immunogenicity often hinder their potential usage. Attachment of PEG to proteins has been used to improve their solubility, immunogenicity, pharmacokinetics, and pharmacodynamics. Antibodies against unmodified proteins do not recognize the PEG-modified protein as an antigen. Increased intravascular half-life is the result of decreased degradation of the protein by enzymes in the blood stream, and reduced renal clearance. (Greenwald et al., 2000a, 2003; Harris and Chess, 2003).

ADAGEN, a PEGylated bovine enzyme adenosine deaminase (ADA) for treatment of ADA-deficient severe combined immunodeficiency disease, and ONCASPAR, a PEGylated L-asparaginase used for treatment of lymphoblastic leukemia were commercial products approved for use in the early 1990s. Now, several PEGylated proteins have been cleared for clinical use (Table 2).

PEG is commercially available with either one or two attachment points. It can be produced as a linear or branched polymer. The –OH groups at the chain termini can be conjugated with drugs or other functional groups. PEG was used intensively as a carrier of low molecular weight anticancer drugs. For example, PEG–camptothecin conjugate (PEG–CPT) demonstrated improved pharmacokinetics, enhanced solid tumor accumulation, and increased anticancer activity when compared to free CPT (Yu et al., 2005).

The lack of reacting functional groups limits the loading capacity and the potential use for PEG in polymer–drug conjugates. To overcome this limitation, reactive groups on PEG chains can be multiplied by reaction with multifunctional compounds. For example, Greenwald and coworkers coupled PEG with aspartic acid (Asp) and AspAsp dendrons to introduce –COOH groups in PEG. These carboxyl groups were conjugated with cytosine arabinoside (Ara-C) via spacers (Choe et al., 2002a,b). Compared to the native drug, the tetrameric and octameric PEG-ara-C amide prodrugs were much more effective in the treatment of cancer.

Schiavon et al. (2004) presented another approach to increase PEG drug loading. The hydroxyl groups were repeatedly modified with adipic acid yielding four or eight functional derivatives capable of binding Ara-C (Fig. 2). The conjugates possessed increased stability of conjugated Ara-C, improved blood residence time in mice, and reduced cytotoxicity with respect to the free Ara-C form. A similar approach to multiply PEG functional groups was published by Pasut et al. (2005).

Pechar et al. (2005) synthesized biodegradable multiblock polymers of PEG by reacting *N*-hydroxysuccinimide-activated PEG (mw 2,000) with an enzymatically degradable tripeptide derivative, N^2,N^6-diglutamyllysine triethyl ester. The ethyl ester groups of the block copolymer were converted to hydrazide by hydrazinolysis. Doxorubicin (Dox) was attached to the polymer carrier via a hydrazone bond susceptible to acid hydrolysis at pH 5.0 (Fig. 3). Human immunoglobulin (IgG) was modified with 2-iminothiolane to introduce –SH groups and then coupled to the polymer–Dox conjugate by using succinimidyl 3-(2-pyridyldisulanyl)propanoate (SPDP). These polymer conjugates have shown antiproliferative and antitumor activities in vitro and in vivo.

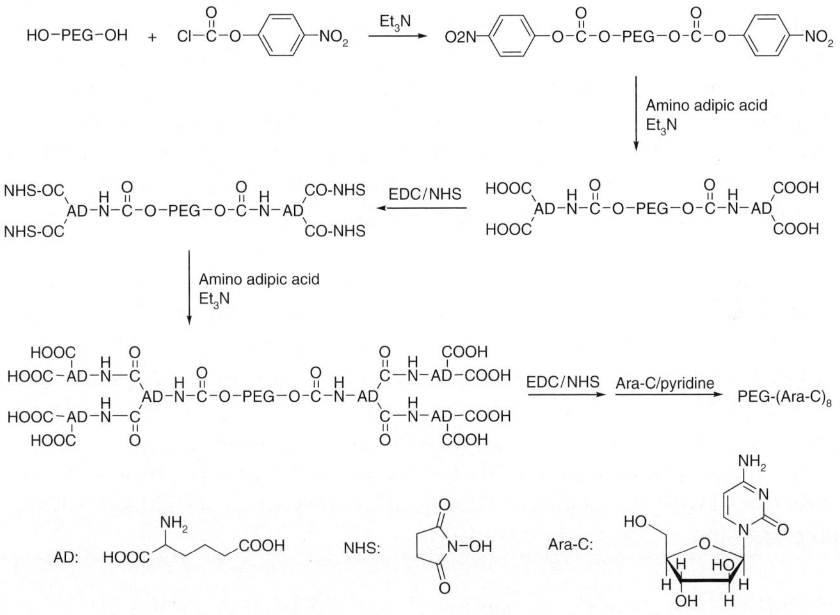

Fig. 2 Introduction of functional groups into poly(ethylene glycol) by consecutive modification with amino adipic acid. Up to eight drug (Ara-C) molecules can be bound to one modified PEG chain (adapted from Schiavon et al., 2004)

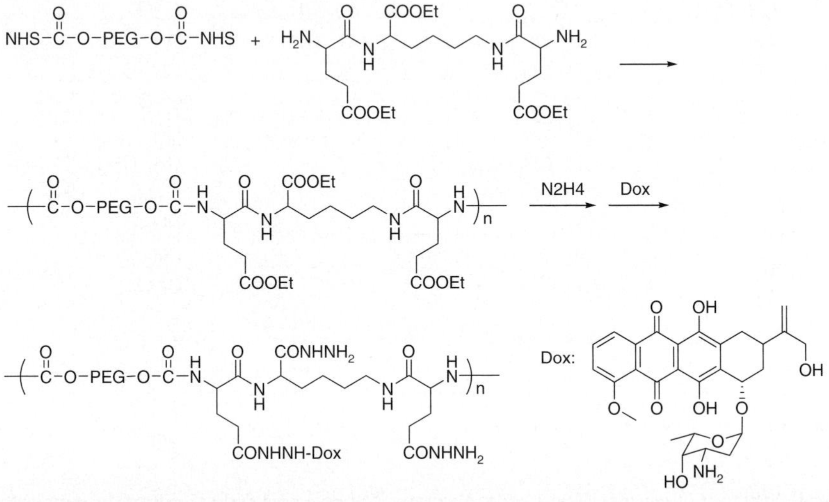

Fig. 3 Synthesis of biodegradable multiblock PEG copolymers containing doxorubicin (adapted from Pechar et al., 2005)

N-(2-hydroxypropyl)methacrylamide (HPMA) copolymers. HPMA copolymers are one of the most studied platforms for polymer–drug conjugates. It is a water-soluble, neutral, nonimmunogenic copolymer. The bulk of the work with the polymer has been focused on the delivery of anticancer drugs (Kopeček, 1977; Kopeček et al., 2000; Kunath et al., 2000; Putnam and Kopeček, 1995a,b), tumor-specific antisense oligonucleotides (Jensen et al., 2002), and site-specific delivery to the gastrointestinal (GI) tract (Gao et al., 2006a,b, 2007). Biologically active HPMA copolymers were synthesized by free-radical polymerization of HPMA monomer with other functional comonomers. These comonomers included drug-containing monomers (Obereigner et al., 1979), monomers with targeting moieties (Rathi et al., 1991), or reactive monomers for subsequent polymer analogous reactions (Fig. 4) (Rejmanová et al., 1977) and attachment of biologically active compounds (Chytrý et al., 1978; Lääne et al., 1985; Solovskij et al., 1983). The molecular weight of the copolymers was controlled by the concentration of monomers, initiator, or by the addition of chain transfer agents (Strohalm and Kopeček, 1978). Using functionalized initiators or chain transfer agents, functional groups may be introduced at one end of the polymer chain to create semitelechelic (*co*)polymers (Kamei and Kopeček, 1995; Lu et al., 1998).

In recent years, living free radical polymerization methods, including atom transfer radical polymerization (ATRP) and reversible addition-fragmentation chain transfer (RAFT) polymerization, have been used for the synthesis of HPMA copolymers with narrow molecular weight distributions.

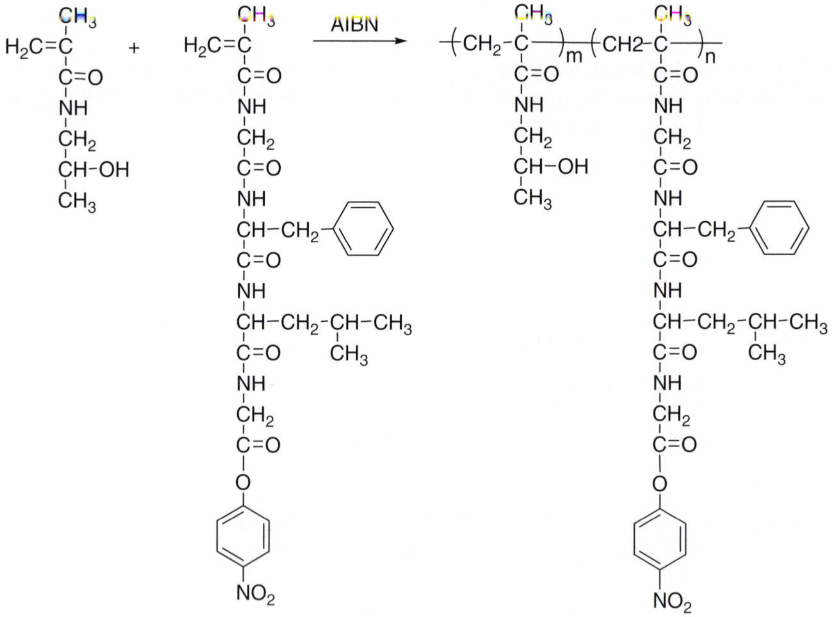

Fig. 4 Synthesis of reactive HPMA copolymer precursor containing GFLG side chains terminated in *p*-nitrophenyl ester groups (Rejmanová et al., 1977)

Polymerization of HPMA by the ATRP method was first reported by the Matyjaszewski laboratory (Teodorescu and Matyjaszewski, 1999, 2000). Here, methyl 2-bromopropionate/CuBr/1,4,8,11-tetramethyl-1,4,8,11-tetraazacyclotetradecane (Me4Cyclam) and 2-chloropropionate/CuCl/tris(2-dimethylaminoethyl)amine were used as initiators in organic solvent. It was found, however, that the polymerization was difficult to control and the yield low in some cases. Using the ATRP method, block copolymers of HPMA with butyl acrylate were synthesized (Fig. 5) (Koňák et al., 2002).

McCormick and coworkers first reported the polymerization of HPMA using the RAFT method (Scales et al., 2005). Here, 4-cyanopentanoic acid dithiobenzoate was the chain transfer agent (CTA) and 4,4′-azobis(4-cyanopentanoic acid) the initiating species. Using an aqueous acetic buffer, polymer was obtained at high yields and a low polydispersity. RAFT polymerization was also used for the synthesis of different HPMA copolymers (Fig. 6). Using the RAFT process, poly(HPMA-*block*-*N*-[3-(dimethylamino)propyl]methacrylamide) (DMAPMA) was synthesized with well-defined block lengths and charge distributions. These block copolymers were used as block ionomer complexes (BICs) for stabilization and the delivery of small interfering RNAs (siRNA), to silence specific genes (Scales et al., 2006). Hong et al. synthesized biotinylated trithiocarbonate as a new RAFT agent, and used RAFT polymerization to prepare temperature-responsive, biotin-terminating HPMA and poly(*N*-isopropylacrylamide) (PNIPAAm) block copolymers (Hong and Pan, 2006). It is known that a large difference in the reactivity ratios of monomers always results in significant variations in copolymer composition with increasing conversion during batch copolymerization. Because it is a "living" polymerization, the RAFT method allows to add additional amounts of monomers during the reaction.

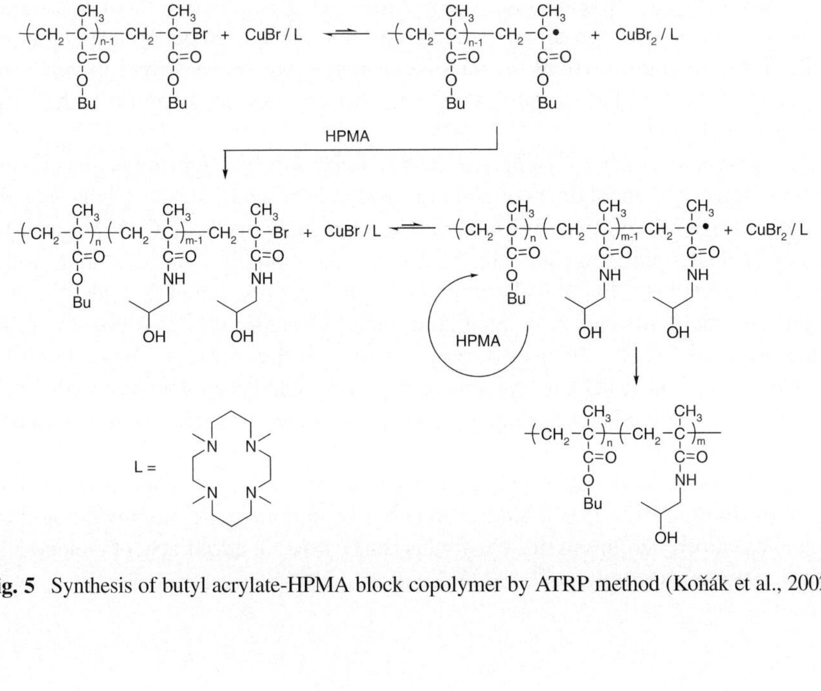

Fig. 5 Synthesis of butyl acrylate-HPMA block copolymer by ATRP method (Koňák et al., 2002)

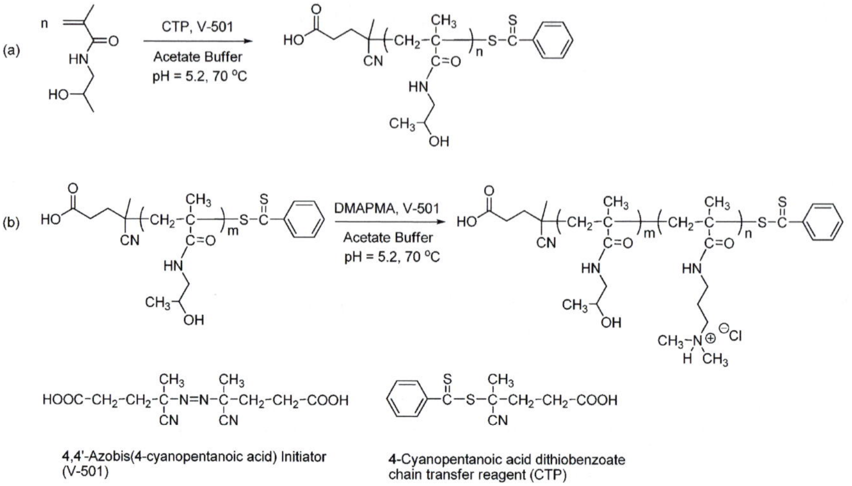

Fig. 6 Synthesis of (**a**) polyHPMA and (**b**) copolymer of HPMA and *N*-(3-dimethylaminopropyl) methacrylamide (DMAPMA) by the RAFT polymerization method (Scales et al., 2005, 2006)

Kane et al. used the RAFT method along with semibatch copolymerization by the gradual addition of the more reactive monomer. They obtained homogeneous HPMA copolymers containing active ester functional groups (Yanjarappa et al., 2006).

Poly(styrene-co-maleic anhydride) (SMA). Styrene-maleic anhydride copolymer (SMA) was synthesized by the copolymerization of styrene and maleic anhydride in cumene using dicumyl peroxide as the initiator at 152–153 °C. Partial half-esterification or partial hydrolysis of SMA was performed in dioxane at 90 °C using lithium acetate as a catalyst (Fig. 7) (Maeda et al., 1985). SMA was used for the quantitative modification of the amino group of neocarzinostatin (NCS). Neocarzinostatin is an inhibitor tumor cell growth at the nanomolar range, but shows severe toxicity and very short half-life. The conjugate (SMANCS) demonstrated improved half-life and bioavailability.

Poly(glutamic acid) (PG) and poly(β-L-malic acid). High-molecular-weight homopolymers and copolymers of glutamic acid may be synthesized by ring-opening (*co*)polymerization of the corresponding *N*-carboxyanhydrides (NCA), initiated by amines or nucleophilic agents (Fig. 8). PG is water-soluble, nontoxic, and biodegradable. It has been found to be more susceptible to lysosomal degradation than poly(L-aspartic acid) and poly(D-glutamic acid) (Kishore et al., 1990). Cysteine proteases, particularly cathepsin B, play key roles in the lysosomal degradation of PG (Chiu et al., 1997). PG has a γ-carboxyl group in each repeating unit of L-glutamic acid that offers potential drug attachment points. These properties make PG a good drug carrier (Li, 2002).

Poly(β-L-malic acid) can be prepared in the myxomycete *Physarum polycephalum* (Cheremisinoff, 1997). It is a biodegradable, nonimmunogenic, nontoxic polymer, containing carboxy groups in side chains that can be used for attachment of biologically active compounds (Fujita et al., 2006; Lee et al., 2006).

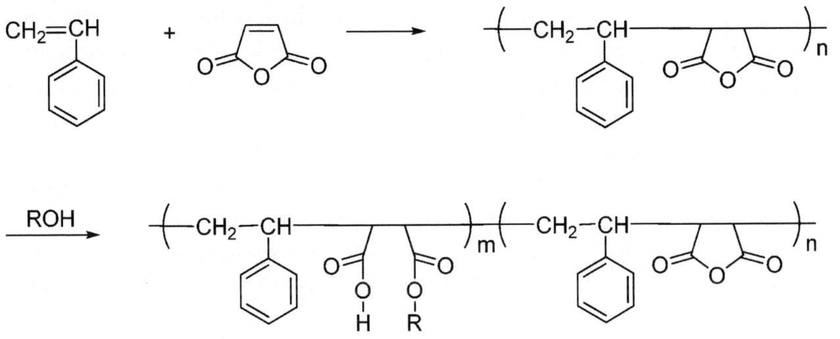

Fig. 7 Synthesis and partial esterification (hydrolysis) of styrene-maleic anhydride copolymer

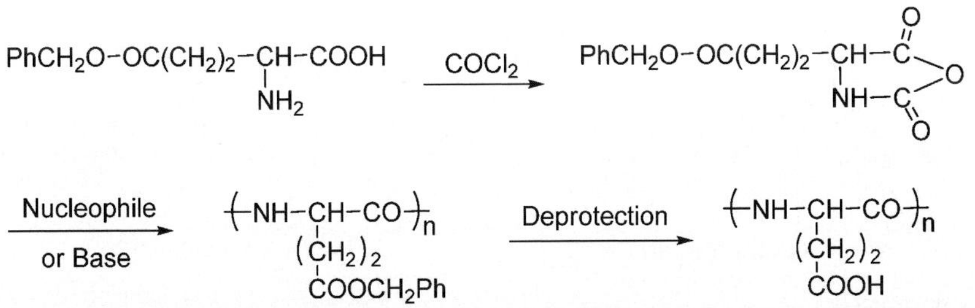

Fig. 8 Synthesis of poly(glutamic acid) by ring-opening polymerization of *N*-carboxyanhydride

Cationic polymers – polyethyleneimine (PEI), polyamidoamine, and poly(amino acid)s. Multifunctional PEIs, PEI derivatives, polymers containing amino acids, including poly(L-lysine) (PLL), and polyamidoamine have been used for gene delivery (Brown et al., 2003; Brownlie et al., 2004; Esfand and Tomalia, 2001; Gebhart et al., 2002; Malik et al., 2000; Thomas and Klibanov, 2002; Wiwattanapatapee et al., 2000). The structures of these polymers are shown in Fig. 9. For gene delivery, the polymer should have a controllable quantity and density of positive charges. These bind with DNA via electrostatic interactions and release DNA after cell uptake.

DNA is reversibly condensed into tightly packed complexes after binding to cationic polymers. Studies have shown that DNA binding to linear or low molecular weight PEI created large complexes, but DNA binding with higher molecular weight or branched PEIs lead to more compact complexes (Godbey et al., 1999a,b). The size and shape of the final complex also depend on other conditions, such as temperature (Bloomfield, 1996).

Dextran. Dextran is a natural polysaccharide obtained by fermenting sucrose using *Leuconostoc mesenteroides* or *Streptococcus mutans*. It is an α-D-1,6-glucose-linked glucan with 1–2 glucose units long side chains 1–3 linked to the backbone of the biopolymer. The degree of branching is typically about 5% (Fig. 10).

Dextran is a water-soluble biopolymer, but it can also dissolve in some organic solvents, such as DMSO, DMF, formamide, ethylene glycol, and glycerol. Dextran is biocompatible and biodegradable in blood and in the GI tract (Vercauteren et al., 1992), however, it is not degraded in lysosomes (Chiu et al., 1994). It has been used extensively for prodrug conjugation research. Dextran possesses multiple primary and secondary hydroxyl groups that can be used for binding drugs or proteins directly or via spacers (Fig. 11). For example, the hydroxyl groups of dextran can directly link the carboxylates of drugs using dicyclohexyl carbodiimide (DCC) as a coupling agent (Arefjev et al., 1999), or by using a functional linker (Mehvar and Hoganson, 2000; Rensberger et al., 2000). The dextran chain contains vicinal diol structure. Consequently, it can be oxidized by periodic acid or salts into two aldehyde groups (Fig. 12), which can then be used to react with amino groups

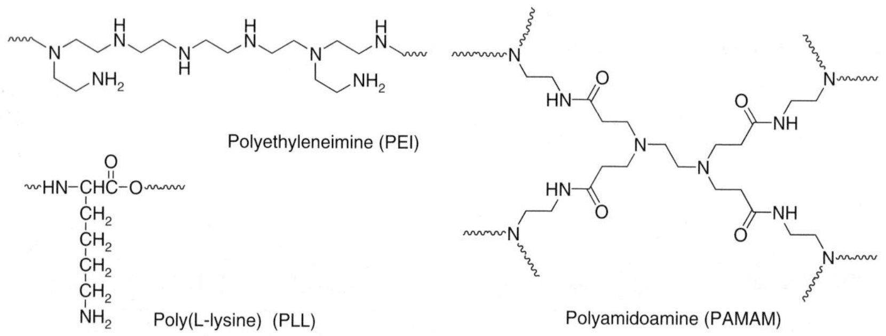

Fig. 9 Structures of some polycations for drug and gene delivery

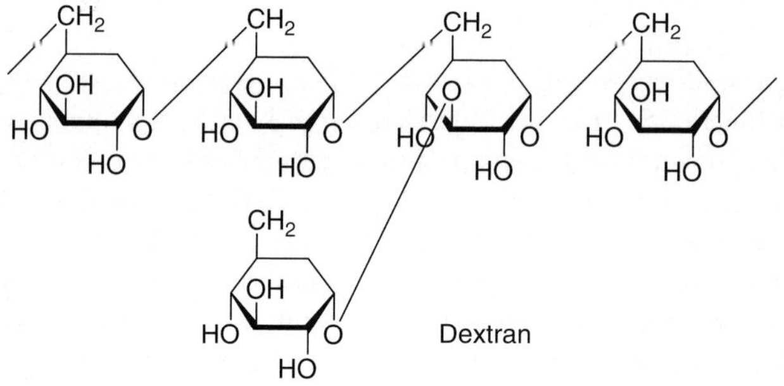

Fig. 10 Structure of dextran

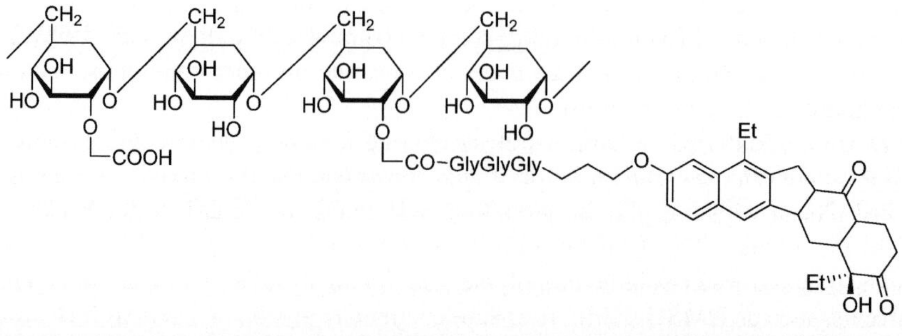

Fig. 11 Carboxymethyl dextran–camptothecin analogue conjugate (Okuno et al., 2000)

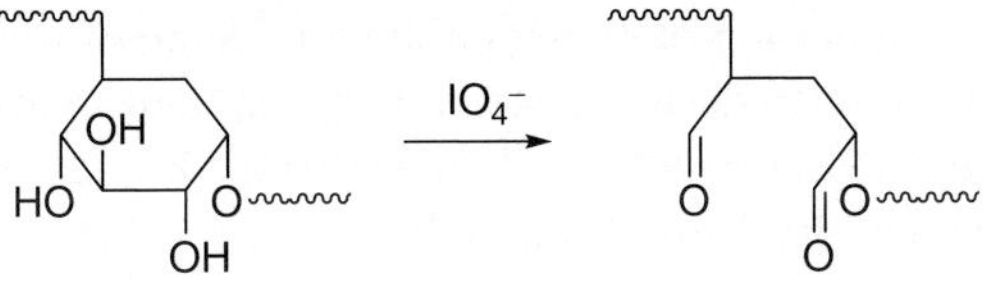

Fig. 12 Oxidation of dextran by periodic acid

(Mehvar, 2000). The application of dextran in polymer–drug conjugates has been reviewed by Khandare and Minko (2006).

Environmentally sensitive polymers. Smart or stimuli-responsive polymers exhibit sharp changes in behavior in response to an external stimulus, such as pH, temperature, solvents, salts, electrical field, light, and chemical or biochemical agents. Linear (*co*)polymers containing stimuli-sensitive groups change their conformation as a result of environmental change. Covalent attachment of stimuli-sensitive polymers to proteins may alter their bioactivity and/or permit an on–off switch of their biological activity in response to the changes of the conformation of the attached macromolecules. Similarly, the biorecognition of a ligand bound to a stimuli-sensitive macromolecule can be controlled by conformational changes induced by external stimuli (Kopeček, 2003). In a living organism, polymer-based drug delivery systems may be exposed to a stimulus after moving from one compartment to another. For example, the pH near and inside the tumor tissue is slightly acidic; the glucose concentration is higher in diabetes patients than in healthy people. During subcellular trafficking, polymer–drug conjugates are exposed to a pH change, from 7.4 in the extracellular fluid to 5.5–6.5 in endosomes and 4.5–5 in lysosomes. Smart polymers have been designed which disrupt the endosomal membrane and permit the escape of biologically active compounds before fusion of endosomes with lysosomes. This is of utmost importance for the delivery of (lysosomally) labile drugs, such as antisense oligodeoxynucleotides, and short interfering RNA (siRNA). Hoffman et al. synthesized a series of pH sensitive poly(2-alkylacrylic acid)s that can selectively disrupt endosomes and release their contents into the cytosol. Poly(propylacrylic acid) (PPAA) enhanced translocation of a biotinylated PPAA complex containing a biotinylated anti-CD3 antibody and streptavidin to the cytoplasm (Bulmus et al., 2003; Lackey et al., 2002). The same research group also found that propylamine, butylamine, and pentylamine derivatives of poly(styrene-*alt*-maleic anhydride) (PSMA) copolymers also show an endosome membrane disability property. These polymers are hydrophilic and membrane inactive at physiological pH; however, they become hydrophobic and membrane disruptive in response to endosomal pH, and the pH-dependent membrane-destabilizing activity of PSMA derivatives can be controlled by varying the length of the alkylamine group, the degree of modification of the copolymer, and the molecular weight of the PSMA copolymer backbone (Henry et al., 2006).

Copolymers. Random, graft, and block copolymers provide the opportunity to combine properties of the individual polymers and beyond. For example, copolymers containing hydrophilic and hydrophobic polymer segments may form polymeric micelles and polymersomes (Dalheimer et al., 2004; Discher and Ahmed, 2006; Lin et al., 2004); the phase transition point of a particular polymer may be adjusted by the incorporation of comonomer units. Synthesis of block copolymers provides an opportunity to design systems responsive to multiple stimuli. For example, poly (*N*-isopropylacrylamide-*block*-acrylic acid) possesses both pH- and temperature-sensitive properties (Kulkarni et al., 2006).

2.2   Design of Spacers

Spacers between the polymer backbone and the drug play an important role in polymer–drug conjugates. The stability of the conjugate, the drug release rate, and the drug release site often depend on its specific structure. A spacer should have high stability in the bloodstream and the extracellular interstitium, but be cleavable at the target site (i.e., in lysosomes). If the spacer is designed to be cleaved by an enzyme at a particular site, then the spacer can confer some targeting ability. The spacers typically used in polymer–drug conjugates are either enzymatically cleavable, pH-sensitive, or degradable by bacteria (Fig. 13).

Enzymatically cleavable oligopeptide spacers. Kopeček and coworkers thoroughly studied the enzymatic cleavage of oligopeptide spacers in the 1970s and 1980s (Drobník et al., 1976; Duncan et al., 1980; Kopeček and Rejmanová, 1983; Kopeček et al., 1981a,b; Putnam and Kopeček, 1995b; Rejmanová et al., 1981, 1983, 1985; Ulbrich et al., 1980, 1981). Using the substrate–enzyme interaction model (Schechter and Berger, 1967), they demonstrated the relationship between the structure of oligopeptide and cleavage rate.

In contrast to low-molecular-weight drugs that enter the cell by diffusion, water-soluble polymer–drug conjugates enter the cell via endocytosis and are then trafficked to the lysosomes. Many different classes of enzymes including nucleases, proteases, phosphatases, lipases, etc. are present in the lysosomes, and cathepsin B, a cysteine proteinase, is one of the major lysosomal enzymes.

HPMA copolymers containing oligopeptide spacers (Kopeček, 1984; Kopeček and Rejmanová, 1983) terminated with drug and susceptible to enzymatically catalyzed hydrolysis (drug release) in the lysosomes were utilized in numerous drug delivery systems (Kopeček et al., 2000). The cathepsin B-cleavable tetrapeptide Gly-Phe-Leu-Gly was used in HPMA copolymer conjugates for delivery of anticancer drugs, for example in HPMA copolymer–doxorubicin (Dox) conjugates. This spacer is stable in blood (Rejmanová et al., 1985), but can be cleaved by cathepsin B in the lysosomal compartment of cells (Duncan, 2003; Rejmanová et al., 1983).

The Gly-Gly-Pro-Nle spacer has been used in polymer–drug conjugates to deliver drugs to bone for the treatment of bone diseases. The Gly-Gly-Pro-Nle linker was designed to be cleaved by cathepsin K. Cathepsin K, a cysteine proteinase of the papain superfamily (Brömme and Okamoto, 1995; Kafienah et al., 1998), is expressed at high levels in osteoclasts, while cathepsins B, L, and S are expressed at relatively low or undetectable levels (Drake et al., 1996).

Acid-labile spacers. The environment in endosomal and lysosomal compartments is acidic. The extracellular pH surrounding tumor tissue is also slightly lower

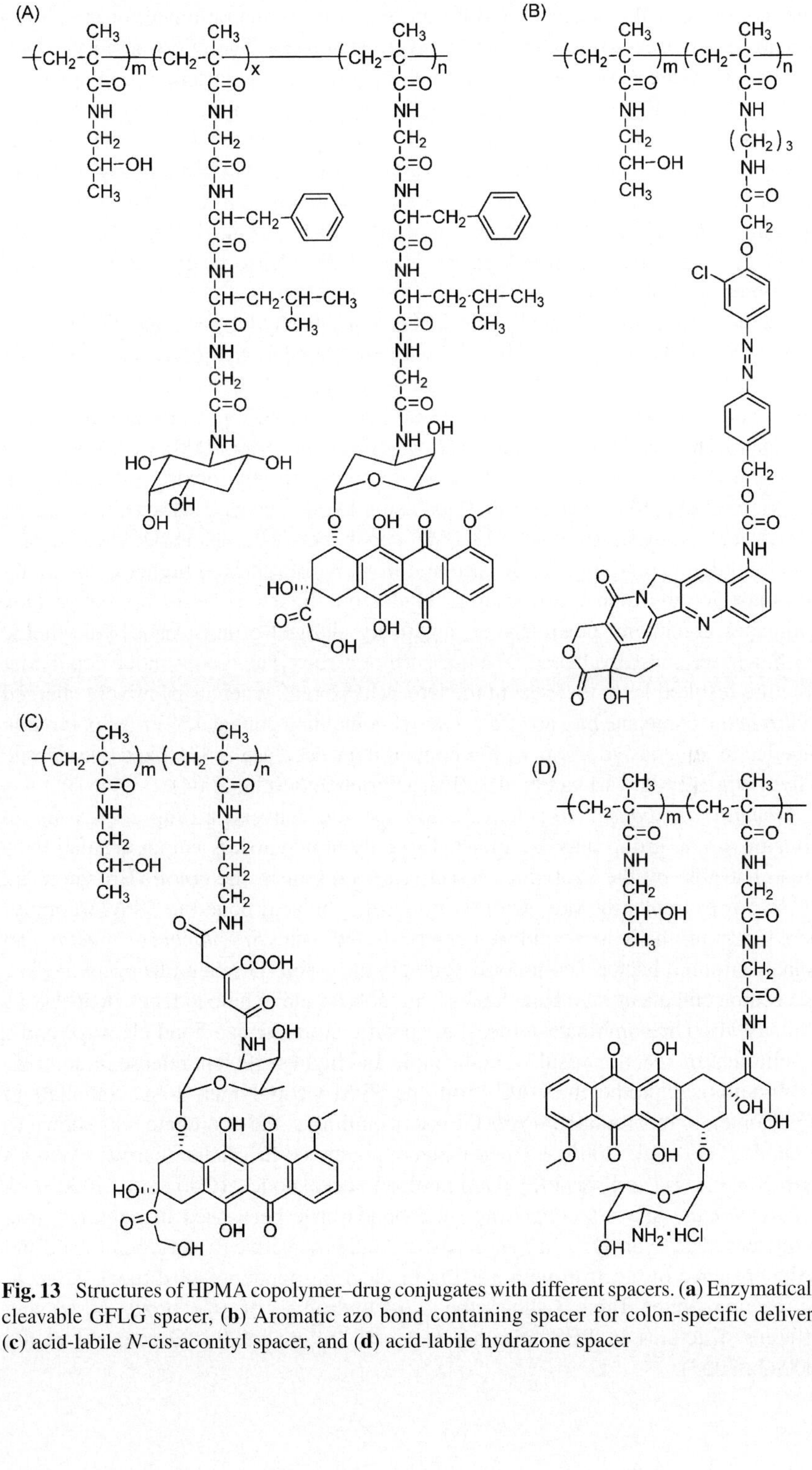

Fig. 13 Structures of HPMA copolymer–drug conjugates with different spacers. (**a**) Enzymatically cleavable GFLG spacer, (**b**) Aromatic azo bond containing spacer for colon-specific delivery, (**c**) acid-labile *N*-cis-aconityl spacer, and (**d**) acid-labile hydrazone spacer

than that of normal tissue. These differences in pH are the rationale for the design of acid-sensitive *N-cis*-aconityl and hydrazone linkers. The *N-cis*-aconityl linker was first introduced into polymer–drug conjugates in daunorubicin conjugates in 1981 (Shen and Ryser, 1981). The *N-cis*-aconityl spacer in the conjugate has considerable stability at pH 6, but was readily hydrolyzed at pH 4. Since then, many polymeric prodrugs used the *cis*-aconityl amide linkage; poly(aminopropyl)dextran-daunorubicin (Mann et al., 1992), alginate-daunomycin (Al-Shamkhani and Duncan, 1995), HPMA copolymer–doxorubicin (Choi et al., 1999; Ulbrich et al., 2003), etc., have been synthesized targeting drug release in the endosomes and lysosomes of tumor cells.

Another frequently used acid-labile attachment is the hydrazone linker (Kratz et al., 1999). The aliphatic aldehyde-based hydrazone bond is hydrolyzed at pH 5, but relatively stable at pH 7.4, whereas the aromatic aldehyde-based hydrazone bond is highly stable at both pH values (Kale and Torchilin, 2007). Many polymer–drug conjugates, such as PEG–doxorubicin (Rodrigues et al., 1999, 2006), PEG–paclitaxel (Rodrigues et al., 2003), polyglutamine–streptomycin, and dextran–streptomycin (Coessens et al., 1996) were synthesized using hydrazone as a pH-sensitive spacer. Ulbrich et al. synthesized a series of HPMA copolymer–doxorubicin (Dox) conjugates containing hydrazone linkers. In vitro and in vivo tests showed higher cytotoxicity antitumor activity than the free drug. Hydrazone-linked HPMA copolymer–Dox conjugates containing positively or negatively charged groups, or a hydrophobic substituent were also evaluated. The presence of carboxylate groups in the copolymer structure resulted in an increase in the Dox release rate, whereas positively charged groups in the conjugate had no effect. Oleoyl-containing conjugate formed polymeric micelles in aqueous solution with a concomitant decrease in the Dox release rate (Chytil et al., 2006; Mrkvan et al., 2005; Ulbrich and Šubr, 2004).

Aromatic azo bonds. To achieve colon-specific delivery, a drug possessing an aromatic amine group may be attached to polymeric carriers via an aromatic azo bond cleavable by the azoreductase activities present in the colon (Brown et al., 1983). For example, the release of 5-aminosalicylic acid bound to HPMA copolymers via an aromatic azo bond was demonstrated using *Streptococcus faecium*, an isolated strain of bacteria commonly found in the colon (Grim and Kopeček, 1991), the cecum contents of rats, guinea pigs, and rabbits, and in human feces (Kopečková et al., 1994). The combination of colon-specific aromatic azo bond cleavage and a 1,6-elimination reaction resulted in the rapid and highly efficient release of unmodified 9-aminocamptothecin (9-AC) from the HPMA copolymer–9-AC conjugate in cecal contents. In simulated upper GI tract conditions, the conjugate was shown to be stable (Gao et al., 2006a). The conjugate possessed favorable pharmacokinetics (Gao et al., 2007) and was effective in colon cancer models (Gao et al., 2006b).

Ester bonds. Spacers containing ester bonds have been used in polymer–drug conjugates. The hydrolysis rate of the ester bond is sensitive to steric hindrance and to the presence of electron-withdrawing or electron-donating substituents close to the reaction center. Table 3 shows the $t_{1/2}$ of hydrolysis of ester bonds possessing different structures in PBS, plasma, and cell culture media (Greenwald et al., 2000a,b, 2003).

Disulfide bonds. Spacers containing disulfide bonds (–S–S–) have been used to control subcellular cleavage. Disulfide bond reduction and oxidation are mediated by small redox molecules such as glutathione or redox enzymes. The extracellular space is usually an oxidizing environment that favors the retention of disulfide bonds. Inside cells, due to the presence of reduced glutathione (GSH) and thioredoxin reductase, the cytosol is a reducing environment (Saito et al., 2003). In polymer conjugates internalized via endocytosis and localized in lysosomes, disulfide bonds may be cleaved by γ-interferon-inducible lysosomal thiol reductase (GILT). Cysteine is actively transported into lysosomes and acts as a reducing buffer to maintain the activity of the enzyme (Arunachalam et al., 2000; Phan et al., 2000).

Table 3 *In vitro* and *in vivo* hydrolysis of the prodrugs with ester bond (Greenwald et al., 2000b)

Compound	$t_{1/2}$ (h) PBS pH 7.4	$t_{1/2}$ (h) Rat plasma	$t_{1/2}$ (h) Cell media
	>24	0.2	80
	>24	1.9	14
	>24	17	32
	>24	8	94
	>24	21	36
	>24	1.1	8
	>24	>24	53

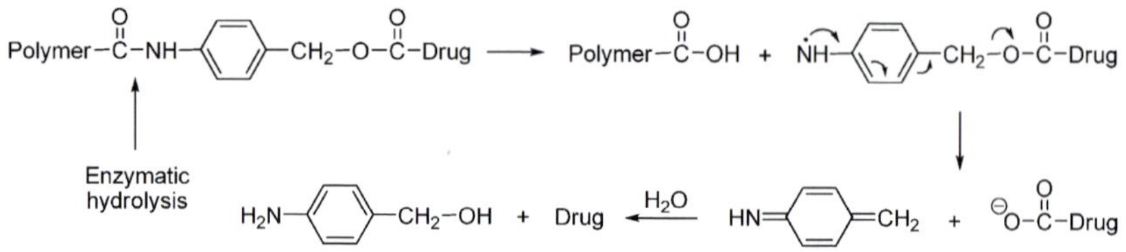

Fig. 14 Mechanism of drug release by enzymatic hydrolysis followed by 1,6-elimination reaction

Self-eliminating linkers. If a drug cannot be directly linked to an enzymatically cleavable spacer, a self-elimination group can be introduced between the spacer and the drug (Fig. 14). Elongated spacers that separate the enzymatically cleaved bond from the drug by a self-eliminating group have been designed by several laboratories (Carl et al., 1981; de Groot et al., 2001; Toki et al., 2002). The most predominant example of an electronic cascade spacer is the 1,6-elimination spacer (Carl et al., 1981). One example is bifunctional *p*-aminobenzyl alcohol group linked to an enzymatically cleavable group through an amine moiety. Amine-containing drugs can be bound through the benzylic hydroxyl group forming a carbamate functionality. After enzymatic cleavage, the strong electron-donating amine group of the 1,6-elimination spacer is unmasked and immediately initiates an electronic cascade that leads to cleavage of the benzyl-carbamate bond and the release of carbamic acid. The unstable carbamic acid rapidly releases carbon dioxide to yield the unmodified drug (de Groot et al., 2001; Toki et al., 2002). Such an approach was used for the design of oral drug delivery systems based on HPMA copolymer–9-amino-camptothecin conjugates (Gao et al., 2006a).

2.3 Targeting Groups

Passive targeting. Enhanced permeability and retention (EPR) effect was first proposed by Maeda (Matsumura and Maeda, 1986) to describe the observation that macromolecular conjugates tend to accumulate in tumor tissues more compared to normal tissues. The EPR principle is shown in Fig. 15. The phenomenon has been attributed to the high vascular density of the tumor, the increased permeability of tumor vessels, defects in tumor vasculature, and decreased lymphatic drainage in the tumor interstitium (Fang et al., 2003; Greish et al., 2003; Maeda, 2001a). Other factors, however, may have an opposite effect. For example, a high intratumoral pressure may result in a convective fluid flow from the center of the tumor to the periphery (Jain, 1989). Nevertheless, a number of studies have shown increased accumulation of macromolecules in tumors as compared to normal tissue (Maeda, 2001b; Shiah et al., 2001a).

Negative targeting. Sometimes the term "negative targeting" is used to describe the fact that by binding a drug to a water-soluble polymer carrier, the site of the nonspecific toxicity of the drug can be avoided. For example, DOX is cardiotoxic,

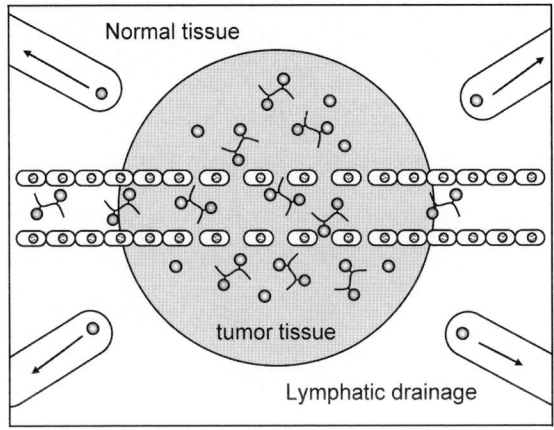

Fig. 15 The enhanced permeability and retention (EPR) effect (Matsumura and Maeda, 1986)

and its MTD (maximum tolerated dose) in humans is 60–80 mg/m². The MTD of HPMA copolymer–DOX conjugate was 320 mg/m², which was attributable to the fact that endocytosis is attentuated in heart tissue (Vasey et al., 1999).

Active targeting. Anticancer chemotherapy drugs are limited by a number of serious side effects that arise from toxicities to normal cells because the therapies are typically only selective toward proliferating cells. The EPR effect and specific spacers can favorably modify the biodistribution polymer–drug conjugates when compared to low molecular weight drugs; but how can selectivity be improved further? One strategy is to use targeting moieties, i.e., to couple macromolecular therapeutics with antibodies or other ligands that recognize tumor-associated antigens or recognition sites.

For different purposes, antibodies, ligands, including carbohydrates, small molecular hormones, oligopeptides, and charged compounds can be used for biorecognition. The choice of targeting groups rests on the fact that cells express unique proteins specific to the cell type, and that the binding of some ligands can trigger receptor-mediated endocytosis (Mukherjee et al., 1997). Antibodies (including monoclonal and polyclonal antibodies, and their Fab fragments) (Fig. 16) and many non-antibody ligands can be selected for this purpose. A review of the antibodies and other ligands that have been used are discussed by Allen (2002). Table 4 lists selected receptors that can be targeted by specific ligands.

Lectins are carbohydrate-specific cell surface receptors. They can be targeted by specific carbohydrates or their derivatives. Galactose was often used as a targeting group for hepatocytes, because of the large number of asialoglycoprotein receptors (ASGP-R) on their surface. HPMA copolymer conjugates containing galactose moieties have been shown to rapidly accumulate into hepatocytes (Duncan et al., 1983, 1986). Other carbohydrates such as mannose and fucose were studied as targeting moieties for macrophages.

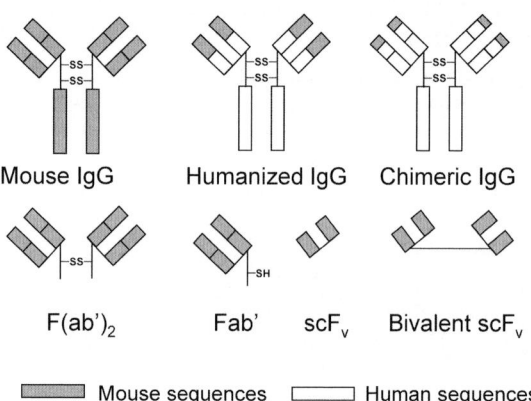

Mouse IgG   Humanized IgG   Chimeric IgG

F(ab')₂   Fab'   scF_v   Bivalent scF_v

Mouse sequences   Human sequences

Fig. 16 Antibodies and antibody fragments (Allen, 2002)

Table 4 Selected examples of receptors and their ligands

Receptor	Ligand	References
Lectins	Carbohydrates	Gref et al. (2003)
Vitamin receptor	Vitamins	Gabizon et al. (1999)
Hormone receptor	Hormones	Akhlynina et al. (1995)
Growth-factor receptor	Growth factors	Gijsens and De Witte (2000), Lanciotti et al. (2003), and Vega et al. (2003)
$\alpha_v\beta_3$ Integrin	RGD peptides	Hersel et al. (2003) and Hynes (1992)
Transferrin receptor	Transferrin	Derycke and De Witte (2002)
GM-CSF receptor	GM-CSF	Frankel et al. (2002)
Neurotensin receptor	Neurotensin	Martinez-Fong et al. (1999)
Acetylcholine receptor	RVG peptide	Kumar et al. (2001)

$\alpha_v\beta_3$ integrins are expressed on newly formed endothelial cells and on various tumor cells. Arginine-glycine-aspartic acid (RGD) peptides preferentially bind to $\alpha_v\beta_3$ integrin. The binding constant may be increased when conformational constrain is introduced into the ligand. To increase the efficiency of binding, RGD-containing targeting groups were often designed as cyclic peptides (Mitra et al., 2006). Another integrin-binding peptide, PLAEIDGIELTY, discovered by phage display is specific for the $\alpha_9\beta_1$-integrin receptor (Schneider et al., 1998), which is highly expressed in human airway epithelia.

A 29-amino-acid peptide derived from rabies virus glycoprotein (RVG) can specifically bind to the nicotinic acetylcholine receptor (AchR) on neuronal cells. Nonamer arginine derivative of RVG (RVG-9R) enables binding and transvascular delivery of small interfering RNA (siRNA) to the brain. The peptide RVG-9R provides a safe way for the delivery of siRNA or other therapeutic molecules across the blood-brain barrier (Kumar et al., 2007).

Monoclonal antibodies (mAb) and antibody fragments often have a higher degree of specificity than non-antibody ligands for target cells. They are frequently being used in polymer–drug conjugates for targeting; however, they are expensive and prone to immunological reactions. Monoclonal antibodies are usually of murine origin (Köhler and Milstein, 1975) and are prone to generate immune responses by the production of human antimouse antibodies (HAMA) when administered in humans. Chimeric mAbs and humanized mAbs may reduce this immunological problem, but it may still lead to the mAb binding to normal tissues through Fc receptors. Antibody fragments (F(ab')$_2$, Fab' and scFv) lack the Fc domain and the complement-activating region, and this has been shown to reduce to immunogenicity. Binding affinity may also be increased by using bivalent or multivalent fragments.

Bone targeting moieties. The most frequently used bone targeting groups (Wang et al., 2005) are tetracyclines, bisphosphonates, and oligopeptides of aspartic acid (Fig. 17). HPMA copolymers containing bisphosphonates or octapeptide of D-aspartic acid (D-Asp$_8$) showed efficient adsorption on a hydroxyapatite bone model (Hruby et al., 2006; Wang et al., 2003). An in vivo biodistribution study of I^{125}-labeled HPMA copolymers containing D-Asp$_8$ in mice showed a high accumulation of the conjugates to bone (Wang et al., 2006).

Protein transduction domains (PTD). Some transcription factors, including the TAT protein of human immunodeficiency virus (HIV) (Frankel and Pabo, 1988) and VP22 protein of herpes simplex virus (HSV) (Elliott and O'Hare, 1997), have been shown to possess regions that allow them to efficiently transduce to the cytoplasm. These PTDs consist of large number of basic amino acid residues (arginine and lysine). HIV-1 TAT is an 86 amino-acid protein, and TAT transdution peptide Y$_{47}$GRKKRRQRRR$_{57}$ possesses a high net positive charge at physiological pH.

TAT-mediated transduction can direct the uptake of proteins, nanoparticles, and liposomes of large sizes (Wadia and Dowdy, 2005). In vitro studies on HPMA copolymer–Dox conjugates containing the TAT peptide also showed that the intracellular delivery of polymer-bound Dox was improved, and that the TAT peptide can also transport synthetic macromolecules into cytosol and nucleus (Nori et al., 2003a,b). The mechanisms of TAT peptide-mediated cell uptake and the use of the TAT peptide in polymer-drug/gene delivery systems was reviewed by Nori and Kopeček (2005).

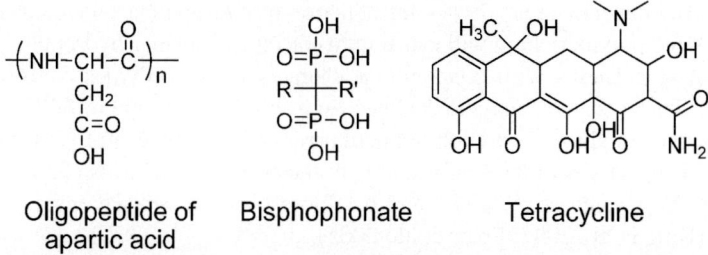

Oligopeptide of apartic acid   Bisphophonate   Tetracycline

Fig. 17 Structures of some bone affinity compounds

2.4   Coupling Reactions in the Synthesis of Conjugates

Drugs, targeting moieties, or other functional groups can be incorporated into copolymer conjugates either by copolymerization using functionalized monomers (e.g., polymerizable drug derivative), or by coupling reactions with copolymers having reactive groups (polymeranalogous reactions). Design of functional monomers and the choice of a polymerization method are important for the direct synthesis of polymer–drug conjugates. The selection of coupling method is important for the polymeranalogous route of conjugate synthesis. Table 5 summarizes coupling reactions usually used in the synthesis of polymer–drug conjugates and some bifunctional commercial coupling (crosslinking) reagents.

The thiazolidine-2-thione reactive group can be used to highlight a recent development in the synthesis of conjugates. This reactive amide group showed low susceptibility to hydrolysis and a high rate of aminolysis in aqueous solutions (Šubr and Ulbrich, 2006; Šubr et al., 2006). It has been very useful as a coupling reaction in aqueous solutions.

3   Physical Properties and Drug Delivery

Polymeric drug carriers containing multiple components, such as targeting moieites, drug releasing modules, and endosome-disrupting modules, have showed the potential to perform multiple functions within a single structure (Lee et al., 2006). Ideally, each component within the delivery system should function independently, without affecting the functionality of the other components. In reality, however, the physical and bio-logical properties of each component can be expected to exert some influence on the other components (Pan et al., 2006; Ulbrich et al., 1987). Therefore, awareness of the complexities caused by the introduction of each component (like polymer self-asso-ciation) is required to design multicomponent drug carriers.

3.1   Solution Properties of Polymer–Drug Conjugates

Water-soluble polymers typically assume a random coil conformation in aqueous solution (Bohdanecký et al., 1974). Introduction of hydrophobic side chains into a water-soluble polymer chain will cause intra- or intermolecular hydrophobic asso-ciation of side chains with concomitant changes in the polymer conformation (Fig. 18). These changes may be detected by a shift in the shape of GPC profiles, by differences in phase separation temperature, and by changes in the viscometric behavior. The changes of the conformation of the polymer can be also demonstrated by NMR (Nagayama et al., 2002) and by fluorescence resonance energy transfer (FRET) (Ding et al., 2007; Sparr et al., 2005).

Polymer associations may have a strong influence on the biological functions of the polymer conjugates. Associations are the result of inter- and/or intramolecular

Table 5 Examples of some coupling reactions and commercial coupling reagents

Coupling reaction

(a) R_1-COOH + H_2N-R_2 $\xrightarrow{\text{Coupling agent}}$ $R_1-CO-NH-R_2$

(b) $R_1-\overset{O}{\overset{\|}{C}}-X$ + H_2N-R_2 $\longrightarrow$ $R_1-CO-NH-R_2$

(c) $R_1-\overset{O}{\overset{\|}{C}}-CH_2-X$ + H_2N-R_2 $\longrightarrow$ $R_1-CO-CH_2-NH-R_2$

(d) $R_1-N=C=S$ + H_2N-R_2 $\longrightarrow$ $R_1-NH-\overset{S}{\overset{\|}{C}}-NH-R_2$

(e) $R_1-CH=O$ + H_2N-R_2 $\longrightarrow$ $R_1-CH=N-R_2$

(f) + $HS-R_2$ $\longrightarrow$

(g) $R_1-\overset{O}{\overset{\|}{C}}-CH_2-X$ + $HS-R_2$ $\longrightarrow$ $R_1-CO-CH_2-S-R_2$

(h) $R_1-\overset{\ominus}{N}-\overset{\oplus}{N}{\equiv}N$ + $\equiv-R_2$ $\longrightarrow$

Some useful coupling reagents

(i) N-Succinimidyl 4-(maleimidomethyl) cyclohexanecarboxylate (SMCC)   ACS No. 64987-85-5

(j) N-Succinimidyl 3-(2-pyridyldithio)propionate (SPDP)   ACS No. 68181-17-9

(k) (+)-Biotin N-hydroxysuccinimide ester (NHS-Biotin)   ACS No. 35013-72-0

(a) Coupling reaction between carbonic acid and amine mediated by coupling agents; (b) aminolysis of active ester or active amide; (c) nucleophilic substitution; (d) reaction of isothiocyanate with amine; (e) condensation of carbonyl compound with amine; (f) addition of thiol to maleimide group; (g) nucleophilic substitution; (h) Huisgen 1,3-dipolar cycloaddition; (i) coupling reagent to link an amino group with a thiol group; (j) coupling reagent to link an amino group with a thiol group; (k) reagent for biotinylation of an amino group containing compound or protein to provide recognizability by avidin.

Fig. 18 Intramolecular association of a polymer chain containing hydrophobic side chains in aqueous solution – formation of a unimolecular micelle

interactions, and are typically hydrophobic interactions. The preference for either types of association is intrinsically determined by polymer concentration, and structural parameters such as the size of the polymer chain, or the type, content, and arrangement of hydrophobes in the polymer (Chang and McCormick, 1993). It can also be influenced by extrinsic factors, such as temperature, solvent, ionic strength, and pH (Chang and McCormick, 1994).

Ulbrich et al. (1987) systematically studied this phenomenon using HPMA copolymers containing different amounts of p-nitroaniline covalently bound via an enzymatic cleavable oligopeptide spacer. Light scattering, GPC, and sedimentation studies have shown that these conjugates associate and form micelles in water. In this model, micelles were arranged with hydrophobic p-nitoaniline groups inside and hydrophilic polymer chain outside. The number and shape of the micelles depended on the density of the hydrophobic side chain in the conjugate, the conjugate concentration, and the temperature.

Polyelectrolytes bearing hydrophobic grafts may also aggregate in aqueous solution due to hydrophobic association. Poujol et al. (2000) have shown that the association of polymers containing carboxyl groups and alkyl grafts is considerably more stable than micelles formed from amphiphilic diblock copolymers. These processes may be used for the solubilization of hydrophobic drugs, e.g., paclitaxel. However, the amount of solubilized drug is much lower than in the micelles formed from amphiphilic diblock copolymers (Gautier et al., 1997; Poujol et al., 2000). Uchegbu et al. studied hydrophobic grafts with PLL and PEI and found that PEI containing low levels of hydrophobic grafts (cetyl groups < 23 mol%) forms micelles in water. Increasing the content of hydrophobic grafts, the polymers formed vesicles (cetyl groups = 23–42 mol%) and dense nanoparticles (cetyl groups < 49 mol%) (Wang et al., 2004a).

Block copolymers composed from hydrophilic and hydrophobic polymer segments form polymeric micelles in aqueous solutions via hydrophobic association. They have been used as versatile vehicles for drug delivery (Alakhov and Kabanov, 1998; Nishiyama and Kataoka, 2006).

The distinction between the inter- and intramolecular association of side chains in multifunctional polymers can be made using a combination of physicochemical methods. HPMA copolymers containing side chains terminated in chlorin e_6 (a photosensitizer) were evaluated by determination of the quantum yield of singlet oxygen formation and by light scattering. The decrease of the quantum yield indicated polymer association, while minimal changes in the hydrodynamic volumes of the conjugates (as observed by dynamic light scattering) indicated that the association was intramolecular (Shiah et al., 1997, 1998).

Ding et al. (2007) evaluated the relationship between structure and self-association of HPMA copolymer–heptapeptide (YILIHRN) conjugates, where the peptide was a ligand for the CD21 receptor. Using fluorescence energy resonance transfer (FRET), their conformation in solution was evaluated. In addition to heptapeptide (HP), HPMA copolymers contained side chains terminating in tryptophan (energy donor) and dansyl (energy acceptor). Conjugate solutions were evaluated using an excitation wavelength of 295 nm (ratio of emission intensity 510 nm/370 nm indicated energy transfer efficiency). It was found that higher HP content correlated with higher FRET efficiency, indicating the formation of compact coils (Fig. 19). Modification of the HPMA copolymer backbone by the incorporation of acrylic acid comonomers resulted in decreased FRET efficiency, presumably due to the expansion of the polymer coils as a result of electrostatic repulsion. The dependence of FRET efficiency on pH was in agreement with the ionization profile of the acrylic acid residues (Ding et al., 2007).

The use of small-angle neutron scattering (SANS) is also a method for direct study of the conformation of the polymer–drug conjugates in the solution (Paul et al., 2007).

3.2 Impact of Conformational Change on the Biological Properties of the Conjugates

Conformation can also influence the biological properties of polymer conjugates, and may affect the rate of enzymatic release of drugs or ligands (Putnam and Kopeček, 1995a,b). Increase in the content of hydrophobic heptapeptide in HPMA copolymer–heptapeptide–Dox conjugate (Ding et al., 2007), and increase of

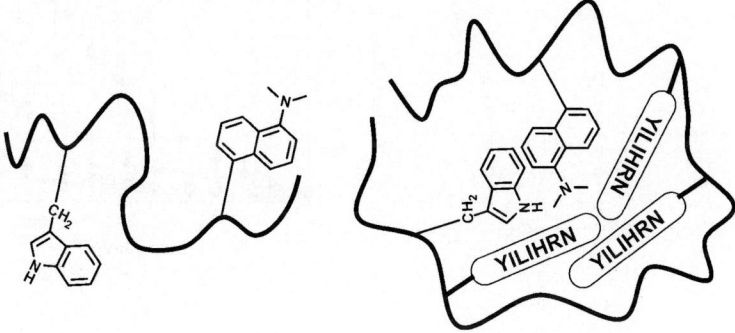

Fig. 19 Incorporation of tryptophan (energy donor) and dansyl (energy acceptor) groups permits to follow the formation of compact polymer coils by fluorescence resonance energy transfer (FRET) (Ding et al., 2007)

prostaglandin E_1 (PGE$_1$) in HPMA copolymer–PGE$_1$ conjugates (Pan et al., 2006) caused decreases in the drug release rate. Apparently, the hydrophobic association resulted in compact coil formation, which thereby impaired the formation of the enzyme–substrate complexes (Ulbrich et al., 1987).

3.3 Molecular Weight, Shape, and Efficiency

The retention time of the polymer–drug conjugates in the body is related to the hydrodynamic radius or molecular weight of the conjugate. The hydrodynamic radius of the renal threshold is about 45 Å. If the hydrodynamic radius of conjugate molecules is above the renal threshold, elimination of the conjugate via glomerular filtration is slow. At this threshold, the molecular weight for polyHPMA is 45 kDa, PEG is 30 kDa, and dextran is 40 kDa (D'Souza and Topp, 2004). One strategy to extend the intravascular half-life of polymer–drug conjugates without impairing the elimination of the carrier from the organism is to include biodegradable linkers between polymer chains (Dvořák et al., 1999). Such conjugates are long circulating before degradation, but are quickly eliminated after degradation. Shiah et al. (2001a) used lysosomal degradable linker as cross-linking agent and prepared four different molecular weights (22, 160, 895, 1230 kDa) HPMA copolymer–Dox conjugates. Results showed that the half-life in circulation was five times longer, and the elimination rate from the tumor was 25 times slower, when the molecular weight increased from 22 to 1230 kDa. The antitumor efficiency was much higher for conjugates with molecular weights higher than 160 kDa (Fig. 20). Similarly, the

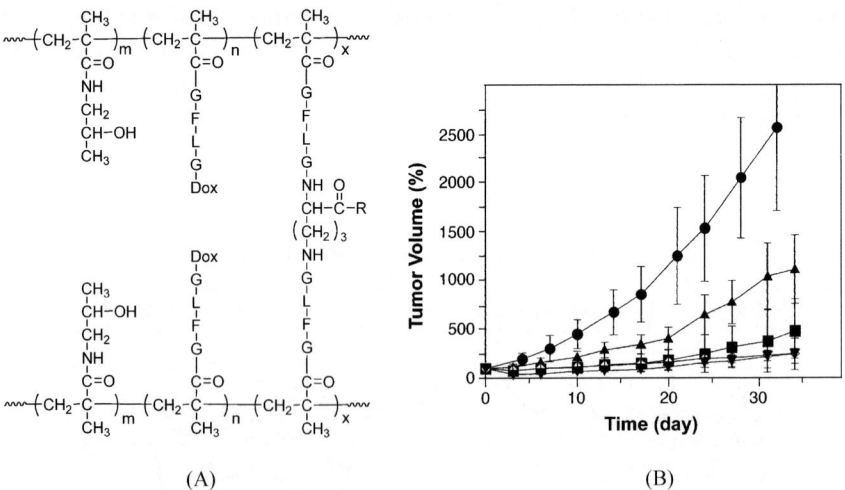

(A) (B)

Fig. 20 Structure (**a**) and therapeutic efficacy (**b**) of HPMA copolymer–Dox conjugates with different molecular weight toward subcutaneous human OVCAR-3 carcinoma xenografts in nu/nu mice. *Filled circle*, control; *Filled triangle*, 22 kDa; *Filled square*, 160 kDa; *Open diamond*, 859 kDa; *Inverted triangle*, 1,230 kDa (Shiah et al., 2001a)

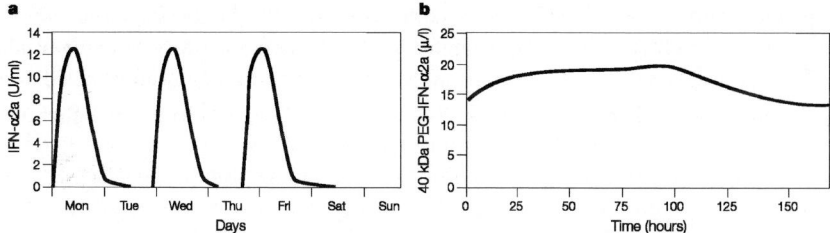

Fig. 21 Human blood level profiles for (**a**) interferon (IFN)-α2a and (**b**) 40 kDa PEG-IFN-α2a administered by subcutaneous injection (adapted from Harris and Chess, 2003)

circulating half-life of PEG–protein conjugate increased dramatically when the molecular weight of PEG was increased from 5 kDa to 40 kDa (Greenwald et al., 2000a) (Fig. 21).

The shape of the conjugate may also affect the activity of the conjugates. Star-shaped antibody containing polymer–drug conjugates had a higher antitumor effect than linear ones (Jelínková et al., 2003; Kovář et al., 2002a). Dendritic polymer–drug conjugates also are efficient macromolecular therapeutics (Khandare et al., 2006b).

3.4 Physicochemical Properties: Biodistribution

Physicochemical properties such as molecular weight, type, and amount of functional groups affect the biodistribution of the polymer–drug conjugates. Lammers et al. (2005) studied the circulation kinetics, tissue distribution, and tumor accumulation of HPMA copolymer conjugates. The results have shown that the relative accumulation levels of HPMA copolymers were significantly higher in tumor than in liver, testis, heart, skin, ileum, and muscle. Accumulation in the spleen was always found to be more selective than accumulation in tumor, and relative levels in lung were generally comparable to levels in tumor. Increasing the average molecular weight of HPMA copolymers resulted in prolonged circulation times and in increased tumor and organ concentrations. The higher molecular weight also increased the tumor-to-organ ratios. Because of the influence of molecular weight, the authors considered the molecular weight of currently used HPMA copolymer–drug conjugates to be suboptimal. Introduction of functional groups, such as carboxyl (COOH) and hydrazide (NHNH$_2$), into HPMA copolymers caused rapid elimination of the conjugate from circulation and lowered the levels in tumors in all organs other than kidney, which had increased accumulation. Seymour et al. studied the effect of molecular weight on body distribution and rate of excretion of HPMA copolymers. They found that the molecular weight threshold limiting glomerular filtration is approximately 45 kDa for intravenous administration. Molecular weight did not influence the movement of copolymers from the peritoneal compartment to the bloodstream after intraperitoneal injection, which may have been due to bulk phase lymphatic drainage. By subcutaneous administration, the large molecular weight copolymer (778 kDa) showed high retention at the injection site (Seymour et al., 1987). When galactosamine targeting moieties were incorporated into the HPMA copolymer, the copolymer was cleared more rapidly than the unsubstituted copolymer from circulation after intravenous administration, and selectively accumulated in the liver. Targeting to the liver was also observed after intraperitoneal and subcutaneous administrations, but the level of liver targeting was lower compared to intravenous administration. There was no significant transfer of HPMA copolymer into tissues following oral administration (Cartlidge et al., 1987a,b).

4   Internalization and Subcellular Fate of Polymer–Drug Conjugates

4.1   Cellular Binding and Internalization

Binding drugs to a polymer carrier results in a change of the mechanism of cell entry. Whereas low molecular weight lipophilic drugs enter the cell by diffusion, water-soluble macromolecular therapeutics enter the cell via endocytosis. Endocytosis refers to the cell uptake of extracellular solutes by engulfing them and trapping them in membrane-bound intracellular vesicles. Endocytosis may be classified into two categories, pinocytosis (cell drinking) and phagocytosis (cell eating) (Conner and Schmid, 2003). Pinocytosis takes place in all cells, whereas phagocytosis is a triggered phenomenon performed by specialized cells (Fig. 22). Kinetically, endocytosis can be classified as fluid-phase, adsorptive, and receptor-mediated endocytosis (Amyere et al., 2002; Khalil et al., 2006). The terms endocytosis and pinocytosis are frequently used synonymously. At least four different pathways may occur: clathrin-mediated endocytosis, caveolae-mediated endocytosis, macropinocytosis, and clathrin/caveolae-independent endocytosis (Lamaze and Schmid, 1995; Mukherjee et al., 1997; Soldati and Schliwa, 2006).

The structure of macromolecules has a strong impact on the kinetics of internalization. Hydrophilic macromolecules, such as homopolymeric HPMA, are internalized via

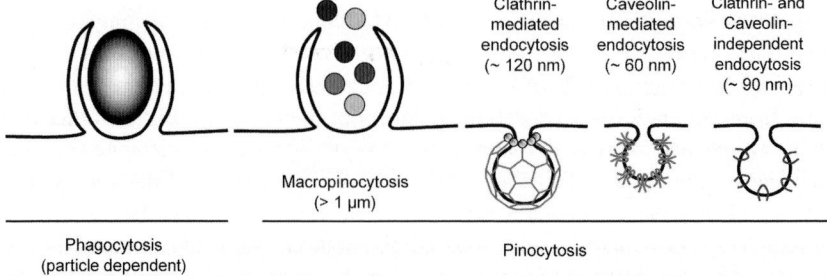

Fig. 22 Endocytosis of mammalian cells (adapted from Conner and Schmid, 2003)

fluid-phase pinocytosis, a slow process. Modification of the structure with hydrophobic moieties or introduction of positively charged groups favors internalization by adsorptive pinocytosis (Duncan et al., 1984), a signficiantly faster process, but nonspecific in all cells. Specific interaction with a subset of cells and internalization via receptor-mediated pinocytosis is achieved by the introduction of specific ligands, or targeting moieties, into macromolecular therapeutics. Introduction of a galactose moiety can produce liver-parenchymal cell-specific targeting (Duncan et al., 1983). Mannose targets macrophages, which overexpress mannose receptors on their surface (Kawakami et al., 2004). Transferrin (Tf), an iron-binding glycoprotein, has been used as a tumor-targeting ligand (Kakudo et al., 2004; Wagner et al., 1992). The folate receptor is also overexpressed in tumor cells, and has been used for tumor targeting (Cho et al., 2005; Lee and Huang, 1996).

Let us briefly describe the potential subcellular fate of macromolecular therapeutics:

Clathrin-mediated endocytosis (CME). CME occurs constitutively in all mammalian cells. It is typically initiated by the formation of clathrin-coated pits in the plasma membrane. The coated pits invaginate and assisted by dynamin, a multidomain GTPase, then pinch off from the plasma membrane to form intracellular clathrin-coated vesicles (CCVs) ranging in size of 100–150 nm in diameter. The clathrin coat depolymerizes forming early endosomes. These progress to late endosomes and fuse with lysosomes where the contents are digested. During the internalization process, the pH drops from neutral to pH about 6 in early endosomes with a further reduction to about 5 in late endosomes and lysosomes (Khalil et al., 2006; Maxfield and McGraw, 2004). The pH drop results in the dissociation of ligands and receptors in early endosomes and the recycling of receptors back to the plasma membrane.

Caveolae-mediated endocytosis. Caveolae are 60–70 nm in diameter, caveolin-coated, flask-shaped invaginations of the plasma membrane enriched in cholesterol and glycosphingolipids (Harris et al., 2002a,b; Matveev et al., 2001). The shape and structure of caveolae depend on caveolin, a dimeric cholesterol-binding protein (Conner and Schmid, 2003). Caveolae are present in many cell types, and are especially common in endothelial cells.

Caveolae have several functions including cholesterol homeostasis, glycosphingolipid transport, and negative regulation of endothelial nitric oxide synthase. Experiments with green fluorescent protein (GFP)-conjugated caveolin show a resting state of caveolae in plasma membranes, but ligands binding to the receptors in caveolae can trigger a signal cascade and cause their internalization (Pelkmans and Helenius, 2002; Pelkmans et al., 2002; Schousboe et al., 2004; Thomsen et al., 2002). Some cationic polymer–DNA complexes and TAT peptide have been hypothesized to be internalized by caveolae (Ferrari et al., 2003; Fittipaldi et al., 2003; Rejman et al., 2005).

Caveolae-mediated endocytosis is a slow internalization process, uptake is nonacidic, and is unlikely to result in lysosomal degradation (Ferrari et al., 2003; Harris et al., 2002a).

Macropinocytosis. Macropinocytosis refers to the formation of large endocytic vesicles mediated by actin-driven invagination of the plasma membrane (Khalil et al., 2006). It is usually associated with a cell surface ruffle formed by a linear band of outward-directed polymerized actin near the plasma membrane resulting from stimulation by growth factors or other signals. After stimulation by mitogenic factors, the ruffles become longer and broader and frequently close into large macropinosomes (Swanson and Watts, 1995). In cells, after the loss of actin, the macropinosomes may shrink, acidify, and fuse with lysosomes (in macrophages), or alternately, recycle their contents back to the cell surface (Meier and Greber, 2003).

Macropinocytosis is an efficient route for the nonselective endocytosis of solute macromolecules (Conner and Schmid, 2003). Some pathogens trigger macropinocytosis to facilitate their own uptake. The TAT peptide and its cargos may be taken up by macropinocytosis (Kaplan et al., 2005; Nakase et al., 2004; Wadia et al., 2004).

Clathrin/caveolae-independent endocytosis. Beside caveolae, there are other microdomains, referred to as "rafts," with unique lipid composition and diffusing freely on cell surface (Edidin, 2001a,b; Kirkham and Parton, 2005). Their specific membrane proteins and glycolipids provide the pathway for clathrin/caveolae-independent endocytosis. For example, the interleukin-2 (IL-2) receptor on lymphocytes is internalized associated with lipid microdomains, but in a clathrin- and caveolin-independent manner (Lamaze et al., 2001).

Phagocytosis. Phagocytosis in mammalian cells is conducted primarily by macrophages, monocytes, and neutrophils. It functions to clear pathogens ($>0.5 \mu m$) such as bacteria and yeast, or large debris such as dead cells and arterial fat deposits (Allen and Aderem, 1996). It is regulated by specific receptors and signaling cascades mediated by Rho-family GTPases. Interaction of specific receptors on the phagocyte with ligands on the surface of the particle triggers the assembly of actin, resulting in the formation of cell surface extensions around the particle and the engulfment of the particle. After internalization, phagosomes lose actin and mature to phagolysosomes after a series of fusion and fission processes whereby internalized particles are degraded (Allen and Aderem, 1996).

Nonendocytic delivery. Some cationic peptides, the PTDs, such as the TAT, penetratin, and VP22 peptides, may have the ability to directly penetrate cell membranes by an energy-independent route without endocytic events (Brooks et al., 2005; Gupta and Torchilin, 2006; Gupta et al., 2005; Vives et al., 1997), but this point was challenged by later studies. Duchardt et al. (2007) showed that antennapedia-homeodomain-derived antennapedia (Antp) peptide, nona-arginine, and the HIV-1 Tat-protein-derived TAT peptide may also simultaneously use three endocytic pathways: macropinocytosis, clathrin-mediated endocytosis, and caveolae/lipid-raft-mediated endocytosis. At higher concentrations of peptide, an endocytosis-independent mode of uptake may be observed.

Microinjection, permeabilization, and electroporation may be also used to deliver drugs or maromolecules into cells, although these are restricted to in vitro experimentation.

Multidrug resistance. The acquired resistance of malignant tumors to chemotherapeutic agents remains the major cause of cancer therapy failure. One reason is the increased efflux of low molecular weight cytotoxic agents by ATP-driven efflux pumps, such as P-glycoprotein. Polymer–drug conjugates, due to their mechanism of cell entry, have the potential to bypass this resistance mechanism (Gottesman et al., 2002; Kopeček and Kopečková, 2003; Larsen et al., 2000; Omelyanenko et al.,

1998). Numerous studies have demonstrated enhanced efficacy of macromolecular therapeutics in resistant tumors (Kabanov et al., 2002; Minko et al., 1999, 2000; Okhawa et al., 1993; Štastný et al., 1999).

Techniques used for the evaluation of the endocytotic uptake pathways. Endocytic uptake pathways can be inhibited by treatment with specific inhibitors. In general, endocytic uptake is an energy-dependent mechanism; therefore, it can be inhibited by the use of metabolic inhibitors to deplete the ATP pool or by lowering the temperature (Duncan et al., 1986; Saraste et al., 1986).

The assembly of the clathrin lattice on the plasma membrane is essential for CME. Treatments that cause the dissociation of clathrin, such as potassium depletion, cytosol acidification, hypertonicity, and chlorpromazine treatment, can specifically inhibit CME (Lamaze and Schmid, 1995).

Cholesterol is required for the caveolae pathway. Drugs that specifically bind, sequester, or deplete cholesterol such as filipin, nystatin, and methyl-β-cyclodextrin, respectively, perturb internalization through the caveolae (Lamaze and Schmid, 1995). Caveolae also depend on the actin cytoskeleton, and drugs that cause the depolymerization of the actin cytoskeleton, such as cytochalasin D, can inhibit caveolae uptake (Parton et al., 1994). For internalization of poly(N,N-(dimethylamino)ethyl methacrylate) (pDMAEMA)- and PEI-complexed DNA, it was found that both complexes were colocalized with fluorescently labeled transferrin and cholera toxin in COS-7 cells, which indicates uptake via the clathrin- and caveolae-dependent pathways. By using specific inhibitors, blocking the caveolae-mediated uptake route resulted in an almost complete loss of polyplex-mediated gene expression, whereas gene expression was not negatively affected by blocking the clathrin-dependent route of uptake. That shows the importance of caveolae-mediated uptake for pDMAEMA and PEI-mediated gene transfection (van der Aa, 2007).

Ruffling is dependent on actin cytoskeleton. Drugs that disrupt the actin cytoskeleton can inhibit macropinocytosis. The ruffling response is also dependent on protein kinase C (Conner and Schmid, 2003). Wortmannin, amiloride, and its analogs also inhibit macropinocytosis (Arcaro and Wymann, 1993; Hewlett et al., 1994).

Akita et al. used sequential Z-series images captured by confocal laser scanning microscopy (confocal image-assisted three-dimensionally integrated quantification (CIDIQ)) to quantitatively analyze intracellular trafficking of plasmid DNA transfected by a nonviral gene delivery system. The DNA was labeled with rhodamine and different intracellular compartments were labeled with different fluorescent markers. Cells were scanned three-dimensionally and images of different sections were recorded. The amount of DNA in each organelle could be obtained by analysis of the data (Akita et al., 2004; Hama et al., 2006).

Although confocal microscopy is typically used to monitor intracellular fate, subcellular fractionation may give more quantitative results to measure the organelles concentration over time (Tijerina et al., 2003a,b). Seib et al. established a differential centrifugation method for B16F10 murine melanoma cells and used it to define the intracellular trafficking of HPMA copolymer–Dox conjugate (PK1). This method demonstrated lysosomotropism of PK1, subsequent Dox liberation, and nuclear localization (Seib et al., 2006).

4.2   Endosomal/Lysosomal Escape

After internalization via endocytosis, uptaken molecules are entrapped in endosomes. These endosomes may fuse with lysosomes for degradation or they may recycle their contents back to the cell surface. Conjugated drugs need to escape the endosomal/lysosomal compartment to reach their targeted organelles (Fig. 23). Low molecular weight drugs linked to the polymer via an enzymatic cleavable or acid-labile spacer can be released from the carrier and then diffuse into the cytosol. Large hydrophilic drugs such as proteins and DNA, however, need a different mechanism for endosomal/lysosomal escape lest they be trapped and degraded in the lysosomes.

In one strategy, PEI and its derivatives are used to release macromolecules from endosomes by disrupting its membrane structure. PEIs contain numerous secondary amino groups that can be protonated in the acidic environment of the endosome/lysosomes. PEI absorbs the protons, which results in osmotic swelling of endosomes/lysosomes and destabilization of their membranes (Boussif et al., 1995; Merdan et al., 2002; Thomas and Klibanov, 2002).

Polylysine cannot efficiently escape the endosome without the addition of endosomolytic agents such as chloroquine or fusogenic peptides (Read et al., 2005). Chloroquine can diffuse into the endosomes/lysosomes and protonate in its acidic environment and produce swelling and destabilization of endosomal/lysosomal membranes. Chloroquine also inhibits the acidification and maturation of endosomes, which retards the lysosomal degradation of genes. Histidine can be introduced into polylysine to provide an endosomal/lysosomal escape route without the addition of endosomolytic agents (Midoux and Monsigny, 1999).

Other polycations containing imidazole groups have also been synthesized. They also act as a proton sponge and mediate endosomal/lysosomal escape. Furthermore, their cytotoxicity is much lower than that of PEI and polylysine (Pichon et al., 2001; Putnam et al., 2001, 2003).

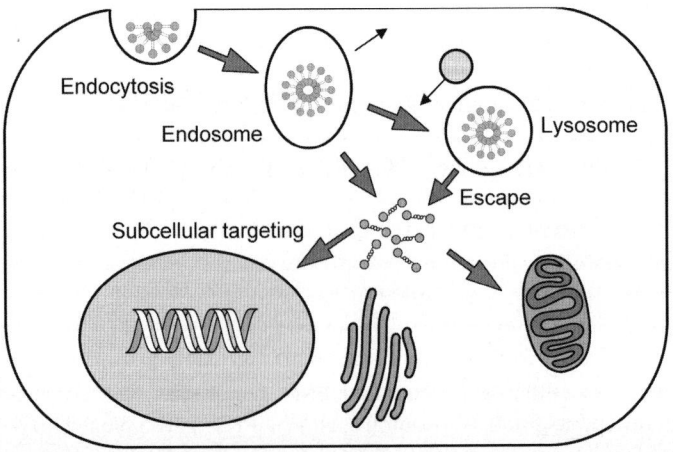

Fig. 23 Uptake and subcellular trafficking of polymer–drug conjugates

pH-sensitive fusogenic peptides are used by bacteria and viruses to facilitate the escape of their cargo from lysosomes to host cells. Peptides derived from these viruses, such as the peptide derived from the *N*-terminal sequence of the influenza virus hemagglutinin subunit HA-2 (GLFGAIAGFIENGWEGMIDGWYG), or synthetic analogues, such as GALA, KALA, or JTS-1 (Gottschalk et al., 1996; Simoes et al., 1999; Wagner et al., 1992), can also perturb the endosomal membrane and affect release. GALA (WEAALAEALAEALAEHLAEALAEALEALAA) is a 30 amino acid peptide that undergoes a conformation transition from random coil to α-helix when the pH changes from 7.4 to 5 in the lysosomal compartment. The α-helical GALA can form multimer aggregates and insert into the lysosomal membrane to form pores. KALA (WEAKLAKALAKALAKHLAKALAKALKACEA) is a cationic counterpart of GALA. KALA also undergoes a pH-dependent conformation change and induces membrane leakage. JTS-1 (GLFEALLELLESLWELLLEA) is an amphipathic peptide able to form an α-helix with nonpolar amino acids on the hydrophobic face and glutamic acid residues on the opposite hydrophilic face. It produced lysis in phosphatidylcholine liposomes and in erythrocytes at pH 5.

4.3   Organelle Delivery

To increase the pharmaceutical efficiency of polymer–drug conjugates, a double-targeted approach is desirable. The conjugates should not only target the specific cells, but also direct the drug into a specific subcellular compartment. For example, photosensitizers are significantly more effective when they are localized in mitochondria or nuclei.

Transport into the nucleus generally occurs through nuclear pore complexes (NPCs), but the inner diameter of an NPC is only ~9 nm, too small for efficient gene or other types of macromolecular delivery. Callahan et al. (2008) studied nuclear entry kinetics of a series of HPMA copolymers containing different functional groups after microinjecting them into the cytoplasm of ovarian carcinoma cells. HPMA copolymers of molecular weight <20 kDa can quickly enter the nuclei, whereas HPMA copolymers with a molecular weight >70 kDa were excluded. The rate of entry of HPMA copolymers with molecular weight between 20–70 kDa depended on both their structure and molecular weight.

Conjugation with subcellular targeting peptide sequences is one method of trafficking macromolecules (reviewed by Cuchelkar and Kopeček, 2006; Nori and Kopeček, 2005). To ensure nuclear targeting, many nuclear localization signal (NLS) peptides have been developed to allow entry into the nucleus through the NPCs by active transport (Boulikas, 1993; Christophe et al., 2000; Goldfarb et al., 1986; Hodel et al., 2001; Mahat et al., 1999; Nigg, 1997). NLSs are short basic monopartite or bipartite amino acid sequences that can bind to cytoplasmic receptors known as importins (Bremner et al., 2001; Gorlich and Mattaj, 1996; Zanta et al., 1999). The most frequently used monopartite NLS is derived from the large tumor antigen of simian virus 40 (SV40). A 7 amino acid sequence of PKKKRKV is required for activity. It has been shown to increase the rate of nuclear uptake of plasmid DNA (Collas and Alestrom, 1996; Collas et al., 1996; Lanford et al., 1986), and enhance the chemotherapeutic activity of HPMA copolymer–Mce$_6$ conjugate in human ovarian carcinoma cells (Tijerina et al., 2003a,b). Peptide fragments M1 from human c-*myc* protein P$_{320}$AAKRVKLD$_{328}$, and M2 R$_{364}$QRRNELKRSP$_{374}$ are both able to target their cargo to the nucleus (Dang and Lee, 1988).

Bipartite NLS sequences contain two basic amino acids domains separated by an oligopeptide spacer (Mahat et al., 1999). *Xenopus* protein nucleoplasmin is one of the most characterized bipartite NLSs. Nucleoplasmin has two basic amino acid sequences separated by a 10 amino acid spacer. The overall sequence is KRPAATKKAGQAKKKK (Robbins et al., 1991). Another bipartite NLS, also derived from *Xenopus* N1 protein, is a 24 amino acid sequence composed of two nuclear uptake sequences linked by a 10 amino acid spacer (VRKKRKTEEESPLK DKDAKKSKQE) (Kleinschmidt and Seiter, 1988). For efficient nuclear targeting, the full length of these bipartite NLSs is needed.

Different low molecular weight hormones can also be used as potential nuclear targeting groups such as 5α-dihydrotestosterone (DHT) and cortisol (Cuchelkar et al., 2008).

Mitochondria represent another important target. In eukaryotic cells, mitochondria produce ATP, are involved in citrate cycle, in the synthesis of steroid hormones and gluconeogenesis. Mitochondrial dysfunction contributes to a large number of human disorders, such as neurodegenerative diseases, obesity, diabetes, etc. Mitochondria play a key role in the complex apoptosis mechanism. Many drugs, including paclitaxel, directly act on mitochondria triggering apoptosis (Torchilin, 2006b; Weissig, 2005). Some cationic compounds can penetrate the cell membrane and specifically target mitochondria, such as rhodamine-123 (Teicher et al., 1986, 1991), oligoguanidinium (Fernandez-Carneado et al., 2005), and triphenylphospho-nium (TPP) (Coulter et al., 2000; Kelso et al., 2002; Smith et al., 1999, 2004). Callahan and Kopeček (2006) synthesized HPMA copolymer containing a TPP moiety bound to the end of the copolymer chain (Fig. 24). TPP-HPMA conjugate was shown to be specifically absorbed by isolated mitochondria; however, in cultured cells, it was shown that the cellular uptake was by endocytosis only. Furthermore, microinjected TPP-HPMAs failed to specifically localize to mitochondria. In another

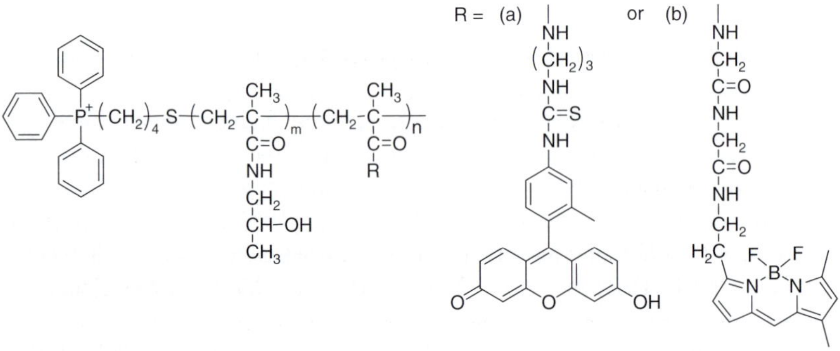

Fig. 24 Structure of triphenylphosphonium (TPP)-functionalized semitelechelic HPMA copolymer for mitochondrial targeting labeled with (**a**) FITC and (**b**) BOPIPY (Callahan and Kopeček, 2006)

approach, a low molecular weight drug was directly linked to TPP (drug-TPP), and then bound to the HPMA copolymer via a lysosomally cleavable spacer. After uptake by cultured cells, the drug–TPP conjugate was released from the lysosomes and specifically trafficked into mitochondria (Cuchelkar et al., 2008).

5   Recent Progress in Polymeric Drug Delivery Systems

5.1   *Polymer–Antitumor Drug Conjugates*

Many drugs, including Dox, daunorubicin, 1-β-D-arabinofuranosylcytosine (Ara-C), paclitaxel, and camptothecin, have been attached to various polymer carriers such as HPMA copolymers, PEG, and PG for the purpose of improving efficiency and lowering side effects. The properties of many preclinically evaluated and clinically tested polymer–drug conjugates have been reviewed by Cuchelkar and Kopeček (2006), Duncan (2006), Hoste et al. (2004), Khandare and Minko (2006), Satchi-Fainaro et al. (2006), and Twaites et al. (2005). Here, we will briefly mention some clinical results of selected polymer–drug conjugates and discuss the implications for the improved design of polymer–drug conjugates.

HPMA copolymer–GFLG–doxorubicin conjugate (PK1) was the first polymer–drug conjugate to enter phase I clinical trials (Bilim, 2003; Thomson et al., 1999; Vasey et al., 1999). Clinical pharmacokinetics showed a prolonged plasma circulation and uptake of the conjugate, and positive responses by some tumors. By using HPMA copolymer–doxorubicin conjugate, high doses (four–fivefold more than free drug) could be administered with reduced immunogenic and toxic responses (Duncan, 2006). PK2 is a HPMA copolymer–GFLG–doxorubicin conjugate containing galactosamine as a targeting moiety (for structures of PK1 and PK2 see Fig. 13). Introduction of galactosamine moiety was designed for specific targeting of the hepatocyte asialoglycoprotein receptor. Clinical trials for the treatment of primary hepatocellular carcinoma showed that PK2 efficiently targeted the liver and achieved 12–50-fold higher drug levels in tumors, compared to nongalactosylated polymers or free drug. Both PK1 and PK2 showed antitumor activities and were two–fivefold less toxic than free Dox (Paul et al., 2007).

Styrene-maleic anhydride copolymer–neocarzinostatin conjugate (SMANCS) was the first clinically approved polymer–drug conjugate (Fig. 25). NCS is released from SMANCS when mixed with tissue homogenates (Maeda et al., 1985). SMANCS showed significant antitumor activity and very high tumor/blood ratio in a number of animal models. Maeda described the phenomenon as "EPR effect." During phase I and II clinical trials, 80% of patients with primary hepatomas showed regression of tumor size. SMANCS has been marketed in Japan since 1990 for the treatment of hepatocellular carcinoma (Maeda, 2001b).

PG–paclitaxel conjugate (CT-2103, XYOTAX) (Fig. 26) has been evaluated in phase III clinical trials. Coupling with PG greatly improved the water solubility of paclitaxel. In in vitro and in vivo tests, PG–paclitaxel demonstrated significantly

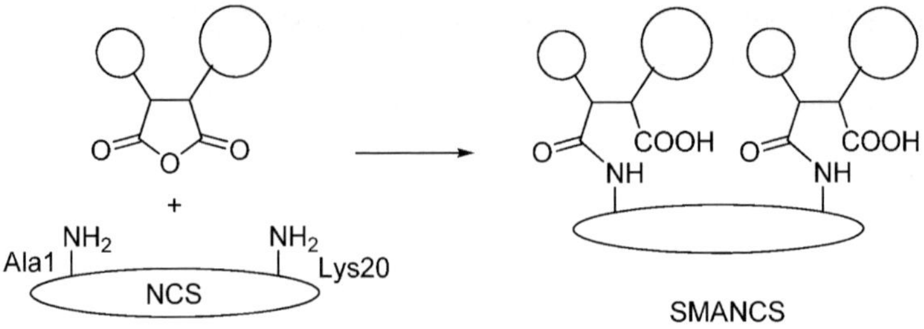

Fig. 25 Structure of styrene-maleic anhydride copolymer–neocarzinostatin conjugate (SMANCS)

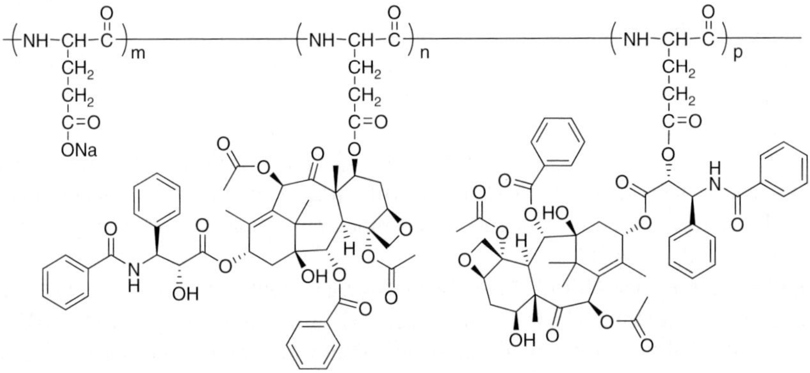

Fig. 26 Structure of poly(glutamic acid)–paclitaxel conjugate (CT-2103, XYOTAX)

reduced systemic toxicity and improved antitumor efficacy (Li et al., 1998). Phase I/II trials in Europe and the USA (Langer, 2004a,b; Markman, 2004; Sabbatini et al., 2004) confirmed several advantages, including ease of administration, improved antitumor activity, and manageable toxicity. XYOTAX provides a superior first-line treatment for performance status 2 (PS2) patients, compared to paclitaxel or vinorelbine. XYOTAX has been shown to significantly prolong survival in PS2 nonsmall cell lung cancer patients (Bonomi et al., 2006; Langer et al., 2005, 2006; Singer et al., 2005). Another PG conjugate, PG–camptothecin conjugate (CT-2106), is also in clinical testing.

Many methods have been explored for further improving the antitumor efficiency of polymer–drug conjugates. New drugs are being developed that interfere with the discrete transduction pathways that cause tumorigenesis. HPMA conjugates have been developed using new drugs such as 9-aminocamptothecin (9-AC) (Gao et al., 2006a, 2007; Sakuma et al., 2001), geldanamycin (Kasuya et al., 2001, 2002; Nishiyama et al., 2003), and TNP-470, an antiangiogenesis drug (Satchi-Fainaro et al., 2004, 2005).

Improved targeting is also an efficient way to enhance the efficiency of drug conjugates. Introduction of an antibody to target the conjugates to specific tumor cells is a direct way to improve efficacy. Antibodies, such as antimouse CD71

monoclonal antibody (mAb) for 38C13 tumors (Kovář et al., 2002b) and B1 monoclonal antibody (mAb) for BCL1 leukemia cells (Kovář et al., 2003) were used in
HPMA copolymer–Dox conjugates. In vitro and in vivo data showed that, compared
with the conjugates without antibody or with nonspecific ligands, introduction of
the antibody significantly improved antitumor activity.

Because of the high molecular weight of the antibody, the penetration into tumor
tissue may be limited. The use of mouse monoclonal antibodies (mAbs) in humans
proved to be problematic due to immune responses that produced human antimouse
antibody (HAMA). This issue was solved by the fusion of mouse variable regions
to human constant regions (chimeric antibodies), removal of T cell epitopes
(de-immunization), or grafting mouse antigen binding regions onto human antibody
frameworks (humanized antibodies) (Hudson and Souriau, 2003).

The use of antibody fragments decreases immunogenecity and improves tissue
diffusivity when compared to whole antibodies. Monovalent Fab fragments, single-
chain variable fragments (scFv), and bivalent scFv exhibit better pharmacokinetics
for tissue penetration with full binding specificity, but show poor reduced intravascular times (Hegazy et al., 2006; Vasir and Labhasetwar, 2005; Vasir et al., 2005).
Binding the antibody fragment in polymer conjugates may prolong the retention
time of Fab or scFv fragments.

Compared with the entire antibody, keeping the recognition domain and remov-
ing the Fc domain, Fab may improve selectivity and reduce immunogenicity.
Studies on the cytotoxicity of free mesochlorin e_6 (Mce$_6$), HPMA copolymer–Mce$_6$
conjugate, and HPMA copolymer–Mce$_6$–Fab conjugate toward OVCAR-3 cells
showed IC$_{50}$ doses of 7.9, 230, and 2.6 µM, respectively. The HPMA–Mce$_6$–Fab
conjugate more efficiently inhibited the growth of OVCAR-3 cells (Lu et al.,
1999b, 2001, 2003). Confocal microscopy showed that rapid internalization of the
HPMA–Mce$_6$–Fab conjugate in OVCAR-3 cells was by receptor-mediated endocytosis
and the conjugate was localized in the lysosomal compartment. The internalization
of the nontargeted conjugate was much slower.

Using receptor-binding peptide epitopes is another way to improve targeting to
malignant cells (Nan et al., 2005). Compared with antibody and Fab, more epitopes
can be introduced onto one polymer chain. With multivalent binding, multiple ligands bind to multiple receptors on one cell, and can greatly enhance the binding
constant of the polymer–drug conjugate (Fig. 27). The small size of the epitopes

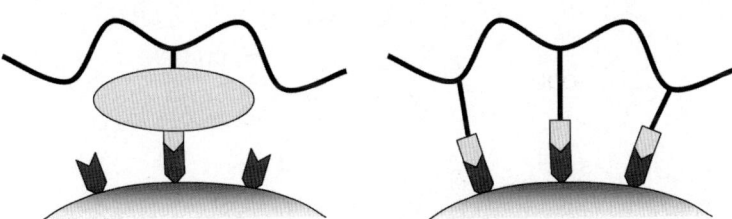

Fig. 27 Multivalent binding between polymer–drug conjugate and target cell

also makes them more stable during coupling reactions and other applications. The CD21 receptor has been found to be overexpressed on lymphoblastoid cells, and has been used as a target for lymphomas (Rask et al., 1988; Tang and Kopeček, 2002; Tang et al., 2003). For example, a nonapeptide (EDPGFFNVE) epitope was used in HPMA copolymer–Dox conjugate to target the cells with CD21 receptors. Multivalent interactions between HPMA copolymer–peptide conjugates and target receptors played an important role in the biorecognition of HPMA copolymer–peptide–drug conjugates (Ding et al., 2006a,b; Tang et al., 2003). As a result, enhanced cytotoxicity of HPMA copolymer–Dox conjugates containing multiple copies of targeting peptide toward cancer cells was observed (Tang et al., 2003). Bis(2-carboxyethyl) polyethylene glycol containing three copies of the targeting luteinizing hormone-releasing hormone (LHRH) peptide and camptothecin showed a dramatically enhanced antitumor activity compared with the analogous nontargeted prodrug and prodrugs containing one or two copies of active components (Khandare et al., 2006a).

Introduction of TAT peptide has been used to allow intracellular targeting of polymer–drug conjugates (Fig. 28). Targeting the intracellular organelles can

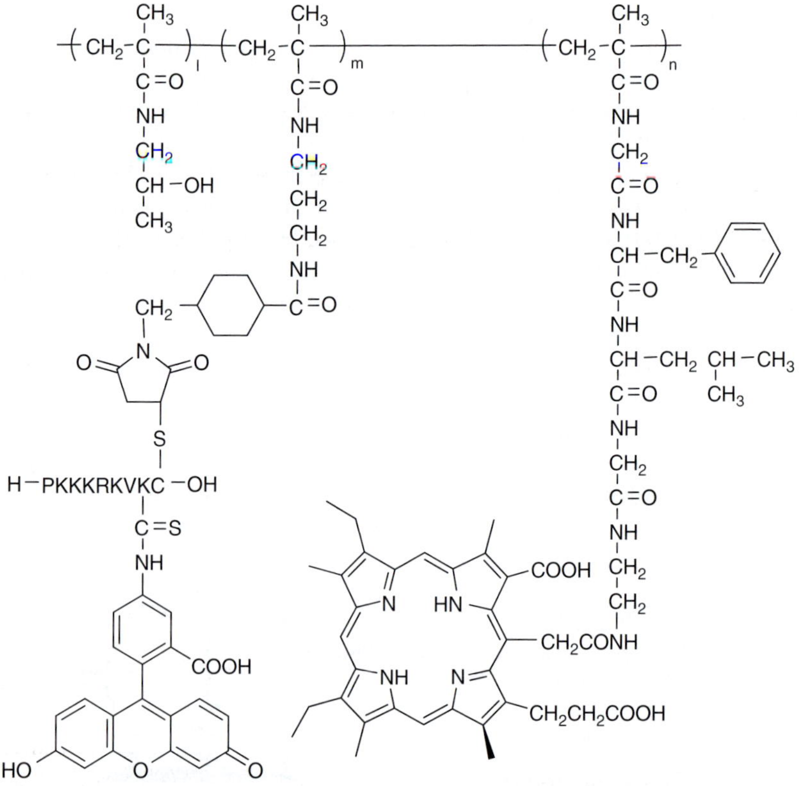

Fig. 28 Structure of HPMA copolymer–Mce$_6$ conjugate containing NLS peptides (Tijerina et al., 2003b)

improve the efficiency of the polymer–drug conjugates (Tijerina et al., 2003a,b). TAT peptide was used in an HMPA copolymer–Mce$_6$ conjugate for improving the photodynamic therapy efficiency. The result was a threefold increase in cytotoxicity compared with the non-TAT conjugate (Nori et al., 2003a,b).

5.2   Gene Delivery

There are two strategies to use the multifunctional polymer carriers in gene delivery systems. One is to synthesize new polymers suitable for gene delivery and the other is to modify viruses with functionalized polymers.

Synthetic polymers, primarily cationic polymers, have been developed to form polyplexes with DNAs to deliver genes to cells. The most commonly used have been the cationic polymers, PEI and PLL, despite toxicity issues associated with both. For screening of new polymers for gene delivery with reduced toxicities and improved efficacies, combinatorial methods have been used. Langer et al. used parallel synthetic methods for the preparation of 140 poly(β-amino ester)s from a series of seven diacrylates and 20 primary or secondary diamines. Two polymers from this library mediated gene transfection rates comparable to branched polyethylenimine and lipofectamine 2000 (Akinc et al., 2003; Lynn et al., 2001). By the same method, 2,350 poly(β-amino ester) structures were synthesized from 94 amine and 25 diacrylate monomers. In these polymers, 46 new structures were discovered with transfection efficiencies that exceeded optimized polyethylenimine (Anderson et al., 2003).

Combinatorial methods were also used to modify the side chains of well-defined polymers. Brocchini and coworkers synthesized by ATRP an active ester homopolymer precursor and two blocked copolymer active ester precursors to prepare a library of 16 cationic functionalized polymethacrylamides. The 16 polycations were prepared by the coupling of two amines from the group of (2-aminoethyl)tri methylammonium chloride hydrochloride (TMA), 3-(dimethylamino)propylamine (DMA), histamine (His), and 1-amino-2-propanol (AP). Consequently, the conjugates contained different amounts of tertiary amine, quaternary ammonium, imidazole, and hydroxypropyl pendent moieties, and possessed varying degrees of charge density along the polymer main chain (Pedone et al., 2003).

Viral-based gene-delivery vehicles are highly efficient, but they can be immunogenic. PEGylation is a stealth technology that can reduce the immunogenicity of viral vectors. Multifuctional polyHPMA has also been used for the modification of viral vectors. PolyHPMA with multiple reactive side chains forms a hydrophilic coat around the virus (Green et al., 2004). Both PEG and polyHPMA modifications mask the capsid domains of the virus and increase the blood circulation time of viral vectors. Incorporation of targeting ligands such as basic fibroblast growth factor (FGF2) and vascular endothelial growth factor on to the polymer-coated virus produces ligand-mediated, coxsackie- and adenovirus receptor (CAR)-independent binding and uptake into cells bearing appropriate receptors (Fisher et al., 2001; Lanciotti et al., 2003).

5.3   Bone Targeting Polymer–Drug Conjugate Therapies

Most of the studies in the polymer–drug conjugate area are concentrated on anti-cancer therapies, although there are many other diseases that seriously affect human health. Osteoporosis is a significant public health problem that affects an estimated 44 million Americans, or 55% of people over the age of 50. It contributes to an estimated 1.5 million bone fractures per year. Osteoporosis is a common metabolic bone disease. Many therapeutic agents are being used for the treatment of osteoporosis; however, none of them are very effective in the rebuilding of new skeletal mass, and many have significant side effects (Lacey et al., 2002a,b; Willhite et al., 1998; Wang et al., 2005). Recently, a novel, bone-targeted, water-soluble HPMA copolymer conjugate was developed using with a well-established bone anabolic agent, prostaglandin E_1 (PGE_1) (Pan et al., 2006; Wang et al., 2003, 2006). Biorecognition of the conjugates was mediated by an octapeptide of D-aspartic acid (D-Asp_8) or alendronate (Wang et al., 2003, 2007). The therapeutic activity of the conjugate both in vitro and in vivo was established.

6   Future Prospects

The living organisms emerged 3.5 billion years ago (Schopf et al., 2002) and after billions of years of evolution, the complexity of life emerged. On the molecular level, biomacromolecules exhibit precise chemical structures and spatial architecture, highly specific organization, signaling, high efficiency, and precisely concerted actions. At present, the level of sophistication of polymer–drug conjugates is quite inferior to biological systems. Design of new synthetic conjugates, based on the understanding of the structure–property–function relationship of biomacromolecules, would be a great step forward.

6.1   Control of Molecular Weight and Architecture

A deep understanding of the interaction of polymer–drug conjugates with cells, organelles, and individual biomacromolecules is necessary for the design of new macromolecular therapies. For a detailed study of the biological properties of polymer–drug conjugates at the cellular and molecular levels, polymer–drug conjugates should have specific molecular mass and a controlled architecture. However, the heterogeneity in the chemical composition and molecular weights of synthetic polymers has many implications to their interactions with cells. These issues can affect self-association states, diffusion rates, binding strengths, and transmembrane transport. Also, analysis of results may be a challenge. One way to improve the design of the conjugates is to take advantage of new methods that permit better control of the molecular weight of the polymer, such as ATRP and RAFT polymerizations. These techniques can be used to control the distribution of functional groups along the polymer chain and to control molecular weight distributions.

6.2   Click Chemistry

The term "click chemistry" was first proposed by Sharpless and coworkers (Kolb et al., 2001). To fit the category, the reaction must be modular, wide in scope, give very high yields, generate only inoffensive byproducts that can be easily removed, and be stereospecific (but not necessarily enantioselective). An example is the widely used copper-catalyzed Huisgen 1,3-dipolar cycloaddition of azides and terminal alkynes. In the presence of Cu(I), cycloaddition reactions are fast, regioselective, and efficient. It can be performed in various solvents, even in water, and the presence of other functional groups does not affect the reaction (Fig. 29).

For biomaterials, click reactions have been used for the synthesis of well-defined PEG hydrogels (Malkoch et al., 2006), the preparation of glycopolymers (Ladmiral et al., 2006), grafting PEG and oligopeptide moieties to aliphatic polyesters (Parrish et al., 2005), for example. Parrish and Emrick (2007) also modified alkyne-functionalized aliphatic polyesters with azide-functionalized camptothecin derivatives and azide-terminated PEG to produce water-soluble polyester–camptothecin conjugates. Click reactions are becoming an attractive method for the functionalization of polymers and for the preparation of new polymer–drug conjugates.

6.3   New Ligand Exploration

The advent of combinatorial chemistry has dramatically expedited the screening and identification of new chemical identities with desired properties (Service, 1996). Among the classes of combinatorial libraries, oligonucleotide, oligosaccharide, peptide, and chemically synthesized small molecules, peptide libraries have been most widely used for pharmaceutical applications, such as epitope mapping, and the identification of inhibitors and targeting ligands.

The methods to prepare combinatorial peptide libraries can be generally divided into two categories: chemically prepared libraries including methods such as one-bead one-compound (OBOC) (Lam et al., 2003) and spot synthesis (Frank, 2002); and biologically prepared libraries with methods such as phage

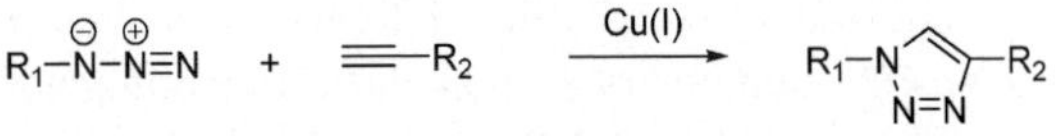

Fig. 29 Copper-catalyzed Huisgen 1,3-dipolar cycloaddition of azide and terminal alkynes, one of the click reactions

display (Smith and Petrenko, 1997) or bacterial display (Westerlund-Wikstrom, 2000). The biologically prepared library is less limited with respect to peptide length; oligopeptides, polypeptides, and even whole proteins can be displayed on the surface of phage or bacteria. However, chemically prepared libraries are more flexible for the introduction of linkers or the incorporation of unnatural amino acids. Both the OBOC method and phage display have been extensively used for selection and identification of oligopeptide ligands for tumor targeting (Wang et al., 2004b).

Phage display (Ding et al., 2006a) and OBOC (Ding et al., 2006b) were successfully used for the identification of peptides specific toward the CD21 receptor. Numerous reports described multivalent interactions with cells of polymer conjugates containing multiple peptide moieties (see, e.g., Ding et al., 2006a; Nan et al., 2005; Tang et al., 2003).

Aptamers are RNA or DNA oligonucleotides. Like antibodies, they can fold into unique 3D conformations capable of binding to target antigens. Aptamers are generally selected using the SELEX (Systematic Evolution of Ligands by EXponential enrichment) process (Eaton, 2005; Tuerk and Gold, 1990). In contrast to antibodies, the synthesis of aptamers is an entirely chemical process that can decrease batch-to-batch variability (Farokhzad et al., 2004, 2006). The dissociation constant of aptamers is in the picomolar to nanomolar range. Binding is highly specific and can discriminate between related proteins that share common sets of structural domains (Lee and Sullenger, 1997; Rusconi et al., 2002). Aptamers show almost no immunogenicity (Eyetech Study Group, 2002, 2003) and exhibit remarkable stability at pHs between 4 and 9 and a variety of temperatures and organic solvents, without loss of activity (Nimjee et al., 2005; Wilson and Szostak, 1998).

Unmodified aptamers are highly susceptible to nuclease degradation, however. A way to improve their resistance to nuclease degradation is to substitute the $2'$ hydroxyl of the ribose moiety of pyrimidines with fluorine (F) or an amino group (NH_2) (Pieken et al., 1991). A further modification that improves RNA oligonucleotide stability against nucleases is the substitution of $2'$-O-methyl ($2'$-O-Me) at the $2'$-hydroxyl position of purine (Burmeister et al., 2005). Aptamers are often capped at their $3'$ end with a deoxythymidine to further decrease exonuclease degradation (Beigelman et al., 1995).

The molecular weight of aptamers truncated to their core-binding domain is about 5–15 kDa (15–40 nucleotides), which is below the fast renal clearance threshold. A modification by site-specific attachment of PEG or other macromolecules would be one method to improve their retention time (Tucker et al., 1999; Willis et al., 1998).

Aptamers are potentially useful for the targeting of drug delivery systems. Langer et al. first reported targeted drug delivery with nanoparticle–aptamer bioconjugates. The results demonstrated that these bioconjugates can efficiently target and be taken up by prostate LNCaP epithelial cells that express the prostate-specific membrane antigen (PSMA) protein. The uptake of these particles was not enhanced in cells that do not express PSMA (Farokhzad et al., 2004).

6.4   Double Targeting

As discussed above, macromolecular therapeutics are internalized by endocytosis and are trafficked to the lysosomes. Numerous conjugates have been synthesized and evaluated based on this biological pathway. Recently, however, research has been focusing on the identification of different routes of cell entry with the aim to deliver drugs into subcellular compartments other than lysosomes. Selective localization of therapeutic agents in specific organelles can serve to enhance therapeutic efficacy (Fig. 30). The so-called double targeting refers to the combination of cell targeting and subcellular targeting. Nuclear targeting is important in the delivery of genes and many low molecular weight drugs (Torchilin, 2006b). For example, the activity of photosensitizers is considerably higher in nucleus when compared to other subcellular locations. Mitochondria are another important target for drugs. The destabilizatin of the mitochondrial membrane and release of cytochrome c are essential in triggering of apoptosis.

6.5   Combination Therapy

It is well established that the clinical treatment of cancer is optimized using combination therapy. A similar concept has been developed for the treatment with polymer-bound drugs. Photodynamic therapy is a newer paradigm in anticancer therapy that involves activation of specific compounds called photosensitizers with specific wavelengths of light to induce cell death. Illumination of these compounds results in the generation of singlet oxygen and free radicals, which cause cell damage and death. A combination of chemotherapy and photodynamic therapy may result in a synergistic response, resulting in better cure rate than monotherapy. In two cancer

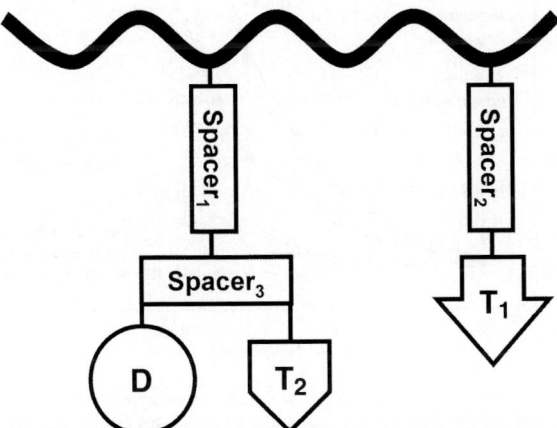

Fig. 30 Double-targeted polymer–drug conjugate

models, Neuro 2A neuroblastoma (Krinick et al., 1994) and human ovarian carcinoma in mice (Peterson et al., 1996), it was demonstrated that combination therapy with HPMA copolymer–anticancer drugs [Dox and meso chlorin e$_6$ mono(*N*-2-aminoethylamide) {Mce$_6$}] conjugates showed tumor cures that could not be obtained with either chemotherapy or photodynamic therapy alone. Cooperativity of action of both drugs contributed to the observed effect (Lu et al., 1999a). Based on biodistribution data (Shiah et al., 1999), it was hypothesized that combination therapies of s.c. human ovarian carcinoma OVCAR-3 xenografts in nude mice using *multiple doses* of P(GFLG)-Mce$_6$ (P is the HPMA copolymer backbone) and P(GFLG)-Dox may allow low effective doses without sacrificing the therapeutic efficacy. Indeed, 10 out of 12 tumors exhibited complete responses in the cohort of mice receiving multiple PDT plus multiple chemotherapy (Shiah et al., 2000).

Finally, additional enhancement of therapeutic efficacy may be reached using *targeted combination chemotherapy and photodynamic therapy* with OV-TL16-mAb-HPMA copolymer–Dox and HPMA copolymer–mesochlorin e$_6$ conjugates. OV-TL16 antibodies are specific for the OA-3 (CD47) antigen present on the majority of ovarian cancers. The immunoconjugates preferentially accumulated in human ovarian carcinoma OVCAR-3 xenografts in nude mice with a concomitant increase in therapeutic efficacy when compared with nontargeted conjugates (Fig. 31). The targeted conjugates suppressed tumor growth for the entire length of the experiment (Shiah et al., 2001b).

An interesting result was the observation that HPMA copolymer conjugates carrying both an aromatase inhibitor (aminoglutethimide) and doxorubicin on one polymer chain showed markedly enhanced in vitro cytotoxicity to MCF-7 breast

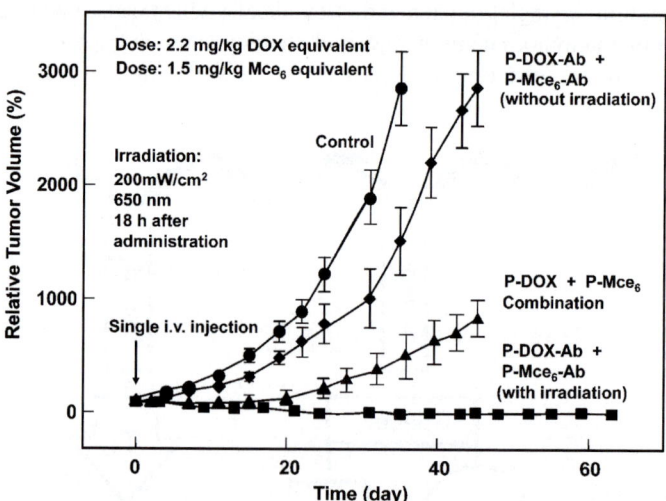

Fig. 31 Eficacy of combination chemotherapy and photodynamic therapy of OVCAR-3 xenografts in nude mice with nontargeted and OV-TL 16 antibody-targeted HPMA copolymer–doxorubicin (P-Dox) and HPMA copolymer–mesochlorin e$_6$ (P-Mce$_6$) conjugates (Shiah et al., 2001b)

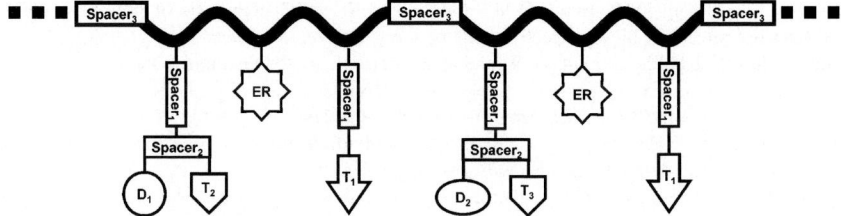

Fig. 32 Multifunctional dream conjugate

cancer cells compared to a mixture of conjugates with only one drug (Greco et al., 2005, 2007).

Polymer-directed enzyme prodrug therapy (PDEPT) is a new two-step antitumor approach including administration of a polymer–drug conjugate containing an enzymatically cleavable linker followed by the administration of a polymer–enzyme conjugate to generate a cytotoxic drug rapidly and selectively at the tumor site. PDEPT did not activate the prodrug in the circulation, and polymer–enzyme conjugates have reduced immunogenicity. It appears to be advantageous compared to antibody-directed enzyme prodrug therapy (ADEPT) and gene-directed enzyme prodrug therapy (GDEPT) (Duncan et al., 2001; Satchi-Fainaro et al., 2001, 2003).

6.6 Dream Conjugate

An efficient conjugate for the future must have an increased intravascular half-life. Consequently, for the long-circulating conjugate to remain biocompatible and to be eliminated from the organism, labile bonds, cleavable either enzymatically or pH sensitive should be incorporated into the polymer backbone. Two targeting moieties should be attached: one for cellular targeting to direct the conjugate to a subset of cells and the other for subcellular targeting to direct the released drug into a particular orgenelle. Two biologically active moieties might be attached to the same macromolecule to provide the conjugate with combination therapy property. In addition, environmentally responsive groups (ER) might be also attached to the macromolecule, so the conjugates can adjust the drug release rate in response to the microenvironment (Fig. 32).

Acknowledgments The research in authors' laboratory was supported in part by NIH grants CA51578, GM069847, and EB005288, and the Department of Defense grant W81XWH-04-1-0900. We thank Jon Callahan for carefully editing the manuscript.

References

Akhlynina, T.V., Rosenkranz, A.A., Jans, D.A., and Sobolev, A.S. 1995. Insulin-mediated intracellular targeting enhances the photodynamic activity of chlorin e_6. *Cancer Res.* **55:** 1014–1019.

Akinc, A., Anderson, D.G., Lynn, D.M., and Langer, R. 2003. Synthesis of poly(beta-amino ester)s optimized for highly effective gene delivery. *Bioconjug. Chem.* **14:** 979–988.

Akita, H., Ito, R., Khalil, I.A., Futaki, S., and Harashima, H. 2004. Quantitative three-dimensional analysis of the intracellular trafficking of plasmid DNA transfected by a nonviral gene delivery system using confocal laser scanning microscopy. *Mol. Ther.* **9:** 443–451.

Alakhov, V.Y. and Kabanov, A.V. 1998. Block copolymeric biotransport carriers as versatile vehicles for drug delivery. *Expert Opin. Investig. Drugs* **7:** 1453–1473.

Allen, T.M. 2002. Ligand-targeted therapeutics in anticancer therapy. *Nat. Rev. Cancer* **2:** 750–763.

Allen, L.A. and Aderem, A. 1996. Mechanisms of phagocytosis. *Curr. Opin. Immunol.* **8:** 36–40.

Al-Shamkhani, A. and Duncan, R. 1995. Synthesis, controlled release properties and antitumour activity of alginate-cis-aconityl-daunomycin conjugates. *Int. J. Pharm.* **122:** 107–119.

Amyere, M., Mettlen, M., Van Der Smissen, P., Platek, A., Payrastre, B., Veithen, A., and Courtoy, P.J. 2002. Origin, originality, functions, subversions and molecular signalling of macropinocytosis. *Int. J. Med. Microbiol.* **291:** 487–494.

Anderson, D.G., Lynn, D.M., and Langer, R. 2003. Semi-automated synthesis and screening of a large library of degradable cationic polymers for gene delivery. *Angew. Chem. Int. Ed. Engl.* **42:** 3153–3158.

Arcaro, A. and Wymann, M.P. 1993. Wortmannin is a potent phosphatidylinositol 3-kinase inhibitor: the role of phosphatidylinositol 3,4,5-trisphosphate in neutrophil responses. *Biochem. J.* **296 (Pt 2):** 297–301.

Arefjev, D.V., Domnina, N.S., Komarova, E.A., and Bilibin, A.Y. 1999. Sterically hindered phenol-dextran conjugates: synthesis and radical scavenging activity. *Eur. Polym. J.* **35:** 279–284.

Arunachalam, B., Phan, U.T., Geuze, H.J., and Creswell, P. 2000. Enzymatic reduction of disulfide bonds in lysosomes: characterization of a gamma-interferon-inducible lysosomal thiol reductase (GILT). *Proc. Natl. Acad. Sci. USA* **97:** 745–750.

Beigelman, L., McSwiggen, J.A., Draper, K.G., Gonzalez, C., Jensen, K., Karpeisky, A.M., Modak, A.S., Matulic-Adamic, J., DiRenzo, A.B., Haeberli, P., Sweedler, D., Tracz, D., Grimm, S., Wincott, F.E., Thackray, V.G., and Usman, N. 1995. Chemical modification of hammerhead ribozymes. Catalytic activity and nuclease resistance. *J. Biol. Chem.* **270:** 25702–25708.

Bhatt, R., de Vries, P., Tulinsky, J., Bellamy, G., Baker, B., Singer, J.W., and Klein, P. 2003. Synthesis and in vivo antitumor activity of poly(L-glutamic acid) conjugates of 20S-camptothecin. *J. Med. Chem.* **46:** 190–193.

Bilim, V. 2003. Technology evaluation: PK1, Pfizer/Cancer Research UK. *Curr. Opin. Mol. Ther.* **5:** 326–330.

Bissett, D., Cassidy, J., de Bono, J.S., Muirhead, F., Main, M., Robson, L., Fraier, D., Magne, M.L., Pellizzoni, C., Porro, M.G., Spinelli, R., Speed, W., and Twelves, C. 2004. Phase I and pharmacokinetic (PK) study of MAG-CPT (PNU 166148): a polymeric derivative of camptothecin (CPT). *Br. J. Cancer* **91:** 50–55.

Bloomfield, V.A. 1996. DNA condensation. *Curr. Opin. Struct. Biol.* **6:** 334–341.

Bohdanecký, M., Bažilová, H., and Kopeček, J. 1974. Poly[N-(2-hydroxypropyl)methacrylamide]. II. Hydrodynamic properties of diluted polymer solutions. *Eur. Polym. J.* **10:** 405–410.

Bonomi, P., Paz-Ares, L., and Langer, C.J. 2006. XYOTAX™ vs. docetaxel for the second-line treatment of non-small cell lung canceer: the STELLAR 2 phase III study. *Lung Cancer* S35.

Boulikas, T. 1993. Nuclear localization signals (NLS). *Crit. Rev. Eukaryot. Gene Expr.* **3:** 193–227.

Boussif, O., Lezoualc'h, F., Zanta, M.A., Mergny, M.D., Scherman, D., Demeneix, B., and Behr, J.P. 1995. A versatile vector for gene and oligonucleotide transfer into cells in culture and in vivo: polyethylenimine. *Proc. Natl. Acad. Sci. USA* **92:** 7297–7301.

Bremner, K.H., Seymour, L.W., and Pouton, C.W. 2001. Harnessing nuclear localization pathways for transgene delivery. *Curr. Opin. Mol. Ther.* **3:** 170–177.

Brömme, D. and Okamoto, K. 1995. Human cathepsin O2, a novel cysteine protease highly expressed in osteoclastomas and ovary molecular cloning, sequencing and tissue distribution. *Biol. Chem. Hoppe-Seyler* **376**: 379–384.

Brooks, H., Lebleu, B., and Vives, E. 2005. Tat peptide-mediated cellular delivery: back to basics. *Adv. Drug. Deliv. Rev.* **57**: 559–577.

Brown, J.P., McGarraugh, G.V., Parkinson, T.M., Wingard, R.E., Jr., and Onderdonk, A.B. 1983. A polymeric drug for treatment of inflammatory bowel disease. *J. Med. Chem.* **26**: 1300–1307.

Brown, M.D., Gray, A.I., Tetley, L., Santovena, A., Rene, J., Schätzlein, A.G., and Uchegbu, I.F. 2003. In vitro and in vivo gene transfer with poly(amino acid) vesicles. *J. Control. Release* **93**: 193–211.

Brownlie, A., Uchegbu, I.F., and Schätzlein, A.G. 2004. PEI-based vesicle-polymer hybrid gene delivery system with improved biocompatibility. *Int. J. Pharm.* **274**: 41–52.

Bulmus, V., Woodward, M., Lin, L., Murthy, N., Stayton, P., and Hoffman, A. 2003. A new pH-responsive and glutathione-reactive, endosomal membrane-disruptive polymeric carrier for intracellular delivery of biomolecular drugs. *J. Control. Release* **93**: 105–120.

Burmeister, P.E., Lewis, S.D., Silva, R.F., Preiss, J.R., Horwitz, L.R., Pendergrast, P.S., McCauley, T.G., Kurz, J.C., Epstein, D.M., Wilson, C., and Keefe, A.D. 2005. Direct in vitro selection of a 2'-O-methyl aptamer to VEGF. *Chem. Biol.* **12**: 25–33.

Callahan, J. and Kopeček, J. 2006. Semitelechelic HPMA copolymers functionalized with triphenylphosphonium as drug carriers for membrane transduction and mitochondrial localization. *Biomacromolecules* **7**: 2347–2356.

Callahan, J., Kopečková, P., and Kopeček, J. 2008. To be submitted.

Campone, M., Rademaker-Lakhai, J.M., Bennouna, J., Howell, S.B., Nowotnik, D.P., Beijnen, J.H., Schellens, J.H. 2007. Phase I and pharmacokinetic trial of AP5346, a DACH-platinum-polymer conjugate, administered weekly for three out of every 4 weeks to advanced solid tumor patients. *Cancer Chemotherapy Pharmacol.* **60**: 523–533.

Carl, P.L., Chakravarty, P.K., and Katzenellenbogen, J.A. 1981. A novel connector linkage applicable in prodrug design. *J. Med. Chem.* **24**: 479–480.

Cartlidge, S.A., Duncan, R., Lloyd, J.B., Kopečková-Rejmanova, P., and Kopeček, J. 1987a. Soluble crosslinked N-(2-hydroxypropyl)methacrylamide copolymers as potential drug carriers. 2. Effect of molecular weight on blood clearance and body distribution in the rat after intravenous administration. Distribution of unfractionated copolymer after intraperitoneal, subcutaneous or oral administration. *J. Control. Release* **4**: 253–264.

Cartlidge, S.A., Duncan, R., Lloyd, J.B., Kopečková-Rejmanova, P., and Kopeček, J. 1987b. Soluble, crosslinked N-(2-hydroxypropyl)methacrylamide copolymers as potential drug carriers. 3. Targeting by incorporation of galactosamine residues. Effect of route of administration. *J. Control. Release* **4**: 265–278.

Chang, Y. and McCormick, C.L. 1993. Water-soluble copolymers. 49. Effect of the distribution of the hydrophobic cationic monomer dimethyldodecyl (2-acrylamidoethyl)ammonium bromide on the solution behavior of associating acrylamide copolymers. *Macromolecules* **26**: 6121–6126.

Chang, Y. and Mccormick, C.L. 1994. Water-soluble copolymers. 57. Amphiphilic cyclocopolymers of diallylalkoxybenzyl-methylammonium chloride and diallyldimethylammonium chloride. *Polymer* **35**: 3503–3512.

Cheremisinoff, N. 1997. *Handbook of Engineering Polymeric Materials*. Marcel Dekker, New York.

Chiu, H.-C., Koňák, Č., Kopečková, P., and Kopeček, J. 1994. Enzymatic degradation of poly(ethylene glycol) modified dextrans. *J. Bioact. Compat. Polym.* **9**: 388–410.

Chiu, H.C., Kopečková, P., Deshmane, S.S., and Kopeček, J. 1997. Lysosomal degradability of poly(alpha-amino acids). *J. Biomed. Mater. Res.* **34**: 381–392.

Cho, K.C., Kim, S.H., Jeong, J.H., and Park, T.G. 2005. Folate receptor-mediated gene delivery using folate-poly(ethylene glycol)-poly(L-lysine) conjugate. *Macromol. Biosci.* **5**: 512–519.

Choe, Y.H., Conover, C.D., Wu, D., Royzen, M., Gervacio, Y., Borowski, V., Mehlig, M., and Greenwald, R.B. 2002a. Anticancer drug delivery systems: multi-loaded N4-acyl poly(ethylene glycol) prodrugs of ara-C. II. Efficacy in ascites and solid tumors. *J. Control. Release* **79**: 55–70.

Choe, Y.H., Conover, C.D., Wu, D., Royzen, M., and Greenwald, R.B. 2002b. Anticancer drug delivery systems: N4-acyl poly(ethyleneglycol) prodrugs of ara-C. I. Efficacy in solid tumors. *J. Control. Release* **79**: 41–53.

Choi, W.-M., Kopečková, P., Minko, T., and Kopeček, J. 1999. Synthesis of HPMA copolymer containing adriamycin bound via an acid-labile spacer and its activity toward human ovarian carcinoma cells. *J. Bioact. Compat. Polym.* **14**: 447–456.

Christophe, D., Christophe-Hobertus, C., and Pichon, B. 2000. Nuclear targeting of proteins: how many different signals? *Cell Signal* **12**: 337–341.

Chytil, P., Etrych, T., Koňák, Č., Šírová, M., Mrkvan, T., Říhová, B., and Ulbrich, K. 2006. Properties of HPMA copolymer-doxorubicin conjugates with pH-controlled activation: Effect of polymer chain modification. *J. Control. Release* **115**: 26–36.

Chytrý, V., Vrána, A., and Kopeček, J. 1978. Synthesis and activity of a polymer which contains insulin covalently bound on a copolymer of N-(2-hydroxypropyl)methacrylamide and N-methacryloylglycylglycine 4-nitrophenyl ester. *Makromol. Chem.* **179**: 329–336.

Coessens, V., Schacht, E., and Domurado, D. 1996. Synthesis of polyglutamine and dextran conjugates of streptomycin with an acid-sensitive drug-carrier linkage. *J. Control. Release* **38**: 141–150.

Collas, P. and Alestrom, P. 1996. Nuclear localization signal of SV40 T antigen directs import of plasmid DNA into sea urchin male pronuclei in vitro. *Mol. Reprod. Dev.* **45**: 431–438.

Collas, P., Husebye, H., and Alestrom, P. 1996. The nuclear localization sequence of the SV40 T antigen promotes transgene uptake and expression in zebrafish embryo nuclei. *Transgenic Res.* **5**: 451–458.

Conner, S.D. and Schmid, S.L. 2003. Regulated portals of entry into the cell. *Nature* **422**: 37–44.

Coulter, C.V., Kelso, G.F., Lin, T.K., Smith, R.A., and Murphy, M.P. 2000. Mitochondrially targeted antioxidants and thiol reagents. *Free Radic. Biol. Med.* **28**: 1547–1554.

Cuchelkar, V. and Kopeček, J. 2006. Polymer-drug conjugates. In: *Polymers in Drug Delivery*, I.F. Uchegbu, A.G. Schätzlein, Eds., CRC Press, Boca Raton, Florida, 2006, pp. 155–182.

Cuchelkar, V., Kopečková, P., and Kopeček, J. 2008. To be submitted.

Dahlheimer, P., Engler, A.J., Parthasarathy, R., and Discher, D.E. 2004. Targeted worm micelles. *Biomacromolecules* **5**: 1714–1719.

Dang, C.V. and Lee, W.M. 1988. Identification of the human c-myc protein nuclear translocation signal. *Mol. Cell Biol.* **8**: 4048–4054.

Danhauser-Riedl, S., Hausmann, E., Schick, H.D., Bender, R., Dietzfelbinger, H., Rastetter, J., and Hanauske, A.R. 1993. Phase I clinical and pharmacokinetic trial of dextran conjugated doxorubicin (AD-70, DOX-OXD). *Invest. New Drugs* **11**: 187–195.

de Groot, F.M., Loos, W.J., Koekkoek, R., van Berkom, L.W., Busscher, G.F., Seelen, A.E., Albrecht, C., de Bruijn, P., and Scheeren, H.W. 2001. Elongated multiple electronic cascade and cyclization spacer systems in activatible anticancer prodrugs for enhanced drug release. *J. Org. Chem.* **66**: 8815–8830.

Derycke, A.S. and De Witte, P.A. 2002. Transferrin-mediated targeting of hypericin embedded in sterically stabilized PEG-liposomes. *Int. J. Oncol.* **20**: 181–187.

Ding, H., Prodinger, W.M., and Kopeček, J. 2006a. Identification of CD21-binding peptides with phage display and investigation of binding properties of HPMA copolymer-peptide conjugates. *Bioconjug. Chem.* **17**: 514–523.

Ding, H., Prodinger, W.M., and Kopeček, J. 2006b. Two-step fluorescence screening of CD21-binding peptides with one-bead one-compound library and investigation of binding properties of N-(2-hydroxypropyl)methacrylamide copolymer-peptide conjugates. *Biomacromolecules* **7**: 3037–3046.

Ding, H., Kopečková, P., and Kopeček, J. 2007. Self-association properties of HPMA copolymers containing an amphipathic heptapeptide. *J. Drug Target.* **15**: 465–475.

Discher, D.E. and Ahmed, F. 2006. Polymersomes. *Annu. Rev. Biomed. Eng.* **8:** 323–341.

Drake, F.H., Dodds, R.A., James, I.E., Connor, J.R., Debouck, C., Richardson, S., Lee-Rykaczewski, E., Coleman, L., Rieman, D., Barthlow, R., Hastings, G., and Gowen, M. 1996. Cathepsin K, but not cathepsins B, L, or S, is abundantly expressed in human osteoclasts. *J. Biol. Chem.* **271:** 12511–12516.

Drobník, J., Kopeček, J., Labský, J., Rejmanová, P., Exner, J., Saudek, V., and Kálal, J. 1976. Enzymatic cleavage of side-chains of synthetic water-soluble polymers. *Makromol. Chem.* **177:** 2833–2848.

D'Souza, A.J. and Topp, E.M. 2004. Release from polymeric prodrugs: linkages and their degradation. *J. Pharm. Sci.* **93:** 1962–1979.

Duchardt, F., Fotin-Mleczek, M., Schwarz, H., Fisher, R., Brock, R. 2007. A comprehensive model for the cellular uptake of cationic cell-penetrating peptides. *Traffic* **8:** 848–866.

Duncan, R. 2003. The dawning era of polymer therapeutics. *Nat. Rev. Drug Discov.* **2:** 347–360.

Duncan, R. 2006. Polymer conjugates as anticancer nanomedicines. *Nat. Rev. Cancer* **6:** 688–701.

Duncan, R., Lloyd, J.B., and Kopeček, J. 1980. Degradation of side chains of N-(2-hydroxypropyl) methacrylamide copolymers by lysosomal enzymes. *Biochem. Biophys. Res. Commun.* **94:** 284–290.

Duncan, R., Kopeček, J., Rejmanová, P., and Lloyd, J.B. 1983. Targeting of N-(2-hydroxypropyl) methacrylamide copolymers to liver by incorporation of galactose residues. *Biochim. Biophys. Acta* **755:** 518–521.

Duncan, R., Cable, H.C., Rejmanová, P., Kopeček, J., and Lloyd, J.B. 1984. Tyrosinamide residues enhance pinocytic capture of N-(2-hydroxypropyl)methacrylamide copolymers. *Biochim. Biophys. Acta* **799:** 1–8.

Duncan, R., Seymour, L.C., Scarlett, L., Lloyd, J.B., Rejmanová, P., and Kopeček, J. 1986. Fate of N-(2-hydroxypropyl)methacrylamide copolymers with pendent galactosamine residues after intravenous administration to rats. *Biochim. Biophys. Acta* **880:** 62–71.

Duncan, R., Gac-Breton, S., Keane, R., Musila, R., Sat, Y.N., Satchi, R., and Searle, F. 2001. Polymer-drug conjugates, PDEPT and PELT: basic principles for design and transfer from the laboratory to clinic. *J. Control. Release* **74:** 135–146.

Dvořák, M., Kopečková, P., and Kopeček, J. 1999. High-molecular weight HPMA copolymer-adriamycin conjugates. *J. Control. Release* **60:** 321–332.

Eaton, B. 2005. The joys of in vitro selection: chemically dressing oligonucleotides to satiate protein targets. *Curr. Opin. Chem. Biol.* **1:** 10–16.

Edidin, M. 2001a. Membrane cholesterol, protein phosphorylation, and lipid rafts. *Sci. STKE* **2001:** PE1.

Edidin, M. 2001b. Shrinking patches and slippery rafts: scales of domains in the plasma membrane. *Trends Cell. Biol.* **11:** 492–496.

Elliott, G. and O'Hare, P. 1997. Intercellular trafficking and protein delivery by a herpesvirus structural protein. *Cell* **88:** 223–233.

Esfand, R. and Tomalia, D.A. 2001. Poly(amidoamine) (PAMAM) dendrimers: from biomimicry to drug delivery and biomedical applications. *Drug Discov. Today* **6:** 427–436.

Eyetech Study Group. 2002. Preclinical and phase 1A clinical evaluation of an anti-VEGF pegylated aptamer (EYE001) for the treatment of exudative age-related macular degeneration. *Retina* **22:** 143–152.

Eyetech Study Group. 2003. Anti-vascular endothelial growth factor therapy for subfoveal choroidal neovascularization secondary to age-related macular degeneration: phase II study results. *Ophthalmology* **110:** 979–986.

Fang, J., Sawa, T., and Maeda, H. 2003. Factors and mechanism of "EPR" effect and the enhanced antitumor effects of macromolecular drugs including SMANCS. *Adv. Exp. Med. Biol.* **519:** 29–49.

Farokhzad, O.C., Jon, S., Khademhosseini, A., Tran, T.N., Lavan, D.A., and Langer, R. 2004. Nanoparticle-aptamer bioconjugates: a new approach for targeting prostate cancer cells. *Clin. Cancer Res.* **64:** 7668–7672.

Farokhzad, O.C., Cheng, J., Teply, B.A., Sherifi, I., Jon, S., Kantoff, P.W., Richie, J.P., and Langer, R. 2006. Targeted nanoparticle-aptamer bioconjugates for cancer chemotherapy in vivo. *Proc. Natl. Acad. Sci. USA* **103:** 6315–6320.

Fernandez-Carneado, J., Van Gool, M., Martos, V., Castel, S., Prados, P., de Mendoza, J., and Giralt, E. 2005. Highly efficient, nonpeptidic oligoguanidinium vectors that selectively internalize into mitochondria. *J. Am. Chem. Soc.* **127:** 869–874.

Ferrari, A., Pellegrini, V., Arcangeli, C., Fittipaldi, A., Giacca, M., and Beltram, F. 2003. Caveolae-mediated internalization of extracellular HIV-1 tat fusion proteins visualized in real time. *Mol. Ther.* **8:** 284–294.

Fisher, K.D., Stallwood, Y., Green, N.K., Ulbrich, K., Mautner, V., and Seymour, L.W. 2001. Polymer-coated adenovirus permits efficient retargeting and evades neutralising antibodies. *Gene Ther.* **8:** 341–348.

Fittipaldi, A., Ferrari, A., Zoppe, M., Arcangeli, C., Pellegrini, V., Beltram, F., and Giacca, M. 2003. Cell membrane lipid rafts mediate caveolar endocytosis of HIV-1 Tat fusion proteins. *J. Biol. Chem.* **278:** 34141–34149.

Frank, R. 2002. The SPOT-synthesis technique. Synthetic peptide arrays on membrane supports – principles and applications. *J. Immunol. Methods* **267:** 13–26.

Frankel, A.D. and Pabo, C.O. 1988. Cellular uptake of the tat protein from human immunodeficiency virus. *Cell* **55:** 1189–1193.

Frankel, A.E., Powell, B.L., Hall, P.D., Case, L.D., and Kreitman, R.J. 2002. Phase I trial of a novel diphtheria toxin/granulocyte macrophage colony-stimulating factor fusion protein (DT388GMCSF) for refractory or relapsed acute myeloid leukemia. *Clin. Cancer Res.* **8:** 1004–1013.

Fujita, M., Khazenzon, N.M., Ljubimov, A.V., Lee, B.S., Virtanen, I., Holler, E., Black, K.L., and Ljubimova, J.Y. 2006. Inhibition of laminin-8 in vivo using a novel poly(malic acid)-based carrier reduces glioma angiogenesis. *Angiogenesis* **9:** 183–191.

Gabizon, A., Horowitz, A.T., Goren, D., Tzemach, D., Mandelbaum-Shavit, F., Qazen, M.M., and Zalipsky, S. 1999. Targeting folate receptor with folate linked to extremities of poly(ethylene glycol)-grafted liposomes: in vitro studies. *Bioconjug. Chem.* **10:** 289–298.

Gao, S.Q., Lu, Z.R., Petri, B., Kopečková, P., and Kopeček, J. 2006a. Colon-specific 9-aminoc-amptothecin-HPMA copolymer conjugates containing a 1,6-elimination spacer. *J. Control. Release* **110;** 323–331.

Gao S.Q., Kopečková P., Sun Y., Lu Z.-R., Peterson C.M., and Kopeček J. 2006b. Activity of a polymer-bound 9-aminocamptothecin in a colon cancer model. *Eur. J. Clin. Invest.* **36**(Suppl 1): 52–53.

Gao, S.Q., Lu, Z.R., Kopečková, P., and Kopeček, J. 2007. Biodistribution and pharmacokinetics of colon-specific HPMA copolymer–9-aminocamptothecin conjugate in mice. *J. Control. Release* **117:** 179–185.

Garnett, M.C. 2001. Targeted drug conjugates: principles and progress. *Adv. Drug Deliv. Rev.* **53:** 171–216.

Gautier, S., Boustta, M., and Vert, M. 1997. Poly (L-Lysine Citramide), a water-soluble bioresorbable carrier for drug delivery: Aqueous solution properties of hydrophobized derivatives. *J. Bioact. Compat. Polym.* **12:** 77–98.

Gebhart, C.L., Sriadibhatla, S., Vinogradov, S., Lemieux, P., Alakhov, V., and Kabanov, A.V. 2002. Design and formulation of polyplexes based on pluronic-polyethyleneimine conjugates for gene transfer. *Bioconjug. Chem.* **13:** 937–944.

Gijsens, A. and De Witte, P. 2000. [Targeting of chlorine E6 by EGF increasing its photodynamic activity in selective ways]. *Verh. K. Acad. Geneeskd. Belg.* **62:** 329–352.

Godbey, W.T., Wu, K.K., and Mikos, A.G. 1999a. Poly(ethylenimine) and its role in gene delivery. *J. Control. Release* **60:** 149–160.

Godbey, W.T., Wu, K.K., and Mikos, A.G. 1999b. Size matters: molecular weight affects the efficiency of poly(ethylenimine) as a gene delivery vehicle. *J. Biomed. Mater. Res.* **45:** 268–275.

Goldfarb, D.S., Gariépy, J., Schoolnik, G., and Kornberg, R.D. 1986. Synthetic peptides as nuclear localization signals. *Nature* **322:** 641–644.

Görlich, D. and Mattaj, I.W. 1996. Nucleocytoplasmic transport. *Science* **271:** 1513–1518.

Gottesman, M.M., Fojo, T., and Bates, S.E. 2002. Multidrug resistance in cancer: role of ATP-dependent transporters. *Nat. Rev. Cancer* **2:** 48–58.

Gottschalk, S., Sparrow, J.T., Hauer, J., Mims, M.P., Leland, F.E., Woo, S.L., and Smith, L.C. 1996. A novel DNA-peptide complex for efficient gene transfer and expression in mammalian cells. *Gene Ther.* **3:** 448–457.

Graham, M.L. 2003. Pegaspargase: a review of clinical studies. *Adv. Drug Deliv. Rev.* **55:** 1293–1302.

Greco, F., Vicent, M.J., Penning, N.A., Nicholson, R.I., and Duncan, R. 2005. HPMA copolymer-aminoglutethimide conjugates inhibit aromatase in MCF-7 cell lines. *J. Drug Target.* **13:** 459–470.

Greco, F., Vicent, M.J., Gee, S., Jones, A.T., Gee, J., Nicholson, R.I., and Duncan, R. 2007. Investigating the mechanism of enhanced cytotoxicity of HPMA copolymer-Dox-AGM in breast cancer cells. *J. Control. Release* **117:** 28–39.

Green, N.K., Herbert, C.W., Hale, S.J., Hale, A.B., Mautner, V., Harkins, R., Hermiston, T., Ulbrich, K., Fisher, K.D., and Seymour, L.W. 2004. Extended plasma circulation time and decreased toxicity of polymer-coated adenovirus. *Gene Ther.* **11:** 1256–1263.

Greenwald, R.B., Conover, C.D., and Choe, Y.H. 2000a. Poly(ethylene glycol) conjugated drugs and prodrugs: a comprehensive review. *Crit. Rev. Ther. Drug Carrier Syst.* **17:** 101–161.

Greenwald, R.B., Choe, Y.H., Conover, C.D., Shum, K., Wu, D., and Royzen, M. 2000b. Drug delivery systems based on trimethyl lock lactonization: poly(ethylene glycol) prodrugs of amino-containing compounds. *J. Med. Chem.* **43:** 475–487.

Greenwald, R.B., Choe, Y.H., McGuire, J., and Conover, C.D. 2003. Effective drug delivery by PEGylated drug conjugates. *Adv. Drug Deliv. Rev.* **55:** 217–250.

Gref, R., Couvreur, P., Barratt, G., and Mysiakine, E. 2003. Surface-engineered nanoparticles for multiple ligand coupling. *Biomaterials* **24:** 4529–4537.

Greish, K., Fang, J., Inutsuka, T., Nagamitsu, A., and Maeda, H. 2003. Macromolecular therapeutics: advantages and prospects with special emphasis on solid tumour targeting. *Clin. Pharmacokinet.* **42:** 1089–1105.

Grim, Y. and Kopeček, J. 1991. Bioadhesive water-soluble polymeric drug carriers for site-specific oral drug delivery. Synthesis, characterization and release of 5-aminosalicylic acid by Streptococcum faecium in vitro. *New Polym. Mater.* **3:** 49–59.

Gupta, B. and Torchilin, V.P. 2006. Transactivating transcriptional activator-mediated drug delivery. *Expert Opin. Drug Deliv.* **3:** 177–190.

Gupta, B., Levchenko, T.S., and Torchilin, V.P. 2005. Intracellular delivery of large molecules and small particles by cell-penetrating proteins and peptides. *Adv. Drug. Deliv. Rev.* **57:** 637–651.

Hama, S., Akita, H., Ito, R., Mizuguchi, H., Hayakawa, T., and Harashima, H. 2006. Quantitative comparison of intracellular trafficking and nuclear transcription between adenoviral and lipoplex systems. *Mol. Ther.* **13:** 786–794.

Harris, J.M. and Chess, R.B. 2003. Effect of pegylation on pharmaceuticals. *Nat. Rev. Drug Discov.* **2:** 214–221.

Harris, J., Werling, D., Hope, J.C., Taylor, G., and Howard, C.J. 2002a. Caveolae and caveolin in immune cells: distribution and functions. *Trends Immunol.* **23:** 158–164.

Harris, J., Werling, D., Koss, M., Monaghan, P., Taylor, G., and Howard, C.J. 2002b. Expression of caveolin by bovine lymphocytes and antigen-presenting cells. *Immunology* **105:** 190–195.

Hegazy, A.K., Barakat, H.N., and Kabiel, H.F. 2006. Anatomical significance of the hygrochastic movement in Anastatica hierochuntica. *Ann. Bot. (Lond.)* **97:** 47–55.

Henry, S.M., El-Sayed, M.E., Pirie, C.M., Hoffman, A.S., and Stayton, P.S. 2006. pH-responsive poly(styrene-*alt*-maleic anhydride) alkylamide copolymers for intracellular drug delivery. *Biomacromolecules* **7:** 2407–2414.

Hersel, U., Dahmen, C., and Kessler, H. 2003. RGD modified polymers: biomaterials for stimulated cell adhesion and beyond. *Biomaterials* **24:** 4385–4415.

Hewlett, L.J., Prescott, A.R., and Watts, C. 1994. The coated pit and macropinocytic pathways serve distinct endosome populations. *J. Cell Biol.* **124:** 689–703.

Hodel, M.R., Corbett, A.H., and Hodel, A.E. 2001. Dissection of a nuclear localization signal. *J. Biol. Chem.* **276:** 1317–1325.

Hong, C.Y. and Pan, C.Y. 2006. Direct synthesis of biotinylated stimuli-responsive polymer and diblock copolymer by RAFT polymerization using biotinylated trithiocarbonate as RAFT agent. *Macromolecules* **39**: 3517–3524.

Hoste, K., De Winne, K., and Schacht, E. 2004. Polymeric prodrugs. *Int. J. Pharm.* **277**: 119–131.

Hrubý, M., Etrych, T., Kučka, J., Forsterová, M., and Ulbrich, K. 2006. Hydroxybisphosphonate-containing polymeric drug-delivery systems designed for targeting into bone tissue. *J. Appl. Polym. Sci.* **101**: 3192–3201.

Hudson, P.J. and Souriau, C. 2003. Engineered antibodies. *Nat. Med.* **9**: 129–134.

Hynes, R.O. 1992. Integrins: versatility, modulation, and signaling in cell adhesion. *Cell* **69**: 11–25.

Jain, R.K. 1989. Delivery of novel therapeutic agents in tumors: physiological barriers and strategies. *J. Natl. Cancer Inst.* **81**: 570–576.

Jatzkewitz, H. 1955. Peptamin (glycyl-L-leucyl-mescaline) bound to blood plasma expander (polyvinylpyrrolidone) as a new depot form of a biologically active primary amine (mescaline). *Z. Naturforsch* **10b**: 27–31.

Jelínková, M., Strohalm, J., Etrych, T., Ulbrich, K., and Říhová, B. 2003. Starlike vs. classic macromolecular prodrugs: two different antibody-targeted HPMA copolymers of doxorubicin studied in vitro and in vivo as potential anticancer drugs. *Pharm. Res.* **20**: 1558–1564.

Jensen, K.D., Kopečková, P., and Kopeček, J. 2002. Antisense oligonucleotides delivered to the lysosome escape and actively inhibit the hepatitis B virus. *Bioconjug. Chem.* **13**: 975–984.

Kabanov, A.V., Batrakova, E.V., and Alakhov, V.Y. 2002. Pluronic block copolymers for overcoming drug resistance in cancer. *Adv. Drug Deliv. Rev.* **54**: 759–779.

Kafienah, W., Brömme, D., Buttle, D.J., Croucher, L.J., and Hollander, A.P. 1998. Human cathepsin K cleaves native type I and II collagens at the N-terminal end of the triple helix. *Biochem. J.* **331 (Pt 3):** 727–732.

Kakudo, T., Chaki, S., Futaki, S., Nakase, I., Akaji, K., Kawakami, T., Maruyama, K., Kamiya, H., and Harashima, H. 2004. Transferrin-modified liposomes equipped with a pH-sensitive fusogenic peptide: an artifical viral-like delivery system. *Biochemistry* **43**: 5618–5628.

Kale, A.A. and Torchilin, V.P. 2007. Design, synthesis, and characterization of pH-sensitive PEG-PE conjugates for stimuli-sensitive pharmaceutical nanocarriers: the effect of substitutes at the hydrazone linkage on the ph stability of PEG-PE conjugates. *Bioconjug. Chem.* **18**: 363–370.

Kamei, S. and Kopeček, J. 1995. Prolonged blood circulation in rats of nanospheres surface-modified with semitelechelic poly[N-(2-hydroxypropyl)methacrylamide]. *Pharm. Res.* **12**: 663–668.

Kaplan, I.M., Wadia, J.S., and Dowdy, S.F. 2005. Cationic TAT peptide transduction domain enters cells by macropinocytosis. *J. Control. Release* **102**: 247–253.

Kasuya, Y., Lu, Z.R., Kopečková, P., Minko, T., Tabibi, S.E., and Kopeček, J. 2001. Synthesis and characterization of HPMA copolymer-aminopropylgeldanamycin conjugates. *J. Control. Release* **74**: 203–211.

Kasuya, Y., Lu, Z.R., Kopečková, P., Tabibi, S.E., and Kopeček, J. 2002. Influence of the structure of drug moieties on the in vitro efficacy of HPMA copolymer-geldanamycin derivative conjugates. *Pharm. Res.* **19**: 115–123.

Kawakami, S., Hattori, Y., Lu, Y., Higuchi, Y., Yamashita, F., and Hashida, M. 2004. Effect of cationic charge on receptor-mediated transfection using mannosylated cationic liposome/plasmid DNA complexes following the intravenous administration in mice. *Pharmazie* **59**: 405–408.

Kelso, G.F., Porteous, C.M., Hughes, G., Ledgerwood, E.C., Gane, A.M., Smith, R.A., and Murphy, M.P. 2002. Prevention of mitochondrial oxidative damage using targeted antioxidants. *Ann. N. Y. Acad. Sci.* **959**: 263–274.

Khalil, I.A., Kogure, K., Akita, H., and Harashima, H. 2006. Uptake pathways and subsequent intracellular trafficking in nonviral gene delivery. *Pharmacol. Rev.* **58**: 32–45.

Khandare, J. and Minko, T. 2006. Polymer-drug conjugates: progress in polymeric prodrugs. *Prog. Polym. Sci.* **31**: 359–397.

Khandare, J.J., Chandna, P., Wang, Y., Pozharov, V.P., and Minko, T. 2006a. Novel polymeric prodrug with multivalent components for cancer therapy. *J. Pharmacol. Exp. Ther.* **317**: 929–937.

Khandare, J.J., Jayant, S., Singh, A., Chandna, P., Wang, Y., Vorsa, N., and Minko, T. 2006b. Dendrimer versus linear conjugate: influence of polymeric architecture on the delivery and anticancer effect of paclitaxel. *Bioconjug. Chem.* **17:** 1464–1472.

Kirkham, M. and Parton, P.G. 2005. Clathrin-independent endocytosis: new insight into caveolae and non-caveolar lipid raft carriers. *Biochim. Biophys. Acta* **1746:** 350–363.

Kishore, B.K., Lambricht, P., Laurent, G., Maldague, P., Wagner, R., and Tulkens, P.M. 1990. Mechanism of protection afforded by polyaspartic acid against gentamicin-induced phospholipidosis. II. Comparative in vitro and in vivo studies with poly-L-aspartic, poly-L-glutamic and poly-D-glutamic acids. *J. Pharmacol. Exp. Ther.* **255:** 875–885.

Kleinschmidt, J.A. and Seiter, A. 1988. Identification of domains involved in nuclear uptake and histone binding of protein N1 of Xenopus laevis. *EMBO J.* **7:** 1605–1614.

Köhler, G. and Milstein, C. 1975. Continuous cultures of fused cells secreting antibody of predefined specificity. *Nature* **256:** 495–497.

Kolb, H.C., Finn, M.G., and Sharpless, K.B. 2001. Click chemistry: diverse chemical function from a few good reactions. *Angew. Chem. Int. Ed. Engl.* **40:** 2004–2021.

Koňák, Č., Ganchev, B., Teodorescu, M., Matyjaszewski, K., Kopečková, P., and Kopeček, J. 2002. Poly[N-(2-hydroxypropyl)methacrylamide-*block*-n-butyl acrylate] micelles in water/DMF mixed solvents. *Polymer* **43:** 3735–3741.

Kopeček, J. 1977. Soluble biomedical polymers. *Polim. Med.* **7:** 191–221.

Kopeček, J. 1984. Controlled biodegradability of polymers – a key to drug delivery systems. *Biomaterials* **5:** 19–25.

Kopeček, J. 2003. Smart and genetically engineered biomaterials and drug delivery systems. *Eur. J. Pharm. Sci.* **20:** 1–16.

Kopeček, J. and Kopečková, P. 2003. Macromolecular therapeutics: state-of-the-art and future potential. *Bull. Tech. Gattefossé* **96:** 9–21.

Kopeček, J. and Rejmanová, P. 1983. Enzymatically degradable bonds in synthetic polymers. In: *Controlled Drug Delivery*, S.D. Bruck, Ed., CRC Press, Boca Raton, Florida, pp. 81–124.

Kopeček, J., Cífková, I., Rejmanová, P., Strohalm, J., Obereigner, B., and Ulbrich, K. 1981a. Polymers containing enzymatically degradable bonds. 4. Preliminary experiments in vivo. *Makromol. Chem.* **182:** 2941–2949.

Kopeček, J., Rejmanová, P., and Chytrý, V. 1981b. Polymers containing enzymatically degradable bonds 1. Chymotrypsin catalyzed hydrolysis of p-nitroanilides of phenylalanine and tyrosine attached to side-chains of copolymers of N-(2-hydroxypropyl)methacrylamide. *Makromol. Chem.* **182:** 799–809.

Kopeček, J., Kopečková, P., Minko, T., and Lu, Z. 2000. HPMA copolymer-anticancer drug conjugates: design, activity, and mechanism of action. *Eur. J. Pharm. Biopharm.* **50:** 61–81.

Kopečková, P., Rathi, R.C., Takada, S., Říhová, B., Berenson, M.M., and Kopeček, J. 1994. Bioadhesive N-(2-hydroxypropyl)methacrylamide copolymers for colon-specific drug delivery. *J. Control. Release* **28:** 211–222.

Kovář, M., Strohalm, J., Etrych, T., Ulbrich, K., and Říhová, B. 2002a. Star structure of antibody-targeted HPMA copolymer-bound doxorubicin: a novel type of polymeric conjugate for targeted drug delivery with potent antitumor effect. *Bioconjug. Chem.* **13:** 206–215.

Kovář, M., Strohalm, J., Ulbrich, K., and Říhová, B. 2002b. In vitro and in vivo effect of HPMA copolymer-bound doxorubicin targeted to transferrin receptor of B-cell lymphoma 38C13. *J. Drug Target.* **10:** 23–30.

Kovář, M., Mrkvan, T., Strohalm, J., Etrych, T., Ulbrich, K., Šťastný, M., and Říhová, B. 2003. HPMA copolymer-bound doxorubicin targeted to tumor-specific antigen of BCL1 mouse B cell leukemia. *J. Control. Release* **92:** 315–330.

Kratz, F., Beyer, U., and Schutte, M.T. 1999. Drug-polymer conjugates containing acid-cleavable bonds. *Crit. Rev. Ther. Drug Carrier Syst.* **16:** 245–288.

Krinick, N.L., Sun, Y., Joyner, D., Spikes, J.D., Straight, R.C., and Kopeček, J. 1994. A polymeric drug delivery system for the simultaneous delivery of drugs activatable by enzymes and/or light. *J. Biomat. Sci., Polym. Ed.* **5:** 303–324.

Kulkarni, S., Schilli, C., Grin, B., Muller, A.H., Hoffman, A.S., and Stayton, P.S. 2006. Controlling the aggregation of conjugates of streptavidin with smart block copolymers prepared via the RAFT copolymerization technique. *Biomacromolecules* **7:** 2736–2741.

Kumar, P., Wu, H., McBride, J.L., Jung, K.E., Kim, M.H., Davidson, B.L., Lee, S.K., Shankar, P., Manjunath, N., 2007. Transvascular delivery of small interfering RNA to the central nervous system. *Nature* **448:** 39–43.

Kumazawa, E. and Ochi, Y. 2004. DE-310, a novel macromolecular carrier system for the camptothecin analog DX-8951f: potent antitumor activities in various murine tumor models. *Cancer Sci.* **95:** 168–175.

Kunath, K., Kopečková, P., Minko, T., and Kopeček, J. 2000. HPMA copolymer-anticancerdrug-OV-TL16 antibody conjugates. 3. The effect of free and polymer-bound adriamycin on the expression of some genes in the OVCAR-3 human ovarian carcinoma cell line. *Eur. J. Pharm. Biopharm.* **49:** 11–15.

Lääne, A., Aaviksaar, A., Haga, M, Chytrý, V., and Kopeček, J. 1985. Preparation of polymer-modified enzymes of prolonged circulation times. Poly[N-(2-hydroxypropyl)methacrylamide] bound acetylcholinesterase. *Makromol. Chem., Suppl.* **9:** 35–42.

Lacey, J.V., Mink, P.J., Lubin, J.H., Sherman, M.E., Troisi, R., Hartge, P., Schatzkin, A., and Schairer, C. 2002a. Menopausal hormone replacement therapy and risk of ovarian cancer. *JAMA* **288:** 334–341.

Lacey, J.V., Mink, R., Lubin, J.H., Sherman, M.E., Troisi, R., Hartge, P., Schatzkin, A., and Schairer, C. 2002b. Estrogen replacement therapy and risk of ovarian cancer in postmenopausal women. *JAMA* **288:** 2539–2539.

Lackey, C.A., Press, O.W., Hoffman, A.S., and Stayton, P.S. 2002. A biomimetic pH-responsive polymer directs endosomal release and intracellular delivery of an endocytosed antibody complex. *Bioconjug. Chem.* **13:** 996–1001.

Ladmiral, V., Mantovani, G., Clarkson, G.J., Cauet, S., Irwin, J.L., and Haddleton, D.M. 2006. Synthesis of neoglycopolymers by a combination of "click chemistry" and living radical polymerization. *J. Am. Chem. Soc.* **128:** 4823–4830.

Lam, K.S., Lehman, A.L., Song, A., Doan, N., Enstrom, A.M., Maxwell, J., and Liu, R. 2003. Synthesis and screening of "one-bead one-compound" combinatorial peptide libraries. *Methods Enzymol.* **369:** 298–322.

Lamaze, C. and Schmid, S.L. 1995. The emergence of clathrin-independent pinocytic pathways. *Curr. Opin. Cell Biol.* **7:** 573–580.

Lamaze, C., Dujeancourt, A., Baba, T., Lo, C.G., Benmerah, A., and Dautry-Varsat, A. 2001. Interleukin 2 receptors and detergent-resistant membrane domains define a clathrin-independent endocytic pathway. *Mol. Cell* **7:** 661–671.

Lammers, T., Kuhnlein, R., Kissel, M., Šubr, V., Etrych, T., Pola, R., Pechar, M., Ulbrich, K., Storm, G., Huber, P., and Peschke, P. 2005. Effect of physicochemical modification on the biodistribution and tumor accumulation of HPMA copolymers. *J. Control. Release* **110:** 103–118.

Lanciotti, J., Song, A., Doukas, J., Sosnowski, B., Pierce, G., Gregory, R., Wadsworth, S., and O'Riordan, C. 2003. Targeting adenoviral vectors using heterofunctional polyethylene glycol FGF2 conjugates. *Mol. Ther.* **8:** 99–107.

Lanford, R.E., Kanda, P., and Kennedy, R.C. 1986. Induction of nuclear transport with a synthetic peptide homologous to the SV40 T antigen transport signal. *Cell* **46:** 575–582.

Langer, C.J. 2004a. CT-2103: a novel macromolecular taxane with potential advantages compared with conventional taxanes. *Clin. Lung Cancer* **6 (Suppl 2):** S85–S88.

Langer, C.J. 2004b. CT-2103: emerging utility and therapy for solid tumours. *Expert. Opin. Investig. Drugs* **13:** 1501–1508.

Langer, C.J., Socinski, M.A., Ross, H., and O'Byrne, K.J. 2005. Paclitaxel poliglumex (PPX)/carboplatin vs. paclitaxel/carboplatin for the treatment of PS2 patients with chemotherapy-naïve advanced non-small cell lung cancer (NSCLC): a phase III study. *J. Clin. Oncol.* **23:** 7011, *2005 ASCO Annual Meeting Proceedings.*

Langer, C.J., Socinski, M.A., and O'Byrne, K.J. 2006. Paclitaxel poliglumex (PPX/carboplatin vs. paclitaxel/carboplatin for the treatment of PS2 patients with chemotherapy-naïve advanced non-small cell lung cancer: a phase III study. *Proc. Am. Soc. Clin. Oncol.* **23:** 623s.

Larsen, A.K., Escargueil, A.E., and Skladanowski, A. 2000. Resistance mechanisms associated with altered intracellular distribution of anticancer agents. *Pharmacol. Ther.* **85:** 217–229.

Lee, B.S., Fujita, M., Khazenzon, N.M., Wawrowsky, K.A., Wachsmann-Hogiu, S., Farkas, D.L., Black, K.L., Ljubimova, J.Y., and Holler, E. 2006. Polycefin, a new prototype of a multifunctional nanoconjugate based on poly(beta-L-malic acid) for drug delivery. *Bioconjug. Chem.* **17:** 317–326.

Lee, R.J. and Huang, L. 1996. Folate-targeted, anionic liposome-entrapped polylysine-condensed DNA for tumor cell-specific gene transfer. *J. Biol. Chem.* **271:** 8481–8487.

Lee, S.W. and Sullenger, B.A. 1997. Isolation of a nuclease-resistant decoy RNA that can protect human acetylcholine receptors from myasthenic antibodies. *Nat. Biotechnol.* **15:** 41–45.

Levy, Y., Hershfield, M.S., Fernandez-Mejia, C., Polmar, S.H., Scudiery, D., Berger, M., and Sorensen, R.U. 1988. Adenosine deaminase deficiency with late onset of recurrent infections: response to treatment with polyethylene glycol-modified adenosine deaminase. *J. Pediatr.* **113:** 312–317.

Li, C. 2002. Poly(L-glutamic acid)–anticancer drug conjugates. *Adv. Drug Deliv. Rev.* **54:** 695–713.

Li, C., Yu, D.F., Newman, R.A., Cabral, F., Stephens, L.C., Hunter, N., Milas, L., and Wallace, S. 1998. Complete regression of well-established tumors using a novel water-soluble poly(L-glutamic acid)-paclitaxel conjugate. *Clin. Cancer Res.* **58:** 2404–2409.

Lin, J.J., Silas, J.A., Bermudez, H., Milam, V.T., Bates, F.S., and Hammer, D.A. 2004. The effect of polymer chain length and surface density on the adhesiveness of functionalized polymersomes. *Langmuir* **20:** 5493–5500.

Lu, Z.R., Kopečková, P., Wu, Z., and Kopeček, J. 1998. Functionalized semitelechelic poly[N-(2-hydroxypropyl)methacrylamide] for protein modification. *Bioconjug. Chem.* **9:** 793–804.

Lu, J.M., Peterson, C.M., Guo-Shiah, J., Gu, Z.W., Peterson, C.A., Straight, R.C., and Kopeček, J. 1999a. Cooperativity between free and N-(2-hydroxypropyl) methacrylamide copolymer bound adriamycin and meso-chlorin e_6 monoethylene diamine induced photodynamic therapy in human epithelial ovarian carcinoma in vitro. *Int. J. Oncol.* **15:** 5–16.

Lu, Z.R., Kopečková, P., and Kopeček, J. 1999b. Polymerizable Fab' antibody fragments for targeting of anticancer drugs. *Nat. Biotechnol.* **17:** 1101–1104.

Lu, Z.R., Shiah, J.G., Kopečková, P., and Kopeček, J. 2001. Preparation and biological evaluation of polymerizable antibody Fab' fragment targeted polymeric drug delivery system. *J. Control. Release* **74:** 263–268.

Lu, Z.R., Shiah, J.G., Kopečková, P., and Kopeček, J. 2003. Polymerizable Fab' antibody fragment targeted photodynamic cancer therapy in nude mice. *STP Pharm. Sci.* **13:** 69–75.

Lynn, D.M., Anderson, D.G., Putnam, D., and Langer, R. 2001. Accelerated discovery of synthetic transfection vectors: parallel synthesis and screening of a degradable polymer library. *J. Am. Chem. Soc.* **123:** 8155–8156.

Maeda, H. 2001a. The enhanced permeability and retention (EPR) effect in tumor vasculature: the key role of tumor-selective macromolecular drug targeting. *Adv. Enzyme Regul.* **41:** 189–207.

Maeda, H. 2001b. SMANCS and polymer-conjugated macromolecular drugs: advantages in cancer chemotherapy. *Adv. Drug Deliv. Rev.* **46:** 169–185.

Maeda, H., Ueda, M., Morinaga, T., and Matsumoto, T. 1985. Conjugation of poly(styrene-comaleic acid) derivatives to the antitumor protein neocarzinostatin: pronounced improvements in pharmacological properties. *J. Med. Chem.* **28:** 455–461.

Mahat, R.I., Monera, O.D., Smith, L.C., and Rolland, A. 1999. Peptide-based gene delivery. *Curr. Opin. Mol. Ther.* **1:** 226–243.

Malik, N., Wiwattanapatapee, R., Klopsch, R., Lorenz, K., Frey, H., Weener, J.W., Meijer, E.W., Paulus, W., and Duncan, R. 2000. Dendrimers: relationship between structure and biocompatibility in vitro, and preliminary studies on the biodistribution of 125I-labelled polyamidoamine dendrimers in vivo. *J. Control. Release* **65:** 133–148.

Malkoch, M., Vestberg, R., Gupta, N., Mespouille, L., Dubois, P., Mason, A.F., Hedrick, J.L., Liao, Q., Frank, C.W., Kingsbury, K., and Hawker, C.J. 2006. Synthesis of well-defined hydrogel networks using click chemistry. *Chem. Commun.* **26:** 2774–2776.

Mann, J.S., Huang, J.C., and Keana, J.F. 1992. Molecular amplifiers: synthesis and functionalization of a poly(aminopropyl)dextran bearing a uniquely reactive terminus for univalent attachment to biomolecules. *Bioconjug. Chem.* **3:** 154–159.

Markman, M. 2004. Improving the toxicity profile of chemotherapy for advanced ovarian cancer: a potential role for CT-2103. *J. Exp. Ther. Oncol.* **4:** 131–136.

Martinez-Fong, D., Navarro-Quiroga, I., Ochoa, I., Alvarez-Maya, I., Meraz, M.A., Luna, J., and Arias-Montano, J.A. 1999. Neurotensin-SPDP-poly-L-lysine conjugate: a nonviral vector for targeted gene delivery to neural cells. *Brain Res. Mol. Brain Res.* **69:** 249–262.

Matsumura, Y. and Maeda, H. 1986. A new concept for macromolecular therapeutics in cancer chemotherapy: mechanism of tumoritropic accumulation of proteins and the antitumor agent smancs. *Cancer Res.* **46:** 6387–6392.

Matveev, S., Li, X., Everson, W., and Smart, E.J. 2001. The role of caveolae and caveolin in vesicle-dependent and vesicle-independent trafficking. *Adv. Drug Deliv. Rev.* **49:** 237–250.

Maxfield, F.R. and McGraw, T.E. 2004. Endocytic recycling. *Nat. Rev. Mol. Cell Biol.* **5:** 121–132.

Meerum Terwogt, J.M., ten Bokkel Huinink, W.W., Schellens, J.H., Schot, M., Mandjes, I.A., Zurlo, M.G., Rocchetti, M., Rosing, H., Koopman, F.J., and Beijnen, J.H. 2001. Phase I clinical and pharmacokinetic study of PNU166945, a novel water-soluble polymer-conjugated prodrug of paclitaxel. *Anticancer Drugs* **12:** 315–323.

Mehvar, R. 2000. Dextrans for targeted and sustained delivery of therapeutic and imaging agents. *J. Control. Release* **69:** 1–25.

Mehvar, R. and Hoganson, D.A. 2000. Dextran-methylprednisolone succinate as a prodrug of methylprednisolone: immunosuppressive effects after in vivo administration to rats. *Pharm. Res.* **17:** 1402–1407.

Meier, O. and Greber, U.F. 2003. Adenovirus endocytosis. *J. Gene Med.* **5:** 451–462.

Merdan, T., Kunath, K., Fischer, D., Kopeček, J., and Kissel, T. 2002. Intracellular processing of poly(ethylene imine)/ribozyme complexes can be observed in living cells by using confocal laser scanning microscopy and inhibitor experiments. *Pharm. Res.* **19:** 140–146.

Midoux, P. and Monsigny, M. 1999. Efficient gene transfer by histidylated polylysine/pDNA complexes. *Bioconjug. Chem.* **10:** 406–411.

Minko, T., Kopečková, P., and Kopeček, J. 1999. Chronic exposure to HPMA copolymer-bound adriamycin does not induce multidrug resistance in a human ovarian carcinoma cell line. *J. Control. Release* **59:** 133–148.

Minko, T., Kopečková, P., and Kopeček, J. 2000. Efficacy of the chemotherapeutic action of HPMA copolymer-bound doxorubicin in a solid tumor model of ovarian carcinoma. *Int. J. Cancer* **86:** 108–117.

Mitra, A., Coleman, T., Borgman, M., Nan, A., Ghandehari, H., and Line, B.R. 2006. Polymeric conjugates of mono- and bi-cyclic alphaVbeta3 binding peptides for tumor targeting. *J. Control. Release* **114:** 175–183.

Molineux, G. 2004. The design and development of pegfilgrastim (PEG-rmetHuG-CSF, Neulasta). *Curr. Pharm. Des.* **10:** 1235–1244.

Mrkvan, T., Šírová, M., Etrych, T., Chytil, P., Strohalm, J., Plocová, D., Ulbrich, K., and Říhová, B. 2005. Chemotherapy based on HPMA copolymer conjugates with pH-controlled release of doxorubicin triggers anti-tumor immunity. *J. Control. Release* **110:** 119–129.

Mukherjee, A., Monson, J.P., Jonsson, P.J., Trainer, P.J., and Shalet, S.M. 2003. Seeking the optimal target range for insulin-like growth factor I during the treatment of adult growth hormone disorders. *J. Clin. Endocrinol. Metab.* **88:** 5865–5870.

Mukherjee, S., Ghosh, R.N., and Maxfield, F.R. 1997. Endocytosis. *Physiol. Rev.* **77:** 759–803.

Nagayama, T., Hashidzume, A., and Morishima, Y. 2002. Characterization of self-association in water of polycations hydrophobically modified with hydrocarbon and siloxane chains. *Langmuir* **18:** 6775–6782.

Nakanishi, T., Fukushima, S., Okamoto, K., Suzuki, M., Matsumura, Y., Yokoyama, M., Okano, T., Sakurai, Y., and Kataoka, K. 2001. Development of the polymer micelle carrier system for doxorubicin. *J. Control. Release* **74:** 295–302.

Nakase, I., Niwa, M., Takeuchi, T., Sonomura, K., Kawabata, N., Koike, Y., Takehashi, M., Tanaka, S., Ueda, K., Simpson, J.C., Jones, A.T., Sugiura, Y., and Futaki, S. 2004. Cellular uptake of arginine-rich peptides: roles for macropinocytosis and actin rearrangement. *Mol. Ther.* **10:** 1011–1022.

Nan, A., Ghandehari, H., Hebert, C., Siavash, H., Nikitakis, N., Reynolds, M., and Sauk, J.J. 2005. Water-soluble polymers for targeted drug delivery to human squamous carcinoma of head and neck. *J. Drug Target.* **13:** 189–197.

Nigg, E.A. 1997. Nucleocytoplasmic transport: signals, mechanisms and regulation. *Nature* **386:** 779–787.

Nimjee, S.M., Rusconi, C.P., and Sullenger, B.A. 2005. Aptamers: an emerging class of therapeutics. *Annu. Rev. Med.* **56:** 555–583.

Nishiyama, N. and Kataoka, K. 2006. Current state, achievements, and future prospects of polymeric micelles as nanocarriers for drug and gene delivery. *Pharmacol Ther* **112:** 630–648.

Nishiyama, N., Nori, A., Malugin, A., Kasuya, Y., Kopečková, P., and Kopeček, J. 2003. Free and N-(2-hydroxypropyl)methacrylamide copolymer-bound geldanamycin derivative induce different stress responses in A2780 human ovarian carcinoma cells. *Cancer Res.* **63:** 7876–7882.

Nori, A. and Kopeček, J. 2005. Intracellular targeting of polymer-bound drugs for cancer chemotherapy. *Adv. Drug. Deliv. Rev.* **57:** 609–636.

Nori, A., Jensen, K.D., Tijerina, M., Kopečková, P., and Kopeček, J. 2003a. Subcellular trafficking of HPMA copolymer-Tat conjugates in human ovarian carcinoma cells. *J. Control. Release* **91:** 53–59.

Nori, A., Jensen, K.D., Tijerina, M., Kopečková, P., and Kopeček, J. 2003b. Tat-conjugated synthetic macromolecules facilitate cytoplasmic drug delivery to human ovarian carcinoma cells. *Bioconjug. Chem.* **14:** 44–50.

Obereigner, B., Burešová, M., Vrána, A., and Kopeček, J. 1979. Preparation of polymerizable derivatives of N-(4-aminobenzenesulfonyl)-N'-butylurea. *J. Polym. Sci., Polym. Symp.* **66:** 41–52.

Ohkawa, K., Hatano, T., Yamada, K., Joh, K., Takada, K., Tsukada, Y., and Matsuda, M. 1993. Bovine serum albumin-doxorubicin conjugate overcomes multidrug resistance in a rat hepatoma. *Cancer Res.* **53:** 4238–4242.

Okuno, S., Harada, M., Yano, T., Yano, S., Kiuchi, S., Tsuda, N., Sakamura, Y., Imai, J., Kawaguchi, T., and Tsujihara, K. 2000. Complete regression of xenografted human carcinomas by camptothecin analogue-carboxymethyl dextran conjugate (T-0128). *Cancer Res.* **60:** 2988–2995.

Omelyanenko, V., Gentry, C., Kopečková, P., and Kopeček, J. 1998. HPMA copolymer-anticancer drug-OV-TL16 antibody conjugates. II. Processing in epithelial ovarian carcinoma cells in vitro. *Int. J. Cancer* **75:** 600–608.

Pan, H., Kopečková, P., Wang, D., Yang, J., Miller, S., and Kopeček, J. 2006. Water-soluble HPMA copolymer–prostaglandin E1 conjugates containing a cathepsin K sensitive spacer. *J. Drug Target.* **14:** 425–435.

Panarin, E.F. and Ushakov, S.N. 1968. Synthesis of polymer salts and amidopenicillines (in Russian). *Khim. Pharm. Zhur.* **2:** 28–31.

Parrish, B. and Emrick, T. 2007. Soluble camptothecin derivatives prepared by click cycloaddition chemistry on functional aliphatic polyesters. *Bioconjug. Chem.* **18:** 263–267.

Parrish, B., Breitenkamp, R.B., and Emrick, T. 2005. PEG- and peptide-grafted aliphatic polyesters by click chemistry. *J. Am. Chem. Soc.* **127:** 7404–7410.

Parton, R.G., Joggerst, B., and Simons, K. 1994. Regulated internalization of caveolae. *J. Cell Biol.* **127:** 1199–1215.

Pasut, G., Scaramuzza, S., Schiavon, O., Mendichi, R., and Veronese, F.M. 2005. PEG-epirubicin conjugates with high drug loading. *J. Bioact. Compat. Polym.* **20:** 213–230.

Paul, A., Vicent, M.J., and Duncan, R. 2007. Using small-angle neutron scattering to study the solution conformation of N-(2-hydroxypropyl)methacrylamide copolymer-doxorubicin conjugates. *Biomacromolecules* **8**: 1573–1579.

Pechar, M., Braunová, A., Ulbrich, K., Jelínková, M., and Říhová, B. 2005. Poly(ethylene glycol) – Doxorubicin conjugates with pH-controlled activation. *J. Bioact. Compat. Polym.* **20**: 319–341.

Pedone, E., Li, X.W., Koseva, N., Alpar, O., and Brocchini, S. 2003. An information rich biomedical polymer library. *J. Mater. Chem.* **13**: 2825–2837.

Pelkmans, L. and Helenius, A. 2002. Endocytosis via caveolae. *Traffic* **3**: 311–320.

Pelkmans, L., Puntener, D., and Helenius, A. 2002. Local actin polymerization and dynamin recruitment in SV40-induced internalization of caveolae. *Science* **296**: 535–539.

Peterson, C.M., Lu, J.M., Sun, Y., Peterson, C.A., Shiah, J.G., Straight, R.C., and Kopeček, J. 1996. Combination chemotherapy and photodynamic therapy with N-(2-hydroxypropyl) methacrylamide copolymer-bound anticancer drugs inhibit human ovarian carcinoma heterotransplanted in nude mice. *Cancer Res.* **56**: 3980–3985.

Phan, U.T., Arunachalam, B., and Creswell, P. 2000. Gamma-interferon-inducible lysosomal thiol reductase (GILT). Maturation, activity, and mechanism of action. *J. Biol. Chem.* **275**: 25907–25914.

Pichon, C., Goncalves, C., and Midoux, P. 2001. Histidine-rich peptides and polymers for nucleic acids delivery. *Adv. Drug Deliv. Rev.* **53**: 75–94.

Pieken, W.A., Olsen, D.B., Benseler, F., Aurup, H., and Eckstein, F. 1991. Kinetic characterization of ribonuclease-resistant 2′-modified hammerhead ribozymes. *Science* **253**: 314–317.

Poujol, S., Pinguet, F., Bressole, F., Boustta, M., and Vert, M. 2000. Molecular microencapsulation: paclitaxel formations in aqueous medium using hydrophobized poly(L-lysine citramide imide). *J. Bioact. Compat. Polym.* **15**: 99–114.

Putnam, D. and Kopeček, J. 1995a. Polymer conjugates with anticancer activity. *Adv. Polym. Sci.* **122**: 55–123.

Putnam, D. and Kopeček, J. 1995b. Enantioselective release of 5-fluorouracil from HPMA based copolymers via lysosomal enzymes. *Bioconjug. Chem.* **6**: 483–492.

Putnam, D., Gentry, C.A., Pack, D.W., and Langer, R. 2001. Polymer-based gene delivery with low cytotoxicity by a unique balance of side-chain termini. *Proc. Natl. Acad. Sci. USA.* **98**: 1200–1205.

Putnam, D., Zelikin, A.N., Izumrudov, V.A., and Langer, R. 2003. Polyhistidine-PEG:DNA nanocomposites for gene delivery. *Biomaterials* **24**: 4425–4433.

Qiu, L.Y. and Bae, Y.H. 2006. Polymer architecture and drug delivery. *Pharm. Res.* **23**: 1–30.

Rademaker-Lakhai, J.M., Terret, C., Howell, S.B., Baud, C.M., De Boer, R.F., Pluim, D., Beijnen, J.H., Schellens, J.H., and Droz, J.P. 2004. A phase I and pharmacological study of the platinum polymer AP5280 given as an intravenous infusion once every 3 weeks in patients with solid tumors. *Clin. Cancer Res.* **10**: 3386–3395.

Rajender Reddy, K., Modi, M.W., and Pedder, S. 2002. Use of peginterferon alfa-2a (40 KD) (Pegasys) for the treatment of hepatitis C. *Adv. Drug Deliv. Rev.* **54**: 571–586.

Rask, R., Rasmussen, J.M., Hansen, H.V., Bysted, P., and Svehag, S.E. 1988. Complement C3d,g/ Epstein-Barr virus receptor density on human B-lymphocytes estimated by immunoenzymatic assay and immunocytochemistry. *J. Clin. Lab. Immunol.* **25**: 153–156.

Rathi, R.C., Kopečková, P., Říhová, B., and Kopeček, J. 1991. N-(2-Hydroxypropyl)methacrylam ide copolymers containing pendant saccharide moieties. Synthesis and bioadhesive properties. *J. Polym. Sci. [A1]* **29**: 1895–1902.

Read, M.L., Singh, S., Ahmed, Z., Stevenson, M., Briggs, S.S., Oupický, D., Barrett, L.B., Spice, R., Kendall, M., Berry, M., Preece, J.A., Logan, A., and Seymour, L.W. 2005. A versatile reducible polycation-based system for efficient delivery of a broad range of nucleic acids. *Nucleic Acids Res.* **33**: e86.

Rejman, J., Bragonzi, A., and Conese, M. 2005. Role of clathrin- and caveolae-mediated endocytosis in gene transfer mediated by lipo- and polyplexes. *Mol. Ther.* **12**: 468–474.

Rejmanová, P., Labský, J., and Kopeček, J. 1977. Aminolyses of monomeric and polymeric p-nitrophenyl esters of methacryloylated amino acids. *Makromol. Chem.* **178**: 2159–2168.

Rejmanová, P., Obereigner, B., and Kopeček, J. 1981. Polymers containing enzymatically degradable bonds. 2. Poly[N-(2-hydroxypropyl)methacrylamide] chains connected by oligopeptide sequences cleavable by chymotrypsin. *Makromol. Chem.* **182:** 1899–1915.

Rejmanová, P., Pohl, J., Baudyš, M., Kostka, V., and Kopeček, J. 1983. Polymers containing enzymatically degradable bonds. 8. Degradation of oligopeptide sequences in N-(2-hydroxy propyl)methacrylamide copolymers by bovine spleen cathepsin B. *Makromol. Chem.* **184:** 2009–2020.

Rejmanová, P., Kopeček, J., Duncan, R., and Lloyd, J.B. 1985. Stability in rat plasma and serum of lysosomally degradable oligopeptide sequences in N-(2-hydroxypropyl)methacrylamide copolymers. *Biomaterials* **6:** 45–48.

Rensberger, K.L., Hoganson, D.A., and Mehvar, R. 2000. Dextran-methylprednisolone succinate as a prodrug of methylprednisolone: in vitro immunosuppressive effects on rat blood and spleen lymphocytes. *Int. J. Pharm.* **207:** 71–76.

Ringsdorf, H. 1975. Structure and properties of pharmacologically active polymers. *J. Polym. Sci. Symp.* **51:** 135–153.

Robbins, J., Dilworth, S.M., Laskey, R.A., and Dingwall, C. 1991. Two interdependent basic domains in nucleoplasmin nuclear targeting sequence: identification of a class of bipartite nuclear targeting sequence. *Cell* **64:** 615–623.

Rodrigues, P.C., Beyer, U., Schumacher, P., Roth, T., Fiebig, H.H., Unger, C., Messori, L., Orioli, P., Paper, D.H., Mülhaupt, R., Kratz, F. 1999. Acid-sensitive polyethylene glycol conjugates of doxorubicin: preparation, in vitro efficacy and intracellular distribution. *Bioorg. Med. Chem.* **7:** 2517–2524.

Rodrigues, P.C, Scheuermann, K., Stockmar, C., Maier, G., Fiebig, H., Unger, C., Mülhaupt, R., and Kratz, F. 2003. Synthesis and in vitro efficacy of acid-sensitive poly(ethylene glycol) paclitaxel conjugates. *Bioorg. Med. Chem. Lett.* **13:** 355–360.

Rodrigues, P.C., Roth, T., Fiebig, H.H., Unger, C., Mülhaupt, R., and Kratz, F. 2006. Correlation of the acid-sensitivity of polyethylene glycol daunorubicin conjugates with their in vitro antiproliferative activity. *Bioorg. Med. Chem.* **14:** 4110–4117.

Rowinsky, E.K., Rizzo, J., Ochoa, L., Takimoto, C.H., Forouzesh, B., Schwartz, G., Hammond, L.A., Patnaik, A., Kwiatek, J., Goetz, A., Denis, L., McGuire, J., Tolcher, A.W. 2003. A phase I and pharmacokinetic study of pegylated camptothecin as a 1-hour infusion every 3 weeks in patients with advanced solid malignancies. *J. Clin. Oncol.* **21:** 148–157.

Rusconi, C.P., Scardino, E., Layzer, J., Pitoc, G.A., Ortel, T.L., Monroe, D., and Sullenger, B.A. 2002. RNA aptamers as reversible antagonists of coagulation factor IXa. *Nature* **419:** 90–94.

Sabbatini, P., Aghajanian, C., Dizon, D., Anderson, S., Dupont, J., Brown, J.V., Peters, W.A., Jacobs, A., Mehdi, A., Rivkin, S., Eisenfeld, A.J., Spriggs, D. 2004. Phase II study of CT-2103 in patients with recurrent epithelial ovarian, fallopian tube, or primary peritoneal carcinoma. *J. Clin. Oncol.* **22:** 4523–4531.

Saito, G., Swanson, J.A., and Lee, K.D. 2003. Drug delivery strategy utilizing conjugation via reversible disulfide linkages: role and site of cellular reducing activities. *Adv. Drug Deliv. Rev.* **55:** 199–215.

Sakuma, S., Lu, Z.R., Kopečková, P., and Kopeček, J. 2001. Biorecognizable HPMA copolymer-drug conjugates for colon-specific delivery of 9-aminocamptothecin. *J. Control. Release* **75:** 365–379.

Sarapa, N., Britto, M.R., Speed, W., Jannuzzo, M., Breda, M., James, C.A., Porro, M., Rocchetti, M., Wanders, A., Mahteme, H., and Nygren, P. 2003. Assessment of normal and tumor tissue uptake of MAG-CPT, a polymer-bound prodrug of camptothecin, in patients undergoing elective surgery for colorectal carcinoma. *Cancer Chemother. Pharmacol.* **52:** 424–430.

Saraste, J., Palade, G.E., and Farquhar, M.G. 1986. Temperature-sensitive steps in the transport of secretory proteins through the Golgi complex in exocrine pancreatic cells. *Proc. Natl. Acad. Sci. USA* **83:** 6425–6429.

Satchi-Fainaro, R., Connors, T.A., and Duncan, R. 2001. PDEPT: polymer-directed enzyme prodrug therapy. I. HPMA copolymer-cathepsin B and PK1 as a model combination. *Br. J. Cancer* **85:** 1070–1076.

Satchi-Fainaro, R., Hailu, H., Davies, J.W., Summerford, C., and Duncan, R. 2003. PDEPT: poly-mer-directed enzyme prodrug therapy. 2. HPMA copolymer-beta-lactamase and HPMA copolymer-C-Dox as a model combination. *Bioconjug. Chem.* **14:** 797–804.

Satchi-Fainaro, R., Puder, M., Davies, J.W., Tran, H.T., Sampson, D.A., Greene, A.K., Corfas, G., and Folkman, J. 2004. Targeting angiogenesis with a conjugate of HPMA copolymer and TNP-470. *Nat. Med.* **10:** 255–261.

Satchi-Fainaro, R., Mamluk, R., Wang, L., Short, S.M., Nagy, J.A., Feng, D., Dvorak, A.M., Dvorak, H.F., Puder, M., Mukhopadhyay, D., and Folkman, J. 2005. Inhibition of vessel per-meability by TNP-470 and its polymer conjugate, caplostatin. *Cancer Cell* **7:** 251–261.

Satchi-Fainaro, R., Duncan, R., and Barnes, C.M. 2006. Polymer therapeutics for cancer: current status and future challenges. *Adv. Polym. Sci.* **193:** 1–65.

Scales, C.W., Vasilieva, Y.A., Convertine, A.J., Lowe, A.B., and McCormick, C.L. 2005. Direct, controlled synthesis of the nonimmunogenic, hydrophilic polymer, poly(N-(2-hydroxypropyl) methacrylamide) via RAFT in aqueous media. *Biomacromolecules* **6:** 1846–1850.

Scales, C.W., Huang, F.Q., Li, N., Vasilieva, Y.A., Ray, J., Convertine, A.J., and McCormick, C.L. 2006. Corona-stabilized interpolyelectrolyte complexes of SiRNA with nonimmuno-genic, hydrophilic/cationic block copolymers prepared by aqueous RAFT polymerization. *Macromolecules* **39:** 6871–6881.

Schechter, I. and Berger, A. 1967. On the size of the active site in proteases. I. Papain. *Biochem. Biophys. Res. Commun.* **27:** 157–162.

Schiavon, O., Pasut, G., Moro, S., Orsolini, P., Guiotto, A., and Veronese, F.M. 2004. PEG-Ara-C conjugates for controlled release. *Eur. J. Med. Chem.* **39:** 123–133.

Schneider, H., Harbottle, R.P., Yokosaki, Y., Kunde, J., Sheppard, D., and Coutelle, C. 1998. A novel peptide, PLAEIDGIELTY, for the targeting of alpha9betal-integrins. *FEBS Lett.* **429:** 269–273.

Schopf, J.W., Kudryavtsev, A.B., Agresti, D.G., Wdowiak, T.J., and Czaja, A.D. 2002. Laser–Raman imagery of Earth's earliest fossils. *Nature* **416:** 73–76.

Schousboe, I., Thomsen, P., and van Deurs, B. 2004. Factor XII binding to endothelial cells depends on caveolae. *Eur. J. Biochem.* **271:** 2998–3005.

Seib, F.P., Jones, A.T., and Duncan, R. 2006. Establishment of subcellular fractionation tech-niques to monitor the intracellular fate of polymer therapeutics I. Differential centrifugation fractionation B16F10 cells and use to study the intracellular fate of HPMA copolymer–doxorubicin. *J. Drug Target.* **14:** 375–390.

Service, R.F. 1996. Combinatorial chemistry hits the drug market. *Science* **272:** 1266–1268.

Seymour, L.W., Duncan, R., Strohalm, J., and Kopeček, J. 1987. Effect of molecular weight (Mw) of N-(2-hydroxypropyl)methacrylamide copolymers on body distribution and rate of excretion after subcutaneous, intraperitoneal, and intravenous administration to rats. *J. Biomed. Mater. Res.* **21:** 1341–1358.

Seymour, L.W., Ferry, D.R., Anderson, D., Hesslewood, S., Julyan, P.J., Poyner, R., Doran, J., Young, A.M., Burtles, S., and Kerr, D.J. 2002. Hepatic drug targeting: phase I evaluation of polymer-bound doxorubicin. *J. Clin. Oncol.* **20:** 1668–1676.

Shen, W.C. and Ryser, H.J. 1981. cis-Aconityl spacer between daunomycin and macromolecular carriers: a model of pH-sensitive linkage releasing drug from a lysosomotropic conjugate. *Biochem. Biophys. Res. Commun.* **102:** 1048–1054.

Shiah, J.-G., Koňák, Č., Spikes, J.D., and Kopeček, J. 1997. Solution and photoproperties of N-(2-hydroxypropyl)methacrylamide copolymer–meso-chlorin e_6 conjugates. *J. Phys. Chem. B* **101:** 6803–6809.

Shiah, J.-G., Koňák, Č., Spikes, J.D., and Kopeček, J. 1998. Influence of pH on solubility and photoproperties of N-(2-hydroxypropyl)methacrylamide copolymer–meso-chlorin e_6 conju-gates. *Drug Deliv.* **5:** 119–126.

Shiah, J.-G., Sun, Y., Peterson, C.M., Kopeček, J. 1999. Biodistribution of free and N-(2-hydroxy-propyl)methacrylamide copolymer-bound meso chlorin e_6 and adriamycin in nude mice bear-ing human ovarian carcinoma OVCAR-3 xenografts. *J. Control. Release* **61**: 145–157.

Shiah, J.G., Sun, Y., Peterson, C.M., Straight, R.C., and Kopeček, J. 2000. Antitumor activity of N-(2-hydroxypropyl) methacrylamide copolymer-mesochlorin e_6 and adriamycin conjugates in combination treatments. *Clin. Cancer Res.* **6:** 1008–1015.

Shiah, J.G., Dvořák, M., Kopečková, P., Sun, Y., Peterson, C.M., and Kopeček, J. 2001a. Biodistribution and antitumour efficacy of long-circulating N-(2-hydroxypropyl)methacrylamide copolymer–doxorubicin conjugates in nude mice. *Eur. J. Cancer* **37:** 131–139.

Shiah, J.G., Sun, Y., Kopečková, P., Peterson, C.M., Straight, R.C., and Kopeček, J. 2001b. Combination chemotherapy and photodynamic therapy of targetable N-(2-hydroxypropyl)methacrylamide copolymer-doxorubicin/mesochlorin e_6-OV-TL 16 antibody immunoconjugates. *J. Control. Release* **74:** 249–253.

Simoes, S., Slepushkin, V., Pires, P., Gaspar, R., de Lima, M.P., and Duzgunes, N. 1999. Mechanisms of gene transfer mediated by lipoplexes associated with targeting ligands or pH-sensitive peptides. *Gene Ther.* **6:** 1798–1807.

Singer, J.W., Shaffer, S., Baker, B., Bernareggi, A., Stromatt, S., Nienstedt, D., and Besman, M. 2005. Paclitaxel poliglumex (XYOTAX; CT-2103): an intracellularly targeted taxane. *Anticancer Drugs* **16:** 243–254.

Smith, G.P. and Petrenko, V.A. 1997. Phage display. *Chem. Rev.* **97:** 391–410.

Smith, R.A., Porteous, C.M., Coulter, C.V., and Murphy, M.P. 1999. Selective targeting of an antioxidant to mitochondria. *Eur. J. Biochem.* **263:** 709–716.

Smith, R.A., Kelso, G.F., James, A.M., and Murphy, M.P. 2004. Targeting coenzyme Q derivatives to mitochondria. *Methods Enzymol.* **382:** 45–67.

Soldati, T. and Schliwa, M. 2006. Powering membrane traffic in endocytosis and recycling. *Nat. Rev. Mol. Cell Biol.* **7:** 897–908.

Solovskij, M.V., Ulbrich, K., and Kopeček, J. 1983. Synthesis of N-(2-hydroxypropyl)methacrylamide copolymers with antimicrobial activity. *Biomaterials* **4:** 44–48.

Sparr, E., Ash, W.L., Nazarov, P.V., Rijkers, D.T.S., Hemminga, M.A., Tieleman, D.P., and Killian, J.A. 2005. Self-association of transmembrane alpha-helices in model membranes – Importance of helix orientation and role of hydrophobic mismatch. *J. Biol. Chem.* **280:** 39324–39331.

Šťastný, M., Strohalm, J., Plocová, D., Ulbrich, K., and Říhová, B. 1999. A possibility to overcome P-glycoprotein (PGP)-mediated multidrug resistance by antibody-targeted drugs conjugated to N-(2-hydroxypropyl)methacrylamide (HPMA) copolymer carrier. *Eur. J. Cancer* **35:** 459–466.

Strohalm, J. and Kopeček, J. 1978. Poly[N-(2-hydroxypropyl)methacrylamide]. IV. Heterogeneous polymerization. *Angew. Makromol. Chem.* **70:** 109–118.

Šubr, V. and Ulbrich, K. 2006. Synthesis and properties of new N-(2-hydroxypropyl)methacrylamide copolymers containing thiazolidine-2-thione reactive groups. *React. Funct. Polym.* **66:** 1525–1538.

Šubr, V., Koňák, Č., Laga, R., and Ulbrich, K. 2006. Coating of DNA/poly(L-lysine) complexes by covalent attachment of poly[N-(2-hydroxypropyl)methacrylamide]. *Biomacromolecules* **7:** 122–130.

Swanson, J.A. and Watts, C. 1995. Macropinocytosis. *Trends Cell. Biol.* **5:** 424–428.

Tang, A. and Kopeček, J. 2002. Presentation of epitopes on genetically engineered peptides and selection of lymphoma-targeting moieties based on epitope biorecognition. *Biomacromolecules* **3:** 421–431.

Tang, A., Kopečková, P., and Kopeček, J. 2003. Binding and cytotoxicity of HPMA copolymer conjugates to lymphocytes mediated by receptor-binding epitopes. *Pharm. Res.* **20:** 360–367.

Teicher, B.A., Holden, S.A., Jacobs, J.L., Abrams, M.J., and Jones, A.G. 1986. Intracellular distribution of a platinum-rhodamine 123 complex in cis-platinum sensitive and resistant human squamous carcinoma cell lines. *Biochem. Pharmacol.* **35:** 3365–3369.

Teicher, B.A., Varshney, A., Khandekar, V., and Herman, T.S. 1991. Effect of hypoxia and acidosis on the cytotoxicity of six metal(ligand)4(rhodamine-123)2 complexes at normal and hyperthermic temperatures. *Int. J. Hyperthermia* **7:** 857–868.

Teodorescu, M. and Matyjaszewski, K. 1999. Atom transfer radical polymerization of (meth)acrylamides. *Macromolecules* **32:** 4826–4831.

Teodorescu, M. and Matyjaszewski, K. 2000. Controlled polymerization of (meth)acrylamides by atom transfer radical polymerization. *Macromol. Rapid Commun.* **21:** 190–194.

Thomas, M. and Klibanov, A.M. 2002. Enhancing polyethylenimine's delivery of plasmid DNA into mammalian cells. *Proc. Natl. Acad. Sci. USA* **99:** 14640–14645.

Thomsen, P., Roepstorff, K., Stahlhut, M., and van Deurs, B. 2002. Caveolae are highly immobile plasma membrane microdomains, which are not involved in constitutive endocytic trafficking. *Mol. Biol. Cell* **13**: 238–250.

Thomson, A.H., Vasey, P.A., Murray, L.S., Cassidy, J., Fraier, D., Frigerio, E., and Twelves, C. 1999. Population pharmacokinetics in phase I drug development: a phase I study of PK1 in patients with solid tumours. *Br. J. Cancer* **81**: 99–107.

Tijerina, M., Kopečková, P., and Kopeček, J. 2003a. Correlation of subcellular compartmentalization of HPMA copolymer-Mce$_6$ conjugates with chemotherapeutic activity in human ovarian carcinoma cells. *Pharm. Res.* **20**: 728–737.

Tijerina, M., Kopečková, P., and Kopeček, J. 2003b. Mechanisms of cytotoxicity in human ovarian carcinoma cells exposed to free Mce$_6$ or HPMA copolymer-Mce$_6$ conjugates. *Photochem. Photobiol.* **77**: 645–652.

Toki, B.E., Cerveny, C.G., Wahl, A.F., and Senter, P.D. 2002. Protease-mediated fragmentation of p-amidobenzyl ethers: a new strategy for the activation of anticancer prodrugs. *J. Org. Chem.* **67**: 1866–1872.

Torchilin, V.P. 2006a. Multifunctional nanocarriers. *Adv. Drug. Deliv. Rev.* **58**: 1532–1555.

Torchilin, V.P. 2006b. Recent approaches to intracellular deličvery of drugs and DNA and organelle targeting. *Annu. Rev. Biomed. Eng.* **8**: 343–375.

Tsukioka, Y., Matsumura, Y., Hamaguchi, T., Koike, H., Moriyasu, F., and Kakizoe, T. 2002. Pharmaceutical and biomedical differences between micellar doxorubicin (NK911) and liposomal doxorubicin (Doxil). *Jpn. J. Cancer Res.* **93**: 1145–1153.

Tucker, C.E., Chen, L.S., Judkins, M.B., Farmer, J.A., Gill, S.C., and Drolet, D.W. 1999. Detection and plasma pharmacokinetics of an anti-vascular endothelial growth factor oligonucleotide-aptamer (NX1838) in rhesus monkeys. *J. Chromatogr. B Biomed. Sci. Appl.* **732**: 203–212.

Tuerk, C. and Gold, L. 1990. Systematic evolution of ligands by exponential enrichment: RNA ligands to bacteriophage T4 DNA polymerase. *Science* **249**: 505–510.

Twaites, B., Alarcon, C.D., and Alexander, C. 2005. Synthetic polymers as drugs and therapeutics. *J. Mater. Chem.* **15**: 441–455.

Ulbrich, K. and Šubr, V. 2004. Polymeric anticancer drugs with pH-controlled activation. *Adv. Drug Deliv. Rev.* **56**: 1023–1050.

Ulbrich, K., Zacharieva, E.I., Obereigner, B., and Kopeček, J. 1980. Polymers containing enzymatically degradable bonds. 5. Hydrophilic polymers degradable by papain. *Biomaterials* **1**:199–204.

Ulbrich, K., Strohalm, J., and Kopeček, J. 1981. Polymers containing enzymatically degradable bonds. 3. Poly[N-(2-hydroxypropyl)methacrylamide] chains connected by oligopeptide sequences cleavable by trypsin. *Makromol. Chem.* **182**: 1917–1928.

Ulbrich, K., Koňák, Č., Tuzar, Z., and Kopeček, J. 1987. Solution properties of drug carriers based on poly[N-(2-hydroxypropyl)methacrylamide] containing biodegradable bonds. *Makromol. Chem.* **188**: 1261–1272.

Ulbrich, K., Etrych, T., Chytil, P., Jelínková, M., and Říhová, B. 2003. HPMA copolymers with pH-controlled release of doxorubicin: in vitro cytotoxicity and in vivo antitumor activity. *J. Control. Release* **87**: 33–47.

van der Aa, M.A., Huth, U.S., Hafele, S.Y., Schubert, R., Oosting, R.S., Mastrobattista, E., Hennink, W.E., Peschka-Suss, R., Koning, G.A., and Crommelin, D.J. 2007. Cellular uptake of cationic polymer-DNA complexes via caveolae plays a pivotal role in gene transfection in COS-7 cells. *Pharm. Res.* **24**: 1590–1598.

Vasey, P.A., Kaye, S.B., Morrison, R., Twelves, C., Wilson, P., Duncan, R., Thomson, A.H., Murray, L.S., Hilditch, T.E., Murray, T., Burtles, S., Fraier, D., Frigerio, E., Cassidy, J., and on behalf of the Cancer Research Campaign Phase I/II Committee. 1999. Phase I clinical and pharmacokinetic study of PK1 [N-(2-hydroxypropyl)methacrylamide copolymer doxorubicin]: first member of a new class of chemotherapeutic agents-drug-polymer conjugates. *Clin. Cancer Res.* **5**: 83–94.

Vasir, J.K. and Labhasetwar, V. 2005. Targeted drug delivery in cancer therapy. *Technol. Cancer Res. Treat.* **4**: 363–374.

Vasir, J.K., Reddy, M.K., and Labhasetwar, V.D. 2005. Nanosystems in drug targeting: opportunities and challenges. *Curr. Nanosci.* **1**: 47–64.

Vega, J., Ke, S., Fan, Z., Wallace, S., Charsangavej, C., and Li, C. 2003. Targeting doxorubicin to epidermal growth factor receptors by site-specific conjugation of C225 to poly(L-glutamic acid) through a polyethylene glycol spacer. *Pharm. Res.* **20**: 826–832.

Vercauteren, R., Schacht, E., and Duncan, R. 1992. Effect of the chemical modification of dextran on the degradation by rat liver lysosomal enzymes. *J. Bioact. Compat. Polym.* **7**: 346–357.

Vives, E., Brodin, P., and Lebleu, B. 1997. A truncated HIV-1 Tat protein basic domain rapidly translocates through the plasma membrane and accumulates in the cell nucleus. *J. Biol. Chem.* **272**: 16010–16017.

Wachters, F.M., Groen, H.J., Maring, J.G., Gietema, J.A., Porro, M., Dumez, H., de Vries, E.G., and van Oosterom, A.T. 2004. A phase I study with MAG-camptothecin intravenously administered weekly for 3 weeks in a 4-week cycle in adult patients with solid tumours. *Br. J. Cancer* **90**: 2261–2267.

Wadia, J.S. and Dowdy, S.F. 2005. Transmembrane delivery of protein and peptide drugs by TAT-mediated transduction in the treatment of cancer. *Adv. Drug Deliv. Rev.* **57**: 579–596.

Wadia, J.S., Stan, R.V., and Dowdy, S.F. 2004. Transducible TAT-HA fusogenic peptide enhances escape of TAT-fusion proteins after lipid raft macropinocytosis. *Nat. Med.* **10**: 310–315.

Wagner, E., Plank, C., Zatloukal, K., Cotten, M., and Birnstiel, M.L. 1992. Influenza virus hemagglutinin HA-2 N-terminal fusogenic peptides augment gene transfer by transferrin-polylysine-DNA complexes: toward a synthetic virus-like gene-transfer vehicle. *Proc. Natl. Acad. Sci. USA* **89**: 7934–7938.

Wang, Y.S., Youngster, S., Grace, M., Bausch, J., Bordens, R., and Wyss, D.F. 2002. Structural and biological characterization of pegylated recombinant interferon alpha-2b and its therapeutic implications. *Adv. Drug Deliv. Rev.* **54**: 547–570.

Wang, D., Miller, S., Sima, M., Kopečková, P., and Kopeček, J. 2003. Synthesis and evaluation of water-soluble polymeric bone-targeted drug delivery systems. *Bioconjug. Chem.* **14**: 853–859.

Wang, W., Qu, X., Gray, A.I., Tetley, L., and Uchegbu, I.F. 2004a. Self-assembly of cetyl linear polyethylenimine to give micelles, vesicles, and dense nanoparticles. *Macromolecules* **37**: 9114–9122.

Wang, X., Zhang, J., Song, A., Lebrilla, C.B., and Lam, K.S. 2004b. Encoding method for OBOC small molecule libraries using a biphasic approach for ladder-synthesis of coding tags. *J. Am. Chem. Soc.* **126**: 5740–5749.

Wang, D., Miller, S.C., Kopečková, P., and Kopeček, J. 2005. Bone-targeting macromolecular therapeutics. *Adv. Drug Deliv. Rev.* **57**: 1049–1076.

Wang, D., Miller, S.C., Shlyakhtenko, L.S., Portillo, A.M., Liu, X.M., Papangkorn, K., Kopečková, P., Lyubchenko, Y., Higuchi, W.I., Kopeček, J., 2007. Osteotropic Peptide that differentiates functional domains of the skeleton. *Bioconjug. Chem.* **18**: 1375–1378.

Wang, D., Sima, M., Mosley, R.L., Davda, J.P., Tietze, N., Miller, S.C., Gwilt, P.R., Kopečková, P., and Kopeček, J. 2006. Pharmacokinetic and biodistribution studies of a bone-targeting drug delivery system based on N-(2-hydroxypropyl)methacrylamide copolymers. *Mol. Pharm.* **3**: 717–725.

Weissig, V. 2005. Targeted drug delivery to mammalian mitochondria in living cells. *Expert Opin. Drug Deliv.* **2**: 89–102.

Westerlund-Wikstrom, B. 2000. Peptide display on bacterial flagella: principles and applications. *Int. J. Med. Microbiol.* **290**: 223–230.

Willhite, S.L., Goebel, S.R., and Scoggin, J.A. 1998. Raloxifene provides an alternative for osteoporosis prevention. *Ann. Pharmacother.* **32**: 834–837.

Willis, M.C., Collins, B.D., Zhang, T., Green, L.S., Sebesta, D.P., Bell, C., Kellogg, E., Gill, S.C., Magallanez, A., Knauer, S., Bendele, R.A., Gill, P.S., Janjic, N. 1998. Liposome-anchored vascular endothelial growth factor aptamers. *Bioconjug. Chem.* **9**: 573–582 (erratum, 633).

Wilson, C. and Szostak, J.W. 1998. Isolation of a fluorophore-specific DNA aptamer with weak redox activity. *Chem. Biol.* **5:** 609–617.

Wiwattanapatapee, R., Carreno-Gomez, B., Malik, N., and Duncan, R. 2000. Anionic PAMAM dendrimers rapidly cross adult rat intestine in vitro: a potential oral delivery system? *Pharm. Res.* **17:** 991–998.

Yanjarappa, M.J., Gujraty, K.V., Joshi, A., Saraph, A., and Kane, R.S. 2006. Synthesis of copolymers containing an active ester of methacrylic acid by RAFT: controlled molecular weight scaffolds for biofunctionalization. *Biomacromolecules* **7:** 1665–1670.

Yu, D., Peng, P., Dharap, S.S., Wang, Y., Mehlig, M., Chandna, P., Zhao, H., Filpula, D., Yang, K., Borowski, V., Zhang, Z., and Minko, T. 2005. Antitumor activity of poly(ethylene glycol)-camptothecin conjugate: the inhibition of tumor growth in vivo. *J. Control. Release* **110:** 90–102.

Zanta, M.A., Belguise-Valladier, P., and Behr, J.P. 1999. Gene delivery: a single nuclear localization signal peptide is sufficient to carry DNA to the cell nucleus. *Proc. Natl. Acad. Sci. USA* **96:** 91–96.

"Smart" pH-Responsive Carriers for Intracellular Delivery of Biomolecular Drugs

P.S. Stayton and A.S. Hoffman

1   Introduction

A large number of therapeutic drugs that utilize biological molecules, i.e., DNA, RNA, and proteins, are under current development in the biotechnology and pharmaceutical industries. Their potential is widely recognized, but bringing them into medical practice remains a major challenge. For proteins such as antibodies that act at the extracellular membrane face, considerable progress has been made in bringing them into medical practice. However, for biomolecules that function at intracellular locations (e.g., immunotoxins, antisense oligonucleotides, siRNA, antigens for vaccines), there is the additional difficult barrier of cytoplasmic entry (Kyriakides et al., 1999, 2001) in addition to the general challenges of drug stability, tissue penetration and transport, and therapeutic targeting. The predominant fates of internalized biomolecules are enzymatic degradation in the lysosome or recycling and extracellular clearance. In this chapter, we review the development of synthetic polymeric carriers that mimic the highly efficient intracellular delivery systems found in pathogenic viruses and organisms. Their most important property ties together the sensing of pH changes to membrane-destabilizing activity. The carriers are applicable to a wide range of biotherapeutics, and might additionally open up new families of protein or nucleic acid candidates that attack intracellular targets.

The smart polymeric drug delivery systems are specific examples of the more general class of "polymer therapeutics" that have generated considerable interest in the pharmaceutical field (Maeda and Matsumura, 1989). Polymer therapeutics exhibit several interesting properties that distinguish them from the drug alone and from prior delivery strategies. One of the most important discoveries was that the biodistribution and pharmaco-kinetic properties of polymeric drug conjugates are distinct from the drug itself. Maeda, Duncan, and others have thus pointed out that polymer–drug conjugates should be considered a new class of therapeutics, rather than a simple drug delivery system (Maeda, 1991; Maeda and Matsumura, 1989; Vasey et al., 1999). Polymer–drug conjugates show higher plasma levels (Takakura and Hashida, 1996) and lower renal clearance and longer circulation half-lives than small MW drugs (Noguchi

V. Torchilin (ed.), *Multifunctional Pharmaceutical Nanocarriers*,
© Springer Science+Business Media, LLC 2008

et al., 1998). The polymer also improves the stability of therapeutic molecules by shielding against degrading enzymes (Maeda et al., 1979, 1984), reduces immunogenicity (Maeda et al., 1984), and allows the modification of solubility properties. A representative example of the altered biodistribution properties of drugs delivered by polymeric carriers is the "enhanced permeation and retention" (EPR) effect. The EPR effect was first demonstrated by Maeda and coworkers, who noted that polymer therapeutics passively accumulate at tumors because of the leaky vasculature around tumor sites (Maeda and Matsumura, 1989). Numerous polymer–drug conjugates have been applied clinically to exploit the EPR effect, including poly(ethylene glycol) (PEG)-modified adenosine deaminase (Hershfield, 1997), L-asparginase (Holle, 1997), and SMANCS, a poly(styrene-*co*-maleic acid)-conjugated neocarzinostatin (Tsuchiya et al., 2000). PEG-bound camptothecin (Conover et al., 1998) and *N*-(2-hydroxypropyl) methacrylamide (HPMA) copolymer conjugates with doxorubicin (Julyan et al., 1999) and camptothecin (Caiolfa et al., 2000) are being evaluated in clinical trials. Polymeric drug carriers with intrinsic endosomal-disruptive properties have been previously proposed as potential drug carriers, and this remains a generally active area of research in many groups around the world. The cationic carrier poly(ethyleneimine) (PEI) has been hypothesized to act as an endosome-disrupting agent by functioning as a proton sponge (Behr, 1997; Boussif et al., 1995). Amido-amine cationic carriers have also been shown to act as endosomal-disrupting agents through the proton sponge effect (Richardson et al., 1996).

Several groups have developed polymeric carriers that are responsive to biologically relevant stimuli. Kabanov and coworkers have developed nanoscale polymeric delivery vehicles that assemble via polyelectrolyte interactions, and the complexes can display responsiveness to environmental factors such as temperature, ionic strength, and pH (Kyung et al., 2006; Oh et al., 2006; Alakhov et al., 2001; Gebhart et al., 2001; Kabanov et al., 2001). The Kataoka group has developed new polymeric delivery systems for nucleic acids, including pH-responsive systems for antisense and siRNA drugs, as well as for plasmid delivery systems (Fukushima et al., 2005; Kakizawa et al., 2006; Nishiyama and Kataoka, 2006). The Kim group was also one of the earliest developers of polymeric gene carriers, and they have reported a variety of PEGylated cationic carriers that are currently in clinical trials (Ahn et al., 2002; Kim et al., 2006; Park et al., 2006). A novel new pH-responsive polymeric carrier has been reported recently by the Bae group that is based on a sharply pH-responsive poly(methacryloyl sulfadimethoxine) composition (Sethuraman et al., 2006). The Davis group has described a carrier system for nucleic acids and also has proposed physical design parameters for drug complexes and particles (Davis et al., 2004; Popielarski et al., 2005). An interesting stimuli-responsive (bio)polymer system based on elastin peptide sequences has been developed for small molecule drugs by the Chilkoti group (Chilkoti et al., 2002). Szoka and coworkers have recently utilized acid-labile orthoesters to incorporate poly(ethylene glycol) (PEG) chains to stabilize the particles and to prolong circulation time in vivo (Guo and Sozka, 2001; Guo et al., 2003; Huang et al., 2006; Li et al., 2005).

The PEG chains are cleaved in the endosome, promoting destabilization of the NLP and release of the condensed DNA with low cytotoxicity.

2   General Design of pH-Responsive, Membrane-Destabilizing Polymeric Carriers

The polymeric carriers developed in our group offer the general polymeric carrier advantages (such as the EPR effect), but are distinguished (with potential advantages and disadvantages) by a new approach to pH-dependent, membrane-destabilizing activity. These carriers were designed to mimic the highly efficient intracellular delivery systems utilized by viruses and pathogenic organisms that have faced a similar "delivery" challenge. The pathogens have evolved remarkable molecular machines that enhance transport of DNA or proteins across the endosomal membrane to the cytosolic compartment. A key functional property of many of these proteins is their membrane-destabilizing activity that is closely coupled to a pH-sensing mechanism. At physiological pH (7.4), viral proteins such as hemagglutinin (Wiley and Skehel, 1987) or pathogenic proteins such as diphtheria toxin (Hughson, 1995) are in a "stealth" conformation, but as the pH of these compartments drops during endosomal development to values of 5.5 or lower, a conformational change is triggered to expose a membrane-destabilizing domain. Many of these proteins share a common pH-sensing strategy. When the pH has dropped sufficiently to protonate key carboxylate residues, the conformational equilibria of the proteins are shifted toward a membrane-active state that results in endosomal membrane destabilization. The destabilization allows enhanced delivery of the DNA, RNA, or protein cargo to the cytoplasm. The design of synthetic peptides with these types of pH-dependent membrane-disrupting characteristics has also been reported by Szoka and coworkers (Parente et al., 1990; Plank et al., 1994; Subbarao et al., 1987).

We have designed synthetic polymers that incorporate a pH-sensing carboxylate moiety that triggers membrane-destabilizing activity at a defined lower pH value (Albarran et al., 2005; Bulmus et al., 2003; El-Sayed et al., 2005; Henry et al., 2006; Murthy et al., 1999, 2003; Yin et al., 2006) (Fig. 1). These polymeric carrier systems are designed like viruses and pathogenic proteins to have modular components that possess different functional properties. The carriers can incorporate a targeting element that directs uptake into specific cells, a versatile conjugation or complexation element that allows covalent conjugation or ionic complexation of biomolecular drugs, a pH-responsive component that enhances membrane transport selectively in the low pH environment of the endosome, and a masking component as necessary to optimize circulation stability. The carrier can be designed to release the drug in the low pH environment of the endosome, "unmasking" the membrane-disruptive element and enhancing the delivery of the free drug into the cytoplasm (in other designs, the polymer–drug conjugate or complex may escape the endosome without first releasing the drug, after which it can then release the drug by disulfide bond reduction in the reducing environment of the cytoplasm).

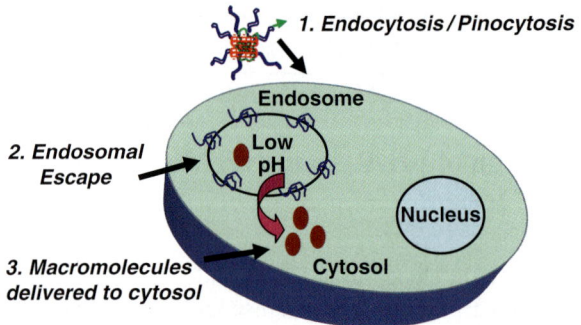

Fig. 1 General design of the pH-responsive, membrane-destabilizing drug carriers. The low pH gradient that is created in the endosomal compartment triggers a switch to the membrane-destabilizing state of the carrier, enhancing cytosolic delivery of macromolecular drugs.

3   Development of pH-Responsive PDSA Polymer Carriers

Tirrell and coworkers previously described the pH-dependent disruptive properties of poly(ethylacrylic acid) (PEAA) with lipid vesicle membranes (Thomas and Tirrell, 1992; Thomas et al., 1994). We have synthesized new members of the alkyl(acrylic acid) family where the carboxylate group was connected to longer alkyl segments. Poly(propylacrylic) acid (PPAA) exhibited a sharply pH-dependent membrane-destabilizing activity below 7.4, and a surprising order of magnitude increase over PEAA in hemolytic efficiency. PPAA is significantly more potent than even the best reported hemolytic peptides and proteins. PPAA also exhibited a shift to the membrane-active state at higher pH values than PEAA. This shift is reflected in the shift of the pK of the polymeric acid to higher pH values as the polymer becomes more hydrophobic. The addition of another methylene unit to make poly(butyl acrylic acid) shifted the pH profile for membrane disruption even further toward physiological pH. The concentration dependencies and pH profiles are also dependent on the polymer molecular weight, with the profiles (and pKs) shifted to higher pH transitions as molecular weight increases.

To create true carrier systems that provide a flexible conjugation route for therapeutic molecules, complexing agents, and/or targeting moieties, a new family termed the "PDSA" carriers has been developed. These carriers consist of an alkyl(acrylic acid) monomer, an alkyl(acrylate) monomer, and a functionalized monomer for conjugating targeting elements, the drug, and/or nucleic acid-condensing segments. As a first example of a functionalized monomer, a new pyridyl disulfide acrylate (PDSA) monomer was synthesized. The PDSA monomer allows efficient conjugation of targeting moieties, radiolabels, or biomolecular drugs through disulfide linkages that can be reduced in the cytoplasm after endosomal translocation of the therapeutics.

The general design of the PDSA family allows molecular tuning of the pH profile and membrane-destabilizing activity. The pH profile is controlled by the choice of the alkylacrylic acid monomer and by the ratio of the carboxylate-containing alkylacrylic acid monomer to the alkylacrylate monomer. Similarly, the membrane-destabilizing activity is controlled by the lengths of the alkyl segment on the alkylacrylic acid monomer and the alkylacrylate monomer, as well as by their ratio in the final polymer chains. We have recently shown that there is a significant increase in polymer backbone hydrophobic character upon butylacrylate (BA) incorporation that is directly tied to membrane-destabilizing activity (El-Sayed et al., 2005). Both poly(EAA-*co*-BA-*co*-PDSA) and poly(PAA-*co*-BA-*co*-PDSA) polymers exhibited high hemolytic/membrane-destabilizing activity at the low molecular weights of 9 and 12 kDa, respectively (Fig. 2). This finding suggests that carriers below the renal excretion size limit can be designed to have potent pH-responsive, membrane-destabilizing activity. This finding is also expected to be important in regard to the issue of tumor penetration, where smaller chains could be desirable. The ability to tune the pH profile and membrane-destabilizing activity is particularly important because the biomolecular therapeutics themselves are hydrophilic macromolecules that can significantly shift the pH profile and membrane-destabilizing activity of the polymer. The precise pH profile and activity will depend on the nature of the macromolecular drug (e.g., DNA vs. protein) and the quantity of drug loading, so that the ability to tune the intrinsic pH profile and membrane-destabilizing activity is critical. There is thus a rich polymer engineering opportunity to optimize the properties of these carriers for nucleic acid delivery.

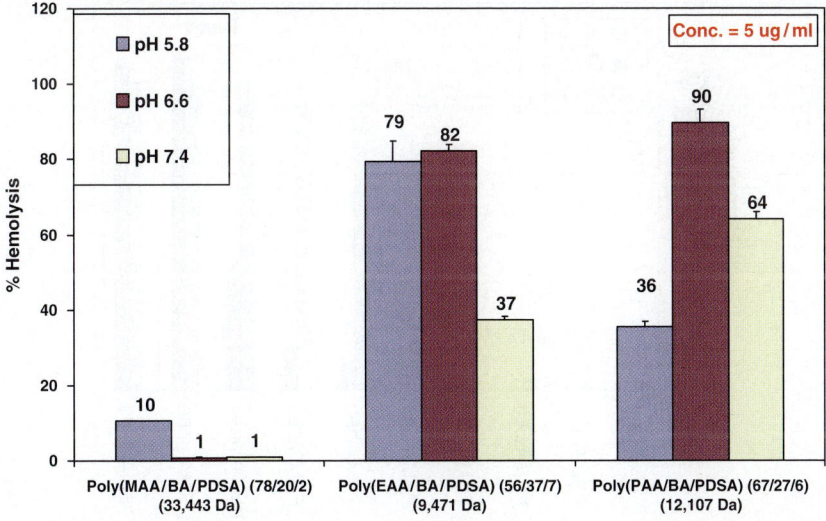

Fig. 2 Activity of the PDSA ter-copolymer carriers as a function of the alkyl(acrylic acid) composition. The percent hemolysis represents the fraction of red blood cells (RBCs) lysed compared to the positive control of Triton-X detergent. 10^8 RBCs per 1 ml of phosphate buffer were incubated with the polymers in a 37°C water bath for 1 h at the specified pH values. The tubes were centrifuged for 5 min at 13,500 g to separate intact RBCs and disrupted membranes from the solution. The supernatant, containing released Hb, was collected and transferred to 96 well plates, and the absorbance measured on a Saphire 2 (Tecan, Austria) plate reader at 541 nm.

4 Development of a pH-Responsive Poly(styrene-*alt*-maleic anhydride) (PSMA) Family

A new family of pH-responsive polymers based on poly(styrene-*alt*-maleic anhydride) copolymers have been recently developed (Fig. 3). PSMA has been in previous clinical use in Japan as the anticancer drug SMANCS (Maeda, 2001; Maeda et al., 1985, 2001) providing this polymer with an *in vivo* track record of biocompatibility. We modified the PSMA polymers by reacting the maleic anhydride group with primary alkylamines, to obtain alkyl amide/carboxylic acid derivatives capable of membrane disruption at endosomal pHs. In previous work, Maeda and coworkers prepared butyl "partial half esters" of PSMA in which 70% of the maleic anhydride groups were opened using butanol (Maeda, 2001; Maeda et al., 1985, 2001). The primary alkylamines react with anhydrides faster, and amide bonds have greater *in vitro* and *in vivo* stability in the resulting PSMA copolymer derivative.

By adjusting the degree and type of alkylamine modification, as well as the molecular weight of the PSMA backbone, we have been able to "molecularly engineer" the membrane-disrupting activity of these polymers to act only within specific pH ranges (Henry et al., 2006) (Fig. 3). The PSMA-alkyl amide derivatives are hydrophilic and membrane-inactive at physiological pH, and become membrane-disruptive in sharp

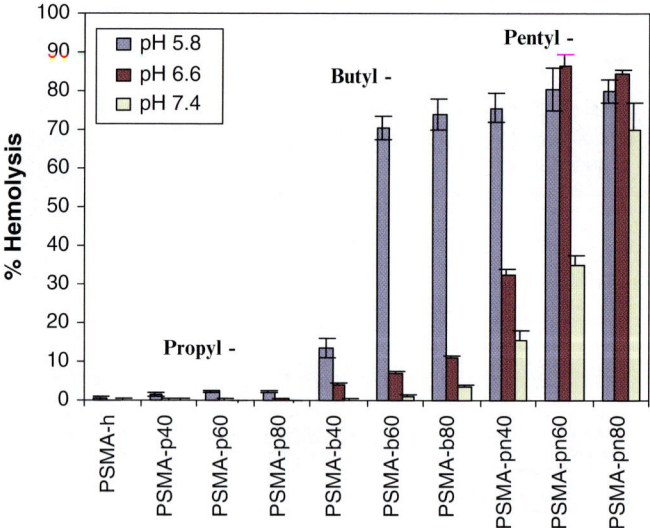

Fig. 3 Hemolysis properties of the alkyl-substituted PSMA carriers. Hemolysis characterization was conducted as in Fig. 2 at a polymer concentration of 20μg/ml. Propylamine derivatives and unmodified PSMA copolymers have insignificant hemolytic activity at all pH values. Butylamine derivatives exhibit higher pH-sensitive membrane disruption, particularly at higher substitution ratios. The pentylamine derivatives displayed even higher pH-dependent membrane disruption, with a favorable 70–75% hemolytic activity at pH 5.8. These derivatives also produced 57–65% hemolytic activity at pH 6.6, but remained inactive at pH 7.4.

response to lower endosomal pH values. The pH-dependent, membrane-destabilizing activity can be controlled by varying the length of the alkylamine group, the degree of modification of the copolymer, and the MW of the PSMA copolymer backbone. Anhydride moieties that remain in the polymer backbone after alkylamine modification can be readily used for further functionalization, such as the conjugation of cell-targeting ligands. After all the functionalization has been completed, the remaining anhydride groups can be readily hydrolyzed or derivatized with inert moieties.

5 Encrypted pH-Responsive Carriers

A related family of "smart" carriers termed "encrypted" polymers has also been recently developed. These polymers were termed "encrypted" by analogy to encrypted protein domains present in nature, where active domains of several extracellular and matricellular proteins are initially masked but become exposed and activated by the action of proteolytic enzymes at controlled time points. Similarly, the encrypted polymers contain a "masked" membrane-destabilizing backbone that becomes activated in response to acidic endosomal pH gradients (Murthy et al., 1999, 2003). Our first encrypted polymer (E1) utilized a hydrophobic backbone composed of a styrene acetal (SA) monomer with butyl methacrylate (BMA) and dimethylaminoethyl methacrylate (DMAEMA) monomers (Murthy et al., 2003). The hydrophobic backbone was "masked" by the direct conjugation of hydrophilic PEG chains through acid-sensitive acetal linkages, which was designed to improve the solubility and serum stability of the polymer backbone. Endocytosis of encrypted polymers would result in their accumulation in the endosomes where the acid-sensitive acetal linkages between the PEG chains and the hydrophobic backbone become hydrolyzed. The "unmasking" of the hydrophobic backbone in turn causes destabilization of the endosomal membrane and release of drugs into the cytoplasm (Fig. 4).

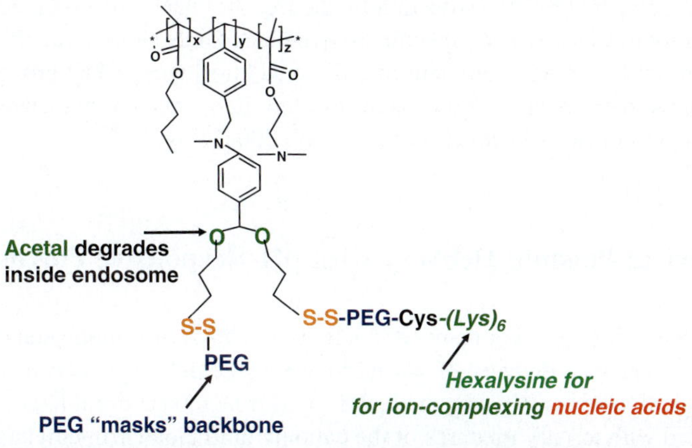

Fig. 4 The design and general structure of the pegylated "encrypted" polymer family.

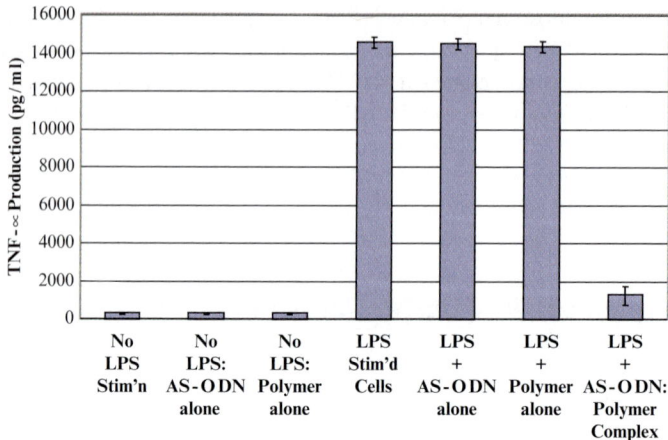

Fig. 5 Functional analysis of anti-IRAK delivery activity using a pH-responsive polyplex carrier. An antisense oligonucleotide directed against the RAW cell IRAK sequence was delivered with the encrypted polymeric carrier of Fig. 4. The RAW cells were then challenged with lipopolysaccharide (LPS) and the quantity of TNF-α was measured by ELISA assay. Only the antisense oligonucleotide delivered with the encrypted polymer was shown to knock down the responsiveness of the RAW cells to LPS stimulation.

The ability of the encrypted polymers carrying heterobifunctional PEG grafts modified with $(Lys)_3$-$(mannose)_3$ targeting groups to enhance the cytoplasmic delivery of antisense oligonucleotides against inducible nitric oxide synthase (iNOS) was investigated in model RAW macrophage-like cells. The macrophage and associated cell models represent a particularly challenging cell model to deliver nucleic acids because of the active phagosomal compartment that contains relatively high levels of degrading enzymes. The ionic complexes of polymer-$(Lys)_3$-$(mannose)_3$ conjugate with the antisense oligonucleotide for iNOS produced 80% reduction in NO production compared to 25% reduction by the free ASODN. The observed increase in ODN activity was sequence specific as scrambled ODN showed no effect on the production of NO. In addition, antisense-directed knockouts of key pro-inflammatory proteins have been demonstrated for the IL-1 receptor-associated kinase (IRAK) target in macrophages (Cuschieri et al., 2004) (Fig. 5).

6   Nonviral Plasmid Delivery with pH-Responsive Polymers

The excellent pH-dependent hemolytic activity of PPAA also motivated studies of its ability to enhance delivery of plasmid-based gene delivery when incorporated into cationic lipoplexes (Cheung et al., 2001). Murine mouse fibroblasts (NIH3T3) were treated with ternary mixtures of the cationic lipid dioleyltrimethylammonium propane (DOTAP), the pCMVβ plasmid DNA, and PPAA and compared against binary DOTAP/DNA mixtures. In serum-free media, ternary DOTAP/DNA/PPAA particles exhibited higher transfection efficiencies compared to the binary DOTAP/DNA particles at all +/− charge ratios reaching maximum β-galactosidase gene expression at a +/− charge ratio of 1.3. In addition, the PPAA increased the stability of ternary lipoplexes against serum inactivation. The specific effects of serum proteins, the stability, cellular uptake, and transfection efficiency of DOTAP/DNA lipoplexes with/without PPAA were studied at concentrations matching their average level in whole blood (Cheung et al., 2004). DNA condensation, cell uptake, and transfection results collectively showed that incorporation of PPAA greatly improved the complex stability of these formulations.

The favorable cell transfection results encouraged further *in vivo* evaluation. A mouse model of wound healing was utilized for the evaluation of cationic lipoplexes with and without PPAA (Kyriakides et al., 2002). This model was based on previous studies demonstrating that excisional wound healing is accelerated in thrombospondin-2 (TSP2)-null knockout mice (Kyriakides et al., 1998). In the absence of TSP2 in the knockout mouse, excisional wounds exhibit irregular deposition of extracellular matrix and enhanced vascularization that is associated with a significantly accelerated rate of wound healing. These results suggested that delivery of a plasmid encoding an antisense oligonucleotide to inhibit TSP2 expression could enhance healing in the wild-type mouse.

The deposition of TSP2 in wild-type mice wounds was found to be absent during the early inflammatory phase and peaking on day 10, coinciding with the period of maximal vascular regression (Kyriakides et al., 1999, 2001). The ternary DOTAP/DNA/PPAA formulations were therefore injected into the wound on days 4, 8, and 12 followed by wound evaluation on day 14 (Fig. 6). The delivery of PPAA-containing lipoplexes resulted in significantly enhanced disorganization of the wound extracellular matrix, resembling that seen with the TSP2-null wound-healing process. Immuno-histochemical staining of wound sections for the endothelial marker PECAM1 also showed that PPAA addition resulted in significantly greater vascularization at the wound site, again similar to that seen in TSP2-null wounds. These experiments collectively demonstrated that inclusion of PPAA in cationic lipoplexes formulations greatly enhanced transfection and resulted in the localized modulation of the wound-healing response.

7   siRNA Delivery Results

There is considerable current excitement over the development of small interfering RNA (siRNA) for the silencing of specific gene products. Because the siRNA must reach the cytosolic compartment to exert their biological effect, the pH-responsive polymer carriers have been studied recently as a method for shuttling the RNA through the endosomal membrane and into the cytoplasm of targeted cells. Initial studies have utilized a new cationic condensing agent, poly(dimethylaminoethylmeth acrylate)-*co*-poly(dimethylaminoethylmethacrylate-*co*-butyl methacrylate-*co*-methyl

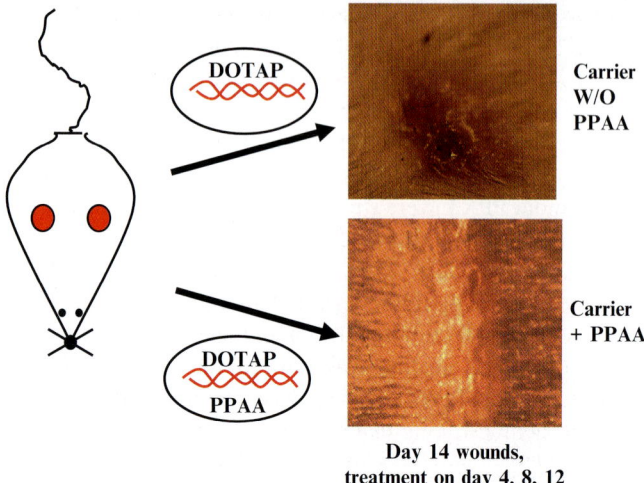

Fig. 6 A mouse model of excisional skin wound healing was used to evaluate the activity of PPAA in enhancing plasmid delivery in vivo. The delivery of a plasmid encoding an antisense oligonucleotide sequence that knocks down TSP2 expression was tested by injecting ternary or binary DOTAP/DNA lipoplexes that contained PPAA or not, respectively. The formulations were injected into the wound on days 4, 8, and 12 followed by wound evaluation on day 14. The wounds treated with PPAA-containing lipoplexes exhibited a significantly enhanced disorganization of the wound extracellular matrix and vascularization, resembling that seen with the TSP2-null wound-healing process. The results demonstrated that inclusion of PPAA in cationic lipoplexes greatly enhanced transfection efficiency and resulted in the localized modulation of the wound-healing response.

methacrylate)(pDMAEMA-b-p(DMAEMA-*co*-BMA-*co*-MMA) block copolymer (block 1 = 7,600 Mn, block 2 = 20,000 Mn, block 2: DMAEMA 60%, BMA 20%, MMA 20%), that was synthesized by the reversible addition fragmentation chain transfer (RAFT) technique (Johas et al., 2008). A hydrophobic derivative of pDMAEMA was previously studied by our group for the intracellular delivery of antisense oligonucleotides (Cuschieri et al., 2004; Murthy et al., 1999, 2003) motivating the polymeric design used in this study. A commercial polymer with the trade name Eudragit® E100 (Röhm Pharma, Germany) has a similar composition to the second block and has been previously shown to exhibit interaction with liposomes and red blood cells (Alasino et al., 2005) as well as pH-responsive hemolysis (Kusonwiriyawong et al., 2003).

This cationic diblock copolymer condenses siRNA into particles of less than 100 nm, and ternary complexes of the diblock copolymer, siRNA, and PPAA display sharply pH-responsive hemolysis. Delivery efficiencies were characterized using a glyceraldehyde phosphate dehydrogenase (GAPDH) siRNA sequence in the RAW 264.7 murine macrophage cell line (Fig. 7). The negative control siRNA is a scramble sequence selected by Ambion to have minimal background genomic complementarity or nonspecific silencing activity. The ternary complexes displayed highly active and specific anti-GAPDH activities, with a dose-dependent

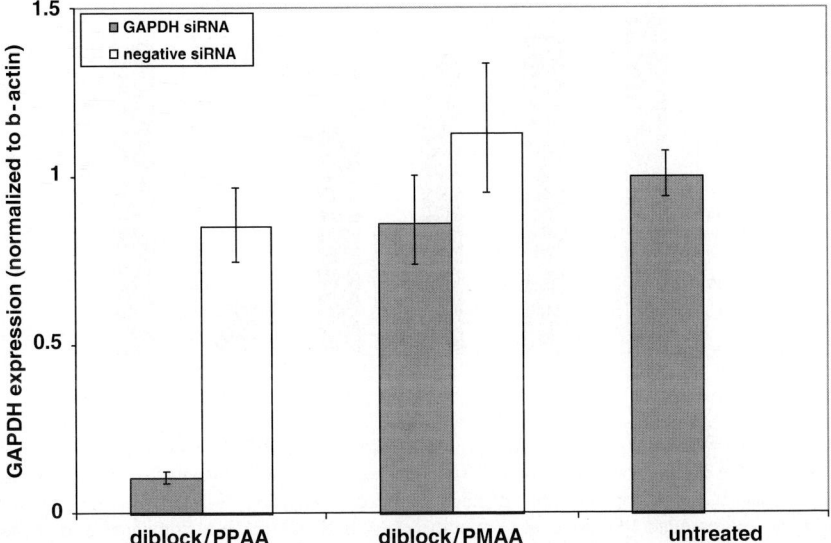

Fig. 7 GAPDH expression in RAW 264.7 cells treated with ternary complexes of poly(DMAEMA-b-DMAEMA-*co*-BMA-*co*-MMA), siRNA, and either PPAA or PMAA of equal mass. The polyplexes were formulated with a +1.4 charge ratio. The relative GAPDH expression in cells treated with GAPDH siRNA (100 nM) is shown in *gray bars*, and the *white bars* show cells treated with a negative control siRNA sequence that shows minimal background hybridization to the mouse genome. The knockdown of GAPDH messenger RNA levels was dependent on PPAA presence, and the polyplexes formed from the PMAA (which does not exhibit membrane-destabilizing activity) do not show an enhanced gene knockdown activity.

silencing efficiency of ca. 90% at 25 nM siRNA and 75% at 10 nM. These results demonstrate the synergistic effect of the diblock cationic carrier with the pH-responsive PPAA anionic polymer for siRNA delivery. Similar particles formulated with the control polymer PMAA (which does not exhibit pH-dependent membrane destabilization) did not show reduced gene expression. These results show that PPAA efficiently enhances siRNA activity with this ternary delivery system and opens the possibility that other related carrier compositions could find utility in siRNA delivery applications.

8 Intracellular Protein Delivery

Intracellular protein delivery also represents an important delivery frontier for macromolecular drug carrier development. We have recently investigated the ability of PPAA to enhance the cytosolic delivery of proteins and peptides. In one study, a genetically engineered TAT-streptavidin (TAT-SA) fusion protein was used to deliver biotinylated protein/peptide cargo to HeLa cells (Albarran et al.,

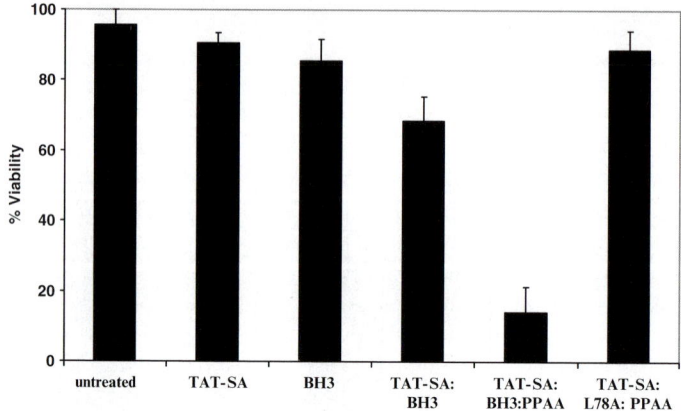

Fig. 8 Delivery of a pro-apoptotic petide with a TAT-streptavidin/PPAA bioconjugate. The degree of HeLa cell apoptosis was determined by staining fixed cells with DAPI nuclear stain, followed by fluorescence microscopic analysis of nuclear morphology. The TAT-SA:BH3 complexes displayed an approximate 30% increase in apoptosis ($p < 0.05$) after 24 h as compared to untreated cells (Fig. 2d). The number of cells displaying condensed nuclei was further increased to ~85% ($p < 0.001$) when the endosomal-releasing polymer PPAA was incorporated into TAT-SA:BH3 complexes, but not when cells were treated with PPAA alone. The null peptide mutant BH3(L78A) was delivered to test the peptide specificity, and as expected, the delivery of this peptide did not induce apoptosis.

2005). The TAT-SA is a model protein that is efficiently taken up by a variety of cultured cells, and may represent a useful *in vitro* cell transfection reagent. TAT-SA was used to deliver the pro-apoptotic Bak BH3 peptide (Fig. 8), which requires cytosolic delivery to induce its cell-killing activity (Albarran et al., submitted). The Bak BH3 peptide was shown previously to exhibit a cell-killing effect when fused to the Antennapedia PTD (Holinger et al., 1999). Bak BH3 peptides bind the death-suppressing proteins Bcl-2 and Bcl-x_L, and these binding events lead to apoptotic cell death through the release of the pro-apoptotic factor cytochrome c and the activation of caspase proteases (Shangary et al., 2004). Bcl-2 and Bcl-x_L are expressed in outer mitochondrial membranes (Walsh et al., 2002) and thus serve as targets for evaluating the endosomal release of peptides carried into cells via the TAT-SA carrier protein.

The TAT-streptavidin containing an N-terminal TAT peptide sequence was used to optimize the pinocytotic cell uptake of biotinylated BH3 peptide and end-biotinylated PPAA. Approximately 30% of cells treated with TAT-SA:BH3 complexes revealed morphologically distinct nuclear condensation. The incorporation of biotinylated PPAA strongly enhanced the cell-killing effect of BH3 peptides by an additional 55% ($p < 0.001$) to a total cell-killing efficiency of 85%. Caspase-3 activity was upregulated in a TAT-SA:BH3:PPAA dose-dependent manner. The induction of apoptosis with the TAT-SA:BH3:PPAA complex was abrogated with a L78A BH3 peptide, that had been previously shown to abrogate antagonization activity (Shangary et al., 2004). The caspase and L78A peptide results demonstrate that the delivered BH3 peptide is indeed working through the biologically relevant apoptosis signaling pathway. These studies establish the ability of PPAA to strongly enhance the intracellular delivery of a functional pro-apoptotic peptide. Together with the PPAA, the TAT-SA adaptor complex could prove useful as a carrier of peptide/protein cargo to cultured cells.

Previous studies have also established the ability of PPAA to enhance intracellular antibody delivery. Immunotoxins are a class of therapeutics that exploit monoclonal antibody specificity for the delivery of toxins to cellular and tissue targets. A model protein complex composed of a biotinylated anti-CD3 antibody/streptavidin bioconjugate was constructed to test the activity of PPAA (Lackey et al., 2002). The monoclonal anti-CD3 antibody 64.1 (MoAb) was previously shown to traffick to the lysosome with minimal translocation to the cytoplasm, which allowed a rigorous evaluation of PPAA's enhancement of the cytoplasmic release (Press et al., 1988). Changes in the trafficking of Ab/streptavidin conjugates with and without complexed PPAA-biotin were evaluated by visualizing the intracellular distribution of different protein complexes in Jurkat cells using confocal microscopy and western blotting techniques. Incorporation of PPAA-biotin resulted in a diffuse intracellular staining of the cells indicating the cytoplasmic delivery of the protein complex. In contrast, protein complexes with no PPAA-biotin showed a punctuate fluorescence indicating entrapment of the protein complexes in the endosomal/lysosomal vesicles of the cells. The quantity of the protein complex present in the cytoplasmic fraction was compared to that present in the total cell homogenate using quantitative western blotting techniques (Fig. 9). Results showed that only the protein complexes with PPAA-biotin were detected (~73%) in the cytoplasmic fraction of cell homogenate. The physical mixture of PPAA polymer with the MoAb-biotin/streptavidin complex yielded some cytoplasmic release (~29%), indicating that PPAA can function *in trans* to enhance release. These results collectively demonstrate that PPAA can enhance the cytoplasmic delivery of antibody-targeted protein conjugates that are internalized through receptor-mediated endocytosis.

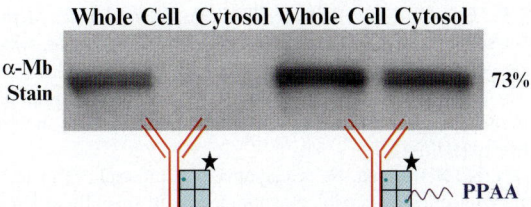

Fig. 9 The intracellular delivery of an IgG antibody with PPAA was studied with a biotinylated anti-CD3 antibody/streptavidin complex (Popielarski et al., 2005). The monoclonal anti-CD3 antibody 64.1 (MoAb) was co-complexed with streptavidin and biotin-PPAA. Changes in the trafficking of antibody–protein complexes with/without PPAA-biotin were evaluated by quantitative western blotting. The streptavidin complexes with biotin-PPAA showed ~73% localization in the cytosolic cell fraction compared to no detectable cytosolic delivery without PPAA, while a physical mixture of PPAA and the MoAb-biotin/streptavidin complex yielded a low but significant level of ~29.

References

Ahn, C. H., Chae, S. Y., Bae, Y. H., and Kim, S. W. 2002. Biodegradable poly(ethylenimine) for plasmid DNA delivery. J. Control. Release 80:273–282.

Alakhov, V., Klinski, E., Lemieux, P., Pietrzynski, G., and Kabanov, A. V. 2001. Block copolymeric biotransport carriers as versatile vehicles for drug delivery. Expert Opin. Biol. Ther. 1:583–602.

Alasino, R. V., Ausar, S. F., Bianco, I. D., Castagna, L. F., Contigiani, M., and Beltramo, D. M. 2005. Amphipathic and membrane-destabilizing properties of the cationic acrylate polymer Eudragit(R E100). Macromol. Biosci. 5:207–213.

Albarran, B., To, R., and Stayton, P. S. 2005. A TAT-streptavidin fusion protein directs uptake of biotinylated cargo into mammalian cells. Protein Eng. Des. Sel. 18:147–152.

Albarran, B., Hoffman, A. S., and Stayton, P. S. 2007. Efficient intracellular delivery of a pro-apoptotic peptide with a pH-responsive carrier. Manuscript submitted.

Behr, J. 1997. The proton sponge: a trick to enter cells the viruses did not exploit. Chimia 51:34–36.

Boussif, O., Lezoualch, F., Zanta, M., Mergny, M., Scherman, D., Demeneix,B., and Behr, J. 1995. A versatile vector for gene and oligonucleotide transfer into cells in culture and in vivo: polyethyleneimine. Proc. Natl. Acad. Sci. USA 92:7297–7301.

Bulmus, V., Woodward, M., Lin, L., Murthy, N., Stayton, P. S., and Hoffman, A. S. 2003. A new pH-responsive and glutathione-reactive, endosomal membrane-disruptive polymeric carrier for intracellular delivery of biomolecular drugs. J. Control. Release 93:105–120.

Caiolfa, V., Zamai, A., Fiorino, A., Frigerio, E., Pellizoni, C., d'Argy, R., Ghiglieri, A., Castelli, M., Farao, M., Pesenti, E., Gigli, M., Angelucci, F., and Suarato, A. 2000. Polymer-bound camptothecin: initial biodistribution and antitumor activity studies. J. Control. Release 65:105–119.

Cheung, C. Y., Murthy, N., Stayton, P. S., and Hoffman, A. S. 2001. A pH-sensitive polymer that enhances cationic lipid-mediated gene transfer. Bioconjug. Chem. 12:906–910.

Cheung, C. Y., Stayton, P. S., and Hoffman, A. S. 2004. Poly(propylacrylic acid) protects cationic lipoplexes against serum inactivation. J. Biomater. Sci. 16:163–179.

Chilkoti, A., Dreher, M. R., and Meyer, D. E. 2002. Design of thermally responsive, recombinant polypeptide carriers for targeted drug delivery. Adv. Drug Deliv. Rev. 54:1093–1111.

Conover, C., Greenwald, R., Pendri, A., Gilbert, C., and Shum, K. 1998. Camptothecin delivery systems: enhanced efficacy and tumor accumulation of campotothecin to polyethylene glycol via a glycine linker. Cancer Chemother. Pharmacol. 42:407–414.

Cuschieri, J., Bulmus, V., Gourlay, D., Garcia, I., Stayton, P., and Maier, R. 2004. Modulation of macrophage responsiveness to lipopolysaccharide by IRAK-1 manipulation. Shock 21:182–188.

Davis, M. E., Pun, S. H., Bellocq, N. C., Reineke, T. M., Popielarski, S. R., Mishra, S., and Heidel, J. D. 2004. Self-assembling nucleic acid delivery vehicles via linear, water-soluble, cyclodextrin-containing polymers. Curr. Med. Chem. 11:179–197.

El-Sayed, M. E. H., Hoffman, A. S., and Stayton, P. S. 2005. Rational design of composition and activity correlations for pH-responsive and glutathione-reactive polymer therapeutics. J. Control. Release 104:417–427.

Fukushima, S., Miyata, K., Nishiyama, N., Kanayama, N., Yamasaki, Y., and Kataoka, K. 2005. PEGylated polyplex micelles from triblock catiomers with spatially ordered layering of condensed pDNA and buffering units for enhanced intracellular gene delivery. J. Am. Chem. Soc. 127:2810–2811.

Gebhart, C. L. and Kabanov, A. V. 2001. Evaluation of polyplexes as gene transfer agents. J. Control. Release 73:401–416.

Guo, X. and Sozka, F. C. 2001. Steric stabilization of fusogenic liposomes by a low-pH sensitive PEG-diortho ester-lipid conjugate. Bioconjug. Chem. 12:291–300.

Guo, X., MacKay, J. A., and Szoka, F. C. 2003. Mechanism of pH-triggered collapse of phosphatidylethanolamine liposomes stabilized by an ortho ester polyethyleneglycol lipid. Biophys. J. 84:1784–1795.

Henry, S. M., El-Sayed, M. E. H., Pirie, C. M., Hoffman, A. S., and Stayton, P. S. 2006. pH-responsive poly(styrene-alt-maleic anhydride) alkylamide copolymers for intracellular drug delivery. Biomacromolecules 7:2407–2414.

Hershfield, M. 1997. Biochemistry and immunology of poly(ethylene glycol)-midifie adenosine deaminase (PEG-ADA). In: Harris, J., Zalipsky, S., eds. ACS Symposium. Poly(Ethylene Glycol) Biological Applications: American Chemical Society. pp. 145–154.

Holinger, E. P., Chittenden, T., and Lutz, R. J. 1999. Bak BH3 peptides antagonize Bcl-xL function and induce apoptosis through cytochrome c-independent activation of caspases. J. Biol. Chem. 274:13298–13304.

ŠHolle, L. 1997. Pegaspargase: an alternative? Ann. Pharmacother. 31:616–624.

Huang, Z., Guo, X., Li, W., MacKay, A., and Szoka, F. C. 2006. Acid-triggered transformation of diortho ester phosphocholine liposome. J. Am. Chem. Soc. 128:60–61.

Hughson, F. M. 1995. Structural characterization of viral fusion proteins. Curr. Biol. 5:265–274.

Johns, R. E., Convertine, A. J., Hoffman, A. S., and Stayton, P. S. 2007. Synergistic delivery of siRNA with the pH-responsive poly(propylacrylic acid) and a new cationic diblock copolymer. Manuscript submitted.

Julyan, P., Seymour, L., Ferry, D., Daryani, S., Boivin, C., Doran, J., David, M., Anderson, D., Christodolou, C., Young, A., Hesselwood, S., and Kerr, D. 1999. Preliminary clinical study of the distribution of HPMA copolymers bearing doxorubicin and galactosamine. J. Control. Release 57:281–290.

Kakizawa, Y., Furukawa, S., Ishii, A., and Kataoka, K. 2006. Organic-inorganic hybrid-nanocarrier of siRNA constructing through the self-assembly of calcium phosphate and PEG-based block aniomer. J. Control. Release 111:368–370.

Kim, W. J., Yockman, J. W., Jeong, J. H., Christensen, L. V., Lee, M., Kim, Y. H., and Kim, S. W. 2006. Anti-angiogenic inhibition of tumor growth by systemic delivery of PEI-g-PEG-RGD/pCMV-sFlt-1 complexes in tumor-bearing mice. J. Control. Release 114:381–388.

Kusonwiriyawong, C., van de Wetering, P., Hubbell, J. A., Merkle, H. P., and Walter, E. 2003. Evaluation of pH-dependent membrane-disruptive properties of poly(acrylic acid) derived polymers. Eur. J. Pharm. Biopharm. 56:237–246.

Kyriakides, T. R., Zhu, Y. N., Smith, L. T., Bain, S. D, Yang, Z., Lin, M. T., Danielson, K. G., Iozzo, R. V., LaMarca, M., McKinney, C. E., Ginns, E. I., and Bornstein, P. 1998. Mice that lack thrombospondin 2 display connective tissue abnormalities that are associated with disordered collagen fibrillogenesis, an increased vascular density, and a bleeding diathesis. J. Cell Biol. 140:419–430.

Kyriakides, T. R., Tam, J. W., and Bornstein, P. 1999. Accelerated wound healing in mice with a disruption of the thrombosponding 2 gene. J. Invest. Dermatol. 113:782–787.

Kyriakides, T. R., Hartzel, T., Huyen, G., and Bornstein, P. 2001. Modulation of angiogenesis and matrix remodeling by localized, matrix-mediated, antisense gene delivery. Mol. Ther. 3:842–849.

Kyriakides, T. R., Cheung, C. Y., Murthy, N., Bornstein, P., Stayton, P. S., and Hoffman, A. S. 2002. pH-Sensitive polymers that enhance intracellular drug delivery in vivo. J. Control. Release 78:295–303.

Kyung, T., Oh, T. K. B., Bronberg, L., Hatton, T. A., and Kabanov, A. V. 2006. Block ionomer complexes as prospective nanocontainers for drug delivery. J. Control. Release 115:9–17.

Lackey, C. A., Press, O. W., Hoffman, A. S., and Stayton, P. S. 2002. A biomimetic pH-responsive polymer directs endosomal release and intracellular delivery of an endocytosed antibody complex. Bioconjug. Chem. 13:996–1001.

Li, W., Huang, Z., MacKay, J. A., Grube, S., and Szoka, F. C. 2005. Low pH-sensitive poly(ethylene glycol) (PEG)-stabilized plasmid nanolipioparticles: effects of PEG chain length, lipid composition and assembly conditions on gene delivery. J. Gene Med. 7:67–79.

Maeda, H. 1991. SMANCS and polymer-conjugated macromolecular drugs: advantages in cancer chemotherapy. Adv. Drug Deliv. Rev. 6:181–202.

Maeda, H. 2001. SMANCS and polymer-conjugated macromolecular drugs: advantages in cancer chemotherapy. Adv. Drug Deliv. Rev. 46:169–185.

Maeda, H. and Matsumura, Y. 1989. Tumoritropic and lymphotropic principles of macromolecular drugs. Crit. Rev. Ther. Drug Carrier Syst. 6:193–210.

Maeda, H., Takeshita, J., and Kanamaru, R. 1979. A lipophilic derivative of neocarzinostatin. A polymer conjugation of an antitumor protein antibiotic. Int. J. Pept. Protein Res. 14:81–87.

Maeda, H., Matsumoto, T., Konno, T., Iwaqi, K., and Ueda, M. 1984. Tailor-making of protein drugs by polymer conjugation for tumor targeting: a brief review on SMANCS. J. Protein Chem. 3:181–193.

Maeda, H., Ueda, M., Morinaga, T., and Matsumoto, T. 1985. Conjugation of poly(styrene-co-maleic acid) derivatives to the antitumor protein neocarzionstatin: pronounced improvements in pharmacological properties. J. Med. Chem. 28:455–461.

Maeda, H., Sawa, T., and Konno, T. 2001. Mechanism of tumor-targeted delivery of macromolecular drugs, including the EPR effect in solid tumor and clinical overview of the prototype polymeric drug SMANCS. J. Control. Release 74:47–61.

Murthy, N., Robichaud, J. R., Tirrell, D. T., Stayton, P. S., and Hoffman, A. S. 1999. The design and synthesis of polymers for eukaryotic membrane disruption. J. Control. Release 61:137–143.

Murthy, N., Campbell, J., Fausto, N., Hoffman, A. S., and Stayton, P. S. 2003. Design and synthesis of pH-responsive polymeric carriers that target uptake and enhance the intracellular delivery of oligonucleotides. J. Control. Release 89:365–374.

Nishiyama, N. and Kataoka, K. 2006. Current state, achievements, and future prospects of polymeric micelles as nanocarriers for drug and gene delivery. Pharmacol. Ther. 112:630–648.

Noguchi, Y., Wu, J., Duncan, R., Strohalm, J., Ulbrich, K., Akaike, T., and Maeda, H. 1998. Early phase tumor accumulation of macromolecules: a great difference between the tumor vs normal tissue in their clearance rate. Jpn. J. Cancer Res. 89:307–314.

Parente, R. A., Nir, S., and Szoka, F. C. 1990. Mechanism of leakage of phospholipid vesicle contents induced by the peptide GALA. Biochemistry 29:8720–8728.

Park,T. G., Jeong, J. H., and Kim, S.W. 2006. Current status of polymeric gene delivery systems. Adv. Drug Deliv. Rev. 58:467–486.

Plank, C., Oberhauser, B., Mechtler, K., Koch, C., and Wagner, E. 1994. The influence of endosome-disruptive peptides on gene transfer using synthetic virus-like gene transfer systems. J. Biol. Chem. 269:12918–12924.

Popielarski, S. R., Hu-Lieskovan, S., French, S. W., Triche, T. J., and Davis, M. E. 2005. A nanoparticle-based model delivery system to guide the rational design of gene delivery to the liver. 2. In vitro and in vivo uptake results. Bioconjug. Chem. 16:1071–1080.

Press, O. W., Hansen, J. A., Farr, A., and Martin, P. J. 1988. Endocytosis and degradation of murine anti-human CD3 monoclonal antibodies by normal and malignant T-lymphocytes. Cancer Res. 48:2249–2257.

Richardson, S., Ferruti, S., and Duncan, R. 1996. Poly(amidoamine)s as potential endosomolytic polymers: evaluation in vitro and body distribution in normal and tumour bearing animals. J. Drug Target. 6:391–394.

Sethuraman, V. A., Na, K., and Bae, Y. H. 2006. pH-responsive sulfonamide/PEI system for tumor specific gene delivery: an in vitro study. Biomacromolecules 7:64–70.

Shangary, S., Oliver, C. L., Tillman, T. S., Cascio, M., and Johnson, D. E. 2004. Sequence and helicity requirements for the proapoptotic activity of Bax BH3 peptides. Mol. Cancer Ther. 3:1343–1354.

Subbarao, N. K., Parente, R. A., Szoka, F. C., Nadasdi, L., and Pongracz, K. 1987. pH-dependent bilayer destabilization by an amphipathic peptide. Biochemistry 26:2964–2972.

Takakura, Y. and Hashida, M. 1996. Macromolecular carrier systems for targeted drug delivery: pharmacokinetic considerations on biodistribution. Pharm. Res. 13:820–831.

Thomas, J. L. and Tirrell, D. A. 1992. Polyelectrolyte-sensitized phospholipid vesicles. Acc. Chem. Res. 25:336–342.

Thomas, J. L., Barton, S. W., and Tirrell, D. A. 1994. Membrane solubilization by a hydrophobic polyelectrolyte: surface activity and membrane binding. Biophys. J. 67:1101–1106.

Tsuchiya, K., Uchida, T., Kobayashi, M., Maeda, H., Konno, T., and Yamanaka, H. 2000. Tumor-targeted chemotherapy with SMANCS in lipiodol for renal cell carcinoma: longer survival with larger size tumors. Urology 55:495–500.

Vasey, P. A., Kaye, S. B., Morrison, R., Twelves, C., Wilson, P., Duncan, R., Thomson, A. H., Murray, L. S., Hilditch, T. E., Murray, T., Burtles, S., Fraier, D., Frigerio, E., and Cassidy, J. 1999. Phase I clinical and pharmacokinetic study of PK1 [N-(2-hydroxypropyl)methacrylamide copolymer doxorubicin]: first member of a new class of chemotherapeutic agents-drug-polymer conjugates. Cancer Research Campaign Phase I/II Committee. Clin. Cancer Res. 5:83–94.

Walsh, M., Lutz, R. J., Cotter, T. G., and O'Connor, R. 2002. Erythrocyte survival is promoted by plasma and suppressed by a Bak-derived BH3 peptide that interacts with membrane-associated Bcl-XL. Blood 99:3439–3448.

Wiley, D. C. and Skehel, J. J. 1987. The structure and function of the hemagglutinin membrane glycoprotein of influenza virus. Ann. Rev. Biochem. 56:365–394.

Yin, X., Hoffman, A. S., and Stayton, P. S. 2006. Poly(N-isopropylacrylamide-co-propylacrylic acid) copolymers that respond sharply to temperature and pH. Biomacromolecules 7:1381–1385.

Stimuli-Sensitive Nanosystems: For Drug and Gene Delivery

Han Chang Kang, Eun Seong Lee, Kun Na, and You Han Bae

1 Introduction

Apart from its previous history in pharmaceutics, nanotechnology has recently become a major paradigm for the delivery of anticancer drugs, imaging agents, and genetic material. Pharmaceutical nanosystems have shown beneficial therapeutic efficacy with reduced side effects in treating diseases when compared to traditional dosage forms. For example, delivery of high doses of therapeutic and/or diagnostic agents to target cancer sites has been achieved using nano-sized carrier systems. This effect is primarily attributed to passive accumulation in solid tumors and inflamed regions by the EPR effect and the size (20–200 nm) of the carriers, followed by passive diffusional release of the drug in the extracellular space and/or active internalization into the cells via various entry mechanisms.

To further improve local high-dose therapy, nanosystems need to be inert, meaning minimal interactions with biological components and negligible drug release while circulating in the blood stream. However, upon reaching their target sites, the nanosystems should switch their nature to induce aggressive cellular interaction, rapid localization at their intracellular destination, and enhanced drug-release kinetics. The switching property could be endowed by employing stimuli-sensitive components when constructing nanocarriers. It is anticipated that the responsive systems would reach their target sites more effectively by overcoming biological barriers such as drug-resistant mechanisms and entrapment in harsh lysosomal compartments.

A summary of recent stimuli and stimuli-sensitive nanosystems for effective drug/gene delivery are presented in this chapter, with particular emphasis on tumor treatment. While pharmaceutical stimuli-sensitive nanocarriers for drug/gene delivery include various carriers constructed from polymers, phospholipids, lipids, and their hybrids, this chapter will focus on polymeric nanocarriers.

2 Stimuli and Stimuli-Sensitive Polymers

Nanocarriers constructed by self-assembling processes can have stimuli-sensitive properties induced by a variety of environmental changes covering a broad range of stimuli. Within the body, stimuli are limited to physiological signals such as

V. Torchilin (ed.), *Multifunctional Pharmaceutical Nanocarriers*,
© Springer Science+Business Media, LLC 2008

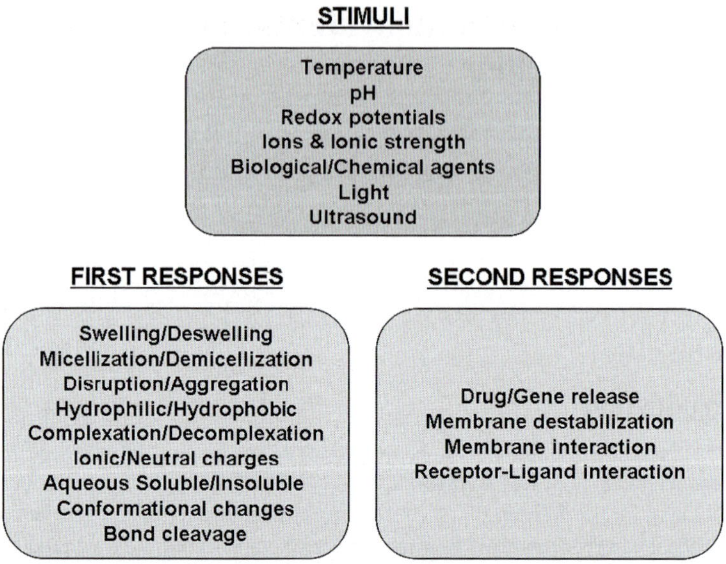

STIMULI

Temperature
pH
Redox potentials
Ions & Ionic strength
Biological/Chemical agents
Light
Ultrasound

FIRST RESPONSES   **SECOND RESPONSES**

Swelling/Deswelling
Micellization/Demicellization
Disruption/Aggregation   Drug/Gene release
Hydrophilic/Hydrophobic   Membrane destabilization
Complexation/Decomplexation   Membrane interaction
Ionic/Neutral charges   Receptor-Ligand interaction
Aqueous Soluble/Insoluble
Conformational changes
Bond cleavage

Fig. 1 Stimuli for drug/gene delivery-induced physical/chemical changes of pharmaceutical nanosystems as "first responses," successively leading to "second responses"

temperature, pH, redox gradients, ionic species, and concentration, which are often linked to biological and pathological events. External physical signals such as light, temperature, and ultrasound can also be employed. As shown in Fig. 1, internal and external stimuli affect the physical/chemical properties of materials or nanosystems causing "first responses." This includes swelling/deswelling (see Cammas et al., 1997; Chung et al., 2000), disruption/aggregation (see Lee et al., 2003a,b, 2005a,b; Na et al., 2003, 2004), chain cleavage (see Kaneko et al., 1991; Sawant et al., 2006), and complexation/decomplexation (see Sethuraman and Bae, 2007; Sethuraman et al., 2006). The "first responses" successively activate "second responses" such as drug/gene release (see Takeda et al., 2004), cell interactions (see Sethuraman et al., 2006), cell membrane destabilization (see Kang and Bae, 2007), and ligand–receptor interactions (see Lee et al., 2005b) for effective drug/gene delivery.

2.1   Temperature and Thermo-Sensitive Polymers

Although changes in local body temperature are occasionally induced by pathological conditions (i.e., fever and irregular metabolism), external control of local body temperature by physical means offers a more consistent signal and provides broader application of thermo-sensitive pharmaceutical nanosystems. Hyperthermic therapy is one physical process that is known to kill or weaken cancer cells and/or facilitate radiation and anticancer drug treatment. To heat target sites, microwave (see Yokoyama, 2002), ultrasound (see Tacker and Anderson, 1982), and magnetic devices (see Alexiou et al., 2006) have been routinely applied, and most internal organs and tissues can be accessed from outside of the body. As shown in Fig. 2, locally heating solid tumors to 40–45°C increases blood flow and vascular permeability, causing increased accumulation of pharmaceutical nanosystems and improved antitumor therapeutic effects (see Ponce et al., 2006). In addition, normal cells are unaffected by local hyperthermic conditions, which greatly influence the biological functions of cancer cells. Cancer cells show decreased DNA synthesis, heat shock protein expression, microtubule disruption, alteration in receptor expression and changed cell morphology (see Gerlowski and Jain, 1985; Jain, 1987; Ponce et al., 2006; Song, 1978). In contrast to heating, results obtained by cooling specific sites are rather limited. Using catheters with a cooling mechanism could be one option because catheters can access most internal organs and tissues via blood vessels for short-term applications. Prolonged cooling may reduce various biological functions of tissues including protein synthesis and gene regulation (see Yokoyama, 2002).

Thermo-sensitive nanosystems or polymers often utilize thermal phase transitions (i.e., coil-to-globule transition). Most thermo-responsive polymers are water-soluble below their lower critical solution temperature (LCST) and become water-insoluble (hydrophobically collapsed or aggregated) upon raising the temperature above the LCST. Poly(N-isopropylacrylamide) (poly(NIPAAm)) is a representative thermo-sensitive polymer. This polymer forms hydrogen bonds between its amine groups and water molecules, hydrating of the N-isopropyl groups below the LCST, giving the polymer a hydrophilic character. However, above the LCST,

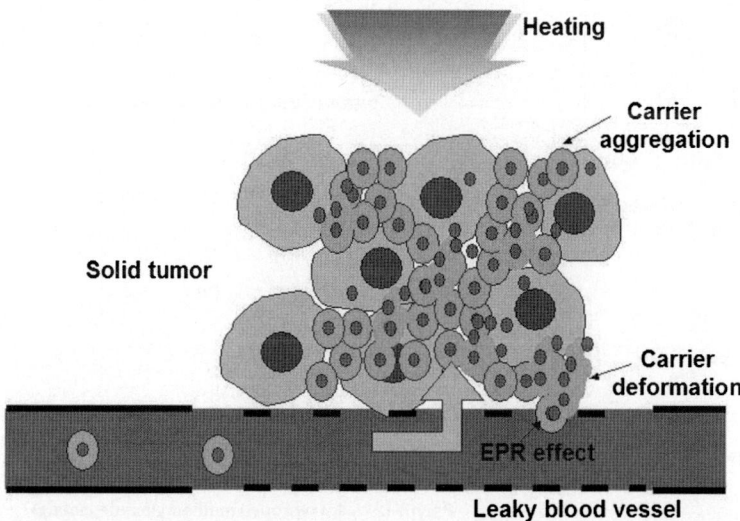

Fig. 2 Tumor-specific accumulation of thermo-sensitive nanocarriers under local hyperthermia

poly(NIPAAm) is dehydrated because of decreased interaction with water mole-cules (see Yuk and Bae, 1999). Polymers with LCST characteristics include poly(NIPAAm) (see Chung et al., 2000), polyester block copolymers (see Jeong et al., 1999; Na et al., 2006), and elastin-like polypeptides (ELP) (see Dreher et al., 2003; Furgeson et al., 2006; Matsumura and Maeda, 1986; Rodriguez-Cabello et al., 2006; Urry, 1997). Their chemical structures are presented in Fig. 3. The poly-mers and their copolymers showed LCST in the range of 30–50°C (poly(NIPAAm) and its copolymers), 20–100°C (polyester block copolymers), and 27–40°C (ELP). For specific applications, phase transition temperatures can be tuned by controlling comonomer composition, hydrophilic/hydrophobic balance, stereochemistry (see Chung et al., 1998, 1999, 2000; Kikuchi and Okano, 2002) and additives (i.e., salts and surfactants) (see Makhaeva et al., 1998).

Drugs or genes can be incorporated into the thermo-responsive nanosystems, which can be constructed from thermo-sensitive polymers only or polymers modi-fied to include drug/gene-interacting segments via chemical linkages, electrostatic attraction, and hydrophobic interactions. For example, block copolymers composed of a thermo-sensitive block and a hydrophobic block are used for water-insoluble drug loading, while positively charged blocks are used for loading negatively charged genetic material.

For water-insoluble anticancer drug delivery, typical hydrophobic blocks include methacrylic acid stearoyl ester (see Cammas et al., 1997), stearoyl chloride (see Kikuchi and Okano, 2002), poly(styrene (St)) (see Chung et al., 2000; Gaucher et al., 2005), poly(n-butyl methacrylate (BMA)) (see Chung et al., 2000; Kikuchi

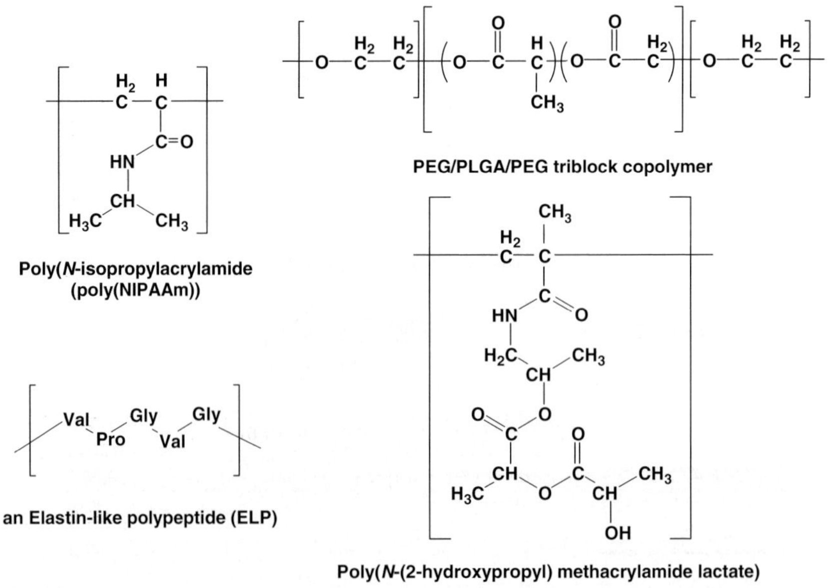

Fig. 3 Typical thermo-sensitive polymers having LCST characteristics

and Okano, 2002), and poly(D,L-lactide-*co*-glycolide) (PLGA) (see Liu et al., 2005). Below the LCST, amphiphilic block copolymers form nano-sized micelles or nanoparticles depending on the copolymer architecture (di-, tri-, and multi-blocks, random, and graft copolymers). Diblock copolymers (i.e., poly(NIPAAm)-*b*-stearoyl chloride and poly(NIPAAm)-*b*-polySt) formed a typical core-shell micellar structure, and their structural transition temperatures were close to the LCST of the thermo-sensitive blocks (i.e., poly(NIPAAm)) (see Kikuchi and Okano, 2002). Micelle formation might be caused if there is no interfering interactions between the hydrophilic poly(NIPAAm) chains and the hydrophobic core (see Heskins and Guillet, 1968). However, unlike diblock types, poly(NIPAAm-*co*-methacrylic acid stearoyl ester) random copolymers showed aggregation at temperatures lower than the LCST of poly(NIPAAm) (see Cammas et al., 1997). This might be caused by incomplete phase separation between the hydrophobic and hydrophilic segments.

For genetic material, thermo-sensitive polymers have been linked to cationic segments such as poly(L-lysine) (PLL) (see Oupicky et al., 2003), polyethylene-imine (PEI) (see Bisht et al., 2006; Lavigne et al., 2007; Türk et al., 2004), and dimethylaminoethyl methacrylate (DMAEMA) (see Kurisawa et al., 2000; Takeda et al., 2004). Genes can also be chemically conjugated to thermo-responsive polymers (see Murata et al., 2003a,b). The block copolymers form nanocomplexes with genes via electrostatic interactions or self-assembly. Above the LCST of polymers, the loaded genes were protected from nucleases because of the hydrophobic thermo-sensitive blocks, whereas temperatures lower than the LCST made the thermo-sensitive blocks more hydrophilic and caused gene release from the swollen nanocomplexes.

Thus, combining the applications of thermo-responsive materials and temperature modulation (i.e., heating and cooling) has the potential to effectively deliver chemical drugs and therapeutic genes. Specific examples will be introduced in Sects. 3.1. and 4.1.

2.2   pH and pH-Sensitive Polymers

Table 1 compiles the specific pH of various organs, intracellular compartments, and body fluids in normal and pathological conditions (see Diessemond et al., 2003; Kang et al., 2005; Na and Bae, 2005; Okada and Hillery, 2001; Owen and Katz, 2005; Schmaljohann, 2006; Tannock and Rotin, 1989). Under nonpathological conditions, the gastrointestinal tract (i.e., stomach, duodenum, and colon) and intracellular compartments (i.e., early endosomes, late endosomes, lysosomes, cytosol, and Golgi) within a single cell have their own specific ranges of pH. pH can be a signal to trigger the release of a drug/gene from carriers at a particular location of the GI track or in select intracellular compartments. Specific pHs are also found in wounds (~pH 5.5–8.7) (see Diessemond et al., 2003), semen (~pH 7.5) (see Owen and Katz, 2005), and vaginal fluid (~pH 4–5) (see Okada and Hillery, 2001).

Table 1 pH Specificity of organs, tissues, cells, intracellular compartments, and body fluids in normal biological environments and pathological environments

Normal biological environments		pH
Organ	Stomach	1–2[a]
	Duodenum	4.0–5.5[a]
	Jejunum	5.5–7[a]
	Ileum	7–8[a]
	Colon and rectum	7–7.5[a]
Intracellular compartment	Early endosomes	5.5–7[b]
	Late endosomes	5–6[b,c]
	Lysosomes	4–5[b]
	Golgi	6.4[c]
	Cytosol	~7.2[a]
Body fluid	Blood	~7.4[c]
	Vaginal fluid	4–5[d]
	Semen	~7.5[e]
Pathological environments		pH
Body fluid	Extratumoral blood	6.5–7.2[f]
	Wound	5.5–8.7[g]

Cited from [a]Na and Bae, 2005
[b]Kang et al., 2005
[c]Schmaljohann, 2006
[d]Okada and Hillery, 2001
[e]Owen and Katz, 2005
[f]Tannock and Rotin, 1989
[g]Diessemond et al., 2003

The extracellular tumor pH (pH_e) presents a high potential for targeting nanocarrier therapeutics. More than 80% of measured pH_e values in human and animal solid tumors showed pHs lower (~6.5–7.2) than normal blood pH (7.4) (see Engin et al., 1995; Leeper et al., 1994; Ojugo et al., 1999; van Sluis et al., 1999). Acidic pH_e values were affected by tumor histology, location, size, and glycolysis rates (see Hobbs et al., 1998; Stubbs et al., 2000; Tannock and Rotin, 1989). In addition, acidic pH is regarded to be a phenotype of tumor cells (see Yamagata et al., 1998). Inadequate blood supply, poor lymphatic drainage, and high interstitial pressure in tumor tissues contributed to the development of acidic environments (see Engin et al., 1995; Leeper et al., 1994; Ojugo et al., 1999). Acidic tumor pH has prompted the development of pH-sensitive antitumor nanosystems to deliver chemical drugs (see Drummond et al., 2000; Gerasimov et al., 1999; Lee et al., 2003a,b, 2005a,b; Sethuraman and Bae, 2007) and genes (see Sethuraman et al., 2006). Nanosystems that were sensitive to tumor acidity demonstrated accelerated release of chemical drugs (see Lee et al., 2003a,b, 2005a,b) and exposed shielded functions (i.e., ligands (see Lee et al., 2005b; Sethuraman and Bae, 2007) or positive charges (see Sethuraman et al., 2006)) to enhance cellular internalization.

The more acidic endosomal/lysosomal compartments are also sites of pH targeting. Sequestration of pharmaceutical nanosystems in these compartments should be avoided because drugs/genes that are exposed to lysosomal enzymes are subject to extensive degradation or metabolism. Avoiding sequestration increases the bioavailability of delivered agents to target subcellular organelles such as cytoplasm, cytoskeleton, mitochondria, and nucleus. Two properties associated with pH-sensitive materials that aid this pursuit are osmotic shock (the so-called proton sponge) effect and/or fusogenicity. The "proton sponge" effect first suggested by Behr's group (see Boussif et al., 1995) has been supported by Sonawane et al. (2003) who monitored changes in endosomal volume, pH, and chloride concentration using representative polycations such as PEI and PAMAM. The sponge effect's mechanism is related to the acidification of endosomal compartments. Vacuolar ATPase-H^+ pumps which exist in endosomal membranes induce proton transfer from the cytoplasm to endosomal compartments. If certain materials with protonable groups are entrapped in these acidic compartments, the transferred protons will be captured by protonable moieties. This proton buffering leads to a smaller decrease in endosomal pH. This further increases the influx of protons into the endosomes simultaneously with their counter ions (Cl^-), causing osmotic imbalances between the cytosol and the endosomal compartments. This causes endosomal swelling caused by continuous water influx and finally leads to endosomal rupture (see Cho et al., 2003; Kang et al., 2005; Pack et al., 2005). Fusogenicity is related to the physical interaction between pH-sensitive membrane destabilizers and endosomal membranes. Positively charged amino acid-containing materials can interact with negatively charged phospholipid bilayers in the endosomal membrane and cause destabilization (see Cho et al., 2003). For materials that have pH-induced hydrophilic-to-hydrophobic transitions (i.e., anionic polyelectrolytes), mechanisms of membrane destabilization have been investigated using lipid bilayers (i.e., liposomes). Typically, hydrophobic segments can adsorb to the outer phospholipid bilayers membrane (see Tirrell et al., 1985) or negatively charged polymers may interact with dipoles contained in the headgroups of the lipid bilayers via electrostatic attractions (see Xie and Granick, 2002). Polymers with increased hydrophobicity via pH modulation may penetrate into the hydrophobic tail region of the bilayers (see Xie and Granick, 2002), increasing lateral compression and causing pore formation in the bilayers (see Thomas and Tirrell, 2000). These processes induced by hydrophobic polymer chains ultimately cause membrane disruption (see Eum et al., 1989; Yessine and Leroux, 2004). As shown in Fig. 4, representative endosome-disrupting polymers/ oligomers can contain several functional groups (i.e., amines (see Boussif et al., 1995; Lee et al., 2005b; Park et al., 2006; Pichon et al., 2001; Yang et al., 2006), carboxylic acids (see Jones et al., 2003; Kiang et al., 2004; Yessine and Leroux, 2004; Yessine et al., 2003), and sulfonamides (see Kang and Bae, 2007)) which enable protonation/deprotonation transition in the endosomal compartments. These polymers/oligomers start functioning after being physically or chemically introduced into pharmaceutical nanosystems.

Acid-labile degradable bonds such as acetal bonds (see Gillies et al., 2004a; Murthy et al., 2003), hydrazone bonds (see Bae et al., 2005a,b; Hruby et al., 2005; Sawant et al., 2006; Walker et al., 2005), Schiff's base (or azomethine) bonds (see Kim et al., 2005b), and vinyl ether bonds (see Shin et al., 2003) (see Fig. 5) are often employed in designing pH-sensitive systems. These bonds are cleaved upon

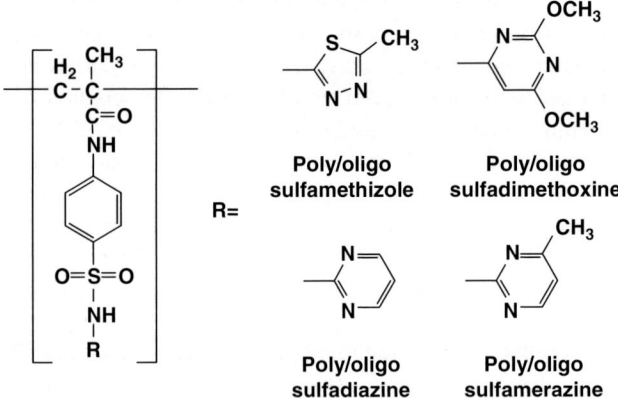

Fig. 4 Examples of protonable oligomers/polymers that destabilize the endosomal membrane

contacting acidic milieu in solid tumors and endosomal/lysosomal compartments, ultimately resulting in the accelerated release of delivered drug/gene.

Thus, drug/gene-carrying nanopharmaceuticals which recognize minute differences in pH at specific sites such as acidic tumors and endosomal/lysosomal compartments can be very beneficial in increasing drug/gene bioavailability at the sites of drug action while reducing unwanted effects. The systems will be examined further in Sect. 3.2 for drugs and Sect. 4.2 for genes.

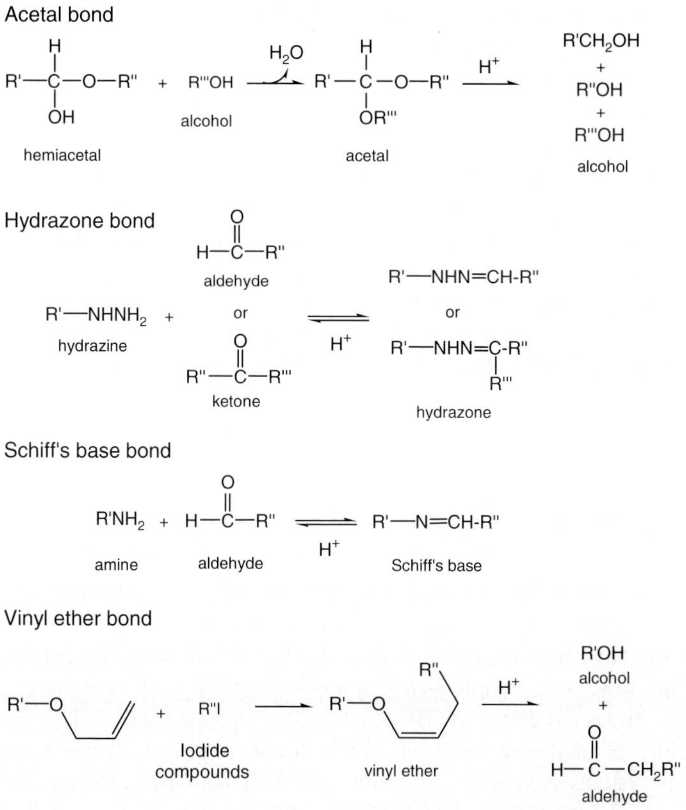

Acetal bond

$$R'-\overset{\overset{\displaystyle H}{|}}{\underset{\underset{\displaystyle OH}{|}}{C}}-O-R'' \quad + \quad R'''OH \xrightarrow{\quad H_2O \quad} R'-\overset{\overset{\displaystyle H}{|}}{\underset{\underset{\displaystyle OR'''}{|}}{C}}-O-R'' \xrightarrow{\quad H^+ \quad} \begin{array}{l} R'CH_2OH \\ + \\ R''OH \\ + \\ R'''OH \end{array}$$

hemiacetal · alcohol · acetal · alcohol

Hydrazone bond

$$R'-NHNH_2 \quad + \quad \begin{array}{c} \overset{\displaystyle O}{\underset{\displaystyle \parallel}{}} \\ H-C-R'' \\ \text{aldehyde} \\ \text{or} \\ \overset{\displaystyle O}{\underset{\displaystyle \parallel}{}} \\ R''-C-R''' \\ \text{ketone} \end{array} \underset{H^+}{\rightleftharpoons} \begin{array}{c} R'-NHN{=}CH\text{-}R'' \\ \text{or} \\ R'-NHN{=}\overset{\displaystyle C}{\underset{\displaystyle R'''}{}}\text{-}R'' \\ \text{hydrazone} \end{array}$$

hydrazine

Schiff's base bond

$$R'NH_2 \quad + \quad \overset{\overset{\displaystyle O}{\parallel}}{H-C-R''} \underset{H^+}{\rightleftharpoons} R'-N{=}CH\text{-}R''$$

amine · aldehyde · Schiff's base

Vinyl ether bond

$$R'-O \diagup\!\!\!\diagdown \quad + \quad R''I \longrightarrow R'-O \diagdown_{R''} \xrightarrow{\quad H^+ \quad} \begin{array}{l} R'OH \\ \text{alcohol} \\ + \\ \overset{\displaystyle O}{\underset{\displaystyle \parallel}{}} \\ H-C-CH_2R'' \\ \text{aldehyde} \end{array}$$

Iodide compounds · vinyl ether

Fig. 5 Typical pH-triggered hydrolytic linkages: formation and acidic hydrolysis

2.3 Redox Potentials and Thiol-Sensitive Polymers

Reduction/oxidation reactions (redox) are chemical reactions where molecules, atoms, or ions gain (for reduction) or lose (for oxidation) electrons. Site-dependent local concentrations of reduced and oxidized chemical species generate a redox potential. In most mammalian cells, intracellular compartments (i.e., cytosol and nuclei) have higher populations of reduced forms than extracellular milieu (i.e., body fluid, blood). Most known biological redox potentials are induced by balancing reduced glutathione, thioredoxin, peroxiredoxins, and nicotinamide adenine dinucleotides (NADH and NADPH) with their oxidized forms (see Hansen et al., 2006). A representative redox pair is reduced glutathione (GSH) and oxidized glutathione (GSSG). Concentrations of GSH in the plasma and cytosol are reported to be 2–20 μM and 0.5–10 mM, respectively (see Wu et al., 2004). In addition, GSH is increasingly preferred over GSSG intracellularly: mitochondria > nuclei > cytoplasm > endoplasmic reticulum > extracellular spaces (see Hansen et al., 2006).

This site-specific preference of GSH or GSSH has prompted the design of site-specific triggered drug/gene delivery.

Thiol-triggered degradable bonds can be positioned either in the polymer backbone or between the polymer backbone and drug/gene. The former system traps the drug/gene via electrostatic or hydrophobic interactions. Upon encountering high thiol concentrations, disulfide bonds in the polymer backbone are cleaved to quickly release drug/gene. In the latter case, drug/gene is chemically introduced into the polymer backbone via disulfide bonds. Redox preference in the cytosol changes disulfide bonds into thiols, inducing prompt release of drug/gene.

These thiol-specific degradable nanosystems enhance the bioavailability of delivered drug/gene at the targeted sites (i.e., mitochondria, cytosol, and nucleus). Bond cleavage also offers better biocompatibility since the short degraded fragments are easily removed from the body.

3   Stimuli-Sensitive Drug Delivery

Most anticancer drugs demonstrate a powerful capability in killing tumor cell lines in in vitro experiments. However, in animals and in clinical settings, the same drugs are often less effective than expected. These results are linked to nonspecific toxicity which limits dose size, inappropriate tumor-selectivity, and multidrug resistance (MDR) (see Brannon-Peppas and Blanchette, 2004; Szakacs et al., 2006). Poor aqueous solubility of antitumor drugs is also a strong limiting factor in formulations. Nanocarrier systems such as micelles, liposomes, and nanoparticles have been intensively investigated to improve drug solubility, enhance passive accumulation in solid tumors, improve stability, and increase circulation time, tumor-targetability (when a specific ligand is equipped), and biodegradability in vivo. Despite these advantages, there is still room for improving drug accumulation in tumor sites and intracellular compartments.

When nanocarriers for antitumor drugs are designed, controlling the release kinetics of the incorporated drugs should be considered. During circulation in the blood stream, released drugs reduce bioavailability at the target sites while increasing systemic side effects. At the tumor site, Gardner (2000) suggested that slow drug release in tumors might ameliorate the drug efficacy, especially for drug-resistant cells. However, Gottesman et al. (1996) reported that slow drug release could increase the probability of developing MDR, finally inducing weak antitumor effects. Despite this controversy, accelerated drug release at tumors has been investigated. Temperature and pH, which are linked with antitumor therapy (i.e., hyperthermia (see Ponce et al., 2006)) and tumor milieu (i.e., acidic pHs (see Engin et al., 1995)), have been candidate signals for triggered release. Polymer-based nanocarriers for triggered release by temperature and/or pH are the focus here. (External photodynamic and ultrasonic stimulation are not covered in this part because these stimuli are not directly affecting polymer properties.)

3.1 Thermo-Responsive Polymeric Drug Carriers

Thermo-sensitive polymeric nanocarriers in the forms of micelles and nanoparticles have been investigated. The self-assembly of such nanosystems is related to the polymer architecture and composition using temperature-sensitive and hydrophobic blocks. Thermal stability of the hydrophobic core affects drug release profiles (see Chung et al., 2000; Kikuchi and Okano, 2002). Drug loading and release kinetics can be tailored by polymer design.

3.1.1 Thermosensitive Nondegradable Nanocarriers

Micelles composed of poly(NIPAAm)-b-polyBMA or poly(NIPAAm)-b-polySt represent nondegradable and thermo-sensitive nanocarriers (see Chung et al., 2000). The AB-type block copolymers showed LCST (about 34°C) similar to poly(NIPAAm). As shown in Fig. 6, at 40°C (above the LCST) doxorubicin (DOX, an antitumor drug) release was enhanced due to structural deformation, whereas release was inhibited

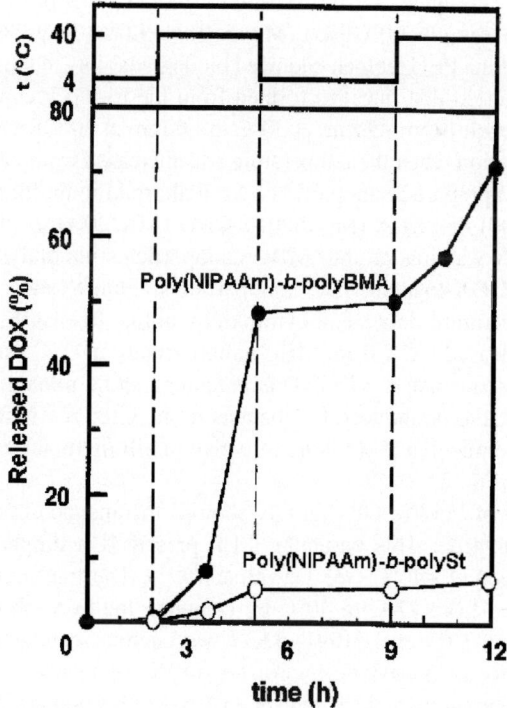

Fig. 6 DOX release from poly(NIPAAm)-b-polyBMA and poly(NIPAAm)-b-polySt micelles in response to temperature switching between 4 and 40°C (reproduced with permission; Chung et al., 2000)

regardless of the hydrophobic core materials at 4°C (below the LCST). However, poly(NIPAAm)-*b*-polyBMA micelles showed more accelerated drug release than poly(NIPAAm)-*b*-polySt micelles at 40°C. This difference in DOX release profiles was interpreted to be caused by differences in the stability of the hydrophobic cores, which held the antitumor drugs. PolySt core is more stable than polyBMA against thermal change (heating–cooling cycle), resulting in lower DOX release rates. Liu et al. (2005) also reported consistent results using poly(NIPAAm-*co*-N,N-dimethylacrylamide)-*b*-PLGA copolymer (LCST 39.1°C). Micelles about 200 nm in diameter were stable at 37°C but became deformed and triggered DOX release at 39.5°C. Above the LCST, DOX release from the micelles was more accelerated than below the LCST, causing greater antitumor effects (cytotoxicity against cancer cell lines).

These in vitro results may support the benefit of using temperature-responsive micelles coupled with local hyperthermic treatment (40–45°C). However, the non-degradable nature of poly(NIPAAm), if it is used as the thermo-sensitive functionality, may limit its clinical applications.

3.1.2 Thermo-Sensitive Degradable Nanocarriers

Jeong et al. (1999) designed thermo-sensitive degradable tri-block polymers composed of poly(ethylene glycol) (PEG) (Mw 550)-*b*-PLGA (Mw 2810)-*b*-PEG (Mw 550). Hydrolysis of the PLGA block endowed biodegradability to the thermo-sensitive polymer. They reported that micelles formed from the tri-block copolymers showed abrupt micellar growth (from ~30 nm at 30°C to ~60 nm at 40°C) and increased polymer–polymer attraction when the temperature was increased from 20 to 50°C.

Na et al. (2006) prepared nanoparticles from alternating multiblock copolymers (MBC, Mw 30,600) of PLLA (Mw 800)-*b*-PEG (Mw 2000)-*b*-PLLA (Mw 800). Hydrophobic DOX was loaded into MBC nanoparticles via dialysis. In vitro anti-tumor studies of DOX-loaded MBC nanoparticles against Lewis lung carcinoma cells showed temperature-dependent cytotoxicity unlike free DOX. After 1 day, the total amount of released DOX from MBC nanoparticles at 42°C was approximately two times higher than that at 37°C. This enhanced DOX release rate from MBC nanoparticles seems to be induced by changes in the interior structure of the nano-particles. Thus, the micelles were more effective in killing tumor cells at 42°C than 37°C as shown in Fig. 7.

ELP composed of Val-Pro-Gly-Val-Gly showed therapeutic effects similar to the synthetic MBC at 40°C. This particular ELP presented hydrophobic folding and assembling transition at ~40°C (see Dewhirst, 1995). The transition temperature of ELP was controlled from 27°C to 40°C by changing the hydrophobicity via amino acid substitution (see Urry et al., 1991). DOX was chemically conjugated to ELP via hydrazone bonds for acid catalytic hydrolysis. DOX-conjugated ELP showed more accumulation in solid tumor at 42°C than at 34°C (see Meyer et al., 2001). Also, heat-treated cells showed two- to threefold higher cellular DOX uptake than normo-thermic cells (see Kaneko et al., 1991). ELP-drug conjugate seems to be a promising candidate for tumor targeting under hyperthermic conditions (see Kong and Dewhirst, 1999).

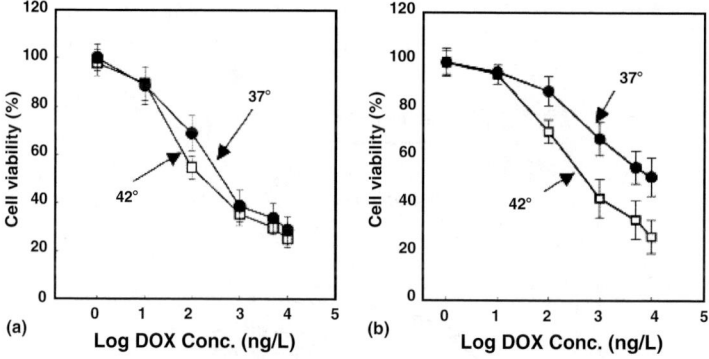

Fig. 7 In vitro cytotoxicity of free DOX (**a**) and MBC1 nanoparticles containing DOX (**b**) against Lewis Lung Carcinoma cells as a function of drug concentration at 37 and 42°C. Values are mean ± the standard deviation (S.D.) ($n = 7$) (reproduced with permission; Na et al., 2006)

3.2 pH-Sensitive Polymeric Drug Carriers

Tumor extracellular pH and endo/lysosomal pH have been used to cleave chemical bonds and cause physical transitions of carrier properties by ionization, hydrophobic interaction, and solubility. In turn, the switching property caused by pH changes influences the rate of drug release, carrier stability, and cellular interactions.

3.2.1 Acid-Induced Cleavage of Chemical Bonds

Fréchet's group has used acetal bonds for acid-triggered drug release. Their system released 5-fluorouridine (5-FU) at pH 5.0 via acidic hydrolysis of acetal bonds between 5-FU and PEG, whereas neutral pH caused minimal breakage in the acetal linkages (see Gillies et al., 2004a). Similarly, as shown in Fig. 8, acetal bonds could be introduced to link hydrophilic copolymers (linear PEG-*b*-dendritic polylysine or polyester) and hydrophobic chemicals (see Gillies and Fréchet, 2005; Gillies et al., 2004b). These dendrites with acetal bonds formed micellar structures at neutral pH. Under acidic conditions, the bond was cleaved, resulting in DOX release and demicellization. Further, lowering the pH caused more accelerated DOX release (pH 4.0 > pH 5.0 > pH 6.0 > pH 7.4).

Hydrazone bonds were used to link PEG-*b*-poly(allyl glycidyl ether) and DOX (Hruby et al., 2005). The polymeric drug formed micelles approximately 100 nm in diameter in pH 7.4 buffer solution at 37°C. These micelles released about 43% of DOX at pH 5.0 after 24-h incubation whereas at pH 7.4 only 16% of DOX was released. Similarly, Kataoka's group has designed PEG-*b*-poly(aspartate)-based micelles for cytosolic drug delivery (see Bae et al., 2003, 2005a,b). DOX was conjugated to poly(aspartate) via hydrazone bonds (see Fig. 9). In vitro experiments with the micelles showed pH-dependent DOX release; 5–20% and 40–60% of the

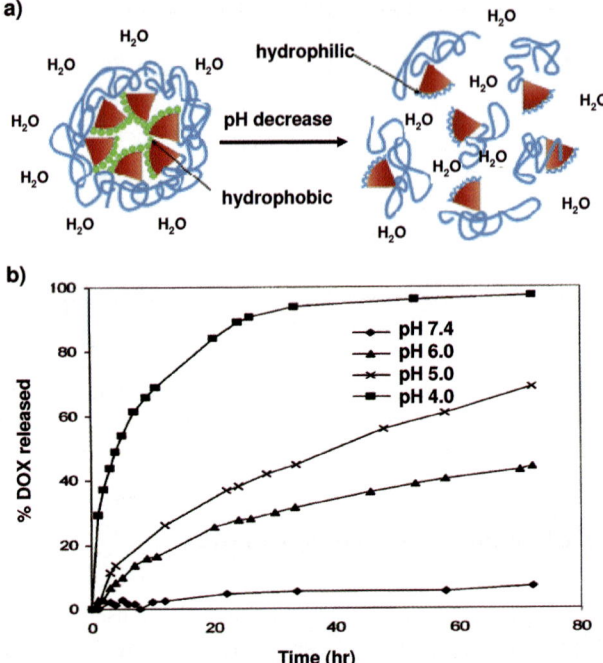

Fig. 8 pH-Sensitive supramolecular micelles having acetal linkages between hydrophilic copo-lymers (linear PEG-*b*-dendritic polylysine or polyester) and hydrophobic chemicals. (**a**) Schematic concepts (reproduced with permission; Gillies et al., 2004b) and (**b**) pH-dependent DOX release of DOX-loaded micelles (reproduced with permission; Gillies and Fréchet, 2005)

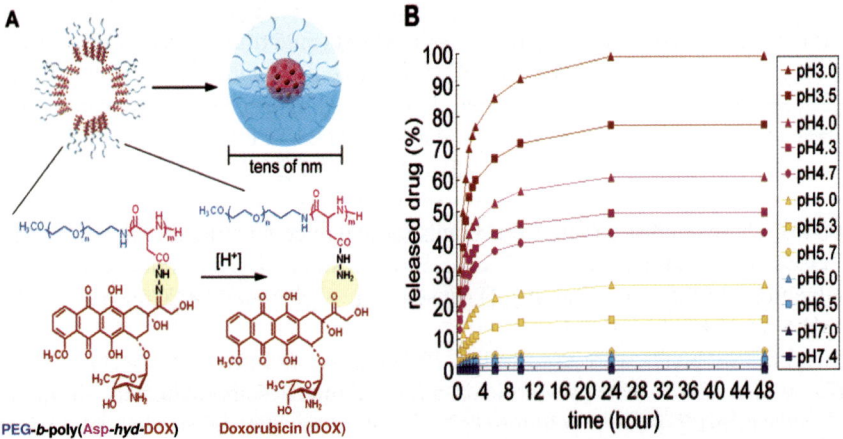

Fig. 9 Tumor-infiltrating polymeric micelles with intracellular pH-sensitivity. (**a**) Micelles with diameters of tens of nm were prepared from self-assembling amphiphilic block copolymers, PEG-*b*-poly(Asp-*hyd*-DOX), where the anticancer drug doxorubicin (DOX) was conjugated through acid-sensitive hydrazone linkers. (**b**) The micelles released the loaded drugs under acidic conditions below pH 6.0 corresponding to the intracellular space, but remained stable under the conditions of vascular and extracellular space (pH 7.4–7.0) (reproduced with permission; Bae et al., 2005b)

loaded DOX released in the pH ranges of pH 5.0–6.5 and pH 4.0–4.7, respectively. Besides, more acidic pH 3–4 induced 70–90% release of total DOX. These studies demonstrate that the acid-sensitive hydrazone linkage is sensitive to lysosomal pH.

3.2.2   Acidic pH-Induced Structural Change

Polymeric micelles having a pH-sensitive shell experience structural changes when the pH is lowered, which promotes protonation of the ionized acidic block. This makes hydrophilic surfaces hydrophobic.

Na et al. (2003, 2004) prepared pH-sensitive nanogels using hydrophobically modified pullulan (pullulan acetate (PA)) coupled with oligomeric sulfadimethoxine (OSDM) via ester linkages. These nanogels were formed by self-assembly of the polymer dissolved in dimethylsulfoxide during dialysis against pH 8.5. At pH 7.4, the pH-sensitive component OSDM was fully ionized and existed on the surface of the nanogel, acting as a hydrophilic shell. As the pH decreased to 6.8, OSDM became hydrophobic due to its deionization (protonation), causing perturbation/reorganization of the nanogel structure as depicted in Fig. 10. These pH-dependent structural

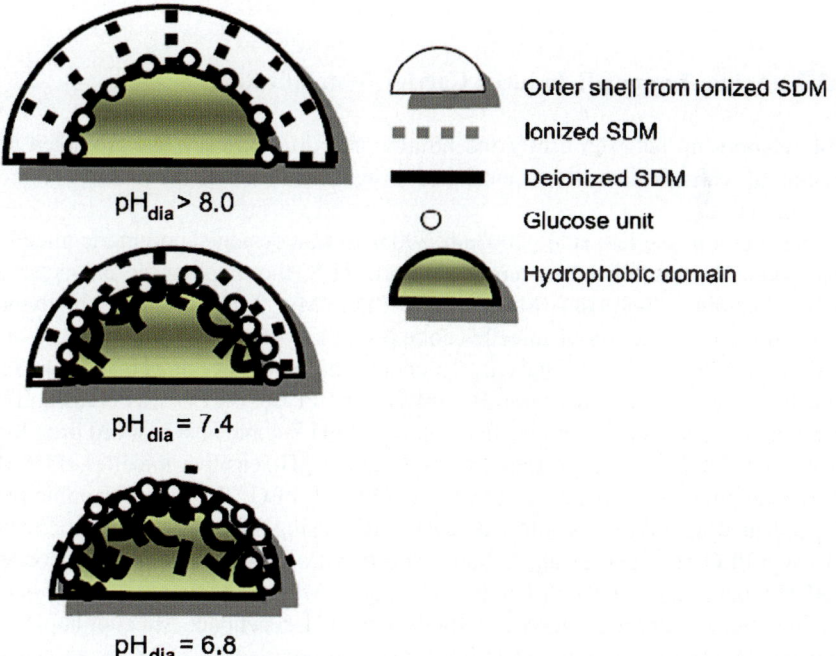

Fig. 10 Nanoparticles whose outer shell width is controllable and pH-sensitive. At higher pH_{dia} (above 8.0), OSDM on the surface of the particle may be entirely elongated because it was fully ionized by the adequate energy supply, forming a broader outer shell, while at lower pH 7.4, OSDM may form relatively narrow shells due to the hydrophobic interactions between deionized SDM groups and/or acetyl groups. Ionized (*dashed line*), deionized (*solid line*) OSDM, and hydroxyl group (*open circle*) (reproduced with permission; Na et al., 2004)

changes of the nanogels accelerated the release rate of DOX. Acidic pH (pH < 6.8) promoted more DOX release than neutral pH. In addition, the nanogels aggressively adhered to and internalized into tumor cells at pH 6.8, probably due to hydrophobic interactions between cellular membranes and the deionized OSDM. The combined effects resulted in significantly higher cytotoxicity (antitumor effect) against MCF-7 tumor cells at pH 6.8 than at pH 7.4.

Liu et al. (2007) designed polymers dually sensitive to pH and temperature. The amphiphilic polymer, poly(NIPAAm-co-N,N-dimethylacrylamide-co-2-aminoethyl methacrylate)-b-poly(10-undecenoic acid), showed pH-dependant LCSTs: 38.0°C at pH 7.4 and 36.2°C at pH 6.6. The hydrophilic nature of poly(10-undecenoic acid) caused by the higher degree of ionization of carboxylic groups at pH 7.4 resulted in self-assembled core-shell micelles. At pH 6.6 and 37°C, a decrease in the number of ionized carboxylic groups made poly(10-undecenoic acid) slightly more hydrophobic, shifting the LCST to a lower temperature and leading to deformed/precipitated micelles. The polymer also showed pH-dependent DOX release patterns (about 40 and 85% of the loaded DOX at pH 7.4 and pH 6.8, respectively). In vitro studies and tumor-bearing animal models showed that acidic conditions improved cytotoxicity and caused more DOX accumulation in tumors. However, the nondegradable nature of the polymer may be a drawback of this system.

3.2.3 Acidic Tumor pH-Induced Carrier Destabilization

The relationship between drug concentration and drug efficacy has prompted the design of carriers that show enhanced drug release in slightly acidic tumor environments.

Bae's group (see Lee et al., 2003a,b, 2005a,b) first designed polymeric micelles that could be destabilized in tumor pH_e. At pH 8, the amphiphilic character of poly(L-histidine) (polyHis) (Mw 5,000)-b-PEG (Mw 2,000) diblock copolymer allowed the construction of micelles composed of hydrophobic polyHis core and hydrophilic PEG shell via dialysis. To endow more stability at pH 7.4, a mixed micelle system composed of polyHis-b-PEG and PLLA-b-PEG was created. The resulting micelles improved micellar stability at pH 7.4 and also reduced drug loss at pH 7.4. DOX release profiles from the mixed pH-sensitive micelles (PHSM) were manipulated by changing the wt% of PLLA-b-PEG. The most favorable pH-dependent drug release profile was seen when using PHSM containing 25wt% PLLA-b-PEG (see Lee et al., 2003b). The in vitro cytotoxicity of DOX-loaded PHSM (containing 25wt% PLLA-b-PEG) against MCF-7 (human breast adenocarcinoma) tumor cells after incubation for 48 h was pH-dependent. Micelles at pH 6.8 showed DOX-dose-dependent cytotoxicity similar to free DOX as shown in Fig. 11a (see Lee et al., 2003b). However, as shown in Fig. 11b, DOX-loaded PHSM (containing 25wt% PLLA-b-PEG) showed potent inhibition of tumor (s.c. MCF-7 xenografts) growth compared to other non-pH-sensitive controls (i.e., PLLA-b-PEG micelles, free DOX, and saline) at the same dose of DOX. The tumor growth inhibition of PHSM was maintained for about 6 weeks (see Lee et al., 2005a).

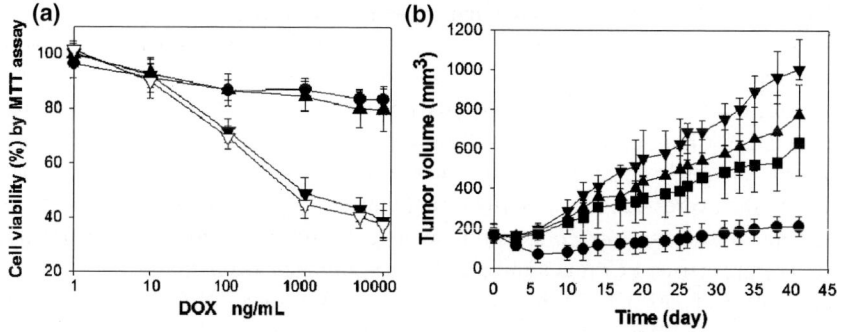

Fig. 11 (**a**) In vitro pH-dependent cytotoxicity of DOX-loaded PHSM (containing 25 wt% PLL-*b*-PEG; pH 7.4(*filled circle*), 7.0(*filled triangle*), and 6.8(*filled down triangle*)) and free DOX (6.8(*open square*)) after 48-h incubation (reproduced with permission; Lee et al., 2003b). (**b**) Tumor growth inhibition after treatment with PHSM (*filled circle*), PLLA-*b*-PEG micelles (*filled square*), free DOX (*filled triangle*), and saline (*filled down triangle*) in MCF-7 tumor-bearing BALB/c nude mice. Two *i.v.* injections of 10 mg/kg DOX equivalent dose were made on days 0 and 3. Values are the mean ± the standard deviation (S.D.) (*n* = 5) (reproduced with permission; Lee et al., 2005a)

Similar pH-sensitivity was seen in other polymers such as poly(L-histidine-*co*-L-phenylalanine)-*b*-PEG diblock copolymer (see Kim et al., 2005a) and [2-(methacryloyloxy)ethyl phosphorylcholine (MPC)]-*b*-[2-(diisopropyl amino) ethyl methacrylate (DPA)] (see Licciardi et al., 2006). Poly(L-histidine-*co*-L-phenylalanine) and DPA blocks showed pH-induced hydrophobic–hydrophilic transition in each polymer when the pH was lowered from the pH used for micelle formation to more acidic pHs. For the former polymers, their p*K*a values were controlled by adjusting the ratio of histidine/phenylalanine in the poly(amino acid) block. Their micelles showed micellar destabilization below pH 6.7 (see Kim et al., 2005a). In the case of PTX-loaded micelles made from MPC-DPA polymers, PTX was completely released at pH 5.0 after 20 h, whereas at pH 7.4 after 50 h, the amount of PTX released was 70% (see Licciardi et al., 2006). Leroux's group (see Drummond et al., 2000; Leroux et al., 2001; Zignani et al., 2000) has also studied lipid/polymer hybridized nanocarriers composed of lipid bilayers and hydrophobically modified copolymers of *N*-isopropylacrylamide bearing a pH-sensitive methacrylic acid. Those nanosystems showed pH- and copolymer/lipid ratio-dependent drug release. The amount of DOX released sharply increased when the pH dropped from 6.5 to 5 and the loaded DOX was almost completely released at pH 5, while minimal DOX release was seen in the range of pH 6.5–8.0 (about 10% of the loaded DOX) (see Leroux et al., 2001). Drug carriers that showed abrupt drug release at pHs lower than tumoral pH$_e$ 6.5–7.2 have been justified as cytosolic drug delivery carriers using the endocytosis mechanism. These nanosystems may be more beneficial to endosomal/lysosomal pH targeting than tumoral pH targeting.

3.2.4 pH-Induced Endosomal Membrane Destabilization

Although abrupt and accelerated drug release in the extratumoral space showed improved antitumor effects for wild-type sensitive tumors, the approach was not convincing for multidrug resistance (MDR) tumors. The drug resistance of certain tumors to cytotoxic agents observed from the beginning of chemotherapy was often linked to the expression of multidrug efflux pumps (P-glycoprotein (Pgp) or MDR-associated proteins), altered glutathione metabolism, reduced topoisomerase II activity, and various changes in cellular proteins and mechanisms (see Boesch et al., 1991; Cataldo et al., 2003; Lazzarino et al., 1998; Sehested et al., 1987; Simon and Schindler, 1994). Drug sequestration in the endosomal/lysosomal compartments also decreases antitumor effects because the predominant form (ionized form) of weakly basic antitumor drugs limits its passage through cellular and intracellular membranes. Moreover, MDR tumor cells actively remove weakly acidic drugs through exocytosis (see Hager et al., 1991; Hooijberg et al., 2003; van Adelsberg and Al-Awqati, 1998; Warren et al., 1991). Thus, cytosolic drug delivery would be a more effective method for treating MDR cells.

For endosomal pH-triggered drug release, Lee et al. (2005b) induced the active cellular internalization of nanocarriers using PHSM decorated with folate (PHSM/f). PHSM/f targeting folate receptors on sensitive MCF-7 cells caused potent cytotoxicity (~16% cell viability at DOX 10 µg/ml) (see Lee et al. 2003b, 2005a). However, PLLA-*b*-PEG–folate micelles (pH-insensitive micelles having folate; PHIM/f) showed less cytotoxicity (~80% cell viability at DOX 10 µg/ml). This difference in cytotoxicity between DOX-loaded PHSM/f and PHIM/f was caused by differences in the intracellular localization of DOX. DOX-loaded PHSM/f resulted in drug distribution in the cytosol as well as the nucleus of MCF-7 cells, while DOX carried by PHIM/f was significantly localized within the endosomal/lysosomal compartments (see Lee et al., 2005b). The result was explained by the ability to escape from the endosomes conferred by polyHis in PHSM/f. This endosomolytic effect of polyHis has stimulated the design of histidine-containing polymers such as histidine-conjugated poly(2-hydroxyethyl aspartamide) (PHEA) or octade-cylamine-*g*-PHEA (see Yang et al., 2006) and *N*-acetyl histidine-conjugated glycol chitosan (see Park et al., 2006).

Bae's group tested the effects of PHSM/f against MDR-tumor cells. DOX-loaded PHSM/f using Pgp-expressing MCF-7/DOX[R] cells showed cytotoxicity levels similar to those obtained when using free DOX against wild cells (MCF-7) (see Lee et al., 2005a). However, DOX-loaded pH-insensitive micelles (PHIM/f) were not significantly effective against MCF-7/DOX[R] cells, although they were still effective using the sensitive MCF-7 cell line. This difference in anticancer effects might be due to the endosomal pH-triggered DOX release and the endosomal escaping activity of polyHis-*b*-PEG-folate unimers released after micellar destabilization in the endosomal compartments. Cellular retention of PHSM/f in MCF-7/DOX[R] cells was about twofold higher than that of PHIM/f (see Mohajer et al., 2007). This supported a decrease in MDR-related exocytosis by disturbing the drug sequestration mechanism in MDR cells (Fig. 12). In in vivo tests using

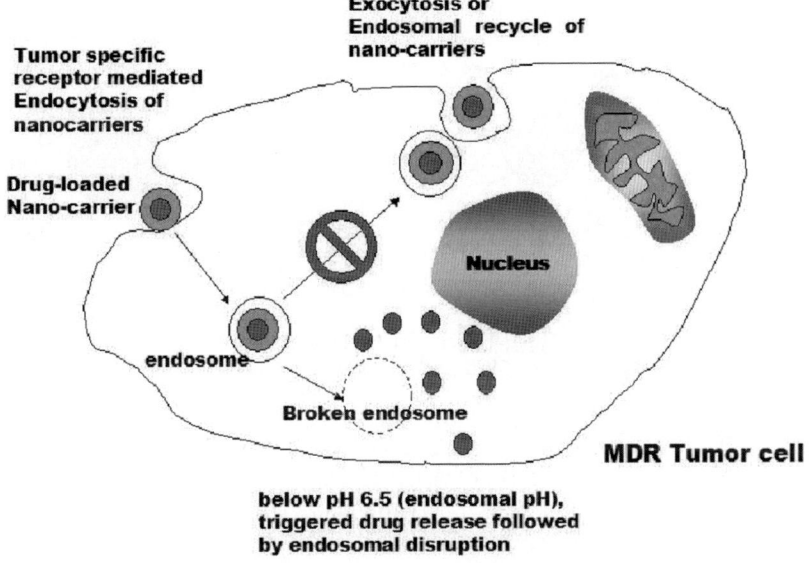

Fig. 12 The proposed design concept of PHSM/f for effective antitumor drug delivery to MDR tumor cells. Carriers are equipped with folate for receptor-mediated endocytosis. PolyHis controls drug-release kinetics via pH-activated switching mechanism and represents fusogenic activity in the endosomal compartments (reproduced with permission; Mohajer et al., 2007)

MCF-7/DOX^R xenografts in nude mice, PHSM/f significantly inhibited tumor growth and reduced tumor size for 2 weeks (see Lee et al., 2005a).

3.2.5 Acid-Induced TAT Deshielding

Effective cellular internalization using the cell-penetrating peptide TAT (transactivating transcriptional activator) may improve antitumor effects if TAT is conjugated to drug-carrying nanosystems. However, the positively charged character of TAT causes it to interact nonspecifically with the cellular membrane. Torchilin's and Bae's groups have investigated the effectiveness of using TAT expressing nanocarriers in the acidic tumor extracellular space.

Sawant et al. (2006) devised dual-targeting pH-responsive pharmaceutical micelles. Micelles were composed of long PEG-*b*-phosphatidylethanolamine (PE) (long PEG-*b*-PE) with tumor-specific monoclonal antimyosin antibody (mAb 2G4) and short PEG-*b*-PE (short PEG-*b*-PE) with TAT. Acid-labile hydrazone bonds were introduced between the long PEG and PE. At neural pH or in the blood, the PEG corona temporarily shielded TAT, leading to long-circulation and reduced nonspecific interactions of TAT with cells. However, at acidic pH 5.0, the long PEG detached by hydrazone bond hydrolysis, resulting in the exposure of TAT coupled to the short PEG/PE on the micelles. These micelles could be potentially used for acidic tumor targeting if the system works at pH 7.2–6.5 rather than at pH 5.0.

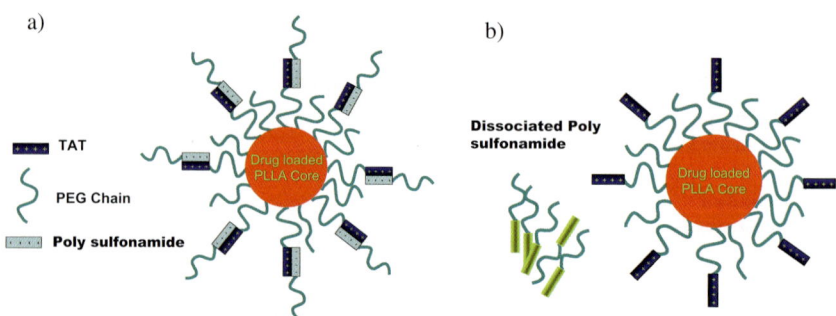

Fig. 13 Concept for a proposed drug delivery system: the carrier system consists of two components, PLLA-*b*-PEG-TAT micelles (TAT-micelles) and pH-sensitive PEG-*b*-PSD. (**a**) At normal blood pH, PSD is negatively charged, and when mixed with TAT, PSD shields the TAT by electrostatic interaction. Only PEG is exposed to the outside which could make the carrier long circulating; (**b**) when the system experiences a decrease in pH (near tumor), PSD loses charge and detaches, exposing TAT for interaction with tumor cells (reproduced with permission; Sethuraman and Bae, 2007)

Very recently, Bae's group developed PHSM consisting of PLLA-*b*-PEG-*b*-TAT and PEG-*b*-poly(methacryloyl sulfadimethoxine) (PSD) for tumor pH$_e$-induced pop-up of the active internalization moiety (see Sethuraman and Bae, 2007). The micelles utilized the unique pH-dependent characteristics of PSD (pKa 6.8). At neutral pH (> pKa), PSD was negatively charged and could be coupled with positively charged TAT via electrostatic attraction. At pH 6.6 (<pKa; tumor pH$_e$), PSD lost its charge, leading to the detachment of PEG-*b*-PSD from the micelles and exposure of the hidden TAT in tumoral environments (see Fig. 13). Unlike the experiments conducted by Sawant et al. (2006), these designed nanocarriers targeted realistic tumoral pH. The in vivo results of this approach are promising (unpublished data). With the biodegradable nature and improved design of PEG-*b*-polysulfonamide, this system was able to distinguish pH 7.0 from pH 7.4 (unpublished data).

3.2.6   Acidic pH-Induced Ligand Pop-Up and Micellar Destabilization

Similar to the acidic pH-induced deshielding, ligand pop-up systems at extracellular tumor pH are hypothesized to reduce undesired nonspecific interactions with biological components. While the deshielding strategy has adopted the use of detaching the shell part of nanocarriers by breaking physical and/or chemical linkages, the ligand pop-up strategy has utilized the positional change of ligands from the middle of the hydrophilic shell to the surface of the shell. As an example, Bae's group investigated an ultra-smart, polymeric micellar carrier that had tumor specificity using nonspecific ligands on the carrier (see Lee et al., 2005b). The micellar systems were constituted from two block copolymer components: polyHis-*b*-PEG

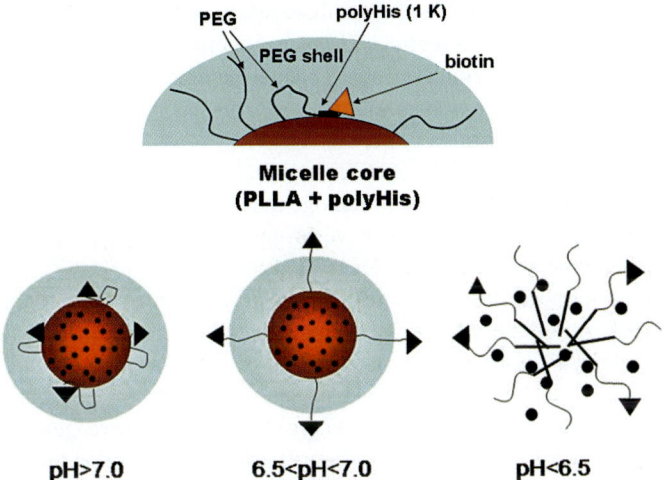

Fig. 14 The central concept of pH-induced vitamin repositioning on the micelles. The location of biotin anchored to the micelle core is controlled by a pH-sensitive molecular chain actuator (polyHis). Above pH 7.0, biotin is shielded by the PEG shell of the micelles. However, 6.5 < pH < 7.0 induces exposure of biotin on the micelle surface, leading to biotin receptor-mediated endocytosis and further acidification (pH < 6.5) stimulates micellar destabilization, resulting in enhanced drug release and membranes disruption such as the endosomal membrane (reproduced with permission; Lee et al., 2005b)

and PLLA-*b*-PEG-*b*-polyHis-biotin. The unique pH-dependent hydrophilic/hydrophobic characteristics of polyHis hid a nonspecific ligand (i.e., biotin) within the hydrophilic shell of the micelles during blood circulation due to the hydrophobic nature of polyHis at neutral pH. However, upon encountering the acidic tumor environment, the hydrophilic shift of polyHis induced the pop-up of biotins on the micellar surface, resulting in biotin receptor-mediated endocytosis of the micellar systems (Fig. 14). After endocytosis of the micelles, endosomal acidic pH further increased the hydrophilic nature of polyHis, and caused the successive dissociation of micelles, disrupted the endosomal membrane, and released DOX (see Lee et al., 2003a,b, 2005a).

The same working principle can be applied to various low molecular weight endogenous ligands and cell-penetrating peptides for tumor pH targeting.

4 Stimuli-Sensitive Gene Delivery

The delivery of genetic materials such as plasmid DNA, short-interfering RNA (siRNA), and oligonucleotides (ODN) to target cells/tissues for repairing/replacing defective genes, modulating irregular gene expression, or imparting new functions is a promising strategy. The potential targets in clinical applications are very broad including cancers, vascular diseases, liver diseases, muscles diseases, heart diseases, neurodegenerative diseases, infectious diseases, inherited monogenic diseases, and so on. However, unlike chemical drugs, genetic materials are easily degraded by nucleases in the blood and extravascular spaces. This has prompted the use of various gene carriers such as viral vectors, nonviral vectors, or their hybrids. Viral vectors have shown higher transfection efficiency than nonviral systems, although the former has safety concerns such as immunogenicity and wild-type viral production, unlike the latter. These safety issues have prompted the development of effective and safer nonviral gene carriers (see Kang et al., 2005, 2007).

To reach the intracellular target sites of genes (i.e., cytosol and nucleus), nonviral vectors such as polymers, liposomes, and their hybrids face several extracellular and intracellular barriers, leading to decreased gene bioavailability and reduced therapeutic effects at the target sites (i.e., gene expression for plasmid DNA and gene silencing for siRNA and ODN). For example, polymeric gene complexes (polyplexes) interact nonspecifically with serum proteins, erythrocytes, and nontarget cells during blood circulation. These problems have been decreased using hydrophilic corona (i.e., PEG) and ligands targeted to specific cells. However, once within the cells, polyplexes still need to overcome endosomal sequestration, cytosolic transport, nuclear import, and decomplexation as shown in Fig. 15 (see Kang et al., 2005, 2007). Each barrier should be overcome to improve transfection efficiency using nonviral carrier systems. To this end, temperature, pH, and redox potential may add some beneficial aspects to gene delivery carrier design.

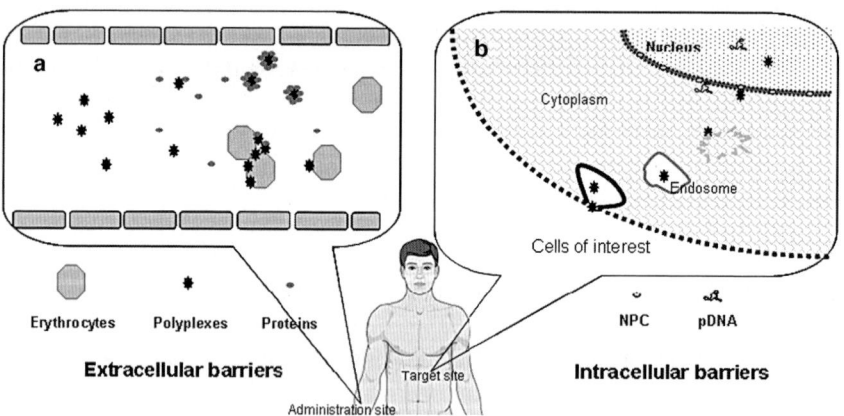

Fig. 15 Extracellular and intracellular transfection barriers. (**a**) Polyplexes face serum components (i.e., serum proteins and erythrocytes) and nontarget cells during blood circulation. Polyplexes may interact nonspecifically with each other, leading to reduced gene bioavailability at the sites of interest. (**b**) After reaching the target cells/tissues, polyplexes need to be internalized into the cells and then moved to the nucleus for gene expression. During the intracellular trafficking process, polyplexes have to overcome the endosomal membrane, cytosolic transport, and the nuclear membrane, and gene release may be aggressively limited (reproduced with permission; Kang et al., 2005)

4.1 Temperature-Modulated Gene Carriers

Thermo-sensitive gene carriers have been used to protect the genetic material cargo from nucleases and to quickly release genes upon reaching the target sites. In the carriers, thermo-sensitive materials trap genes by physical means or by chemical linkages. Unlike the hyperthermia strategy of antitumor therapies, effective gene release at the site of interest has been executed using cooling processes.

4.1.1 Physical Trapping of Genes in Thermosensitive Carriers

To form complexes with genes, thermo-sensitive polymers were conjugated to positively charged polymers. Gene complexes showed decreased positive surface charges due to charge-shielding by the thermo-sensitive polymer, leading to reduced cytotoxicity (see Bisht et al., 2006). The thermo-sensitive gene complexes were prepared above the LCST, whereas gene release from the complexes was induced below the LCST. Okano's group was the first to utilize this "cold-shock" release method using temperatures lower than the LCST (see Kurisawa et al., 2000; Takeda et al., 2004). Takeda et al. (2004) synthesized random copolymers composed of NIPAAm, (dimethylamino)ethylmethacrylate (DMAEMA), and BMA. IP-20D-10B copolymer (NIPAAm:DMAEMA:BMA = 70:20:10 in feed; LCST 28.4°C) complexed with genes showed incubation temperature-dependent transfection. In the experiments, it was intended that the initial temperature of 37°C would cause cellular internalization of the polyplexes, the second lower temperature were applied to cause quick release of the gene in the cell for a short period, and a third incubation temperature allowed for gene expression. Using 37–20–37°C and 37–27–37°C incubation schedules caused about twofold and 1.7-fold higher transfection efficiencies than just 37°C incubation. Also, these incubation processes demonstrated that 37–20–37°C condition showed slightly better transfection efficiency than 37–27–37°C condition. This result may be related to better gene release at 20°C than 27°C or 37°C since the thermo-sensitive components were more hydrophilic below the LCST than above the LCST.

Similarly, Lavigne et al. (2007) synthesized PEI-g-poly(NIPAAm) having an LCST at 34–37°C (polymer name: P3) or 40–44°C (polymer name: P4). Complex formation and transfection experiments were executed at various temperatures. The complexation temperature of 45°C (above the LCST) was more favorable than 25°C (below the LCST). As shown in Fig. 16, the incubation temperature used in the transfection experiments influenced gene expression. P3/gene complexes prepared at 45°C showed that 37–32–37°C incubation schedule induced 1.4-fold higher gene expression than the 37–39–37°C schedule because gene release was easier at 32°C than at 37°C or 39°C.

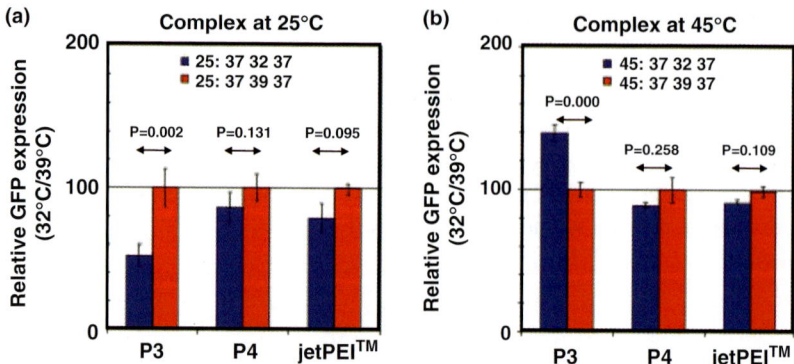

Fig. 16 In vitro GFP expression in C2C12 and COS-7 cells following thermo-controlled transfection using thermoresponsive polymers. Complexes were prepared at 25 or 45°C and the cells were incubated at 37°C for 1 h (to induce cellular uptake), transferred to 32°C (cold shock) or 39°C (heat shock) for 2 h, and then transferred back to 37°C for the rest of the experiment. P3 and P4 were PEI-*g*-poly(NIPAAm) having 34–37°C and 40–44°C as LCST, respectively (reproduced with permission; Lavigne et al., 2007)

4.1.2 Gene Conjugates with Thermo-Sensitive Carriers

Murata et al. (2003a,b) investigated whether antisense ODN conjugated poly(NIPAAm)s improved nuclease resistance. As shown in Fig. 17, the system was prepared by radical copolymerization of NIPAAm and methacryloyl-terminated ODN and showed coil-to-globule transition at about 33°C in a physiological pH buffer. The ODN-poly(NIPAAm) conjugate showed more resistance to nuclease attack at 37°C (above the LCST) than at 27°C (below the LCST) due to the temperature-sensitive hydrophobicity/hydrophilicity of poly(NIPAAm) (Murata et al., 2003a). This nuclease-resistance strategy will be advantageous to improve conventional ODN delivery. However, after cellular internalization of the ODN-poly(NIPAAm) nanoparticles, incubation below the LCST was better than above the LCST since below the LCST the ODN appeared to interact more easily with the target mRNA in the cytosol. This fact was confirmed because the ODN-poly(NIPAAm) conjugate showed 60% gene silencing at 27°C, whereas at 37°C there was no gene silencing (see Murata et al., 2003b). These nanoparticles may need to improve their cell-targetability and endosomal escaping-capability so that they can more effectively target the cells of interest and then the cytoplasm, respectively.

4.2 pH-Modulating Gene Carriers

A major approach in gene delivery is to prevent endosomal/lysosomal sequestration of delivered genes using pH modulation, because lysosomal enzymes degrade the delivered genes and reduce gene expression. To pursue this approach, pH-sensitive polymeric gene carriers target endosomal/lysosomal pHs to cause endosomal

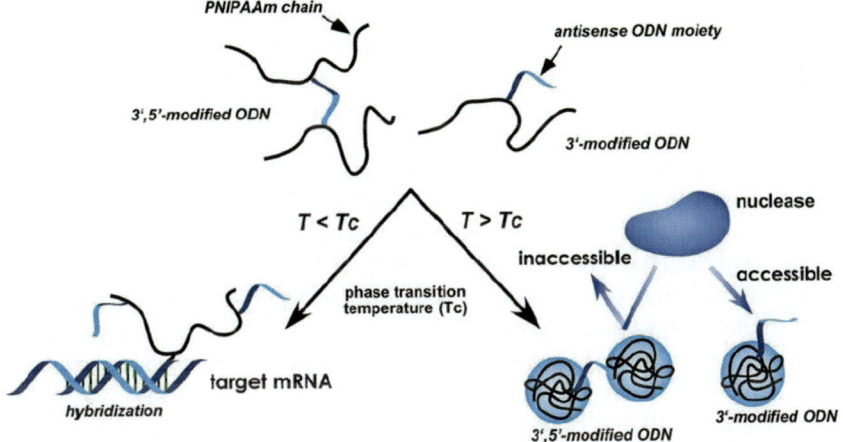

Fig. 17 Thermoresponsive antisense systems composed of oligonucleotides (ODN) and poly(*N*-isopropylacrylamide) (poly(NIPAAm)). Above the transition temperature (T_c = LCST), self-assembled gene complexes protect ODN from nuclease attack, whereas below T_c, hydrophilic poly(NIPAAm) allow ODN to interact with target mRNA (reproduced with permission; Murata et al., 2003a)

destabilization and endosomal pH-induced deshielding. Also, pH-sensitive nano-complexes have been designed for tumor pH-induced deshielding in extratumoral spaces to maximize tumor-specific gene expression.

4.2.1 Endosomal Membrane Destabilization

Proton buffering and/or the fusogenic character (or conformation transition, hydrophilicity/hydrophobicity transition) of the carrier materials can induce endosomal disruption, aiding the endosomal escape of the genes to be delivered. These materials have been chemically or physically introduced into the polymeric gene carriers depending on their charge character. Amine-containing materials directly interact with genetic materials due to their charge character, whereas negatively charged materials (carboxylic acid and sulfonamides) at neutral pH are physically mixed with genes and polycations.

Amine-containing polycations such as PEI, polyamidoamine (PAMAM), poly-His, and their derivatives have proton-buffering capacity due to secondary or tertiary amines or imines (see Kang et al., 2005, 2007). Generally PEI-based or PAMAM-based polyplexes showed higher transfection efficiency than PLL-based polyplexes due to proton-buffering capacity of PEI and PAMAM, unlike PLL (see Sonawane et al., 2003). In addition, histidine has been frequently utilized as well, due to its p*K*a of 6 of its imidazole ring. This amino acid was introduced into cationic polymers via several methods (i.e., homopolymer, graft polymer, block copolymer, and random copolymer) to endow proton buffering to the system, for effective endosomal escape of the delivered gene and/or polyplexes (see Kang et al., 2005, 2007; Midoux and

Mosigny, 1999; Pichon et al., 2001). As an example, Midoux and Mosigny (1999) synthesized histidine-grafted PLL (His-PLL), which showed higher transfection efficiency than unmodified PLL. Optimum transfection results were obtained when histidine was substituted for 38% of the amino groups in PLL.

Carboxylic acid-containing polyanions such as poly(ethylacrylic acid) (PEAA) (see Jones et al., 2003), poly(propylacrylic acid) (PPAA) (see Jones et al., 2003), and methacrylic acid copolymers (see Yessine et al., 2003) were used as additives when preparing either polyplexes or liposomal gene complexes (lipoplexes). These polymers had a random-coil structure at neutral pH, but in the weak acidic pH range (5–7) their conformation transitioned to a hydrophobic helical structure, resulting in endosomal rupture (see Yessine and Leroux, 2004). This membrane-destabiliz-ing ability meant that PPAA-containing chitosan/DNA complexes enhanced trans-fection efficiency twofold and tenfold compared to chitosan/DNA complexes (control) in 293 cells and HeLa cells, respectively (see Kiang et al., 2004).

However, some polycations showed no noticeable transfection enhancement due to their inability to cause endosomal breakage, even though the materials had a pro-ton-buffering capacity at pH 5–7 (see Funhoff et al., 2004). Also, PPAA, a representa-tive protonable polyanion, demonstrated hemolytic activity at endosomal/lysosomal pH as well as at pH 7.4 (see Jones et al., 2003). Yessine et al. (2003) found that the endosomolytic ability of methacrylic acid copolymers was affected by endosomal membrane composition. Furthermore, endosomal characteristics such as endosomal pH (see Rybak and Murphy, 1998; Rybak et al., 1997), membrane composition (see Alberts et al., 2002; Evans and Hardison, 1985), and acidification rate are cell-spe-cific. These facts have stimulated the development of cell-customized endosomolytic agents for more effective cell transfection. Bae's group designed sulfonamide-based endosomolytic oligomers using agents from sulfonamide pool having broad-range pKa values (3–11) conferred by various substituted groups, R (see Fig. 4) (see Kang and Bae, 2007). Kang and Bae (2007) focused on selectable pKa and hydrophobicity from the molecular pool of sulfonamides and expected cell-specific endosomolytic agents to induce the highest transfection efficiency. For feasibility, the pKa values of four sulfonamides were selected within endosomal/lysosomal pHs (sulfamethizole (SMT; pKa 5.45), sulfadimethoxine (SDM; pKa 6.1), sulfadiazine (SDZ; pKa 6.4), and sulfamerazine (SMZ; pKa 7.0)) and their oligomers (OSMT, OSDM, OSDZ, and OSMZ) had M_n = 1.8–2.5 kDa. Synthesized oligomeric sulfonamides (OSAs) showed different proton buffering and aqueous solubility transition within the endosomal/ lysosomal pH ranges. As shown in Fig. 18a, OSMT and OSDZ displayed broad pro-ton buffering in the range of pH 5.0–6.4 and 5.7–7.3, respectively, whereas proton buffering of OSDM and OSMZ occurred at a particular pH, 6.5 and 7.3, respectively. Also, in solubility transition studies of the OSAs (see Fig 18b), OSMT, OSDM, and OSDZ showed relatively sharp transmittance changes within narrow pH ranges unlike OSMZ which showed solubility transitions within a broad pH range. Apparent pKa values of the OSAs were 5.7 (OSMT), 6.5 (OSDM and OSDZ), and 7.3 (OSMZ). Using three different cell lines (i.e., HepG2, HEK293, and RINm5F cells), OSA-containing PLL/DNA complexes (OSA-polyplexes) showed 4–55-fold better transfection efficiency than control polyplexes (PLL/DNA). Interestingly, OSDM and

OSDZ were more favorable for transfecting HEK293 cells (human embryonic kidney cells) whereas OSMZ was the best for transfecting RINm5F cells (rat insulinoma cells). This study verified the need for cell-customized endosomolytic agents to achieve clinically effective gene delivery. In addition, these anionic materials could provide an opportunity to revive biocompatible but less transfection-efficient gene carriers.

4.2.2 Tumor pH-Induced Deshielding

This strategy is similar to the acid-induced deshielding technologies for delivering antitumor drugs. The approach excludes nonspecific cellular internalization of cationic polyplexes to minimize gene expression in nontarget cells, while inducing

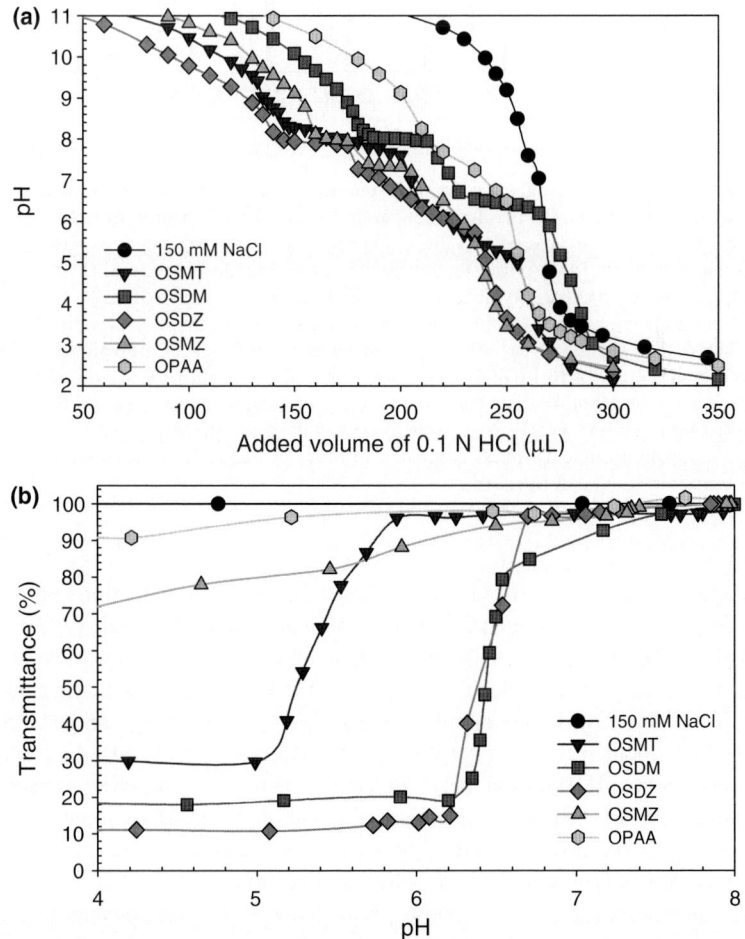

Fig. 18 (continued)

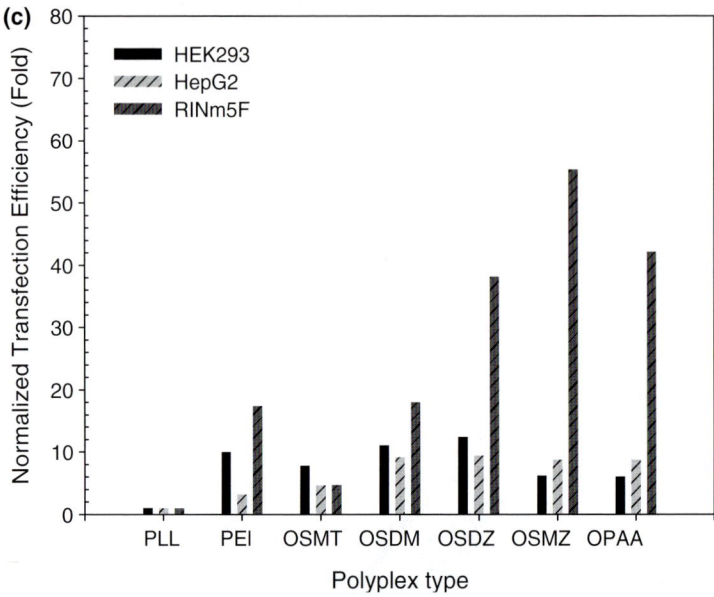

Fig. 18 (continued) (**a**) Acid-base titration curves of OSA and OPAA (oligomeric pro-pylacrylic acid) solutions. (**b**) pH-dependent aqueous solubility transition of OSA and OPAA solutions. For (**a**) and (**b**), means of values obtained from two independent experiments were used as data points. (**c**) In vitro transfection studies using a luciferase gene of OSA-containing PLL/DNA complexes (OSA-polyplexes) to HepG2, HEK293, and RINm5F cells. Dose of OSA and OPPA was 5 nmol (based on their monomeric units) per 1 µg DNA. Normalized transfection efficiency was defined as (Absolute transfection efficiency of polyplexes)/(Absolute transfection efficiency of PLL/DNA complexes) for a specific cell. Unit of absolute transfection efficiency is RLU/mg Protein. Charge ratio (+/−) of polyplexes was 3 except for PEI/DNA (+/− = 5). (Mean ± SEM; $n \geq 4$). All results of OSA-polyplexes showed statistical significance ($p < 0.05$; unpaired Student t-tests) compared to those of PLL/DNA complexes (reproduced with permission; Kang and Bae, 2007)

tumor-specific endocytosis of the polyplexes to maximize transfection at the target cells. To illustrate this concept, Sethuraman et al. (2006) designed PEG-*b*-PSD-shielded PEI/DNA complexes via electrostatic attraction. As shown in Fig. 19, the polyplexes had limited nonspecific interactions because of PEG shielding at normal blood pH and during blood circulation because the negatively charged PSD in PEG-*b*-PSD shielded the positively charged surface of the PEI/DNA complexes. However, acidic pH (tumoral pH) induced a charge loss in PSD, causing the deshielding PEG-*b*-PSD and exposing the cationic PEI/DNA complex. This approach decreased positive charge-induced cytotoxicity at pH 7.4. Interestingly, the deshielded polyplexes showed better transfection efficiency at pH 6.6 than shielded polyplexes at pH 7.4. These pH-dependent shielding/deshielding smart polyplexes may minimize unwanted gene expression in nontumor cells and maximize transfection efficiency at target tumors.

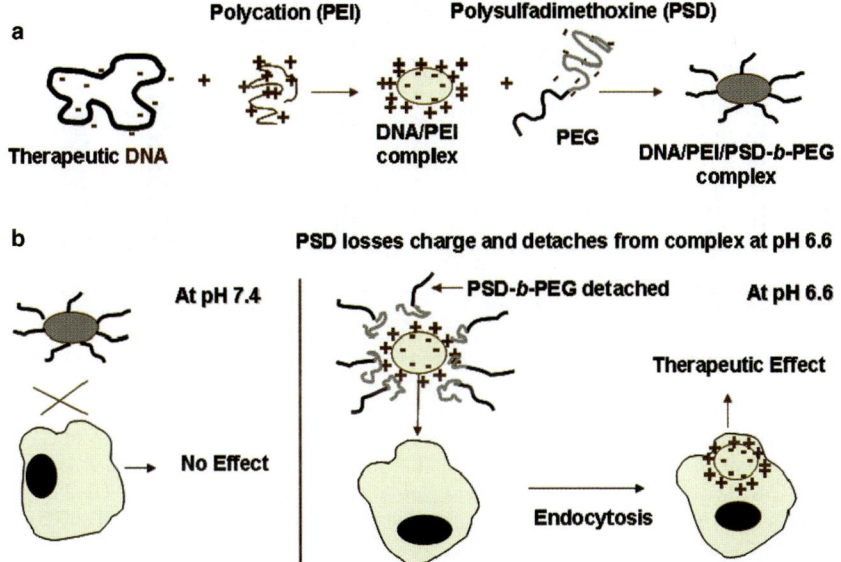

Fig. 19 Targeting based on differences in pH: (**a**) shows formation of the nanoparticle complex through charge–charge interaction between DNA, cationic polymer (PEI), and PSD-*b*-PEG; (**b**) shows the complex shielded at physiological pH and deshielded at cancer pH (reproduced with permission; Sethuraman et al., 2006)

4.2.3 Endosomal pH-Induced Cleavage of Chemical Bonds

As discussed before, chemical bonds such as acetal bonds, hydrazone bonds, Schiff's base bonds, and vinyl ether bonds (see Fig. 5) are easily hydrolyzed at endosomal/lysosomal pHs. These bonds have been introduced into polyplexes and/or polymer backbones for detaching the PEG corona or to enhance biocompatibility.

It is well known that adding a PEG corona to polyplexes is beneficial for increasing blood circulation by inhibiting nonspecific interactions. However, studies by Walker et al. (2005) showed that after cellular internalization, PEGylated polyplexes can negatively impact intracellular trafficking due to the presence of PEG. In these studies, PLL-*hydrazone*-PEG showed 90% hydrolysis within 10 min at pH 5 and 37°C, but at pH 7.4 50% hydrolysis was seen after about 1.5 h. These acid-induced hydrolytic linkages were utilized to prepare reversibly shielded polyplexes. The polyplexes were compared to nonshielded polyplexes and stably shielded polyplexes to examine the effect of PEG in endosomes/lysosomes. In in vitro experiments using cell lines (i.e., K562, Neuro2A, HUH-7, and EGFR-Renca cells), PEGylated polyplexes having hydrazone linkages showed up to two orders higher gene expression than PEGylated polyplexes with nondegradable linkages as shown in Fig. 20. In addition, in vivo studies confirmed that detachable PEGylated polyplexes showed better gene expression in the tumors than nondetachable PEGylated polyplexes. These studies indicated that different functionalities used to

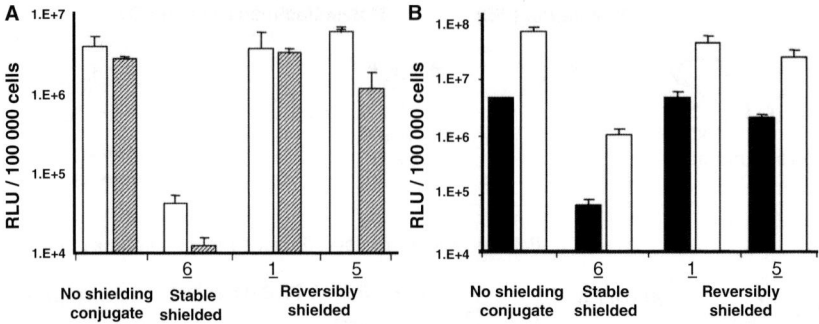

Fig. 20 Gene transfection efficiency of targeted polyplexes with stable or acid-labile conjugates with shielding and without shielding. (**a**) Transfection of K562 (*white bars*) and Neuro2A (*shaded bars*) cells using transferrin (Tf) receptor-targeted polyplexes containing Tf–PEI conjugate (at *N/P* 0.6) with unmodified PLL (no shielding conjugate), PLL–PEG (**6**), or PLL–*hydrazone*–PEG conjugates (**1**, **5**) at *N/P* = 0.2 and PEI (*N/P* = 4.8). (**b**) Transfection of HUH-7 cells (*black bars*) and EGFR-Renca cells (*white bars*) using EGF receptor-targeted polyplexes containing EGF–PEI plus other conjugates at ratios as indicated in (**a**) (reproduced with permission; Walker et al., 2005)

overcome a specific transfection barrier in the transfection process could interfere with each other. Functionalities which have played their roles should dissociate or be detached before the complexes proceed further through downstream extracellular and intracellular trafficking processes (see Kang et al., 2007; Wagner, 2004).

Generally, high molecular weight (MW) polycations are more cytotoxic than low MW polycations (see Kunath et al., 2003). However, low MW polymers loosely complexed with genes, and their complexes showed less transfection efficiency than the gene complexes prepared with high MW polycations. Thus, to improve transfection efficiency and to reduce cytotoxicity caused by polycations, acid-labile bonds were used to link low (MW) cationic blocks together to produce high MW polycations. As an example, Kim et al. (2005b) synthesized cross-linked polymers about 7–23 kDa using branched PEI (1.8 kDa) and Schiff's base bonds. The acid-labile bonds in the polymer had a half-life of 1.1 h at pH 4.5 but 118 h at pH 7.4. Degradable cross-linked PEI (23 kDa) had similar cytotoxicity to branched PEI (1.8 kDa) but showed better cell viability than branched PEI (25 kDa). Furthermore, degradable PEI (23 kDa)/gene complexes showed slight lower luciferase expression than branched PEI (25 kDa)/gene complexes and induced much higher transfection than branched PEI (1.8 kDa)/gene complexes. However, this system may need to improve its transfection efficiency before it can be considered for clinical applications.

4.3   Redox Potential-Triggered Gene Release

To express proteins from delivered genes, polyplexes need to release the genes at appropriate sites such as in the cytosol and/or in the nucleus. The cytosol and nucleus prefer the reduced form of thiol containing components to the oxidized form (disulfide bond). This suggests that thiol-triggered degradable polyplexes have the potential to enhance polymeric transfection by quick DNA release (or decomplexation). Many research groups have synthesized reducible polymeric gene carriers (see Balakirev et al., 2000; Blessing et al., 1998; Dauty et al., 2001; Manickam and Oupicky, 2006; Manickam et al., 2005; Miyata et al., 2004; Oupicky et al., 2001, 2002; Read et al., 2005). As shown in Fig. 21, thiol-triggered degradable bonds (disulfide bonds) were included in or grafted into the polymeric backbone and were disrupted when facing thiol-rich environments, leading to effective DNA release. For instance, peptide-based thiol-triggered degradable polymers were prepared by the polymerization of Cys-X$_n$-Cys (see Manickam and Oupicky, 2006; Oupicky et al., 2002; Read et al., 2005) or Cys-Y-Cys (see Manickam and Oupicky, 2006; Manickam et al., 2005), where X was lysine, histidine, or a mixture, n was 9–15, and Y was TAT or NLS. Overall transfection efficacies were MW and composition

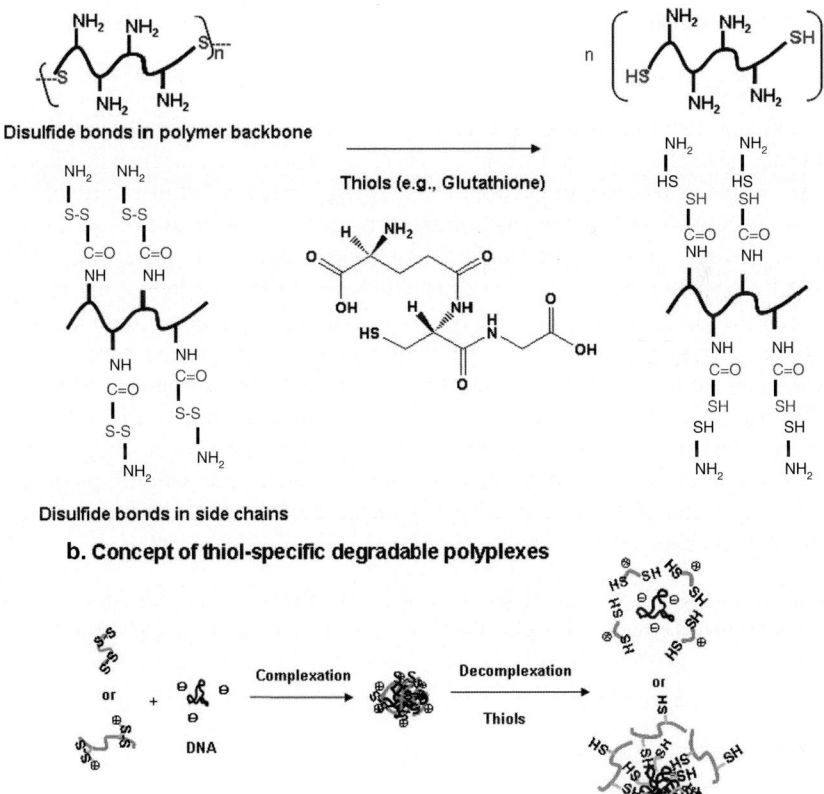

Fig. 21 Thiol-specific degradable gene carriers (reproduced with permission; Kang et al., 2007)

dependent. These polymeric gene carriers have shown controversial results in transfection efficiency. That is, poly(Cys-Lys$_{10}$-Cys) (187 kDa) had higher transfection efficiency (more than 20 times) than PLL (205 kDa) (see Oupicky et al., 2002), whereas poly(Cys-Lys$_{10}$-Cys) (65 kDa) showed less GFP expression than PLL (56 kDa) (see Read et al., 2005). The reasons for this discrepancy are unclear although it appears to be related to the rate of decomplexation (see Schaffer et al., 2000), which leads to differences in the proximity of the DNA to the nucleus. Probably, poly(Cys-Lys$_{10}$-Cys) (65 kDa) released DNA earlier than poly(Cys-Lys$_{10}$-Cys) (187 kDa), meaning the DNA was further from the nucleus. Reducible polymers prepared from arginine-containing peptides such as TAT and NLS showed less transfection efficiency than branched PEI (25 kDa) due to the lower endosomolytic activity of TAT (see Manickam et al., 2005) and NLS (see Manickam and Oupicky, 2006). This problem was overcome by introducing histidine into the peptide backbone (see Manickam and Oupicky, 2006; Read et al., 2005).

5   Conclusion

Many investigators in drug/gene delivery have pursued effective strategies for maximizing the therapeutic effects of delivered drug/gene while minimizing unwanted side effects. Therefore, pharmaceutical nanocarriers that can intelligently sense minute differences in environmental changes (such as temperature, pH, and redox potentials) and effectively release drug/gene at target sites such as solid tumors, tissues, cells, or intracellular compartments have been aggressively investigated. Depending on the target sites and disease, stimuli-sensitive drug/gene carriers need to be designed that undergo transitions in response to stimuli. In most research approaches, tailor-made materials (especially, polymers) have been utilized to create nanosystems that meet these needs. Thus, this chapter introduced polymeric nanocarriers that could sense heating/cooling, tumoral pH, endosomal pH, and intracellular redox potentials. Depending on the therapeutic drugs used, the genes of interest, and the target diseases, well-organized multistimuli-modulating nanopharmaceuticals will be beneficial to achieve effective therapeutic results in clinical applications in the future.

Acknowledgments This work was partially supported by NIH grants (CA 101850, CA 122356, and DK 56884). The authors appreciate the critical reading of the manuscript by Deepa Mishra.

References

Alberts, B., Johnson, A., Lewis, J., Raff, M., Roberts, K., and Walter, P. 2002. Molecular Biology of the Cell. 4th Edition. New York: Garland Science.
Alexiou, C., Jurgons, R., Seliger, C., and Iro, H. 2006. Medical applications of magnetic nanoparticles. J. Nanosci. Nanotechnol. 6:2762–2768.

Bae, Y., Fukyshima, S., Harada, A., and Kataoka, K. 2003. Design of environment-sensitive supramolecular assemblies for intracellular drug delivery: polymeric micelles that are responsive to intracellular pH change. Angew. Chem. Int. Ed. Engl. 42:4640–4643.

Bae, Y., Jang, W. D., Nishiyama, N., Fukushima, S., and Kataoka, K. 2005a. Multifunctional polymeric micelles with folate-mediated cancer cell targeting and pH-triggered drug releasing properties for active intracellular drug delivery. Mol. BioSyst. 1:242–250.

Bae, Y., Nishiyama, N., Fukushima, S., Koyama, H., Yasuhiro, M., and Kataoka, K. 2005b. Preparation and biological characterization of polymeric micelle drug carriers with intracellular pH-triggered drug release property: tumor permeability, controlled subcellular drug distribution, and enhanced in vivo antitumor efficacy. Bioconjug. Chem. 16:122–130.

Balakirev, M., Schoehn, G., and Chroboczek, J. 2000. Lipoic acid-derived amphiphiles for redox-controlled DNA delivery. Chem. Biol. 7:813–819.

Bisht, H. S., Manickam, D. S., You, Y., and Oupicky, D. 2006. Temperature-controlled properties of DNA complexes with poly(ethyleneimine)-*graft*-poly(N-isopropylacrylamide). Biomacromolecules 7:1169–1178.

Blessing, T., Remy, J. S., and Behr, J. P. 1998. Template oligomerization of DNA-bound cations produces calibrated nanometric particles J. Am. Chem. Soc. 120:8519–8520.

Boesch, D., Gaveriaux, C., Jachez, B., Pourtier-Manzanedo, A., Bollinger, P., and Loor, F. 1991. In vivo circumvention of P-glycoprotein-mediated multidrug resistance of tumor cells with SDZ PSC 833. Cancer Res. 51:4226–4233.

Boussif, O., Lezoualc'h, F., Zanta, M. A., Mergny, M. D., Scherman, D., Demeneix, B., and Behr, J. P. 1995. A versatile vector for gene and oligonucleotides transfer into cells in culture and in vivo: polyethyleneimine. Proc. Natl. Acad. Sci. USA. 92:7297–7301.

Brannon-Peppas, L. and Blanchette, J. O. 2004. Nanoparticle and targeted systems for cancer therapy. Adv. Drug Deliv. Rev. 56:1649–1659.

Cammas, S., Suzuki, K., Sone, C., Sakuri, Y., Kataoka, K., and Okano, T. 1997. Thermo-responsive polymer nanoparticles with a core-shell micelle structure as site-specific drug carriers. J. Control. Release 48:157–164.

Cataldo, A. M., Petanceska, S., Peterhoff, C. M., Terio, N. B., Epstein, C. J., Villar, A., Carlson, E. J., Staufenbiel, M., and Nixon, R. A. 2003. App gene dosage modulates endosomal abnormalities of Alzheimer's disease in a segmental trisomy 16 mouse model of Down syndrome. J. Neurosci. 23:6788–6792.

Cho, Y. W., Kim, J. D., and Park, K. 2003. Polycation gene delivery systems: escape from endosomes to cytosol. J. Pharm. Pharmacol. 55:721–734.

Chung, J. E., Yokoyama, M., Aoyagi, Y., Sakurai, Y., and Okano, T. 1998. Effect of molecular architecture of hydrophobically modified poly(N-isopropylacrylamide) on the formation of thermoresponsive core-shell micellar drug carriers. J. Control. Release 53:119–130.

Chung, J. E., Yokoyama, M., Yamato, T., Aoyagi, Y., Sakurai, Y., and Okano, T. 1999. Thermo-responsive drug delivery from polymeric micelles constructed using block copolymers of poly(N-isopropylacrylamide) and poly(butylmethacrylate). J. Control. Release 62:115–127.

Chung, J. E., Yokoyama, M., and Okano, T. 2000. Inner core segment design for drug delivery control of thermo-responsive polymeric micelles. J. Control. Release 65:93–103.

Dauty, E., Remy, J. S., Blessing, T., and Behr, J. P. 2001. Dimerizable cationic detergents with a low cmc condense plasmid DNA into nanometric particles and transfect cells in culture. J. Am. Chem. Soc. 123:9227–9234.

Dewhirst, M. W. 1995. Thermal dosimetry. In Principles and Practice of Thermoradiotherapy and Thermochemotherapy, ed. M. H. Seegenschmiedt, P. Fessenden, and C. C. Vernon, pp. 123–136. Berlin: Springer.

Diessemond, J., Witthoff, M., Brauns, T. C., Harberer, D., and Gros, M. 2003. pH Values on chronic wounds: evaluation during modern wound therapy. Hautarzt 54:959–965.

Dreher, M. R., Raucher, D., Balu, N., Colvin, O. M., Ludeman, S. M., and Chilkoti, A. 2003. Evaluation of an elastin-like polypeptide-doxorubicin conjugate for cancer therapy. J. Control. Release 91:31–43.

Drummond, D. C., Zignani, M., and Leroux, J. C. 2000. Current status of pH-sensitive liposomes in drug delivery. Prog. Lipid Res. 39:409–460.

Engin, K., Leeper, D. B., Cater, J. R., Thistlethwaite, A. J. Tupchong, L., and McFarlane, J. D. 1995. Extracellular pH distribution in human tumors. Int. J. Hyperthermia 11:211–216.

Eum, K. M., Langley, K. H., and Tirrell, D. A. 1989. Quasi-elastic and electrophoretic light-scattering studies of the reorganization of dioleoylphosphatidylcholine vesicle membrane by poly(2-ethylacrylic acid). Macromolecules 22:2755–2760.

Evans, W. H. and Hardison, W. G. 1985. Phospholipid, cholesterol, polypeptide and glycoprotein composition of hepatic endosome subfractions. Biochem. J. 232:33–36.

Funhoff, A. M., van Nostrum, C. F., Koning, G. A., Schuurmans-Nieuwenbroek, N. M., Crommelin, D. J., and Hennink, W. E. 2004. Endosomal escape of polymeric gene delivery complexes is not always enhanced by polymers buffering at low pH. Biomacromolecules 5:32–39.

Furgeson, D. Y., Dreher, M. R., and Chilkoti, A. 2006. Structural optimization of a "smart" doxo-rubicin-polypeptide conjugate for thermally targeted delivery to solid tumors. J. Control. Release 110:362–369.

Gardner, S. N. 2000. A mechanistic, predictive model of dose-response curves for cell cycle phase-specific and -nonspecific drugs. Cancer Res. 60:1417–1425.

Gaucher, G., Dufresne, M. H., Sant, V. P., Kang, N., Maysinger, D., and Leroux, J. C. 2005. Block copolymer micelles: preparation, characterization and application in drug delivery. J. Control. Release 109:169–188.

Gerasimov, O. V., Boomer, J. A., Qualls, M. M., and Thompson, D. H. 1999. Cytosolic drug delivery using pH- and light-sensitive liposomes. Adv. Drug Deliv. Rev. 38:317–338.

Gerlowski, L. E. and Jain, R. K. 1985. Effect of hyperthermia on microvascular permeability to macromolecules in normal and tumor tissues. Int. J. Microcirc. Clin. Exp. 4:363–372.

Gillies, E. R. and Fréchet, J. M. 2005. pH-Responsive copolymer assemblies for controlled release of doxorubicin. Bioconjug. Chem. 16:361–368.

Gillies, E. R., Goodwin, A. P., and Fréchet, M. J. 2004a. Acetals as pH-sensitve linkages for drug delivery. Bioconjug. Chem. 15:1254–1263.

Gillies, E. R., Jonsson, T. B., and Fréchet, M. J. 2004b. Stimuli-responsive supramolecular assemblies of linear-dendritic copolymers. J. Am. Chem. Soc. 126:11936–11943.

Gottesman, M. M., Pastan, I., and Ambudkar, S. V. 1996. P-glycoprotein and multi drug resistance. Curr. Opin. Genet. Dev. 6:610–617.

Hager, A., Debus, G., Edel, H. G., Stransky, H., and Serrano, R. 1991. Auxin induces exocytosis and the rapid synthesis of a high-turnover pool of plasma-membrane H^+ ATPase. Planta 185:527–537.

Hansen, J. M., Go, Y. M., and Jones, D. P. 2006. Nuclear and mitochondrial compartmentation of oxidative stress and redox signaling. Annu. Rev. Pharmacol. Toxicol. 46:215–234.

Heskins, M. and Guillet, J. E. 1968. Solution properties of poly(N-isopropylacrylamide). J. Macromol. Sci. Chem. A. 2:1441–1455.

Hobbs, S. K., Monsky, W. L., Yuan, F., Roberts, W. G., Griffith, L., Torchilin, V. P., and Jain, R. K., 1998. Regulation of transport pathways in tumor vessels: role of tumor type and microenvironment. Proc. Natl. Acad. Sci. USA. 95:4607–4612.

Hooijberg, J. H., Peters, G. J., Assaraf, Y. G., Kathmann, I., Priest, D. G., Bunni, M. A., Veerman, A. J., Scheffer, G. L., Kaspers, G. J., and Jansen, G. 2003.The role of multidrug resistance proteins MRP1, MRP2 and MRP3 in cellular folate homeostasis. Biochem. Pharmacol. 65:765–771.

Hruby, M., Konak, C., and Ulbrich, K. 2005. Polymeric micellar pH-sensitive drug delivery system for doxorubicin. J. Control. Release 103:137–148.

Jain, R. K. 1987. Transport of molecules across tumor vasculature. Cancer Metastasis Rev. 6:559–593.

Jeong, B., Bae, Y. H., and Kim, S. W. 1999. Biodegradable thermosensitive micelles of PEG–PLGA–PEG triblock copolymers. Colloids Surf. B. Biointerfaces 16:185–193.

Jones, R. A., Cheung, C. Y., Black, F. E., Zia, J. K., Stayton, P. S., Hoffman, A. S., and Wilson., M. R. 2003. Poly(2-alkylacrylic acid) polymers deliver molecules to the cytosol by pH-sensitive disruption of endosomal vesicles. Biochem. J. 372:65–75.

Kaneko, T., Willner, D., Monkovic, I., Knipe, J. O., Braslawsky, G. R., Greenfield, R. S., and Vyas, D. M. 1991. New hydrazone derivatives of adriamycin and their immunoconjugates: a correlation between acid stability and cytotoxicity. Bioconjug. Chem. 2:133–141.

Kang, H. C. and Bae, Y. H. 2007. pH-Tunable endosomolytic oligomers for enhanced nucleic acid delivery. Adv. Funct. Mater. 17:1263–1272.

Kang, H. C., Lee, M., and Bae, Y. H. 2005. Polymeric gene carriers. Crit. Rev. Eukaryot. Gene Expr. 15:317–342.

Kang, H. C., Lee, M., and Bae, Y. H. 2007. Polymeric gene delivery vectors. In Nanotechnology in Therapeutics: Current Technology and Application, ed. N. A. Peppas, J. Z. Hilt, and J. B. Thomas, pp. 131–161. Wymondham: Horizon Bioscience.

Kiang, T., Bright, C., Cheung, C. Y., Stayton, P. S., Hoffman, A. S., and Leong, K. W. 2004. Formulation of chitosan-DNA nanoparticles with poly(propyl acrylic acid) enhances gene expression. J. Biomater. Sci. Polym. Ed. 15:1405–1421.

Kikuchi, A. and Okano, T. 2002. Intelligent thermoresponsive polymeric stationary phases for aqueous chromatography of biological compounds. Prog. Polym. Sci. 27:1165–1193.

Kim, G. M., Bae, Y. H., and Jo, W. H. 2005a. pH-Induced micelle formation of poly(histidine-cophenylalanine)-block-poly(ethylene glycol) in aqueous media. Macromol. Biosci. 5:1118–1124.

Kim, Y. H., Park, J. H., Lee, M., Kim, Y. H., Park, T. G., and Kim, S. W. 2005b. Polyethylenimine with acid-labile linkages as a biodegradable gene carrier. J. Control. Release 103:209–219.

Kong, G. H. and Dewhirst, M. W. 1999. Hyperthermia and liposomes. Int. J. Hyperthermia 15:345–370.

Kunath, K., von Harpe, A., Fischer, D., Petersen, H., Bickel, U., Voigt, K., and Kissel, T. 2003. Low-molecular-weight polyethylenimine as a non-viral vector for DNA delivery: comparison of physicochemical properties, transfection efficiency and in vivo distribution with high-molecular-weight polyethylenimine. J. Control. Release 89:113–125.

Kurisawa, M., Yokoyama, M., and Okano, T. 2000. Gene expression control by temperature with thermo-responsive polymeric gene carriers. J. Control. Release 69:127–137.

Lavigne, M. D., Pennadam, S. S., Ellis, J., Yates, L. L., Alexander, C., and Górecki, D. C. 2007. Enhanced gene expression through temperature profile-induced variations in molecular architecture of thermoresponsive polymer vectors. J. Gene Med. 9:44–54.

Lazzarino, D. A., Blier, P., and Mellman, I. 1998. The monomeric guanosine triphosphatase rab4 controls an essential step on the pathway of receptor-mediated antigen processing in B cells. J. Exp. Med. 188:1769–1774.

Lee, E. S., Shin, H. J., Na, K., and Bae, Y. H. 2003a. Poly(L-histidine)-PEG block copolymer micelles and pH-induced destabilization. J. Control. Release 90:363–374.

Lee, E. S., Na, K., and Bae, Y. H. 2003b. Polymeric micelle for tumor pH and folate-mediated targeting. J. Control. Release 91:103–113.

Lee, E. S., Na, K., and Bae, Y. H. 2005a. Doxorubicin loaded pH-sensitive polymeric micelles for reversal of resistant MCF-7 tumor. J. Control. Release 103:405–418.

Lee, E. S., Na, K., and Bae, Y. H. 2005b. Super pH-sensitive multifunctional polymeric micelle. Nano Lett. 5:325–329.

Leeper, D. B., Engin, K., Thistlethwaite, A. J., Hitchon, H. D., Dover, J. D., Li, D. J., and Tupchong, L. 1994. Human tumor extracellular pH as a function of blood glucose concentration. Int. J. Radiat. Oncol. Biol. Phys. 28:935–943.

Leroux, J., Roux, E., Le Garrec, D., Hong, K., and Drummond, D. C. 2001. N-isopropylacrylamide copolymers for the preparation of pH-sensitive liposomes and polymeric micelles. J. Control. Release 72:71–84.

Licciardi, M., Giammona, G., Du, J., Armes, S. P., Tang, Y., and Lewis, A. L. 2006. New folate-functionalized biocompatible block copolymer micelles as potential anti-cancer drug delivery systems. Polymer 47:2946–2955.

Liu, S. Q., Tong, Y. W., and Yang, Y. Y. 2005. Incorporation and in vitro release of doxorubicin in thermally sensitive micelles made from poly(N-isopropylacrylamide-co-N,N-dimethylacry-lamide)-b-poly(D,L-lactide-co-glycolide) with varying compositions. Biomaterials 26:5064–5074.

Liu, S. Q., Wiradharma, N., Gao, S. J., Tong, Y. W., and Yang, Y. Y. 2007. Bio-functional micelles self-assembled from a folate-conjugated block copolymer for targeted intracellular delivery of anticancer drugs. Biomaterials 28:1423–1433.

Makhaeva, E. E., Tenhu, H., and Khokhlov, A. R. 1998. Conformational changes of poly(vinylcaprolactam) macromolecules and their complexes with ionic surfactants in aqueous solution. Macromolecules 31:6112–6118.

Manickam, D. S. and Oupicky, D. 2006. Multiblock reducible copolypeptides containing histidine-rich and nuclear localization sequences for gene delivery. Bioconjug. Chem. 17:1395–1403.

Manickam, D. S., Bisht, H. S., Wan, L., Mao, G., and Oupicky, D. 2005. Influence of TAT-peptide polymerization on properties and transfection activity of TAT/DNA polyplexes. J. Control. Release 102:293–306.

Matsumura, Y. and Maeda, H. 1986. A new concept for macromolecular therapeutics in cancer-chemotherapy—mechanism of tumoritropic accumulation of proteins and the antitumor agent SMANCS. Cancer Res. 46:6387–6392.

Meyer, D. E., Shin, B. C., Kong, G. A., Dewhirst, M. W., and Chilkoti, A. 2001. Drug targeting using thermally responsive polymers and local hyperthermia. J. Control. Release 74:213–224.

Midoux, P. and Monsigny, M. 1999. Efficient gene transfer by histidylated polylysine/pDNA complexes. Bioconjug. Chem. 10:406–411.

Miyata, K., Kakizawa, Y., Nishiyama, N., Harada, A., Yamasaki, Y., Koyama, H., and Kataoka, K. 2004. Block catiomer polyplexes with regulated densities of charge and disulfide cross-linking directed to enhance gene expression. J. Am. Chem. Soc. 126:2355–2361.

Mohajer, G., Lee, E. S., and Bae, Y. H. 2007. Enhanced intercellular retention activity of novel pH-sensitive polymeric micelles in wild and multidrug resistant MCF-7 Cells. Pharm. Res. 24:1618–1627.

Murthy, N., Campbell, J., Fausto, N., Hoffman, A. S., and Stayton, P. S. 2003. Bioinspired pH-responsive polymers for the intracellular delivery of biomolecular drugs. Bioconjug. Chem. 14:412–419.

Murata, M., Kaku, W., Anada, T., Sato, Y., Kano, T., Maeda, M., and Katayama, Y. 2003a. Novel DNA/polymer conjugate for intelligent antisense reagent with improved nuclease resistance. Bioorg. Med. Chem. Lett. 13:3967–3970.

Murata, M., Kaku, W., Anada, T., Sato, Y., Maeda, M., and Katayama, Y. 2003b. Temperature-dependent regulation of antisense activity using a DNA/poly(N-isopropylacrylamide) conjugate. Chem. Lett. 32:986–987.

Na, K. and Bae, Y. H. 2005. pH-Sensitive polymers for drug delivery. In Polymeric Drug Delivery Systems, ed. G. S. Kwon, pp. 129–194. Boca Raton: Taylor & Francis Group.

Na, K., Lee, E. S., and Bae, Y. H. 2003. Adriamycin loaded pullulan acetate/sulfonamide conjugate nanoparticles responding to tumor pH: pH-dependent cell interaction, internalization and cytotoxicity in vitro. J. Control. Release 87:3–13.

Na, K., Lee, K. H., and Bae, Y. H. 2004. pH-Sensitivity and pH-dependent interior structural change of self-assembled hydrogel nanoparticles of pullulan acetate/oligo-sulfonamide conjugate. J. Control. Release 97:513–525.

Na, K., Lee, K. H., Lee, D. H., and Bae, Y. H. 2006. Biodegradable thermo-sensitive nanoparticles from poly(L-lactic acid)/poly(ethylene glycol) alternating multi-block copolymer for potential anti-cancer drug carrier. Eur. J. Pharm. Sci. 27:115–122.

Ojugo, A. S. E., Mcsheehy, P. M. J., Mcintyre, D. J. O., Mccoy, C., Stubbs, M., Leach, M. O., Judson, I. R., and Griffiths, J. R. 1999. Measurement of the extraceullar pH of solid tumours in mice by magnetic resonance spectroscopy: a comparison of exogenous ^{19}F and ^{31}P probes. NMR Biomed. 12:495–504.

Okada, H. and Hillery, A. M. 2001. Vaginal Drug Delivery. New York: Taylor and Francis.

Oupicky, D., Carlisle, R. C., and Seymour, L. W. 2001. Triggered intracellular activation of disulfide crosslinked polyelectrolyte gene delivery complexes with extended systemic circulation in vivo. Gene Ther. 8:713–724.

Oupicky, D., Parker, A. L., and. Seymour, L. W. 2002. Laterally stabilized complexes of DNA with linear reducible polycations: strategy for triggered intracellular activation of DNA delivery vectors. J. Am. Chem. Soc. 124:8–9.

Oupicky, D., Reschel, T., Konak, C., and Oupicka, L. 2003. Temperature-controlled behavior of self-assembly gene delivery vectors based on complexes of DNA with poly(L-lysine)-*graft*-poly(*N*-isopropylacrylamide). Macromolecules 36:6863–6872.

Owen, D. H. and Katz, D. F. 2005. A review of the physical and chemical properties of human semen and the formulation of a semen stimulant. J. Androl. 26:459–469.

Pack, D. W., Hoffman, A. S., Pun, S., and Stayton, P. S. 2005. Design and development of polymers for gene delivery. Nat. Rev. Drug Discov. 4:581–593.

Park, J. S., Han, T. H., Lee, K. Y., Han, S. S., Hwang, J. J., Moon, D. H., Kim, S. Y., and Cho, Y. W. 2006. N-acetyl histidine-conjugated glycol chitosan self-assembled nanoparticles for intracytoplasmic delivery of drugs: endocytosis, exocytosis and drug release. J. Control. Release 115:37–45.

Pichon, C., Goncalves, C., and Midoux, P. 2001. Histidine-rich peptides and polymers for nucleic acids delivery. Adv. Drug Deliv. Rev. 53:75–94.

Ponce, A. M., Vujaskovic, Z., Yuan, F., Needham, D., and Dewhirst, M. W. 2006. Hyperthermia mediated liposomal drug delivery. Int. J. Hyperthermia 22:205–213.

Read, M. L., Singh, S., Ahmed, Z., Stevenson, M., Briggs, S. S., Oupicky, D., Barrett, L. B., Spice, R., Kendall, M., Berry, M., Preece, J. A., Logan, A., and Seymour, L. W. 2005. A versatile reducible polycation-based system for efficient delivery of a broad range of nucleic acids. Nucleic Acids Res. 33:e86.

Rodriguez-Cabello, J. C., Reguera, J., Girotti, A., Arias, F. J., and Alonso, M. 2006. Genetic engineering of protein-based polymers: the example of elastin like polymers. Adv. Polym. Sci. 200:119–167.

Rybak, S. L. and Murphy, R. F. 1998. Primary cell cultures from murine kidney and heart differ in endosomal pH. J. Cell. Physiol. 176:216–222.

Rybak, S. L., Lanni, F., and Murphy, R. F. 1997. Theoretical considerations on the role of membrane potential in the regulation of endosomal pH. Biophys. J. 73:674–687.

Sawant, R. M., Hurley, J. P., Salmaso, S., Kale, A., Tolcheva, E., Levchenko, T. S., and Torchilin, V. P. 2006. "Smart" drug delivery systems: double-targeting pH-responsive pharmaceutical nanocarriers. Bioconjug. Chem. 17:943–949.

Schaffer, D. V., Fidelman, N. A., Dan, N., and Lauffenburger, D. A. 2000. Vector unpacking as a potential barrier for receptor-mediated polyplex gene delivery. Biotechnol. Bioeng. 67:598–606.

Schmaljohann, D. 2006. Thermo- and pH-responsive polymers in drug delivery. Adv. Drug Deliv. Rev. 58:1655–1670.

Sehested, M., Skovsgarrd, T., van Deurs, B., and Winther-Nielsen, H. 1987. Increase in nonspecific adsorptive endocytosis in anthracycline- and vinca alkaloid-resistant Ehrlich ascites tumor cell lines. J. Natl. Cancer Inst. 78:171–179.

Sethuraman, V. A. and Bae, Y. H. 2007. TAT peptide-based micelle system for potential active targeting of anti-cancer agents to acidic solid tumors. J. Control. Release 118:216–224.

Sethuraman, V. A., Na, K., and Bae, Y. H. 2006. pH-Responsive sulfonamide/PEI system for tumor specific gene delivery: in vitro study. Biomacromolecules 7:64–70.

Shin, J., Shum, P., and Thompson, D. H. 2003. Acid-triggered release via dePEGylation of DOPE liposomes containing acid-labile vinyl ether PEG-lipids. J. Control. Release 91:187–200.

Simon, S. M. and Schindler, M. 1994. Cell biological mechanisms of multidrug resistance in tumors. Proc. Natl. Acad. Sci. USA. 91:3497–3504.

Sonawane, N. D., Szoka, F. C., Jr., and Verkman, A. S. 2003. Chloride accumulation and swelling in endosomes enhances DNA transfer by polyamine-DNA polyplexes. J. Biol. Chem. 278:44826–44831.

Song, C. W. 1978. Effect of hyperthermia on vascular functions of normal tissues and experimental tumors: brief communication. J. Natl. Cancer Inst. 60:711–713.

Stubbs, M., Mcsheehy, R. M. J., Griffiths, J. R., and Bashford, L. 2000. Causes and consequences of tumour acidity and implications for treatment. Opinion 6:15–19.

Szakacs, G., Paterson, J. K., Ludwig, J. A., Booth-Genthe, C., and Gottesman, M. M. 2006. Targeting multidrug resistance in cancer. Nat. Rev. Drug Discov. 5:219–234.

Tacker, J. R. and Anderson, R. U. 1982. Delivery of antitumor drug to bladder cancer by use of phase transition liposomes and hyperthermia. J. Urology 127:1211–1214.

Takeda, N., Nakamura, E., Yokoyama, M., and Okano, T. 2004. Temperature-responsive polymeric carriers incorporating hydrophobic monomers for effective transfection in small doses. J. Control. Release 95:343–355.

Tannock, I. F. and Rotin, D. 1989. Acid pH in tumors and its potential for therapeutic exploitation. Cancer Res. 49:4373–4384.

Thomas, J. L. and Tirrell, D. A. 2000. Polymer-induced leakage of cations from dioleoyl phosphatidylcholine and phosphatidylglycerol liposomes. J. Control. Release 67:203–209.

Tirrell, D. A., Takigawa, D. Y., and Seki, K. 1985. pH Sensitization of phospholipid vesicles via complexation with synthetic poly(carboxylic acid)s. Ann. N. Y. Acad. Sci. 446:237–248.

Türk, M., Dinçer, S., Yuluğ, I. G., and Piskin, E. 2004. In vitro transfection of HeLa cells with temperature sensitive polycationic copolymers. J. Control. Release 96:325–340.

Urry, D. W. 1997. Physical chemistry of biological free energy transduction as demonstrated by elastic protein-based polymers. J. Phys. Chem. B 101:11007–11028.

Urry, D. W., Luan, C. H., Parker, T. M., Gowda, D. C., Prasad, K. U., Reid, M. C., and Safavy, A. 1991. Temperature of polypeptide inverse temperature transition depends on mean residue hydrophobicity. J. Am. Chem. Soc. 113:4346–4348.

van Adelsberg, J. and Al-Awqati, Q. 1998. Regulation of cell pH by Ca^{2+} -mediated exocytotic insertion of H-ATPase. J. Cell Biol. 102:1638–1645.

van Sluis, R., Bhujwalla, Z. M., Raghunand, N., Ballesteros, P., Alvarez, J. Cerdán, S. Galons, J. P., and Gillies, R.J. 1999. In vivo imaging of extracellular pH using ^{1}H MRSI. Magn. Reson. Med. 41:743–750.

Wagner, E. 2004. Strategies to improve DNA polyplexes for in vivo gene transfer: will "artificial viruses" be the answer? Pharm. Res. 21:8–14.

Walker, G. F., Fella, C., Pelisek, J., Fahrmeir, J., Boeckle, S., Ogris, M., and Wagner, E. 2005. Toward synthetic viruses: endosomal pH-triggered deshielding of targeted polyplexes greatly enhances gene transfer in vitro and in vivo. Mol. Ther. 11:418–425.

Warren, L., Jardillier, J. C., and Ordentlich, P. 1991. Secretion of lysosomal enzymes by drug-sensitive and multiple drug-resistant cells. Cancer Res. 51:1996–2001.

Wu, G., Fang, Y. Z., Yang, S., Lupton, J. R., and Turner, N. D. 2004. Glutathione metabolism and its implications for health. J. Nutr. 134:489–492.

Xie, A. F. and Granick, S. 2002. Phospholipid membranes as substrates for polymer adsorption. Nat. Mater. 1:129–133.

Yamagata, M., Hasuda, K., Stamato, T., and Tannock, I. F. 1998. The contribution of lactic acid to acidification of tumours: studies of variant cells lacking lactate dehydrogenase. Br. J. Cancer 77:1726–1731.

Yang, S. R., Lee, H. J., and Kim, J. D. 2006. Histidine-conjugated poly(amino acid) derivatives for the novel endosomolytic delivery carrier of doxorubicin. J. Control. Release 114:60–68.

Yessine, M. A. and Leroux, J. C. 2004. Membrane-destabilizing polyanions: interaction with lipid bilayers and endosomal escape of biomacromolecules. Adv. Drug Deliv. Rev. 56:999–1021.

Yessine, M. A., Lafleur, M., Meier, C., Petereit, H. U., and Leroux, J. C. 2003. Characterization of the membrane-destabilizing properties of different pH-sensitive methacrylic acid copolymers. Biochim. Biophys. Acta 1613:28–38.

Yokoyama, M. 2002. Gene delivery using temperature-responsive polymeric carriers. Drug Discov. Today 7:426–432.

Yuk, S. H. and Bae, Y. H. 1999. Phase-transition polymers for drug delivery. Crit. Rev. Ther. Drug Carrier Syst. 16:385–423.

Zignani, M., Drummond, D. C., Meyer, O., Hong, K., and Leroux, J. C. 2000. In vitro characterization of a novel polymeric-based pH-sensitive liposome system. Biochim. Biophys. Acta 1463:383–394.

Functionalized Dendrimers as Nanoscale Drug Carriers

Rohit Kolhatkar, Deborah Sweet, and Hamidreza Ghandehari

1 Introduction

Dendrimers represent a unique class of nanostructures, playing an important role in the field of nanobiotechnology. The term dendrimer is derived from Greek (dendra means tree and meros means part) and describes highly branched three-dimensional structures. The dendritic architecture was first reported in the late 1970s and the early 1980s by the research groups of Vogtle, Denkwalter, Tomalia, and Newkome (Lee 2005). Poly(amidoamine) (PAMAM) dendrimers were the first to be synthesized and developed in Dow Laboratories between 1979 and 1985. After patents on this new technology had been filed, the dendritic architecture was presented to the public by Tomalia in 1983. Although met with initial skepticism as most new scientific inventions are, dendrimers were soon accepted, and by 1991 the number of dendrimer-related publications and presentations began to climb rapidly (Tomalia and Frechet 2001).

Since then, many research groups from diverse fields have investigated the unique attributes of dendritic architectures in various applications. More than 100 different types of dendrimers with over 1,000 types of surface modifications have been developed to date (Svenson and Tomalia 2005). Dendrimers have been explored as light-harvesting agents (Ahn 2006; Nantalaksakul 2006; Wang 2006), chemical sensors (Gong et al. 2001; Pugh 2001; Svobodova 2004), catalysts (Delort 2006; Muller 2004; Reek 2002; Wu et al. 2006b), and cross-linking agents. Their use in controlled chemical delivery has been explored for drug delivery (Chandrasekar 2007; Gupta 2006; Majoros 2006; Wu et al. 2006a,b), gene therapy (Dufes et al. 2005; Huang 2007; Manunta 2004; Wada 2005), and delivery of contrast agents (Kobayashi 2004; Koyama 2007; Langereis 2006). There are various reviews published describing the utility of dendrimers in biomedical applications (Svenson and Tomalia 2005), including but not limited to gene delivery (Dufes et al. 2005), development of MRI contrast agents (Kobayashi and Brechbiel 2005), and oral drug delivery systems (Kitchens et al. 2005). Duncan et al. have recently reviewed the biocompatibility and toxicity of these nanostructures (Duncan and Izzo 2005). Structural and physical properties of dendrimers have also been described in detail elsewhere (Bosman et al. 1999). Several dendrimer-based products are now approved for use in biomedical applications or in human clinical

V. Torchilin (ed.), *Multifunctional Pharmaceutical Nanocarriers*,
© Springer Science+Business Media, LLC 2008

trials. VivaGel™ (Starpharma) was granted *Fast Track* status in January 2006 and is designed as a topical microbicide for prevention of HIV (McCarthy 2005). SuperFect®, developed by Qiagen, is used for gene transfection of a broad range of cell lines (Crampton and Simanek 2007; Eichman 2003; Tang et al. 1996). Alert Ticket™, developed by U.S. Army Research Laboratory, is used for anthrax detection (Yin 2001). Finally, Stratus® CS, a cardiac marker diagnostic system, commercialized by Dade Behring, is also based on dendrimers (Couck 2005).

2 Dendrimers as Multifunctional Nanocarriers

Dendrimers have been identified as versatile, compositionally and structurally controlled nanoscale building blocks, providing critically needed starting constructs in the nanometer range for the development of highly specialized materials. They are generally monodisperse, providing structural and chemical uniformity, which is critical for drug delivery applications. For PAMAM dendrimers, the most commonly studied dendrimer, the polydispersity values range from 1.000002 to 1.005, with higher generations showing more defects because of purification difficulties (Svenson and Tomalia 2005). In addition, the branched nature of dendrimers allows a high number of surface functional groups, while still remaining nanoscale structures. As dendrimer generation increases, the number of terminal branches increases exponentially, while the diameter increases linearly by ~1 nm/generation (Svenson and Tomalia 2005). This compactness gives dendrimers a significant advantage over traditional linear polymers, which are restricted to low density drug loading. The densely packed attachment sites on the surface can be exploited to attach high densities of drugs or to create multifunctional systems involving drugs, imaging agents, targeting moieties, and other ligands. Drugs or other payloads can also be encapsulated in the dendrimer interior, having the potential to increase solubility and reduce toxicity of pharmaceutical compounds. The potential of dendrimers as multifunctional nanocarriers has been exploited by conjugation of multiple functionalities, including imaging agents, drugs, and targeting moieties, along with surface modification to reduce toxicity and optimize their in vivo behavior (Scheme 1) (Majoros 2005, 2006; Wu et al. 2006a,b, 2007; Huang 2007; Konda 2000, 2001; Quintana 2002; Thomas 2005).

Majoros et al. (2006) synthesized multifunctional G5 PAMAM dendrimers for targeted anticancer therapy. Folic acid, which targets overexpressed folate receptors on cancer cells, paclitaxel, a chemotherapeutic agent that induces apopotosis by mitotic arrest, and fluorescein isothiocyanate (FITC), a fluorescent label, were successfully conjugated to the dendrimer surface, creating a trifunctional, nanoscale therapeutic agent. The surface of the dendrimer was also partially acetylated, providing enhanced solubility of the dendrimers when conjugated to FITC. Cytotoxicity and cellular internalization of this multifunctional nanocarrier were evaluated using KB cells grown in folic acid deficient media (folate receptor up-regulated) and KB cells grown in complete media (folate receptor down regulated). The authors demonstrated that modified dendrimers, with or without conjugated paclitaxel, were internalized by

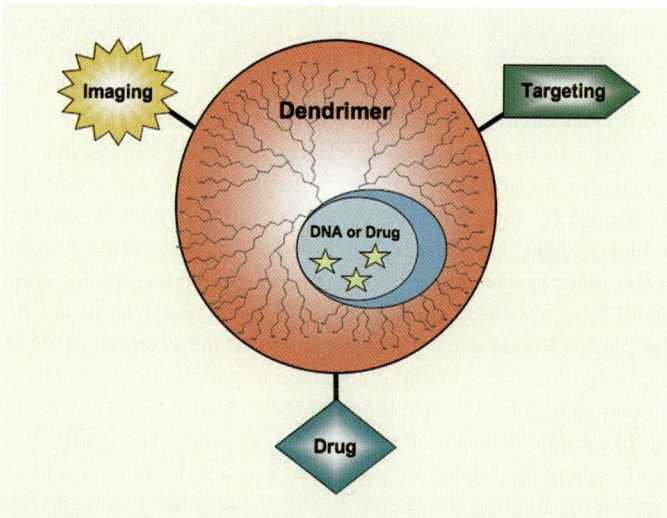

Scheme 1 Schematic of dendrimers as multifunctional nanocarriers. Targeting moieties can be attached to the dendrimer surface groups while drugs and imaging agents can be conjugated to the surface, encapsulated in the dendrimer core or complexed (such as therapeutic DNA) with the dendrimer structure. Additional surface functionalization not shown here can improve biocompatibility (e.g., carboxyl vs. amine terminated) (Duncan and Izzo 2005), half-life (e.g., conjugation of linear polymers such as PEO (Gillies 2005; Gillies and Frechet 2002) or PEG (Okuda et al. 2006a, b)) or cellular uptake (e.g., opening of tight junctions) (Kitchens 2006)

folate-receptor-expressing KB cells, with only the drug-containing conjugates showing cytotoxicity. In contrast, the folate-receptor-negative KB cells did not internalize the dendrimer conjugates and even those conjugates containing paclitaxel were nontoxic, indicating a strong targeting ability of the nanocarrier. This work demonstrates in vitro that multifunctional dendrimers can be used effectively without cross-functionality interference and sets the stage for further in vivo characterization. It also illustrates the importance of successively testing variations of multifunctional conjugates to assess the contribution of each factor to the biological response.

Multifunctional dendrimers can also be used for specialized delivery, such as to the brain. Huang et al. (2007) investigated the potential of PAMAM dendrimers modified with poly(ethylene glycol) (PEG)–transferrin(Tf) for gene delivery across the blood–brain barrier. The transferrin receptor is expressed throughout the blood–brain barrier and can be exploited for transport of drugs from the blood circulation into the brain. Dendrimer–DNA complexes were studied in vitro and in vivo to assess transfection efficiency. In vitro, PAMAM–PEG–Tf–DNA complexes showed a twofold higher transfection into brain capillary endothelial cells, compared with PAMAM–PEG–DNA and PAMAM–DNA complexes. In vivo, the PAMAM–PEG–Tf–DNA complexes showed the highest transfection in the brain. Modifying PAMAM dendrimers with PEG alone did not significantly impact biodistribution, while PAMAM–PEG–Tf conjugates showed a significantly higher accumulation in brain, heart, liver, spleen, lungs, and kidneys.

Dendrimers have also been investigated for treatment of brain tumors with boron neutron capture therapy. Wu et al. (2007) investigated the efficacy of a highly boronated G5 PAMAM dendrimer modified with cetuximab, an antibody that targets epidermal growth factor (EGF) receptors that are overexpressed on gliomas. The bioconjugate was selectively taken up into EGFR-positive F98 cells than in cells that did not express this receptor. In vivo, the conjugate selectively accumulated in the brain, as evidenced by high boron levels. In conjunction with boronphenylalanine (BPA), the bioconjugate significantly increased mean survival time, indicating the viability of this delivery system for boron neutron capture therapy for brain tumors.

Dendrimers have also been used as targeted imaging agents to aid in the early diagnosis of cancer. Konda et al. (2001) investigated the potential of PAMAM dendrimers conjugated to folic acid and complexed to gadolinium (Gd) chelates as targeted MRI contrast agents. The specificity of these dendrimers was tested in vitro in mouse erythroleukemia cells and in vivo in ovarian cancer xenografts. In each case, cells with and without the folate receptors were tested and free folic acid was administered to determine binding inhibition. In vitro, the folate-conjugated dendrimers bound over 2,700% more to receptor-positive cells than to receptor-negative cells. In vivo, a 33% contrast enhancement was observed for the folate–dendrimer chelating agent that showed results far superior to those of a nonspecific Gd agent. Compared to traditional, nonspecific contrast agents, these conjugates were unique because multiple Gd chelates were complexed to each dendrimer, and folic acid was conjugated as a targeting moiety. These results illustrate the potential of multifunctional dendrimers as contrast agents. Coupled with the ability to target drugs to the site of action, PAMAM dendrimers can be used as image-guided drug delivery systems.

These studies are only a few examples in which the unique branched architecture of dendrimers in the nanoscale is useful as multifunctional carriers for targeted delivery of drugs and imaging agents. A comprehensive account of the use of multifunctional dendrimers for targeted delivery is beyond the scope of this chapter and was recently reviewed by a number of investigators in the field (Florence 2005). In this chapter we will focus briefly on dendrimer chemistry, the synthesis of multifunctional carriers, strategies for complexation and conjugation of bioactive agents, and surface modification as they pertain to drug delivery. Finally we will focus on recent results from our laboratory on transport of PAMAM dendrimers across the intestinal epithelial barrier.

3   Dendrimer Chemistry

3.1   Synthesis

A typical dendrimer molecule has an interior core, several layers composed of repeating units, and multiple active terminal groups. Historically, divergent synthesis of dendrimers emerged first, pioneered by Tomalia, followed by convergent synthesis, initiated by Hawker and Frechet (Tomalia and Frechet 2001). In divergent synthesis, monomeric branches are repeatedly added to a central dendrimer core, increasing the dendrimer generation with every addition. PAMAM dendrimers were the first to be commercially produced by this method and use acrylate Michael addition and amidation chemistry (Tomalia and Frechet 2001). In convergent synthesis, reactive dendrons are synthesized and then attached to a multifunctional core to generate the final product (Hawker and Frechet 1990). Convergent synthesis is generally used for lower generation dendrimers, while divergent synthesis can produce higher generation polymers (Scheme 2).

After the initial conception of the dendritic architecture, the focus was shifted to developing synthetic procedures that could be scaled up to produce dendrimers commercially. Central to this process was having high yields of intermediary and final reaction products, minimum volumes of solvent, nontoxic by-products, and reproducibility. Dendrimer synthetic strategies have been improved by "lego chemistry," which uses branched monomers, thus reducing the number of steps to produce higher generation dendrimers and minimizing difficult purification steps. In addition, "click chemistry" has been used to produce dendrimers with specific surface chemistries using a Cu catalyst. Both of these techniques reduce production of toxic by-products to water and nitrogen and sodium chloride respectively (Svenson and Tomalia 2005).

There are many different types of dendrimer families, including poly(amidoamine), poly(propylene imine), and polyether dendrimers (Fig. 1). Polyether dendrons, or "Frechet type" dendrons, are synthesized by the convergent method and are derived from 3,5-dihydroxybenzyl alcohols. They have been used for light harvesting and catalysis, and as molecular machines (Tomalia and Frechet 2001). Poly(propylene imine) (PPI) dendrimers were first developed by Vogtle and are synthesized by the divergent method with repetition of a double Michael addition of amine to acetonitrile, followed by reduction of the end groups into primary amines. Although the initial synthetic process yielded products that were difficult to purify and could

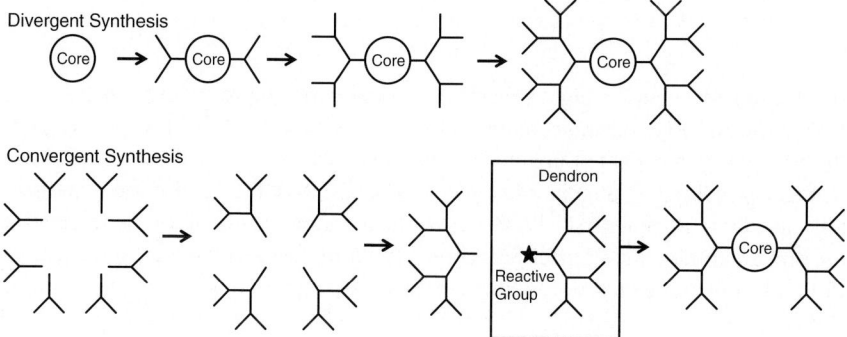

Scheme 2 Scheme for divergent and convergent synthesis of dendrimers. In divergent synthesis, monomeric branches are repeatedly added to a central dendrimer core, increasing the dendrimer generation with every addition. In convergent synthesis, reactive dendrons are synthesized and then attached to a multifunctional core

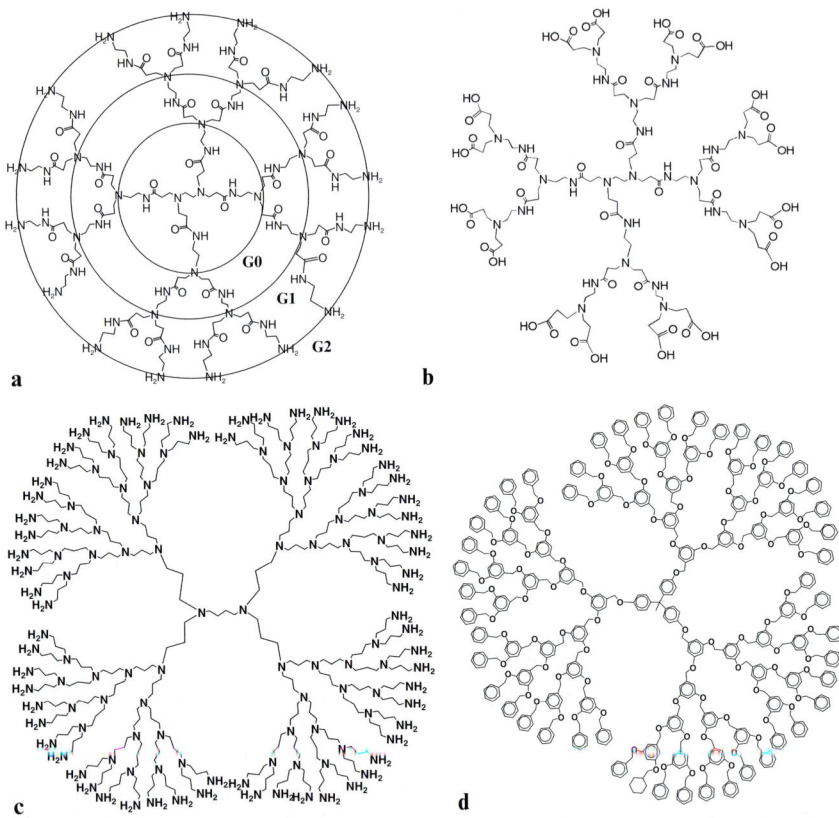

Fig. 1 Schematic of (**a**) G2 amine-terminated and (**b**) G1.5 carboxyl PAMAM dendrimers with ethylenediamine as core, (**c**) G4 PPI dendrimer with 1,4-diaminobutane as core, and (**d**) G4 polyether dendrimer

not be used to generate high generation dendrimers, improvements in the early 1990s allowed PPI dendrimers to be synthesized commercially. PPI dendrimers can either be amine terminated or nitrile terminated, which affects their physical and chemical properties (Tomalia and Frechet 2001). PAMAM dendrimers, commercially available as Starburst® PAMAM dendrimers based on an ethylene diamine core and an amidoamine repeat branching structure, have a diameter ranging from 1.5 to 14.5 nm. As generation number is increased, the number of active terminal groups doubles. For example, G3 dendrimers contain 32 terminal groups and G4 dendrimers contain 64 terminal groups, while the diameter increases by ~1 nm. In PAMAM dendrimers, full generations (G1, G2, G3, etc.) have terminal amine groups while half generation dendrimers (G1.5, G2.5, G3.5, etc.) have carboxylic acid terminal groups (Esfand and Tomalia 2001).

3.2   Drug Complexation and Conjugation

To use dendrimers as drug delivery systems, active pharmaceuticals must be incorporated into the dendritic structure. Complexation of drugs with dendrimers can be achieved by simply mixing the two compounds together. Complexation relies mostly on electrostatic interactions between the drug and dendrimer and will be discussed in more detail later in this chapter. Covalent conjugation of drugs to dendrimers is more challenging and few reactions have been reported.

In many instances amine-terminated PAMAM dendrimers (unmodified or surface modified) have been used for conjugation. Different strategies used for drug conjugation are illustrated in Scheme 3. Drugs or linkers with acidic groups can be attached to PAMAM dendrimer using EDC (1-ethyl-3-(3-dimethylaminopropyl)

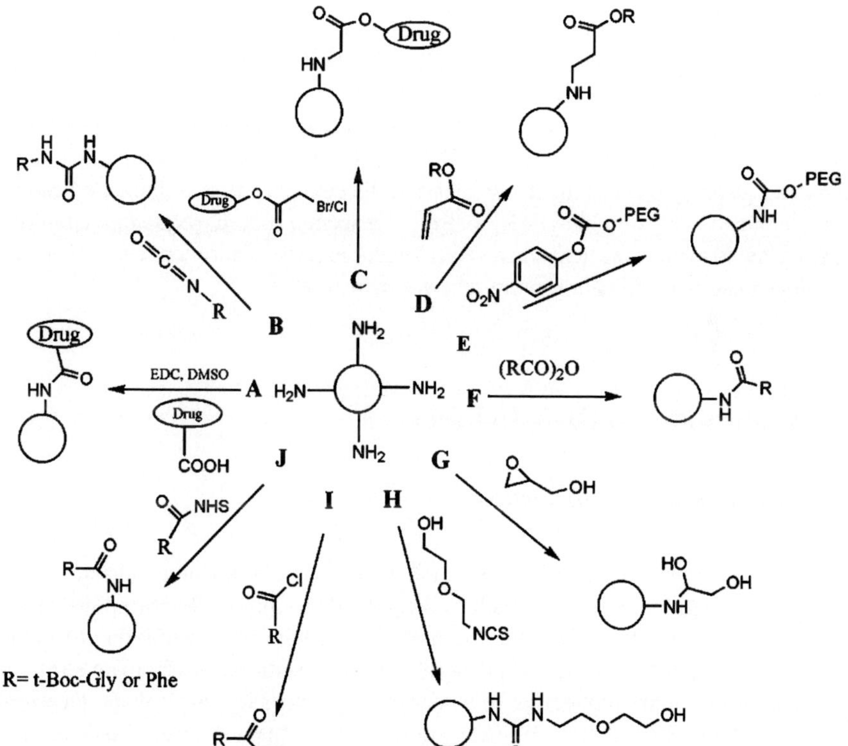

Scheme 3 Scheme for functionalization of amine terminated PAMAM dendrimers with drugs or targeting moieties having acidic group (**a**), isocynate (**b**), alkyl halide (**c**), alkene (**d**), activated acidic group (**e,j**), acidic anhydride (**f**), epoxide (**g**), isothiocynate (**h**) and acid chloride (**i**)

carbodiimide hydrochloride) (Quintana 2002) as a coupling agent or by activation of acidic group to *N*-hydroxysuccinimide ester (Bhadra 2003), followed by conjugation to PAMAM dendrimers to obtain an amide linkage (Fahmy et al. 2007). Folic acid, succinic acid, and methotrexate were conjugated to amine-terminated PAMAM dendrimers following this strategy (Quintana 2002; Islam et al. 2005; Hong 2007). Anhydrides or acid chlorides have also been used to produce amide linkage (Quintana 2002; Hong 2007). Partial or complete functionalization can be achieved by controlling the stoichiometry of reagents in these conjugation reactions.

Drugs with hydroxyl groups can be attached to dendrimers using various linker chemistries. Propranolol was attached to a two carbon linker, followed by conjugation to PAMAM dendrimers (D'Emanuele 2004), while succinic acid was used as a linker for paclitaxel. Linear poly(ethylene glycol) was functionalized with 4-nitrophenyl chloroformate to form an active ester, which was then reacted with the terminal amine groups of the PAMAM dendrimer by a nucleophilic substitution reaction. Fluorescent labeling and radiolabeling were also achieved by nucleophilic substitution reaction of FITC and radiolabeled acetic anhydride respectively with PAMAM dendrimers (El-Sayed 2002; Kolhatkar 2007). Newkome et al. (1998) have reported the reaction of mixtures of branched isocyanates with polypropylene(imine) and 12-cascade acid dendrimers and have shown that different ratios of functional groups can be added to the dendrimer surface. Isothiocyanates of comparable reactivity can also be added to the dendrimer and the degree of functionalization can be easily controlled by stoichiometric control of reagents. Indeed, sequential additions can be used to synthesize heterogeneously functionalized dendrimers even when one isothiocyanate is significantly more reactive than another.

4   Applications in Drug Delivery

4.1   *Encapsulation of Drugs*

Since the inception of dendritic structure, numerous classes of dendrimers have been synthesized, yet fewer efforts have been made to physically entrap drug molecules within them. As a consequence, there is a paucity of information on chemical features desirable for controlling the release of guest molecules from dendrimers. The macromolecular architecture of a dendrimer governs its physical and chemical properties. Dendrimer architecture can significantly influence performance as drug carriers, in particular, encapsulation efficiency and release. The presence of internal voids in the dendritic architecture has led several groups to investigate the possibility of encapsulating drug molecules within the branches of a dendrimer (Svenson and Tomalia 2005). In addition, dendrimers have high surface charge density because of the presence of multiple ionizable groups at the periphery, which can be exploited to attach the drugs electrostatically. This offers dendrimers the potential to interact with labile or poorly soluble drugs, as each macromolecule could attach numerous drug molecules when compared to classical complexing agents, where in many cases only a 1:1 stoichiometry can be expected. The resultant solubility enhancement allows the presentation of the drug to the biological membrane and subsequent internalization and transport. Such systems may also enhance drug stability.

Encapsulation of a drug within a dendrimer may also be used to provide a means of controlling its release. Several types of dendrimers have been investigated for the encapsulation of drugs, including systems designed for triggered release (Paleos 2004). The nature of drug encapsulation within a dendrimer may be simple physical entrapment, or can involve nonbonding interactions with specific structures within the dendrimer. Much of the work in this area has been based on the assumption that dendrimers possess a hollow core and a dense shell, and indeed much of the literature on drug encapsulation supports this hypothesis (Tomalia and Frechet 2001). The number of molecules incorporated into dendrimers is thought to be dependent on the number of surface groups available. There is a desire to use higher generation dendrimers since with each increase in generation the number of groups available for complexation doubles. However, not all the surface groups may be available for interaction, because of either steric hindrance or back folding of chains into the dendrimer. As the generation number increases, more and more back folding occurs, giving a spherical shape to higher generation dendrimers when compared to an extended conformation of lower generation dendrimers (Bosman et al. 1999). Factors that affect complexation, such as surface charge density and ionic strength, have been studied by Newkome and coworkers, for the interaction of small carboxyl-terminated dendrimers with no interior titratable groups, and the strong polycation poly(dimethyldiallylammonium chloride). Complexes formed abruptly at a critical surface charge density that was lower for G3 than for G1 dendrimers, i.e., binding occurred more easily for the larger dendrimers (Miura 1999; Zhang 1999). Kabanov et al. (1998) reported differential complexation behavior for positively and negatively charged dendrimers. The studies reveal that flexible linear polyanions can interact with positively charged amine groups at the surface as well as in the interior of the dendrimer. However, linear polycations were unable to penetrate into the interior and were able to interact with only surface carboxyl groups of dendrimer.

Most of the work done so far with biologically active molecules uses PAMAM dendrimers with amine surface groups, with a few studies using negatively charged carboxyl-terminated and neutral hydroxyl-terminated PAMAM dendrimers. Amine-terminated PAMAM dendrimers generally showed enhanced solubility for various compounds, compared with carboxyl-terminated and hydroxyl-terminated dendrimers (Chauhan 2003; Devarakonda et al. 2005; Markatou 2007). This is thought to be due to the ionic interactions involving positively charged amine groups at physiological pH, as solubility was found to decrease at lower pH (Chauhan 2003; Yiyun et al. 2005). Few studies have suggested other phenomena such as physical entrapment, dendrimer architecture, and nonspecific noncovalent interaction (Chauhan 2003; Yiyun et al. 2005) such as hydrophobic interactions (Jansen et al. 1994; Liu et al. 2000) and weak hydrogen bonding (Chauhan 2003; Prieto 2006) for encapsulation of drugs in the crevices of dendritic structure and observed increase in the solubility. Dhanikula and Hildgen (2007) have also reported the influence of dendrimer architecture on encapsulation of methotrexate. A series of polyether-*co*-polyester (PEPE) dendrimers with different architectures were synthesized. It was observed that an increase in the branching as well as an interior void volume enhanced encapsulation. Physical entrapment, weak hydrogen bonding, and hydrophobic interactions were postulated to be mechanisms of encapsulation. A decrease in encapsulation in dendrimers lacking aromatic rings as branching unit further demonstrated the importance of hydrophobic interactions (Dhanikula and Hildgen 2007). A correlation with hydrophilic character of a guest molecule and solubility in the presence of a dendrimer has been reported by Yiyun and Tongwen (2005a). Kojima et al. (2000) have reported an increase in the loading of methotrexate associated with grafting of PEG on the surface of dendrimer. The loading of methotrexate was dependant on the molecular weight of PEG grafted on the surface. Table 1 includes a partial list of pharmacologically active compounds with increased solubility when complexed with dendrimers. Several such complexes are described here.

Table 1 Partial list of dendrimer–drug complexes

Type of dendrimer	Terminal group (generation)	Mol. wt. (Da)	Drug	Drug/ dendrimer ratio	Reference
PAMAM	NH_2 (3)	6,909	Dimethoxy- curcumin	1.9–5/1	Markatou (2007)
	COOH (3.5)	12,931		1.8–4.2/1	
PAMAM	NH_2 (4)	14,215	Indomethacin	13.9/1	Chauhan (2003)
	COOH (4.5)	26,258		4.8/1	
	OH (4)	14,215		12.5/1	
PAMAM	NH_2 (2–5)	3,256	Ketoprofen	–	Yiyun et al. (2005) and Yiyun and Tongwen (2005)
PAMAM	NH_2 (4)	14,215	Sulfadiazine	30/1	Prieto (2006)
	COOH (4.5)	26,258			
Tris (poly glycerol)	OH (3)	1,690	Paclitaxel	1.87	Ooya et al. (2003)
	OH (4)	3,508		1.81	
	OH (5)	7,087		2.30	
PPI	NH_2 (5)	14,476	Lamivudine	54% (w/w)	Dutta and Jain (2007)
	Mannosamine			69% (w/w)	

(continued)

Table 1 (continued)

Type of dendrimer	Terminal group (generation)	Mol. wt. (Da)	Drug	Drug/ dendrimer ratio	Reference
PPI	NH_2 (5)	14,476	Efivirenz	47% (w/w)	Dutta (2007)
	NH_2, Mannosamine			32% (w/w)	
	NH_2, t-boc			23% (w/w)	
PAMAM	NH_2 (3)	6,909	Niclosamide	32/1	Devarakonda et al. (2005)
PAMAM	NH_2 (2–4)	3,256– 14,215	Nocitinic acid	–	Yiyun and Tongwen (2005)
			Naproxen		Yiyun and Tongwen (2005)
			Diflunisal		Yiyun and Tongwen (2005)
			Pilocarpine		Vandamme and Brobeck (2005)
			Tropicamide		Vandamme and Brobeck (2005)
			Sulfame- thoxazole		Ma (2007)
PAMAM	NH_2 (6) PEG, NH_2	58,046	Doxorubicin	–	Lee (2006)
PAMAM	NH_2 (3)	6,909	Ibuprofen	33	Milhem et al. (2000)
	(4)	14,215		41/1, 78/1	
Polylysine with PEG as core	NH_2 (3)	2,792	Chloroquine	2.2/1	Agrawal et al. (2007)
	Galactose/NH_2 (3)			6.1/1	
	NH_2 (4)	4,840		4.3	
	Galctose/NH_2 (4)			15.27	

PAMAM poly(amidoamine), *PPI* poly(propyleneimine), PEG poly(ethylene glycol)

Malik et al. (1999) have prepared chelates of PAMAM dendrimers with cisplatin, a potent anticancer drug with nonspecific toxicity and poor water solubility. The chelates showed increased solubility, decreased systemic toxicity, and selective accumulation in solid tumors. The dendrimer–platinum complexes showed increased efficacy relative to cisplatin in the treatment of subcutaneous B16F10 melanoma. Dimethoxycurcumin had higher water solubility when complexed with G3.5 or G4 PAMAM dendrimers (Markatou 2007). Increased solubility of indomethacin in the presence of PAMAM dendrimers increased the flux of indomethacin across the skin (Chauhan 2003). The rank order for this increase was $G4NH_2$ > G4OH > G4.5COOH. Similarly, a linear increase in solubility of nonsteroidal anti-inflammatory drugs and nicotinic acid in the presence of amine-terminated PAMAM dendrimers at a fixed pH is reported by Yiyun and Tongwen (2005b). Increased aqueous solubility was in the order of Naproxen > Ketoprofen > Ibuprofen > Diflunisal. Higher generation PAMAM dendrimers showed an enhanced effect on solubility (G5 > G4 > G3 > G2) whereas lowering the pH decreased solubility, suggesting the involvement of electrostatic interactions. Further studies by the same group reported an increase in solubility of ketoprofen using G5 PAMAM dendrimers (Yiyun et al. 2005; Na 2006). It was postulated that the increased solubility of ketoprofen was due to electrostatic interactions between the acidic group of ketoprofen and amine groups of PAMAM dendrimers. G5-ketoprofen conjugates were stable in distilled water as drug release from the complex was slower when compared with free ketoprofen. Similar electrostatic interactions between the weak acid sulfadiazine and G4 PAMAM dendrimers were reported, whereas hydrogen-bond interactions between tertiary amine groups of G4.5 PAMAM dendrimer and sulfadiazine led to the formation of water-soluble complexes with 30 molecules of sulfadiazine per PAMAM dendrimer (Prieto 2006). These complexes were stable in water with less than 15% drug released in 24 h.

Milhem et al. (2000) reported the solubility enhancement of ibuprofen in the presence of PAMAM dendrimers. The number of molecules associated with G4 PAMAM dendrimer was calculated to be 41. Kolhe et al. (2004) later showed that 78 molecules of ibuprofen can be associated with $G4NH_2$ PAMAM dendrimers and that these complexes enter A549 cells much more rapidly than free ibuprofen. Later studies from the same group reported conjugation of ibuprofen with G4OH PAMAM dendrimers via ester bonds (Kolhe et al. 2006).

Devarakonda et al. (2005) reported a significant enhancement in solubility of niclosamide in the presence of G3 PAMAM dendrimers. The authors have suggested that dendrimer–niclosamide complexes were more stable than cyclodextrin–niclosamide complexes. Interestingly, niclosamide is only weakly acidic (pK_a, 7.3) and shows complexation with amine-terminated PAMAM dendrimers through ionic interactions. Half generation PAMAM dendrimers did not show any increased aqueous solubility for niclosamide. Both G3 and G2.5 PAMAM dendrimers significantly enhanced the aqueous solubility of piroxicam at pH values 6 and 8 (D'Emanuele and Attwood 2005). In the case of the G3 dendrimer, this enhancement was attributed to electrostatic complexation or hydrogen bonding, or both. The solubility increase noted in the presence of carboxylated dendrimer (G2.5) was thought to be simply due to an increase in solution pH by the highly basic dendrimer, since there is no opportunity for electrostatic interaction with these similarly charged compounds.

The interaction of unsubstituted and substituted benzoic acid at neutral pH (virtually water-insoluble) with PAMAM dendrimers formed by conversion of the terminal groups to hydroxyls was reported by Beezer et al. (2003). Stable water-soluble complexes were formed in which the benzoic acid was thought to be bound to the internal tertiary nitrogens of the dendrimer by simple ion-pairing. Paclitaxel solubilities in polyglycerol G3, G4, and G5 dendrimers at a 10-wt% concentration were 270-, 370-, and 430-fold higher, respectively, than that in water (Ooya et al. 2003). A further increase in solubility was observed in the presence of 80 wt% of polyglycerol dendrimer. Solubility in the presence of dendrimer was enhanced to a higher extent, compared with that in the presence of other linear and star-shaped poly(ethylene glycol)-based polymers. The authors suggested that the high density of ethylene glycol units in the dendritic structure contributed to an increase in paclitaxel solubility. Recently, Dutta and Jain showed an increase in solubility of efivirenz and lamivudine by using PPI dendrimers (Dutta 2007; Dutta and Jain 2007). The same group reported the use of polylysine dendrimers for increasing the solubility of chloroquine phosphate (Agrawal et al. 2007). The surface of these dendrimers was modified with mannose, galactose, and *t*-Boc to reduce the toxicity associated with the amine terminal groups. These examples illustrate that dendrimer–drug complexation is a powerful tool for enhancing drug solubility, which is often a limiting factor in the development of novel therapeutics.

4.2 Dendrimer–Drug Conjugates

Noncovalent complexation of drugs to dendrimers suffers from potential nonspecific release. Grafting linear polymers on the surface has expanded the scope of encapsulation of drugs such as 5-fluorouracil, methotrexate, and doxorubicin and can slow the drug release rates in these systems to some extent. This method, however, has yet to be demonstrated as a general strategy. Patri et al. (2005) evaluated the release profile of the complexed and conjugated form of methotrexate with surface-modified PAMAM dendrimers. The dendrimer–methotrexate complex was stable in water but the drug was released rapidly from the complex in buffer. In contrast, the conjugates were stable in aqueous environment.

Dendrimers have been used for delivery of various drugs by conjugation (Table 2). Covalent attachment allows for the conjugation of several different moieties with a high degree of control to create multifunctional drug delivery nanodevices. Hong et al. (2007) conjugated folate residues to G5 PAMAM dendrimers along with fluorescein isothiocyanate (FITC; Fluorochrome) for targeting as well as detection of such conjugates in tumor cells. Methotrexate was attached to G5-FITC–FA (folic acid) conjugates via amide and ester linkages. The remaining amine groups were glycidylated, to minimize nonspecific binding to the receptors. Both types of conjugates were internalized in the KB cell line of human epidermoid carcinoma, which overexpresses

Table 2 Partial list of dendrimer–drug conjugates

Type of dendrimer	Terminal group (generation)	Mol. wt. (Da)	Drug	Dendrimer/ drug ratio	Linker (bond)	Reference
PAMAM	COOH (3.5)	12,931	Strepto-kinase	1:1–1:20	– (Amide)	Wang (2007)
1,4,7,10-Tetraazacyclodododacane	NH$_2$	7,021	5-Fluorou-racil	1:15	Bromoacetyl (ester)	Zhuo et al. (1999)
		14,326				
PAMAM	NH$_2$ (4)	14,215	Avidin–biotin/ DNA	100–2.1–1	– (Amide)	Mamede (2004)
PAMAM	NH$_2$ (3)	6,904	Biotin	1:7	Succinic acid (amide)	Sato (2003)
	NH$_2$ (succinate)					
PAMAM	NH$_2$ (3)	6,904	5-Amino salicylic acid	3–9/1	PABA (amide)	Wiwattana-patapee et al. (2003)
					PAH	
PAMAM	NH$_2$ (0)	517	Naproxen	1/1	Lactic acid (amide) Diethylene glycol (ester)	Najlah (2007)
PAMAM	NH$_2$ (4)	14,214	Ibuprofen	58/1	– (Ester)	Kolhe et al. (2004)
	OH	14,214				

PAMAM poly(amidoamine), *PABA* p-aminobenzoic acid, *PAH* p-aminohippuric acid

the folate receptor. In vitro internalization of these conjugates was as efficient as conjugates of dendrimers to FA without drug (Hong 2007). Using PEG as a linker, Yang and Lopina (2003) conjugated penicillin V to G2.5 and G3 PAMAM dendrimers via ester and amide linkages respectively. Less susceptibility of the release of free penicillin V from the amide linkage prompted the authors to evaluate the conjugates with only ester linkages. Assessment of antimicrobial activity against the *Staphylococcus aureus* strain of bacteria revealed that penicillin-V-conjugated PEG–PAMAM (G2.5) dendrimers were as effective as an unmodified penicillin V in inhibiting bacterial growth. Zhuo et al. (1999) described the preparation of PAMAM dendrimers from a cyclic tetraamine core and the subsequent attachment of 5-fluorouracil to the dendrimer periphery via a two-carbon linker by reacting with bromoacetyl chloride. The amount of 5-fluorouracil released due to hydrolysis of the conjugates upon incubation in phosphate-buffered saline varied with G5 dendrimer, showing higher release than with G4.

The use of dendrimers in pH-triggered colon-specific delivery of sulfasalazine was suggested by Wiwattanapatapee et al. (2003). G3 PAMAM dendrimers were conjugated with two types of spacers, namely, *p*-aminobenzoic acid (PABA) and *p*-aminohippuric acid (PAH), via amide linkage. Both spacers are azo linkers, and the drug 5-amino salicylic acid (5-ASA) was bound to both conjugates via azo linkage. Conjugates using PAH as the spacer carry three times more 5-ASA than do conjugates using PABA as the spacer. Both dendrimer conjugates (PAMAM–PABA–SA and PAMAM–PAH–SA) were incubated in homogenates of small intestinal and cecal content of albino rats in vitro for 12 h. It was found that in cecal content, about 28 and 38% of a dose of 5-ASA was released from PAMAM–PABA–SA and PAMAM–PAH–SA, respectively, which increased up to 24 hours. In contrast, in the case of small intestine homogenate the release of 5-ASA was about 4.5 and 7.2% for PAMAM–PABA–SA and PAMAM–PAH–SA, respectively. In the case of small intestinal homogenate, significant amounts of PABA–SA (3.8%) and PAH–SA (12.5%) were also released. Release of 5-ASA in cecal content was due to the activity of azo reductase, which led to the breakdown of the azo bond between the spacer and drug molecule. About 45 and 57% of 5-ASA was released from PAMAM–PABA–SA and PAMAM–PAH–SA conjugates, respectively, in 24 h. This release was much slower (80% in 6 h) than that of 5-ASA from sulfasalazine presumably due to the highly branched structure potentiality limiting the cleavability of the azo bond.

4.3   Dendrimer Surface Modification

In addition to covalent conjugation of drugs to the dendrimer surface, the dendritic ends can also serve as a scaffold to custom-tune surface chemistry for particular applications. A variety of molecules have been conjugated to dendrimers to improve their biological characteristics, including reduction in cytotoxicity, generating multivalency, controlling encapsulated drug release, and increasing biological half-life. Surface modification strategies to address each of these critical issues in drug delivery are described here.

4.3.1   Surface Modification to Reduce Cytotoxicity

Despite the extensive interest in the pharmaceutical applications of dendrimers, there is conflicting evidence regarding their biological safety, and indeed, cationic PAMAM dendrimers have been shown to be hemolytic (Malik 2000). Previous studies in our laboratory examined the effect of charge, generation, and concentration on cytotoxicity of Caco-2 cells by using lactate dehydrogenase and water-soluble tetrazolium (WST-1) assays (Kitchens et al. 2005; El-Sayed 2002; Kitchens 2006; E1-Sayed 2003). Positively charged dendrimers were more cytotoxic compared with negatively charged and neutral dendrimers. Also, higher generation and concentration of dendrimers led to higher cytotoxicity. To further explore the effect of charge, generation, and concentration, we used transmission electron microscopy (TEM) to examine the effect of PAMAM dendrimers on Caco-2 cell monolayer integrity (Kitchens 2007). Untreated, control Caco-2 cells displayed well-differentiated monolayers with intact microvilli (Fig. 2a). This morphology did not change upon treatment with 1 mM G2NH$_2$, G1.5COOH, and G3.5COOH (Fig. 2b, c, and e, respectively). However, cells treated with 1 mM G4NH$_2$ dendrimers displayed distortion in monolayer morphology as well as noticeable cell membrane disruption, as evidenced by a loss of microvilli (Fig. 2d). The effect of G4NH$_2$ dendrimers on microvilli was concentration-dependent, with well-defined microvilli in control cells and at low G4NH$_2$ concentrations (0.01 mM), and an escalation in the disruption and loss of microvilli with increase in G4NH$_2$ concentration (Fig. 2f–i). The greater number of cationic surface groups on these dendrimers increases the interaction with the negatively charged cell membrane when compared to anionic or neutral dendrimers. This, in turn, contributes to the enhanced cytotoxicity of G4NH$_2$ dendrimers.

Many functionalization strategies have been used to mask the terminal amine groups, thereby reducing the positive charge. For example, acetylation of the surface amine reduces cytotoxicity (Kolhatkar 2007). In our laboratory, the effect of dendrimer concentration and the relative degree of surface amine group acetylation on cytotoxicity was evaluated by performing WST-1 assay, which quantifies cell

Top panel

Bottom panel

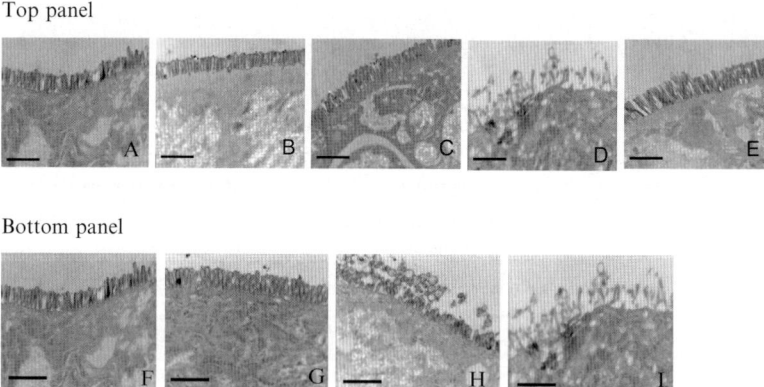

Fig. 2 Transmission electron microscopy (TEM) images of Caco-2 cell monolayers in the presence and absence of PAMAM dendrimers. *Top panel:* Generation and charge-dependent effect of PAMAM dendrimers on Caco-2 microvilli after treatment with 1 mM PAMAM dendrimers for 2 h visualized by TEM. (**a**) control cells, (**b**) G2NH$_2$, (**c**) G1.5COOH, (**d**) G4NH$_2$, and (**e**) G3.5COOH. *Bottom panel:* Concentration-dependent effect of G4NH$_2$ on Caco-2 microvilli after treatment with G4NH$_2$ dendrimers for 2 h is shown in **f–i**: (**f**) control cells, (**g**) 0.01 mM G4NH$_2$, (**h**) 0.1 mM G4NH$_2$, (**i**) 1.0 mM G4NH$_2$. ×12,500. Scale bars = 1 μm. Adapted from Kitchens (2007), with permission from Springer

proliferation and cell viability based on the cleavage of the tetrazolium salt WST-1 by mitochondrial dehydrogenase in viable cells. Studies at 0.01 and 0.1 mM concentrations of G2 and G4 PAMAM dendrimers suggest that partial surface acetylation can reduce cytotoxicity. At 1 mM concentration cytotoxicity was absent in cells incubated with G2 and G4 PAMAM dendrimers exhibiting the highest degree of acetylation, whereas an increase in cytotoxicity was seen with a decrease in acetylation, suggesting a linear relationship between the number of amine groups and toxicity. A simple plot of cell viability vs. the number of surface amine groups (Fig. 3a) suggests that an increase in the number of surface amine groups resulted in increased toxicity; however, the slope of this plot is different for $G2NH_2$ vs. $G4NH_2$. When cell viability was plotted as a function of surface density (SD_a), a linear correlation was observed irrespective of dendrimer generation (Fig. 3b). These results suggest that in correlating polymer structure with function, not only the sheer number of surface functionalities but also their spatial orientation must be considered. The unique three-dimensional branched architecture of PAMAM dendrimers (compared with, e.g., linear multifunctional polymers) provides opportunities for spatial control at the nanoscale. Evaluating how such control influences the interaction of multifunctional carriers with biological environments (e.g., micrometer-sized cells) would be intriguing and yet unexplored.

Jevprasesphant et al. (2003b) compared the cytotoxicity of unmodified amine-terminated (G2, G3, G4) and differentially modified PAMAM dendrimers on Caco-2 cells. The surface amine groups were masked with 6, 9, and 15 long-chain lauroyl groups and with 2 and 4 linear PEG of molecular weight 2,000. The half maximal inhibitory concentration (IC_{50}) values of G3 and G4 PAMAM dendrimers with 6 attached lipid chains were sevenfold higher than those of unmodified dendrimers. Conjugation of G4 dendrimers with 4 PEG chains resulted in marked decrease in cytotoxicity.

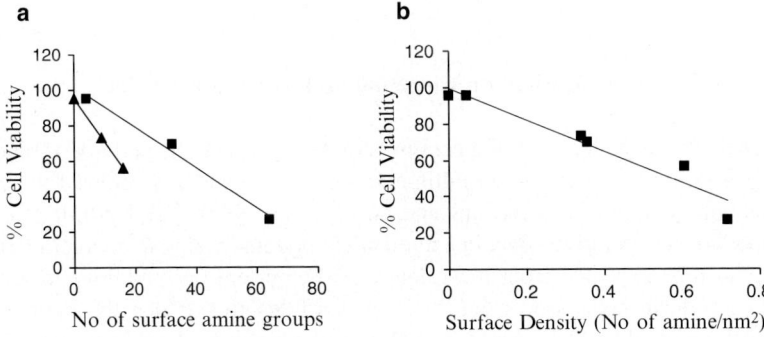

Fig. 3 a Relationship between cell viability and number of surface amine groups for (*filled square*) G4 ($R^2 = 0.99$) and (*filled triangle*) G2 analogs ($R^2 = 1$). Increasing the number of surface amine groups causes a linear increase in toxicity with a generation-dependent slope. **b** Relationship between cell viability and surface density ($R^2 = 0.93$). When surface density of amine groups is accounted for, toxicity is linear and generation-independent. Reprinted from Kolhatkar (2007), with permission from ACS

Meijer's group proposed PPI dendrimer surface modification with Boc protected amino acids to control the release of encapsulated drug, and the synthesis of carbohydrate-coated PPI dendrimers was first proposed by the same group for targeting to lectin-rich organs (Jansen et al. 1994; Ashton 1997). Following a similar synthetic approach, Agashe et al. (2006) functionalized PPI dendrimers using various Boc-protected amino acids and carbohydrates. Hemolytic toxicity using a 2% red blood cell (RBC) suspension, and cytotoxicity by performing MTT (3-(4, 5-Dimethylthiazole-2-yl)-2,5-diphenyltetrazolium bromide) assay using human hepatoma (HEPG2) and adherent SV 40 virus transformed African green monkey (COS-7) cell lines were assessed. Hematological studies were conducted to determine white blood cells (WBCs), red blood cells (RBCs), hemoglobin (Hb), hematocrit (HCT), and mean corpuscular hemoglobin (MCH). All the modified dendrimers were nontoxic irrespective of the concentration (0.001–1 mg mL^{-1}) and various incubation periods (1,2, and 4 h for hemolytic studies and 24, 48, and 72 h for MTT assay). Compared to control, saline-treated animals, a significant increase in WBC count and a decrease in RBC, Hb, and HCT were observed after treatment of rats with unmodified G5 PPI dendrimer. All surface-modified dendrimers did not show any significant changes in the values of hematological parameters when compared to control. When evaluated for immunogenicity in BALB/C mice, the polymers did not elicit detected antibody(IgG) titre, suggesting nonimmumogenicity of these polymers under the experimental conditions. The authors suggest that masking of positively charged surface amine groups presumably reduced the toxicity and improved biocompatibility of the surface-modified G5 PPI dendrimers. Although, clearly surface chemistry is significant, the toxicity of dendrimers is not solely determined by the nature of their surface groups. Dendrimers based on an aromatic polyether skeleton having anionic carboxylate groups on the surface have been shown to be hemolytic in a solution of rat blood cells after 24 h. It is suggested that the aromatic interior of the dendrimer may cause hemolysis through hydrophobic membrane contact.

4.3.2 Surface Modification for Targeting and Multivalent Effect

The synergistic enhancement of a certain activity, e.g., catalytic activity or binding affinity from a monomeric to a multimeric system, is generally referred to as the "cluster" or "dendritic" effect (Zeng and Zimmerman 1997). The dendritic effect is attributed to a cooperative effect in a multivalent system leading to a larger increase in activity than expected from the valency of the system (i.e., additive increase). Dendrimers have been used extensively as scaffolds to produce the multivalent effect (Page et al. 1996; Vrasidas 2001). In particular, glycodendrimers have emerged from the need for having multivalent effect for carbohydrate-based drugs (Ashton 1997; Zeng and Zimmerman 1997). Carbohydrates constitute an important class of biological recognition molecules. Carbohydrate-binding proteins are called lectins, and lectin–carbohydrate interactions have been described in numerous cases in the immune system (cellular activation events), in bacterial and viral infections, and in relation to cancer and cell growth. Carbohydrate-based drugs are of interest as microbial antiadhesins, microbial toxin antagonists, anti-inflammatory, antiviral, and anticancer drugs. However, there are synthetic difficulties in obtaining such bioactive carbohydrate ligands. A strategy to overcome this uses the multivalency/cluster effect (*vide supra*) obtainable by dendrimer presentation, to create carbohydrate ligands with adequate binding affinity from simple mono- or oligosaccharides (Zeng and Zimmerman 1997).

Many examples of saccharide-functionalized dendrimers have been reported. Andre et al. (1999) reported the lactose functionalization of a G5 amine-terminated PAMAM dendrimer to study how the topology of binding site presentation and ligand display affect binding selectivity (Andre et al. 1999). This group has also demonstrated that dendrons having eight sialoside residues on the surface were 1,000-fold more effective than monomeric sialic acid for the inhibition of binding of influenza virus to human erythrocytes (Roy et al. 1993). Tsutsumiuchi et al. (1999) reported the galactose and *N*-acetylglucose functionalization of G6–G8 PAMAM to study the effect that sugar density had on the ability of the dendrimer to recognize and encapsulate a naphthalenesulfonate salt, as well as the ability of these dendrimers to inhibit wheat germ agglutinin. Woller et al. (2003) described the complete functionalization of G1–G6 PAMAM dendrimers with mannose. Relative to methyl mannose, significant enhancements in binding toward Concanavalin A (Con A) were observed with the larger dendrimers because they were capable of spanning multiple binding sites on Con A while the smaller dendrimers could not. A significant effect of the degree of mannose loading on the activity of G3–G6 amine-terminated PAMAM dendrimers toward Con A was also reported (Woller 2003).

Quintana et al. (2002) modified the surface of G5 amine-terminated PAMAM dendrimers with acetamide, 2,3-dihydroxy propyl and succinic acid groups to reduce nonspecific binding associated with the positively charged amine groups. Among these surface-modified dendrimers, apparent affinity and binding capacity of acetamide dendrimers was highest with hydroxyl-terminated dendrimers, showing half the affinity of acetamide, whereas carboxyl-terminated dendrimers showed the least affinity toward KB (human epidermoid carcinoma) cells. A systematic study of folic acid targeted dendritic nanodevices revealed marked enhancement of binding to KB receptors due to multivalency effect (Quintana 2002). However, no increase in cellular internalization was observed, suggesting the possibility that reduced tumor volume associated with these nanodevices is due to increased residence time, followed by internalization, rather than increased binding, followed by endocytosis.

4.3.3 Surface Modification to Influence Drug Encapsulation and Release

The influence of dendrimer generation (G3 and G4) and PEG molecular weight (550 or 2,000) on the ability of PEG-grafted dendrimers to encapsulate the hydrophobic drugs adriamycin and methotrexate was examined by Kojima et al. (2000). Nuclear magnetic resonance data indicated that essentially every terminal amino group on the dendrimer reacted with a PEG chain. Drug loading increased with dendrimer size and increasing chain length of PEG grafts, with up to 6.5 adriamycin or 26 methotrexate molecules incorporated per dendrimer (G4) molecule. While there was evidence of the sustained release of methotrexate from a dendrimer carrier in an aqueous solution of low ionic strength, drug release was not controlled in isotonic solutions (Kojima et al. 2000). In order to overcome this problem and improve the retention of guest molecules, the shell structure on the dendrimer surface was introduced (Haba 2005). G4 PAMAM dendrimers were prepared with PEG 2000 conjugated to the surface and a methacryloyl group at every chain end of the dendrimer through L-lysine residue. The methacryloyl groups were polymerized using a free radical initiator to form a nanocapsule that could retain small molecules within the PAMAM environment. When methacryloyl groups were polymerized in the presence of Bengal Rose, the guest molecules were tightly associated with the dendrimer. However, encapsulation efficiency was low, with an average of only 0.4 Bengal Rose molecules associated per dendrimer (Haba 2005). The influence of the degree of PEG substitution on encapsulation efficiency and release characteristics of PEGylated G3 PAMAM dendrimers was examined by Pan et al. (2005). Using methotrexate as a model drug, it was found that the degree of substitution had little effect on encapsulation efficiency, suggesting that the drug was localized within the dendrimer rather than the surrounding PEG chains. The degree of substitution was reported to have an effect on release characteristics, but the effect was not dramatic.

The synthesis of PEGylated dendritic systems as nanoparticulate depots for drug delivery was described by Bhadra et al. (2003). Amine-terminated G4 PAMAM dendrimers were synthesized and PEGylated using methoxy poly(ethylene glycol) (MPEG)-5000. A comparison was made between the properties of G4 PAMAM dendrimers and PEGylated dendrimers. The PEGylated systems had a higher drug-loading capacity (12-fold for 5-fluorouracil), a slower drug release rate (1/6th), and a decreased toxicity, compared with the non-PEGylated dendrimers. Sideratou et al. investigated the solubilization and release properties of PEGylated diaminobutane (DAB)–PPI dendrimers using pyrene, betamethasone valerate (BV), and bethamethasone dipropionate (BD) (Paleos 2004). Two types of PEGylated dendrimers – weakly PEGylated (DAB64–4PEG) and densely PEGylated (DAB64–8PEG) dendrimers – along with DAB64 dendrimers, were used for the study. Pyrenes as well as BD and BV were successfully solubilized within the interiors of the dendrimer but only partially in PEG coat. Densely PEGylated derivatives solubilized higher concentrations of pyrene. For DAB64–8PEG the loading was 13 wt% and 7 wt% for BV and BD, respectively, whereas for DAB64–4PEG it was 6 wt% and 4 wt%, respectively. The enhanced solubilization of these drugs in PEGylated dendrimers suggests their application as promising controlled-release drug carriers.

4.3.4 Surface Modification to Increase Biological Half-Life

Low accumulation of dendrimers in nontarget organs is a desirable feature for many drug delivery approaches. A longer half-life is generally required to obtain passive tumor targeting via the enhanced permeability and retention (Maeda 2000) effect. Because the rate of renal filtration is based on hydrodynamic volume, one approach to increase the half-life is to make the dendrimer larger. However, the synthesis of well-defined, high-generation dendrimers is time-consuming and, as a result of the globular architectures of the dendrimers, the increase in hydrodynamic volume is modest. In an alternative approach, hybrids of polyester dendrimers and poly(ethylene oxide) star polymers have been prepared with an increase in molecular weight up to 22,000 (Gillies 2005; Gillies and Frechet 2002). Poly(ethylene oxide) was chosen because it is biocompatible and available in polydispersity index of as low as 1.02, thereby providing hybrids with similar polydispersity. Hybrid polymers conjugated to doxorubicin via a hydrazone linkage were prepared. The hydrazone linkage is stable at the physiological pH of 7.4, but is designed to undergo hydrolysis upon endocytosis and subsequent trafficking to mildly acidic subcellular organelles such as endosomes and lysosomes. These drug conjugates show an increased circulation time of more than 1 h and are toxic to a range of tumor cell lines.

PEGylation of dendrimers has been shown to enhance biodistribution and half-life. Okuda et al. (2006a) reported the comparison of biodistribution after intravenous injection of amino-acid-based dendrimers (dendritic poly(L-lysine)s (DPKs) and dendritic poly(L-ornithine)s (DPOs) with PEGylated generation 6 DPKs. Different generations DPKs and DPOs showed similar biodistribution profiles with rapid elimination from the blood stream and significant accumulation in the liver and kidney. In contrast, PEGylation of generation 6 DPKs, showed increased blood retention, decreased hepatic accumulation, and almost no renal accumulation. The authors that hypothesized PEGylation reduced the nonspecific interaction between the dendrimers and biomolecules, thus improving blood retention and biodistribution. The same group proceeded to study the impact of PEGylation of generation 6 DPK dendrimers on their biodistribution in tumor-bearing mice (Okuda 2006b). Generation 6 DPK dendrimers with 76 PEG chains showed significantly more tumor accumulation (14.6% of dose per g tumor), compared with unmodified dendrimers (2.2% of dose per g tumor), because of the enhanced permeability and retention effect, again illustrating the ability of PEGylation to improve biodistribution for drug delivery.

5 Dendrimers as Oral Drug Delivery Agents

5.1 Transepithelial Transport of PAMAM Dendrimers

Few reports are available on the ability of dendrimers to cross gastrointestinal (GI) membranes. Wiwattanapatapee et al. (2000) first suggested the use of PAMAM dendrimers for oral drug delivery. It was observed that transport of dendrimers across epithelial barriers was much faster when compared with that of other linear polymers such as poly(vinylpyrrolidone), poly(N-vinylpyrollidone-co-maleic anhydride), and N-(2-hydroxypropyl)methacrylamide copolymers (Wiwattanapatapee et al. 2000). Transport of cationic (G3 and G4) and anionic (G2.5, G3.5, and G5.5) PAMAM dendrimers was further investigated using everted rat intestinal sacs. For cationic dendrimers tissue uptake was higher than serosal transport, whereas for anionic dendrimers, the serosal transfer rate was higher than tissue uptake. These findings suggested that dendrimers could effectively transverse the epithelial layer of the gut, but that their transport was size and charge dependent. Specifically, G2.5 and G3.5 carboxyl-terminated dendrimers showed the highest serosal transfer rates. PAMAM dendrimers have also been used for improving the oral bioavailability of drugs. D'Emanuele et al. (2004) used G3 PAMAM dendrimers to decrease the efflux and improve the transport of propranolol, which is a substrate for P-glycoprotein efflux pump. Conjugates with 2, 4, and 6 molecules of propranolol and with propranol and lauroyl chains attached to G3 dendrimers were synthesized. A significant increase in apical-to-basolateral (AB) and decrease in basolateral-to-apical transport were observed when propranolol was conjugated to G3 dendrimer. Transport was unaffected when a physical mixture of propranolol and G3 dendrimer was used. This suggested that when covalently attached to G3, propranolol bypasses the P-glycoprotein efflux pump. The extent of AB enhancement was not dependent on the number of propranolol molecules attached. A further 3.5-fold enhancement of AB permeability was observed when six lauroyl chains were attached to the dendrimer. The lauroyl chains likely enhanced permeability by a surfactant effect. The mechanisms and relative contributions of the dendrimer cores vs. lauroyl chains to penetration enhancement need to be further evaluated.

Using Caco-2 monolayers, as a model of intestinal cells, we evaluated the permeability of fluorescently labeled cationic PAMAM–NH_2 dendrimers as a function of incubation time, generation number, and concentration (El-Sayed 2002). Smaller dendrimers (G0–G2) had similar AB permeability despite their different molecular weights. PAMAM permeability increased with time (from 90 to 150 min) and concentration (from 1.0 to 10.0 mM). These smaller dendrimers exhibited higher permeability than did G3 while exerting no toxicity (evident by lactate dehydrogenase assay) at 1.0 mM toward Caco-2 cell monolayers. Subsequent studies were carried out to evaluate the effect of surface charge on permeability characteristics of dendrimers (Kitchens 2006). Permeability of FITC-labeled PAMAM dendrimers G2NH$_2$, G1.5-COOH, and G2OH with similar molecular weight but different surface charges was evaluated at nontoxic concentrations using Caco-2 cell monolayers. An increase in permeability of amine-terminated PAMAM, compared to hydroxyl- and carboxyl-terminated dendrimers, was attributed to their interaction with the negatively charged cell membrane. A similar trend was also observed with the effect of PAMAM dendrimers on the integrity of Caco-2 cell monolayers. The effect of PAMAM dendrimer charge, size, and incubation time on transepithelial electrical resistance (TEER), as well as permeability of known paracellular permeability marker (^{14}C mannitol), was evaluated. An increase in permeability of mannitol and a decrease in TEER values with increase in concentration and generation of amine-terminated PAMAM dendrimers were observed, while neutral PAMAM dendrimers did not produce any effect. A decline in TEER value and an increase in permeability of mannitol were observed in the presence of G2.5 and G3.5 PAMAM dendrimers but not in the presence of higher (4.5) or lower (0.5, 1.5, 2.5) generation dendrimers with negatively charged terminal acidic groups. Together these initial studies (El-Sayed 2002, 2003; Kitchens 2006) provided information about the influences of surface charge, generation, incubation time, and concentration of PAMAM dendrimers on toxicity and their transport across the epithelial barrier of the gut.

5.2 Effect of Surface Modification on Transport

Recently, we evaluated the influence of surface modification of amine-terminated G4 PAMAM dendrimers on permeability across Caco-2 cell monolayers (Kolhatkar 2007). The surface amine groups (50 and 94%) of G4 were functionalized to acetamide to yield G4A32 and G4A60, respectively. Permeability and cellular uptake of native and surface acetylated PAMAM dendrimers were evaluated. Permeability enhancement was found to be a function of both dendrimer surface acetylation and donor concentration. At 0.01 mM, a significant increase in the permeability of G4A32 was observed when compared with unmodified G4 dendrimer. However, an additional increase in acetylation levels (e.g., G4A60) did not further enhance permeability. Although unmodified G4 is toxic at 0.1 mM, a 1.5- and twofold enhancement of cellular permeability was achieved by increasing the G4A32 and G4A60 concentration from 0.01 to 0.1 mM, respectively. The observed permeability enhancement of surface-modified PAMAM dendrimers could be the result of a decrease in hydrophilicity of the polymers leading to conformational changes and/or reduction in nonspecific binding. To further investigate their cellular entry mechanism, the uptake into Caco-2 cells of G4 surface-modified dendrimers was determined at nontoxic concentrations (0.01 mM). A significant decrease in initial uptake for acetylated G4 dendrimer was observed when compared with native dendrimer. In simultaneous studies, G4NH$_2$ dendrimers were conjugated with FITC at feed molar ratios of 1:1, 1:4, and 1:8 (G4NH$_2$/FITC) to investigate how an incremental increase in conjugation would modify the toxicity and permeability of the dendrimers (Kitchens 2006). The toxicity of G4NH$_2$ dendrimers reduced with an increase in FITC content. G4NH$_2$–FITC (1:8) displayed the highest relative permeability with an increase in mannitol flux, most likely due to a higher degree of tight junctional modulation.

Other studies have reported the influence of surface modification of cationic PAMAM dendrimers (G2 and G3) with longer molecules such as lauroyl chains or PEG on their permeability (Jevprasesphant et al. 2003a). PAMAM permeability generally increased with an increase in the number of lipid chains and concentration. The only exception was an observed decrease in permeability when the number of lauroyl chains attached to G2 dendrimers increased from 6 to 9. This was explained by the self-association of the conjugate, since aggregation was facilitated by the small size of G2 and the increased number of hydrophobic chains. In summary, native (unmodified) as well as surface-modified (e.g., acetyl or lauroyl) dendrimers can permeate intestinal barriers. This permeation is a function of size, charge, and the characteristics of the functional group attached. By carefully tuning these properties, it is possible to tailor-make dendritic structures useful for the oral delivery of poorly bioavailable and highly potent drugs.

5.3   Dendrimer Internalization

Limited information is available about the mechanism(s) of transport and trafficking of dendrimers across cellular barriers. Jevprasesphant et al. (2004) studied the transport mechanisms of PAMAM dendrimers across Caco-2 cells. Transport across these cells is particularly important because of its implications for oral drug delivery, in which dendrimers must traverse the intestinal barrier. The cellular transport mechanisms of G3 PAMAM dendrimers with and without surface-modifying lauroyl chains across Caco-2 cell monolayers were evaluated using flow cytometry, confocal microscopy, and transmission electron microscopy. A significant degree of FITC-labeled G3 dendrimer internalization was observed, with little loss of fluorescence after extracellular quenching. A higher cell fluorescence intensity was observed with the lauroyl-modified dendrimers, further confirming the enhancing effect of these functional groups. TEM studies demonstrated that dendrimer–gold composites were found on the apical membrane, in the membrane invaginations, and in multivessicular bodies, indicating internalization by endocytosis.

Recently, Seib et al. (2007) investigated the endocytosis of linear and branched PEI polymers and G2 G4 cationic PAMAM dendrimers into B16f10 melanoma cells. The objective of this work was to investigate the effect of polymer architecture on endocytosis and intracellular trafficking. Polymers were fluorescently labeled using oregon green. All polymers studied showed internalization as well as extracellular binding, but little evidence of exocytosis within the 60-min experimental period was present. PAMAM G4 dendrimers had the highest rate of internalization, followed by branched PEI, linear PEI, G3 and G2 dendrimers. MβCD, an inhibitor of cholesterol-dependent endocytosis, significantly reduced the internalization of PEI branched polymers and G4 dendrimers, while it led to an increase in uptake of linear PEI, showing that these polymers are endocytosed by different mechanisms.

Previous studies in our laboratory suggested the contribution of an energy-dependent process to PAMAM transport across Caco-2 cells (El-Sayed 2003). Temperature-dependent studies demonstrated that permeability of cationic and anionic dendrimers was lower at 4°C than at 37°C. Subsequent studies further investigated the specific mechanism(s) of endocytosis of PAMAM dendrimers by fluorescently labeling dendrimers with FITC and visualizing colocalization with endocytosis markers by confocal microscopy (Kitchens 2007). Clathrin was used as a classical endocytosis marker involved in coated pit formation upon membrane invagiation, and showed significant colocalization with the transferrin which was used as a positive control. Dendrimers showed significant colocalization with clathrin, implying that they were internalized through clathrin-dependent endocytosis (Fig. 4). LAMP-1 (lysosome-associated membrane protein 1), a lysosomal marker,

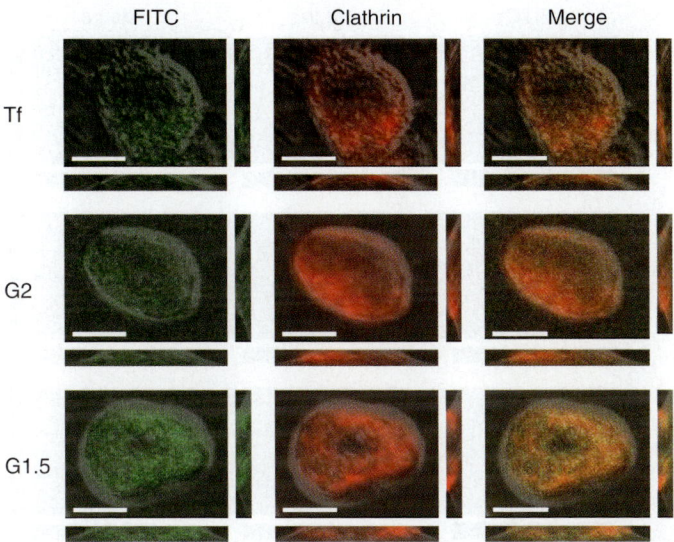

Fig. 4 Internalization of fluorescently labeled transferrin (250 µg mL⁻¹) and PAMAM dendrimers (100 nM) in Caco-2 cells. The orange color in merged panels indicates colocalization with clathrin heavy chain after 20 min. *Main panels* illustrate the *xy* plane; *vertical panels* illustrate the *yz* plane; *horizontal panels* illustrate the *xz* plane. Scale bars = 5 µm. Reprinted from Kitchens (2007), with permission from Springer

showed much lower colocalization with transferrin and also low colocalization with dendrimers (Fig. 5). Interestingly, dendrimers showed high colocalization with clathrin regardless of incubation time, indicating the constant presence of dendrimers in early endosomes, while colocalization with LAMP-1 increased proportionally with time, suggesting time-dependent lysosomal trafficking of dendrimers (Table 3) (Kitchens 2007). When examining the effect of dendrimer surface charge, we found that G1.5 carboxyl-terminated dendrimers had much greater colocalization with LAMP-1 at early incubation times, while G2 amine-terminated dendrimers showed increased lysosomal localization after longer incubation times, indicating time and charge dependence of dendrimer trafficking. Therefore, it appears that PAMAM generation and surface functionality influence not only their toxicity and permeability, but also the mechanisms and extent of intracellular trafficking across Caco-2 cells.

To confirm the route of endocytosis, we investigated the uptake and apparent permeability of G4 PAMAM dendrimers in Caco-2 cells in the presence of four endocytosis inhibitors (Kitchens 2008). Specifically, brefeldin A, colchicine, filipin, and sucrose were investigated to elucidate the endocytosis mechanisms that transport PAMAM dendrimers across Caco-2 cell monolayers. Brefeldin A and colchicines are known to interfere with microtubule tracking, while filipin regulates caveolae-mediated endocytosis and sucrose inhibits clathrin recycling. [³H]riboflavin was used as a positive control for endocytosis, as it is known to be internalized by

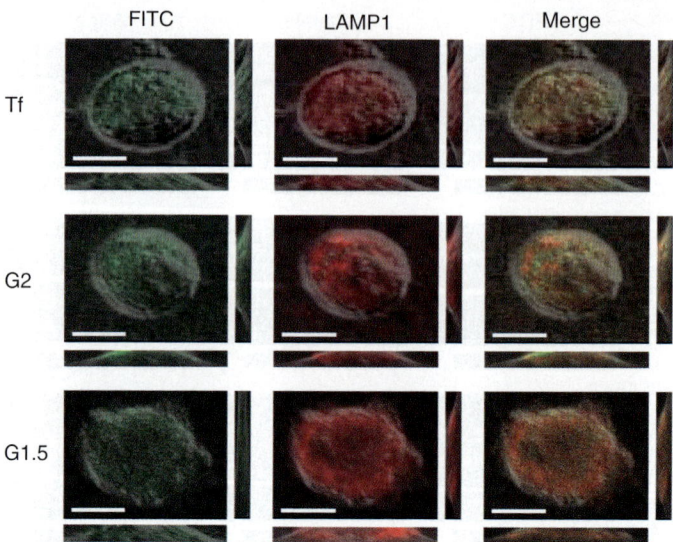

Fig. 5 Internalization of fluorescently labeled transferrin (250 µg mL⁻¹) and PAMAM dendrimers (100 nM) in Caco-2 cells. The orange color in merged panels indicates colocalization with lysosome-associated membrane protein 1 (LAMP-1) after 20 min. *Main panels* illustrate the *xy* plane; *vertical panels* illustrate the *yz* plane; *horizontal panels* illustrate the *xz* plane. Scale bars = 5 µm. Reprinted from Kitchens (2007), with permission from Springer

Table 3 Extent of colocalization between transferrin and PAMAM dendrimers with endocytosis markers

Incubation time (min)	Clathrin (%)	EEA-1 (%)	LAMP-1 (%)
Transferrin			
20	80.3 ± 3.1	70.4 ± 7.0	63.9 ± 7.3
60	83.4 ± 0.2	79.6 ± 0.4	57.0 ± 8.6
G2NH₂			
20	74.3 ± 5.2	76.7 ± 3.4	37.4 ± 5.9
60	73.7 ± 3.8	72.1 ± 0.4	59.0 ± 5.2
G1.5COOH			
20	70.8 ± 3.9	60.1 ± 6.8	58.7 ± 2.0
60	75.3 ± 1.2	53.8 ± 9.7	48.9 ± 1.3

Colocalization coefficients (M_x) are given as mean ± standard error of the mean ($n = 6$). Reprinted from Kitchens (2007), with permission from Springer
LAMP-1 lysosome-associated membrane protein 1

receptor-mediated endocytosis. The internalization of both [³H]riboflavin and [³H]-G4NH₂ dendrimers increased with increasing concentration. Significant reductions in adsorption rates were observed for all four endocytosis inhibitors with high concentrations of [³H]riboflavin. While all the inhibitors decreased [³H]riboflavin apparent permeability almost equally, they affected the permeability of [³H]-G4NH₂ to different degrees (Fig. 6). Based on this data and the specific endocytosis mecha-

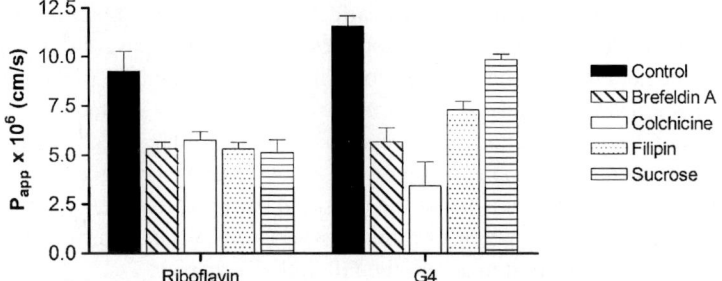

Fig. 6 Reduced apparent permeability ($P_{app} \times 10^{-6}$ cm s^{-1}) of [^{3}H]riboflavin (500 nM) and [^{3}H]-G4NH$_2$ (1 µM) across Caco-2 cell monolayers in the absence (control) and presence of various endocytosis inhibitors: brefeldin A (5 µm), colchicine (10 µm), filipin (1 µg mL^{-1}) and sucrose (200 mM). Results are given as mean ±SD (n = 3). Reprinted from Kitchens et al. (2007) with permission from ACS

nisms regulated by the inhibitors, it is unlikely that PAMAM dendrimers are transported by receptor-mediated endocytosis; instead, G4NH$_2$ is probably nonspecifically internalized partly by clathrin vesicles. Understanding the transport mechanisms of these potential drug carriers in Caco-2 cells paves the way for future oral drug delivery applications of PAMAM dendrimers. Taken together, these studies show that dendrimers are in part internalized by endocytic mechanisms and their trafficking is affected by size and charge. Using this knowledge, we can custom-tailor dendrimers to a given drug delivery application.

6 Conclusions and Future Directions

Dendrimers' unique branched architecture, monodispersity, and nanosized structure make them particularly useful for delivery of bioactive agents. There are a number of ways that dendrimers can be surface-functionalized depending on the desired application. It is critical to systematically investigate how these modifications of chemistry and structure influence the dendrimers' physiocochemical properties and subsequent biological behavior. Studies are emerging about the influence of generation number, surface charge, and surface modification on the interaction of dendrimers with cell surfaces and their subcellular trafficking. A unique advantage of dendritic structures over other random linear polymers is that surface topology and functionality can be controlled to a higher extent. As a result, intriguing studies can be conducted to systematically evaluate the influence of such control over interaction with cell surfaces and subsequent effects on subcellular pathways. Limited in vivo studies have been performed to investigate how dendrimer chemistry and architecture affect biodistribution, pharmacokinetics, biological half-life and toxicity. These studies are needed to pave the way for the use of dendrimers for systemic delivery of drugs in human clinical trials. Although the potential of PAMAM and other dendrimers in oral drug delivery has been studied, to date there is no report of the utility of such approach for the effective delivery of therapeutic agents. Successful transport of stable polymer–drug conjugates (or complexes) in the GI tract and across the intestinal barrier at therapeutically relevant dosages are prerequisites. More exciting prospects are to target drugs to specific sites once they are transported across the GI epithelial barrier.

Acknowledgments Support was provided in part by a Multidisciplinary Postdoctoral Fellowship from the Department of Defense to Rohit Kolhatkar (W81XWH-06-1-0698), a Graduate Research Fellowship to Deborah Sweet from National Science Foundation, a University System of Maryland Integrated Nanobio seed grant from the Maryland Department of Business and Economic Development and NIHR01Eb07470.

References

Agashe, H.B., et al., 2006. *Investigations of the toxicological profile of functionalized fifth-generation poly(propylene imine) dendrimer.* J Pharm Pharmacol, **58**(11): 1491–8.

Agrawal, P., U. Gupta, and N.K. Jain, 2007. *Glycoconjugated peptide dendrimers-based nanoparticulate system for the delivery of chloroquine phosphate.* Biomaterials, **28**(22): 3349–59.

Ahn, T.S., et al., 2006. *Light-harvesting in carbonyl-terminated phenylacetylene dendrimers: the role of delocalized excited states and the scaling of light-harvesting efficiency with dendrimer size.* J Phys Chem B Condens Matter Mater Surf Interfaces Biophys, **110**(40): 19810–9.

Andre, S., et al., 1999. *Lactose-containing Starburst dendrimers: influence of dendrimer generation and binding-site orientation of receptors (plant/animal lectins and immunoglobulins) on binding properties.* Glycobiology, **9**(11): 1253–61.

Ashton, P.R., et al., 1997. *Synthesis of glycodendrimers by modification of poly(propylene imine) dendrimers.* Chem – Eur J, **3**(6): 974–84.

Beezer, A.E., et al., 2003. *Dendrimers as potential drug carriers: encapsulation of acidic hydrophobes within water soluble PAMAM derivatives.* Tetrahedron, **59**: 3873–80.

Bhadra, D., et al., 2003. *A PEGylated dendritic nanoparticulate carrier of fluorouracil.* Int J Pharm, **257**(1/2): 111–24.

Bosman, A.W., H.M. Janssen, and E.W. Meijer, 1999. *About dendrimers: structure, physical properties, and applications.* Chem Rev, **99**(7): 1665–88.

Chandrasekar, D., et al., 2007. *Folate coupled poly(ethyleneglycol) conjugates of anionic poly(amidoamine) dendrimer for inflammatory tissue specific drug delivery.* J Biomed Mater Res A, **82**(1): 92–103.

Chauhan, A.S., et al., 2003. *Dendrimer-mediated transdermal delivery: enhanced bioavailability of indomethacin.* J Control Release, **90**(3): 335–43.

Couck, P., et al., 2005. *Evaluation of the stratus CS fluorometer for the determination of plasma myoglobin.* Acta Clin Belg, **60**(2): 75–8.

Crampton, H.L. and E.E. Simanek, 2007. *Dendrimers as drug delivery vehicles: non-covalent interactions of bioactive compounds with dendrimers.* Polym Int, **56**: 489–96.

D'Emanuele, A. and D. Attwood, 2005. *Dendrimer–drug interactions.* Adv Drug Deliv Rev, **57**(15): 2147–62.

D'Emanuele, A., et al., 2004. *The use of a dendrimer–propranolol prodrug to bypass efflux transporters and enhance oral bioavailability.* J Control Release, **95**(3): 447–53.

Delort, E., et al., 2006. *Synthesis and activity of histidine-containing catalytic peptide dendrimers.* J Org Chem, **71**(12): 4468–80.

Devarakonda, B., et al., 2005. *Comparison of the aqueous solubilization of practically insoluble niclosamide by polyamidoamine (PAMAM) dendrimers and cyclodextrins.* Int J Pharm, **304**(1/2): 193–209.

Dhanikula, R.S. and P. Hildgen, 2007. *Influence of molecular architecture of polyether-co-polyester dendrimers on the encapsulation and release of methotrexate.* Biomaterials, **28**(20): 3140–52.

Dufes, C., I.F. Uchegbu, and A.G. Schatzlein, 2005. *Dendrimers in gene delivery.* Adv Drug Deliv Rev, **57**(15): 2177–202.

Duncan, R. and L. Izzo, 2005. *Dendrimer biocompatibility and toxicity.* Adv Drug Deliv Rev, **57**(15): 2215–37.

Dutta, T. and N.K. Jain, 2007. *Targeting potential and anti-HIV activity of lamivudine loaded mannosylated poly(propyleneimine) dendrimer.* Biochim Biophys Acta, **1770**(4): 681–6.

Dutta, T., et al., 2007. *Poly(propyleneimine) dendrimer based nanocontainers for targeting of efavirenz to human monocytes/macrophages in vitro.* J Drug Target, **15**(1): 89–98.

Eichman, J.D., et al., 2003. *The use of PAMAM dendrimers in the efficient transfer of genetic material into cells.* Pharm Sci Technol Today, **3**(7): 232–45.

El-Sayed, M., et al., 2002. *Transepithelial transport of poly(amidoamine) dendrimers across Caco-2 cell monolayers.* J Control Release, **81**(3): 355–65.

El-Sayed, M., et al., 2003. *Influence of surface chemistry of poly(amidoamine) dendrimers on Caco-2 cell monolayers.* J Bioactive Compat Polym, **18**: 7–21.

El-Sayed, M., et al., 2003. *Transport mechanism(s) of poly(amidoamine) dendrimers across Caco-2 cell monolayers.* Int J Pharm, **265**(1/2): 151–7.

Esfand, R. and D.A. Tomalia, 2001. *Poly(amidoamine) (PAMAM) dendrimers: from biomimicry to drug delivery and biomedical applications.* Drug Discov Today, **6**(8): 427–36.

Fahmy, T.M., J.P. 2007. Schneck, and W.M. Saltzman, *A nanoscopic multivalent antigen-presenting carrier for sensitive detection and drug delivery to T cells.* Nanomedicine, **3**(1): 75–85.

Florence, A. (Ed.), 2005. *Dendrimers: a versatile targeting platform.* Adv Drug Deliv Rev, **57**(15): 2101–286.

Gillies, E.R. and J.M. Frechet, 2002. *Designing macromolecules for therapeutic applications: polyester dendrimer–poly(ethylene oxide) "bow-tie" hybrids with tunable molecular weight and architecture.* J Am Chem Soc, **124**(47): 14137–46.

Gillies, E.R., et al., 2005. *Biological evaluation of polyester dendrimer: poly(ethylene oxide) "bow-tie" hybrids with tunable molecular weight and architecture.* Mol Pharm, **2**(2): 129–38.

Gong, L.Z., Q.S. Hu, and L. Pu, 2001. *Optically active dendrimers with a binaphthyl core and phenylene dendrons: light harvesting and enantioselective fluorescent sensing.* J Org Chem, **66**(7): 2358–67.

Gupta, U., et al., 2006. *A review of in vitro–in vivo investigations on dendrimers: the novel nanoscopic drug carriers.* Nanomedicine, **2**(2): 66–73.

Haba, Y., et al., 2005. *Synthesis of biocompatible dendrimers with a peripheral network formed by linking of polymerizable groups.* Polymer, **46**: 1813–20.

Hawker, C.J. and J.M. Frechet, 1990. *Preparation of polymers with controlled molecular architecture.* A new convergent approach to dendritic macromolecules. J Am Chem Soc, **112**: 7638–47.

Hong, S., et al., 2007. *The binding avidity of a nanoparticle-based multivalent targeted drug delivery platform.* Chem Biol, **14**(1): 107–15.

Huang, R.Q., et al., 2007. *Efficient gene delivery targeted to the brain using a transferrin-conjugated polyethyleneglycol-modified polyamidoamine dendrimer.* FASEB J, **21**(4): 1117–25.

Islam, M.T., I.J. Majoros, and J.R. Baker Jr., 2005. *HPLC analysis of PAMAM dendrimer based multifunctional devices.* J Chromatogr B Anal Technol Biomed Life Sci, **822**(1/2): 21–6.

Jansen, J.F.G.A., E.M.M. 1994. Debrabandervandenberg, and E.W. Meijer, *Encapsulation of guest molecules into a dendritic box.* Science, **266**: 1226–9.

Jevprasesphant, R., et al., 2003a. *Engineering of dendrimer surfaces to enhance transepithelial transport and reduce cytotoxicity.* Pharm Res, **20**(10): 1543–50.

Jevprasesphant, R., et al., 2003b. *The influence of surface modification on the cytotoxicity of PAMAM dendrimers.* Int J Pharm, **252**(1/2): 263–6.

Jevprasesphant, R., et al., 2004. *Transport of dendrimer nanocarriers through epithelial cells via the transcellular route.* J Control Release, **97**(2): 259–67.

Kabanov, V.A., et al., 1998. *Polyelectrolyte behavior of astramol poly(propyleneimine) dendrimers.* Macromolecules, **31**(15): 5142–4.

Kitchens, K.M., M.E. El-Sayed, and H. Ghandehari, 2005. *Transepithelial and endothelial transport of poly(amidoamine) dendrimers.* Adv Drug Deliv Rev, **57**(15): 2163–76.

Kitchens, K.M., et al., 2006. *Transport of poly(amidoamine) dendrimers across Caco-2 cell monolayers: influence of size, charge and fluorescent labeling.* Pharm Res, **23**(12): 2818–26.

Kitchens, K.M., et al., 2005. *Endocytosis and interaction of poly(amidoamine) dendrimers with caco-2 cells.* Pharm Res, **24**(11): 2138–2145.

Kitchens, K.M., et al., *Endocytosis inhibitors prevent poly(amidoamine) dendrimer internalization and permeability across Caco-2 cells.* Mol Pharm, in press.

Kobayashi, H. and M.W. Brechbiel, 2005. *Nano-sized MRI contrast agents with dendrimer cores.* Adv Drug Deliv Rev, **57**(15): 2271–86.

Kobayashi, H., et al., 2004. *Application of a macromolecular contrast agent for detection of alterations of tumor vessel permeability induced by radiation.* Clin Cancer Res, **10**(22): 7712–20.

Kojima, C., et al., 2000. *Synthesis of polyamidoamine dendrimers having poly(ethylene glycol) grafts and their ability to encapsulate anticancer drugs.* Bioconjug Chem, **11**(6): 910–7.

Kolhatkar, R.B., et al., 2007. *Surface acetylation of poly(amidoamine) (PAMAM) dendrimers decreases cytotoxicity while maintaining membrane permeability.* Bioconjug Chem, **18**(6): 2054–2060.

Kolhe, P., et al., 2004. *Hyperbranched polymer–drug conjugates with high drug payload for enhanced cellular delivery.* Pharm Res, **21**(12): 2185–95.

Kolhe, P., et al., 2006. *Preparation, cellular transport, and activity of polyamidoamine-based dendritic nanodevices with a high drug payload.* Biomaterials, **27**(4): 660–9.

Konda, S.D., et al., 2000. *Development of a tumor-targeting MR contrast agent using the high-affinity folate receptor: work in progress.* Invest Radiol, **35**(1): 50–7.

Konda, S.D., et al., 2001. *Specific targeting of folate–dendrimer MRI contrast agents to the high affinity folate receptor expressed in ovarian tumor xenografts.* Magma, **12**(2/3): 104–13.

Koyama, Y., et al., 2007. *A dendrimer-based nanosized contrast agent dual-labeled for magnetic resonance and optical fluorescence imaging to localize the sentinel lymph node in mice.* J Magn Reson Imaging, **25**(4): 866–71.

Langereis, S., et al., 2006. *Evaluation of Gd(III)DTPA-terminated poly(propylene imine) dendrimers as contrast agents for MR imaging.* NMR Biomed, **19**(1): 133–41.

Lee, C.C., et al., 2005. *Designing dendrimers for biological applications.* Nat Biotechnol, **23**(12): 1517–26.

Lee, C.C., et al., 2006. *A single dose of doxorubicin-functionalized bow-tie dendrimer cures mice bearing C-26 colon carcinomas.* Proc Natl Acad Sci USA, **103**(45): 16649–54.

Liu, M., K. Kono, and J.M. Frechet, 2000. *Water-soluble dendritic unimolecular micelles: their potential as drug delivery agents.* J Control Release, **65**(1/2): 121–31.

Ma, M., et al., 2007. *Evaluation of polyamidoamine (PAMAM) dendrimers as drug carriers of anti-bacterial drugs using sulfamethoxazole (SMZ) as a model drug.* Eur J Med Chem, **42**(1): 93–8.

Maeda, H., et al., 2000. *Tumor vascular permeability and the EPR effect in macromolecular therapeutics: a review.* J Control Release, **65**(1/2): 271–84.

Majoros, I.J., et al., 2005. *Poly(amidoamine) dendrimer-based multifunctional engineered nanodevice for cancer therapy.* J Med Chem, **48**(19): 5892–9.

Majoros, I.J., et al., 2006. *PAMAM dendrimer-based multifunctional conjugate for cancer therapy: synthesis, characterization, and functionality.* Biomacromolecules, **7**(2): 572–9.

Malik, N., E.G. Evagorou, and R. Duncan, 1999. *Dendrimer-platinate: a novel approach to cancer chemotherapy.* Anticancer Drugs, **10**(8): 767–76.

Malik, N., et al., 2000. *Dendrimers: relationship between structure and biocompatibility in vitro, and preliminary studies on the biodistribution of ^{125}I-labelled polyamidoamine dendrimers in vivo.* J Control Release, **65**(1/2): 133–48.

Mamede, M., et al., 2004. *Hepatocyte targeting of ^{111}In-labeled oligo-DNA with avidin or avidin-dendrimer complex*. J Control Release, **95**(1): 133–41.

Manunta, M., et al., 2004. *Gene delivery by dendrimers operates via a cholesterol dependent pathway*. Nucleic Acids Res, **32**(9): 2730–9.

Markatou, E., et al., 2007. *Molecular interactions between dimethoxycurcumin and PAMAM dendrimer carriers*. Int J Pharm, **339**(1/2): 231–6.

McCarthy, T.D., et al., 2005. *Dendrimers as drugs: discovery and preclinical and clinical development of dendrimer-based microbicides for HIV and STI prevention*. Mol Pharm, **2**(4): 312–8.

Milhem, O.M., et al., 2000. *Polyamidoamine Starburst dendrimers as solubility enhancers*. Int J Pharm, **197**(1/2): 239–41.

Miura, N., et al., 1999. *Complex formation by electrostatic interaction between carboxyl-terminated dendrimers and oppositely charged polyelectrolytes*. Langmuir, **15**(12): 4245–50.

Muller, C., et al., 2004. *Site-isolation effects in a dendritic nickel catalyst for the oligomerization of ethylene*. J Am Chem Soc, **126**(45): 14960–3.

Na, M., et al., 2006. *Dendrimers as potential drug carriers. II. Prolonged delivery of ketoprofen by in vitro and in vivo studies*. Eur J Med Chem, **41**(5): 670–4.

Najlah, M., et al., 2007. *In vitro evaluation of dendrimer prodrugs for oral drug delivery*. Int J Pharm, **336**(1): 183–90.

Nantalaksakul, A., et al., 2006. *Light harvesting dendrimers*. Photosynth Res, **87**(1): 133–50.

Newkome, G.R., et al., 1998. *Isocyanate-based dendritic building blocks: combinatorial tier construction and macromolecular-property modification*. Angew Chem Int Ed, **37**(3): 307–10.

Okuda, T., et al., 2006a. *Biodistribution characteristics of amino acid dendrimers and their PEGylated derivatives after intravenous administration*. J Control Release, **114**(1): 69–77.

Okuda, T., et al., 2006b. *PEGylated lysine dendrimers for tumor-selective targeting after intravenous injection in tumor-bearing mice*. J Control Rel, **116**(3): 330–6.

Ooya, T., J. Lee, and K. Park, 2003. *Effects of ethylene glycol-based graft, star-shaped, and dendritic polymers on solubilization and controlled release of paclitaxel*. J Control Release, **93**(2): 121–7.

Page, D., D. Zanini, and R. Roy, 1996. *Macromolecular recognition: effect of multivalency on the inhibition of binding of yeast mannan to concanavalin A and pea lectins by mannosylated dendrimers*. Bioorg Med Chem, **4**(11): 1949–61.

Paleos, C.M., et al., 2004. *Acid- and salt-triggered multifunctional poly(propylene imine) dendrimer as a prospective drug delivery system*. Biomacromolecules, **5**(2): 524–9.

Pan, G., et al., 2005. *Studies on PEGylated and drug-loaded PAMAM dendrimers*. J Bioactive Compat Polym, **20**(1): 113–28.

Patri, A.K., J.F. Kukowska-Latallo, and J.R. Baker Jr., 2005. *Targeted drug delivery with dendrimers: comparison of the release kinetics of covalently conjugated drug and non-covalent drug inclusion complex*. Adv Drug Deliv Rev, **57**(15): 2203–14.

Prieto, M.J., et al., 2006. *Nanomolar cationic dendrimeric sulfadiazine as potential antitoxoplasmic agent*. Int J Pharm, **326**(1/2): 160–8.

Pugh, V.J., et al., 2001. *Optically active BINOL core-based phenyleneethynylene dendrimers for the enantioselective fluorescent recognition of amino alcohols*. J Org Chem, **66**(18): 6136–40.

Quintana, A., et al., 2002. *Design and function of a dendrimer-based therapeutic nanodevice targeted to tumor cells through the folate receptor*. Pharm Res, **19**(9): 1310–6.

Reek, J.N., et al., 2002. *Core and periphery functionalized dendrimers for transition metal catalysis: a covalent and a non-covalent approach*. J Biotechnol, **90**(3/4): 159–81.

Roy, R., et al., 1993. *Solid-phase synthesis of dendritic sialoside inhibitors of influenza A virus haemagglutinin*. J Chem Soc Chem Commun, (24): 1869–72.

Sato, N., et al., 2003. *Synthesis of dendrimer-based biotin radiopharmaceuticals to enhance whole-body clearance*. Nucl Med Biol, **30**(6): 617–25.

Seib, F.P., A.T. Jones, and R. Duncan, 2007. *Comparison of the endocytic properties of linear and branched PEIs, and cationic PAMAM dendrimers in B16f10 melanoma cells*. J Control Release, **117**(3): 291–300.

Svenson, S. and D.A. Tomalia, 2005. *Dendrimers in biomedical applications – reflections on the field*. Adv Drug Deliv Rev, **57**(15): 2106–29.

Svobodova, L., et al., 2004. *Properties of mixed alkanethiol–dendrimer layers and their applications in biosensing.* Bioelectrochemistry, **63**(1/2): 285–9.

Tang, M.X., C.T. Redemann, and F.C. Szoka Jr., 1996. *In vitro gene delivery by degraded polyamidoamine dendrimers.* Bioconjug Chem, **7**(6): 703–14.

Thomas, T.P., et al., 2005. *Targeting and inhibition of cell growth by an engineered dendritic nanodevice.* J Med Chem, **48**(11): 3729–35.

Tomalia, D.A. and J.M. Frechet, *Dendrimers and other dendritic polymers.* Wiley, 2002.

Tsutsumiuchi, K., K. Aoi, and M. Okada, 1999. *Globular carbohydrate macromolecule "Sugar Balls" IV. Synthesis of dendritic nanocapsules with molecular recognition sites on periphery.* Polym J, **31**(11-1): 935–41.

Vandamme, T.F. and L. Brobeck, 2005. *Poly(amidoamine) dendrimers as ophthalmic vehicles for ocular delivery of pilocarpine nitrate and tropicamide.* J Control Release, **102**(1): 23–38.

Vrasidas, I., et al., 2001. *Synthesis of lactose dendrimers and multivalency effects in binding to the cholera toxin B subunit.* Eur J Org Chem, **2001**(24): 4685–92.

Wada, K., et al., 2005. *Improvement of gene delivery mediated by mannosylated dendrimer/ alpha-cyclodextrin conjugates.* J Control Release, **104**(2): 397–413.

Wang, J.L., et al., 2006. *Nanosized gradient pi-conjugated thienylethynylene dendrimers for light harvesting: synthesis and properties.* Org Lett, **8**(11): 2281–4.

Wang, X., et al., 2007. *Synthesis, characterization, and in vitro activity of dendrimer–streptokinase conjugates.* Bioconjug Chem, **18**(3): 791–9.

Wiwattanapatapee, R., et al., 2000. *Anionic PAMAM dendrimers rapidly cross adult rat intestine in vitro: a potential oral delivery system?* Pharm Res, **17**(8): 991–8.

Wiwattanapatapee, R., L. Lomlim, and K. Saramunee, 2003. *Dendrimer conjugates for colonic delivery of 5-aminosalicylic acid.* J Control Release, **88**(1): 1–9.

Woller, E.K., et al., 2003. *Altering the strength of lectin binding interactions and controlling the amount of lectin clustering using mannose/hydroxyl-functionalized dendrimers.* J Am Chem Soc, **125**(29): 8820–6.

Wu, G., et al., 2006a. *Targeted delivery of methotrexate to epidermal growth factor receptor-positive brain tumors by means of cetuximab (IMC-C225) dendrimer bioconjugates.* Mol Cancer Ther, **5**(1): 52–9.

Wu, G., et al., 2007. *Molecular targeting and treatment of an epidermal growth factor receptor-positive glioma using boronated cetuximab.* Clin Cancer Res, **13**(4): 1260–8.

Wu, L., et al., 2006b. *Phosphine dendrimer-stabilized palladium nanoparticles, a highly active and recyclable catalyst for the Suzuki–Miyaura reaction and hydrogenation.* Org Lett, **8**(16): 3605–8.

Yang, H. and S.T. Lopina, 2003. *Penicillin V-conjugated PEG–PAMAM star polymers.* J Biomater Sci Polym Ed, **14**(10): 1043–56.

Yin, R., et al., 2001. *Dendrimer-based alert ticket: a novel nanodevice for bio-agent detection.* Polym Mater Sci Eng, **84**: 856.

Yiyun, C. and X. Tongwen, 2005a. *Dendrimers as potential drug carriers. I. Solubilization of non-steroidal anti-inflammatory drugs in the presence of polyamidoamine dendrimers.* Eur J Med Chem, **40**(11): 1188–92.

Yiyun, C. and X. Tongwen, 2005b. *Solubility of nicotinic acid in polyamidoamine dendrimer solutions.* Eur J Med Chem, **40**(12): 1384–9.

Yiyun, C., X. Tongwen, and F. Rongqiang, 2005. *Polyamidoamine dendrimers used as solubility enhancers of ketoprofen.* Eur J Med Chem, **40**(12): 1390–3.

Zeng, F. and S.C. Zimmerman, 1997. *Dendrimers in supramolecular chemistry: from molecular recognition to self-assembly.* Chem Rev, **97**(5): 1681–712.

Zhang, H., et al., 1999. *Interaction of a polycation with small oppositely charged dendrimers.* J Phys Chem B, **103**: 2347–54.

Zhuo, R.X., B. Du, and Z.R. Lu, 1999. *In vitro release of 5-fluorouracil with cyclic core dendritic polymer.* J Control Release, **57**(3): 249–57.

Biodegradable Multitargeting Nanoconjugates for Drug Delivery

Julia Y. Ljubimova, Keith L. Black, Alexander V. Ljubimov, and Eggehard Holler

1 Introduction

Combination of several treatment regimens such as chemotherapy, radiotherapy, and surgery was always more beneficial for cancer patients than monotherapy. Advances in combination chemotherapy in the 1970s brought improvement of life quality and prolonged patient survival. New types of radiotherapy and progress in noninvasive surgery in the mid-1990s were other milestones in cancer treatment. With human genome sequencing and rapid development of individual medicine approaches, new technologies are needed for successful treatment of cancer and many other pathological conditions. The road map for cancer treatment for the first two decades of twenty-first century is likely to focus on simultaneous inhibition of several cancer-specific molecular markers and/or several altered pathways or on conventional chemotherapy in combination with prevention of synthesis of tumor-specific genes/proteins. The new drug engineering strategies will be based on achievements of genomics and proteomics, production of monoclonal antibodies, nanotechnology, and bioinformatics.

They will combine conventional chemotherapy by agents damaging DNA synthesis with compounds targeting cell division and programmed cell death machinery, tumor invasion, and angiogenesis by blocking mRNA and/or protein synthesis. This complex approach has become possible due to modern technologies that allow overcoming problems with specific tumor targeting and drug delivery step by step.

One of these rapidly emerging and innovative technologies concerns polymeric nanoconjugates. They bear numerous chemically functional groups, which allow covalent attachment of a variety of biochemically active groups to precisely target a malignant tumor and achieve a highly efficient treatment. An obvious advantage of polymeric nanoconjugates is that they are capable of bearing functional drugs acting on several tumor targets (e.g., mRNA and/or protein) at the same time. Multitargeting will provide effective inhibition of several tumor pathways, optimal drug concentrations at the site of treatment, minimal adverse effects on healthy tissue, and a choice of more than one prodrug bound to each single conjugate molecule. This is likely to result in simultaneous delivery and synergistic effects in the recipient tumor cell.

To be highly effective, a nanoconjugate should be synthesized around a polymer with pendant functional groups like –OH, –COOH, or –NH$_2$. Synthesis will require

V. Torchilin (ed.), *Multifunctional Pharmaceutical Nanocarriers*,
© Springer Science+Business Media, LLC 2008

a hierarchy of stepwise chemical conjugation reactions, carefully avoiding uncontrollable side chemistry. An ideal nanoconjugate drug should be biodegradable, without prolonged accumulation of possibly harmful deposits and metabolites in nontargeted tissue and organs. Nonbiodegradable polymer platforms typically contain long chain all-carbon backbones without heteroatoms (O, N, S). Besides the prodrug, which may often be synthetic, biodegradable modules or their components can be obtained from natural sources or by appropriate chemical synthesis. Biodegradability should be understood to involve metabolic routes and final decomposition to water and carbon dioxide. Although it seems logical that multifunctional and biodegradable drugs should be of highest priority in cancer treatment, the availability of multifunctional drug-delivery systems remained scarce until now. We will discuss in more detail the Polycefin system of nanoconjugates with a biodegradable natural-derived polymeric platform. Its main advantage is related to the possibility of attaching various inhibitors of multiple molecular targets to the same nanoconjugate platform providing combination therapy with one "superdrug." This system offers an easily accessible route for developing highly efficient multitargeted drug carriers tailored to the need of individual patients.

2 Nanoconjugate Drug Delivery

Multitargeting is a powerful and successful strategy in modern chemotherapy aimed at mounting a multipronged attack on several molecular routes specifically altered in diseased tissue, in particular, cancer (Faivre et al., 2006; Konstantinopoulos et al., 2006; Montemurro et al., 2007; Zinner and Herbst, 2004). Targeting may be provided by the drug itself and/or by other molecules being part of the drug-delivery entity. Multitargeting may be achieved by a single targeting molecule aiming at parallel pathways, or by the simultaneous action of several targeting molecules, or by a combination of both (Faivre et al., 2006; Zinner and Herbst, 2004). Here, we will mainly address multiple targeting structures forming a single chemical entity with the prodrug and releasing free drug at the site of targeted tissue. The sizes of such chemical entities are of the order of 1–200 nm, qualifying them as nanoconjugates and nanodelivery systems. Nanoconjugate delivery systems are distinguished from nonconjugated nanodelivery vehicles (micelles, liposomes, etc.), which are physical entities of drug, targeting, and/or other functional molecules, but that do not form a chemical entity. We shall focus here on nanoconjugate drug-delivery systems with emphasis on the recently developed Polycefin family delivery system.

2.1 Nanoconjugate Concept

Particularly relevant to anticancer therapy is a need to reduce side effects that arise from toxicity to normal cells and to minimize cancer drug resistance. Nanoconjugates may employ several features to overcome this drawback of classical chemotherapy:

sustained release of drug, passive enhanced permeability and retention (EPR; Maeda et al., 2000) effect-based targeting of macromolecules to tumor tissue, ligand-based targeting of cell surface antigens, modules active in endosomal uptake, endosomal membrane disruption, drug release in the cytoplasm, protection from enzymatic degradation. One or more drug molecules and several or all of these modules are covalently bound to a high molecular mass platform thus forming the nanoconjugate. Polymers, as platforms able to deliver inhibitory agents to tumor cells, increasingly gain importance because they are less immunogenic than anti-body platforms or viral vectors, and minimize multidrug resistance (MDR) effect (Kabanov et al., 2002; Luo and Prestwich, 2002), thus being very useful for repetitive tumor treatment, and prevention of recurrences and metastases.

3 Targets and Targeting Concepts

Glivec®/Gleevec® (imatinib mesylate) was developed by Novartis as a break-through drug for chronic myeloid leukemia (CML) (see Heaney and Holyoake, 2007 for review). It was the first medicine that has been tailor-made for a specific pathogenetic gene product. Glivec inhibits the Bcr-Abl tyrosine kinase, which is a constitutively altered tyrosine kinase created by the Philadelphia chromosome abnormality in CML. It inhibits proliferation and induces apoptosis in Bcr-Abl-positive leukemic cells from Philadelphia chromosome-positive CML. Glivec has also dramatically impacted the management of gastro-intestinal stromal tumors (GIST) (Bickenbach et al., 2007). This therapeutic strategy known as "targeted therapy" represents the future of anticancer therapy. It paves the way for new principles of cancer drug development and manufacture.

The use of Glivec therapy for CML and GIST has brought new challenges including optimizing disease monitoring, drug resistance, and teratogenicity (Bickenbach et al., 2007; Heaney and Holyoake, 2007). They may necessitate further quest for novel, more potent, and/or broad-spectrum tyrosine kinase inhibitors. Thus, there was a need to establish new best practices for CML management in the post-imatinib era. This was confirmed by the responses of 956 eligible physicians from the US and Europe to an Internet-based questionnaire, consisting of 26 multiple-choice questions, distributed between November 2005 and January 2006 (Kantarjian et al., 2007).

Current progress of CML treatment is based on the development of drugs inhibiting several targets. One example is dasatinib, an orally active small molecule and a dual inhibitor of both Src and Abl kinases, which was developed by Bristol-Myers Squibb for the treatment of CML patients with imatinib-acquired resistance/intolerance. Whereas imatinib remains a frontline therapy for CML, patients with advanced disease frequently develop resistance to it through multiple mechanisms (Jabbour et al., 2007). One of the possible mechanisms of imatinib-acquired resistance is associated with increased expression of the Src-related kinase Lyn and loss of Bcr-Abl dependence arising from sequence mutations. Dasatinib is also undergoing preclinical evaluation for its potential as a therapy against multiple myeloma.

Novartis oncology is actively developing agents that block signaling by simultaneously targeting several different protein kinases by a drug AEE788 and other proteins involved in cell regulation by a drug RAD001 (Everolimus). AEE788 is an oral multiple-receptor tyrosine kinase inhibitor of EGFR, Her-2, and VEGFR. In preclinical studies, AEE788 showed target specificity, antiproliferative and antiangiogenic activity (Younes et al., 2006). RAD001 is blocking mTOR, a key controller of cell proliferation and angiogenesis. mTOR is a critical component of the PI3K/Akt pathway, the main survival pathway that is deregulated in many cancers and mediates the action of IGF, EGF, PDGF, and VEGF (Yao, 2007).

All the above drugs belong to the family of "small molecules" that are fast cleared through kidneys and demand high therapeutic concentrations, with cardio- or other toxicity as a side effect. Tumor specificity is also questionable, because the majority of targets are present in fast dividing nontumor tissues.

Drug targeting follows an onion skin principle in the order tissue, cell surface, cellular compartment, molecule (Fig. 1). General macromolecular targeting of

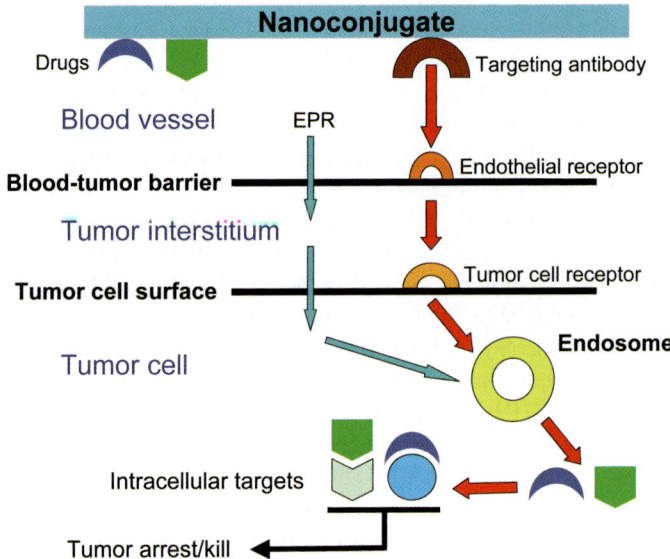

Fig. 1 Multitargeting nanoconjugate. A structure and mechanisms of action of a multitargeting polymeric nanoconjugate are shown. Nanoconjugate bears two drugs to different targets in the tumor cell. Targeting to tumor tissue is achieved by passive (EPR) effect, and by actively targeting module. In case of BTB, active targeting is achieved by an antibody to e.g., TfR. Another antibody ensures binding of the polymer to a tumor cell surface receptor, which can be TfR, insulin receptor, EGF receptor. Nanoconjugate arrives to the endosome inside tumor cells and the active drugs are released to the cytoplasm. Targeting of intracellular tumor targets may be either nonspecific (for example, by doxorubicin or cisplatin) or specific (for example, tumor markers are blocked by inhibitors of kinases or phosphatases, by antisense RNA, siRNA). Several tumor cell surface antigens together with tumor markers may be targeted. Targeting of several tumor cell surface antigens and tumor markers by active moieties on the same carrier molecule has a high probability of provoking favorable additive and/or synergistic effects

tumor tissue may be called "passive," whereas targeting of cell surface, compartments, and molecules, "active" (Leary et al., 2006). Specific targeting of tumor tissue is the advantage of modern macromolecular nanodelivery systems, resulting in high drug concentration only in the tissue of delivery with negligible toxicity for healthy tissue. Nanoconjugates may also be distinguished on the basis of their passive targeting through only EPR effect to tumor tissue and their additional ability of targeting tumor tissue cell surface and intracellular tumor markers.

Cell surface-specific targeting agents are used to direct the drug carrier exclusively to the desired tissue/tumor cell (Khandare and Minko, 2006). Specific cell binding of the carrier system allows delivering toxic drug that acts, for instance, on DNA (cisplatin, doxorubicin) and eliminates the targeted cell by a mechanism that would also kill healthy cells in the absence of cell specificity. In this case, selectivity is achieved by cell surface targeting. For tumor targeting, systemically i.e., intravenously administered drug has to pass one or more barriers before arriving at the tumor cell surface (Fig. 1). Important cases include blood–tumor barrier (BTB) and blood–brain barrier (BBB). Attachment of the drug-delivery molecule on vascular endothelium surface receptor, e.g., transferrin receptor (TfR), allows traversing BTB/BBB by transcytosis (Pardridge, 2002, 2005). Success of a brain tumor targeting drug-delivery vehicle depends on binding to both the endothelial TfR and tumor cell surface. Both receptor-binding activities have to reside on the same vehicle molecule. The role of cell surface targeting moiety on a nanodelivery vehicle is not only to bind selectively to the tumor tissue/cells, but also to provide necessary concentration of the delivered antitumor drug.

Enhanced targeting specificity is achieved if the delivery molecule carries and releases at once several kinds of drugs active exclusively on tumor-produced molecules, i.e., on tumor-specific proteins and mRNAs. Simultaneous delivery of several drugs targeting different tumor markers provokes a statistically higher chance for simultaneous delivery than a mixture of nanoconjugates carrying each drug separately. Typical drugs with high specificity are antisense RNA and small interfering RNA (siRNA) to tumor specific mRNA (tumor marker RNA). Experience has shown that targeting of a single tumor marker at a time may not be necessarily fatal for tumor cell/tissue. Targeting of several different markers at the same time potentially has a more powerful additive/synergistic impact. The delivery of several such drugs by the same delivery molecule ensures the simultaneous action against different tumor markers and thus the best antitumor effect. This concept of combined cell surface targeting agents and multiple tumor marker blocking agents on same drug-delivery molecule should prove the most powerful antitumor treatment. Nanoconjugate drug-delivery systems of this kind are already emerging.

4 History of Targeted Nanoconjugate Drug Delivery

Tissue targeting and sustained delivery were the primary motivations for the design of nanoconjugate drug delivery. Ehrlich (1906) founded the concept of macromolecular drug delivery targeting defined biological sites. Conjugation of drugs to synthetic and biological macromolecules was already used a long time ago (Jatzkewitz, 1955; Panarin and Ushakov, 1968), in particular, targeted delivery by conjugation to immunoglobulins (Mathe et al., 1958), before Ringsdorf (1975) formulated the principle of polymer as a platform for targeted drug delivery.

Antibodies are not only successful therapeutic agents in today's disease treatment (Brekke and Sandlie, 2003; Carter, 2006), but are also suitable agents for the purpose of biodegradable built-in cell surface targeting platforms for conjugated drug delivery (Schrama et al., 2006). Degradation of the polypeptide at the site of delivery provides release of the free active drug. Immunoconjugates and monoclonal antibodies (mAbs), such as doxorubicin (Doxil®), trastuzumab (Herceptin®), and bevacizumab (Avastin® and Lucentis®), are among FDA-approved drugs, and some others are in clinical trials (Schrama et al., 2006). Because of the immunogenicity of antibodies, unless adapted to human immune system (humanized antibodies), structurally simple polymers seem to be more appropriate as nanoconjugate platforms. However, up to now they lacked the ability for active targeting except for the passive EPR-mediated accumulation in tumor tissue.

The pioneering introduction by Kopecek of synthetic N-(2-hydroxypropyl) methacrylamide (HPMA) copolymers as carriers of various drugs taking advantage of pinocytosis and endosomal cleavage of a polymer-conjugated oligopeptide spacer was a milestone in the development of targeted drug delivery (Duncan, 1992; Kopecek et al., 1991, 2000, 2001). HPMA copolymers were synthesized by radical polymerization of methacryloyl derivatives (Kopecek and Baziliva, 1973). The development of HPMA copolymers as anticancer drug carriers was preceded by thorough investigations of their structure–property relationship and biocompatibility (Kopecek et al., 2000 and references therein). In first HPMA conjugates of anticancer drugs, such as doxorubicin, camptothecin, cis-platinate, tumor-EPR effect was the general mode of passive macromolecular tumor specificity. Only those nanoconjugate constructs were active that contained an endosomal cleavage site (peptide spacer recognized by endosomal peptidases or spontaneous cleavage of pH-sensitive hydrazone bonds (Etrych et al., 2001)) for drug release from the carrier. Thus the cleavable peptidyl spacer may be considered as a targeting device for drug release in endosomes. Additional targeting routes were developed by introducing HPMA copolymer-anticancer drug-antibody (Kunath et al., 2000), N-acylated galactosamine (Duncan et al., 1986), or short peptides (Omelyanenko et al., 1999) conjugate that targeted specific cells by recognizing their unique surface antigens or receptors, respectively. The reader is referred to a recent review on HPMA conjugates that target specifically cell surface and intracellular components (Nori and Kopecek, 2005). The importance of polymer architecture for drug delivery has been reviewed by Qiu and Bae (2006). Macromolecular components of the drug-delivery vehicle may also indirectly target and decrease tumor multidrug resistance in a structure-dependent fashion, thus overcoming limitations of chemotherapy with unconjugated free drugs (Nori and Kopecek, 2005). The success of nanoconjugate drug delivery, as manifested by FDA approval and clinical trials, as well as the emergence of "second generation" of nanoconjugate multidrug (cocktail) delivery systems have been reviewed recently (Duncan et al., 2005).

Whereas membrane-penetrating hydrophobic drugs passively diffuse from endosomes into cytoplasm after being released from nanoconjugates, the targeted delivery for hydrophilic drugs and especially for antisense oligonucleotides (AON) or siRNA required further development of the delivery technique. A notable exemption is HPMA copolymer derivatives, which were found to deliver AON in vitro to cell nuclei via the endosomal pathway without the help of additional membrane-penetration moiety (Jensen et al., 2002; Wang et al., 1998). A built-in endosomal membrane disruption device was recently successfully introduced for the endosomal escape and release of free morpholino AON from Polycefin nanoconjugate (Fujita et al., 2006; Lee et al., 2006). This nanoconjugate drug carrier device was designed to contain various functional structures (modules) active in multiple targeting.

Direct access to the cytoplasm is mediated by certain peptides, particularly TAT-derived peptides (Nori and Kopecek, 2005). In contrast to cellular uptake via the endocytotic pathway, however, it is not clear how specificity in cellular targeting is rigorously achieved in this case.

To preclude deposition in the host organism, the nanoconjugate platform should be biodegradable, i.e., be metabolized to water and carbon dioxide. The HPMA drug-delivery platform is not biodegraded and is not even cleared by the urinary pathway in cases of high molecular mass (>40,000) above renal threshold. By cross-linking short-sized HPMA-chains (molecular masses below the threshold of 40,000) with peptides, efficient sustained drug delivery could be achieved, whereas long-term systemic accumulation was avoided (Sprincl et al., 1976). This approach, however, does not account for the fate of persisting nondegradable polymer chains at the cellular level. Alternative biodegradable platforms have been described in nanoconjugate drug delivery, such as amino acid-derived polymers (Haider et al., 2004), especially poly(L-glutamic acid)s (Li, 2002), poly(malic acid) (Abdellaouri et al., 1998; Cammas et al., 1999; Fujita et al., 2006; Lee et al., 2006; Ouchi et al., 1990), polysaccharides (Sugahara et al., 2001), and polyethylene glycol (PEG) (Fleming et al., 2004; Tomlinson et al., 2003; Andersson et al., 2005). Polysaccharides conjugated with the drug have been used for targeting colon cancer after oral administration (Sinha and Kumria, 2001). A remarkable new development is the use of RNA oligonucleotide aptamer-siRNA chimeras that target both cell surface antigen and mRNA to inhibit synthesis of tumor-related proteins (McNamara et al., 2006). Whether they also profit from passive (EPR effect) tumor targeting remains to be investigated.

Farokhzad and co-workers (2006) conjugated docetaxel (Dtxl)-encapsulated nanoparticles (NP) of poly(D,L-lactic-co-glycolic acid)-block-poly(ethylene glycol) (PLGA-b-PEG) copolymer with the A10 2'-fluoropyrimidine RNA aptamers (Apt) that recognize prostate-specific membrane antigen (PSMA). After a single intratumoral injection of Dtxl-NP-Apt bioconjugates, complete tumor reduction was observed in five of seven LNCaP tumor cell xenografted nude mice and they all survived until day 109. In contrast, only two of seven mice in the Dtxl-NP group had complete tumor regression with 109-days survival rate of 57%. Dtxl alone had a survivability of only 14%. This report demonstrates the potential utility of nanoparticle-Apt bioconjugates for cancer therapy.

5   Building of Multifunctional Nanoconjugates

5.1   *Biodegradable Platform*

The use of biodegradable, also naturally occurring polymers, such as poly(aspartic acid), poly(glutamic acid)s, poly(malic acid), and polysaccharides, is preferred when the prevention of vehicle accumulation in treated patients is considered. As a rule of thumb, polymers that contain –O–, –NH–, –Phosphate, –S–, and –S–S– groups as main chain constituents are considered degradable. Such polymers may be hydrolytically cleaved and metabolized to water and carbon dioxide. Ideally, the same principle should hold for nanoconjugates. Furthermore, they should be nontoxic, nonimmunogenic and lack initiation of adverse reactions of the host such as complement response or agglutination. To minimize immunogenicity, a minimum of different kinds and amounts of amino acids are considered, and the conjugate-residing time in the patient should be minimal, a compromise between antitumor efficacy and initiation of adverse systemic responses.

Frequently considered biodegradable polymers of synthetic and biological origin, such as poly(aspartic acid), poly(glutamic acid), and poly(malic acid), have pendant carboxyl groups available for the chemical conjugation of drug and auxiliary delivery modules for targeting, protection, membrane disruption, and drug release. These platforms are used for systemic treatment after injection into the blood circulation system. Other biopolymers such as starch or chitosan and its derivatives, have been used as carriers for controlled oral drug delivery (Golenser et al., 1999; Deres et al., 2004, Roldo et al., 2004).

When choosing a biopolymer as carrier platform, very important properties to be considered are synthetic accessibility, stability, toxicity, or immunogenicity. For instance, both poly(L-glutamic acid) and β-poly(L-malic acid) each contain a pendant, chemically functional, carboxyl group per monomeric unit and a similarly high amount of such groups on a per gram basis. However, polypeptide [poly(L-glutamic acid)] has several disadvantages compared to polyester [β-poly(L-malic acid)]. Polypeptide is insoluble in particular organic solvents, which makes conjugation of groups that are functional in drug delivery more difficult and purification circumstantial. On the contrary, polyester is free from these shortcomings. Further, the poly(L-glutamic acid) chain is rather stiff due to hindered rotation around the peptide bonds. The β-poly(L-malic acid) is superior to poly(L-glutamic acid) in this respect because it is almost freely rotatable and allows the platform molecule to easily relax after conjugation of sterically demanding groups. Enzymatic cleavage of the peptide bonds in poly(L-glutamic acid) or poly(L-aspartic acid) during circulation is more likely than scission of poly(malic acid) ester bonds, which are not recognized by human enzymes. Poly(malic acid) at high doses either as free polymer or conjugated to protein did not provoke polymer-specific antibodies when injected into mice or rabbits (Lee et al., 2002, and unpublished data), whereas the production of antibodies against poly(γ-glutamate) that also recognized poly(α-glutamate), as well as those against poly-L-aspartate, has been reported (Murphy and Sage, 1970; Wang et al., 1993). Antibodies formed against $Glu_{60}Ala_{40}$ and $Glu_{60}Ala_{30}Tyr_{10}$ had dominant specificity against glutamyl residues (Maurer et al., 1964). Moreover, glutamate produced by enzymatic cleavage is known to be cytotoxic (Kumar et al., 2005; Ruiz et al., 2000) inducing apoptosis in neuronal cells (Zhang and Bhavnani, 2006) and may contribute to glaucoma (Gupta and Yucel, 2007). Poly(D-glutamate) caused lysosomal storage disorder (Kishore et al., 1990, 1996). Poly(γ-D-glutamate) is one of the toxic components in the capsule of *Bacillus anthracis* and induces IgG antibodies (Schneerson et al., 2003). Thus the immunogenic properties of the biopolymers and toxicity of the monomeric unit after cleavage of the polymer have to be seriously taken into account, especially in the case of repeated treatment of patients.

5.2 Functional Modules

Nanoconjugates delivering drugs in vivo are equipped with a variety of chemically conjugated modules. The term "modules" refers to functional units of a nanoconjugate that carries out a particular function in drug delivery, such as endosomal escape or drug release from the nanoconjugate. A module not necessary refers to a structural unit as is depicted by the endosomal escape unit of Polycefin (Lee et al., 2006) that comprises leucine ethylester and carboxyl groups distributed over the polymer platform.

A promising nanoconjugate for drug delivery contains several modules with different functions sequentially active during delivery (Fig. 2). A module has one of the following activities: (1) The platform assembles the entity of module. (2) The cell surface targeting modules, i.e., mAbs or peptides against typically tumor overexpressed antigens, such as TfR (Pardridge, 1999; Qian et al., 2002) or folate receptor (Bae et al., 2005; Chytil et al., 2006; Leamon and Reddy, 2004), chaperone the nanoconjugate through cellular barriers (BBB and BTB) and to the tumor cell surfaces. Cell surface targeting modules may or may not be structurally identical. (3) PEG is the module active in protection against degradation and rapid clearance from the circulation (Arpicco et al., 2002; Greenwald et al., 2003; Maruyama et al., 1997; Otsuka et al., 2003). (4) The drug-containing module carries the prodrug. The drug-releasing module connects the drug module to the platform. To release the drug within the endosome, the drug is connected with the platform by an enzymatically hydrolyzable peptide bond (Kopecek et al., 2000) or by a pH-dependent, spontaneously hydrolyzing, hydrazone bond (Bae et al., 2005; Chytil et al., 2006; West and Otto, 2005). To release the drug into the cytoplasm, a disulfide bridge between the platform and the drug is introduced that is cleaved by reduction with glutathione after the nanoconjugate escapes from the endosome (Lee et al., 2006; Saito et al., 2003; West and Otto, 2005). (5) The endosome escape module that is activated during maturation of endosomes to lysosomes. This module consists either of a combination of protonated and thus neutralized carboxylates and hydrophobic groups (Lee et al., 2006; Murthy et al., 1999; Philippova et al., 1997; Rozema et al., 2003; Turk et al., 2002), of ionizable, preferentially imidazole, moieties that become protonated during maturation (Merdan et al., 2002), or peptides that become lysogenic during endosome acidification (Cho et al., 2003; Mastrobattista et al., 2002; Merdan et al., 2002). (6) An optional fluorophore for nanoconjugate localization and imaging after administration (Lee et al., 2006).

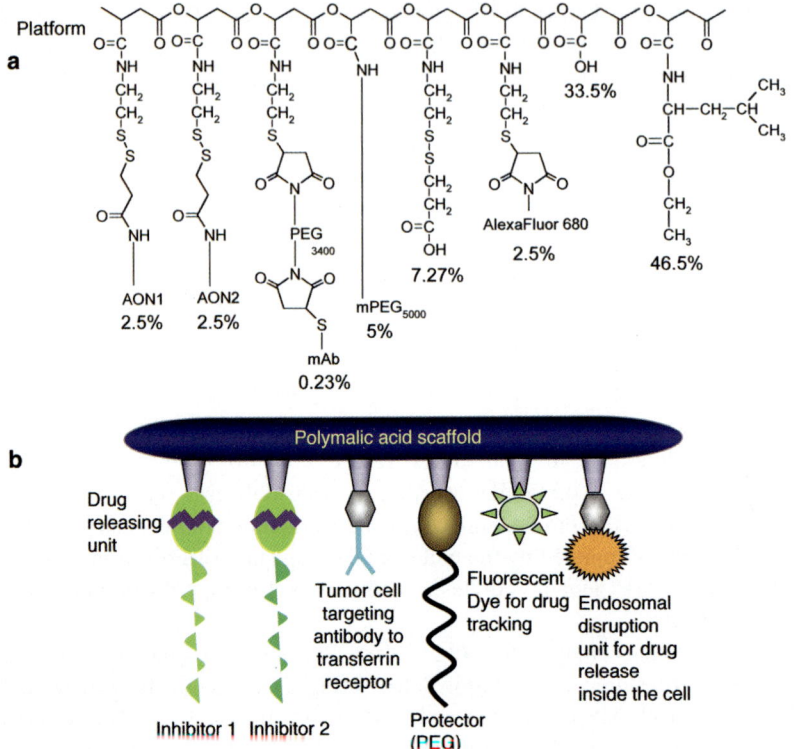

Fig. 2 Polycefin as an example of a multifunctional multitargeting nanoconjugate. **a** Drug-delivery vehicle structure. The percentage values refer to the fraction of pendant carboxyl groups of poly(malic acid). **b** Correlation of the modular structures of Polycefin with their functions. Drugs are morpholino AON to different subunits of laminin 411 (formerly, laminin-8), a marker of tumor angiogenesis. They inhibit the synthesis of α4 and β1 chains that together with γ1 chain form the complex protein laminin 411. In panel (**a**), the antibodies to transferrin receptor target either mouse BTB or human tumor cell surface and cause drug internalization by endocytosis. The drug-releasing module refers to a disulfide bond that is cleaved in the target cell cytoplasm by glutathione. The endosomal disruption module involves both carboxyl groups and hydrophobic leucine ethylester residues, allowing escape of Polycefin from the endosome at a point when during acidification of maturating endosome protonation of carboxyl groups and membrane insertion of hydrophobic patches formed by leucyl ethylester residues occur. Conjugation of a fluorescent dye to Polycefin is optional and allows following the fate of the nanoconjugate after administration in vitro or in vivo. From Ljubimova et al. 2007 (Chem. Biol. Interact. In press). *Reproduced by permission of Elsevier Ireland Ltd ©2007*

5.3 Multitargeting Property

In the simplest form, a polymer carries the drug bound via a drug-releasing module, thus providing tumor tissue targeting by the EPR effect. Nanoconjugate longevity due to absence or slow degradation and renal clearing provides sustained drug delivery to the tumor. An actively targeting nanoconjugate binds to surface antigens of tumor cells after passive delivery to the tumor by the EPR effect, or by combined EPR and active receptor-mediated permeation of the BTB or BBB (see Lee et al., 2006). In this case, the nanoconjugate contains, besides the polymer platform, the drug module and the drug-releasing module, one or more targeting modules that recognize receptors on the surfaces of BTB and tumor cells. The multitargeting is further elaborated if the drug itself targets specifically a tumor-functional molecule, for instance a cell surface tumor marker protein such as folate receptor (Reddy et al., 2005; Yoshida et al., 2006), TfR (Singh 1999; Thorstensen and Romslo 1993; Daniels et al., 2006b; Pardridge, 2002), EGFR and Her-2 (Solit and Rosen, 2007 and references therein), an intracellular tumor marker protein such as certain kinases and phosphatases, or a tumor marker-encoding mRNA by virtue of a carrier platform attached AON or siRNA.

6   Pathways of Nanoconjugate Drug Delivery

To understand the module functions in the context of multiple targeting, the pathway of a water-soluble nanoconjugate from the site of intravasal administration, e. g., of intravenous injection, along its route to the drug targeting effector site(s) is briefly considered. Various biomedical aspects of drug targeting have been extensively reviewed (Muzykantov and Torchilin, 2003). Once administered, the nanoconjugate lifetime is limited by at least three factors: (1) soluble and cellular degradative activities in the vascular system; (2) clearance by renal glomerular filtration; (3) uptake by the reticuloendothelial system (RES), particularly liver and spleen. A solution to the stability problem is the PEG module. Its task is to sterically diminish access of degrading proteases and nucleases, and to minimize recognition and elimination by blood cells (Papisov, 1998). To avoid fast kidney clearance, the delivery vehicle size should be above a certain threshold. For a polymer-based nanoconjugate, the size of the platform can be decisive. For HPMA nanoconjugates, the polymer molecular mass is tailored to higher than 40,000–50,000 (Sprincl et al., 1976). In general, however, the size is a function of the composition, reflecting the sizes of the platform and the modules, depending on their spatial arrangement and other parameters (Seymour et al., 1987); the overall size is less than proportional to the sum of molecular masses of the constituents.

One has to keep in mind that whereas increased sizes preclude rapid clearance by glomerular filtration, they favor the uptake and clearance by the reticuloendothelial system (Gaur et al., 2000). Various mechanisms and clearance-affecting factors have been reviewed (Moghimi et al., 2005). To access the tumor, the drug-delivery vehicle has to pass through the vascular endothelium into the interstitial fluid to target tumor cells. In the brain, the BBB is an impermeable interface. The endothelial cells lining the brain–blood vessels form the principle barrier, and their unique phenotype is a consequence of interactions with perivascular cell types including glia present in the brain parenchyma. The reactivity of efflux transporters in the BBB prevents many administered drugs from entering the brain (Taylor, 2002). Recognition of a tumor from within the blood vessels would be advantageous for targeted drug delivery, allowing rapid access to the tumor and avoiding encounter of toxic agents with healthy tissue. Targeting and transcytosis of the drug carrier via binding of its conjugated mAb to TfR or insulin receptor, overexpressed at BTB, are experimentally probed routes to cross BTB (Coloma et al., 2000; Lee et al., 2006; Pardridge, 2002; Zhang and Pardridge, 2005; Zhang et al., 2004). An overview on BBB drug targeting is presented by Pardridge (Pardridge, 2003). Fortunately, the vascular endothelium surrounding the tumor is imperfect and leaky contributing to the generation of the EPR effect. Once arrived to the interstitium, the drug vehicle has to penetrate through tumor tissue, probably by diffusion, its rate also depending on cell adhesion and cell density (Grantab et al., 2006).

After binding to a surface antigen typically overexpressed by tumor cells, the drug vehicle is internalized into early endosomes, which then, by fusion with primary lysosomes maturate, under concomitant acidification toward pH 5, into lysosomes. If the drug-delivering nanoconjugate is tailored in such a way as to be cleaved by enzymes contained in the maturing endosome, the drug is liberated from the carrier and is free to penetrate through the endosomal membrane into the cytoplasm. Low molecular mass drugs either find their target in the cytoplasm or diffuse passively into the nucleus to react with nucleic acids or nuclear proteins. In cases where drugs cannot penetrate the endosomal membrane, i.e., hydrophilic charged molecules, especially inhibitory RNA (AON or siRNA), the nanoconjugate has to be equipped with a module that disrupts the endosomal membrane for the escape of the delivery vehicle into the cytoplasm. The disrupting module contains residues that penetrate into the membrane and destabilize the lipid bilayer forming escape leakages. These residues are fusogenic peptides or hydrophobic residues and carboxylates, which are inactive at physiological pH, but form large hydrophobic membrane active patches after electrostatic neutralization through carboxylate protonation at decreasing pH of the maturating endosomes. Another kind of module contains ionizing, e.g., imidazole residues that have an appropriate pKa and ionize during endosome acidification. The ions and their counterions thus being accumulated in the endosome raise osmotic pressure and provoke membrane disruption.

The nanoconjugate can be designed to contain a disulfide group as drug-releasing module, which connects the drug with the carrier platform. Typically, the cytoplasm contains abundant glutathione that rapidly cleaves the bond to release free drug. This mechanism works very efficiently for AON or siRNA that become free to target specific mRNA in the cytoplasm for silencing the synthesis of targeted proteins (Lee et al., 2006).

7   Nanoconjugate Synthesis

7.1   Chemical Concept of the Nanoconjugate Synthesis

Chemical synthesis of nanoconjugates makes use of a variety of available synthetic and natural polymers that offer pendant carboxyl (–COOH), hydroxyl (–OH), or amino (–NH$_2$) groups for the conjugation of drugs and auxiliary drug-delivering modules. In some cases for HPMA (Duncan, 1992; Duncan et al., 1986; Etrych et al., 2001; Jensen et al., 2002; Kopecek and Baziliva, 1973; Kopecek et al., 2001; Kunath et al., 2000; Omelyanenko et al., 1999) or poly(malic acid) (Cammas et al., 1999) nanoconjugates, low molecular mass spacers for the attachment of large modules have been introduced by the polymerization or polycondensation of the appropriate monomeric units.

7.2 Biodegradable Nanoconjugate Platform as Exemplified by β-Poly(L-malic acid)

Biodegradable polymers are available as biosynthetic or man-made synthetic material. Examples are poly(glutamic acid), poly(aspartic acid), and poly(malic acid). In general, chemical synthesis allows preparation of large quantities, frequently however suffering from sticky, possibly toxic synthetic impurities or undesired chiral isomers formed during synthesis. An example is β-poly(L-malic acid), a linear polyester of L-malic acid involving the carboxyl group in β-position, while the carboxyl group in α-position remains pendant and available for conjugation with functional groups in the nanoconjugate. The biology and chemistry of the polymer have been investigated in some detail (Lee et al., 2002). This polycation is synthesized by the slime mold *Physarum polycephalum*, which uses the polymer for stockpiling and trafficking of nucleic acid-binding proteins across the giant polynucleate amoeba-like cell, the plasmodium. Large amounts are secreted from the plasmodium and eventually enzymatically cleaved to L-malate by a hydrolase (Korherr et al., 1995). β-Poly(L-malic acid) with molecular mass M_w = 50,000 (polydispersity M_w/M_n = 1.3) from the culture broth of *P. polycephalum* was highly purified and size-fractionated on Sephadex G25 (Lee and Holler, 1999; Ljubimova et al., 2007). The lyophilized polyacid was devoid of material absorbing at 260-nm and 280-nm wavelengths. This polymer was used as a backbone for the construction of an anticancer drug-delivery system, which we termed Polycefin (Lee et al., 2006).

7.3 Currently Used Drugs Conjugated to the β-Poly(L-malic acid) Nanoplatform

In current Polycefin variants, the drug module was represented by morpholino AON to α4 and β1 chains of tumor-specific protein, laminin-8 (currently termed laminin 411), which is overexpressed in vessel walls of highly vascular glial tumors, invasive breast cancers, and their metastases (Fujita et al., 2005; Ljubimova et al., 2001). This type of AON, unlike others, is very resistant against nuclease degradation and is highly specific at low concentrations (Summerton and Weller, 1997).

Laminins are a family of major trimeric structural proteins of basement membranes that are important for cell differentiation, migration, and proliferation

(Hallmann et al., 2005). They participate in tumor invasion as barriers for tumor cell penetration of surrounding tissues. At the same time, some laminins produced by tumor cells facilitate their migration via integrin receptors (Fujiwara et al., 2004; Hallmann et al., 2005). Laminin 411 ($\alpha 4\beta 1\gamma 1$; formerly, laminin-8), a vascular basement membrane component, plays important roles in angiogenesis (capillary maturation) and cell migration. We have documented overexpression of laminin 411 in grade IV human glioma (glioblastoma multiforme, GBM) and ductal breast carcinoma (Fujita et al., 2005; Ljubimova et al., 2001, 2004). Because laminins are trimeric proteins, inhibition of synthesis of more than one chain provides a more extensive blocking of its production than inhibition of only one chain (Khazenzon et al., 2003). Therefore, our strategy was to block two laminin 411 chains rather than one, in a multitargeting approach. We further showed that AON inhibition of two laminin 411 chains ($\alpha 4 + \beta 1$) was able to block glioma invasion in vitro (Khazenzon et al., 2003). Laminin 411 involvement in vessel formation and its overexpression in tumors suggested that its inhibition could reduce tumor neovascularization in vivo. In fact, this was recently shown for intracranial GBM implanted in rat brain (Fujita et al., 2006).

7.4   Active Tumor-Targeting Delivery Module

Transferrin receptor antibody was chosen to be the active targeting device that could carry Polycefin across the BTB in brain and to the GBM target cells by promoting binding of the nanoconjugate to pertinent cell surface receptor and its endosomal uptake (Jefferies et al., 1985; Skarlatos et al., 1995; Broadwell et al., 1996; Daniels et al., 2006a, b; Lee et al., 2006). Some nanoconjugate variants contained other targeting antibodies as discussed below (see also Fujita et al., 2007).

7.5   Synthesis of the Polycefin Nanoconjugate Family from β-Poly(L-malic acid) Platform

Poly(malic acid) is a platform of choice for a number of different nanoconjugates. A general protocol for the synthesis of Polycefin variants has been established (Lee et al., 2006), which involves a limited number of chemical reactions and techniques, depending on the chemical nature, i.e., solubility and reactivity of the nanoconjugate intermediates and the number and chemical nature of the conjugated modules. Synthetic reactions are carried out in organic solvents or aqueous solutions. Because purification of macromolecular intermediates and products is tedious, the experimental conditions have been designed to favor completion of chemical reactions. If the number of different modules is higher than two or three, this can be achieved by establishing a hierarchy of sequential conjugations along chemically different reacting groups typically in the synthesis of Polycefin (Lee et al., 2006). A simplified reaction scheme is shown in Fig. 3. Here, in a first set of amide forming reactions, the protection module (PEG), the endosome escape

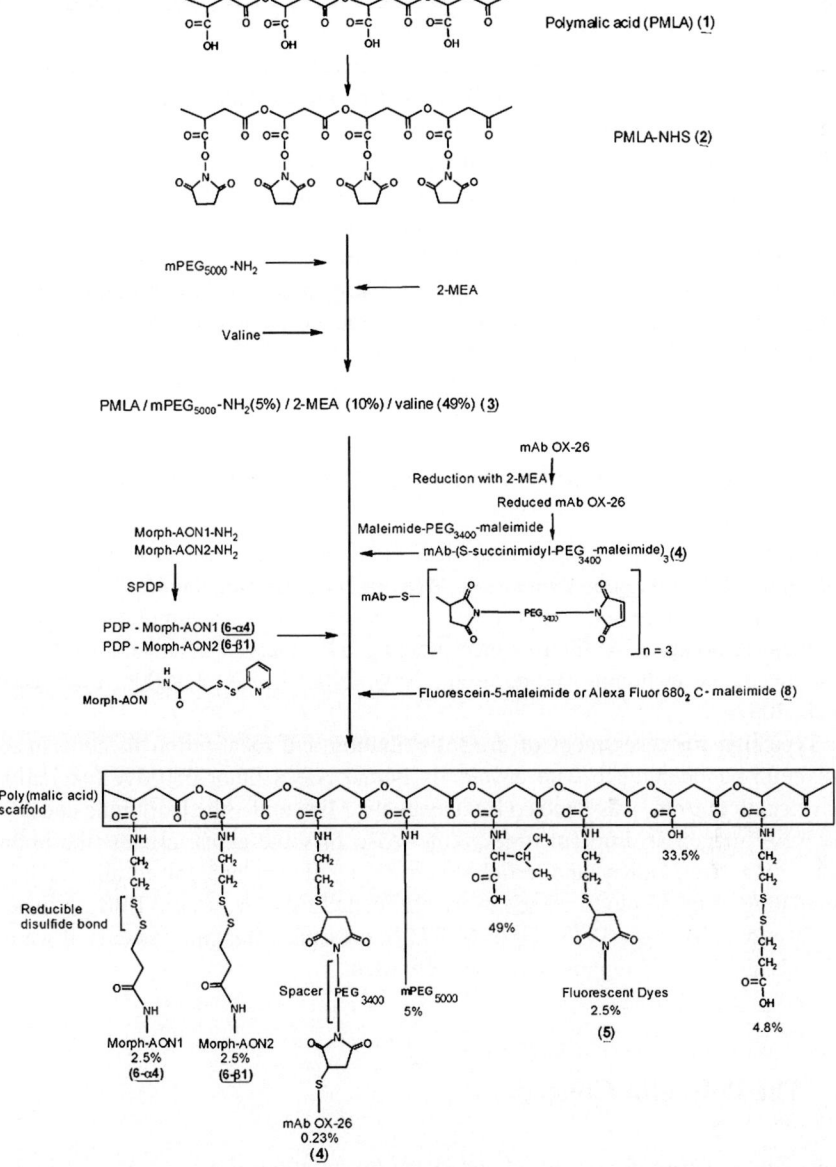

Fig. 3 Synthesis of Polycefin nanoconjugate carrying a variety of functional modules. Note the hierarchy in the choice of the conjugation reactions. The platform consists of biodegradable β-poly(L-malic acid) highly purified from the slime mold *Physarum polycephalum*. The carboxyl groups are at first chemically activated by their esterification with *N*-hydroxysuccinimide. The hierarchy involves a first series of conjugation reactions with more durable residues, i.e., with PEG, leucine ethylester, and the thiol-containing spacer, 2-thioethane-1-amine. In the second series of conjugation reactions, the biochemically fragile molecules, AONs and mAbs, are conjugated by forming thioether and disulfide bonds, respectively. An optional fluorescent reporter molecule can be similarly conjugated. Unreacted free thiol groups are then chemically blocked to prevent uncontrolled reactions. From Lee et al. 2006 (Bioconjug. Chem. 17:317–326). Reproduced by permission of American Chemical Society ©2006

module (leucine ethylester), and a reactive spacer (2-amino-1-mercaptoethanol) are conjugated with the NHS-ester activated poly(malic acid) platform. In a second set, the targeting antibody and the AON drugs are conjugated via a thioether bond and a disulfide bond, respectively. If a multiplicity of different drugs or antibodies is chosen, they will be conjugated together in a mixture with any specific stoichiometry. Purification follows the principle of separation of excess small reactants from macromolecular products by molecular sieving methods, either membrane filtration or size exclusion chromatography. By this technique, nanoconjugates can be tailored with predictable composition and stoichiometry (Fujita et al., 2006, 2007; Lee et al., 2006; Ljubimova et al., 2007).

8 Physical and Chemical Methods of Delivery Validation

Whereas a variety of classical analytical methods such as confocal microscopy, radioactive labeling, gene expression, Western blotting, and chemical or cell biological methods have been employed to follow the selective accumulation of nanoconjugates, noninvasive fluorescence imaging has emerged as a sensitive, rapid, and convenient technique (Fujita et al., 2006, 2007; Lee et al., 2006; Ljubimova et al., 2007).

Typically, for assessment of drug distribution and localization in nude mice, Polycefin nanoconjugate with covalently bound Alexa Fluor 680 dye (excitation wavelength 660 nm) was injected intravenously. Mice under the Isoflurane anesthesia were imaged at different time points. Also in some cases, organs like brain, heart, lung, liver, kidney, and testicles were prepared from euthanized mice and the circulating drugs in blood vessels were eliminated by intra-arterial PBS perfusion for 20 min. A Xenogen IVIS 200 or MISTI fluorescence-imaging systems of whole animals and isolated organs were routinely used.

9 The Polycefin Concept

Polycefin is a fruitful concept allowing the formulation of a large host of multi-targeting nanoconjugates by following very similar chemical protocols. The platform, poly(malic acid) (Figs. 2 and 3), and technical principles have been published before (Fujita et al., 2006, 2007; Lee et al., 2006; Ljubimova et al., 2007). Drugs acting on both molecular and cellular targets can be varied by introducing nucleic acids, proteins, peptides, and nonprotein molecules. The only requisite is a functional group for conjugation to the carrier platform. Other auxiliary groups like PEG, drug-releasing residues, and endosome escape functions can be varied as well. The degree of platform loading with functional groups is adaptable to specific tasks. Examples of variants have been published (Fujita et al., 2007), and some interesting properties will be presented here.

9.1 Nomenclature

The term Polycefin denotes the drug-delivery device with poly(malic acid) as platform and the various functional groups described previously, specifically a mAb to TfR and leucyl ethylester as the active endosomal escape residue. Other versions of Polycefin that contain substitutions of the originally conjugated groups are indicated by the substituting molecule(s). Newly synthesized versions of Polycefin contain other mAbs instead of anti-TfR mAb. Polycefin(mTfR) denotes the version with mouse antihuman TfR. Polycefin(2C5) is the version with nucleosome-specific monoclonal antinuclear mouse autoantibody 2C5 recognizing cancer cell surface-bound nucleosomes (Fujita et al., 2007; Gupta and Torchilin, 2007; Iakoubov and Torchilin, 1997; Torchilin and Lukyanov, 2003) and Polycefin(mTfR,2C5) is the version with both antibodies.

9.2 Cell-Free Studies: Stoichiometry and Function of Conjugated Antibodies, Drug Release from the Platform, Membrane Disruption

Poly(malic acid) platform did not significantly contribute to molecular mass when measured by reducing sodium dodecyl sulfate polyacrylamide gel electrophoresis, exhibiting Polycefin (single IgG conjugated) with electrophoretic mobility coinciding with that of IgG heavy chain (Lee et al., 2006). The presence of two immunologically functional mAbs with different specificities was documented by ELISA of Polycefin(mTfR,2C5) (Fujita et al., 2007).

AONs conjugated via disulfide bonds to the Polycefin platform were rapidly released in the presence of millimolar concentrations of glutathione prevailing in the cytoplasm (Lee et al., 2006). Employing a hemolytic assay, it was shown that hydrophobic amino acids, when bound to the platform, contributed strongly to membrane rupture necessary for endosomal escape, whereas PEG functioned as a stabilizer (Lee et al., 2006).

9.3 In Vitro Results: Receptor-Dependent Endocytosis, Endosomal Escape, Laminin 411 Protein Synthesis Inhibition in Glioma Tissue

Receptor-dependent endosomal uptake was demonstrated by confocal microscopy (Fig. 4a) for human GBM U87MG cells using fluorescein-labeled Polycefin with conjugated anti-TfR antibody (Lee et al., 2006). The same uptake mechanism was shown in U87MG cells for Oregon Green-labeled Polycefin bearing antinucleosome autoantibody 2C5 or both anti-TfR and 2C5 mAbs in tandem (Fujita et al., 2007).

In vitro endosomal escape could be visualized by fluorescence confocal microscopy and further documented on Western blots by the inhibition of synthesis of glioma-specific laminin 411 chains $\alpha4$ an $\beta1$ by morpholino AON (Fig. 4b) after their release from Polycefin in the cytoplasm (Lee et al., 2006).

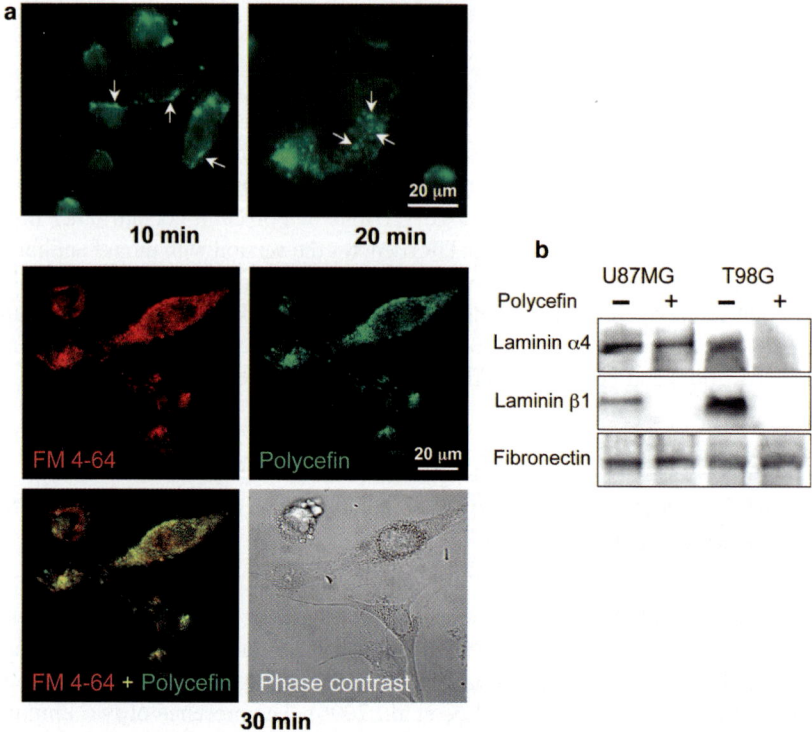

Fig. 4 Polycefin delivery into cultured glioma U87MG cells and in vitro targeting and inhibition of laminin chain synthesis. **a** Kinetics of intracellular accumulation of Polycefin. At 10 min of treatment with fluorescein-labeled Polycefin, it is localized near the cell membrane and early endosomes are beginning to form. At 20 min after treatment with Polycefin, maturing endosomes are visible inside the cells. At 30 min after treatment there is extensive codistribution of endosomal marker FM 4–64 with Polycefin in cultured glioma U87MG cells. FM 4–64 stains endosomes (*red color*), and Polycefin is found in the same place (*green color*). Colocalization is revealed as *yellow color* (*lower left*). Confocal microscopy. **b** Western blot analysis of laminin chain secretion into medium with or without treatment. Conditioned media of both glioma cell lines, U87MG and T98G, contain α4 and β1 chains of laminin 411. Polycefin dramatically inhibited secretion of both laminin chains. Gels were normalized by the content of secreted fibronectin. From Lee et al. 2006 (Bioconjug. Chem. 17:317–326). Reproduced by permission of American Chemical Society ©2006

9.4 In Vivo Results: Imaging of Glioma and Breast Cancer Tumors, Tandem Antibody Targeting of Human Glioma Implants on Mice, Drug-Induced Inhibition of Tumor Angiogenesis, Significant Increase of Survival of Glioma-Bearing Rodents

Targeting of human glioma grown in rat brain was efficient with Polycefin-conjugated rat anti-TfR antibody, which cross-reacted with human TfR on the surface of human GBM cells thus allowing efficient transfer through the BTB and uptake by the tumor

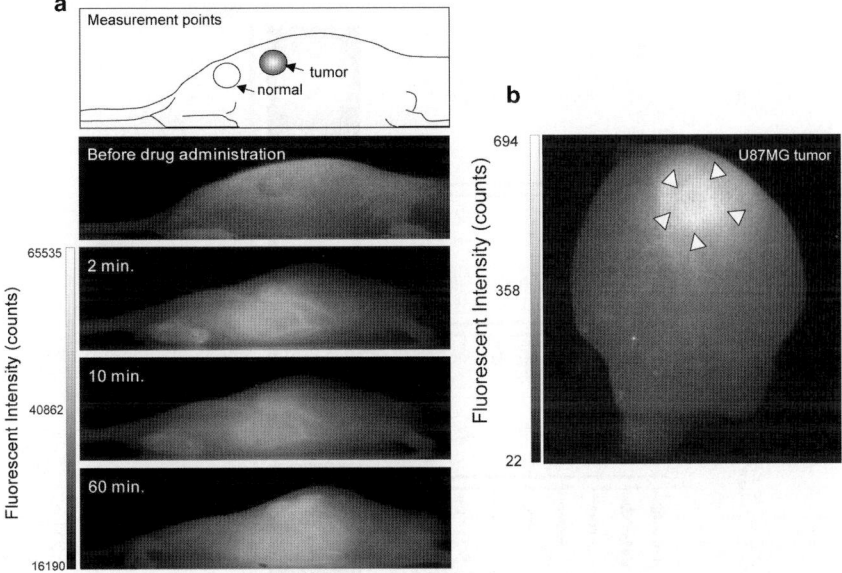

Fig. 5 Polycefin imaging with MISTI system after intravenous administration into tumor-bearing animals. **a** Breast cancer accumulation. In vivo fluorescence images of mice with xenografted human MDA-MB 468 breast tumor before treatment and in 2, 10, and 60 min after intravenous injection of Alexa Fluor 680-labeled Polycefin. Polycefin circulated in blood vessels in early phase (2–10 min), but declined after 60 min. However, the drug kept retaining in tumor tissue even after 60 min. **b** Brain tumor accumulation. In vivo fluorescence imaging of mouse brain with xenografted human U87MG glioma 24 h after intravenous injection of Polycefin. The tumor area (*within arrowheads*) shows distinct fluorescent signal. From Ljubimova et al. 2007 (Chem. Biol. Interact. In press). *Reproduced by permission of Elsevier Ireland Ltd ©2007*

cells themselves (Lee et al., 2006). Efficient targeting was also observed with Polycefin for human breast cancer grown in mice (Ljubimova et al., 2007) (Fig. 5).

Targeting of human tumor in mouse brain was also attempted using Polycefin(mTfR) that facilitated transfer through mouse BTB but did not have the ability to target human glioma cells due to lack of cross-reactivity of antimouse mAb with human TfR (Fujita et al., 2007; Lee et al., 2006). Maximum targeting efficacy was achieved by Polycefin(mTfR,hTfR) that conferred in a tandem configuration targeting of both mouse BTB by antimouse TfR antibody and human glioma cells by antihuman TfR antibody (Fujita et al., 2007). A similarly enhanced tumor targeting was observed with Polycefin(mTfR,2C5) where antihuman TfR antibody was substituted by an antinucleosome mAb 2C5 reacting with nucleosomes on human tumor cell surface (Fig. 6). The accumulation was conveniently followed by whole animal and isolated organ fluorescence imaging using Alexa Fluor 680-labeled Polycefin variants (Fujita et al., 2006, 2007; Lee et al., 2006; Ljubimova et al., 2007).

The drug component of Polycefin was aimed at inhibiting laminin 411 synthesis and thus tumor-specific angiogenesis. Indeed, angiogenesis in treated glioma was substantially reduced (Figs. 7 and 8) after treatment with Polycefin (Fujita et al., 2006).

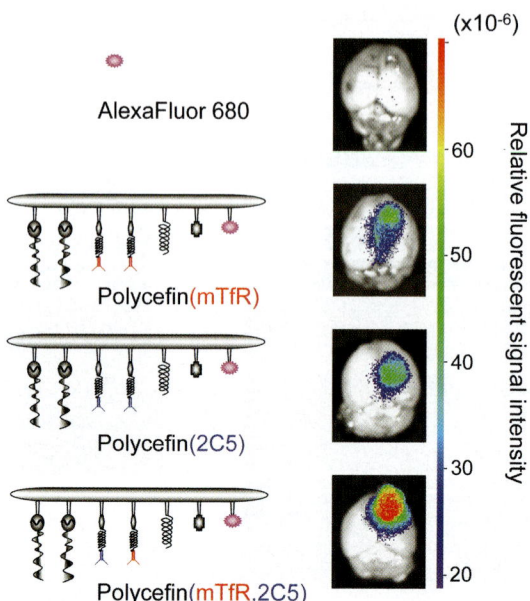

Fig. 6 Fluorescence imaging of brain from nude mice bearing human U87MG glioma. 24 h after intravenous injection of free Alexa Fluor 680 and various Alexa Fluor 680-labeled Polycefin variants, only the tumor contains fluorescent drug. The highest drug accumulation in the tumor is observed for the tandem configuration in Polycefin(mTfR,2C5). Polycefin variants with one antibody (antimouse TfR or antinucleosome 2C5) show less accumulation. 100 μl of Alexa Fluor 680 (0.6 μM) or labeled Polycefin variants at concentrations of 3 μM AONs were injected intravenously. From Fujita et al. 2007 (J. Control. Release 122:356–363). *Reproduced by permission of Elsevier B.V. ©2007*

Inhibition of tumor angiogenesis was accompanied by a significantly prolonged survival of treated rodents (Fujita et al., 2006; Lee et al., 2006) (Fig. 7).

10   Conclusions and Future Directions

The results reported for the Polycefin family of nanoconjugates support the idea that tandem of even more versatile configurations of platform-conjugated antibodies or other entities with appropriate target specificities and a tandem or more elaborated combinations of different molecular drug species are feasible and may significantly enhance the efficacy of tumor treatment. Multiple cell-targeting devices can facilitate penetration of barriers, whereas the function of multiple drug moieties will provoke simultaneous or even synergistic inhibition of tumor cell pathways and eventually tumor cell death.

An alternative approach, to simplify Polycefin structure, may involve the use of single inhibitors that act on multiple pathways simultaneously by blocking the

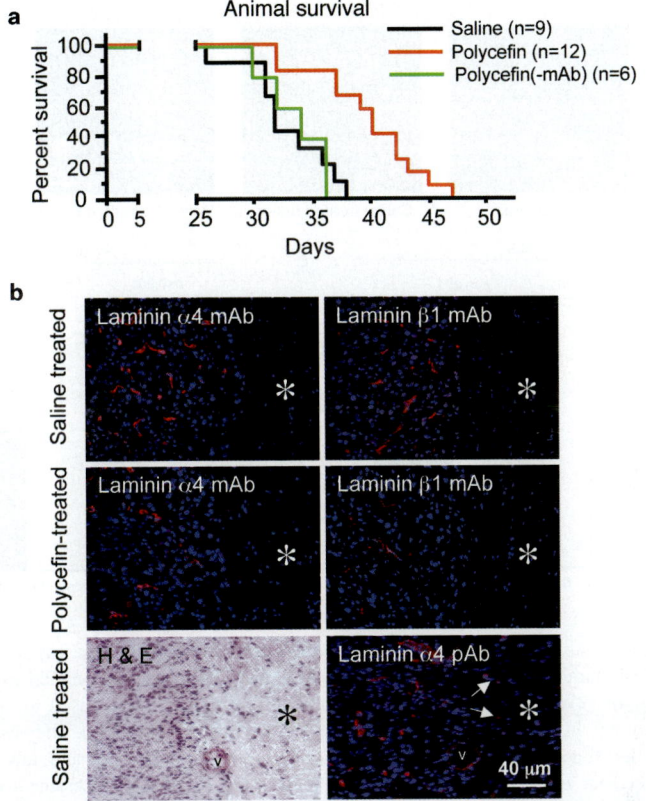

Fig. 7 In vivo target inhibition by Polycefin and animal survival. (**a**) Survival of Polycefin-treated (0.5 mg/kg body weight) and control animals. After intracranial administration of four doses of Polycefin, the animal survival time was significantly increased ($p < 0.0004$) compared to saline-treated or Polycefin(-mAb)-treated rats. (**b**) Immunofluorescent analysis of xenotransplanted brain tumors with antihuman mAbs to laminin $\alpha 4$ or $\beta 1$ chains. After Polycefin treatment, the number of tumor vessels positive for either laminin chain was markedly diminished. Therefore, Polycefin inhibited the expression of both its targets and their incorporation into basement membrane by human tumor cells. *Asterisks* denote tumor-adjacent (normal) brain area. This area has significantly decreased cellularity (revealed by *blue* nuclear staining with DAPI) compared to highly cellular tumor at the *left*. No vascular staining is observed in tumor-adjacent area with both mAbs, because the antibodies only recognized human laminin chains. *Left lower panel*, a hematoxylin-eosin (H&E) stained tumor showing a sharp boundary between highly cellular tumor and surrounding brain parenchyma with significantly fewer cells. *Right lower panel*, staining of a serial section with a polyclonal antibody (pAb) to laminin $\alpha 4$ chain recognizing human and rat protein that reveals all vessels. Note increased vascularity and cellularity of the tumor as opposed to hypocellular surrounding tissue (*asterisk*) that has only scattered vessels (*arrows*). From Fujita et al. 2006 (Angiogenesis 9:183–191). Reproduced by permission of Springer Science + Business Media B.V. ©2006

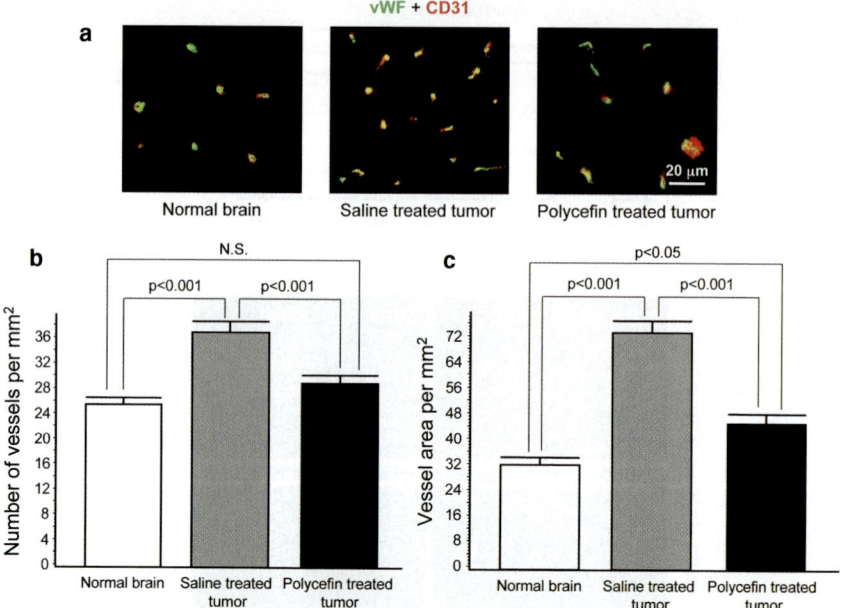

Fig. 8 Decreased vascular density and area in tumors treated with Polycefin. (**a**) Double immunohistochemical staining of rat brain vessels using antibodies to two endothelial markers, von Willebrand factor (vWF, *green*) and CD31 (*red*). The two markers were used to optimize the screening accuracy. In most vessels both markers codistribute (*yellow color*). The vessel number is higher in the xenotransplanted U87MG tumor than in normal brain, and this number is decreased after four intracranial Polycefin treatments. (**b**) Quantitative assessment of vascular density in treated and untreated tumors compared to normal brain. Vessels were revealed by either marker (**a**) and their number was quantitated at ×200 direct magnification. Images were analyzed using ImageJ software. Statistical significance was determined by ANOVA. Microvascular density in xenotransplanted U87MG human glioma is significantly increased compared to normal brain ($p < 0.001$). After four intracranial treatments with Polycefin, tumor vessel density was significantly decreased ($p < 0.001$) and became similar to normal brain tissue (NS, not significant with $p > 0.05$). (**c**) Quantitative assessment of vascular area in treated and untreated tumors compared to normal brain. Vessels were revealed by either marker (**a**) and their relative area quantitated as for vessel density. Vessel area in xenotransplanted U87MG human glioma is significantly increased compared to normal brain ($p < 0.001$). After four intracranial treatments with Polycefin, tumor vascular area significantly decreased ($p < 0.001$) but remained somewhat higher than in normal brain ($p < 0.05$). From Fujita et al. 2006 (Angiogenesis 9:183–191). Reproduced by permission of Springer Science + Business Media B.V. ©2006

production and/or function of groups of enzymes, such as tyrosine kinases, or master regulatory molecules, such as protein kinase CK2. In the first group of drugs, for instance, SU11248 (Faivre et al., 2006), AEE788 (Younes et al., 2006), and RAD001 (Yao, 2007) are being seriously considered as potent anticancer drugs. Each of them inhibits several targets associated with tyrosine kinase-depending signaling pathways abnormal in many tumors. Serine-threonine protein kinase CK2, as a representative of the second group, acts upon more than 300 substrates inside the cell and its single inhibitors can block cell proliferation, migration, tumor growth, and angiogenesis, and increase cancer cell apoptosis (Ahmad et al., 2005; Kramerov et al., 2006; Ljubimov et al., 2004; Pagano et al., 2006; Wang et al., 2006). The pertinent inhibitor drugs are potential candidates for future Polycefins or similar nanoconjugates.

The future Polycefins compounds may be directed not only against cancer but also against other pathological conditions, where specific delivery may be necessary. Such conditions may include neovascularization in the eye as exemplified by the most vision-threatening diseases to date like age-related macular degeneration (AMD) and proliferative diabetic retinopathy (PDR). Certain aptamer (pegaptanib/ Macugen® from Pfizer) and antibody-based (ranibizumab/Lucentis® from Genentech) anti-VEGF drugs have been already approved for human use in wet (neovascular) form of AMD, with clear beneficial effects in patients after intravitreal injections (Takeda et al., 2007). It is quite possible that their use as part of neovasculature-targeting nanoconjugate may increase the drug efficacy and specificity and reduce side effects. Drugs against PDR are only being slowly developed, which may be difficult because this disease appears to be more multifactorial than AMD, requiring multitargeted approach. A Polycefin with a conjugated CK2 inhibitor or inhibitors of several angiogenic pathways may prove to be a viable candidate for PDR therapy.

In the cardiovascular field, nanoconjugates of Polycefin type may be used that bear inhibitors of antiangiogenic molecules, such as thrombospondins. These drug-delivery systems may prove beneficial for patients with myocardial infarctions who need enhanced angiogenesis at the site of injury (Chatila et al., 2007). Alternatively, after myocardial infarction there is a need for stimulating angiogenesis that may also be accomplished with nanoconjugates. Some small molecules, such as a synthetic prostacyclin agonist, ONO-1301, have been described that activate angiogenic growth factor signaling pathways and promote angiogenesis in the ischemic heart (Nakamura et al., 2007).

Poly(malic acid) as a polymer with multiple chemically functional pendant groups offers a great potential for future syntheses of tandem configured drug carrier systems. For instance, with the combination of anti-TfR to direct the conjugate across the BTB and 2C5 antibody to target tumor cells, the Polycefin system represents an ideal combination for brain tumor treatment. Further investigation of this system should provide a more profound understanding of the optimal targeting mechanisms. A tandem-type Polycefin might serve in the future as a potential therapeutic intervention for treatment of patients with brain tumor and in experimental cancer studies. Moreover, the versatility, multitargeting abilities, and the biodegradable nature of the poly(malic acid) platform favor further development of the Polycefin concept for treatment of various diseases including cancer, ocular neovascular diseases, and cardiovascular ischemic problems.

Acknowledgments This work was supported by grants from NIH (CA123495 to J. Y.L., EY13431 and RR00425 to A.V.L.), Winnick Family Foundation and Skirball Foundation (to A.V.L.), a grant from Arrogene Inc., and a grant from the Department of Neurosurgery, Cedars-Sinai Medical Center.

References

Abdellaouri, K., Boustta, M., Vert, M., Marjani, H., and Manfait, M. 1998. Metabolic-derived artificial polymers designed for drug targeting, cell penetration and bioresorption. Eur. J. Pharm. Sci. 6:61–73.

Ahmad, K. A., Wang, G., Slaton, J., Unger, G., and Ahmed, K. 2005. Targeting CK2 for cancer therapy. Anticancer Drugs 16:1037–1043.

Andersson L., Davies, J., Duncan, R., Ferruti, P., Ford, J., Kneller, S., Mendichi, R., Pasut, G., Schiavon, O., Summerford, C., Tirk, A., Veronese, F. M., Vincenzi, V., and Wu, G. 2005. Poly(ethylene glycol)-poly(ester-carbonate) block copolymers carrying PEG-peptidyl-doxorubicin pendant side chains: synthesis and evaluation as anticancer conjugates. Biomacromolecules 6:914–926.

Arpicco, S., Dosio, F., Bolognesi, A., Lubelli, C., Brusa, P., Stells, B., Ceruti, M., and Cattel, L. 2002. Novel poly(ethylene glycol) derivatives for preparation of ribosome-inactivating protein conjugates. Bioconjug. Chem. 13:757–765.

Bae, Y., Jang, W. D., Nishiyama, N., Fukushima, S., and Kataoka, K. 2005. Multifunctional polymeric micelles with folate-mediated cancer cell targeting and pH-triggered drug releasing properties for active intracellular drug delivery. Mol. Biosyst. 1:242–250.

Bickenbach, K., Wilcox, R., Veerapong, J., Kindler, H. L., Posner, M. C., Noffsinger, A., and Roggin, K. K. 2007. A review of resistance patterns and phenotypic changes in gastrointestinal stromal tumors following imatinib mesylate therapy. J. Gastrointest. Surg. 11:758–766.

Brekke, O. H. and Sandlie, I. 2003. Therapeutic antibodies for human diseases at the dawn of the twenty-first century. Nat. Rev. Drug Discov. 2:52–62.

Broadwell, R. D., Baker-Cairns, B. J., Frieden, P. M., Oliver, C., and Villegas, J. C. 1996. Transcytosis of protein through the mammalian cerebral epithelium and endothelium III. Receptormediated transcytosis through the blood-brain barrier of blood-borne transferrin and antibody against the transferrin receptor. Exp. Neurol. 142:47–65.

Cammas, S., Béar M.-M., Moine, L., Escalup, R., Ponchel, G., Kataoka, K., and Guérin, P. 1999. Polymers of malic acid and 3-alkylmalic acid as synthetic PHAs in the design of biocompatible hydrolysable devices. Int. J. Biol. Macromol. 25:273–282.

Carter, P. J. 2006. Potent antibody therapeutics by design. Nat. Rev. Immunol. 6:343–357.

Chatila, K., Ren, G., Xia, Y., Huebener, P., Bujak, M., and Frangogiannis, N. G. 2007. The role of the thrombospondins in healing myocardial infarcts. Cardiovasc. Hematol. Agents Med. Chem. 5:21–27.

Cho, Y. W., Kim, J. D., and Park, K. 2003. Polycation gene delivery systems: escape from endosomes to cytosol. J. Pharm. Pharmacol. 55:721–734.

Chytil, P., Etrych, T., Konak, C., Sirova, M., Mrkvan, T., Rihova, B., and Ulbrich, K. 2006. Properties of HPMA copolymer-doxorubicin conjugates with pH-controlled activation: effect of polymer chain modification. J. Control. Release 115:26–36.

Coloma, M. J., Lee, H. J., Kurihara, A., Landaw, E. M., Boado, R.J., Morrison, S. L., and Pardridge, W. M. 2000. Transport across the primate blood-brain barrier of a genetically engineered chimeric monoclonal antibody to the human insulin receptor. Pharm. Res. 17:266–274.

Daniels, T. R., Delgado, T., Rodriguez, J. A., Helguera, G., and Penichet, M. L. 2006a. The transferrin receptor part I: biology and targeting with cytotoxic antibodies for the treatment of cancer. Clin. Immunol. 121:144–158.

Daniels, T. R., Delgado, T., Helguera, G., and Penichet, M. L. 2006b. The transferrin receptor part II: targeted delivery of therapeutic agents into cancer cells. Clin. Immunol. 121:159–176.

Deres, S., Gdalevsky, G. Y., Gilboa, I., Voorspoels, J., Remon, J. P., and Kost, J. 2004. Bioadhesive grafted starch copolymers as platform for peroral drug delivery: a study of theophylline release. J. Control. Release 94:391–399.

Duncan, R. 1992. Drug-polymer conjugates: potential for improved chemotherapy. Anticancer Drugs 3:175–210.

Duncan, R., Seymour, L. C. W., Scarlett, L., Lloyd, J. B., Rejmanová, P., and Kopecek, J. 1986. Fate of N-(2-hydroxypropyl)methacrylamide copolymers with pendent galactosamine residues after intravenous administration to rats. Biochim. Biophys. Acta 880:62–71.

Duncan, R., Vicent, M. J., Greco, F., and Nicholson, R. I. 2005. Polymer-drug conjugates: towards a novel approach for the treatment of endocrine-related cancer. Endocr. Relat. Cancer 12:S189–S199.

Ehrlich, P. 1906. Studies in Immunity, New York: Plenum Press.

Etrych, T., Jelinkova, M., Rihova, B., and Ulrich, K. 2001. New HPMA copolymers containing doxorubicin bound via pH-sensitive linkage: synthesis and preliminary in vitro and in vivo biological properties. J. Control. Release 73:89–102.

Faivre, S., Delbaldo, C., Vera, K., Robert, C., Lozahic, S., Lassau, N., Bello, C., Deprimo, S., Brega, N., Massimini, G., Armand, J. P., Scigalla, P., and Raymond, E. 2006. Safety, pharmacokinetic, and antitumor activity of SU11248, a novel oral multitarget tyrosine kinase inhibitor, in patients with cancer. J. Clin. Oncol. 24:25–35.

Farokhzad, O. C., Cheng, J., Teply, B. A., Sherifi, I., Jon, S., Kantoff, P. W., Richie, J. P., and Langer, R. 2006. Targeted nanoparticle-aptamer bioconjugates for cancer chemotherapy in vivo. Proc. Natl. Acad. Sci. USA 103:6315–6320.

Fleming, A. B., Haverstick, K., and Saltzman, W. M. 2004. In vitro cytotoxicity and in vivo distribution after direct delivery of PEG-Camptothecin conjugates to the rat brain. Bioconjug. Chem. 15:1364–1375.

Fujita, M., Khazenzon, N. M., Bose, S., Sekiguchi, K., Sasaki, T., Carter, W. G., Ljubimov, A. V., Black, K. L., and Ljubimova, J. Y. 2005. Overexpression of β1 chain-containing laminins in capillary basement membranes of human breast cancer and its metastases. Breast Cancer Res. 7:411–421.

Fujita, M., Khazenzon, N. M., Ljubimov, A. V., Lee, B.-S., Virtanen, I., Holler, E., Black, K. L., Ljubimova, J. Y. 2006. Inhibition of laminin-8 in vivo using a novel poly(malic acid)-based carrier reduces glioma angiogenesis. Angiogenesis 9:183–191.

Fujita, M., Lee, B.-S., Khazenzon, M. N., Penichet, M. L., Wawrowsky, K. A., Patil, R., Ding, H., Holler, E., Black, K. L., and Ljubimova, J. L. 2007. Brain tumor tandem targeting using a combination of monoclonal antibodies attached to biopoly(β-L-malic acid). J. Control. Release 122:356–363.

Fujiwara, H., Gu, J., and Sekiguchi, K. 2004. Rac regulates integrin-mediated endothelial cell adhesion and migration on laminin-8. Exp. Cell Res. 292:67–77.

Gaur, U., Sahoo, S. K., De, T. K., Ghosh, P. C., Maitra, A., and Ghosh, P. K. 2000. Biodistribution of fluoresceinated dextran using novel nanoparticles evading reticuloendothelial system. Int. J. Pharm. 202:1–10.

Golenser, J., Frankenburg, S., Ehrenfreund, T., and Domb, A. J. 1999. Efficacious treatment of experimental Leishmaniasis with amphotericin β-arabinogalactan water-soluble derivatives. Antimicrob. Agents Chemother. 43:2209–2214.

Grantab, R., Sivananthan, S., and Tannock, I. F. 2006. The penetration of anticancer drugs through tumor tissue as a function of cellular adhesion and packing density of tumor. Cancer Res. 66:1033–1039.

Greenwald, R. B., Choe, Y. H., McGuire, J., and Conover, C. D. 2003. Effective drug delivery by PEGylated drug conjugates. Adv. Drug Deliv. Rev. 55:217–250.

Gupta, B. and Torchilin, V. P. 2007. Monoclonal antibody 2C5-modified doxorubicin-loaded liposomes with significantly enhanced therapeutic activity against intracranial human brain U-87 MG tumor xenografts in nude mice. Cancer Immunol. Immunother. 56:1215–1223.

Gupta, N. and Yucel, Y. H. 2007. Glaucoma as a neurodegenerative disease. Curr. Opin. Ophthalmol. 18:110–114.

Haider, M., Megeed, Z., and Ghandehari, H. 2004. Genetically engineered polymers: status and prospects for controlled release. J. Contol. Release 95:1–26.

Hallmann, R., Horn, N., Selg, M., Wendler, O., Pausch, F., and Sorokin, L.M. 2005. Expression and function of laminins in the embryonic and mature vasculature. Physiol. Rev. 85:979–1000.

Heaney, N. B. and Holyoake, T. L. 2007. Therapeutic targets in chronic myeloid leukaemia. Hematol. Oncol. 25:66–75.

Iakoubov, L. Z. and Torchilin, V. P. 1997. A novel class of antitumor antibodies: nucleosome-restricted antinuclear autoantibodies (ANA) from healthy aged nonautoimmune mice. Oncol. Res. 9:439–446.

Jabbour, E., Cortes, J., and Kantarjian, H. 2007. Dasatinib for the treatment of Philadelphia chromosome-positive leukaemias. Expert Opin. Investig. Drugs 16:679–687.

Jatzkewitz, H. 1955. Peptamin (glycyl-L-leucyl-mescaline) bound to blood plasma expander (polyvinylpyrrolidone) as a new depot form of a biologically active primary amine (Mescaline). Z. Naturforsch. 10b:27–31.

Jefferies, W. A., Brandon, M. R., Williams, A. F., and Hunt, V. 1985. Analysis of lymphopoietic stem cells with a monoclonal antibody to the rat transferrin receptor. Immunology 54:333–341.

Jensen, K., D., Kopeckova, P., and Kopecek, J. 2002. Antisense oligonucleotides delivered to the lysosome escape and actively inhibit hepatitis B virus. Bioconjug. Chem. 13:975–984.

Kabanov, A. V., Batrakova, E. V., and Alakhov, V. Y. 2002. Pluronic block copolymers for overcoming drug resistance in cancer. Adv. Drug Deliv. Rev. 13:759–779.

Kantarjian, H. M., Cortes, J., Guilhot, F., Hochhaus, A., Baccarani, M., and Lokey, L. 2007. Diagnosis and management of chronic myeloid leukemia: a survey of American and European practice patterns. Cancer 109:1365–1375.

Khandare, J. J. and Minko, T. 2006. Antibodies and peptides in cancer therapy. Crit. Rev. Ther. Drug Carrier Syst. 23:401–436.

Khazenzon, N. M., Ljubimov, A. V., Lakhter, A. J., Fujita, M., Fujiwara, H., Sekiguchi, K., Sorokin, L. M., Petajaniemi, N., Virtanen, I., Black, K. L., and Ljubimova, J. Y. 2003. Antisense inhibition of laminin-8 expression reduces invasion of human gliomas in vitro. Mol. Cancer Ther. 2:985–994.

Kishore, B. K., Lambricht, P., Laurent, G., Maldague, P., Wagner, R., and Tulkens, P. M. 1990. Mechanism of protection afforded by polyaspartic acid against gentamicin-induced phospholipidosis. II. Comparative in vitro and in vivo studies with poly-L-aspartic, poly-L-glutamic and poly-D-glutamic acids. J. Pharmacol. Exp. Ther. 255:875–885.

Kishore, B. K., Maldague, P., Tulkens, P. M., and Courtoy, P. J. 1996. Pol-D-glutamic acid induces an acute lysosomal thesaurismosis of proximal tubules and a marked proliferation of intrstitium in rat kidney. Lab. Invest. 74:1013–1023.

Konstantinopoulos, P. A., Vandoros, G. P., and Papavassiliou, A. G. 2006. FK228 (depsipeptide): a HDAC inhibitor with pleiotropic antitumor activities. Cancer Chemother. Pharmacol. 58:711–715.

Kopecek, J. and Baziliva, H. 1973. Poly[n-(2-hydroxypropyl)methacrylamide]. 1. Radical polymerization and copolymerization. Eur. Polym. J. 9:7–14.

Kopecek, J., Reijmanová, P, Strohalm, J., Ulbrich, K., Rihová, B., Chytrý, V., Lloyd, J. B., and Duncan, R. 1991. Synthetic polymeric drugs. U.S. Patent. 5,037,883.

Kopecek, J., Kopecková, P., Minko, T., and Lu, Z.-R. 2000. HPMA Copolymer-anticancer drug conjugates: design, activity, and mechanism of action. Eur. J. Pharm. Biopharm. 50:61–81.

Kopecek, J., Kopeckova, P., Minko, T., Lu, Z.-R., and Peterson, C. M. 2001. Water soluble polymers in tumor targeted delivery. J. Control. Release 74:147–158.

Korherr, C., Roth, M., and Holler, E. 1995. Poly(β-L-malate) hydrolase from plasmodia of Physarum polycephalum. Can. J. Microbiol. 41(Suppl. 1):192–199.

Kramerov, A. A., Saghizadeh, M., Pan, H., Kabosova, A., Montenarh, M., Ahmed, K., Penn, J. S., Chan, C. K., Hinton, D. R., Grant, M. B., and Ljubimov, A. V. 2006. Expression of protein kinase CK2 in astroglial cells of normal and neovascularized retina. Am. J. Pathol. 168:1722–1736.

Kumar, D. M., Perez, E., Cai, Z. Y., Aoun, P., Brun-Zinkernagel, A. M., Covey, D. F., Simpkins, J. W., and Agarwal, N. 2005. Role of nonfeminizingestrogen analogues in neuroprotection of rat retinal ganglion cells against glutamate-induced cytotoxicity. Free Radic. Biol. Med. 38:1152–1163.

Kunath, K., Kopceckova, P., Minko, T., and Kopecek, J. 2000. HPMA copolymer-anticancer drug-OV-TL16 antibody conjugates. 3. The effect of free and and polymer-bound adriamycin on the expression of some genes in the OVCAR-3 human ovarian carcinoma cell line. Eur. J. Pharm. Biopharm. 49:11–15.

Leamon, C. P. and Reddy, J. A. 2004. Folate-targeted chemotherapy. Adv. Drug Deliv. Rev. 56:1127–1141.

Leary, S. P., Liu, C.Y., and Apuzzo, M. L. J. 2006. Toward the emergence of nanoneurosurgery: Part III – nanomedicine: targeted nanotherapy, nanosurgery, and progress toward the realization of nanoneurosurgery. Neurosurgery 58:1009–1026.

Lee, B.-S. and Holler, E. 1999. Effects of culture conditions on β-poly(L-malate) production by Physarum polycephalum. Appl. Microbiol. Biotechnol. 51:647–652.

Lee, B.-S., Vert, M., and Holler, E. 2002. Water-soluble aliphatic polyesters: poly(malic acid)s. In Biopolymers. Vol. 3a, eds. V. Doi, and A. Steinbuechel, pp. 75–103. New York: Wiley-VCH.

Lee, B.-S., Fujita, M., Khazenzon, N. M., Wawrowsky, K. A., Wachsmann-Hogiu, S., Farkas, D. L., Black, K. L., Ljubimova, J. Y., and Holler, E. 2006. Polycefin, a new prototype of a multifunctional nanoconjugate based on poly(β-L-malic acid) for drug delivery. Bioconjug. Chem. 17:317–326.

Li, C. 2002. Poly(L-glutamic acid)-anticancer drug conjugates. Adv. Drug Deliv. Rev. 54:695–713.

Ljubimov, A. V., Caballero, S., Aoki, A. M., Pinna, L. A., Grant, M. B., and Castellon, R. 2004. Involvement of protein kinase CK2 in angiogenesis and retinal neovascularization. Invest. Ophthalmol. Vis. Sci. 45:4583–4591.

Ljubimova, J. Y., Lakhter, A. J., Loksh, A., Yong, W. H., Riedinger, M. S., Miner, J. H., Sorokin, L. M., Ljubimov, A. V., and Black, K. L. 2001. Overexpression of α4 chain-containing laminins in human glial tumors identified by gene microarray analysis. Cancer Res. 61:5601–5610.

Ljubimova, J. Y., Fugita, M., Khazenzon, N. M., Das, A., Pikul, B. B., Newman, D., Sekiguchi, K., Sorokin, L. M., Sasaki, T., and Black, K. L. 2004. Association between laminin-8 and glial tumor grade, recurrence, and patient survival. Cancer 101:604–612.

Ljubimova, J. Y., Fujita, M., Khazenzon, N. M., Lee, B.-S., Wachsmann-Hogiu, S., Farkas, D. L., Black, K. L., and Holler, E. 2007. Nanoconjugate based on polymalic acid for tumor targeting. Chem. Biol. Interact. In press.

Luo, Y. and Prestwich, G. D. 2002. Cancer-targeted polymeric drugs. Curr. Cancer Drug Targets 2:209–226.

Maeda, H., Wu, J., Sawa, T., Matsumura, Y., and Hori, K. 2000. Tumor vascular permeability and the EPR effect in macromolecular therapeutics: a review. J. Control. Release 65:271–284.

Maruyama, K., Takahashi, N., Toshiaki, T., Nagaike, K., and Iwatsuru, M. 1997. Immunoliposomes bearing polyethyleneglycol-coupled Fab' fragment show prolonged circulation time and high extravasation into targeted solid tumors in vivo. FEBS Lett. 413:177–180.

Mastrobattista, E., Koning, G. A., van Bloois, L., Filipe, A. C., Jiskoot, W., and Storm, G. 2002. Functional characterization of an endosome-disruptive peptide and its application in cytosolic delivery of immunoliposome-entrapped proteins. J. Biol. Chem. 277:27135–27143.

Mathé, G., Lo, T. B., and Bernard, J. 1958. Effect sur la leucémie L1210 de la souris d'une combinaison par diazotation d'améthoptérine et de γ-globulines de hamsters porteurs de cette leucémie par hétérogreffe. C. R. Hebd. Seances Acad. Sci. 246:1626–1628.

Maurer, P. H., Gerulat, B. F., and Pinchuck, P. 1964. Antigenicity of polypeptides (pol-α-amino acids). XI. Quantitative relationships among polymers and rabbit antisera. J. Biol. Chem. 239:922–929.

McNamara II, J. O., Andrechek, E. R., Wang, Y., Viles, K. D., Rempel, R. E., Gilboa, E., Sullenger, B. A., and Giangrande, P. H. 2006. Cell type-specific delivery of siRNA aptamer-siRNA chimeras. Nat. Biotechnol. 24:1005–1015.

Merdan, T., Kopeček, J., and Kissel, T. 2002. Prospects for cationic polymers in gene and oligonucleotide therapy against cancer. Adv. Drug Deliv. Rev. 54:715–758.

Moghimi, S. M., Hunter, A. C., and Murray, J. C. 2005. Nanomedicine: current status and future prospects. FASEB J. 19:311–330.

Montemurro, F., Valabrega, G., and Aglietta, M. 2007. Lapatinib: a dual inhibitor of EGFR and HER2 tyrosine kinase activity. Expert Opin. Biol. Ther. 7:257–268.

Murphy, P. D. and Sage, H. J. 1970. Variation in the size of antibody sites for the poly-L-aspartate hapten during the immune response. J. Immunol. 105:460–470.

Murthy, N., Robichaud, J. R., Tirrell, D. A., Stayton, P. S., and Hoffman, A. S. 1999. The design and synthesis of polymers for eukaryotic membrane disruption. J. Control. Release 61:137–143.

Muzykantov, V. R. and Torchilin, V. P, eds. 2003. Biomedical Aspects of Drug Targeting. New York: Springer.

Nakamura, K., Sata, M., Iwata, H., Sakai, Y., Hirata, Y., Kugiyama, K., and Nagai, R. 2007. A synthetic small molecule, ONO-1301, enhances endogenous growth factor expression and augments angiogenesis in the ischaemic heart. Clin. Sci. (Lond.) 112:607–616.

Nori, A. and Kopecek, J. 2005. Intracellular targeting of polymer-bound drugs for cancer chemotherapy. Adv. Drug Deliv. Rev. 57:609–636.

Omelyanenko, V., Kopeckova, P., Prakash, R. K., Ebert, C. D., and Kopecek, J. 1999. Biorecognition of HPMA copolymer-adriamycin conjugates by lymphocytes mediated by synthetic receptor binding epitopes. Pharm. Res. 16:1010–1019.

Otsuka, H., Nagasaki, Y., and Kataoka, K. 2003. PEGylated nanoparticles for biological and pharmaceutical applications. Adv. Drug Deliv. Rev. 55:403–419.

Ouchi, T., Fujino, A., Tanaka, K., and Banba, T. 1990. Synthesis and antitumor activity of conjugates of poly(α-malic acid) and 5-fluoroacils bound via ester, amide or carbamoyl bonds. J. Control. Release 12:143–153.

Pagano, M. A., Cesaro, L., Meggio, F., and Pinna, L. A. 2006. Protein kinase CK2: a newcomer in the 'druggable kinome'. Biochem. Soc. Trans. 34(Pt 6):1303–1306.

Panarin, E. F. and Ushakov, S. N. 1968. Synthesis of polymer salts and amidopenicillines. Khim. Pharm. Zhur. 2:28–31.

Papisov, M. I. 1998. Theoretical considerations of RES-avoiding liposomes. molecular mechanics and chemistry of liposome interactions. Adv. Drug Deliv. Rev. 32:119–138.

Pardridge, W. M. 1999. Vector-mediated drug delivery to the brain. Adv. Drug Deliv. Rev. 36:299–321.

Pardridge, W. M. 2002. Drug and gene delivery to the brain: the vascular route. Neuron 36:555–558.

Pardridge, W. M. 2003. Blood-brain-barrier drug targeting: the future of brain drug development. Mol. Interv. 3:90–105.

Pardridge W. M. 2005. Molecular biology of the blood-brain barrier. Mol. Biotechnol. 30:57–70.

Philippova, O. E., Hourdet, D., Audebert, R., and Khokhlov, A. R. 1997. pH-Responsive gels of hydrophobically modified poly(acrylic acid). Macromolecules 30:8278–8285.

Qian, Z. M., Li, H., Sun, H., and Ho, K. 2002. Targeted drug delivery via the transferrin receptor-mediated endocytosis pathway. Pharmacol. Rev. 54:561–587.

Qiu, L. Y. and Bae, Y. H. 2006. Polymer architecture and drug delivery. Pharm. Res. 23:1–30.

Reddy, J. A., Allagadda, V. M., and Leamon, C. P. 2005. Targeting therapeutic and imaging agents to folate receptor positive tumors. Curr. Pharm. Biotechnol. 6:131–150.

Ringsdorf, H. 1975. Structure and properties of pharmacologically active polymers. J. Polym. Sci. Polym. Symp. 51:135–153.

Roldo, M., Hornof, M., Caliceti, P., and Bernkop-Schnurch, A. 2004. Mucoadhesive thiolated chitosans as platforms for oral controlled drug delivery: synthesis and in vitro evaluation. Eur. J. Pharm. Biopharm. 57:115–121.

Rozema, D. B., Ekena, K., Lewis, D. L., Loomis, A. G., and Wolff, J. A. 2003. Endosomolysis by masking of a membrane-active agent (EMMA) for cytoplasmic release of macromolecules. Bioconjug. Chem. 14:51–57.

Ruiz, F., Alvarez, G., Ramos, M., Hernandez, M., Bogonez, E., and Satrustegui, J. 2000. Cyclosporin A targets involved in protection against glutamate excitoxicity. Eur. J. Pharmacol. 404:29–39.

Saito, G., Swanson, J. A., and Lee, K. D. 2003. Drug delivery strategy utilizing conjugation via reversible disulfide linkages: role and site of cellular reducing activities. Adv. Drug Deliv. Rev. 55:199–215.

Schneerson, R., Kubler-Kielb, J., Liu, T. Y., Dai, Z. D., Leppla, S. H., Yergey, A., Backlund, P., Shiloach, J., Majadly, F., and Robbins, J. B. 2003. Poly(γ-D-glutamic acid) protein conjugates induce IgG antibodies in mice to the capsule of *Bacillus anthracis*: a potential addition to the anthrax vaccine. Proc. Natl. Acad. Sci. USA 100:8945–8950.

Schrama, D., Reisfeld, R. A., and Becker, J. C. 2006. Antibody targeted drugs as cancer therapeutics. Nat. Rev. Drug Discov. 5:147–159.

Seymour, L. W., Duncan, R., Strohalm, J., and Kopecek, J. 1987. Effect of molecular weight of N-(2-Hydroxypropyl)methacrylamide copolymers on body distribution and rate of excretion after subcutaneous, intraperitoneal, and intravenous administration to rats. J. Biomed. Mater. Res. 21:1341–1358.

Singh, M. 1999. Transferrin as a targeting ligand for liposomes and anticancer drugs. Curr. Pharm. Res. 5:443–451.

Sinha, V. R. and Kumria, R. 2001. Polysaccharides in colon-specific drug delivery. Int. J. Pharm. 214:19–38.

Skarlatos, S., Yoshikawa, T., and Pardridge, W. M. 1995. Transport of [^{125}I]-transferrin through the rat blood-brain barrier. Brain Res. 683:164–171.

Solit, D. B. and Rosen, N. 2007. Targeting HER2 in prostate cancer: where to next? J. Clin. Oncol. 3:241–243.

Sprincl, L., Exner, J., Sterba, O., and Kopecek, J., 1976. New types of synthetic infusion solutions. III. Elimination and retention of poly[N-(2-hydroxypropyl)methacrylamide] in a test organism. J. Biomed. Mater. Res. 10:953–963.

Sugahara, S., Okuno, S., Yano, T., Hamana, H., and Inoue, K. 2001. Characteristics of tissue distribution of various polysaccharides as drug carriers: influences of molecular weight and anionic charge on tumor targeting. Biol. Pharm. Bull. 24:535–543.

Summerton, J. and Weller, D. 1997. Morpholino antisense oligomers: design, preparation and properties. Antisense Nucleic Acid Drug Dev. 7:187–195.

Takeda, A. L., Colquitt, J. L., Clegg, A. J, and Jones, J. 2007. Pegaptanib and ranibizumab for neovascular age-related macular degeneration: a systematic review. Br. J. Ophthalmol. 91:1177–1182.

Taylor, E. M. 2002. The impact of efflux transporters in the brain on the development of drugs for CNS disorders. Clin. Pharmacokinet. 41:81–92.

Thorstensen, K. and Romslo, I. 1993. The transferrin receptor: its diagnostic value and its potential as therapeutic target. Scand. J. Clin. Lab. Invest. 53:113–120.

Tomlinson, R., Heller, J., Brocchini, S., and Duncan, R. 2003. Polyacetal-doxorubicin conjugates designed for pH-dependent degradation. Bioconjug. Chem. 14:1096–1106.

Torchilin, V. P. and Lukyanov, A. N. 2003. Peptide and protein drug delivery to and into tumors: challenges and solutions. Drug Discov. Today 8:259–266.

Turk, M. J., Reddy, J. A., Chmielewski, J. A., and Low, P. S. 2002. Characterization of a novel pH-sensitive peptide that enhances drug release from folate-targeted liposomes at endosomal pHs. Biochim. Biophys. Acta 1559:56–68.

Wang, Y., Dias, J. A., Nimec, Z., Rotundo, R., O'Conner, B. M., Freisheim, J., and Galivan, J. 1993. The properties and function of gama-glutamyl hydrolase and poly-gamma-glutamate. Adv. Enzyme Regul. 33:207–218.

Wang, L., Kristensen, J., and Ruffner, D. E. 1998. Delivery of antisense oligonucleotides using HPMA polymer: synthesis of a thiol polymer and its conjugation to water-soluble molecules. Bioconjug. Chem. 9:749–757.

Wang, G., Ahmad, K. A., Unger, G., Slaton, J. W., and Ahmed, K. 2006. CK2 signaling in androgen-dependent and -independent prostate cancer. J. Cell Biochem. 99:382–391.

West, K. R. and Otto, S. 2005. Reversible covalent chemistry in drug delivery. Curr. Drug Discov. Technol. 2:123–160.

Yao, J. C. 2007. Molecular targeted therapy for carcinoid and islet-cell carcinoma. Best Pract. Res. Clin. Endocrinol. Metab. 21:163–172.

Yoshida, T., Oide, N., Sakamoto, T., Yotsumoto, S., Negishi, Y., Tsuchiya, S., and Aramaki, Y. 2006. Induction of cancer cell-specific apoptosis by folate-labeled cationic liposomes. J. Control. Release 111:325–332.

Younes, M. N., Park, Y. W., Yazici, Y. D., Gu, M., Santillan, A. A., Nong, X., Kim, S., Jasser, S. A., El-Naggar, A. K., and Myers, J. N. 2006. Concomitant inhibition of epidermal growth factor and vascular endothelial growth factor receptor tyrosine kinases reduces growth and metastasis of human salivary adenoid cystic carcinoma in an orthotopic nude mouse model. Mol. Cancer Ther. 5:2696–2705.

Zhang, Y. and Bhavnani, B. R. 2006. Glutamate-induced apoptosis in neuronal cells is mediated via caspase-dependent and independent mechanisms involving calpain and caspase-3 proteases as well as apoptosis inducing factor (AIF) and this process is inhibited by equine estrogens. BMC Neurosci. 7:49–71.

Zhang, Y. and Pardridge, W. M. 2005. Delivery of β-galactosidase to mouse brain via the blood-brain barrier transferrin receptor. J. Pharmacol. Exp. Ther. 313:1075–1081.

Zhang, Y., Zhang, Y. F., Bryant, J., Charles, A., Boado, R. J., and Pardridge, W. M. 2004. Intravenous RNA interference gene therapy targeting the human epidermal growth factor receptor prolongs survival in intracranial brain cancer. Clin. Cancer Res. 10:3667–3677.

Zinner, R. G. and Herbst, R. S. 2004. Pemetrexed in the treatment of advanced non-small-cell lung cancer: a review of the clinical data. Clin. Lung Cancer 5(Suppl 2):S67–S74.

Pharmaceutical Micelles: Combining Longevity, Stability, and Stimuli Sensitivity

Myrra G. Carstens, Cristianne J.F. Rijcken, Cornelus F. van Nostrum, and Wim E. Hennink

1 Introduction

Targeted drug delivery is of particular importance for the treatment of life-threatening diseases such as cancer, since the adverse effects of cytostatic drugs can be very detrimental (Crommelin et al., 1995; Gros et al., 1981; Moses et al., 2003). Nowadays, polymer micelles are extensively studied as drug delivery systems to fulfill the requirements for selective and tissue-specific drug delivery (Adams et al., 2003; Allen et al., 1999; Gaucher et al., 2005; Jones and Leroux, 1999; Kabanov et al., 1992; Kataoka et al., 2001; Kreuter, 2006; Kwon, 2003; Lavasanifar et al., 2002b; Liu et al., 2007; Moghimi et al., 2001; Nishiyama and Kataoka, 2006; Torchilin, 2001, 2006; Yokoyama et al., 1992). The most attractive feature is their hydrophobic core with a relatively large capacity to accommodate hydrophobic agents, which are normally difficult to formulate. Polymeric micelles have been used to encapsulate a great variety of highly potent but hydrophobic drugs (Avgoustakis et al., 2002; Cheng et al., 2007; Djordjevic et al., 2005; Lee et al., 2007; Lin et al., 2003, 2005; Nishiyama et al., 2001, 2003; Yi et al., 2005; Yokoyama et al., 1998; Zamboni, 2005), such as doxorubicin (DOX) (Gillies and Frechet, 2005; Hruby et al., 2005; Kabanov et al., 2002b; Lee et al., 2005; Nakanishi et al., 2001; Rapoport, 2004; Yokoyama et al., 1992), paclitaxel (PTX) (Cavallaro et al., 2003; Hamaguchi et al., 2005; Huh et al., 2005; Kim et al., 2001, 2004; Krishnadas et al., 2003; Liggins and Burt, 2002; Shuai et al., 2004; Torchilin et al., 2003;), amphotericin B (Lavasanifar et al., 2002), and photosensitizers used for the treatment of cancer (Le Garrec et al., 2004; van Nostrum, 2004; Zhang et al., 2003). Some of these micellar formulations have already entered clinical trials and showed promising results with regard to their therapeutic index in cancer patients (Barratt, 2000; Kim et al., 2004; Matsumura et al., 2004; Torchilin, 2006). A drug delivery system needs to fulfill several (pharmaceutical) requirements such as a significant increase in therapeutic effect with respect to the free drug, good biocompatibility, and the possibility to scale-up the production of the micellar formulation. In addition, the ideal micellar system (1) has long circulating properties and adequate stability in the blood stream, (2) has a high drug-loading capacity, (3) is able to selectively accumulate at the target site, and (4) offers the possibility to control the release of

V. Torchilin (ed.), *Multifunctional Pharmaceutical Nanocarriers*,
© Springer Science+Business Media, LLC 2008

the drug at the target site, for example, by external stimuli (Allen et al., 1999; Gaucher et al., 2005; Jones and Leroux, 1999; Kabanov et al., 1992; Kreuter, 2006; Kwon, 2003; Moghimi et al., 2001; Nishiyama and Kataoka, 2006; Torchilin, 2001, 2006, 2007). Other desirable properties of polymeric micelles are the ability to be (degraded and) excreted from the body after the drug is released and the possibility to track and trace the micellar structure by encapsulating an imaging agent (Torchilin, 2002).

The primary focus of this chapter is the description and discussion of longevity and stability of drug-loaded polymeric micelles after intravenous injection, and the possibility to release their payload in a controlled manner upon local and/or external stimuli.

2   Longevity

Drug delivery systems such as polymeric micelles should deliver their payloads selectively at the target sites, and therefore longevity in the blood circulation is a prerequisite. Provided that the encapsulated drug will remain associated with the nanocarrier, a long circulating nanocarrier will longer maintain the blood level of its loaded drug, thereby enhancing the therapeutic effect of the drug as a result of the prolonged interactions in the target organ (Moghimi et al., 2001; Torchilin, 2001, 2006). In addition, a long circulation time allows the accumulation of polymeric micelles themselves in pathological tissue via the so-called enhanced permeability and retention (EPR) effect. This EPR effect was proposed by Maeda et al. in the eighties, and is attributed to the higher permeability of the vasculature in diseased areas due to discontinuous endothelium, and to impaired lymphatic drainage (Gaucher et al., 2005; Matsumura and Maeda, 1986; Maeda et al., 2000, 2003; Torchilin, 2001). These two features enable extravasation of colloidal particles through the "leaky" endothelial layer into the tumor and inflamed areas, and subsequent retention there.

However, the human immune system rapidly recognizes and eliminates foreign objects via adsorption of opsonic proteins onto their surface. Therefore, a key issue for prolonged circulation of colloidal drug carriers is to reduce the rate and extent of this opsonization, and the recognition by cells of the reticulo-endothelial system (RES). It has been shown that the so-called steric stabilization, i.e., coating of the particle surface with hydrophilic polymers (e.g., poly(ethylene glycol), poloxamer) effectively reduces the interaction with opsonic proteins, and thereby uptake by the RES cells of the liver, spleen, and bone marrow (Illum et al., 1987; Leroux et al., 1995; Moghimi et al., 1991b; Owens and Peppas, 2006; Stolnik et al., 1995; Woodle and Lasic, 1992). Next to surface characteristics, the biodistribution of polymeric micelles depends on many other factors including predominantly particle size (Gaucher et al., 2005; Nishiyama and Kataoka, 2006; Torchilin, 2007; Vonarbourg et al., 2006a) and particle rigidity (Sun et al., 2005; Vonarbourg et al., 2006a) as will be described in this section (vide infra).

2.1   Steric Stabilization

2.1.1   Poly(Ethylene Glycol)

Poly(ethylene glycol) (PEG) (also called poly(ethylene oxide) (PEO)) is the most frequently used hydrophilic segment of amphiphilic micelle-forming copolymers, for example in PEG-b-poly(propyleneglycol) (PPO) (Kabanov et al., 1992), PEG-b-polyesters (Hagan et al., 1996; Carstens et al., 2007a; Luo et al., 2002; Shi et al., 2005; Yamamoto et al., 2001), PEG-b-phospholipid (PL) (Lukyanov and Torchilin, 2004; Weissig et al., 1998), and various PEG-b-poly(meth)acrylamide derivatives (Neradovic et al., 2004; Rijcken et al., 2005; Soga et al., 2004; Topp et al., 1997). The ubiquitous use of PEG results from its low toxicity and immunogenicity, and FDA approval in various pharmaceutical formulations. Moreover, it has unique physicochemical properties, such as excellent water solubility, high flexibility, and a large exclusion volume, resulting in good "stealth" properties (Adams et al., 2003; Allen and Hansen, 1991; Lee et al., 1995; Molineux, 2002; Torchilin and Trubetskoy, 1995; Woodle and Lasic, 1992;). Since the discovery of these "stealth" properties and the positive effect of a PEG coating on the circulation kinetics of colloidal drug delivery systems in the early nineties (Klibanov et al., 1990), the responsible mechanisms and factors that influence this effect have been extensively studied, but are not fully elucidated yet, as reviewed recently by Vonarbourgh et al. (Vonarbourg et al., 2006a). In general, the reduction of opsonization by PEG is ascribed to shielding of surface charge and an increased hydrophilic surface, thereby preventing the two main driving forces for protein adsorption, i.e., electrostatic and hydrophobic interactions. In addition, reduction of the Van der Waals interactions, enhanced repulsive forces, and the formation of an impermeable polymeric layer on the particle surface, and also the binding of dysopsonins (i.e., naturally occurring substances known to inhibit phagocytic ingestion) are considered to attribute to the protective effect (Owens and Peppas, 2006; Soppimath et al., 2001; Torchilin, 2006; Torchilin and Trubetskoy, 1995; Vonarbourg et al., 2006a). Prolonged circulation times as a result of effective blocking of opsonization can only be achieved when the protective polymer layer is sufficiently thick. On the other hand, the PEG chains should retain their flexibility for an optimal protection against recognition by the immune system. Both factors are related to PEG molecular weight, conformation, and the surface chain density (Gaucher et al., 2005; Michel et al., 2005; Owens and Peppas, 2006; Soppimath et al., 2001; Vonarbourg et al., 2006a).

The positive effect of a higher PEG molecular weight to reduce protein adsorption (Fig. 1), and to prolong circulation times of colloidal particles was among others demonstrated for ^{14}C-benzylamine labelled PEG-b-P(Asp) micelles with covalently bound doxorubicin. An increase in the molecular weight of PEG from 5,000 to 12,000 resulted in a fivefold increase of the blood level of the micelles at 4h post injection (13 vs. 68% of the injected dose), and decreased hepatosplenic uptake (Kwon et al., 1993b, 1994). The favorable influence of a longer PEG chain on the

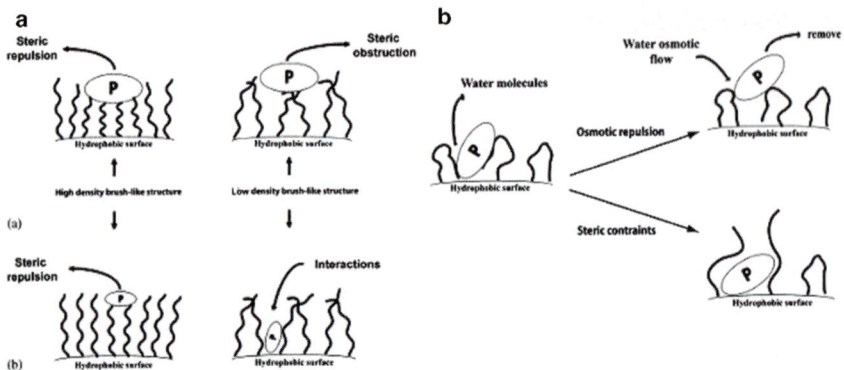

Fig. 1 Effect of chain density (A, large (**a**) and small (**b**) proteins) and conformation (B) on the repulsion of opsonic proteins (P). This figure was published in (Vonarbourg et al., 2006a). Copyright Elsevier (2006)

blood circulation times was also reported for a series of Pluronic® (PEG-PPO-PEG) block copolymers (Kabanov et al., 2002a). Although usually PEG with a molecular weight between 1,000 and 15,000 Da is used to design polymeric micelles for drug delivery, smaller PEGs may also be able to form a protective layer. For instance, coating of lipid nanocapsules with PEG with a molecular weight of 660 Da resulted in steric stabilization and reduced protein opsonization, which was ascribed to a high surface chain density (Vonarbourg et al., 2006b). Remarkably, regarding the PEG chain conformation on the surface, it was shown that the attachment of both ends of the PEG chain onto the surface resulted in better protection than single chain end attachment, despite a lower chain mobility (Hu et al., 2007; Peracchia et al., 1997). It was suggested that single chain end attachment allows easier penetration of the PEG layer by proteins (Fig. 1B), and that it results in a less dense PEG layer, when compared to double chain end attachment (Hu et al., 2007; Peracchia et al., 1997).

Although nowadays PEG is considered to be the golden standard for the steric stabilization of nanoparticles, it is not as inert as generally assumed. Several studies with PEGylated liposomes indicate the binding of blood proteins (Moghimi and Szebeni, 2003), and disappearance of the stealth properties at low lipid doses and/or upon repeated administration (Carstens et al., 2006; Dams et al., 2000; Ishida et al., 2002; Laverman et al., 2000). The repeated administration of drug-loaded polymeric micelles has been reported in several studies. However, these studies focussed on the antitumor effect of the micellar formulation of a cytostatic drug, rather than the biodistribution of the carrier itself, or the loaded drug (Hamaguchi et al., 2005; Kim et al., 2001; Le Garrec et al., 2004; Nishiyama et al., 2003; Yokoyama et al., 1998, 1999;). It is anticipated that the loss of the long circulation behavior upon repeated administration or at low doses is not limited to PEGylated liposomes, and that micelles with a hydrophilic PEG shell may be subject to this phenomenon as well. This warrants the search for alternative protective coatings, as will be described in the next section.

2.1.2   Alternative Coatings

Based on the above-mentioned mechanistical aspects, alternative hydrophilic polymers (Table 1) have to be biocompatible, hydrophilic, water soluble, and highly flexible. Polymeric coatings that successfully prolonged the circulation

Table 1 Various hydrophilic polymers used for the steric stabilization of micelles

Polymer	Abbreviation	Chemical structure	References
Poly(ethylene glycol)	PEG		Hagan et al. (1996); Kabanov et al. (1992); Kwon et al. (1993b); Topp et al. (1997); Weissig et al. (1998); Yokoyama et al. (1998); Yamamoto et al. (2001); Carstens et al. (2007a); Luo et al. (2002); Lukyanov and Torchilin (2004); Neradovic et al. (2004); Rijcken et al. (2005); Shi et al. (2005); Soga et al. (2004)
Poly(N-vinyl pyrrolidone)	PVP		Benahmed et al. (2001); Chung et al. (2004); Le Garrec et al. (2002, 2004); Lee and Lee (2007); Lele and Leroux (2002); Luo et al. (2004)
Poly(vinyl) alcohol	PVA		Zuccari et al. (2005)
Poly(2-ethyl-2-oxazoline)	PEOx		Hsiue et al. (2006); Kim et al. (2000); Lee and Lee (2007); Lee et al. (1999b); Volet et al. (2005)
Poly(N-(2-hydroxy-propyl) methacrylamide)	PHPMA		Lele and Leroux (2002)
Poly(acrylamide)	P(AAm)		Allen et al. (1999); Torchilin and Trubetskoy (1995)
Dextran	Dex		Lemarchand et al. (2005); Rouzes et al. (2000)

times of liposomal carriers are based on poly(oxazoline) (Woodle et al., 1994), poly(glycerol) (Maruyama et al., 1994b), poly(N-vinyl pyrrolidone) (PVP) (Torchilin et al., 1994, 2001), poly(acrylamide) (PAAm) (Torchilin et al., 1994), poly(vinylalcohol) (PVA) (Takeuchi et al., 2001), poly(N-(2-hydroxypropyl) methacrylamide) (PHPMA) (Whiteman et al., 2001), and poly(amino acids) (Metselaar et al., 2003; Romberg et al., 2007). Several of these polymers have also been used as hydrophilic part of amphiphilic block copolymers in the development of polymeric micelles, which will be described in this section.

Poly(N-Vinylpyrrolidone)

An attractive PEG alternative is poly(N-vinylpyrrolidone) (PVP), which is highly hydrophilic, flexible, and biocompatible, similar to PEG. In recent years, micelle formation of several amphiphilic PVP-containing block copolymers has been reported, for example, copolymers with poly(lactic acid) (PLA) (Benahmed et al., 2001; Le Garrec et al., 2004; Luo et al., 2004), PCL (Chung et al., 2004; Lee and Lee, 2007; Lele and Leroux, 2002), and PNIPAAM (Le Garrec et al., 2002), and the effective encapsulation of hydrophobic drugs such as indomethacin (Benahmed et al., 2001), chloro aluminum phthalocyanine (Le Garrec et al., 2002), and paclitaxel (PTX) (Le Garrec et al., 2004) was demonstrated. It has been shown that PVP effectively prolongs the circulation time of liposomes (Torchilin et al., 1994, 2001), but these data are not yet available for micellar systems. In vivo studies with PVP-coated micelles focused on the biodistribution and tumor accumulation of micelles loaded with chloro aluminum phthalocyanine (Le Garrec et al., 2002) or PTX (Le Garrec et al., 2004). For both drugs, no improvement in blood circulation times, tumor accumulation, or therapeutic effect was observed when compared to a similar dose formulated in Cremophor EL®, which may be related to a fast release of the drug from the micelles or disintegration of the micelles themselves. However, the maximum-tolerated dose of PTX-loaded PVP-*b*-PLA micelles in mice was more than five times higher than that of Taxol® (PTX formulated in Cremophor EL). By using a higher dose, a better antitumor activity could be achieved (Le Garrec et al., 2004).

Polysaccharides

Another class of polymers used as PEG alternative is the group of polysaccharides, which play a role in the surface characteristics of several cells, e.g., red blood cells. These cells effectively evade the immune system, which may be related to the presence of oligosaccharide groups on their surface (Lemarchand et al., 2004; Moghimi et al., 2001). It was demonstrated that dextran-*g*-PCL nanoparticles, and dextran-coated PLA nanoparticles showed a lower protein adsorption than bare polyester nanoparticles (Lemarchand et al., 2005; Rouzes et al., 2000). Whereas these studies did not investigate the circulation kinetics of these dextran-coated nanoparticles, prolonged circulation times were reported for poly(methyl methacylate) (PMMA) nanoparticles coated with dextran or heparin, compared to bare PMMA nanoparticles (Passirani et al., 1998). Recently, micelles of hydroxyethyl starch (HES) grafted with acyl chains were reported, which may also evade the RES (Besheer et al., 2007). However, the pharmacokinetics of these HESylated nanoparticles in vivo has not been investigated yet. It was demonstrated that the conformation of polysaccharides is of high importance when minimizing the interactions with plasma proteins (Lemarchand et al., 2006). In contrast to PEG (vide supra, Sect. 2.1.1), a brush-like configuration conferred a more effective protection than the presence of dextran loops at the surface. Although poly- and oligosaccharides present at the nanoparticles surface may protect against RES uptake, saccharide receptors are present in the membranes of several cells (Lemarchand et al., 2004). This enables their use in active targeting approaches (Sect. 2.4), and is illustrated by the recognition of galactose presenting nanoparticles by hepatocytes (Maruyama et al., 1994a).

Other Hydrophilic Blocks

Several other biocompatible hydrophilic polymers have been used as the shell-forming component in polymeric micelles, for example, poly(N,N,dimethylamino-2-ethyl methacrylate)(PDMAEMA) (Bougard et al., 2007), poly(ethylenimine)(PEI) (Nam et al., 2003), poly(acrylic acid) (PAAc) (Allen et al., 1999; Inoue et al., 1998), and poly(asparagine) (Jeong et al., 2005a). However, the longevity of these micelles in vivo is questionable, because of the charges present on the micellar shell surface. In addition, p(NIPAAM-co-N,N-dimethylacrylamide (DMAAm)) was suggested as the shell-forming block (Kohori et al., 1999). Since this copolymer displays thermosensitive behavior, it can be used for the temperature-triggered release of encapsulated drugs, as will be discussed further on in this chapter.

Micelles composed of poly(2-ethyl-2-oxazoline)-*b*-polyester (Hsiue et al., 2006; Kim et al., 2000; Lee and Lee, 2007; Lee et al., 1999b; Volet et al., 2005;), PHPMA-*b*-PCL (Lele and Leroux, 2002), poly(acryl amide) (PAAm)-*b*-palmitate (Allen et al., 1999; Torchilin and Trubetskoy, 1995), and PVA-*b*-oleylamine (Zuccari et al., 2005) may exhibit prolonged circulation times in vivo, since these hydrophilic blocks effectively protected liposomes against rapid RES uptake (vide supra), but evidence is not yet obtained.

2.2   Micellar Size

The size of micelles is another predominant parameter determining its fate after i.v. injection. Nanoparticles larger than 200 nm are removed by mechanical filtration by the interendothelial cell-slits in the spleen, and particles with a molecular weight smaller than 50 kDa (a hydrodynamic diameter of 5–10 nm) are subject to renal excretion (Kabanov et al., 2002a; Kwon, 2003; Moghimi et al., 1991a, 2001). The size of polymeric micelles (10–100 nm) prevents elimination via these routes, but it was shown that the extent of RES uptake and tumor penetration were related to their size as well (Gaucher et al., 2005; Kabanov et al., 2002a; Kwon, 2003; Nishiyama and Kataoka, 2006; Torchilin, 2007). For example, shell-crosslinked poly(*tert*-butylacrylate)-*b*-polystyrene (PBA-*b*-PS) micelles with a size of 20 nm had significantly higher blood residence times than their two times bigger counterparts (50% vs. 5% of the injected dose at 1 h after injection) (Sun et al., 2005), and the same trend was found with PEG-*b*-PHDCA micelles of 80 and 170 nm (Fang et al., 2006). It should be mentioned, however, that in the latter study, the longer circulation times of the 80-nm micelles may be related to a higher surface density of PEG because of the smaller size. As pointed out in Sect. 2.1.1, a higher PEG density results in more effective shielding. Furthermore, it was suggested that the particle size itself is of importance, since small particles have a higher curvature, thereby hampering the adsorption of proteins (Vonarbourg et al., 2006a).

In contrast, Weissig et al. demonstrated that i.v.-injected PEG_{5000}-*b*-distearoyl phosphatidyl ethanolamine (PEG-*b*-DSPE) micelles of 15 nm circulated shorter than 100-nm liposomes in mice with a subcutaneously established Lewis lung carcinoma. Remarkably, the tumor accumulation of the small micelles was much higher than that of the liposomes (Weissig et al., 1998). This is ascribed to the low cutoff size of the tumor vessel wall, which is different in each tumor type, and determines the ability of nanoparticles to penetrate the tumor tissue (Hobbs et al., 1998; Yuan et al., 1995). Especially in solid tumors, such as Lewis lung carcinoma, the small size of micelles is indeed an additional advantage over, e.g., liposomal and other bigger nanoparticulate systems. A different tumor penetration was also reported for doxorubicin-loaded PEG-*b*-poly(aspartate hydrazone adriamycin) micelles of 65 nm and liposomes of 150 nm (Bae et al., 2005; Nishiyama and Kataoka, 2006). The micelles were found inside tumoral spheroids, whereas the 150-nm liposomes were found only in the periphery (Nishiyama and Kataoka, 2006). Thus, small-sized nanoparticles benefit more from the EPR effect as a result of their higher ability to penetrate into tumor tissue.

2.3   Other Strategies to Improve Circulation Times

Next to surface characteristics and particle size, the rigidity of the particle also influences the circulation kinetics. It has been reported that liposomes with a rigid lipid bilayer exhibit long circulation times despite the absence of a protective PEG layer (Oku and Namba, 1994; Senior and Gregoriadis, 1982). Sun et al. compared the biodistribution of poly(*tert*-butylacrylate) (PBA)-*b*-polystyrene (PS) micelles with a high T_g, glassy PS core, to that of PBA-*b*-poly(methyl acrylate) (PMA) micelles with a low T_g, fluid-like PMA core, and demonstrated that a more rigid, glassy micellar core results in a longer blood retention.

In addition to changing particle-related parameters, such as surface characteristics, size, and rigidity, another strategy to improve the circulation times of nanoparticles may be predosing with empty micelles to saturate the elimination mechanisms.

In fact, this was the first strategy to improve the circulation times of liposomes (Abra et al., 1980), but so far it has not yet been investigated for micellar systems.

2.4 Longevity of Actively Targeted Polymeric Micelles

Although long circulation times promote the passive targeting of polymeric micelles and thus the delivery of the entrapped drug at its site of action, their delivery may be further improved by active targeting. Specific ligands such as internalizing antibodies (Gao et al., 2003; Torchilin et al., 2003), sugar moieties (Jeong et al., 2005b), transferrin, RGD, and folate (Park et al,. 2005; Xiong et al., 2007) have been coupled to the micellar shell to promote cellular recognition and internalization of the drug carrier (Lee et al., 2003; Liu et al., 2007b; Torchilin, 2004; Xiong et al., 2007). The delivery of drugs by actively targeted long-circulating micelles is a promising approach to improve its site-specific action, but the properties of such carriers are paradoxal. On the one hand, the presence of a targeting ligand, especially antibodies or other proteins, on the micellar surface may enhance its recognition by the immune system and thereby removal from the circulation. On the other hand, the presence of a protective polymer such as PEG may interfere with the binding of the ligand to its target. A strategy to tackle the above-mentioned paradox is the use of a "sheddable" coating, i.e., a coating that is removed after arrival at the target site. Recently, a pH-sensitive deshielding of a TAT peptide coupled to the surface of PEG-b-PLLA micelles was reported. The cationic TAT ligand was shielded at pH 7.4 by the anionic pH-sensitive poly(methacryloyl sulfadimethoxine) (PSD) block of a PEG-b-PSD copolymer. Upon lowering the pH to 6.6, the complex was disrupted and TAT was exposed, resulting in enhanced cellular uptake and localization at the surface of the nucleus (Sethuraman and Bae, 2007).

3 Micellar Stability

Even with a prolonged circulation, selective drug accumulation can take place only if premature leakage of drug molecules from the micelles is prevented or the release of the drug is slow during the first few hours after administration. Essentially, the core-forming segment determines the micellar stability, its drug-loading capacity, and drug release profile, which explains why so many core-forming, mainly hydrophobic, polymers have been investigated (Table 2) (Allen et al., 1999, 2000; Hagan et al., 1996; Kabanov et al., 2002a; Kwon, 2003; Liggins and Burt, 2002; Torchilin, 2001; van Nostrum, 2004; Yokoyama et al., 1998; Zhang et al., 2006).

The stability of a polymeric micelle can be considered either thermodynamically or kinetically. Polymeric micelles are thermodynamically stable when the polymer concentration in the water is above the critical micelle concentration (CMC), also called the critical aggregation concentration (CAC). Below the CMC, amphiphilic

Table 2 Hydrophobic polymers used as core-forming block in polymeric micelles

Polymer	Abbreviation	Chemical structure	References
Poly (lactic acid)	PLA		Dong and Feng (2004); Le Garrec et al. (2004)
Poly(ε-caprolactone	PCL		Carstens et al. (2007a); Luo et al. (2004); Shi et al. (2005)
Poly(N-iso propylacrylmide)	PNIPAAM	R = CH2 or C2H4	Neradovic et al. (2001); Topp et al. (1997)
Poly(γ-benzyl L-gluta-mate) or poly(γ-benzyl L-aspartate)	PBLG or PBLA		Jeong et al. (2005b); Kwon et al. (1993a); Watanabe et al. (2006); Yokoyama et al. (2004)
Poly(propylene oxide)	PPO	R = H or CH3; n is 1 - 4	Kabanov et al. (1992, 2002b); Rapoport et al. (2002)
Poly(methacrylam ide oligolactates)	pHEMA-Lac$_n$, pHPMA-Lac$_n$	R = H or CH3; n is 1 - 4	Rijcken et al. (2005); Soga et al. (2004)

block copolymers in water are present as single chains in the bulk and at the air–water interface. When the concentration is increased above the CMC, the Gibbs free energy (ΔG) of the system is minimized by the self-assembly of the amphiphiles, as a result of the hydrophobic interactions between the hydrophobic blocks (Attwood and Florence, 1983; Kwon, 2003). Since polymeric micelles are subject to dilution in the circulation upon intravenous injection, it is important to know their CMC, and to administer a sufficiently high dose (Allen et al., 1999; Liu et al., 2007). The kinetic stability of a micellar system is related to the exchange rate of single polymer chains between the micelles and the bulk, and even upon dilution below the CMC, the micellar system may still be kinetically stable (Allen et al., 1999; Attwood and Florence, 1983; Kwon, 2003; Kwon and Okano, 1996). The rate of disassembly is related to the strength of the interactions in the micellar core, which depends on many factors, such as the physical state of the core-forming polymer state (crystalline or amorphous), the presence of solvent (e.g., methanol or dioxane residues due to the preparation procedure) in the micellar core, the ratio between the hydrophilic and the hydrophobic block of the copolymer, and the encapsulation of hydrophobic compounds (Allen et al., 1999; Gaucher et al., 2005; Kwon, 2003; Yokoyama et al., 1998). Preferably, polymeric micelles have a (semi-) crystalline or glassy core at body temperature, and are composed of block copolymers with a low CMC (Allen et al., 1999; Gaucher et al., 2005; Kwon, 2003; Teng et al., 1998). In addition to micellar stability, several other factors influence the release of the loaded drug, such as the length of the core-forming polymer segment and the amount of loaded drug (Kim et al., 1998). Importantly, the compatibility between the core-forming polymer and the drug affects the drug loading and release (Allen et al., 1999; Gaucher et al., 2005). By a proper selection of the block copolymer, the compatibility with the drug was optimized for various micellar systems, which is expected to increase the in vivo drug retention (Forrest et al., 2006; Hamaguchi et al., 2005; Huh et al., 2005; Kwon et al., 1994; Lavasanifar et al., 2002a; Le Garrec et al., 2004; Lee et al., 2003b; Liu et al., 2004; Prompruk et al., 2005; Torchilin, 2001; Wilhelm et al., 1991).

However, the presence of blood components often leads to premature drug release, either by provoking micelle destabilization, or by extraction of encapsulated drug from intact micelles (Konan-Kouakou et al., 2005; Liu et al., 2005; Lo et al., 2007; Opanasopit et al., 2005; Savic et al., 2006). Therefore, much effort is currently undertaken to improve the thermodynamic and the kinetic stability of drug-loaded micelles. Several strategies have been investigated, including modification of the micelle-forming polymers to reduce their CMC (Sect. 3.1), physical and covalent crosslinking (Sects. 3.2 and 3.3), and improving the drug–polymer compatibility (Sect. 3.4).

3.1   Reducing the CMC

The thermodynamic stability of polymeric micelles can be improved by reducing the CMC of the amphiphilic block copolymers. This is easily achieved by adjusting the sizes of the blocks (Allen et al., 1999; Gaucher et al., 2005). Both, a larger hydrophobic block and a smaller hydrophilic block, result in a higher overall hydrophobicity, thereby reducing the CMC (Carstens et al., 2007a; Kabanov et al., 2002a; Kwon, 2003; Le Garrec et al., 2004; Letchford et al., 2004). In addition to size, the nature of the hydrophobic block is an important parameter determining the CMC. Chemical modification of the hydrophobic block, for example by the introduction of aromatic groups, has been demonstrated to effectively reduce the CMC (Carstens et al., 2007a; Mahmud et al., 2006; Opanasopit et al., 2004; Watanabe et al., 2006; Yokoyama et al., 2004). For instance, changing the amount of benzyl groups in the modified poly(β-benzyl L-aspartate) (PBLA$_{mod}$) block of PEG-*b*-PBLA$_{mod}$ copolymer from 44 to 75% resulted in a tenfold reduction of the CMC (Opanasopit et al., 2004). A 60–200-fold reduction was obtained by the end group modification of mPEG$_{750}$-*b*-oligo(ε-caprolactones) with a benzoyl or a naphthoyl moiety (Carstens et al., 2007a). Similarly, an increase in the level of fatty acids (Lavasanifar et al., 2001) or hydrophobic oligolactates (Rijcken et al., 2005) attached to the polymer backbone resulted in a reduced CMC.

3.2   Physical Interactions

The introduction of aromatic groups does not only improve the thermodynamic stability by decreasing the CMC, but may also improve the kinetic stability of the micelles by strengthening the interactions inside the micellar core through π-π-stacking. Mahmud et al. studied the viscosity of the core of PEG-*b*-PCL micelles and PEG-*b*-poly(α-benzyl-ε-caprolactone) micelles with fluorescence spectroscopy and found, besides a decreased CMC (vide supra), an increased rigidity of the micellar core as a result of the introduction of aromatic groups (Mahmud et al., 2006). The introduction of crystallinity or stereocomplex formation was shown to enhance the stability of micelles as compared to the amorphous counterparts (Kang et al., 2005; Zhang et al., 2006a). An increase in physical interactions was also obtained by hydrogen bonding (Yoshida and Kunugi, 2002). A summary of physical interactions that can play a role in the kinetic stability of micelles is illustrated in Fig. 2.

In addition to interactions between the core-forming polymers, the incorporation of a hydrophobic drug may also enhance the micellar stability. For example, Yokoyama et al. demonstrated that the stability of PEG-*b* P(Asp)micelles was not

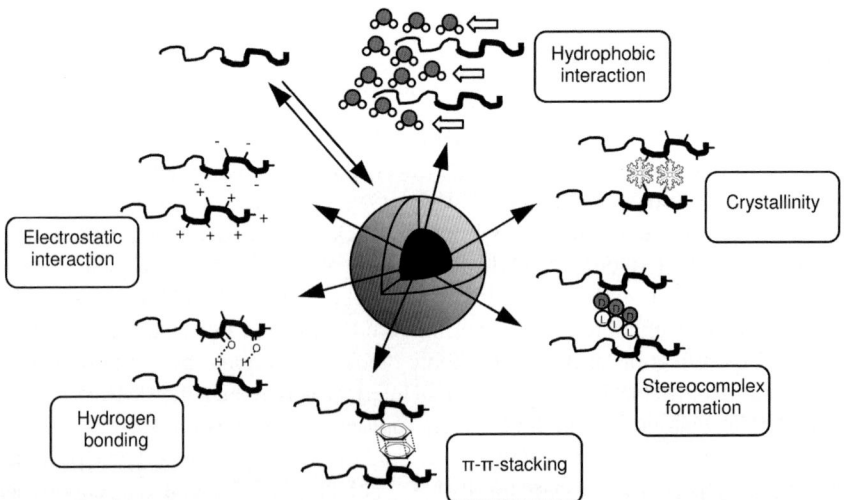

Fig. 2 Interactions in the micellar core that enhance the kinetic stability of polymeric micelles

only increased by the amount of chemically bound doxorubicin (DOX), but also by the amount of physically entrapped DOX (Yokoyama et al., 1998).

Micelle formation may also be driven by the electrostatic ionic interaction forces of oppositely charged block copolymers, to form the so-called polyion complex (PIC) or complex coacervate micelles (Fig. 3) (Cohen Stuart et al., 2005; Harada and Kataoka, 1999; Rodriguez-Hernandez et al., 2005b). Examples of polyion couples are PEG-*b*-poly(L-lysine)(P(Lys)) and PEG-*b*-poly(α, β-aspartic acid) (Harada and Kataoka, 1995, 1999), PEG-*b*-polymethacrylic acid (PMA) and poly(N-ethyl-4-vinylpyridinium)(PEVP)(Kabanov et al., 1996), or PEG-*b*-poly(2-vinylpyridinium) (P2VP) and polystyrene sulfonate (PSS) (Gohy et al., 2000). PIC micelles that comprise thermosensitive shells were described (Cohen Stuart et al., 2005; Park et al., 2007), as well as PIC micelles composed of PEG-PMA and Ca^{2+}, with a crosslinked PMA core (Bronich et al., 2005). Polymers having negatively charged units, such as PMA and P(Asp) (co)polymers, were used to form micelles with cationic drugs or peptides, and polycations such as PEG-P(Lys) were used to form micelles with siRNA/DNA (Harada and Kataoka, 2006). The advantage of PIC micelles is their ease of preparation, i.e., simple mixing of aqueous solutions of drug and polymer. However, their application is limited due to the low stability in physiological saline and the drugs prerequisite to be hydrophilic, although this could be overcome by copolymerizing phenylalanine in the polymer backbone, thereby enhancing the hydrophobic/aromatic interactions (Prompruk et al., 2005). On the other hand, the saline-induced micelle destabilization can be utilized to control the release of the loaded drug. This concept was demonstrated for cisplatin complexed PEG-*b*-p(glutamic acid), as will be discussed in Sect. 4.7 (Nishiyama et al., 2003).

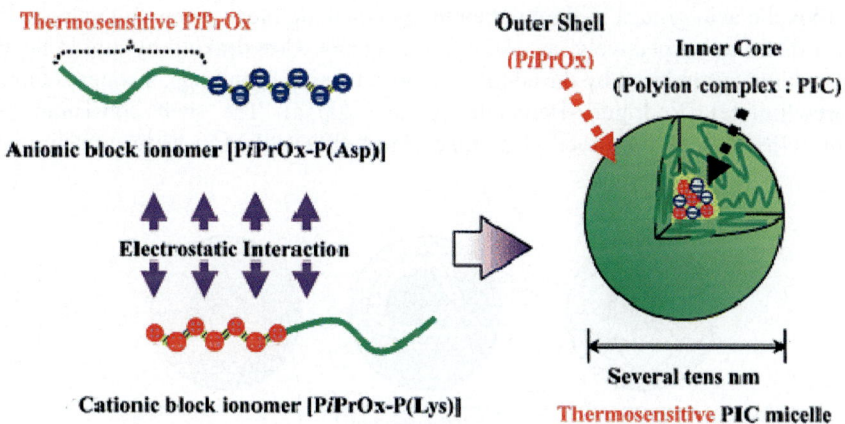

Fig. 3 Formation of PIC micelles with a thermosensitive shell composed of poly(2-isopropyl-2-oxazoline)(PiPrOx) shell and a core composed of the anionic ionomer poly(aspartic acid)P(Asp) and cationic poly(L-lysine) (P(Lys)). Reprinted with permission from (Park et al., 2007). Copyright (2007), American Chemical Society

3.3   Covalent Crosslinking

In addition to the above-mentioned physical means to enhance stability of micelles, chemical crosslinking of either the shell, the interfacial layer, or the core of the micelles has been used to prepare stable particles with a micellar morphology (Fig. 4) (Harada and Kataoka, 2006; O'Reilly et al., 2006; Rodriguez-Hernandez et al., 2005b; Rosler et al., 2001). The increased stability of covalently crosslinked micelles vs. their noncrosslinked counterparts was proven among other techniques by the insensitivity of micelles toward a destabilizing agent (e.g., sodium dodecyl sulphate (SDS)) (Iijima et al., 1999; Kim et al., 1999; Rijcken et al., manuscript submitted). While the micellar morphology was fixed by the crosslinking procedure, drug release could be controlled by the crosslink density (Hu et al., 2006), and stimuli responsiveness was retained (e.g., to pH, temperature, salt concentration) (Bae et al., 2006; Bontha et al., 2006; Liu et al., 2002a; Rijcken et al., manuscript submitted). However, in contrast to physical crosslinking, the covalent crosslinking approach may adversely affect the overall degradability of the micelle and the structural integrity of the encapsulated drug (when the crosslinking procedure is performed in the presence of the drug).

3.3.1   Shell Crosslinking

The hydrophilic shell of polymeric micelles has been covalently crosslinked by chemical or photo-induced reactions (Harada and Kataoka, 2006; Jiang et al., 2006b, 2007; Joralemon et al., 2005; Li et al., 2006, 2006a; O'Reilly et al., 2006; Rodriguez-Hernandez et al., 2005a; Sun et al., 2005). For example, in polypeptide-*b*-polydiene micelles, covalent bonds were formed between either the amine or carboxylic acid groups in the hydrophilic polypeptide block using glutaraldehyde or a diamine, respectively, as crosslinking agents. Crosslinking by amide bond formation was induced by the addition of an activating agent (e.g., a water-soluble carbodiimide) (Rodriguez-Hernandez et al., 2005a). The shell consisting of poly(N,N-dimethylaminoethyl methacrylate) (PDMAEMA) were crosslinked

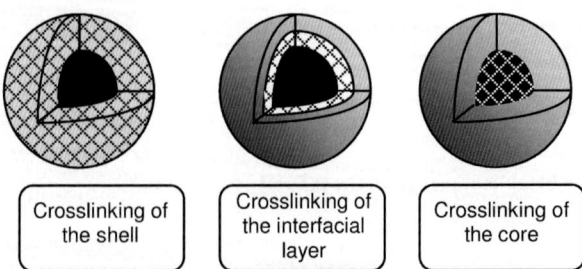

Crosslinking of the shell

Crosslinking of the interfacial layer

Crosslinking of the core

Fig. 4 Covalent crosslinking of the micellar shell, the interfacial layer, or the micellar core

by alkylation with a bifunctional alkyl iodide (P2ilon et al., 2005) and the shell of poly(4-vinylpyridine) (P4VP)-*b*-polystyrene were crosslinked by addition of a water-soluble radical initiator, followed by UV irradiation at 50 °C (Thurmond et al., 1999). Besides shape fixation, shell crosslinking also provides a tool to control the permeability of the micellar shell for drug molecules (Hu et al., 2006). The surface stabilization can also be applied in stimuli-sensitive micelles to further control the drug release (Bae et al., 2006; Li et al., 2006a; Pilon et al., 2005) (see Sect. 4). A major disadvantage of crosslinking the shell segments is that all reactions have to be performed at high polymer dilution, in order to selectively crosslink the micellar shell while avoiding the formation of intermicellar crosslinking (O'Reilly et al., 2006). Furthermore, the shell fixation may hamper the chain flexibility of the shell-forming polymers, thereby impairing the steric stabilization.

3.3.2 Interfacial Crosslinking

Alternatively, the interfacial layer between the micellar core and shell can be crosslinked by the introduction of a crosslinkable spacer between the hydrophobic and the hydrophilic block (Jiang et al., 2006b; Liu et al., 2002b; Zhang et al., 2006b). This approach will leave both the micellar core and shell, and consequently the loaded drug and steric stabilization, respectively, unaffected, while it may provide a way to control drug release. Examples of spacers used are poly(glycerol monomethacrylate) (PGMA) and poly(2-hydroxyethyl methacrylate) (PHEMA), which can be crosslinked by the addition of divinyl sulfone (Liu et al., 2002a,b), or by derivatizing PGMA with cinnamoyl chloride followed by UV irradiation of the aqueous micellar solution (Jiang et al., 2006b).

3.3.3 Core Crosslinking

Core crosslinked (CCL) micelles can be prepared using functional groups present at the chain end or along the core-forming block. Often, hydroxyl moieties present in the hydrophobic block are functionalized with (meth)acrylate groups (Iijima et al., 1999; Li et al., 2006b; Shuai et al., 2004). After micelle formation, the hydrophobic blocks are crosslinked by thermal (Iijima et al., 1999; Li et al., 2006b) or photo-induced polymerization (Kim et al., 1999; Rheingans et al., 2000). Other strategies to obtain CCL micelles from (meth)acrylate functional block copolymers are Michael addition with multifunctional thiol compounds (Lee et al., 2006), or formation of an interpenetrating network using a hydrophobic polyfunctional acrylate (Petrov et al., 2005). Another example is PIC micelles composed of anionic PEG-*b*-P(Asp) and the cationic protein trypsin, which were crosslinked by Schiff-base formation of gluteraldehyde with the protein. An interpenetrating network of crosslinked trypsin was formed in the core which was stable even at high ionic strength (0.6 M NaCl), indicating that, next to protein–protein crosslinks, also covalent bonds between the primary amino groups at the ω-end of the P(Asp) segments and aldehyde groups in glutaraldehyde were formed. Besides, the protein retained its enzymatic activity (Jaturanpinyo et al., 2004). PIC micelles composed of PEG-*b*-PMA/Ca^{2+} were crosslinked by reacting the carboxylic acid groups in PMA with 1,2 diethylenediamine in the presence of a carbodiimide (Bontha et al., 2006; Bronich et al., 2005).

3.3.4   Cleavable Crosslinks

A drawback of crosslinked micelles may be that covalent linkages in the shell, the core, or the interfacial layer can negatively affect the biodegradability of the polymeric assemblies. The use of reversible or degradable crosslinks may (partly) circumvent this. Reversible crosslinked micelles were formed by the introduction of thiol groups on the lysine units in PIC micelles, followed by their oxidation to disulfide bonds (Kakizawa et al., 1999). In addition, disulfide bonds were used to stabilize the interfacial layer of PEG-*b*-(poly(N,N-dimethylacrylamide)-stat-(N-acryloxysuccinimide))-*b*-PNIPAAM micelles. The resulting particles were prone to reduction by agents such as gluthatione or dithiothreitol. Therefore, the reducing environment in the cytoplasm of cells is also a potential trigger for disintegration and drug release from these particles (Li et al., 2006a).

Recently, hydrolyzable CCL micelles were developed via introduction of methacrylate moieties at degradable oligolactate grafts of a core-forming polymer backbone. These crosslinked micelles showed superior physical stability with respect to their noncrosslinked counterparts and the degradation time could be controlled by their crosslink density (Rijcken et al., manuscript submitted).

3.3.5   Effects of Crosslinking on Drug Loading and Release

The above-described crosslinking strategies of the shell, interfacial layer, or micellar core not only fixed the micellar morphology, but also retarded the release of the loaded drug (Hu et al., 2006; Jiang et al., 2006a; Li et al., 2006a, 2007a). In addition, the crosslinking can influence the drug-loading capacity. An eightfold higher amount of triclosan was encapsulated in PEG lipid micelles after polymerization of the chain ends, which was attributed to a higher stability of core crosslinked compared to unmodified micelles (Tian et al., 2004). Obviously, one should always be aware that the structural integrity of the loaded drug molecules should be preserved upon the chemical crosslinking of the core. To avoid unwanted modification of entrapped drug molecules, the micelles can be crosslinked first and subsequently loaded with drugs. For example, mPEG-*b*-PLA micelles were crosslinked by thermally initiated polymerization of methacrylate groups in the core. Next, via a microemulsion method and subsequent evaporation of the organic solvent, paclitaxel (PTX) was loaded into these CCL micelles. A loading capacity of 3–6 weight percent (% w/w), equal to noncrosslinked micelles, was achieved (Kim et al., 1999). Crosslinked PEG-*b*-poly(methacylic acid) micelles were loaded with cisplatin by 48h of incubation with an aqueous drug solution. Subsequently, the unbound cisplatin was removed by ultracentrifugation, and a drug loading of 22% w/w was obtained (Bontha et al., 2006).

3.4   Micellar Core–Drug Compatibility

Even at high micellar stability, the retention of the loaded drug cannot be guaranteed. Upon contact with blood, extraction and redistribution of the drug between the micellar core and blood components might take place (Konan-Kouakou et al., 2005; Liu et al., 2005). The retention and release of a drug is related to the amount of drug loaded, the size of the core (Kim et al., 1998), the compatibility between the micellar core and the drug, and the effect of external stimuli (vide infra). The compatibility between the polymer and the drug can be quantified and predicted by the Flory–Huggings interaction parameter:

$$\chi_{sp} = (\delta_s - \delta_p)^2 \frac{V_s}{RT},$$

where χ_{sp} is the interaction parameter between the drug (solubilizate, s) and the core-forming polymer (p), δ_s is the Scatchard–Hildebrand solubility parameter of the drug and δ_p that of the polymer, and V_s is the molar volume of the drug (Allen et al., 1999; Torchilin, 2001). To obtain an optimal compatibility between the drug and the core-forming polymer, χ_{sp} should be as low as possible. This means that there is no universal micellar system that can be used for all drugs, but an optimal combination has to be found for each drug to improve its retention (Allen et al., 1999; Forrest et al., 2006; Gaucher et al., 2005; Hamaguchi et al., 2005; Huh et al., 2005; Kwon et al., 1993b, 1994; Lavasanifar et al., 2002a; Le Garrec et al., 2004; Lee et al., 2003b; Liu et al., 2004; Prompruk et al., 2005; Torchilin, 2001; Watanabe et al., 2006; Wilhelm et al., 1991; Yokoyama et al., 1994, 1998, 2004). This approach was proven for example by Liu et al. who compared the interaction parameters of a series of core-forming polymers with a drug (ellipticine). It was demonstrated that in this way a good selection could be made, resulting in high drug-loading capacities and slow release (Liu et al., 2004). Similarly, the core-forming polymer can be modified to get a better compatibility with the drug. For example, the loading efficiency in PEG-b-poly(β-benzyl L-aspartate) (PBLA) micelles and the in vivo therapeutic effect of the aromatic drug camptothecin was improved by increasing the number of aromatic groups on the polymer backbone, which was ascribed to aromatic interactions between the benzyl groups and the drug (Watanabe et al., 2006; Yokoyama et al., 2004). A similar approach was used to design a micellar system for doxorubicin (DOX) (Kataoka et al., 2000) and pacli-taxel (PTX) (Hamaguchi et al., 2005). Chemical modification of the drug can be an alternative way to increase the compatibility with the micellar core. Forrest et al. synthesized prodrugs of the anticancer drug geldanamycin and indeed demon-strated that a higher encapsulation could be obtained when the chemical structure of the prodrug was matched with the core-forming segment (Forrest et al., 2006).

The drug is stably retained in the micellar core when the drug is chemically attached to the micelle-forming polymer. This approach was applied by Yokoyama et al., who covalently bound DOX to the P(Asp) block of PEG-*b*-P(Asp) copolymers. The resulting PEG-*b*-P(Asp)-DOX conjugate formed micelles (Kwon et al., 1994; Yokoyama et al., 1994, 1998). In addition to the bound DOX, large amounts of free DOX could be loaded through π–π stacking in these micelles, and encapsulation efficiency depended on the amount of conjugated drug (Yokoyama et al., 1998). In vivo studies demonstrated that these DOX-loaded micelles had a considerably higher antitumor activity compared to free DOX in C26-bearing mice after i.v. injection (Nakanishi et al., 2001). A phase I clinical trial was conducted with this formulation (NK911, Fig. 5) in 23 patients with metastatic or recurrent solid tumors refractory to conventional chemotherapy. It was found that the toxicity profile of NK911 was similar to free DOX. However, NK911 exhibited longer half-lives, a lower clearance, and a larger AUC, suggesting prolonged circulation times compared to free DOX. A phase II clinical trial is currently going on (Matsumura et al., 2004).

In summary, ideally the micelles should circulate long and reach the target site intact, with the drug still loaded in the micellar core. However, this should not be confused with completely inert, nondegrading, and nonreleasing micelles since this would cause long-term accumulation in the body, especially after repeated administration. Moreover, the drug should eventually be released to interact with

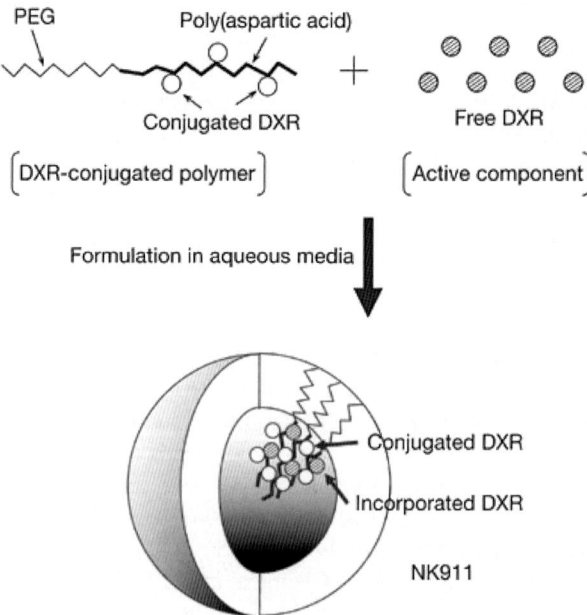

Fig. 5 PEG-*b*-poly(aspartic acid) micelles with covalently bound and physically entrapped doxorubicin, also known as NK911. Reprinted from Macmillan Publishers Ltd: British Journal of Cancer (Matsumura et al., 2004), copyright (2004). Permission request in progress

the therapeutic target. Therefore, ideally, the encapsulated agents should be selectively released at the target site, and the micelles should dissociate into single block copolymer chains or even chain fragments, with a molecular weight less than 50,000 g/mol to enable clearance via the renal pathway (Lavasanifar et al., 2002b; Seymour et al., 1987). At the target site, the drug can be released by degradation of the carrier system, but preferably via a more controlled mechanism, i.e., being the result of specific stimuli as discussed in the next sections of this chapter.

4   Stimuli Sensitivity

As pointed out in the previous sections, in an ideal micellar system, the drug is stably retained in the micelle during circulation, and after accumulation in the targeted tissue, only here the drug is released as a result of environmental triggers. A release mechanism utilizing the locally different conditions in pathological tissue compared to healthy tissue is attractive to achieve a high concentration of the drug in the target tissue. Besides, the loaded drug can be released by an external trigger including temperature, light, or ultrasound. Micelles which are destabilized as a result of either physiological or external triggers are referred to as "stimuli-sensitive micelles." After micelle formation, stimuli-sensitive micelles disassemble only after certain triggers, for example as a result of changed polymers properties (e.g., polarity). Moreover, the originally stably encapsulated drug is expected to be released concomitantly with the disintegration of the micelles. A variety of triggers has been investigated to destabilize drug-loaded polymeric micelles, including temperature (Sect. 4.1), pH (Sect. 4.2), hydrolysis (Sect. 4.3), enzymatic reactions (Sect. 4.4), redox processes (Sect. 4.5), light (Sect. 4.6), other (e.g., ultrasound in Sect. 4.7), as well as combinations thereof (Sect. 4.8) (Fig. 6). These approaches have been described for micelle-forming amphiphilic block copolymers (Rodriguez-Hernandez et al., 2005b), and for peptide amphiphiles (Mart et al., 2006).

A sophisticated stimuli-sensitive release system is obtained by coloading of an imaging agent, which enables tracking of the micelles in vivo as discussed in Sect. 4.9.

4.1   Thermosensitive Polymeric Micelles

An aqueous solution of a thermosensitive polymer is characterized by a so-called cloud point (CP). Below the CP, the polymer is hydrated and intra- and interpolymer interactions are prevented, thus rendering the polymer water soluble. Once the polymer solution is heated above the CP, the hydrogen bonds between the water molecules and the polymer chain are disrupted and water is expelled from the polymer chains. Interactions between the hydrophobic moieties of the polymer chain can now take place, which is associated with the collapse of the polymer and finally results in phase separation (aggregation/precipitation of the polymer).

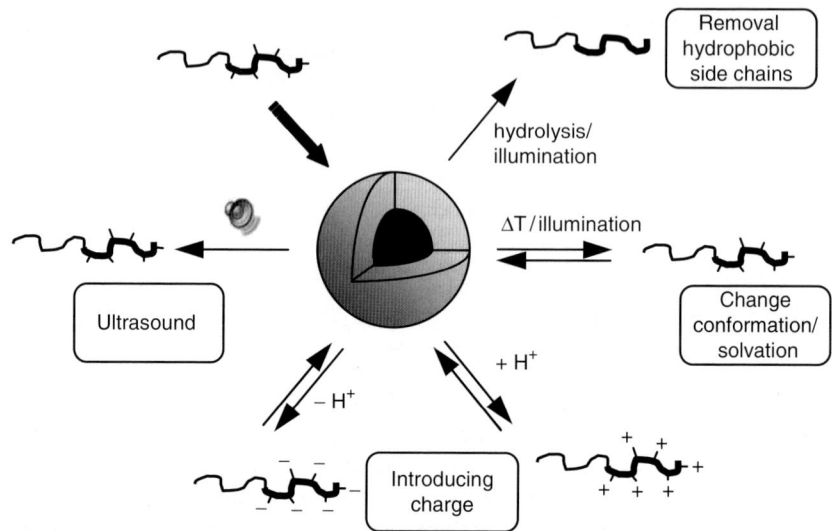

Fig. 6 Examples of stimuli-sensitive polymeric micelle destabilization. Adapted from a figure published in Rijcken et al. (2007). Copyright Elsevier (2007)

Various thermosensitive block polymers are presently under investigation for the development of polymeric micelles for pharmaceutical applications (Chilkoti et al., 2002; Hoffman et al., 2000; Jeong et al., 1997; Kabanov et al., 2002a). Poly(N-isopropylacrylamide) (PNIPAAm) is the most extensively studied thermosensitive polymer with a CP of 32 °C (Heskins and Guillet, 1968; Pelton, 2000; Schild, 1992). The CP of a thermosensitive polymer can be tailored by copolymerization with hydrophobic or hydrophilic comonomers, resulting in a decreased or increased CP, respectively (Feil et al., 1993; Neradovic et al., 2001; Park and Kataoka, 2006; Rijcken et al., 2005; Shibayama et al., 1996; Soga et al., 2004). Via this strategy, polymers with a CP around body temperature were designed to create polymeric micelles, which are suitable for temperature-induced micelle dissociation. Thermosensitive copolymers can be used either as a hydrophilic, shell-forming segment (for example, p(NIPAAM-co-DMAAm) and poly(2-isopropyl-2-oxazoline) (Kohori et al., 1999; Park et al., 2007)) or as a hydrophobic, core-forming segment of block copolymers (for example, pNIPAAM and pHPMA-Lac$_n$ (Soga et al., 2004; Topp et al., 1997)) (Fig. 7).

The advantage of thermosensitive core-forming segments is that micelles are simply prepared by heating an aqueous polymer solution (above the CMC) till above the CP of the thermosensitive part, i.e., no organic solvents are required. The heating rate is a critical factor for the ultimate size of the formed nanoparticles; a fast heating rate resulted in smaller micelles than when a slow heating rate was applied (Neradovic et al., 2004; Qiu and Wu, 1997; Zhu and Napper, 2000). A major drawback for the first generation of thermosensitive polymeric micelles based on nondegradable polymers (e.g., pNIPAAM) is that thermal treatment (hyperthermia

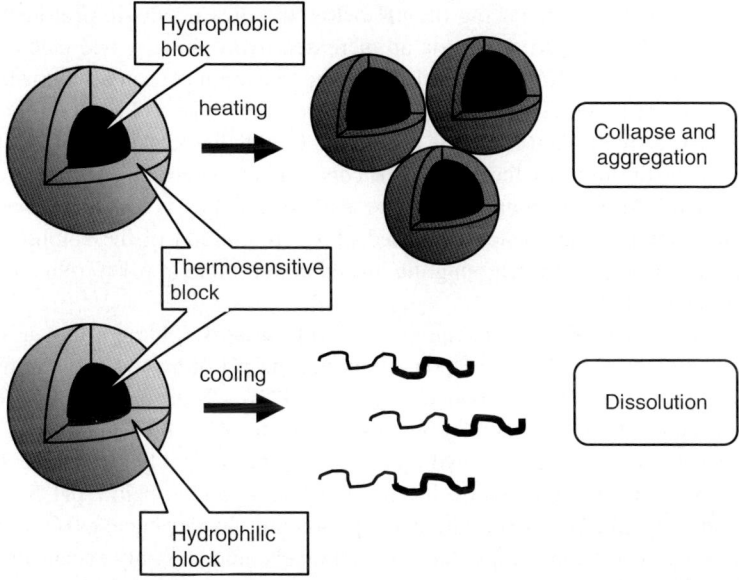

Fig. 7 Drug-loaded block copolymer micelles comprising a thermosensitive block either as the hydrophilic shell below the CP (*top*) or as the hydrophobic core above the CP (*bottom*). Heating or cooling will accomplish distortion of the micellar structures with concomitant release of the loaded drug. This figure was published in Rijcken et al. (2007). Copyright Elsevier (2007)

or hypothermia) is required for their destabilization and concurrent drug release, which is not always feasible in clinical practice. Therefore, thermosensitivity is frequently combined with other stimuli-responsive mechanism, such as pH- or light-sensitivity, and degradability (Sect. 4.8).

4.2   pH-Sensitive Polymeric Micelles

The mildly acidic pH encountered in tumor and inflammatory tissues (pH ~ 6.8) as well as in the endosomal and lysosomal compartments of cells (pH ~ 5–6), provides a potential trigger for destabilization of a pH-sensitive carrier (Engin et al., 1995; Haag, 2004). The major mechanism to induce pH-sensitivity is changes of charges in (polyion complex) micelles; pH-dependent cleavage and destabilization will be discussed in Sect. 4.3.

Typically, block copolymeric micelles that contain basic groups such as L-histidine (His) (Lee et al., 2003, 2003a), pyridine (Martin et al., 1996), and tertiary amine groups (Lee et al., 1999a; Tang et al., 2003) are pH sensitive. The block copolymers assemble into micelles at a pH one unit above the pKa of the amines, where the pH-sensitive block is essentially uncharged and hydrophobic, thereby forming the core of the micelles. Decreasing the pH below the pKa results in protonation of the polymer, which in turn leads via an increased hydrophilicity and electrostatic repulsions to destabilization of the micelles. The transition pH can be controlled by mixing different block copolymers. For instance, a mixture of PEG-*b*-P(His) and PEG-*b*-PLA formed stable micelles at pH 7.4 and dissociated at pH 6.0–7.2, depending on the ratio of the two block copolymers in the micelles (Lee et al., 2003, 2003a). Another example of pH-sensitive micelles is the sulphonamide-containing nanoparticles which collapsed upon protonation of the sulphonamide units (pKa = 6.1), thereby releasing the loaded doxorubicin at pH < 7 (Na and Bae, 2002; Na et al., 2004).

Block copolymers that are in their protonated (= water soluble) form at pH < pKa can be easily loaded with drug. Upon increasing the pH of an aqueous solution of 2-(methacryloyloxy)ethyl phosphorylcholine (PMPC)-*b*-2-(diisopropylamino)ethyl methacrylate (PDPA) block copolymers from pH 3 to pH 7, the PDPA block (pKa 6–7) became deprotonated (i.e., hydrophobic) and self-assembly took place. In the presence of a model compound (dipyridamole, only soluble below pH 5.8), this neutralization resulted in the formation of dipyridamole-loaded micelles (Giacomelli et al., 2006). Furthermore, tamoxifen and paclitaxel could be stably encapsulated in these PMPC-*b*-PDPA micelles at pH 7.4. Lowering of the pH below the pKa of the PDPA block resulted in a fast release of the drugs. This pH-triggered release might be advantageous when loaded micelles permeate into relatively acidic tumor tissue, or when they are taken up via the endocytotic pathway (Licciardi et al., 2006).

4.3   *Chemical Hydrolysis to Induce Micellar Disintegration*

Micellar disintegration and concommittant drug release can be established by chemical hydrolysis, which includes degradation of the polymer backbone (discussed in Sect. 4.3.1), cleavage of side groups (discussed in Sect. 4.3.2), and hydrolyis of covalent bonds between drug and polymer in micelle-forming polymer–drug conjugates (discussed in Sect. 4.3.3).

4.3.1   Chemical Hydrolysis of the Polymeric Backbone

Backbone hydrolysis of the hydrophobic block of an amphiphilic block copolymer is a frequently applied method to destabilize micelles used for drug delivery (Kumar et al., 2001) (Table 3). For example, the chemical degradation of the polyester block in PEG-*b*-poly(DL-lactic-co-glycolic acid) (PLGA) (Zweers et al., 2004), PCL-*b*-PEG-*b*-PCL (Hu et al., 2004), and mPEG-*b*-oligocaprolactones (Carstens et al., 2007b) was associated with changes in particle size, indicating micelle destabilization. Furthermore, transition of PEG-*b*-PCL worm micelles into spherical micelles was observed upon hydrolysis of the ester bonds in the PCL block (Geng and Discher, 2005). The ester hydrolysis is pH dependent and the ester bonds in oligolactates (de Jong et al., 2001) and oligocaprolactones (Carstens et al., 2007b) displayed an optimal stability at pH ~ 4–5. However, even at physiological pH and temperature, chemical hydrolysis of caprolactone-based polymers and oligomers is slow and will hardly play a role in vivo. It is anticipated that in the body these polymers will be mainly cleaved by enzymatic degradation (vide infra) (Carstens et al., 2007b). Moreover, when the drug should be released in the mildly acidic tumor tissue and endosomal compartments of cells and, consequently, degradation of micelles is desired at relatively low pH, other types of polymers such as poly(ortho esters) (POE) have a better degradation profile (Heller and Barr, 2004) (Table 3). Indeed, PEG-*b*-POE micelles displayed a higher stability at pH 7.5 than at pH 5.5 (Heller et al., 2002). The effect of hydrolytic degradation on micelle stability has been extensively studied, and although it is generally believed that micelle destabilization leads to the release of the loaded drugs, experimental data on the relation between degradation and drug release are scarce. One of the few examples was described by Geng et al., who correlated the degradation-induced transition of PEG-*b*-PCL worm-to-sphere

Table 3 Biodegradable moieties embedded in polymeric micelles

Degradable group	Structure	Degradation products	References
Ester			Carstens et al. (2007b); Rijcken et al. (2005)
Orthoester			Heller and Barr (2004)
Acetal			Gillies and Frechet (2003)
Hydrazone			Bae et al. (2003, 2005); Hruby et al. (2005)

micelles with the release of the loaded paclitacel (PTX). PTX release from these micelles was caused by a reduction of the drug-carrying capacity, since spherical micelles have a smaller volume-to-surface ratio than worm micelles (Geng and Discher, 2006).

4.3.2 Cleavable Side Chains

When hydrophobic side chains in the core-forming block, which are contributing to the stability of the micelles, are removed by hydrolysis, the hydrophilicity of the micellar core will increase and micelles undergo destabilization. A good example is PEG-*b*-P(Asp) that is stabilized by cyclic benzylidene acetals in the hydrophobic core via π–π stacking. The micelles were stable at physiological pH, whereas hydrolysis of the acetal bonds at pH 5 generated the more polar diols (Table 3). The overall hydrophilicity of the polymer increased, resulting in micellar dissolution and release of an encapsulated hydrophobic dye (Gillies and Frechet, 2003). A similar mechanism was applied for linear dendritic block copolymers. In vitro, these polymeric micelles displayed an accelerated release of entrapped DOX at acidic pH as a result of micelle disruption (Gillies and Frechet, 2005; Gillies et al., 2004). Thermosensitive (block co) polymers containing biodegradable side chains will be discussed in Sect. 4.8.2.

4.3.3 Cleavage of Polymer–Drug Conjugates

In the case of polymer–drug conjugates, drug release from polymeric micelles can be established by the acid-catalyzed cleavage of a labile linkage by which the drugs are attached to the polymer. For example, the acid-labile hydrazone linkage between DOX and PEG-*b*-P(Asp) resulted in an accelerated release of DOX at acidic pH in vitro (Table 3). In comparison to free DOX, these pH-sensitive DOX-hydrazone-micelles had a 15-fold greater AUC_{blood}, a higher antitumor activity, and a reduced toxicity in vivo. Moreover, micelles in which DOX was bound via a non-degradable amide bond did not exert any antitumor activity (Bae et al., 2003, 2005; Hruby et al., 2005). In a recent study, a triblock copolymer ((PLA-*co*-glycolic acid-*alt*-glutamic acid)-*b*-PEG-*b*-(PLA-*co*-glycolic acid-*alt*-glutamic acid)) was used to couple PTX via an acid labile ester linkage and the resulting micelles displayed a threefold higher release of PTX at pH 4.2 than at 7.4 (Xie et al., 2007).

4.4 Enzymatic Triggered Destabilization of Polymeric Micelles

The abundant presence of certain enzymes in pathological tissues has been applied as an environmental trigger to destabilize (drug-loaded) polymeric micelles. Similar to hydrolytic degradation, enzymes may cleave either the backbone of the hydrophobic block, their side chains, or the bonds between the polymer and drug. It was demonstrated that polyesters are not only degraded hydrolytically, but they are also susceptible to enzymatic degradation, for example by lipases. This lipase-catalyzed degradation was shown for PEG-*b*-poly(3-hydroxybutyrate)(PHB)-*b*-PEG micelles (Chen et al., 2006), PEG-*b*-PCL nanoparticles (Gan et al., 1999), accompanied by the release of encapsulated pyrene, and for PEG-*b*-oligo(ε-caprolactone) micelles (Carstens et al., 2007b). Furthermore, the peptide bonds in poly(amino acid)s, used as hydrophobic blocks, can be cleaved by proteases, as demonstrated for example for poly(γ-glutamic acid)-*g*-L-phenylalanine (PGA-*g*-L-PEA) micelles (Akagi et al., 2006).

4.5   *Oxidation-Sensitive and Reduction-Sensitive Polymeric Micelles*

The reduction of disulphide bonds in polymeric assemblies by intracellular glutathione can be used for micellar decrosslinking (vide supra, Sect. 3.3) or for full destabilization (Ghosh et al., 2006; Kakizawa et al., 1999, 2001; Li et al., 2006a). Furthermore, the reversible redox reactions of organometal compounds, e.g., viologen and ferrocene, are an attractrive trigger to alter the charge density and thus to change the solubility of viologen or ferrocene containing polymers (Anton et al., 1993). Oxidation of redox-active micelles, containing a hydrophobic ferrocenylalkyl moiety in the block copolymer, was demonstrated to shift the hydrophobic/hydrophilic balance, and the micelles disintegrated into water-soluble unimers. The release of a model hydrophobic drug (perylene) from these micelles was precisely controlled by a selective electrochemical oxidation of the ferrocenylalkyl moiety and zero-order kinetics could be realized (Takeoka et al., 1995). Selective drug release at pathogenic sites may be accomplished via externally applied electric current or by taking advantage of the accumulation of activated macrophages in inflamed tissues and certain tumors. These macrophages release oxygen-reactive species, which may also trigger the transformation of redox-sensitive micelles. However, so far, these redox-sensitive polymeric micelles have not yet been investigated in vivo.

4.6   *Light-Induced Micellar Deformation*

Light-responsive polymeric micelles ideally release their entrapped guest molecules only upon either ultraviolet (UV), visible (VIS), or (near-) infrared ((N)IR) light exposure. The use of NIR is of particular interest for biomedical applications because of its deeper tissue penetration and minimal detrimental effects on healthy cells.

4.6.1 Irreversible Reactions upon Illumination (Photolysis)

Light-induced micellar disruption can be applied using UV or IR illumination to cleave photolabile hydrophobic side chains. An amphiphilic block copolymer composed of PEG and a polymethacrylate bearing photolabile pyrene methyl esters in the side chain (PPy) as the hydrophobic core-forming domain was synthesized (Jiang et al., 2005, 2006a). Upon illumination, these ester side groups were split off, thereby transforming the hydrophobic micellar block into a hydrophilic poly(methacrylic acid) (PMA) block which brought about dissociation of the micelles. A controlled release of encapsulated Nile red could be accomplished, since the dissociation kinetics were controlled by the intensity of the light (Jiang et al., 2005). Moreover, upon core crosslinking of these micelles, the photolysis-induced micellar destabilization was prevented by the crosslinks. Nevertheless, the overall hydrophilicity of the polymer increased upon illumination and the micelle swelled, thereby still releasing the loaded hydrophobic guests, although at a lower rate (Jiang et al., 2006a).

In addition, irreversible rearrangements upon illumination are used to eradicate the hydrophobic micellar interaction forces. For example, hydrophobic 2-diazo-1,2-naphthoquinone derivatives were attached to alkyl-PEG chains, which self-assembled into micelles. The so-called Wolff rearrangement (Fig. 8) that takes place upon illumination of these chromophores drastically changed their polarity, destabilized the micelles, and released the encapsulated Nile red (Goodwin et al., 2005).

4.6.2 Light-Induced Reversible Changes

Besides irreversible micellar disintegration, environmental light can induce reversible and nondestructive destabilization. Several photoactive groups that undergo reversible structural changes upon illumination have been attached to amphiphilic block copolymers, thereby mainly shifting the hydrophobic/hydrophilic balance. Chemical entities that display photochemical-induced transitions include azobenzenes (change in dipole moment) (Tong et al., 2005), cinnamoyl (isomerization into a

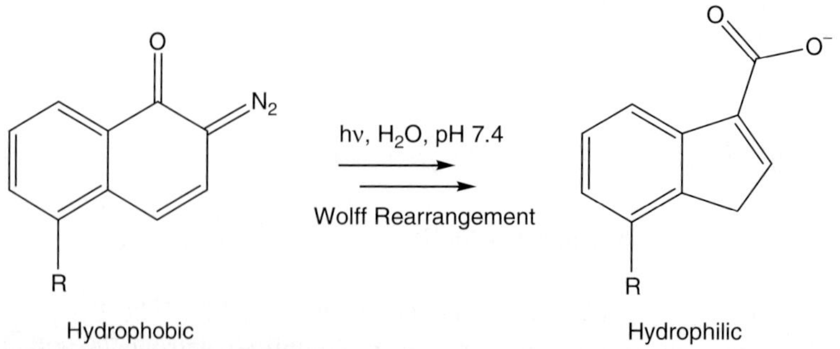

Fig. 8 Solubility change of 2-diazo-1,2-naphthoquinone derivatives after Wolff rearrangement to 3-indenecarboxylate in buffered water

more hydrophilic residue or photodimerization) (Jiang et al., 2007), spyrobenzopyran (formation of zwitterionic species) (Konak et al., 1998), and triphenylmethane leucohydroxide (generation of charges) (Kono et al., 1995), as recently reviewed in detail (Rijcken et al., 2007).

For instance, exposure of azobenzenes-containing methacrylate-*b*-(tert-butyl acrylate-co-acrylic acid) polymers to UV-light results in a trans-*cis* isomerization and a more hydrophilic polymer is generated, causing dissolution of the assemblies (Tong et al., 2005). Another example is the reversible trans-to-*cis* photoisomerization upon UV-irradiation of cinnamoyl-containing PEG-*b*-poly(methacrylate) polymers, which creates compounds with an increased hydrophilicity or leads to reversible photodimerization (Fig. 9) (Jiang et al., 2006b, 2007). Some of these photosensitive moieties were also embedded in thermosensitive block copolymers as will be discussed in Sect. 4.8.3.

4.7 Other Physical Triggers to Destabilize Polymeric Micelles

Besides the above-mentioned temperature, pH, hydrolysis, and light triggers, ultrasound and ion exchange have also been explored to induce drug release. The use of ultrasound as an external trigger to release drug from Pluronic® micellar systems in vitro and in vivo was extensively studied by Rapoport et al. (Gao et al., 2005; Rapoport et al., 2003, 2004). The cellular cytotoxicity of DOX-loaded micelles in combination was 66%, without ultrasound 53% while free DOX without applying

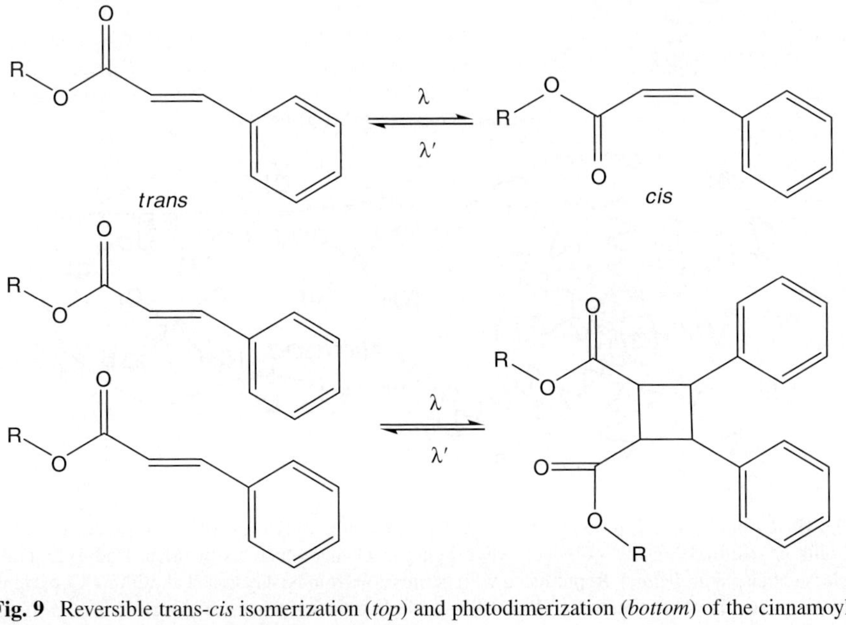

Fig. 9 Reversible trans-*cis* isomerization (*top*) and photodimerization (*bottom*) of the cinnamoyl photoreactive group upon irradiation with UV light

ultrasound resulted only in 15% in cell death (Rapoport et al., 2003). In vivo, an increased uptake of both free drug (in PBS) and DOX-loaded Pluronic® micelles was observed for sonicated tumor cells in vivo (Gao et al., 2005; Rapoport et al., 2004). The mechanisms of the ultrasound effect might be: (1) enhanced permeability of blood vessels results in extravasation of the carriers, (2) dissociation of micelles into unimers with concomitant drug release, (3) accelerated diffusion in the interstitium and tumor, and (4) enhanced membrane permeability which lead to increased cellular uptake of the drug (Gao et al., 2005; Rapoport et al., 2003, 2004).

Ion-sensitive polymer–metal micelles were formed by complexation of cisplatin to the carboxylate groups of PEG-*b*-p(glutamic acid) block copolymer. In 0.15 M NaCl, ion exchange reactions occurred, thereby slowly releasing cisplatin from the micelles, accompanied by the dissociation of the micellar structure (Fig. 10). Intravenous injection of these micelles led to a significantly increased plasma level and tumor accumulation of cisplatin as compared to free drug (Nishiyama et al.,

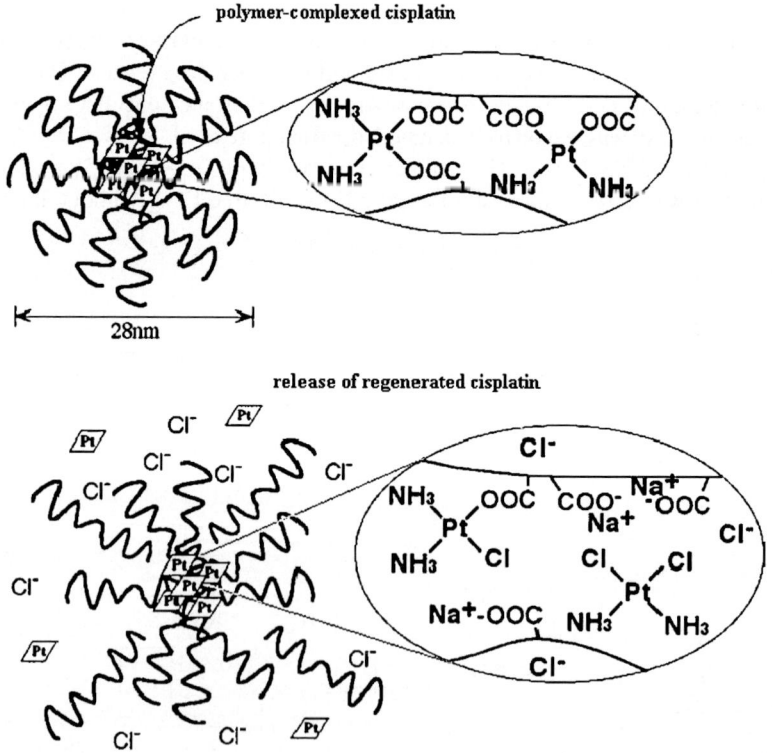

Fig. 10 Cisplatin complexation with carboxylate containing polymers (PEG-*b*-p(glutamic acid)) results in stabilized micelles (*above*), while ligand exchange reactions in saline lead to gradual release of cisplatin (*below*). Reproduced with permission from Nishiyama et al. (2003). Copyright (2003), American Association for Cancer Research. Permission request in progress

2003). Cellular uptake and release of cisplatin was recently also demonstrated with cisplatin-loaded crosslinked PIC micelles (Bontha et al., 2006).

Magnetic micelles (220 ~ 430 nm) were obtained in water by coating iron oxide nanoparticles with peptide-based polymers (polybutadiene-*b*-poly(glutamic acid)). It is anticipated that the micellar shape is manipulable in a magnetic field but experimental data are not available yet. These superparamagnetic self-assembled hybrids might be of interest for future drug delivery systems but also as contrast agents in MRI (Lecommandoux et al., 2006a,b) as described in the Sect. 4.9.

4.8   Polymeric Micelles Sensitive to a Combination of Triggers

4.8.1   pH- and Temperature Sensitivity

Random copolymers based on NIPAAm, N,N-dimethylacrylamide, and 10-undecenoic acid displayed a CP that was not only dependent on the copolymer composition but also on the pH. The polymer was designed in such a way that the polymer was below its CP and thus highly hydrated at pH 7.4 and 20 °C, with the hydrophobic undecenoic acids side chains clustering together to form core-shell morphologies of approximately 200 nm. The hydrophobic drug DOX could be loaded in the unde-cenoic acid core by a dialysis method. Lowering the pH to 6.6 caused protonation of the undecenoic carboxylate group and decreased the CP below 20 °C. Thereby, the micelles disintegrated, which was accompanied by the release of the encapsulated DOX (Liu et al., 2007b; Soppimath et al., 2005). A similar effect was seen at pH 7.4, when increasing the temperature to above the CP of 40 °C (Soppimath et al., 2005). Furthermore, a broad range of other block copolymeric assemblies that use this pH dependency to control the temperature sensitivity have been reported in literature (Liu et al., 2004; Mertoglu et al., 2005; Piskin, 2004; Salgado-Rodriguez et al., 2004; Yin et al., 2006).

4.8.2   Biodegradable Temperature-Sensitive Polymers

Dual-sensitive micellar systems that are based on biodegradable thermosensitive (block co) polymers have been described (Lee and Vernon, 2005; Neradovic et al., 1999, 2001; Rijcken et al., 2005; Shah et al., 1997; Soga et al., 2004). The polymers self-assemble into micelles in aqueous solution above their critical micelle temperature (CMT), which is the temperature above which phase separation of the thermosensitive block takes place. pH-Dependent cleavage of hydrophobic side chains resulted in a "hydrophobic-to-hydrophilic" conversion of the micellar core. Consequently, the CMT gradually increases, which ultimately results in micelle destabilization and polymer dissolution. Our department designed biodegradable thermosensitive polymers that have methacrylamide backbones with oligolactates attached via hydrolytically sensitive ester bonds (e.g., 2-hydroxypropyl methacrylamide lactate

(HPMAm-Lac$_n$, n is the number of lactic acid units in the oligolactate chain) (Lee and Vernon, 2005; Rijcken et al., 2005; Soga et al., 2004). The CMT of these methacrylamide–oligolactate copolymers is precisely tailored by the monomer feed ratio. The CMT of copolymers of HPMAm-Lac$_1$ with HPMAm-Lac$_2$ covers a temperature range of 10–63 °C (corresponding to 0–100% HPMAm-Lac$_1$, respectively). The slightly more hydrophilic homopolymer poly(N-(2-hydroxyethyl)methacrylamide-dilactate) (P(HEMAm-Lac$_2$)) has a CP of 22 °C, which could be lowered by copolymerization with HEMAm-Lac$_4$. Generally, the CMT of the block copolymers can be tuned to be below the temperature at which micelles are wanted (e.g., ambient or body temperature). PEG-b-P(HPMAm-Lac$_2$) (CMT 8 °C) displayed a transient stability at physiological conditions. After one week, a sufficient amount of lactate side chains were hydrolyzed to increase the CMT to above body temperature, which resulted in polymer dissolution and thus the micelles disintegrated (Fig. 11) (Soga et al., 2005). A much shorter destabilization time of 8 h was obtained with PEG-b-P(HEMAm-Lac$_n$) micelles since the hydrolysis of HEMAm-Lac$_n$ was much faster than that of HPMAm-Lac$_n$ (Rijcken et al., 2005)

Recently, other biodegradable thermosensitive polymers with different hydrolyzable groups were reported. A cyclic ester is the degradable moiety in poly(NIPAAm-co-dimethyl-γ-butyrolactone acrylate) (Cui et al., 2007), whereas hydrazone bonds in poly(NIPAAm-hydrazone-alkyl$_n$) (Hruby et al., 2007) or ortho-esters in poly(N-(2-(m)ethoxy-1,3-dioxan-5-yl)methacrylamide) (Huang et al., 2007) have been used as acid-labile groups. It is anticipated that by applying these types of polymers in block copolymer architectures, a second generation of controlled biodegradable thermosensitive micelles can be created.

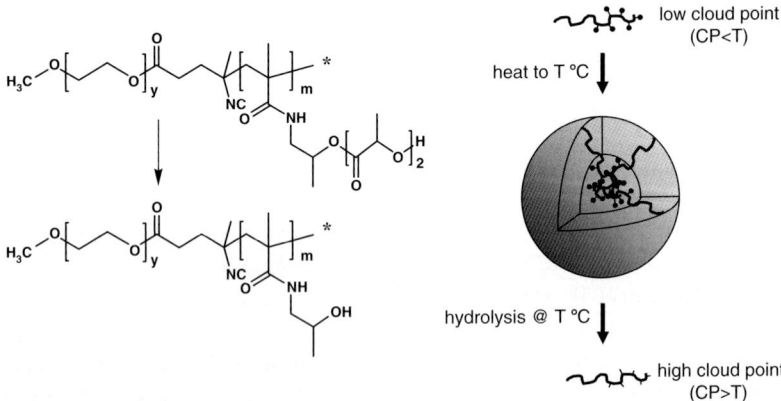

Fig. 11 Hydrolysis of PEG-b-poly(HPMAm-Lac$_2$) (*left*) causes an increase of the critical micelle temperature of the block copolymer by removal of the hydrophobic lactic acid groups (*black dots; right*). Thus, micelles formed above the cloud point of PEG-b-poly(HPMAm-Lac$_2$) destabilize when the CMT passes the incubation temperature. This figure was published in Rijcken et al. (2007). Copyright Elsevier (2007)

4.8.3   Other Dual Sensitive Systems

Photoresponsive thermosensitive copolymers do not only respond to temperature, but also display a photo-induced change in CP. UV illumination of a thermosensitive copolymer that contains light-sensitive compounds (such as those mentioned in Sect. 4.6.2) can result in increased hydrogen-bonding capacity, and consequently an increased CP (Ivanov et al., 2002; Laschewsky and Rekai, 2000; Sugiyama and Sono, 2000). In the case of poly(2-(dimethylamino)ethylmethacrylate)-*b*-poly(6-(4-phenylazo)phenoxy)-hexylmethacrylate) micelles, the application of only light was not sufficient to induce dissociation, since the light-induced trans-to-cis isomerization did not overcome the hydrophobic interactions (Ravi et al., 2005). Additionally, pH sensitivity could be obtained by the introduction of carboxylic acid units (Desponds and Freitag, 2003). The photodimerization of cinnamoyl units was used to design crosslinked polymeric micelles that responded to pH, temperature, and ionic strength (Szczubialka et al., 2004).

The so-called schizophrenic micelles were developed by Armes et al. and are based on the pH-induced micellar inversion of zwitterionic diblock copolymers assemblies. For example, a diblock copolymer composed of 2-(diethylamino)ethyl methacrylate (PDEA) and 2-(N-morpholino)ethyl methacrylate (PMEMA) is fully dissolved at pH 6 and 20 °C, since the PDEA is protonated and the neutral PMEMA is hydrophilic. At pH 8.5, the PDEA block is deprotonated and micelles are formed with PDEA as the core-forming part. Lowering the pH and addition of sufficient electrolyte causes the PMEMA block to be selectively salted out to yield PMEMA-core micelles (Butun et al., 1998). An overview of these types of micelles bearing pH and ionic strength dependency was published recently (Butun et al., 2006).

The combination of enzyme and temperature sensitivity was introduced in a pNI-PAAm-based polymer with peptide side chains. Upon phosphorylation of the peptide by protein kinase A, the CP of the copolymers increased (from 36.7 °C to 40 °C) due to the hydrophilization of the peptide chains (Katayama et al., 2001). At physiological conditions and due to the action of this enzyme, the polymers gradually dissolved in time. However, this concept is not used to design responsive micelles so far.

4.9   Imaging-Guided Drug Delivery

Encapsulation of drugs in stimuli-sensitive polymeric micelles aims at achieving an optimal pharmacotherapeutic effect, e.g., drug release as a result of (external) stimuli after accumulation of the loaded nanocarrier at its site of action. The possibility to detect the presence of the drug-loaded micelles at their aimed site of action by in vivo imaging, and to subsequently trigger the release of the loaded drug would be an important new feature (Wickline and Lanza, 2003).

For imaging purposes, polymeric micelles were loaded with either magnetic resonance imaging (MRI) contrast agents (mainly Fe, Mn, or Gd), γ-emitting radiolabels (such as [111]indium or [99m]technetium) (Trubetskoy et al., 1996), heavy elements (e.g., I, Br, and Ba) for CT imaging, or with quantum dots (QD) (Otsuka et al., 2003; Torchilin, 2002, 2007). Importantly, encapsulation of these agents in micelles is favorable since in that form they are less prone to renal clearance while their accumulation in tumor tissue (via the EPR effect) will enhance the signal strength and specificity, respectively (Ai et al., 2005; Nakamura et al., 2006).

Coloading diacyllipid micelles with Fe_3O_4 and a photosensitizer (a drug that is activated via illumination) enabled to monitor in vitro cellular uptake in real time (Cinteza et al., 2006). Furthermore, applying an external magnetic field to cells incubated with these magnetic nanoparticles resulted in a so-called magnetophoretic control of the cellular uptake (Cinteza et al., 2006; Yoon et al., 2005). Future in vivo administration of iron and drug-loaded micelles in combination with a local magnetic field might thus increase the concentration of the drug at the target site. Another sophisticated carrier for imaging-guided drug delivery was developed by Reddy et al. PEGylated polyacrylamide nanoparticles were loaded with both iron oxide and photofrin (photosensitizer) and also a targeting ligand directing to tumor vasculature was coupled onto the micellar shell. In vivo, a higher photodynamic therapeutic effect (i.e., killing of tumors) was observed when compared to nontargeted nanoparticles or free photofrin. The multifunctional targeted nanoparticles were internalized while the iron oxide enables to monitor the tissue localization of the micelles real time by MRI, and the optimal time for illumination could be chosen (Kopelman et al., 2005; Reddy et al., 2006).

Semiconductor quantum dots (QDs) also have a potential for image-guided drug delivery using polymeric micelles. Several markers can be probed simultaneously since QDs absorb over a very broad spectral range, while the extreme high photostability of QDs enables real-time monitoring over long periods of time (Otsuka et al., 2003; Sahoo et al., 2003). Furthermore, since QDs generate highly reactive free radicals upon illumination, these QDs containing micelles can be used for diagnostic as well as for (photodynamic) therapeutic purposes (Fig. 12) (Maysinger et al., 2007).

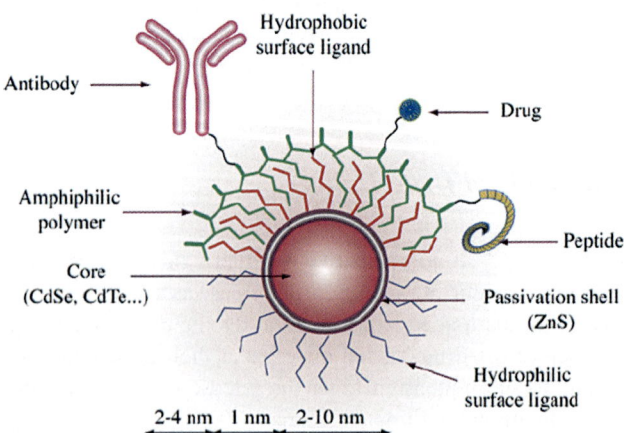

Fig. 12 Multifunctional QD containing micelles. This figure was published in (Maysinger et al., 2007). Copyright Elsevier (2007)

The most extensively studied QD in biology is CdSe, which has been encapsulated in phospholipids (Dubertret et al., 2002) and in antibody-decorated multiblock copolymeric micelles. The latter type of micelles gave a clear imaging signal in vivo and enabled very precise tracking of the active tumor uptake (Gao et al., 2004).

5   Combining Longevity, Stability, and Stimuli Sensitivity

In conclusion, polymeric micelles are very attractive drug delivery carriers for hydrophobic drugs in particular because of their unique morphology, high versatility, and high drug-loading capacity. The ideal micellar system should be able to stably encapsulate drugs, also when circulating in vivo. Several strategies have been investigated to improve the circulation times ("longevity") of the micellar carriers, their stability, and the retention of the loaded drugs in the micellar core. Once the polymeric micelle has reached the aimed target site, the desired release of the entrapped drug poses conflicting requirements on the micellar building blocks. Internal or external triggers create possibilities to develop transiently stable polymeric micelles from which the time and site of release of entrapped drug can be precisely tailored.

The most popular approach to combine longevity and triggered release properties is the use of building blocks that contain a hydrophilic PEG block in combination with a degradable or stimuli-sensitive hydrophobic block, accounting for both desired properties, respectively. However, only a few systems combine additional stabilizing strategies with stimuli-triggered drug release. For example, biodegradable PEG-*b*-PCL micelles were stabilized either by reducing the CMC (Carstens et al., 2007a; Mahmud et al., 2006) or by core crosslinking (Shuai et al., 2004). Furthermore, thermosensitive PEG-*b*-P(HEMA-Lac$_n$) micelles were core crosslinked, and demonstrated transient stability upon pH-dependent degradation (Rijcken et al., manuscript submitted). A third example is PEG-*b*-P(Asp) micelles with covalently coupled doxorubicin, which increased the compatibility of the micellar core with additional physically encapsulated doxorubicin (Yokoyama et al., 1998).

Despite all promising aspects, so far, only few micellar formulations have entered clinical trials; well-known concepts are Genexol-PM® (paclitaxel-loaded mPEG-*b*-PLA micelles) and PEG-*b*-P(Asp) micelles with either encapsulated paclitaxel (NK105) or covalently bound and physically entrapped doxorubicin (NK911) (Kato et al., 2006; Kim et al., 2004; Matsumura et al., 2004). Phase I studies with these formulations indicated prolonged circulation times when compared to the conventional formulations (Kato et al., 2006; Matsumura et al., 2004) or a better toxicity profile, allowing higher dosing (Kim et al., 2004).

In conclusion, it is anticipated that a proper combination of long circulating properties with sophisticated stabilizing strategies will generate highly stable micelles that are able to reach the target site in their intact form. Introduction of stimuli-sensitive building blocks will control the release behavior while further improvement is obtained by attaching targeting ligands. Incorporation of imaging agents allows detection of the drug-loaded micelles at the target site and application of the external trigger at the appropriate site and time. The building blocks of polymeric micelles are synthetic polymers, which offer almost unlimited possibilities to tailor and optimize the micellar structures toward the desired morphology, drug compatibility, and drug release profile. It is therefore expected that within the coming years their favorable properties will be exploited to successfully encapsulate various hydrophobic compounds and that these drug-loaded micelles will display superior performance in the treatment of several diseases.

References

Abra, R. M., Bosworth, M. E., and Hunt, C. A. 1980. Liposome disposition in vivo: effects of predosing with lipsomes. Res. Commun. Chem. Pathol. Pharmacol. 29: 349–360.

Adams, M. L., Lavasanifar, A., and Kwon, G. S. 2003. Amphiphilic block copolymers for drug delivery. J. Pharm. Sci. 92: 1343–1355.

Ai, H., et al. 2005. Magnetite-loaded polymeric micelles as ultrasensitive magnetic-resonance probes. Adv. Mater. 17: 1949–1952.

Akagi, T., et al. 2006. Hydrolytic and enzymatic degradation of nanoparticles based on amphiphilic poly(gamma-glutamic acid)-graft-L-phenylalanine copolymers. Biomacromolecules 7: 297–303.

Allen, T. M. and Hansen, C. 1991. Pharmacokinetics of stealth versus conventional liposomes: effect of dose. Biochim. Biophys. Acta 1068: 133–141.

Allen, C., Maysinger, D., and Eisenberg, A. 1999. Nano-engineering block copolymer aggregates for drug delivery. Colloid Surf B 16: 3–27.

Allen, C., et al. 2000. Polycaprolactone-b-poly(ethylene oxide) copolymer micelles as a delivery vehicle for dihydrotestosterone. J. Control. Release 63: 275–286.

Anton, P., Heinze, J., and Laschewsky, A. 1993. Redox-active monomeric and polymeric surfactants. Langmuir 9: 77–85.

Attwood, D. and Florence, A. T. 1983. In: Surfactant Systems: Their Chemistry, Pharmacy and Biology, p. 108–111. London: Chapman and Hall Ltd.

Avgoustakis, K., et al. 2002. PLGA-mPEG nanoparticles of cisplatin: in vitro nanoparticle degradation, in vitro drug release and in vivo drug residence in blood properties. J. Control. Release 79: 123–135.

Bae, Y., et al. 2003. Design of environment-sensitive supramolecular assemblies for intracellular drug delivery: polymeric micelles that are responsible to intracellular pH change. Angew. Chem. Int. Ed. Engl. 42: 4640–4643.

Bae, Y., et al. 2005. Preparation and biological characterization of polymeric micelle drug carriers with intracellular pH-triggered drug release property: tumor permeability, controlled subcellular drug distribution, and enhanced in vivo antitumor efficacy. Bioconjug. Chem. 16: 122–130.

Bae, K. H., et al. 2006. Thermosensitive pluronic micelles stabilized by shell cross-linking with gold particles. Langmuir 22: 6380–6384.

Barratt, G. M. 2000. Therapeutic applications of colloidal drug carriers. Pharm. Sci. Technol. Today 3: 163–171.

Benahmed, A., Ranger, M., and Leroux, J. C. 2001. Novel polymeric micelles based on the amphiphilic diblock copolymer poly(N-vinyl-2-pyrrolidone)-block-poly(D,L-lactide). Pharm. Res. 18: 323–328.

Besheer, A., et al. 2007. Hydrophobically modified hydroxyethyl starch: synthesis, characterization, and aqueous self-assembly into nano-sized polymeric micelles and vesicles. Biomacromolecules 8: 359–367.

Bontha, S., Kabanov, A. V., and Bronich, T. K. 2006. Polymer micelles with cross-linked ionic cores for delivery of anticancer drugs. J. Control. Release 114: 163–174.

Bougard, F., et al. 2007. Synthesis and supramolecular organization of amphiphilic diblock copolymers combining poly(N,N-dimethylamino-2-ethyl methacrylate) and poly(ε-caprolactone). Langmuir 23: 2339–2345.

Bronich, T. K., et al. 2005. Polymer micelle with cross-linked ionic core. J. Am. Chem. Soc. 127: 8236–8237.

Butun, V., Billingham, N. C., and Armes, S. P. 1998. Unusual aggregation behavior of a novel tertiary amine methacrylate-based diblock copolymer: formation of micelles and reverse micelles in aqueous solution. J. Am. Chem. Soc. 120: 11818–11819.

Butun, V., et al. 2006. A brief review of 'schizophrenic' block copolymers. React. Funct. Polym. 66: 157–165.

Carstens, M. G., et al. 2006. Observations on the disappearance of the stealth property of PEGylated liposomes. Effects of lipid dose and dosing frequency. In: Liposome Technology, ed. G. Gregoriadis, London: CRC press.

Carstens, M. G., et al. 2007a. Small oligomeric micelles based on end group modified mPEG-oligocaprolactone with monodisperse hydrophobic blocks. Macromolecules 40: 116–122.

Carstens, M. G., van Nosrum C. F., Verrijk R., de Leede L. G. J., Crommelin D. J. A., and Hennink W. E. 2008. A mechanistic study of the chemical and enzymatic degradation of PEG-oligo(e-caprolactone) micelles. J. Pharm. Sci. 97: 506–518.

Cavallaro, G., et al. 2003. Poly(hydroxyethylaspartamide) derivatives as colloidal drug carrier systems. J. Control. Release 89: 285–295.

Chen, C., et al. 2006. Biodegradable nanoparticles of amphiphilic triblock copolymers based on poly(3-hydroxybutyrate) and poly(ethylene glycol) as drug carriers. Biomaterials 27: 4804–4814.

Cheng, J., et al. 2007. Formulation of functionalized PLGA-PEG nanoparticles for in vivo targeted drug delivery. Biomaterials 28: 869–876.

Chilkoti, A., et al. 2002. Targeted drug delivery by thermally responsive polymers. Adv. Drug Deliv. Rev. 54: 613–630.

Chung, T. W., et al. 2004. Novel micelle-forming block copolymer composed of poly(e-caprolactone) and poly(vinyl pyrrolidone). Polymer 45: 1591–1597.

Cinteza, L. O., et al. 2006. Diacyllipid micelle-based nanocarrier for magnetically guided delivery of drugs in photodynamic therapy. Mol. Pharm. 3: 415–423.

Cohen Stuart, M. A., et al. 2005. Assembly of polyelectrolyte-containing block copolymers in aqueous media. Curr. Opin. Colloid Interface Sci. 10: 30–36.

Crommelin, D. J. A., Scherphof, G., and Storm, G. 1995. Active targeting with particulate carrier systems in the blood compartment. Adv. Drug Deliv. Rev. 17: 49–60.

Cui, Z., Lee, B. H., and Vernon, B. L. 2007. New hydrolysis-dependent thermosensitive polymer for an injectable degradable system. Biomacromolecules 8: 1280–1286.

Dams, E. T., et al. 2000. Accelerated blood clearance and altered biodistribution of repeated injections of sterically stabilized liposomes. J. Pharmacol. Exp. Ther. 292: 1071–1079.

de Jong, S. J., et al. 2001. New insights into the hydrolytic degradation of poly(lactic acid): participation of the alcohol terminus. Polymer 42: 2795–2802.

Desponds, A. and Freitag, R. 2003. Synthesis and characterization of photoresponsive N-isopropylacrylamide cotelomers. Langmuir 19: 6261–6270.

Djordjevic, J., Barch, M., and Uhrich, K. E. 2005. Polymeric micelles based on amphiphilic scorpion-like macromolecules: novel carriers for water-insoluble drugs. Pharm. Res. 22: 24–32.

Dong, Y. and Feng, S. 2004. Methoxy poly(ethylene glycol)-poly(lactide) (MPEG-PLA) nanoparticles for controlled delivery of anticancer drugs. Biomaterials 25: 2843–2849.

Dubertret, B., et al. 2002. In vivo imaging of quantum dots encapsulated in phospholipid micelles. Science 298: 1759–1762.

Engin, K., et al. 1995. Extracellular pH distribution in human tumours. Int. J. Hyperthermia 11: 211–216.

Fang, C., et al. 2006. In vivo tumor targeting of tumor necrosis factor-alpha-loaded stealth nanoparticles: effect of MePEG molecular weight and particle size. Eur. J. Pharm. Sci. 27: 27–36.

Feil, H., et al. 1993. Effect of comonomer hydrophilicity and ionization on the lower critical solution temperature of N-isopropylacrylamide copolymers. Macromolecules 26: 2496–2500.

Forrest, M. L., et al. 2006. Lipophilic prodrugs of Hsp90 inhibitor geldanamycin for nanoencapsulation in poly(ethylene glycol)-b-poly(epsilon-caprolactone) micelles. J. Control. Release 116: 139–149.

Gan, Z., et al. 1999. Enzymatic biodegradation of poly(ethylene oxide-ε-caprolactone) diblock copolymer and its potential biomedical applications. Macromolecules 32: 590–594.

Gao, Z., et al. 2003. PEG-PE/phosphatidylcholine mixed immunomicelles specifically deliver encapsulated taxol to tumor cells of different origin and promote their efficient killing. J. Drug Target. 11: 87–92.

Gao, X., et al. 2004. In vivo cancer targeting and imaging with semiconductor quantum dots. Nature 22: 969–976.

Gao, Z. -G., Fain, H. D., and Rapoport, N. 2005. Controlled and targeted tumor chemotherapy by micellar-encapsulated drug and ultrasound. J. Control. Release 102: 203–222.

Gaucher, G., et al. 2005. Block copolymer micelles: preparation, characterization and application in drug delivery. J. Control. Release 109: 169–188.

Geng, Y. and Discher, D. E. 2005. Hydrolytic degradation of poly(ethylene oxide)-*block*-polycaprolactone worm micelles. J. Am. Chem. Soc. 127: 12780–12781.

Geng, Y. and Discher, D. E. 2006. Visualization of degradable worm micelle breakdown in relation to drug release. Polymer 47: 2519–2525.

Ghosh, S., Basu, S., and Thayumanavan, S. 2006. Simultaneous and reversible functionalization of copolymers for biological applications. Macromolecules 39: 5595–5597.

Giacomelli, C., et al. 2006. Phosphorylcholine-based pH-responsive diblock copolymer micelles as drug delivery vehicles: light scattering, electron microscopy, and fluorescence experiments. Biomacromolecules 7: 817–828.

Gillies, E. R. and Frechet, J. M. 2003. A new approach towards acid sensitive copolymer micelles for drug delivery. Chem. Commun. 14: 1640–1641.

Gillies, E. R. and Frechet, J. M. 2005. pH-Responsive copolymer assemblies for controlled release of doxorubicin. Bioconjug. Chem. 16: 361–368.

Gillies, E. R., Jonsson, T. B., and Frechet, J. M. J. 2004. Stimuli-responsive supramolecular assemblies of linear-dendritic copolymers. J. Am. Chem. Soc. 126: 11936–11943.

Gohy, J. F., et al. 2000. Water-soluble complexes formed by sodium poly(4-styrenesulfonate) and a poly(2-vinylpyridinium)-block-poly(ethyleneoxide) copolymer. Macromolecules 33: 9298–9305.

Goodwin, A. P., et al. 2005. Synthetic micelle sensitive to IR light via a two-photon process. J. Am. Chem. Soc. 127: 9952–9953.

Gros, L., Ringsdorf, H., and Schupp, H. 1981. Polymeric antitumor agents on a molecular and on a cellular level? Angew. Chem. Int. Ed. Engl. 20: 305–325.

Haag, R. 2004. Supramolecular drug delivery systems based on polymeric core-shell architectures. Angew. Chem. Int. Ed. Engl. 43: 278–282.

Hagan, S. A., et al. 1996. Polylactide-poly(ethylene glycol) copolymers as drug delivery systems. 1. Characterization of water dispersible micelle-forming systems. Langmuir 12: 2153–2161.

Hamaguchi, T., et al. 2005. NK105, a paclitaxel-incorporating micellar nanoparticle formulation, can extend in vivo antitumour activity and reduce the neurotoxicity of paclitaxel. Br. J. Cancer 92: 1240–1246.

Harada, A. and Kataoka, K. 1995. Formation of polyion complex micelles in an aqueous milieu from a pair of oppositely-charged block copolymers with poly(ethylene glycol) segments. Macromolecules 28: 5294–5299.

Harada, A. and Kataoka, K. 1999. Chain length recognition: core-shell supramolecular assembly from oppositely charged block copolymers. Science 283: 65–67.

Harada, A. and Kataoka, K. 2006. Supramolecular assemblies of block copolymers in aqueous media as nanocontainers relevant to biological applications. Prog. Polym. Sci. 31: 949–982.

Heller, J. and Barr, J. 2004. Poly(ortho esters) – from concept to reality. Biomacromolecules 5: 1625–1632.

Heller, J., et al. 2002. Poly(ortho esters): synthesis, characterization, properties and uses. Adv. Drug Deliv. Rev. 54: 1015–1039.

Heskins, M. and Guillet, J. E. 1968. Solution properties of poly (N-isopropylacrylamide). J. Macromol. Sci. A2: 1441–1455.

Hobbs, S. K., et al. 1998. Regulation of transport pathways in tumor vessels: role of tumor type and microenvironment. Proc. Natl. Acad. Sci. USA 95: 4607–4612.

Hoffman, A. S., et al. 2000. Really smart bioconjugates of smart polymers and receptor proteins. J. Biomed. Mater. Res. 52: 577–586.

Hruby, M., Konak, C., and Ulbrich, K. 2005. Polymeric micellar pH-sensitive drug delivery system for doxorubicin. J. Control. Release 103: 137–148.

Hruby, M., et al. 2007. New bioerodable thermoresponsive polymers for possible radiotherapeutic applications. J. Control. Release 119: 25–33.

Hsiue, G. H., et al. 2006. Environmental-sensitive micelles based on poly(2-ethyl-2-oxazoline)-b-poly(L-lactide) diblock copolymer for application in drug delivery. Int. J. Pharm. 317: 69–75.

Hu, F. -Q., et al. 2006. Shell cross-linked stearic acid grafted chitosan oligosaccharide self-aggregated micelles for controlled release of paclitaxel. Colloid Surf. B 50: 97–103.

Hu, Y., et al. 2004. Degradation behavior of poly(ε-caprolactone-b-poly(ethylene glycol)-b-poly(ε-caprolactone) micelles in aqueous solution. Biomacromolecules 5: 1756–1762.

Hu, Y., et al. 2007. Effect of PEG conformation and particle size on the cellular uptake efficiency of nanoparticles with the HepG2 cells. J. Control. Release 118: 7–17.

Huang, X., Rong, F. D., and Li, J. Z. 2007. Novel acid-labile, thermoresponsive poly(methacrylamide)s with pendent ortho ester moieties. Macromol. Rapid Commun. 28: 597–603.

Huh, K. M., et al. 2005. Hydrotropic polymer micelle system for delivery of paclitaxel. J. Control. Release 101: 59–68.

Iijima, M., et al. 1999. Core-polymerized reactive micelles from heterotechelic amphiphilic block copolymers. Macromolecules 32: 1140–1146.

Illum, L., et al. 1987. Surface characteristics and the interaction of colloidal particles with mouse peritoneal macrophages. Biomaterials 8: 113–117.

Inoue, T., et al. 1998. An AB block copolymer of oligo(methyl methacrylate) and poly(acrylic acid) for micellar delivery of hydrophobic drugs. J. Control. Release 51: 221–229.

Ishida, T., Harashima, H., and Kiwada, H. 2002. Liposome clearance. Biosci. Rep. 22: 197–224.

Ivanov, A. E., et al. 2002. Photosensitive copolymer of N-isopropylacrylamide and methacryloyl derivative of spyrobenzopyran. Polymer 43: 3819–3823.

Jaturanpinyo, M., et al. 2004. Preparation of bionanoreactor based on core-shell structured polyion complex micelles entrapping trypsin in the core cross-linked with glutaraldehyde. Bioconjug. Chem. 15: 344–348.

Jeong, B., et al. 1997. Biodegradable block copolymers as injectable drug-delivery systems. Nature 388: 860–862.

Jeong, J. H., et al. 2005a. Biodegradable poly(asparagine) grafted with poly(caprolactone) and the effect of substitution on self-aggregation. Colloid Surf. A 264: 187–194.

Jeong, Y. I., et al. 2005b. Cellular recognition of paclitaxel-loaded polymeric nanoparticles composed of poly(gamma-benzyl L-glutamate) and poly(ethylene glycol) diblock copolymer end-capped with galactose moiety. Int. J. Pharm. 296: 151–161.

Jiang, J., Tong, X., and Zhao, Y. 2005. A new design for light-breakable polymer micelles. J. Am. Chem. Soc. 127: 8290–8291.

Jiang, J., et al. 2006a. Toward photocontrolled release using light-dissociable block copolymer micelles. Macromolecules 39: 4633–4640.

Jiang, X., et al. 2006b. UV irradiation-induced shell cross-linked micelles with pH-responsive cores using ABC triblock copolymers. Macromolecules 39: 5987–5994.

Jiang, J., et al. 2007. Polymer micelles stabilization on demand through reversible photo-cross-linking. Macromolecules 40: 790–792.

Jones, M. -C. and Leroux, J. -C. 1999. Polymeric micelles – a new generation of colloidal drug carriers. Eur. J. Pharm. Biopharm. 48: 101–111.

Joralemon, M. J., et al. 2005. Shell click-crosslinked (SCC) nanoparticles: a new methodology for synthesis and orthogonal functionalization. J. Am. Chem. Soc. 127: 16892–16899.

Kabanov, A. V., et al. 1992. A new class of drug carriers: micelles of poly(oxyethylene)-poly(oxypropylene) block copolymers as microcontainers for drug targeting from blood in brain. J. Control. Release 22: 141–157.

Kabanov, A. V., et al. 1996. Soluble stoichiometric complexes from poly(N-ethyl-4-vinylpyridinium) cations and poly(ethylene oxide)-block-polymethacrylate anions. Macromolecules 29: 6797–6802.

Kabanov, A. V., Batrakova, E. V., and Alakhov, V. Y. 2002a. Pluronic block copolymers as novel polymer therapeutics for drug and gene delivery. J. Control. Release 82: 189–212.

Kabanov, A. V., Batrakova, E. V., and Alakhov, V. Y. 2002b. Pluronic block copolymers for overcoming drug resistance in cancer. Adv. Drug Deliv. Rev. 54: 759–779.

Kakizawa, Y., Harada, A., and Kataoka, K. 1999. Environment-sensitive stabilization of core-shell structured polyion complex micelle by reversible cross-linking of the core through disulfide bond. J. Am. Chem. Soc. 121: 11247–11248.

Kakizawa, Y., Harada, A., and Kataoka, K. 2001. Glutathione-sensitive stabilization of block copolymer micelles composed of antisense DNA and thiolated poly(ethylene glycol)-block-poly(L-lysine): a potential carrier for systemic delivery of antisense DNA. Biomacromolecules 2: 491–497.

Kang, N., et al. 2005. Stereocomplex block copolymer micelles: core-shell nanostructures with enhanced stability. Nano Lett. 5: 315–319.

Kataoka, K., et al. 2000. Doxorubicin-loaded poly(ethylene glycol)-poly([beta]-benzyl-aspartate) copolymer micelles: their pharmaceutical characteristics and biological significance. J. Control. Release 64: 143–153.

Kataoka, K., Harada, A., and Nagasaki, Y. 2001. Block copolymer micelles for drug delivery: design, characterization and biological significance. Adv. Drug Deliv. Rev. 47: 113–131.

Katayama, Y., Sonoda, T., and Maeda, M. 2001. A polymer micelle responding to the protein kinase A signal. Macromolecules 34: 8569–8573.

Kato, K., et al. 2006. Phase I study of NK105, a paclitaxel incorporating micellar nanoparticle, in patients with advanced cancer. J. Clin. Oncol. 24: 2018.

Kim, C., et al. 2000. Amphiphilic diblock copolymers based on poly(2-ethyl-2-oxazoline) and poly(1,3-trimethylene carbonate): Synthesis and micellar characteristics. Macromolecules 33: 7448–7452.

Kim, J., et al. 1999. Core-stabilized polymeric micelle as potential drug carrier: increased solubilization of taxol. Polym. Adv. Technol. 10: 647–654.

Kim, S. C., et al. 2001. In vivo evaluation of polymeric micellar paclitaxel formulation: toxicity and efficacy. J. Control. Release 72: 191–202.

Kim, S. Y., et al. 1998. Methoxy poly(ethylene glycol) and epsilon-caprolactone amphiphilic block copolymeric micelle containing indomethacin. II. Micelle formation and drug release behaviours. J. Control. Release 51: 13–22.

Kim, T. Y., et al. 2004. Phase I and pharmacokinetic study of Genexol-PM, a cremophor-free, polymeric micelle-formulated paclitaxel, in patients with advanced malignancies. Clin. Cancer Res. 10: 3708–3716.

Klibanov, A. L., et al. 1990. Amphipathic polyethyleneglycols effectively prolong the circulation time of liposomes. FEBS Lett. 268: 235–237.

Kohori, F., et al. 1999. Control of adriamycin cytotoxic activity using thermally responsive polymeric micelles composed of poly(N-isopropylacrylamide-co-N,N-dimethylacrylamide)-b-poly(-lactide). Colloid Surf. B 16: 195–205.

Konak, C., et al. 1998. Photoassociation of water-soluble copolymers containing photochromic spirobenzopyran moieties. Polym. Adv. Technol. 9: 641–648.

Konan-Kouakou, Y. N., et al. 2005. In vitro and in vivo activities of verteporfin-loaded nanoparticles. J. Control. Release 103: 83–91.

Kono, K., Nishihara, Y., and Takagishi, T. 1995. Photoresponsive permeability of polyelectrolyte complex capsule membrane containing triphenylmethane leucohydroxide residues. J. Appl. Polym. Sci. 56: 707–713.

Kopelman, R., et al. 2005. Multifunctional nanoparticle platforms for in vivo MRI enhancement and photodynamic therapy of a rat brain cancer. J. Magn. Magn. Mater. 293: 404–410.

Kreuter, J. 2006. Nanoparticles – a historical perspective. Int. J. Pharm. 331: 1–10.

Krishnadas, A., Rubinstein, I., and Önyüksel, H. 2003. Sterically stabilized phospholipid mixed micelles: in vitro evaluation as a novel carrier for water-insoluble drugs. Pharm. Res. 20: 297–302.

Kumar, N., Ravikumar, M. N., and Domb, A. J. 2001. Biodegradable block copolymers. Adv. Drug Deliv. Rev. 53: 23–44.

Kwon, G., et al. 1994. Enhanced tumor accumulation and prolonged circulation times of micelle-forming poly (ethylene oxide-aspartate) block copolymer–adriamycin conjugates. J. Control. Release 29: 17–23.

Kwon, G. S. 2003. Polymeric micelles for delivery of poorly water-soluble compounds. Crit. Rev. Ther. Drug Carrier Syst. 20: 357–403.

Kwon, G. S. and Okano, T. 1996. Polymeric micelles as new drug carriers. Adv. Drug Deliv. Rev. 21: 107–116.

Kwon, G. S., et al. 1993a. Micelles based on AB block copolymers of poly(ethylene oxide) and poly(b-benzyl l-aspartate). Langmuir 9: 945–949.

Kwon, G. S., et al. 1993b. Biodistribution of micelle-forming polymer-drug conjugates. Pharm. Res. 10: 970–974.

Laschewsky, A. and Rekai, E. D. 2000. Photochemical modification of the lower critical solution temperature of cinnamoylated poly(N-2-hydropropylmethacrylamide) in water. Macromol. Rapid Commun. 21: 937–940.

Lavasanifar, A., Samuel, J., and Kwon, G. S. 2001. The effect of alkyl core structure on micellar properties of poly(ethylene oxide)-block-poly(-aspartamide) derivatives. Colloid Surf. B 22: 115–126.

Lavasanifar, A., et al. 2002. Block copolymer micelles for the encapsulation and delivery of amphotericin B. Pharm. Res. 19: 418–422.

Lavasanifar, A., Samuel, J., and Kwon, G. S. 2002a. The effect of fatty acid substitution on the in vitro release of amphotericin B from micelles composed of poly(ethylene oxide)-block-poly(N-hexyl stearate–aspartamide). J. Control. Release 79: 165–172.

Lavasanifar, A., Samuel, J., and Kwon, G. S. 2002b. Poly(ethylene oxide)-block-poly(-amino acid) micelles for drug delivery. Adv. Drug Deliv. Rev. 54: 169–190.

Laverman, P., et al. 2000. Preclinical and clinical evidence for disappearance of long-circulating characteristics of polyethylene glycol liposomes at low lipid dose. J. Pharmacol. Exp. Ther. 293: 996–1001.

Le Garrec, D., et al. 2002. Optimizing pH-responsive polymeric micelles for drug delivery in a cancer photodynamic therapy model. J. Drug Target. 10: 429–437.

Le Garrec, D., et al. 2004. Poly(N-vinylpyrrolidone)-block-poly(D,L-lactide) as a new polymeric solubilizer for hydrophobic anticancer drugs: in vitro and in vivo evaluation. J. Control. Release 99: 83–101.

Lecommandoux, S., et al. 2006a. Smart hybrid magnetic self-assembled micelles and hollow capsules. Prog. Solid State Ch. 34: 171–179.

Lecommandoux, S., et al. 2006b. Self-assemblies of magnetic nanoparticles and di-block copolymers: magnetic micelles and vesicles. J. Magn. Magn. Mater. 300: 71–74.

Lee, A. S., et al. 1999a. Characterizing the structure of pH dependent polyelectrolyte block copolymer micelles. Macromolecules 32: 4302–4310.

Lee, B. H. and Vernon, B. 2005. Copolymers of N-isopropylacrylamide HEMA-lactate and acrylic acid with time-dependent lower critical solution temperature as a bioresorbable carrier. Polym. Int. 54: 418–422.

Lee, B. H., et al. 2006. In-situ injectable physically and chemically gelling NIPAAm-based copolymer system for embolization. Biomacromolecules 7: 2059–2064.

Lee, E. S., Na, K., and Bae, Y. H. 2003. Polymeric micelle for tumor pH and folate-mediated targeting. J. Control. Release 91: 103–113.

Lee, E. S., et al. 2003a. Poly(-histidine)-PEG block copolymer micelles and pH-induced destabilization. J. Control. Release 90: 363–374.

Lee, E. S., Na, K., and Bae, Y. H. 2005. Super pH sensitive multifunctional polymeric micelle. Nano Lett. 5: 325–329.

Lee, J., et al. 2003b. Hydrotropic solubilization of paclitaxel: analysis of chemical structures for hydrotropic property. Pharm. Res. 20: 1022–1030.

Lee, K. Y., Ha, W. S., and Park, W. H. 1995. Blood compatibility and biodegradability of partially N-acylated chitosan derivatives. Biomaterials 16: 1211–1216.

Lee, S. C. and Lee, H. J. 2007. pH-Controlled, polymer-mediated assembly of polymer micelle nanoparticles. Langmuir 23: 488–495.

Lee, S. C., et al. 1999b. Synthesis and micellar characterization of amphiphilic diblock copolymers based on poly(2-ethyl-2-oxazoline) and aliphatic polyesters. Macromolecules 32: 1847–1852.

Lee, S. C., et al. 2007. Hydrotropic polymeric micelles for enhanced paclitaxel solubility: in vitro and in vivo characterization. Biomacromolecules 8: 202–208.

Lele, B. S. and Leroux, J. -C. 2002. Synthesis and micellar characterization of novel amphiphilic A-B-A triblock copolymers of N-(2-hydroxypropyl)methacrylamide or N-vinyl-2-pyrrolidone with poly(ε-caprolactone). Macromolecules 35: 6714–6723.

Lemarchand, C., Gref, R., and Couvreur, P. 2004. Polysaccharide-decorated nanoparticles. Eur. J. Pharm. Biopharm. 58: 327–341.

Lemarchand, C., et al. 2005. Physico-chemical characterization of polysaccharide-coated nanoparticles. J. Control. Release 108: 97–111.

Lemarchand, C., et al. 2006. Influence of polysaccharide coating on the interactions of nanoparticles with biological systems. Biomaterials 27: 108–118.

Leroux, J., et al. 1995. An investigation on the role of plasma and serum opsonins on the internalization of biodegradable poly(D,L-lactic acid) nanoparticles by human monocytes. Life Sci. 57: 695–703.

Letchford, K., et al. 2004. Synthesis and micellar characterization of short block length methoxy poly(ethylene glycol)-block-poly(caprolactone) diblock copolymers. Colloid Surf. B 35: 81–91.

Li, X., Ji, J. and Shen, J. 2006. Synthesis of hydroxyl-capped comb-like poly(ethylene glycol) to develop shell cross-linkable micelles. Polymer 47: 1987–1994.

Li, Y., et al. 2006a. Synthesis of reversible shell cross-linked micelles for controlled release of bioactive agents. Macromolecules 39: 2726–2728.

Li, Z., et al. 2006b. Molecularly imprinted polymeric nanospheres by diblock copolymer self-assembly. Macromolecules 39: 2629–2636.

Licciardi, M., et al. 2006. New folate-functionalized biocompatible block copolymer micelles as potential anti-cancer drug delivery systems. Polymer 47: 2946–2955.

Liggins, R. T. and Burt, H. M. 2002. Polyether-polyester diblock copolymers for the preparation of paclitaxel loaded polymeric micelle formulations. Adv. Drug Deliv. Rev. 54: 191–202.

Lin, W. J., Juang, L. W., and Lin, C. C. 2003. Stability and release performance of a series of pegylated copolymeric micelles. Pharm. Res. 20: 668–673.

Lin, W. J., et al. 2005. Characterization of pegylated copolymeric micelles and in vivo pharmacokinetics and biodistribution studies. J. Biomed. Mater. Res. B 77: 188–194.

Liu, D.- Z., et al. 2007a. Synthesis, characterization and drug delivery behaviors of new PCP polymeric micelles. Carbohydr. Polym. 68: 544–554.

Liu, J., Xiao, Y., and Allen, C. 2004. Polymer-drug compatibility: a guide to the development of delivery systems for the anticancer agent, ellipticine. J. Pharm. Sci. 93: 132–143.

Liu, J., Zeng, F., and Allen, C. 2005. Influence of serum protein on polycarbonate-based copolymer micelles as a delivery system for a hydrophobic anti-cancer agent. J. Control. Release 103: 481–497.

Liu, J., Zeng, F., and Allen, C. 2007. In vivo fate of unimers and micelles of a poly(ethylene glycol)-block-poly(caprolactone) copolymer in mice following intravenous administration. Eur. J. Pharm. Biopharm. 65: 309–319.

Liu, S., et al. 2002a. Synthesis of shell cross-linked micelles with pH-responsive cores using ABC triblock copolymers. Macromolecules 35: 6121–6131.

Liu, S., et al. 2002b. Synthesis of pH-responsive shell crosslinked micelles and their use as nano-reactors for the preparation of gold nanoparticles. Langmuir 18: 8350–8357.

Liu, S. -Q., et al. 2007b. Bio-functional micelles self-assembled from a folate-conjugated block copolymer for targeted intracellular delivery of anticancer drugs. Biomaterials 28: 1423–1433.

Liu, X. -M., et al. 2004. The effect of salt and pH on the phase-transition behaviors of temperature-sensitive copolymers based on N-isopropylacrylamide. Biomaterials 25: 5659–5666.

Lo, C. L., et al. 2007. Mixed micelles formed from graft and diblock copolymers for application in intracellular drug delivery. Biomaterials 28: 1225–1235.

Lukyanov, A. N. and Torchilin, V. P. 2004. Micelles from lipid derivatives of water-soluble polymers as delivery systems for poorly soluble drugs. Adv. Drug Deliv. Rev. 56: 1273–1289.

Luo, L., et al. 2002. Cellular internalization of poly(ethylene oxide)-b-poly(epsilon-caprolactone) diblock copolymer micelles. Bioconjug. Chem. 13: 1259–1265.

Luo, L., et al. 2004. Novel amphiphilic diblock copolymer of low molecular weight poly(N-vinylpyrrolidone)-block-poly(D,L-lactide): synthesis, characterization, and micellization. Macromolecules 37: 4008–4013.

Maeda, H., et al. 2000. Tumor vascular permeability and the EPR effect in macromolecular therapeutics: a review. J. Control. Release 65: 271–284.

Maeda, H., et al. 2003. Vascular permeability enhancement in solid tumor: various factors, mechanisms involved and its implications. Int. Immunopharmacol. 3: 319–328.

Mahmud, A., Xiong, X. B., and Lavasanifar, A. 2006. Novel self-associating poly(ethylene oxide)-block-poly(e-caprolactone) block copolymers with functional side groups on the polyester block for drug delivery. Macromolecules 39: 9419–9428.

Mart, R. J., et al. 2006. Peptide-based stimuli-responsive biomaterials. Soft Matter. 2: 822–835.

Martin, T. J., et al. 1996. pH-dependent micellization of poly(2-vinylpyridine)-block-poly(ethylene oxide). Macromolecules 29: 6071–6073.

Maruyama, A., et al. 1994a. Preparation of nanoparticles bearing high density carbohydrate chains using carbohydrate-carrying polymers as emulsifier. Biomaterials 15: 1035–1042.

Maruyama, K., et al. 1994b. Phosphatidyl polyglycerols prolong liposome circulation in vivo. Int. J. Pharm. 111: 103–107.

Matsumura, Y. and Maeda, H. 1986. A new concept for macromolecular therapeutics in cancer chemotherapy: mechanism of tumoritropic accumulation of proteins and the antitumor agent smancs. Cancer Res. 46: 6387–6392.

Matsumura, Y., et al. 2004. Phase I clinical trial and pharmacokinetic evaluation of NK911, a micelle-encapsulated doxorubicin. Br. J. Cancer 91: 1775–1781.

Maysinger, D., et al. 2007. Fate of micelles and quantum dots in cells. Eur. J. Pharm. Biopharm. 65: 270–281.

Mertoglu, M., et al. 2005. Stimuli responsive amphiphilic block copolymers for aqueous media synthesised via reversible addition fragmentation chain transfer polymerisation (RAFT). Polymer 46: 7726–7740.

Metselaar, J. M., et al. 2003. A novel family of L-amino acid-based biodegradable polymer-lipid conjugates for the development of long-circulating liposomes with effective drug-targeting capacity. Bioconjug. Chem. 14: 1156–1164.

Michel, R., et al. 2005. Influence of PEG architecture on protein adsorption and conformation. Langmuir 21: 12327–12332.

Moghimi, S. M. and Szebeni, J. 2003. Stealth liposomes and long circulating nanoparticles: critical issues in pharmacokinetics, opsonization and protein-binding properties. Prog. Lipid Res. 42: 463–478.

Moghimi, S. M., et al. 1991a. Non-phagocytic uptake of intravenously injected microspheres in rat spleen: influence of particle size and hydrophilic coating. Biochem. Biophys. Res. Commun. 177: 861–866.

Moghimi, S. M., et al. 1991b. The effect of poloxamer-407 on liposome stability and targeting to bone marrow: comparison with polystyrene microspheres. Int. J. Pharm. 68: 121–126.

Moghimi, S. M., Hunter, A. C., and Murray, J. C. 2001. Long-circulating and target-specific nanoparticles: theory to practice. Pharmacol. Rev. 53: 283–318.

Molineux, G. 2002. PEGylation: engineering improved pharmaceuticals for enhanced therapy. Cancer Treat. Rev. 28: 13–16.

Moses, M. A., Brem, H., and Langer, R. 2003. Advancing the field of drug delivery: taking aim at cancer. Cancer Cell 4: 337–341.

Na, K. and Bae, K. H. 2002. Self-assembled hydrogel nanoparticles responsive to tumor extracellular pH from pullulan derivative/sulfonamide conjugate: characterization, aggregation, and adriamycin release in vitro. Pharm. Res. 19: 681–688.

Na, K., Lee, K. H., and Bae, Y. H. 2004. pH-sensitivity and pH-dependent interior structural change of self-assembled hydrogel nanoparticles of pullulan acetate/oligo-sulfonamide conjugate. J. Control. Release 97: 513–525.

Nakamura, E., et al. 2006. A polymeric micelle MRI contrast agent with changeable relaxivity. J. Control. Release 114: 325–333.

Nakanishi, T., et al. 2001. Development of the polymer micelle carrier system for doxorubicin. J. Control. Release 74: 295–302.

Nam, Y. S., et al. 2003. New micelle-like polymer aggregates made from PEI-PLGA diblock copolymers: micellar characteristics and cellular uptake. Biomaterials 24: 2053–2059.

Neradovic, D., et al. 1999. Poly(N-isopropylacrylamide) with hydrolyzable lactic acid ester side groups: a new type of thermosensitive polymer. Macromol. Rapid Commun. 20: 577–581.

Neradovic, D., Van Nostrum, C. F., and Hennink, W. E. 2001. Thermoresponsive polymeric micelles with controlled instability based on hydrolytically sensitive N-isopropylacrylamide copolymers. Macromolecules 34: 7589–7591.

Neradovic, D., et al. 2004. The effect of the processing and formulation parameters on the size of nanoparticles based on block copolymers of poly(ethylene glycol) and poly(N-isopropylacrylamide) with and without hydrolytically sensitive groups. Biomaterials 25: 2409–2418.

Nishiyama, N. and Kataoka, K. 2006. Current state, achievements, and future prospects of polymeric micelles as nanocarriers for drug and gene delivery. Pharmacol. Therapeut. 112: 630–648.

Nishiyama, N., et al. 2001. Cisplatin-loaded polymer-metal complex micelle with time-modulated decaying property as a novel drug delivery system. Pharm. Res. 18: 1035–1041.

Nishiyama, N., et al. 2003. Novel cisplatin-incorporated polymeric micelles can eradicate solid tumors in mice. Cancer Res. 63: 8977–8983.

Oku, N. and Namba, Y. 1994. Long-circulating liposomes. Crit. Rev. Ther. Drug 11: 231–270.

Opanasopit, P., et al. 2004. Block copolymer design for camptothecin incorporation into polymeric micelles for passive tumor targeting. Pharm. Res. 21: 2001–2008.

Opanasopit, P., et al. 2005. Influence of serum and albumins from different species on stability of camptothecin-loaded micelles. J. Control. Release 104: 313–321.

O'Reilly, R. K., Hawker, C. J., and Wooley, K. L. 2006. Cross-linked block copolymer micelles: functional nanostructures of great potential and versatility. Chem. Soc. Rev. 35: 1068–1083.

Otsuka, H., Nagasaki, Y., and Kataoka, K. 2003. PEGylated nanoparticles for biological and pharmaceutical applications. Adv. Drug Deliv. Rev. 55: 403–419.

Owens, D. E., 3rd and Peppas, N. A. 2006. Opsonization, biodistribution, and pharmacokinetics of polymeric nanoparticles. Int. J. Pharm. 307: 93–102.

Park, E. K., et al. 2005. Folate-conjugated methoxy poly(ethylene glycol)/poly(epsilon-caprolactone) amphiphilic block copolymeric micelles for tumor-targeted drug delivery. J. Control. Release 109: 158–168.

Park, J. S. and Kataoka, K. 2006. Precise control of lower critical solution temperature of thermosensitive poly(2-isopropyl-2-oxazoline) via gradient copolymerization with 2-ethyl-2-oxazoline as a hydrophilic comonomer. Macromolecules 39: 6622–6630.

Park, J. S., et al. 2007. Preparation and characterization of polyion complex micelles with a novel thermosensitive poly(2-isopropyl-2-oxazoline) shell via the complexation of oppositely charged block ionomers. Langmuir 23: 138–146.

Passirani, C., et al. 1998. Long-circulating nanoparticles bearing heparin or dextran covalently bound to poly(methyl methacrylate). Pharm. Res. 15: 1046–1050.

Pelton, R. 2000. Temperature-sensitive aqueous microgels. Adv. Colloid Interface 85: 1–33.

Peracchia, M. T., et al. 1997. Complement consumption by poly(ethylene glycol) in different conformations chemically coupled to poly(isobutyl 2-cyanoacrylate) nanoparticles. Life Sci. 61: 749–761.

Petrov, P., Bozukov, M., and Tsvetanov, C. B. 2005. Innovative approach for stabilizing poly(ethylen oxide)-b-poly(propylene oxide)-b-poly(ethylen oxide) micelles by forming nano-sized networks in the micelle. J. Mater. Chem. 15: 1481–1486.

Pilon, L. N., et al. 2005. Synthesis and characterization of shell cross-linked micelles with hydroxy-functional coronas: a pragmatic alternative to dendrimers? Langmuir 21: 3808–3813.

Piskin, E. 2004. Molecularly designed water soluble, intelligent, nanosize polymeric carriers. Int. J. Pharm. 277: 105–118.

Prompruk, K., et al. 2005. Synthesis of a novel PEG-block-poly(aspartic acid-stat-phenylalanine) copolymer shows potential for formation of a micellar drug carrier. Int. J. Pharm. 297: 242–253.

Qiu, X. P. and Wu, C. 1997. Study of the core-shell nanoparticle formed through the "coil-to-globule" transition of poly(N-isopropylacrylamide) grafted with poly(ethylene oxide). Macromolecules 30: 7921–7926.

Rapoport, N. 2004. Combined cancer therapy by micellar-encapsulated drug and ultrasound. Int. J. Pharm. 277: 155–162.

Rapoport, N., et al. 2002. Intracellular uptake and trafficking of pluronic micelles in drug-sensitive and MDR cells: effect on the intracellular drug localization. J. Pharm. Sci. 91: 157–170.

Rapoport, N., et al. 2003. Drug delivery in polymeric micelles: from in vitro to in vivo. J. Control. Release 91: 85–95.

Rapoport, N. Y., et al. 2004. Ultrasound-triggered drug targeting of tumors in vitro and in vivo. Ultrasonics 42: 943–950.

Ravi, P., et al. 2005. New water-soluble azobenzene-containing diblock copolymers: synthesis and aggregation behavior. Polymer 46: 137–146.

Reddy, G. R., et al. 2006. Vascular targeted nanoparticles for imaging and treatment of brain tumors. Clin. Cancer Res. 12: 6677–6686.

Rheingans, O., et al. 2000. Nanoparticles built of cross-linked heterotechelic amphiphilic poly(dimethylsiloxane)-b-poly(ethylene oxide) diblock copolymers. Macromolecules 33: 4780–4790.

Rijcken, C. J. F., et al. 2005. Novel fast degradable thermosensitive polymeric micelles based on PEG-block-poly(N-(2-hydroxyethyl)methacrylamide-oligolactates). Biomacromolecules 6: 2343–2351.

Rijcken, C. J. F., Soga, O., Hennink, W. E., and van Nostrum, C. F. 2007. Triggered destabilisation of polymeric micelles and vesicles by changing polymers polarity: An attractive tool for drug delivery. J. Control. Release 120:131–148.

Rijcken, C. J. F., Snel, C. J., Schiffelers, R. M., van Nostrum, C. F., and Hennink, W. E. (2007). Hydrolyzable core crosslinked polymeric micelles. synthesis, characterisation and in vivo studies. Biomaterials 28:5581–5593.

Rodriguez-Hernandez, J., et al. 2005a. Preparation of shell cross-linked nano-objects from hybrid-peptide block copolymers. Biomacromolecules 6: 2213–2220.

Rodriguez-Hernandez, J., et al. 2005b. Toward 'smart' nano-objects by self-assembly of block copolymers in solution. Prog. Polym. Sci. 30: 691–724.

Romberg, B., et al. 2007. Pharmacokinetics of poly(hydroxyethyl-L-asparagine)-coated liposomes is superior over that of PEG-coated liposomes at low lipid dose and upon repeated administration. Biochim. Biophys. Acta 1768: 737–743.

Rosler, A., Vandermeulen, G. W., and Klok, H. A. 2001. Advanced drug delivery devices via self-assembly of amphiphilic block copolymers. Adv. Drug Deliv. Rev. 53: 95–108.

Rouzes, C., et al. 2000. Surface modification of poly(lactic acid) nanospheres using hydrophobically modified dextrans as stabilizers in an o/w emulsion/evaporation technique. J. Biomed. Mater. Res. 50: 557–565.

Sahoo, S., Ma, and Labhasetwar 2003. Nanotech approaches to drug delivery and imaging. Drug Discov. Today 8: 1112–1120.

Salgado-Rodriguez, R., Licea-Claverie, A., and Arndt, K. F. 2004. Random copolymers of N-isopropylacrylamide and methacrylic acid monomers with hydrophobic spacers: pH-tunable temperature-sensitive materials. Eur. Polym. J. 40: 1931–1946.

Savic, R., et al. 2006. Assessment of the integrity of poly(caprolactone)-b-poly(ethyleneoxide)mi celles under biological conditions: a fluorogenic based approach. Langmuir 22: 3570–3578.

Schild, H. G. 1992. Poly (N-isopropylacrylamide) experiment, theory and application. Prog. Polym. Sci. 17: 163–249.

Senior, J. and Gregoriadis, G. 1982. Stability of small unilamellar liposomes in serum and clearance from the circulation: the effect of the phospholipid and cholesterol components. Life Sci. 30: 2123–2136.

Sethuraman, V. A. and Bae, Y. H. 2007. TAT peptide-based micelle system for potential active targeting of anti-cancer agents to acidic solid tumors. J. Control. Release 118: 216–224.

Seymour, L. W., et al. 1987. Effect of molecular weight (Mw) of N-(2-hydroxypropyl)methacryla mide copolymers on body distribution and rate of excretion after subcutaneous, intraperitoneal, and intravenous administration to rats. J. Biomed. Mater. Res. 21: 1341–1358.

Shah, S. S., et al. 1997. Polymer-drug conjugates: manipulating drug delivery kinetics using model LCST systems. J. Control. Release 45: 95–101.

Shi, B., et al. 2005. Stealth MePEG-PCL micelles: effects of polymer composition on micelle physicochemical characteristics, in vitro drug release, in vivo pharmacokinetics in rats and biodistribution in S180 tumor-bearing mice. Colloid Polym. Sci. 283: 954–967.

Shibayama, M., Mizutani, S., and Nomura, S. 1996. Thermal properties of copolymer gels containing N-isopropylacrylamide. Macromolecules 29: 2019–2024.

Shuai, X., et al. 2004. Core-cross-linked polymeric micelles as paclitaxel carriers. Bioconjug. Chem. 15: 441–448.

Soga, O., Van Nostrum, C. F., and Hennink, W. E. 2004. Poly(N-2-hydroxypropyl) methacrylamide mono/di lactate): a new class of biodegradable polymers with tuneable thermosensitivity. Biomacromolecules 5: 818–821.

Soga, O., et al. 2004. Physicochemical characterization of degradable thermosensitive polymeric micelles. Langmuir 20: 9388–9395.

Soga, O., et al. 2005. Thermosensitive and biodegradable polymeric micelles for paclitaxel delivery. J. Control. Release 103: 341–353.

Soppimath, K. S., et al. 2001. Biodegradable polymeric nanoparticles as drug delivery devices. J. Control. Release 70: 1–20.

Soppimath, K. S., Tan, D. C. -W., and Yang, Y. -Y. 2005. pH-triggered thermally responsive polymer core-shell nanoparticles for drug delivery. Adv. Mater. 17: 318–323.

Stolnik, S., Illum, L., and Davis, S. S. 1995. Long circulating microparticulate drug carriers. Adv. Drug Deliv. Rev. 16: 195–214.

Sugiyama, K. and Sono, K. 2000. Characterization of photo- and thermoresponsible amphiphilic copolymers having azobenzene moieties as side groups. J. Appl. Poly. Sci. 81: 3056–3063.

Sun, X., et al. 2005. An assessment of the effects of shell cross linked nanoparticle size, core composition and surface pegylation on in vivo distribution. Biomacromolecules 6: 2541–2554.

Szczubialka, K., et al. 2004. Photocrosslinkable smart terpolymers responding to pH, temperature and ionic strength. J. Polym. Sci. Pol. Chem. 43: 3879–3886.

Takeoka, Y., et al. 1995. Electrochemical control of drug release from redox-active micelles. J. Control. Release 33: 79–87.

Takeuchi, H., et al. 2001. Evaluation of circulation profiles of liposomes coated with hydrophilic polymers having different molecular weights in rats. J. Control. Release 75: 83–91.

Tang, Y., et al. 2003. Solubilization and controlled release of a hydrophobic drug using novel micelle-forming ABC triblock copolymers. Biomacromolecules 4: 1636–1645.

Teng, Y., et al. 1998. Release kinetics studies of aromatic molecules into water from block polymer micelles. Macromolecules 31: 3578–3587.

Thurmond, K. B. II., et al. 1999. Shell cross-linked polymer micelles: stabilized assemblies with great versatility and potential. Colloid Surface B 16: 45–54.

Tian, L., et al. 2004. Core crosslinkable polymeric micelles from PEG-lipid amphiphiles as drug carriers. J. Mater. Chem. 14: 2317–2324.

Tong, X., et al. 2005. How can azobenzene block copolymer vesicles be dissociated and reformed by light. J. Phys. Chem. B 109: 20281–20287.

Topp, M. D. C., et al. 1997. Thermosensitive micelle-forming block copolymers of poly(ethylene glycol) and poly(N-isopropylacrylamide). Macromolecules 30: 8518–8520.

Torchilin, V. P. 2001. Structure and design of polymeric surfactant-based drug delivery systems. J. Control. Release 73: 137–172.

Torchilin, V. P. 2002. PEG-based micelles as carriers of contrast agents for different imaging modalities. Adv. Drug Deliv. Rev. 54: 235–252.

Torchilin, V. P. 2004. Targeted polymeric micelles for delivery of poorly soluble drugs. Cell. Mol. Life Sci. 61: 2549–2559.

Torchilin, V. P. 2006. Multifunctional nanocarriers. Adv. Drug Deliv. Rev. 58: 1532–1555.

Torchilin, V. P. 2007. Micellar nanocarriers: pharmaceutical perspectives. Pharm. Res. 24: 1–16.

Torchilin, V. P. and Trubetskoy, V. S. 1995. Which polymers can make nanoparticulate drug carriers long-circulating? Adv. Drug Deliv. Rev. 16: 141–155.

Torchilin, V. P., et al. 2001. Amphiphilic poly-N-vinylpyrrolidones: synthesis, properties and liposome surface modification. Biomaterials 22: 3035–3044.

Torchilin, V. P., et al. 2003. Immunomicelles: targeted pharmaceutical carriers for poorly soluble drugs. Proc. Natl. Acad. Sci. USA 100: 6039–6044.

Torchilin, V. P., et al. 1994. Amphiphilic vinyl polymers effectively prolong liposome circulation time in vivo. Biochim. Biophys. Acta 1195: 181–184.

Trubetskoy, V. S., et al. 1996. Stable polymeric micelles: lymphangiographic contrast media for gamma scintigraphy and magnetic resonance imaging. Acad. Radiol. 3: 232–238.

van Nostrum, C. F. 2004. Polymeric micelles to deliver photosensitizers for photodynamic therapy. Adv. Drug Deliv. Rev. 56: 9–16.

Volet, G., et al. 2005. Synthesis of monoalkyl end-capped poly(2-methyl-2-oxazoline) and its micelle formation in aqueous solution. Macromolecules 38: 5190–5197.

Vonarbourg, A., et al. 2006a. Parameters influencing the stealthiness of colloidal drug delivery systems. Biomaterials 27: 4356–4373.

Vonarbourg, A., et al. 2006b. Evaluation of PEGylated lipid nanocapsules versus complement system activation and macrophage uptake. J. Biomed. Mater. Res. 78: 620–628.

Watanabe, M., et al. 2006. Preparation of camptothecin-loaded polymeric micelles and evaluation of their incorporation and circulation stability. Int. J. Pharm. 308: 183–189.

Weissig, V., Whiteman, K. R., and Torchilin, V. P. 1998. Accumulation of protein-loaded long-circulating micelles and liposomes in subcutaneous lewis lung carcinoma in mice. Pharm. Res. 15: 1552–1556.

Whiteman, K., et al. 2001. Poly(HPMA)-coated liposomes demonstrated prolonged circulation in mice. J. Liposome Res. 11: 153–164.

Wickline, S. A. and Lanza, G. M. 2003. Nanotechnology for molecular imaging and targeted therapy. Circulation 107: 1092–1095.

Wilhelm, M., et al. 1991. Poly(styrene-ethylene oxide) block copolymer micelle formation in water: a fluorescence probe study. Macromolecules 24: 1033–1040.

Woodle, M. C. and Lasic, D. D. 1992. Sterically stabilized liposomes. Biochim. Biophys. Acta 1113: 171–199.

Woodle, M. C., Engbers, C. M., and Zalipsky, S. 1994. New amphipatic polymer-lipid conjugates forming long-circulating reticuloendothelial system-evading liposomes. Bioconjug. Chem. 5: 493–496.

Xie, Z., et al. 2007. A novel polymer-paclitaxel conjugate based on amphiphilic triblock copolymer. J. Control. Release 117: 210–216.

Xiong, X. B., et al. 2007. Conjugation of arginine-glycine-aspartic acid peptides to poly(ethylene oxide)-b-poly(epsilon-caprolactone) micelles for enhanced intracellular drug delivery to metastatic tumor cells. Biomacromolecules 8: 874–884.

Yamamoto, Y., et al. 2001. Long-circulating poly(ethylene glycol)-poly(D,L-lactide) block copolymer micelles with modulated surface charge. J. Control. Release 77: 27–38.

Yi, Y., et al. 2005. A polymeric nanoparticle consisting of mPEG-PLA-Toco and PLMA-COONa as a drug carrier: improvements in cellular uptake and biodistribution. Pharm. Res. 22: 200–208.

Yin, X., Hoffman, A. S., and Stayton, P. S. 2006. Poly(N-isopropylacrylamide-co-propylacrylic acid) copolymers that respond sharply to temperature and pH. Biomacromolecules 7: 1381–1385.

Yokoyama, M., et al. 1992. Preparation of micelle-forming polymer-drug conjugates. Bioconjug. Chem. 3: 295–301.

Yokoyama, M., et al. 1994. Improved synthesis of adriamycin-conjugated poly (ethylene oxide)-poly (aspartic acid) block copolymer and formation of unimodal micellar structure with controlled amount of physically entrapped adriamycin. J. Control. Release 32: 269–277.

Yokoyama, M., et al. 1998. Characterization of physical entrapment and chemical conjugation of adriamycin in polymeric micelles and their design for in vivo delivery to a solid tumor. J. Control. Release 50: 79–92.

Yokoyama, M., et al. 1999. Selective delivery of adriamycin to a solid tumor using a polymeric micelle carrier system. J. Drug Target. 7: 171–186.

Yokoyama, M., et al. 2004. Polymer design and incorporation methods for polymeric micelle carrier system containing water-insoluble anti-cancer agent camptothecin. J. Drug Target. 12: 373–384.

Yoon, T. J., et al. 2005. Multifunctional nanoparticles possessing a 'magnetic motor effect' for drug or gene delivery. Angew. Chem. Int. Ed. Engl. 117: 1092–1095.

Yoshida, E. and Kunugi, S. 2002. Micelle formation of poly(vinyl phenol)-block-polystyrene by alfa, omega-diamines. J. Polym. Sci. Pol. Chem. 40: 3063–3067.

Yuan, F., et al. 1995. Vascular permeability in a human tumor xenograft: molecular size dependence and cutoff size. Cancer Res. 55: 3752–3756.

Zamboni, W. C. 2005. Liposomal, nanoparticle and conjugated formulations of anticancer agents. Clin. Cancer Res. 11: 8230 8234.

Zhang, G. -D., et al. 2003. Polyion complex micelles entrapping cationic dendrimer porphyrin: effective photosensitizer for photodynamic therapy of cancer. J. Control. Release 93: 141–150.

Zhang, J., et al. 2006a. Micellization phenomena of amphiphilic block copolymers based on methoxy poly(ethylene glycol) and either crystalline or amorphous poly(caprolactone-b-lactide). Biomacromolecules 7: 2492–2500.

Zhang, L., et al. 2006b. Using the reversible addition-fragmentation chain transfer process to synthesize core-crosslinked micelles. J. Polym. Sci. Pol. Chem. 44: 2177–2194.

Zhang, Z., Grijpma, D. W., and Feijen, J. 2006. Thermo-sensitive transition of monomethoxy poly(ethylene glycol)-block-poly(trimethylene carbonate) films to micellar-like nanoparticles. J. Control. Release 112: 57–63.

Zhu, P. W. and Napper, D. H. 2000. Effect of heating rate on nanoparticle formation of poly(N-isopropylacrylamide)-poly(ethylene glycol) block copolymer microgels. Langmuir 16: 8543–8545.

Zuccari, G., et al. 2005. Modified polyvinylalcohol for encapsulation of all-trans-retinoic acid in polymeric micelles. J. Control. Release 103: 369–380.

Zweers, M. L. T., et al. 2004. In vitro degradation of nanoparticles prepared from polymers based on DL-lactide, glycolide and poly(ethylene oxide). J. Control. Release 100: 347–356.

Multifunctional Nanotherapeutics for Cancer

T. Minko, J.J. Khandare, A.A. Vetcher, V.A. Soldatenkov, O.B. Garbuzenko, M. Saad, and V.P. Pozharov

1 Introduction

Nanotechnology, as a field of applied science, focuses on the development, production, characterization and application of materials, and devices at the level of molecules and atoms with a typical size between 10^{-9} nm and 10^{-6} μm. Nanotherapeutics, a rapidly expanding area of medicine, uses nanotechnology products for highly specific medical interventions at the molecular scale for curing diseases or repairing damaged tissues. Although some nanotechnology products can be applied alone as therapeutic or imaging agents, they are being most often used as pharmaceutical nanocarriers for delivering drugs or imaging agents to the site of the action in desired quantities and releasing therapeutic loads with a specific time profile. Linear and branched polymers, dendrimers, quantum dots, nanoparticles, nanospheres, nanotubes, nanocrystals, nanogels, liposomes, micelles, as well as other types of nanocarriers are being employed in different fields of medicine for diagnostics, imaging, treatment, and prophylaxis of many pathological conditions (Fig. 1)

In contrast to the earlier developed nanotherapeutics, which had a relatively simple two-component drug–carrier composition, modern nanocarriers often include other active ingredients that perform different specific functions for enhancing cellular uptake and efficiency of the main drug, preventing adverse side effects, providing drug release with a predetermined profile in the certain compartment of an organ, tissue, or cell, and preventing the development and/or suppression of the existent drug resistance, etc. The increase in complexity and performed functions of nanocarriers actually converts them into multifunctional nanotherapeutical products. This chapter is mainly focused on reviewing modern multifunctional approaches in nanotherapeutics designed for effective cancer treatment.

2 Requirements for Cancer Nanotherapeutics

Based on literature data and our own research, we conclude that almost all traditional anticancer drugs initiate four main effects in cancer cells: (1) induction of cell death, (2) triggering drug efflux pumps, (3) activation of cellular detoxification

V. Torchilin (ed.), *Multifunctional Pharmaceutical Nanocarriers*,
© Springer Science+Business Media, LLC 2008

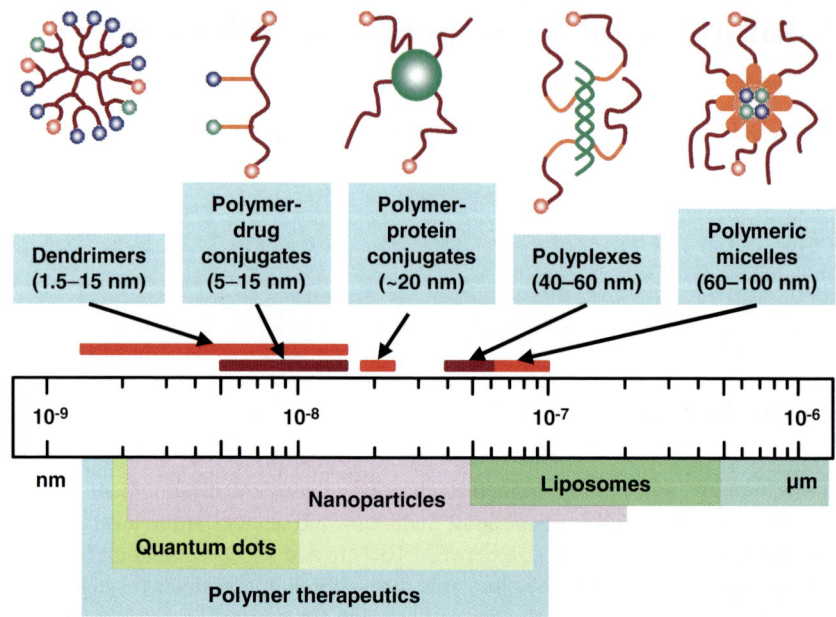

Fig. 1 Classes of nanoscale drug delivery systems (*bottom panel*) and examples of polymeric drug delivery systems

mechanisms, and (4) activation of cell death defensive mechanisms (Fig. 2). The first component – cell death induction – is specific (especially in its initial steps) to a particular anticancer drug. The last three components, particularly after repeated treatment, ultimately lead to an increase in cellular drug resistance limiting the efficacy of cancer chemotherapy. Cellular resistance to chemotherapy in turn might be subdivided into two types: pump and nonpump resistance (Pakunlu et al., 2003). Development of pump resistance, also termed as multidrug resistance, depends on the activation of drug efflux pumps, which pump an anticancer drug out from the cellular cytoplasm to the surrounding medium. In contrast, nonpump resistance includes all cellular defensive mechanisms that are not related to the active efflux of drugs from the cells. The main components of nonpump resistance are cellular detoxification mechanisms and antiapoptotic defense (Fig. 2). Nonpump resistance decreases a specific activity of an anticancer agent limiting cellular damage caused by the drug and prevents the expansion of this initial cellular damage to cell death.

Ideal anticancer therapeutics have to fulfill one main requirement: it should effectively kill cancer cells without harming the surrounding normal cells. Acting like a "magic bullet," an anticancer drug should kill the villain while sparing the victim. Although the concept of "magic bullet" in chemotherapy was first introduced by Paul Ehrlich at the beginning of 20th century, the problem of selective toxicity of a drug only to cancer cells has not yet been completely solved. The most promising way to achieve the desired result is to use a delivery system, which will minimize toxic effects of an anticancer drug by (1) increasing the amount and

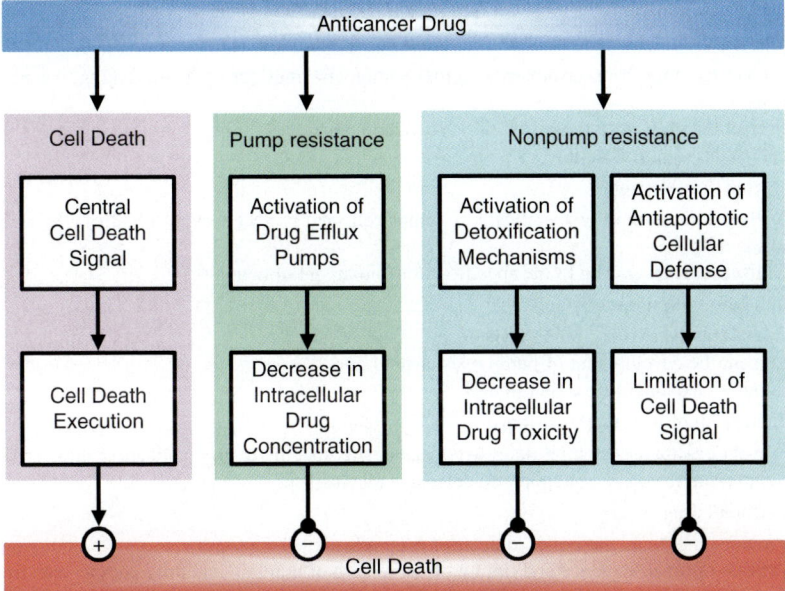

Fig. 2 Cellular effects of anticancer drugs. Most anticancer drugs induce cell death while activating antiapoptotic cellular defense, detoxification mechanisms, and ATP-dependent drug efflux pumps. This limits the efficacy of cancer therapy and develops multidrug resistance

persistence of a drug in the vicinity of "target" (cancer) cells and (2) reducing the drug exposure of "nontarget" (normal) cells. In addition, an advanced drug delivery system (DDS) should promote the therapeutic effects of a drug (an induction of cell death in cancer cells) and minimize its previously discussed undesirable effects (an increase in cellular resistance). Based on the cellular effects of the anticancer drugs discussed earlier, we formulate the main requirements to the advanced anticancer DDSs in general and anticancer nanotherapeutics in particular (Table 1).

2.1 Protected Drug Delivery and Prodrug Approach

The drug and other components of nanotherapeutics should be protected from the action of detoxification enzymes and other types of degradation during their voyage in the bloodstream, organs (stomach, liver, intestines, etc.), and inside the cancer cell. Protected delivery of therapeutics is usually achieved by the so-called "prodrug approach." A prodrug is an inactive form of a drug, which converts to active drug form in the site of its action. Prodrug approach decreases the degradation of the drug during its journey in gastrointestinal tract or systemic circulation and prevents systemic toxicity. In most cases, a type of attachment of a drug to a carrier is designed to prevent substantial degradation during drug transfer to the site of action (organ, tissue, cell, or intracellular organelles) and allows release of

Table 1 Requirements for anticancer multifunctional nanotherapeutics

1. Protected drug delivery
   - Protect the drug from the action of detoxification enzymes during its trafficking to cancer cells
   - Protect the drug from the action of detoxification enzymes during its trafficking inside the cell to the site of its action
2. Targeted drug delivery
   - Deliver anticancer drug specifically to tumor cells and protect normal cells from the action of the drug
   - Deliver anticancer drug to the specific intracellular site of action
3. Modulation of pump resistance
   - Overcome the existing pump resistance
   - Prevent the development of pump resistance
   - Suppress existing drug efflux pumps
4. Suppression of nonpump resistance
   - Prevent or suppress cellular defensive mechanisms specific to used anticancer drug
   - Suppress nonspecific antiapoptotic defensive mechanisms
5. Controlled drug release
   - Provide an optimal pattern of anticancer drug release
   - Provide optimal sequence and timing of the release of nondrug and drug components from the complex system

the drug in its active form in the targeted spot. For a more detail description of different types of prodrugs, the reader is referred to the corresponding reviews and original reports (Mehvar et al., 2000; Duncan et al., 2001; Greenwald et al., 2003; Choe et al., 2004; Chung & Cho, 2004; Greenwald et al., 2004; Minko, 2004; Minko et al., 2005; Naik et al., 2005; Zhang et al., 2005; Khandare & Minko, 2006; Khandare et al., 2006a).

Usually drugs incorporated into special kind of micelles, liposomes, microspheres, nanoparticles, etc, or bound to natural or synthetic polymers, are sufficiently protected from destructive enzymes if two actions occur: (1) the drug carrier does not degrade significantly during its voyage and (2) the carrier–drug linkage is designed to be stable in the bloodstream and interstitial space, but susceptible to hydrolysis or other types of degradation inside the targeted cell compartment (e.g., lysosomes, cytoplasm). This approach also guarantees the targeted delivery of the drug and other components of nanotherapeutics to the specific cellular compartments.

2.2 Targeted Drug Delivery

The term *targeted drug delivery* is typically interpreted as the delivery of a therapeutic agent to the "target" organ or tissue. In our opinion, the definition of this term should be extended to mean "the delivery of the drug to the specific intracellular site of action". For example, DNA damaging agents, such as doxorubicin (DOX), should be delivered to the tumor cells and released in the perinuclear region.

There are two main approaches of delivering drugs specifically to tumors. The first (*passive*) approach is based on the so-called enhanced permeability and retention (EPR) effect (Matsumura & Maeda, 1986; Maeda, 2001; Fang et al., 2003; Greish et al., 2003). The EPR effect is the result of the increased permeability of the tumor vascular endothelium to circulating macromolecules combined with limited lymphatic drainage from the tumor interstitium. The second (*active*) approach is based on the coupling of a drug carrier with a targeted moiety specific to certain or a majority of cancer cells. Many different types of targeting moieties are being used in experiments and clinical trials. For example, substrates to specific receptors expressed on certain types of cancer cells (luteinizing hormone release hormone receptors, somatostatin analogs, folate, vasoactive intestinal peptide, gastrin-releasing peptide, cholecystokinin or gastrin, neurotensin, substance P, neuropeptide Y, etc.), antibodies or their fragments targeted to different proteins (tumor-associated glycoprotein-72, TAG-72; CD20 antigen; tumor necrosis factor, TNF; epidermal growth factor, EGF receptor; etc.) overexpressed in cancer cells. Detailed description of various targeting moieties have been presented in several recent reviews and experimental papers (Torchilin, 2000; Kopecek et al., 2001; Langer & Beck-Sickinger, 2001; Langer, 2001; Dharap et al., 2003; Reubi, 2003; Minko et al., 2004) for the details.

2.3 Modulation of Pump and Nonpump Resistance

One of the very important requirements for advanced anticancer nanotherapeutics is the ability to overcome or suppress drug efflux pumps. In addition, a new generation of nanotherapeutics should simultaneously suppress the nonpump cellular resistance, specifically the antiapoptotic cellular defense – the key component of this type of resistance. It should be stressed, that only simultaneous modulation of multidrug resistance and antiapoptotic cellular defense is capable of significantly increasing the efficacy of traditional anticancer drugs. Therefore, advanced anticancer nanotherapeutics should contain inhibitors of both drug efflux pumps and antiapoptotic cellular defense. Several methods have been recently developed to modulate pump and nonpump resistance. The most promising of them are based on the suppression of the overexpression of P-glycoprotein or multidrug resistance associated proteins and antiapoptotic members of BCL2 family proteins – the key players in pump and nonpump resistance respectively. Two main approaches are currently being used for this purpose. The first approach is based on the use of small molecules that downregulate protein expression by specifically binding to the protein itself. This approach was already applied to the suppression of BCL2 protein (Cosulich et al., 1997; Holinger et al., 1999; Lutz, 2000; Dharap & Minko, 2003; Dharap et al., 2003; Dharap et al., 2006). The second approach is based on the use of antisense oligonucleotides (ASOs) directed to mRNA encoding proteins responsible for both types of cellular defense (Pakunlu et al., 2003; Minko et al., 2004; Pakunlu et al., 2004, 2006).

2.4   Controlled Drug Release

Another requirement for advanced anticancer nanotherapeutics is controlled drug release. Controlled drug release in the context of drug delivery means that the DDS could provide the release of its components through a preprogrammed time profile. This consideration becomes especially important when components of a complex DDS have substantial variations in timing of maximal therapeutic effect. If, for instance, the drug (apoptosis inductor) acts immediately after its release, while a DDS component that suppresses cellular antiapoptotic defense has relatively long lag time of action, the latter should be released from the DDS first, followed by a delay in the release of the apoptosis inductor. The duration of that delay should be sufficient to suppress an antiapoptotic cellular defense. In this case, apoptosis induction will take place on the background of suppressed cellular defense. This should significantly improve the efficacy of the anticancer drug, since the opposite situation is less favorable. Controlled release complex DDS components might be achieved by variations in the type and strength of a bond used to attach the DDS components to the drug carrier. It should be stressed that while many different controlled release dosage forms of drugs are widely being developed, tested, and used, controlled release of drugs from proposed complex DDS containing multiple active ingredients with different mechanisms of action still need to be developed.

3   Novel Multifunctional Nanotherapeutics for Cancer

The modern conception of effective anticancer drugs frequently uses the metaphor "silver bullet" to describe a new anticancer drug or approach that is expected to easily cure one or several cancer types with extreme effectiveness. A narrower meaning of "silver bullet" is used for describing a drug that precisely targets a selected protein or mechanism in specific tumors and demonstrates high effectiveness in killing cancer cells. However, many diseases, including cancer, have very complex etiology and intrinsically possess or rapidly develop defensive mechanisms against anticancer treatment, which prevents an effective cure. In response to this complexity, a new paradigm has recently emerged that challenges the widely held assumption that "silver bullet" agents are superior to "dirty drugs" in therapeutic approaches aimed at the prevention or treatment of cancer, AIDS, and neuropsychiatric and cardiovascular diseases (Van der Schyf et al., 2006). Although such a concept is interesting, we believe that a combination of several "silver bullets" targeted to different mechanisms in one DDS should be used instead of a "dirty drug" with predominately uncontrolled mechanisms of action. In addition to simultaneous targeting of several mechanisms of cell death and drug or cell death resistance, it is extremely important that modern multifunctional anticancer drugs come close to being "magic bullet" – a perfect drug that cures a disease without a risk of side effects. The latter usually is achieved by targeting drugs to the tumor or organ or tissues that bear tumors. Within this paper, we provide some examples of recently developed multifunctional nanotherapeutics, which partially or fully satisfy the earlier-mentioned demands.

3.1   Linear and Branched Water-Soluble Polymers

Several types and designs of water-soluble polymers are extensively used as carriers for multifunctional nanotherapeutics. Some of the designs that are most frequently employed for delivery of anticancer therapeutics are shown in Fig. 3.

The simplest design of a polymeric DDS includes linear polymer with active ingredients conjugated to the distal ends of polymer or polymer backbone directly or through so called "spacers" (Fig. 3a). Such spacers may play several roles in a DDS. First, they provide for binding structurally and chemically different compounds (polymer, drug, targeting moiety, etc.) in one complex relatively stable

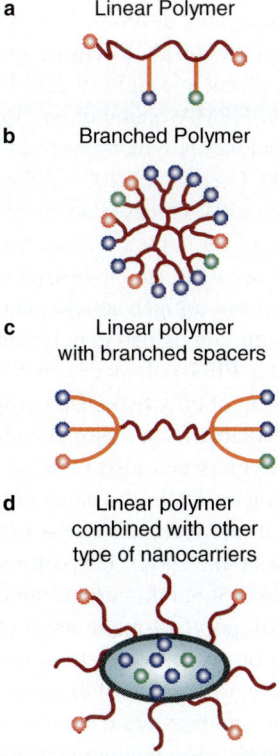

a   Linear Polymer

b   Branched Polymer

c   Linear polymer
            with branched spacers

d   Linear polymer
            combined with other
            type of nanocarriers

Fig. 3 Classification of water-soluble multifunctional anticancer polymeric drug delivery systems. PEGylated liposome is shown as an example of mixed type of drug delivery system

chemical structure. Second, they physically separate the polymer from the active ingredients, preventing their negative influence on the specific activity of the drug(s) used in the DDS. Third, the specific construction of a spacer allows for the regulation of the degree of binding of drugs or other active ingredients to the polymer backbone and therefore predetermines the rate and specific place of drug release (if any). Two polymers are most widely used in these systems – N-(2-Hydroxypropyl)methacryla mide (HPMA) copolymer and poly(ethylene glycol) (PEG).

HPMA copolymer is a water-soluble, biocompatible, nontoxic, nonimunogenic polymer, which is frequently used in different DDSs (Minko et al., 1998; Minko et al., 1999a,b, 2001; Demoy et al., 2000; Kunath et al., 2000; Minko et al., 2000; Wang et al., 2000; Kasuya et al., 2001; Lu et al., 2002; Peterson et al., 2003; David et al., 2004; Funhoff et al., 2004; Kovar et al., 2004; Ulbrich et al., 2004a,b; Luten et al., 2006; Zarabi et al., 2006; Duncan, 2007). HPMA copolymers have been most intensively studied at the Center for Controlled Chemical Delivery (US), at the Institute of Macromolecular Chemistry (Czech Republic), at the Center for Polymer Therapeutics at the University of Cardiff (UK), and at the Utrecht Institute for Pharmaceutical Sciences (Netherlands). The main advantage of HPMA as a drug carrier is the existence of several sites for potential conjugation of drugs, targeting moieties, and other components of a DDS. The HPMA copolymer main chain is not biodegradable, and so all conjugates developed clinically have been limited to a molecular weight of less than 40,000 Da to ensure eventual renal elimination. In most cases, an increase in molecular weight of HPMA copolymers results in the prolonged circulation times of HPMA-bound drugs (Lammers et al., 2005).

PEG is a water-soluble nonionic polymer approved by FDA for pharmaceutical applications (Zhao et al., 2000; Greenwald, 2001, 2004; Minko et al., 2002; Dharap et al., 2003, 2006; Choe et al., 2004; Gunaseelan et al., 2004; Paranjpe et al., 2004; Chang et al., 2005). Because of its nontoxic character, it is widely used in many biochemical, cosmetic, pharmaceutical, and industrial applications. It is also impor- tant that PEG polymers exhibit low antigen activity and, in most cases, decrease the antigenicity of active ingredients conjugated to polymers (Caliceti et al., 2001). The major disadvantage of a linear PEG polymer is that they have only two binding sites. This allows for the binding of only two DDS components directly to the poly- mer. PEG is commercially available from major chemical suppliers.

Several other types of polymers can also be used in linear polymer systems. A new prototype of the polymeric DDS, the nanoconjugate Polycefin, was tested for its ability to accumulate in tumors based on the EPR effect and receptor medi- ated endocytosis (Ljubimova et al., 2007). Polycefin was synthesized for the tar- geted delivery of Morpholino ASOs into certain tumors. It consists of units that are covalently conjugated with poly(β-l-malic acid) (M_w = 50,000, M_w/M_n = 1.3) that is purified from cultures of myxomycete *Physarum polycephalum*. The units are active in endosomal uptake, disruption of endosomal membranes, oligonucle- otide release in the cytoplasm, and protection against enzymatic degradation in the vascular system. The conjugate also contained AlexaFluor 680 C2-maleimide dye for in vivo detection. The polymer is biodegradable, nonimmunogenic, and nontoxic.

An interesting polymeric carrier consisting of folate-poly(ethylene glycol)-grafted chitosan was recently synthesized for targeted plasmid DNA delivery to tumor cells (Chan et al., 2007). The carrier employs PEG polymer to enhance the aqueous solubility of chitosan and folate to improve gene transfection efficiency due to promoted uptake of folate receptor-bearing tumor cells. Such a polymeric construct can be a promising gene carrier because of its solubility in physiological pH, efficiency in condensing DNA, low cytotoxicity, and targeting ability.

A substantial disadvantage of the linear polymeric system is low loading capacity, which is limited to two distal ends or few backbone sites. To overcome this problem, branched polymers (Fig. 3b), linear polymers with branched spacers (Fig. 3c), and combinations of linear polymers with other types of nanocarriers (liposomes, nanoparticles, etc.) (Fig. 3d) have been developed. Dendrimers are routinely synthesized as tunable nanostructures that may be designed and regulated as a function of their size, shape, surface chemistry, and interior void space, and are distinguished by their precise nanoscale scaffolding and nanocontainer properties (Tomalia et al., 2007). As a result of these important properties, they are expected to play an important role in the emerging field of nanomedicine.

A typical branched polymer, dendrimer, is a regularly branched molecule that resembles the branches of a tree (dendron on Greek). A number of surface groups of polymer that can be used for conjugation of active components doubles with the branching of each terminal end of polymer or generation (G) of dendrimer. Typical structures of polyamidoamine (PAMAM) dendrimers with different number of generations are presented in Fig. 4. The increase in G from 0 (native PAMAM) to 4 leads to an increase in number of surface groups from 4 to 64. Such branching slightly enlarges an average diameter of the PAMAM dendrimer (approximately 3 times in most cases), but substantially increases an average molecular weight of the dendrimer (up to 30 times). The solubility of dendrimers and other properties are mainly determined by the functional groups on the molecular surface, e.g., a dendrimer can be water-soluble when its end-groups are hydrophilic groups, such as carboxyl groups. The molecular heterogeneity of complex dendrimer conjugate nanodevices also depends heavily upon the types of terminal functional groups (Islam et al., 2005; Shi et al., 2006). The drug and other active components of DDS can be associated with dendrimers by placing drugs into the inner core of the system or conjugating these components to the outer surface groups directly or through spacers. For instance, we recently used succinic acid as a spacer to synthesize PAMAM succinic acid–paclitaxel conjugate (Khandare et al., 2006b). Conjugation of succinic acid to paclitaxel resulted in the formation of mono carboxylic acid conjugate, which was further conjugated with hydroxyl groups in PAMAM-OH dendrimer. The solubility and cytotoxicity of PAMAM–paclitaxel conjugate were 10 times higher when compared with those of free paclitaxel. It is important that not only several copies of drugs can be associated with one molecule of branched polymers, but other functional groups can be used in multiple copies to enhance the multifunctionality of the carrier. For instance, a dendrimeric system containing five folate molecules was designed and evaluated (Hong et al., 2007). Experimental results showed that dissociation constants K_D between the nanodevices and folate-binding protein were dramatically enhanced through multivalency

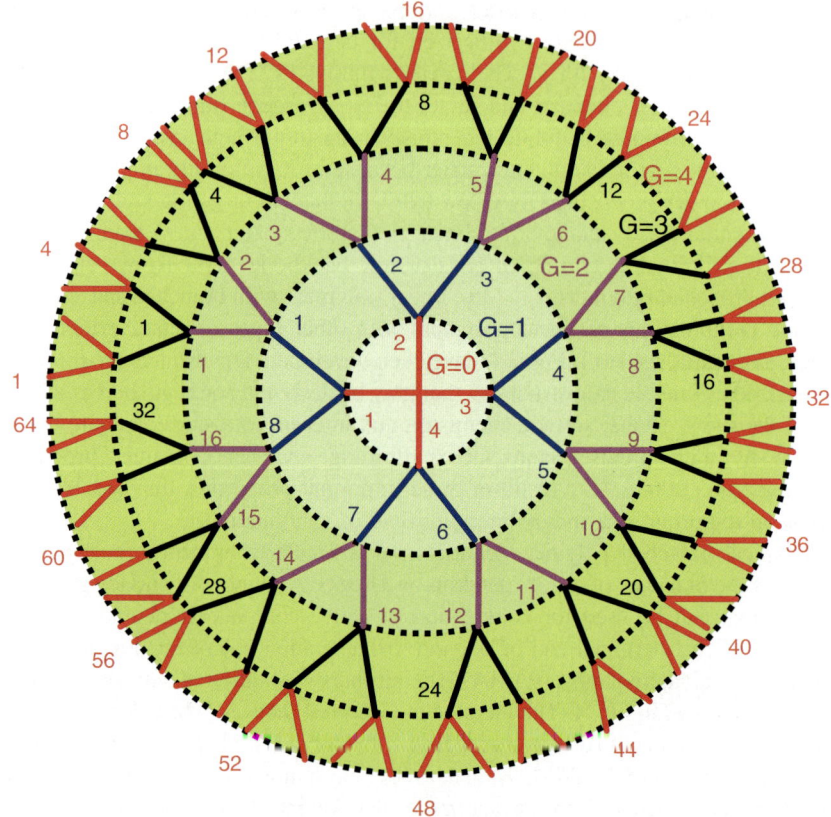

Fig. 4 A schematic of PAMAM dendrimer. G0, G1, G2, G3, G4 – generations of dendrimer

(approximately 2,500–170,000-fold). The data support the hypothesis that multivalent enhancement of K_D, not an enhanced rate of endocytosis, is the key factor resulting in the improved biological targeting by these drug delivery platforms. However, the drug payload and number of copies of other functional groups in one DDS is limited mainly because of the decrease in solubility and steric hindrance. Consequently, a real number of active ingredients can be conjugated to dendrimer substantially lower than the total number of available surface groups and in most cases is close to 10% of total mass of the system. However, covalent conjugation of 58 molecules of a drug to one molecule of G4 PAMAM-OH dendrimer have also been reported (Kolhe et al., 2004).

The type of the bond that conjugates active ingredients to the polymer or spacer is very important for performing their functions. While targeting moieties in most cases conjugated to polymeric carrier by nondegradable in physiological conditions bonds, anticancer drugs should be conjugated via bonds degradable in the tumor environment or inside specific organelles in cancer cells (mainly lysosomes). For example, DOX attached to a biodegradable dendrimer via the pH-sensitive hydra-zone (HZ) linkages was more than 10 times toxic when compared with that of free DOX. At the same time, a dendrimer–DOX complex in which DOX was attached by means of a stable carbamate bond did not show antitumor activity in vivo (Lee et al., 2006).

In addition to targeting moieties and anticancer drugs, several other active ingredients that are able to perform different functions may be conjugated to dendrimers converting them into multifunctional DDSs. For example, Myc et al. (2007) synthesized a PAMAM dendrimer-based nanodevice in which folic acid was used as the targeting molecule, staurosporine was employed as an anticancer drug or apoptosis inducer, and a caspase-hydrolyzable, fluorescence resonance energy transfer (FRET)-based substrates was utilized as the apoptosis-detecting agent. The developed device, in addition to active targeting folic acid-receptor-positive cancer cells and cell death induction, could be used for simultaneously monitoring the apoptotic potential of a delivered drug by a fluorescent detection of caspase activity. Other imaging agents are being used to visualize delivery system and tumors. These agents include rare earth metals (Kobayashi et al., 2006; Langereis et al., 2006), gold (Khan et al., 2005), radionuclides (Mitra et al., 2006), fluorescein isothiocyanate (FITC) (Majoros et al., 2006; Mora et al., 2006), and other fluorescent dyes. Recently, we used near-infrared dye Cy5.5 to visualize PAMAM dendrimers. The main advantage of near-infrared radiation is its ability to penetrate much farther into a sample than do mid-infrared or shorter wavelength rays. This allows for in vivo imaging (Fig. 5)

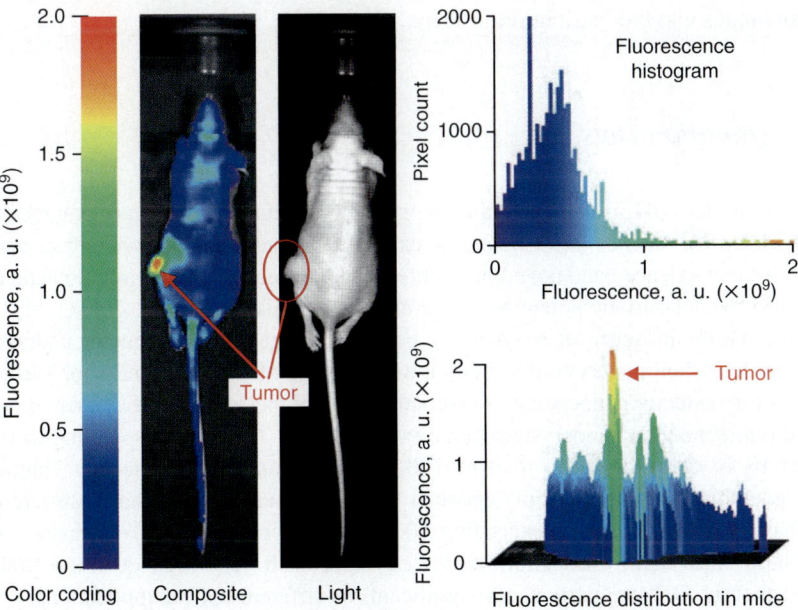

Fig. 5 Typical in vivo images of mouse bearing xenograft of multidrug resistant A2780 human ovarian carcinoma. Mouse was injected with PAMAM generation 4 dendrimer (14.2 kDa, 4.5 nm) labeled with near-infrared fluorophore cyanine dye (Cy 5.5). Images were taken using IVIS Imaging System (Xenogen, Alameda, CA) in anesthetized animals 72 h after injection of dendrimer. Visible light and fluorescent images were overlaid to obtain a composite image. The distribution of dendrimer fluorescence was analyzed using an original software program developed in our laboratory

using commercially available devices, such as the IVIS system (Xenogen Corporation, Alameda, CA). Using this system, we evaluated the distribution of PAMAM generation 4 dendrimers with molecular weight about 14 kDa and size about 4.5 nm in mice nearing xenografts of human ovarian carcinoma. It was found that because of the EPR effect, dendrimer accumulated predominately in the solid tumor.

Instead of using branched polymers to increase the loading capacity of DDS, branched spacer architecture can be utilized (Fig. 3d). Recently, we proposed a novel multifunctional polymeric DDS with linear PEG polymer and branched citric acid spacer (Khandare et al., 2006a). The complex system includes several copies of a targeting moiety (synthetic LHRH peptide, analog of luteinizing hormone-releasing hormone) and anticancer drug (camptothecin). The next generation of such multifunctional DDS with multiple copies of active ingredients per one molecule of PEG carrier has been developed in our laboratory. In addition to specified components (targeting moiety and anticancer drug), the system includes BH3 peptide as a suppressor of cellular antiapoptotic defense.

Several types of more complex polymeric structures, including combinations of dendrimers and linear polymers with liposomes, nanoparticles, oligonucleotides, etc., (Fig. 3d) were developed and evaluated (Torchilin, 2002, 2007; Amirkhanov and Wickstrom, 2005; Choi and Baker, 2005; Gillies et al., 2005; Khan et al., 2005; Papagiannaros et al., 2005; Lee et al., 2006). The main advantages of such complex systems include, but not limit to, an increase in loading capacity and better pharmacokinetics and biodistribution in vivo.

3.2 Quantum Dots

A quantum dot (QD) is a semiconductor crystal that limits the motion of conduction band electrons, valence band holes, or excitons (bound pairs of conduction band electrons and valence band holes) in all three spatial directions. The limitation can be achieved by electrostatic potentials (generated by external electrodes, doping, strain, impurities), the presence of an interface between different semiconductor materials (e.g., in core-shell nanocrystal systems), the presence of the semiconductor surface (e.g., semiconductor nanocrystal), or a combination of these. Small QDs, such as colloidal semiconductor nanocrystals, can be as small as 2–10 nm, corresponding to 10–50 atoms in diameter and a total of 100–100,000 atoms within the QD volume. Self-assembled QDs are typically between 10 and 50 nm in size. QD heterostructures can have lateral dimensions exceeding 100 nm. By being exposed to ultraviolet or near-ultraviolet light, QDs emit fluorescent light with very high quantum yield. Consequently, QDs are particularly significant for different optical applications.

Free QDs, as well as other nanoparticles, injected in the systemic circulation are acively phagocytized by macrophages and are taken up by the liver, spleen, and lymph nodes. It was found that, at low doses, the majority of the QDs are sequestered in the liver, spleen, and lymph nodes. At higher doses, increasing quantities of QDs are noted within the experimental brain tumors. Macrophages and microglia colocalize with glioma cells, carrying the QD and thereby optically outlining the tumor. The optical signal may be detected, allowing for improved identification and visualization of tumors, potentially augmenting brain tumor biopsy and resection (Popescu and Toms, 2006; Jackson et al., 2007). More complex simultaneous two-color in vivo wavelength-resolved spectral fluorescence lymphangiography using two near infrared QDs with different emission spectra was developed for noninvasive and simultaneous visualization of two separate lymphatic flows draining the breast and the upper extremity and variations in the drainage patterns and the water sheds within the axillary node (Hama et al., 2007).

Active targeting of QD to tumor cells should dramatically improve tumor imaging. The simplest tumor targeted structure that utilizes QDs for cancer imaging can be created by linking a targeting moiety with QD. Similar to the targeting of other DDS, antibody or their fragments, cancer-specific peptides, ligands to extracellular receptors, or molecules expressed in specific cancer cells are being used to target QD to cancer. Specific immunofluorescent probes for hepatoma detection in vivo were synthesized by linking QDs to an α-fetoprotein (AFP) antibody (Yu et al., 2007b). AFP is an important marker for hepatocellular carcinoma cell lines. QDs-Anti-AFP probes showed specific fluorescence in tumor and provided for an active tumor targeting and spectroscopic hepatoma imaging. QDs labeled with the monoclonal anti-HER2 antibody were successfully used for targeting of HER2-overexpressing breast cancer cells in mice bearing xenografts of human tumors (Tada et al., 2007). Six stages of the delivery processes were identified: initial circulation within a blood vessel, extravasation, traffic in the extracelullar region, binding to HER2 on the cell membrane, moving from the cell membrane to the perinuclear region, and in the perinuclear region. Other targeting moieties and penetration enhancers that were used to enhance the uptake of different types of DDS by cancer cells, including TAT peptide, matrix metalloprotease, biotin, biotinylated peptides, integrin-targeted arginine–glycine–aspartic acid were also used for tumor targeting of QD (Cai et al., 2006; Li et al., 2006; Young and Rozengurt, 2006; Zhang et al., 2006; Xue et al., 2007).

More complex tumor-targeted delivery systems can be designed using linear or branched polymers as carriers with QDs as imaging agents. All above-mentioned structures (Fig. 3) can be used for carrying QD to tumors and providing their uptake by cancer cells. High-quality nanocrystals formed in organic solvents can be completely solubilized in water using amphiphilic copolymers containing poly(ethylene glycol) or PEG. These copolymers are generated using a maleic anhydride coupling scheme that permits the conjugation of a wide variety of PEG polymers, both unfunctionalized and functionalized, to hydrophobic tails. The composite nanocrystal–polymer assemblies can be targeted to recognize cancer cells with Her2 receptors and are biocompatible if their surface coatings contain PEG. In the particular case of semiconductor nanocrystals (e.g., QDs), the materials in water have the same optical spectra and quantum yield as those formed initially in organic solutions (Yu et al., 2007a).

Multifunctional nanoparticle probes based on semiconductor QDs were developed for simultaneous targeting and imaging of cancer cells in living animals (van Vlerken and Amiji, 2006; Gao et al., 2007; Yu et al., 2007a). The structural design involves encapsulating luminescent QDs with a triblock copolymer, and linking this polymer to tumor-targeting ligands, such as antibodies and drug-delivery functionalities. In vivo studies on mice bearing xenografts of human prostate cancer showed that the QD probes can be delivered to tumor sites by both the EPR effect (passive targeting) and by antibody binding to cancer-specific cell surface biomarkers such as prostate-specific membrane antigen (active targeting). The targeted QD conjugates demonstrated a high sensitivity, providing for multicolor fluorescent imaging of as few as 10–100 cancer cells under in vivo settings.

It was found that CdSe QDs can be efficiently encapsulated in the phospholipid nanoemulsion that mimics the natural lipoprotein core. The QD nanoemulsion had a particle size approximately 80 nm and appears physically stable. On experiments on non–small-cell lung cancer cells, the intensity of cellular fluorescence imaging increased with the cell incubation time while more QDs were taken up by the cells. Two types of fluorescent microscopies confirmed that QDs are primarily localized in the cytoplasm but not in the nucleus of the cells (Liu et al., 2006).

Importantly, cadmium selenide or zinc sulfide (CdSe/ZnS) QDs themselves can induce cell death mediated by cadmium ions (Cd2+) released from the QDs cores (Cho et al., 2007). In addition to Cd2+ ions, cell death induction by QD involved the generation of oxygen-free radicals accompanied by lysosomal enlargement and intracellular redistribution. It was also shown that CdSe-core QDs can induce apoptotic biochemical changes, including JNK activation, loss of mitochondrial membrane potential, mitochondrial release of cytochrome c, and activation of caspase-9 and caspase-3 (Chan et al., 2006).

QD-conjugated probes to specific biomarkers are powerful tools that can be applied in a multiplex manner to single tissue sections of biopsies to measure expression levels of multiple biomarkers (True and Gao, 2007). QDs conjugated to immunoglobulins are commercially available from Invitrogen and can be used for visualizing various biomarkers labeled with primary antibodies. Assessing malignant tumors for the expression of multiple biomarkers provides data that are critical for patient management.

3.3   Nanoparticles

After 100 years of discovery of gold particles by Michael Faraday in 1857, nanoparticles finally entered the field of anticancer therapeutics because of their unique properties and ability to carry out different biological active substances. The term *nanoparticle* in drug delivery generally combines all delivery vehicles with size less then one micrometer (Portney and Ozkan, 2006). However, often polymeric DDS, QDs, and liposomes are excluded from nonviral nanoparticle vectors.

Various types of nanoparticles, including nanospheres, nanotubes, nanocrystals, nanogels, nanowires, nanocantilevers, nanopores, nanoshells, etc., are being applied for cancer treatment (Arayne and Sultana, 2006). Similar to other types of nanocarriers, tumor targeting of nanoparticles is achieved passively by the EPR effect or by adding targeting moieties of different nature (Minko, 2004; Minko et al., 2004). A desired drug release is accomplished by varying the type of spacers, bonds between the drug and nanoparticle material, nanoparticle construction, etc. Different shapes of nanoparticles open a wide road for the selection of desired carrier properties, including pharmacokinetics, biodistribution, and drug release. For instance, recently, we found that soft filament-type nanoparticles circulate longer than spherical particles (Geng et al., in press). The reader is referred to recent reviews for evaluating different types of nanoparticle delivery systems (McNeil, 2005; Arayne & Sultana, 2006; Chen & Zhang, 2006; Corot et al., 2006; Grodzinski et al., 2006; Hillaireau et al., 2006; Mitra et al., 2006; Moghimi, 2006; Sinha et al., 2006; Zhang et al., 2007).

Several designs of multifunctional nanoparticles for cancer treatment were proposed. Sukhorukov et al (Sukhorukov et al., 2005) developed a concept of multifunctional nanoengineered polymer capsules and outlined their applications as new DDSs or supramolecular toolboxes containing enzymes capable of converting nontoxic prodrugs into toxic drugs at a designated location. Such functionalized nanocontainers offer a wide range of applications, including enzymatic catalysis, controlled release, and directed drug delivery, in medicine because of their multifunctionality. The unique advantage of such capsules when compared with other systems is that they can be functionalized or loaded simultaneously with the earlier-mentioned components, thus permitting multifunctional processes in single cells.

Another multifunctional design of nanoparticles permits simultaneous targeting to cancer cells, chemotherapy, and tumor imaging by magnetic resonance imaging (MRI) (Nasongkla et al., 2006). Cancer-targeting capability is achieved via $\alpha_v\beta_3$ integrins targeted to alpha$_v\beta_3$-expressing tumor cells. DOX and a cluster of superparamagnetic iron oxide nanoparticles were loaded inside the micelle core for cell death induction and MRI imaging respectively. A similar targeting and imaging approach was applied to different types of nanoparticles with protein cage architectures such as virus capsids and ferritins (Uchida et al., 2006).

An original approach for fabrication of spontaneous, biologically active molecular networks has been reported (Souza et al., 2006). The assembly consists of bacteriophage (phage) directly assembled with gold nanoparticles (termed *Au-phage*). Each phage particle displays a peptide allowing the cell surface receptor binding and internalization attributes of the displayed peptide. The spontaneous organization of these targeted networks can be manipulated further by the incorporation of imidazole, which induces changes in fractal structure and near-infrared optical properties. The networks can be used as labels for enhanced fluorescence and dark-field microscopy, surface-enhanced Raman scattering detection, and near-infrared photon-to-heat conversion. Together, the physical and biological features within these targeted networks offer convenient multifunctional integration within a single entity with the potential for nanotechnology-based biomedical applications.

Cytotoxicity of anticancer nanomedicine might be maximized if drugs with different activities can be delivered simultaneously to the same cell. However, combination therapy with drugs having distinct properties such as solubility generally requires use of multiple carriers or solvents, limiting the likelihood of simultaneous delivery. Developed in the laboratory of Dr. Discher, novel biodegradable polymersomes can be successfully used for the systemic delivery of an anticancer cocktail. These polymer-based shells exploit a thick hydrophobic membrane and an aqueous lumen to efficiently carry both hydrophobic and hydrophilic drugs, paclitaxel and DOX respectively. Polymersomes are long-circulating in vivo but also degrade and release their drugs on a time scale of about 1 day. In our recent collaborative in vivo experiments we found that a single systemic injection of the dual drug combination shows a higher maximum tolerated dose than did the free drug cocktail. Additionally, it shrinks tumors more effectively and more sustainably than do free drugs: 50% smaller tumors are seen at 5 days with polymersomes. The polymersomes caused two-fold higher cell death in tumors than do free drugs and showed quantitatively similar increases in maximum tolerated dose and drug accumulation within the tumors suggesting promise for multi-drug delivery (Ahmed et al., 2006).

3.4   Liposomes and Micelles

A liposome is a spherical vesicle with a membrane composed of naturally-derived phospholipids with mixed lipid chains (such as egg phosphatidylethanolamine), or of pure surfactant components such as DOPE (dioleolylphosphatidylethanolamine). Liposomes usually contain a core of aqueous solution; lipid spheres that contain no aqueous material are called micelles. However, micelles can be made to encompass an aqueous environment forming so-called reverse micelles. The main advantages of liposomal drugs over the nonencapsulated drugs include: (1) improved pharmacokinetics and drug release, (2) enhanced intracellular penetration, (3) tumor targeting and preventing adverse side effects and (4) ability to include several active ingredients in one complex liposomal DDS (Minko et al., 2006). To prevent rapid clearance of liposomes from the bloodstream by the reticulo-endothelial system, liposomes are often masked by polymeric (mostly by PEG) covering forming STEALTH® or PEGylated liposomes (Fig. 3d). In addition to masking, such a covering provides for additional functional groups, which can be used to conjugate other active elements of DDS, such as targeting moieties. In contrast to the widely spread belief that liposomes, especially PEGylated vesicles, cannot penetrate tumor cells, we recently, using electron, confocal, and fluorescent microscopes, showed that both conventional and PEGylated liposomes are able to penetrate tumor cells both in vitro and in vivo after the systemic delivery (Pakunlu et al., 2006).

Modern methods of liposome and micelle preparation offer a reproducible production of vesicles with relatively uniform size in both traditional and PEGylated configuration. To characterize liposome structure and their size distribution we utilized atomic force microscopy (AFM), a technology with capabilities of imaging wide variety of nano- and microscale objects, including liposomes (Thomson et al., 2000; Kazakov et al., 2003; Ruozi et al., 2005), under physiological conditions.

AFM is preferable for the identification of soft surface water-containing nanoparticles to conventional electron microscopy, which has specific requirements for sample preparation such as complete removal of water from the samples and special procedures of staining to increase the contrast of image. The AFM images of the liposome preparations (Fig. 6) revealed the convex meniscus shape of the lipidic vesicles. To assess effects of PEG-modification on the structure of liposome vehicles, the images were captured in two modes: "height," which demonstrates the topography of the surface, and "dem," which works as a phase contrast and particularly suitable for object edge detection. As expected, the PEGylation of liposome surface leads to the hiding object topology (Fig. 1c and d), although individual particles maintain their convex profile. Importantly, each particle has a structure of a convex meniscus reflecting the flattening of liposomes during deposition. This effect of liposome flattening on the mica surface results in distortion of their actual linear sizes. Therefore, the value of reconstructed diameter (d) was calculated from the liposomes volume measured by AFM (V) under the assumption that liposomes

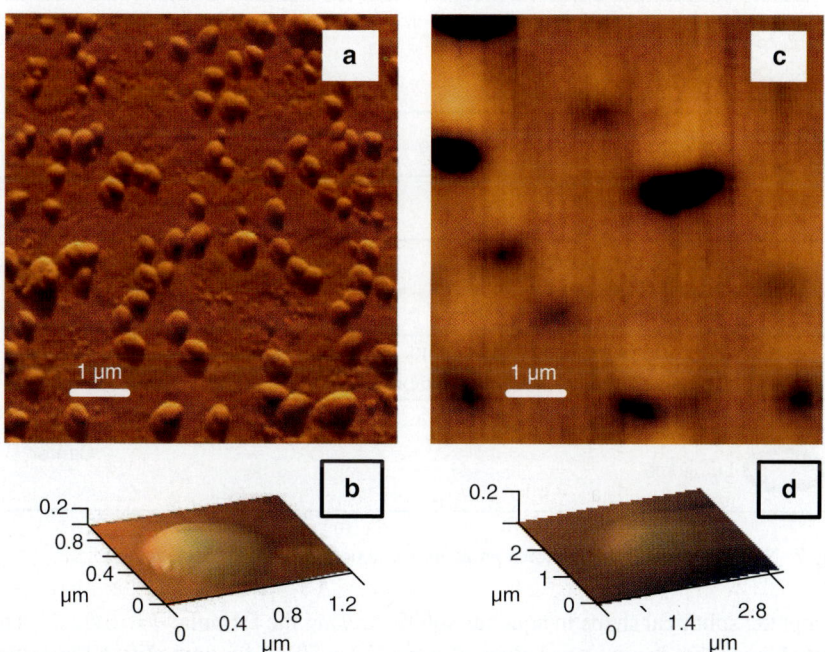

Fig. 6 Typical atomic force microscope (AFM) images of conventional (**a, b**) and PEGylated (**c, d**) liposomes. Liposome suspension was applied on freshly cleaved mica, kept for 10 min at 100% humidity atmosphere to achieve deposition and dried under the flow of nitrogen to remove external water. Images were acquired in dry air with a Nano-R AFM Pacific Nanotechnology instrument (PNI, Santa Clara, CA) in close contact (tapping) mode. Dimensional analysis has been carried out on the images obtained by AFM using Femtoscan Online v. 2.2.85 software (Advanced Technologies Center, Moscow, Russia). Representative images show panoramic view of conventional (**a**) and PEGylated (**d**) liposomes captioned in "dem" mode (phase contrast). Images (**b**) and (**d**) show the three-dimensional zoomed images in "height" mode (topography). The length of the bar is 1 μm

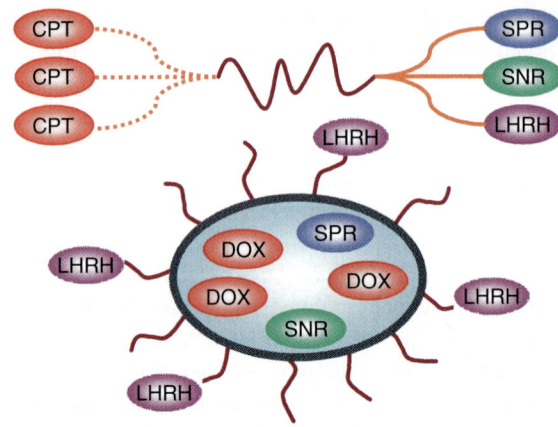

Component		Role
CPT	Camptothecin	Anticancer Drug
DOX	Doxorubicin	
LHRH	Synthetic analog of LHRH peptide	Targeting Moiety
SPR	ASO/siRNA targeted to P-glycoprotein or MRP protein	Suppressor of Pump Resistance
SNR	Synthetic BH3 peptide, ASO/siRNA targeted to BCL2 mRNA	Suppressor of Nonpump Resistance
········	Citric acid with biodegradable bond	Spacer
——	Citric acid with non-biodegradable bond	
∿	PEG polymer	Carrier
⬭	Liposome	

Fig. 7 Novel targeted multifunctional proapoptotic anticancer nanotherapeutics

accept the spherical shape in aqueous solution, using the formula: $d = (6V/\pi)^{1/3}$. The data showed that the average values of d are $177 \pm 50\,\mathrm{nm}$ for unmodified liposomes and $336 \pm 108\,\mathrm{nm}$ for PEGylated liposomes.

Currently, virtually all traditional anticancer drugs have been encapsulated in liposomes using different technologies (Szoka, 1990; Goyal et al., 2005; Hofheinz et al., 2005; Torchilin, 2005; Gabizon et al., 2006; Gabizon et al., 2006; Torchilin, 2006a,b) and many of them are included in clinical trials as cancer imaging agents or anticancer therapeutics (or both). Several liposomal formulations have recently entered different phases of clinical trials. They include liposomal DOX in patients with advanced solid tumors (Kouroussis et al., 2005), liposomal DOX and weekly docetaxel in advanced breast cancer patients (Mrozek et al., 2005), pegylated liposomal DOX and cyclophosphamide as first-line therapy for patients with metastatic or recurrent breast cancer (Overmoyer et al., 2005), liposomal cisplatin analog in patients with malignant pleural mesothelioma (Lu et al., 2005), sphingosomal vincristine in patients with recurrent or refractory adult acute lymphocytic leukemia (Thomas et al., 2006), and liposomal muramyltripeptide phosphatidylethanolamine in the treatment of osteosarcoma (Romet-Lemonne et al., 2005). In addition to the direct treatment of different cancers, liposomal preparations are also being used for tumor imaging (Zielhuis et al., 2005) gene therapy and liposome-based anticancer vaccines (Butts et al., 2005; Chen and Huang, 2005). Recent progress in liposomal technology further opens up the possibility of generating more selectively targeted photosensitizers encapsulated in liposomes for photodynamic therapy of cancer (Chen et al., 2005). Magnetite cationic liposomes (MCLs), one of the groups of cationic magnetic particles, can be used as carriers to introduce magnetite nanoparticles into target cells since their positively charged surface interacts with the negatively charged cell surface. Furthermore, they find applications to hyperthermic treatments (Ito et al., 2003, 2005; Tanaka et al., 2005). Considerable attention is devoted to the development of tumor targeted liposomes and micelles as well as DDS with stimuli-sensitive drug release.

All above-mentioned targeting approaches, which are being used for other types of anticancer nanomedicines, are successfully used for targeting liposomes and micelles (Allen et al., 2005; Brignole et al., 2005; Wang et al., 2005). The monoclonal antibody 2C5 with nucleosome-restricted specificity was studied in the laboratory of Dr. Torchilin for its ability to specifically recognize human brain tumor cells and to serve as a specific ligand for liposome targeting to brain tumor cells in vitro and in vivo (Lukyanov et al., 2004; Gupta et al., 2005). The affinity of 2C5 toward brain tumor cells was tested by flow cytometry. The interaction of 2C5-immunoliposomes with brain tumor cells in vitro was studied by fluorescence microscopy. 2C5 was found to be reactive against several tested brain tumor cell lines. It showed enhanced cell-surface binding with several brain cancer cells in vitro. 2C5-immunoliposomes displayed significantly better accumulation in the subcutaneously grown brain tumor than did nonspecific control. Therefore, 2C5 specifically recognizes brain tumor cells and can serve as a ligand to target drug carriers such as liposomes to brain tumor cells in vivo.

A set of aliphatic and aromatic aldehyde-derived HZ-based acid-sensitive polyethylene glycol–phosphatidylethanolamine (PEG–PE) conjugates was synthesized and evaluated for their hydrolytic stability at neutral and slightly acidic pH values (Kale & Torchilin, 2007). The micelles formed by aliphatic aldehyde-based PEG–HZ–PE conjugates were found to be highly sensitive to mildly acidic pH and reasonably stable at physiologic pH, while those derived from aromatic aldehydes were highly stable at both pH values. The pH-sensitive PEG–PE conjugates with controlled pH sensitivity may find applications in bilogical stimuli-mediated drug targeting for building pharmaceutical nanocarriers capable of specific release of their cargo at certain pathological sites in the body (tumors, infarcts) or intracellular compartments (endosomes, cytoplasm) demonstrating decreased pH.

4   Future Directions

Currently used pharmaceutical nanocarriers, such as liposomes, micelles, dendrimers, nanospheres, nanotubes, nanocrystals, nanogels, nanoemulsions, QDs, polymeric nanoparticles, demonstrate a broad variety of useful properties, such as longevity in the blood allowing for their accumulation in pathological areas with compromised vasculature; specific targeting to certain disease sites due to various targeting ligands attached to the surface of the nanocarriers; enhanced intracellular penetration with the help of surface-attached cell-penetrating molecules; contrast properties due to the carrier loading with various contrast materials allowing for direct carrier visualization in vivo; stimuli-sensitivity allowing for drug release from the carriers under certain physiological conditions, and others (Torchilin, 2006a). However, rare multifunctional nanosized delivery device combine several useful properties in one particle. In terms of anticancer therapeutics, the main aims of multifunctional targeted anticancer proapoptotic delivery system include delivery of payload specifically to tumor cells, effective apoptosis induction, and suppression of cellular defensive mechanisms, including pump and nonpump resistance. Several types of such therapeutics are currently being developed in our laboratory. The systems include (Fig. 6) one or several copies of each of the following: (1) nanosized carrier (PEG polymer with branched spacer or liposomes); (2) targeting moiety (LHRH peptide); (3) anticancer drug; (4) suppressor of pump resistance (ASOs, siRNA targeted to p-glycoprotein or multidrug resistance associated protein); and (5) suppressor of nonpump resistance (BH3 peptide or ASOs or siRNA targeted to BCL2 mRNA). Such delivery systems are being extensively tested in our laboratory in both in vitro and in vivo settings and allow for a multipronged attack on cancer by the simultaneous induction of cell death specifically in tumor cells and the suppression of cell defensive mechanisms (Dharap et al., 2003; Minko et al., 2003; Pakunlu et al., 2003; Minko, 2004; Minko et al., 2004; Pakunlu et al., 2004; Dharap et al., 2005; Dharap et al., 2006; Khandare et al., 2006a; Khandare et al., 2006b; Pakunlu et al., 2006).

Acknowledgements This work was supported in part by National Institutes of Health Grants CA100098, CA111766, and CA074175 from the National Cancer Institute and by American Lung Association.

References

Ahmed, F., Pakunlu, R. I., Brannan, A., Bates, F., Minko, T. and Discher, D. E. 2006. Biodegradable polymersomes loaded with both paclitaxel and doxorubicin permeate and shrink tumors, inducing apoptosis in proportion to accumulated drug. J Contr Rel 116:150–8.

Allen, T. M., Mumbengegwi, D. R. and Charrois, G. J. 2005. Anti-CD19-targeted liposomal doxorubicin improves the therapeutic efficacy in murine B-cell lymphoma and ameliorates the toxicity of liposomes with varying drug release rates. Clin Cancer Res 11:3567–73.

Amirkhanov, N. V. and Wickstrom, E. 2005. Synthesis of novel polydiamidopropanoate dendrimer PNA-peptide chimeras for non-invasive magnetic resonance imaging of cancer. Nucleos Nucleot Nucleic Acids 24:423–6.

Arayne, M. S. and Sultana, N. 2006. Review: nanoparticles in drug delivery for the treatment of cancer. Pak J Pharm Sci 19:258–68.

Brignole, C., Marimpietri, D., Pagnan, G., Di Paolo, D., Zancolli, M., Pistoia, V., Ponzoni, M. and Pastorino, F. 2005. Neuroblastoma targeting by c-myb-selective antisense oligonucleotides entrapped in anti-GD2 immunoliposome: immune cell-mediated anti-tumor activities. Cancer Lett 228:181–6.

Butts, C., Murray, N., Maksymiuk, A., Goss, G., Marshall, E., Soulieres, D., Cormier, Y., Ellis, P., Price, A., Sawhney, R., Davis, M., Mansi, J., Smith, C., Vergidis, D., MacNeil, M. and Palmer, M. 2005. Randomized phase IIB trial of BLP25 liposome vaccine in stage IIIB and IV non-small-cell lung cancer. J Clin Oncol 23:6674–81.

Cai, W., Shin, D. W., Chen, K., Gheysens, O., Cao, Q., Wang, S. X., Gambhir, S. S. and Chen, X. 2006. Peptide-labeled near-infrared quantum dots for imaging tumor vasculature in living subjects. Nano Lett 6:669–76.

Caliceti, P., Schiavon, O. and Veronese, F. M. 2001. Immunological properties of uricase conjugated to neutral soluble polymers. Bioconjug Chem 12:515–22.

Chan, P., Kurisawa, M., Chung, J. E. and Yang, Y. Y. 2007. Synthesis and characterization of chitosan-g-poly(ethylene glycol)-folate as a non-viral carrier for tumor-targeted gene delivery. Biomaterials 28:540–9.

Chan, W. H., Shiao, N. H. and Lu, P. Z. 2006. CdSe quantum dots induce apoptosis in human neuroblastoma cells via mitochondrial-dependent pathways and inhibition of survival signals. Toxicol Lett 167:191–200.

Chang, L. C., Lee, H. F., Chung, M. J. and Yang, V. C. 2005. PEG-modified Protamine with improved pharmacological/pharmaceutical properties as a potential protamine substitute: synthesis and in vitro evaluation. Bioconjug Chem 16:147–55.

Chen, B., Pogue, B. W. and Hasan, T. 2005. Liposomal delivery of photosensitising agents. Expert Opin Drug Deliv 2:477–87.

Chen, W. and Zhang, J. 2006. Using nanoparticles to enable simultaneous radiation and photodynamic therapies for cancer treatment. J Nanosci Nanotechnol 6:1159–66.

Chen, W. C. and Huang, L. 2005. Non-viral vector as vaccine carrier. Adv Genet 54:315–37.

Cho, S. J., Maysinger, D., Jain, M., Roder, B., Hackbarth, S. and Winnik, F. M. 2007. Long-term exposure to CdTe quantum dots causes functional impairments in live cells. Langmuir 23:1974–80.

Choe, Y. H., Greenwald, R. B., Conover, C. D., Zhao, H., Longley, C. B., Guan, S., Zhao, Q. and Xia, J. 2004. PEG prodrugs of 6-mercaptopurine for parenteral administration using benzyl elimination of thiols. Oncol Res 14:455–68.

Choi, Y. and Baker, J. R., Jr. 2005. Targeting cancer cells with DNA-assembled dendrimers: a mix and match strategy for cancer. Cell Cycle 4:669–71.

Chung, Y. and Cho, H. 2004. Preparation of highly water soluble tacrolimus derivatives: poly(ethylene glycol) esters as potential prodrugs. Arch Pharm Res 27:878–83.

Corot, C., Robert, P., Idee, J. M. and Port, M. 2006. Recent advances in iron oxide nanocrystal technology for medical imaging. Adv Drug Deliv Rev 58:1471–504.

Cosulich, S. C., Worrall, V., Hedge, P. J., Green, S. and Clarke, P. R. 1997. Regulation of apoptosis by BH3 domains in a cell-free system. Curr Biol 7:913–20.

David, A., Kopeckova, P., Minko, T., Rubinstein, A. and Kopecek, J. 2004. Design of a multivalent galactoside ligand for selective targeting of HPMA copolymer–doxorubicin conjugates to human colon cancer cells. Eur J Cancer 40:148–57.

Demoy, M., Minko, T., Kopeckova, P. and Kopecek, J. 2000. Time- and concentration-dependent apoptosis and necrosis induced by free and HPMA copolymer-bound doxorubicin in human ovarian carcinoma cells. J Contr Rel 69:185–96.

Dharap, S. S. and Minko, T. 2003. Targeted proapoptotic LHRH-BH3 peptide. Pharm Res 20:889–96.

Dharap, S. S., Qiu, B., Williams, G. C., Sinko, P., Stein, S. and Minko, T. 2003. Molecular targeting of drug delivery systems to ovarian cancer by BH3 and LHRH peptides. J Contr Rel 91:61–73.

Dharap, S. S., Wang, Y., Chandna, P., Khandare, J. J., Qiu, B., Gunaseelan, S., Sinko, P. J., Stein, S., Farmanfarmaian, A. and Minko, T. 2005. Tumor-specific targeting of an anticancer drug delivery system by LHRH peptide. Proc Natl Acad Sci U S A 102:12962–7.

Dharap, S. S., Chandna, P., Wang, Y., Khandare, J. J., Qiu, B., Stein, S. and Minko, T. 2006. Molecular targeting of BCL2 and BCLXL proteins by synthetic BH3 peptide enhances the efficacy of chemotherapy. J Pharmacol Exp Ther 316:992–98.

Duncan, R. 2007. Designing polymer conjugates as lysosomotropic nanomedicines. Biochem Soc Trans 35:56–60.

Duncan, R., Gac-Breton, S., Keane, R., Musila, R., Sat, Y. N., Satchi, R. and Searle, F. 2001. Polymer–drug conjugates, PDEPT and PELT: basic principles for design and transfer from the laboratory to clinic. J Contr Rel 74:135–46.

Fang, J., Sawa, T. and Maeda, H. 2003. Factors and mechanism of "EPR" effect and the enhanced antitumor effects of macromolecular drugs including SMANCS. Adv Exp Med Biol 519:29–49.

Faraday, M. 1857. Experimental relations of gold (and other metals) to light. Philos Trans R Soc London 14:145–81.

Funhoff, A. M., van Nostrum, C. F., Janssen, A. P., Fens, M. H., Crommelin, D. J. and Hennink, W. E. 2004. Polymer side-chain degradation as a tool to control the destabilization of polyplexes. Pharm Res 21:170–6.

Gabizon, A. A., Shmeeda, H. and Zalipsky, S. 2006. Pros and cons of the liposome platform in cancer drug targeting. J Liposome Res 16:175–83.

Gabizon, A. A., Tzemach, D., Horowitz, A. T., Shmeeda, H., Yeh, J. and Zalipsky, S. 2006. Reduced toxicity and superior therapeutic activity of a mitomycin C lipid-based prodrug incorporated in pegylated liposomes. Clin Cancer Res 12:1913–20.

Gao, X., Chung, L. W. and Nie, S. 2007. Quantum dots for in vivo molecular and cellular imaging. Methods Mol Biol 374:135–46.

Geng, Y., Dalhaimer, P., Cai, S., Tsai, R., Tewari, M., Minko, T. and Discher, D. 2007. Soft filaments circulate longer than spherical particles – shape effects in flow and drug delivery. Nat Nanotech 2:249–55.

Gillies, E. R., Dy, E., Frechet, J. M. and Szoka, F. C. 2005. Biological evaluation of polyester dendrimer: poly(ethylene oxide) "bow-tie" hybrids with tunable molecular weight and architecture. Mol Pharm 2:129–38.

Goyal, P., Goyal, K., Vijaya Kumar, S. G., Singh, A., Katare, O. P. and Mishra, D. N. 2005. Liposomal drug delivery systems – clinical applications. Acta Pharm 55:1–25.

Greenwald, R. B. 2001. PEG drugs: an overview. J Contr Rel 74:159–71.

Greenwald, R. B., Zhao, H. and Xia, J. 2003. Tripartate poly(ethylene glycol) prodrugs of the open lactone form of camptothecin. Bioorg Med Chem 11:2635–9.

Greenwald, R. B., Zhao, H., Xia, J., Wu, D., Nervi, S., Stinson, S. F., Majerova, E., Bramhall, C. and Zaharevitz, D. W. 2004. Poly(ethylene glycol) prodrugs of the CDK inhibitor, alsterpaullone (NSC 705701): synthesis and pharmacokinetic studies. Bioconjug Chem 15:1076–83.

Greish, K., Fang, J., Inutsuka, T., Nagamitsu, A. and Maeda, H. 2003. Macromolecular therapeutics: advantages and prospects with special emphasis on solid tumour targeting. Clin Pharmacokinet 42:1089–105.

Grodzinski, P., Silver, M. and Molnar, L. K. 2006. Nanotechnology for cancer diagnostics: promises and challenges. Expert Rev Mol Diagn 6:307–18.

Gunaseelan, S., Debrah, O., Wan, L., Leibowitz, M. J., Rabson, A. B., Stein, S. and Sinko, P. J. 2004. Synthesis of poly(ethylene glycol)-based saquinavir prodrug conjugates and assessment of release and anti-HIV-1 bioactivity using a novel protease inhibition assay. Bioconjug Chem 15:1322–33.

Gupta, B., Levchenko, T. S., Mongayt, D. A. and Torchilin, V. P. 2005. Monoclonal antibody 2C5-mediated binding of liposomes to brain tumor cells in vitro and in subcutaneous tumor model in vivo. J Drug Target 13:337–43.

Hama, Y., Koyama, Y., Urano, Y., Choyke, P. L. and Kobayashi, H. 2007. Simultaneous two-color spectral fluorescence lymphangiography with near infrared quantum dots to map two lymphatic flows from the breast and the upper extremity. Breast Cancer Res Treat 103:23–8.

Hillaireau, H., Le Doan, T. and Couvreur, P. 2006. Polymer-based nanoparticles for the delivery of nucleoside analogues. J Nanosci Nanotechnol 6:2608–17.

Hofheinz, R. D., Gnad-Vogt, S. U., Beyer, U. and Hochhaus, A. 2005. Liposomal encapsulated anti-cancer drugs. Anticanc Drugs 16:691–707.

Holinger, E. P., Chittenden, T. and Lutz, R. J. 1999. Bak BH3 peptides antagonize Bcl-xL function and induce apoptosis through cytochrome c-independent activation of caspases. J Biol Chem 274:13298–304.

Hong, S., Leroueil, P. R., Majoros, I. J., Orr, B. G., Baker, J. R., Jr. and Banaszak Holl, M. M. 2007. The binding avidity of a nanoparticle-based multivalent targeted drug delivery platform. Chem Biol 14:107–15.

Islam, M. T., Majoros, I. J. and Baker, J. R., Jr. 2005. HPLC analysis of PAMAM dendrimer based multifunctional devices. J Chromatogr B Analyt Technol Biomed Life Sci 822:21–6.

Ito, A., Tanaka, K., Honda, H., Abe, S., Yamaguchi, H. and Kobayashi, T. 2003. Complete regression of mouse mammary carcinoma with a size greater than 15 mm by frequent repeated hyperthermia using magnetite nanoparticles. J Biosci Bioeng 96:364–9.

Ito, A., Shinkai, M., Honda, H. and Kobayashi, T. 2005. Medical application of functionalized magnetic nanoparticles. J Biosci Bioeng 100:1–11.

Jackson, H., Muhammad, O., Daneshvar, H., Nelms, J., Popescu, A., Vogelbaum, M. A., Bruchez, M. and Toms, S. A. 2007. Quantum dots are phagocytized by macrophages and colocalize with experimental gliomas. Neurosurgery 60:524–9; discussion 529–30.

Kale, A. A. and Torchilin, V. P. 2007. Design, synthesis, and characterization of pH-sensitive PEG–PE conjugates for stimuli-sensitive pharmaceutical nanocarriers: the effect of substitutes at the hydrazone linkage on the ph stability of PEG-PE conjugates. Bioconjug Chem 18:363–70.

Kasuya, Y., Lu, Z. R., Kopeckova, P., Minko, T., Tabibi, S. E. and Kopecek, J. 2001. Synthesis and characterization of HPMA copolymer–aminopropylgeldanamycin conjugates. J Contr Rel 74:203–11.

Kazakov, S., Kaholek, M., Kudasheva, D., Teraoka, I., Cowman, M. K. and Levon, K. 2003. Poly(N-isopropylacrylamide-co-1-vinylimidazole) hydrogel nanoparticles prepared and hydrophobically modified in liposome reactors: Atomic force microscopy and dynamic light scattering study. Langmuir 19:8086–8093.

Khan, M. K., Nigavekar, S. S., Minc, L. D., Kariapper, M. S., Nair, B. M., Lesniak, W. G. and Balogh, L. P. 2005. In vivo biodistribution of dendrimers and dendrimer nanocomposites – implications for cancer imaging and therapy. Technol Cancer Res Treat 4:603–13.

Khandare, J. and Minko, T. 2006. Polymer–drug conjugates: progress in polymeric prodrugs. Progr Polym Sci 31:359–97.

Khandare, J., Kolhe, P., Pillai, O., Kannan, S., Lieh-Lai, M. and Kannan, R. M. 2005. Synthesis, cellular transport, and activity of polyamidoamine dendrimer–methylprednisolone conjugates. Bioconjug Chem 16:330–7.

Khandare, J. J., Chandna, P., Wang, Y., Pozharov, V. P. and Minko, T. 2006a. Novel polymeric prodrug with multivalent components for cancer therapy. J Pharmacol Exp Ther 317:929–37.

Khandare, J. J., Jayant, S., Singh, A., Chandna, P., Wang, Y., Vorsa, N. and Minko, T. 2006b. Dendrimer versus linear conjugate: influence of polymeric architecture on the delivery and anticancer effect of paclitaxel. Bioconjug Chem 17:1464–72.

Kobayashi, H., Kawamoto, S., Bernardo, M., Brechbiel, M. W., Knopp, M. V. and Choyke, P. L. 2006. Delivery of gadolinium-labeled nanoparticles to the sentinel lymph node: comparison of the sentinel node visualization and estimations of intra-nodal gadolinium concentration by the magnetic resonance imaging. J Contr Rel 111:343–51.

Kolhe, P., Khandare, J., Pillai, O., Kannan, S., Lieh-Lai, M. and Kannan, R. 2004. Hyperbranched polymer–drug conjugates with high drug payload for enhanced cellular delivery. Pharm Res 21:2185–95.

Kopecek, J., Kopeckova, P., Minko, T., Lu, Z. R. and Peterson, C. M. 2001. Water soluble polymers in tumor targeted delivery. J Contr Rel 74:147–58.

Kouroussis, C., Androulakis, N., Vamvakas, L., Kalykaki, A., Spiridonakou, S., Kentepozidis, N., Saridaki, Z., Xiropoulou, E. and Georgoulias, V. 2005. Phase I study of weekly docetaxel and liposomal doxorubicin in patients with advanced solid tumors. Oncology 69:202–7.

Kovar, M., Kovar, L., Subr, V., Etrych, T., Ulbrich, K., Mrkvan, T., Loucka, J. and Rihova, B. 2004. HPMA copolymers containing doxorubicin bound by a proteolytically or hydrolytically cleavable bond: comparison of biological properties in vitro. J Contr Rel 99:301–14.

Kunath, K., Kopeckova, P., Minko, T. and Kopecek, J. 2000. HPMA copolymer-anticancer drug-OV-TL16 antibody conjugates. 3. The effect of free and polymer-bound adriamycin on the expression of some genes in the OVCAR-3 human ovarian carcinoma cell line. Eur J Pharm Biopharm 49:11–15.

Lammers, T., Kuhnlein, R., Kissel, M., Subr, V., Etrych, T., Pola, R., Pechar, M., Ulbrich, K., Storm, G., Huber, P. and Peschke, P. 2005. Effect of physicochemical modification on the biodistribution and tumor accumulation of HPMA copolymers. J Contr Rel 103–18.

Langer, M. and Beck-Sickinger, A. G. 2001. Peptides as carrier for tumor diagnosis and treatment. Curr Med Chem Antican Agents 1:71–93.

Langer, R. 2001. Drug delivery. Drugs on target. Science 293:58–9.

Langereis, S., de Lussanet, Q. G., van Genderen, M. H., Meijer, E. W., Beets-Tan, R. G., Griffioen, A. W., van Engelshoven, J. M. and Backes, W. H. 2006. Evaluation of Gd(III)DTPA-terminated poly(propylene imine) dendrimers as contrast agents for MR imaging. NMR Biomed 19:133–41.

Lee, C. C., Gillies, E. R., Fox, M. E., Guillaudeu, S. J., Frechet, J. M., Dy, E. E. and Szoka, F. C. 2006. A single dose of doxorubicin-functionalized bow-tie dendrimer cures mice bearing C-26 colon carcinomas. Proc Natl Acad Sci U S A 103:16649–54.

Li, Z., Wang, K., Tan, W., Li, J., Fu, Z., Ma, C., Li, H., He, X. and Liu, J. 2006. Immunofluorescent labeling of cancer cells with quantum dots synthesized in aqueous solution. Anal Biochem 354:169–74.

Liu, S., Lee, C. M., Wang, S. and Lu, D. R. 2006. A new bioimaging carrier for fluorescent quantum dots: phospholipid nanoemulsion mimicking natural lipoprotein core. Drug Deliv 13:159–64.

Ljubimova, J. Y., Fujita, M., Khazenzon, N. M., Lee, B. S., Wachsmann-Hogiu, S., Farkas, D. L., Black, K. L. and Holler, E. 2007. Nanoconjugate based on polymalic acid for tumor targeting. Chem Biol Interact Feb 8 [Epub ahead of print].

Lu, C., Perez-Soler, R., Piperdi, B., Walsh, G. L., Swisher, S. G., Smythe, W. R., Shin, H. J., Ro, J. Y., Feng, L., Truong, M., Yalamanchili, A., Lopez-Berestein, G., Hong, W. K., Khokhar, A. R. and Shin, D. M. 2005. Phase II study of a liposome-entrapped cisplatin analog (L-NDDP) administered intrapleurally and pathologic response rates in patients with malignant pleural mesothelioma. J Clin Oncol 23:3495–501.

Lu, Z. R., Shiah, J. G., Sakuma, S., Kopeckova, P. and Kopecek, J. 2002. Design of novel bioconjugates for targeted drug delivery. J Contr Rel 78:165–73.

Lukyanov, A. N., Elbayoumi, T. A., Chakilam, A. R. and Torchilin, V. P. 2004. Tumor-targeted liposomes: doxorubicin-loaded long-circulating liposomes modified with anti-cancer antibody. J Contr Rel 100:135–44.

Luten, J., Akeroyd, N., Funhoff, A., Lok, M. C., Talsma, H. and Hennink, W. E. 2006. Methacrylamide polymers with hydrolysis-sensitive cationic side groups as degradable gene carriers. Bioconjug Chem 17:1077–84.

Lutz, R. J. 2000. Role of the BH3 (Bcl-2 homology 3) domain in the regulation of apoptosis and Bcl-2-related proteins. Biochem Soc Trans 28:51–6.

Maeda, H. 2001. The enhanced permeability and retention (EPR) effect in tumor vasculature: the key role of tumor-selective macromolecular drug targeting. Adv Enzyme Regul 41:189–207.

Majoros, I. J., Myc, A., Thomas, T., Mehta, C. B. and Baker, J. R., Jr. 2006. PAMAM dendrimer-based multifunctional conjugate for cancer therapy: synthesis, characterization, and functionality. Biomacromolecules 7:572–9.

Matsumura, Y. and Maeda, H. 1986. A new concept for macromolecular therapeutics in cancer chemotherapy: mechanism of tumoritropic accumulation of proteins and the antitumor agent smancs. Cancer Res 46:6387–92.

McNeil, S. E. 2005. Nanotechnology for the biologist. J Leukoc Biol 78:585–94.

Mehvar, R., Dann, R. O. and Hoganson, D. A. 2000. Kinetics of hydrolysis of dextran-methylprednisolone succinate, a macromolecular prodrug of methylprednisolone, in rat blood and liver lysosomes. J Contr Rel 68:53–61.

Minko, T. 2004. Drug targeting to the colon with lectins and neoglycoconjugates. Adv Drug Deliv Rev 56:491–509.

Minko, T., Kopeckova, P., Pozharov, V. and Kopecek, J. 1998. HPMA copolymer bound adriamycin overcomes MDRI gene encoded resistance in a human ovarian carcinoma cell line. J Contr Rel 54:223–33.

Minko, T., Kopeckova, P. and Kopecek, J. 1999a. Chronic exposure to HPMA copolymer-bound adriamycin does not induce multidrug resistance in a human ovarian carcinoma cell line. J Contr Rel 59:133–48.

Minko, T., Kopeckova, P. and Kopecek, J. 1999b. Comparison of the anticancer effect of free and HPMA copolymer-bound adriamycin in human ovarian carcinoma cells. Pharm Res 16:986–96.

Minko, T., Kopeckova, P., Pozharov, V., Jensen, K. D. and Kopecek, J. 2000. The influence of cytotoxicity of macromolecules and of VEGF gene modulated vascular permeability on the enhanced permeability and retention effect in resistant solid tumors. Pharm Res 17:505–14.

Minko, T., Kopeckova, P. and Kopecek, J. 2001. Preliminary evaluation of caspases-dependent apoptosis signaling pathways of free and HPMA copolymer-bound doxorubicin in human ovarian carcinoma cells. J Contr Rel 71:227–37.

Minko, T., Paranjpe, P. V., Qiu, B., Lalloo, A., Won, R., Stein, S. and Sinko, P. J. 2002. Enhancing the anticancer efficacy of camptothecin using biotinylated poly(ethylene glycol) conjugates in sensitive and multidrug-resistant human ovarian carcinoma cells. Cancer Chemother Pharmacol 50:143–50.

Minko, T., Dharap, S. S. and Fabbricatore, A. T. 2003. Enhancing the efficacy of chemotherapeutic drugs by the suppression of antiapoptotic cellular defense. Cancer Detect Prev 27:193–202.

Minko, T., Dharap, S. S., Pakunlu, R. I. and Wang, Y. 2004. Molecular targeting of drug delivery systems to cancer. Curr Drug Targets 5:389–406.

Minko, T., Pakunlu, R. I., Wang, Y., Khandare, J. J. and Saad, M. 2006. New generation of liposomal drugs for cancer. Anticanc Agents Med Chem 6:537–52.

Mitra, A., Nan, A., Line, B. R. and Ghandehari, H. 2006. Nanocarriers for nuclear imaging and radiotherapy of cancer. Curr Pharm Des 12:4729–49.

Moghimi, S. M. 2006. Recent developments in polymeric nanoparticle engineering and their applications in experimental and clinical oncology. Anticanc Agents Med Chem 6:553–561.

Mora, J. R., Knoll, J. H., Rogan, P. K., Getts, R. C. and Wilson, G. S. 2006. Dendrimer FISH detection of single-copy intervals in acute promyelocytic leukemia. Mol Cell Probes 20:114–20.

Mrozek, E., Rhoades, C. A., Allen, J., Hade, E. M. and Shapiro, C. L. 2005. Phase I trial of liposomal encapsulated doxorubicin (Myocet; D-99) and weekly docetaxel in advanced breast cancer patients. Ann Oncol 16:1087–93.

Myc, A., Majoros, I. J., Thomas, T. P. and Baker, J. R., Jr. 2007. Dendrimer-based targeted delivery of an apoptotic sensor in cancer cells. Biomacromolecules 8:13–8.

Naik, S. S., Liang, J. F., Park, Y. J., Lee, W. K. and Yang, V. C. 2005. Application of "ATTEMPTS" for drug delivery. J Contr Rel 101:35–45.

Nasongkla, N., Bey, E., Ren, J., Ai, H., Khemtong, C., Guthi, J. S., Chin, S. F., Sherry, A. D., Boothman, D. A. and Gao, J. 2006. Multifunctional polymeric micelles as cancer-targeted, MRI-ultrasensitive drug delivery systems. Nano Lett 6:2427–30.

Overmoyer, B., Silverman, P., Holder, L. W., Tripathy, D. and Henderson, I. C. 2005. Pegylated liposomal doxorubicin and cyclophosphamide as first-line therapy for patients with metastatic or recurrent breast cancer. Clin Breast Cancer 6:150–7.

Pakunlu, R. I., Cook, T. J. and Minko, T. 2003. Simultaneous modulation of multidrug resistance and antiapoptotic cellular defense by MDR1 and BCL-2 targeted antisense oligonucleotides enhances the anticancer efficacy of doxorubicin. Pharm Res 20:351–9.

Pakunlu, R. I., Wang, Y., Tsao, W., Pozharov, V., Cook, T. J. and Minko, T. 2004. Enhancement of the efficacy of chemotherapy for lung cancer by simultaneous suppression of multidrug resistance and antiapoptotic cellular defense: novel multicomponent delivery system. Cancer. Res. 64:6214–24.

Pakunlu, R. I., Wang, Y., Saad, M., Khandare, J. J., Starovoytov, V. and Minko, T. 2006. In vitro and in vivo intracellular liposomal delivery of antisense oligonucleotides and anticancer drug. J Contr Rel 114:153–62.

Papagiannaros, A., Dimas, K., Papaioannou, G. T. and Demetzos, C. 2005. Doxorubicin–PAMAM dendrimer complex attached to liposomes: cytotoxic studies against human cancer cell lines. Int J Pharm 302:29–38.

Paranjpe, P. V., Chen, Y., Kholodovych, V., Welsh, W., Stein, S. and Sinko, P. J. 2004. Tumor-targeted bioconjugate based delivery of camptothecin: design, synthesis and in vitro evaluation. J Contr Rel 100:275–92.

Peterson, C. M., Shiah, J. G., Sun, Y., Kopeckova, P., Minko, T., Straight, R. C. and Kopecek, J. 2003. HPMA copolymer delivery of chemotherapy and photodynamic therapy in ovarian cancer. Adv Exp Med Biol 519:101–23.

Popescu, M. A. and Toms, S. A. 2006. In vivo optical imaging using quantum dots for the management of brain tumors. Expert Rev Mol Diagn 6:879–90.

Portney, N. G. and Ozkan, M. 2006. Nano-oncology: drug delivery, imaging, and sensing. Anal Bioanal Chem 384:620–30.

Reubi, J. C. 2003. Peptide receptors as molecular targets for cancer diagnosis and therapy. Endocr Rev 24:389–427.

Romet-Lemonne, J. L., Mills, B., Fridman, W. H. and Munsell, M. 2005. Prospectively planned analysis of data from a phase III study of liposomal muramyltripeptide phosphatidyleth-anolamine in the treatment of osteosarcoma. J Clin Oncol 23:6437–8.

Ruozi, B., Tosi, G., Forni, F., Fresta, M. and Vandelli, M. A. 2005. Atomic force microscopy and photon correlation spectroscopy: Two techniques for rapid characterization of liposomes. Eur J Pharmaceut Sci 25:81–9.

Shi, X., Majoros, I. J., Patri, A. K., Bi, X., Islam, M. T., Desai, A., Ganser, T. R. and Baker, J. R., Jr. 2006. Molecular heterogeneity analysis of poly(amidoamine) dendrimer-based mono- and multifunctional nanodevices by capillary electrophoresis. Analyst 131:374–81.

Sinha, R., Kim, G. J., Nie, S. and Shin, D. M. 2006. Nanotechnology in cancer therapeutics: bio-conjugated nanoparticles for drug delivery. Mol Cancer Ther 5:1909–17.

Souza, G. R., Christianson, D. R., Staquicini, F. I., Ozawa, M. G., Snyder, E. Y., Sidman, R. L., Miller, J. H., Arap, W. and Pasqualini, R. 2006. Networks of gold nanoparticles and bacteriophage as biological sensors and cell-targeting agents. Proc Natl Acad Sci U S A 103: 1215–20.

Sukhorukov, G. B., Rogach, A. L., Zebli, B., Liedl, T., Skirtach, A. G., Kohler, K., Antipov, A. A., Gaponik, N., Susha, A. S., Winterhalter, M. and Parak, W. J. 2005. Nanoengineered polymer capsules: tools for detection, controlled delivery, and site-specific manipulation. Small 1:194–200.

Szoka, F. C., Jr. 1990. The future of liposomal drug delivery. Biotechnol Appl Biochem 12:496–500.

Tada, H., Higuchi, H., Wanatabe, T. M. and Ohuchi, N. 2007. In vivo real-time tracking of single quantum dots conjugated with monoclonal anti-HER2 antibody in tumors of mice. Cancer Res 67:1138–44.

Tanaka, K., Ito, A., Kobayashi, T., Kawamura, T., Shimada, S., Matsumoto, K., Saida, T. and Honda, H. 2005. Heat immunotherapy using magnetic nanoparticles and dendritic cells for T-lymphoma. J Biosci Bioeng 100:112–5.

Thomas, D. A., Sarris, A. H., Cortes, J., Faderl, S., O'Brien, S., Giles, F. J., Garcia-Manero, G., Rodriguez, M. A., Cabanillas, F. and Kantarjian, H. 2006. Phase II study of sphingosomal vincristine in patients with recurrent or refractory adult acute lymphocytic leukemia. Cancer 106:120–7.

Thomson, N. H., Collin, I., Davies, M. C., Palin, K., Parkins, D., Roberts, C. J., Tendler, S. J. B. and Williams, P. M. 2000. Atomic force microscopy of cationic liposomes. Langmuir 16:4813–18.

Tomalia, D. A., Reyna, L. A. and Svenson, S. 2007. Dendrimers as multi-purpose nanodevices for oncology drug delivery and diagnostic imaging. Biochem Soc Trans 35:61–7.

Torchilin, V. P. 2000. Drug targeting. Eur J Pharm Sci 11 Suppl 2:S81–91.

Torchilin, V. P. 2002. PEG-based micelles as carriers of contrast agents for different imaging modalities. Adv Drug Deliv Rev 54:235–52.

Torchilin, V. P. 2005. Recent advances with liposomes as pharmaceutical carriers. Nat Rev Drug Discov 4:145–60.

Torchilin, V. P. 2006a. Multifunctional nanocarriers. Adv Drug Deliv Rev 58:1532–55.

Torchilin, V. P. 2006b. Recent approaches to intracellular delivery of drugs and DNA and organelle targeting. Annu Rev Biomed Eng 8:343–75.

Torchilin, V. P. 2007. Micellar nanocarriers: pharmaceutical perspectives. Pharm Res 24:1–16.

True, L. D. and Gao, X. 2007. Quantum dots for molecular pathology: their time has arrived. J Mol Diagn 9:7–11.

Uchida, M., Flenniken, M. L., Allen, M., Willits, D. A., Crowley, B. E., Brumfield, S., Willis, A. F., Jackiw, L., Jutila, M., Young, M. J. and Douglas, T. 2006. Targeting of cancer cells with ferrimagnetic ferritin cage nanoparticles. J Am Chem Soc 128:16626–33.

Ulbrich, K., Etrych, T., Chytil, P., Jelinkova, M. and Rihova, B. 2004a. Antibody-targeted polymer–doxorubicin conjugates with pH-controlled activation. J Drug Target 12:477–89.

Ulbrich, K., Etrych, T., Chytil, P., Pechar, M., Jelinkova, M. and Rihova, B. 2004b. Polymeric anticancer drugs with pH-controlled activation. Int J Pharm 277:63–72.

Van der Schyf, C. J., Geldenhuys, W. J. and Youdim, M. B. 2006. Multifunctional drugs with different CNS targets for neuropsychiatric disorders. J Neurochem 99:1033–48.

van Vlerken, L. E. and Amiji, M. M. 2006. Multi-functional polymeric nanoparticles for tumour-targeted drug delivery. Expert Opin Drug Deliv 3:205–16.

Wang, D., Kopeckova, J. P., Minko, T., Nanayakkara, V. and Kopecek, J. 2000. Synthesis of starlike N-(2-hydroxypropyl)methacrylamide copolymers: potential drug carriers. Biomacromolecules 1:313–19.

Wang, Q. W., Lu, H. L., Song, C. C., Liu, H. and Xu, C. G. 2005. Radiosensitivity of human colon cancer cell enhanced by immunoliposomal docetaxel. World J Gastroenterol 11:4003–7.

Xue, F. L., Chen, J. Y., Guo, J., Wang, C. C., Yang, W. L., Wang, P. N. and Lu, D. R. 2007. Enhancement of intracellular delivery of CdTe quantum dots (QDs) to living cells by TAT conjugation. J Fluoresc 17:149–54.

Young, S. H. and Rozengurt, E. 2006. Qdot nanocrystal conjugates conjugated to bombesin or ANG II label the cognate G protein-coupled receptor in living cells. Am J Physiol Cell Physiol 290:C728–32.

Yu, W. W., Chang, E., Falkner, J. C., Zhang, J., Al-Somali, A. M., Sayes, C. M., Johns, J., Drezek, R. and Colvin, V. L. 2007a. Forming biocompatible and nonaggregated nanocrystals in water using amphiphilic polymers. J Am Chem Soc 129:2871–9.

Yu, X., Chen, L., Li, K., Li, Y., Xiao, S., Luo, X., Liu, J., Zhou, L., Deng, Y., Pang, D. and Wang, Q. 2007b. Immunofluorescence detection with quantum dot bioconjugates for hepatoma in vivo. J Biomed Opt 12:014008.

Zarabi, B., Nan, A., Zhuo, J., Gullapalli, R. and Ghandehari, H. 2006. Macrophage targeted N-(2-hydroxypropyl)methacrylamide conjugates for magnetic resonance imaging. Mol Pharm 3:550–7.

Zhang, C., Pei, J., Kumar, D., Sakabe, I., Boudreau, H. E., Gokhale, P. C. and Kasid, U. N. 2007. Antisense oligonucleotides: target validation and development of systemically delivered therapeutic nanoparticles. Methods Mol Biol 361:163–85.

Zhang, Y., So, M. K. and Rao, J. 2006. Protease-modulated cellular uptake of quantum dots. Nano Lett 6:1988–92.

Zhang, Z., Tanabe, K., Hatta, H. and Nishimoto, S. 2005. Bioreduction activated prodrugs of camptothecin: molecular design, synthesis, activation mechanism and hypoxia selective cytotoxicity. Org Biomol Chem 3:1905–10.

Zhao, H., Lee, C., Sai, P., Choe, Y. H., Boro, M., Pendri, A., Guan, S. and Greenwald, R. B. 2000. 20-*O*-acylcamptothecin derivatives: evidence for lactone stabilization. J Org Chem 65:4601–6.

Zielhuis, S. W., Nijsen, J. F., Seppenwoolde, J. H., Zonnenberg, B. A., Bakker, C. J., Hennink, W. E., van Rijk, P. P. and van het Schip, A. D. 2005. Lanthanide bearing microparticulate systems for multi-modality imaging and targeted therapy of cancer. Curr Med Chem Anti-Canc Agents 5:303–13.

Charge Modification of Pharmaceutical Nanocarriers: Biological Implications

N. Nafee, M. Schneider, and C.-M. Lehr

1 Introduction

Multifunctional particles may be necessary to target specific locations and overcome different barriers hampering the delivery of drugs using nanoparticulate matter (Fenart et al. 1999; Kreuter 2004; Lockman et al. 2004). The interaction taking place is based on several properties that are essential for the relation between particle and biological system. For instance, the multifunctional particles will allow to target a certain tissue, deliver an active component, and provide a marker to observe the successful delivery (McNeil 2005).

In this chapter we will focus on the charge as one of the most basic and crucial properties influencing the environmental interaction of nanoparticulate matter. The surface charge of the nanoscale particles is a standard parameter to be characterized and reported. This is attributed to the stability of the suspension, which is achieved by electrostatic repulsion or, less often, by sterical hindrance. Charge is mainly used for complexation. Especially, this holds for the intracellular delivery of genetic material (RNA, DNA, and oligonucleotides), which is typically negatively charged and therefore requires a positively charged carrier (Elouahabi and Ruysschaert 2005).

Charge is a key parameter assumed to play an essential role (Jung et al. 2001; Vila et al. 2002, 2004) and is also known to have an influence on the cytotoxicity (Blau et al. 2000; Borm et al. 2006; Chanana et al. 2005; Goodman et al. 2004; Hardman 2006) and the barrier integrity (Chanana et al. 2005; Lockman et al. 2004). In addition, it is known that a charge facilitates the transfection to cells (Felgner 1999; Huang and Viroonchatapan 1999; Khalil et al. 2007; Nemmar et al. 2001; Sakurai et al. 2000). The transfection efficiency of cationic liposomes, e.g., can be increased dramatically using additional antibodies for cellular targeting (Xu et al. 2002).

To account for the complexity of biological systems, such as environments of different polarities, pH value, viscosities, and the presence of molecular targets, modern nanocarrier systems need to combine a multitude of functionalities. Multicomponent targeting, exploiting the huge available surface of the nanoscale transport vehicles, combining a variety of ligands and surrounding susceptible molecules are considered to be essential for future improvements (Ferrari 2005; Torchilin 2006). All these factors must be considered and adapted to the target area while designing multifunctional nanocarrier systems.

V. Torchilin (ed.), *Multifunctional Pharmaceutical Nanocarriers*,
© Springer Science + Business Media, LLC 2008

In the following, we present various strategies typically used to change or influence the surface charge of possible nanoscale carrier systems. We will also focus on the nanoparticulate systems chosen for the applications and that were successfully connected with the materials mentioned in the first part.

2   Types of Nanocarriers and Their Biological Applications

Currently used pharmaceutical nanocarriers include cationic polymers and lipids, nanoparticles, liposomes, dendrimers, polymeric micelles, and others. These systems have shown a great potential to carry drugs and genetic material to the desired site of action, in a controlled or sustained manner. Figure 1 is a schematic diagram of some of these carriers and their application as nonviral gene delivery system.

2.1   Cationic Polymers and Lipids for Gene Delivery

Since 1987, various cationic polymers and lipids have been introduced as gene carrier systems (Felgner et al. 1987; Wu and Wu 1987). Gene therapy, as promising therapeutics to treat generic or acquired diseases, has achieved exciting development

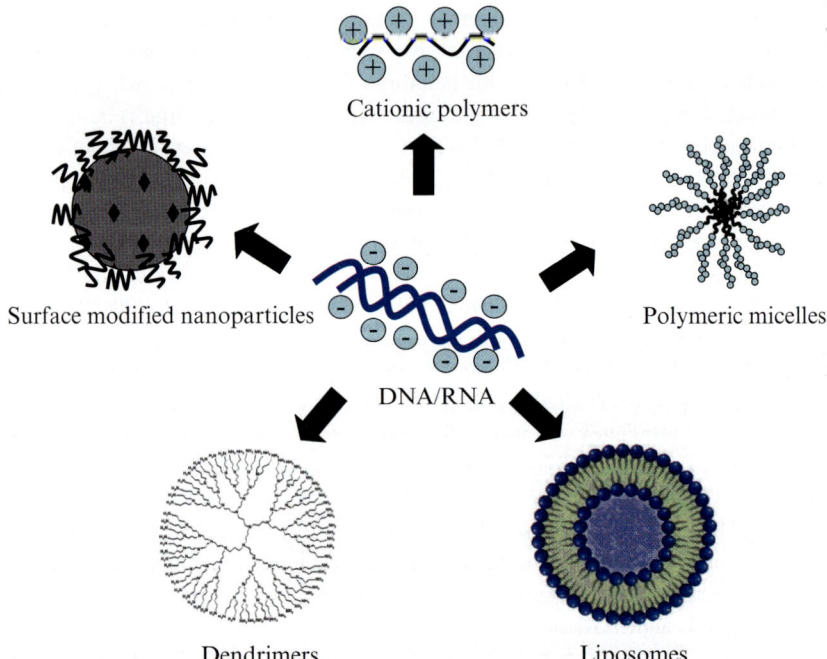

Fig. 1 Schematic of different types of nonviral nanocarriers used for gene delivery

in the past two decades (Brewster et al. 2006; Park et al. 2006; Zhang and Godbey 2006).

Essentially, this group of compounds needs to be delivered intracellularly to exert their action. However, the application of therapies based on antisense oligo-nucleotides is hampered by their instability to cellular nuclease and their weak penetration of the lipophilic cellular membrane. Cationic polymers and lipids, the most important nonviral vectors, have many advantages over viral ones, being non-immunogenic, nononcogenic, and easy to produce (Kaneda and Tabata 2006; Park et al. 2006). In condensing the anionic DNA into compact nanoplexes, the poly-meric gene carrier can provide protection from DNAses and prolong the bioavaila-bility of the incorporated DNA, upon introduction into an enzyme-rich environment. Typically possessing a net positive surface charge, these DNA nanoparticles readily attach to the cell surface via charge–charge interactions with the negatively charged glycocalyx, thereby facilitating internalization by different endocytic mechanisms (Dang and Leong 2006). From this one can deduce the critical role of the polymer charge for both the binding with the plasmid as well as the interaction with the cel-lular membrane. Trials to augment the surface charge are expected to enhance the transfection efficiency of the nanoplexes, but on the other hand might negatively affect the safety aspects of these vectors (Li and Huang 2000; Lv et al. 2006). In addition, most of these polymers contain reactive sites amenable for ligand conju-gation, cross-linking, and other modifications that can render the polymer tailored for a range of clinical applications. Examples of these charged carrier molecules include cationic polysaccharides, polymethacrylates, polyethyleneimine (PEI), poly(L-lysine) (PLL), and cationic lipids. In the following part we will focus on the charge modification of these carriers and its implication on their biological applications.

Chitosan, a naturally derived polysaccharide, is one of the most reported nonvi-ral cationic polymeric gene carriers (Mansouri et al. 2004). It has a strong affinity for DNA and protects against the degradation with DNAse. The transfection effi-ciency of chitosan/DNA complexes is dependent on several factors, including the degree of deacetylation and molecular weight of the chitosan, plasmid concentra-tion, charge ratio of amine (chitosan) to phosphate (DNA), serum concentration, pH, and cell type (Dang and Leong 2006). Owing to their mucoadhesive property (Lehr et al. 1990), chitosan-based gene delivery systems have been successfully applied to oral and nasal route gene therapy systems (Issa et al. 2005). Recently attempts have been made to provide cell-specific targeting and improved transfec-tion efficiency. For instance, methylation, PEGylation, thiolation, and galactosyla-tion of chitosan (Fig. 2) were aimed for fine-tuned adjustments toward targeted delivery of therapeutic genes despite the partial reduction in the net positive charge (Janes et al. 2001; Shi et al. 2006).

Fractional conjugation of poly(ethylene glycol) (PEG) via an amide linkage to soluble chitosan was shown to yield self-aggregates at basic pH that could trap insulin likely due to electrostatic interactions between the unconjugated chitosan monomers and the anionic residues of the protein. The degree of PEGylation influenced the release rate as well as the aggregate size (Ohya et al. 1999). An interesting approach

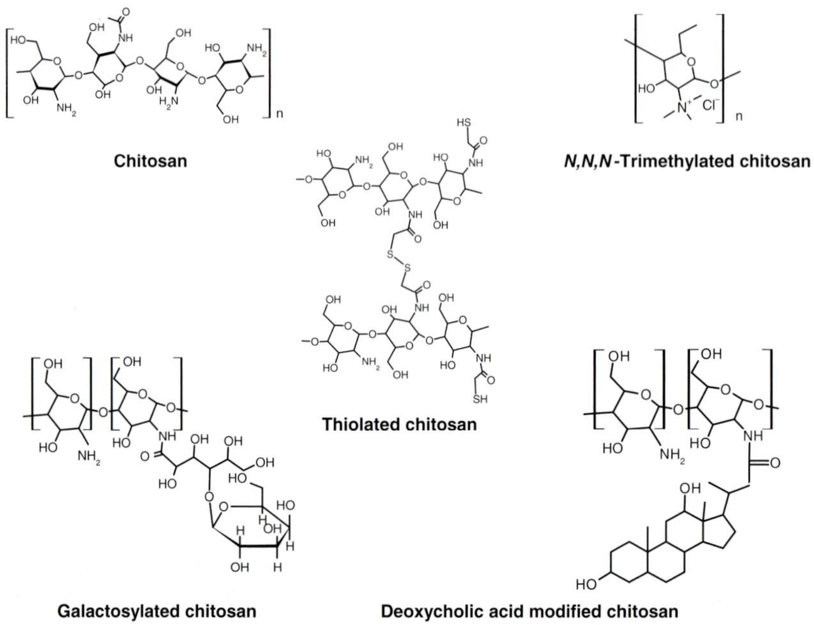

Fig. 2 Chemical structure of chitosan and different modified chitosan derivatives

leading to the formation of chitosan vesicles was developed by Uchegbu et al. (1998). These authors linked palmitic acid to modified glycol chitosan chains, thus producing an amphiphilic polymer, which upon mixing with cholesterol, formed nanovesicles ~300–600 nm in size. These vesicles showed good biocompatibility, hemocompatibility, and stability in serum and bile salts. Moreover, the vesicles were able to encapsulate bleomycin, a chemotherapeutic peptide.

Recently, Kim et al. (2004) described galactosylated chitosan/DNA nano-plexes as a gene carrier. Hepatocytes possess asialoglycoprotein receptors (ASGR) that bind and internalize galactose-terminal (asialo) glycoproteins (Wall et al. 1980). Galactosylated chitosan ensured a higher transfection efficiency of DNA and antisense oligonucleotides into ASGR-bearing cells, HepG2, than into HeLa without ASGR (Gao et al. 2005). Furthermore, lactobionic acid-bearing galactose group was coupled to low molecular weight chitosan to improve hepa-tocytes specificity of plasmid DNA (Gao et al. 2003). MTT assay showed that galactosylated chitosan was less toxic compared with chitosan, which might be due to the comparatively lower positive charge density (Gao et al. 2005). Similarly, mannosylated chitosan was found to induce receptor-mediated endocy-tosis and targeting into antigen-presenting cells, especially dendritic cells having mannose receptors (Kim et al. 2006b). Furthermore, introducing a trisaccharide branch with the N-acetylglucosamine residue at the free end that targets cell-surface lectins to chitosan oligomers was found to increase gene expression levels in the human hepatocytes (HepG2) by tenfold, compared with polyplexes mediated by

PEI (Issa et al. 2005). Despite the partial reduction of surface charge, the colloidal stability was also enhanced with the substituted form, suggesting that the trisaccharide branch sterically stabilized the polyplexes. In vivo studies showed that, 24 h after lung administration to mice, luciferase gene expression was 4-fold higher with the trisaccharide-substituted oligomers than with the corresponding linear chitosan oligomers (Issa et al. 2005). Nevertheless, trials using lactosylated chitosan were characterized by lower affinity to DNA and accordingly lower transfection efficiency (Erbacher et al. 1998).

Another approach to increase transfection rate using chitosan as a vector consists of preparing trimethylated chitosan oligomers through quaternarization of oligomeric chitosan, as demonstrated by Thanou et al. (2002). This process is based on a reductive methylation procedure using methyl iodide in an alkaline environment. Although all derivatives tested were able to effectively transfect MCF-7 cells, cytotoxicity was found to increase with increasing degree of trimethylation (Kean et al. 2005). Further modification was done by Mao et al. (2005) by the synthesis of PEGylated trimethylated chitosan.

Other examples of chitosan modification include dodecylated chitosan (Liu et al. 2001) and deoxycholic acid modified chitosan (Kim et al. 2001).

Polyethyleneimine (PEI) has been one of the most popularly used cationic gene carriers because of its superior transfection efficiency in many types of cells. PEI has primary (25%), secondary (50%), and tertiary amines (25%), of which two-thirds are protonated in physiological media (Fig. 3). The unprotonated amines with different pK_a values confer a buffering effect over a wide range of pH (Garnett 1999). PEI complexes with DNA can be internalized by endocytosis. Inside the cell PEI strongly protonate under the acidic pH in the endosomes and create a charge gradient that provokes a water influx and endosomal swelling and disintegration (Boussif et al. 1995). Besides, the buffering property gives PEI an opportunity to escape from the endosome (proton sponge effect). Similar to chitosan, modified PEI derivatives such as galactosylated PEI (Kunath et al. 2003) and folate-modified PEI (Hwa Kim et al. 2005) were prepared.

The high cation density of PEI contributes to the formation of highly condensed PEI/DNA particles but may confer significant nondesirable cytotoxicity. PEI-grafted PEGs were synthesized to address the cytotoxicity and aggregation problems of PEI. The surface charge of PEI-*g*-PEG/DNA complexes reduced when the molecular weight of PEG was more than 5 kDa, because of the charge shielding effect of PEG, whereas low molecular weight PEG (550 Da) did not show this shielding effect (Petersen et al. 2002). A PEG-conjugated fusogenic KALA peptide (PEG–KALA) was synthesized. KALA is a cationic peptide consisting of 30 amino acids (WEAK–LAKA–LAKA–LAKH–LAKA–LAKA–LKAC–EA). KALA itself was used for both DNA condensation and membrane destabilization, and thus induced marked increase in transfection efficiency (Duguid et al. 1998). When precondensed PEI/DNA complexes, having negatively charged surfaces, were coated with positively charged PEG–KALA to form net positively charged PEG–KALA/DNA/PEI complexes, the complexes maintained the size of 200–400 nm. In contrast, the negatively charged PEI/DNA

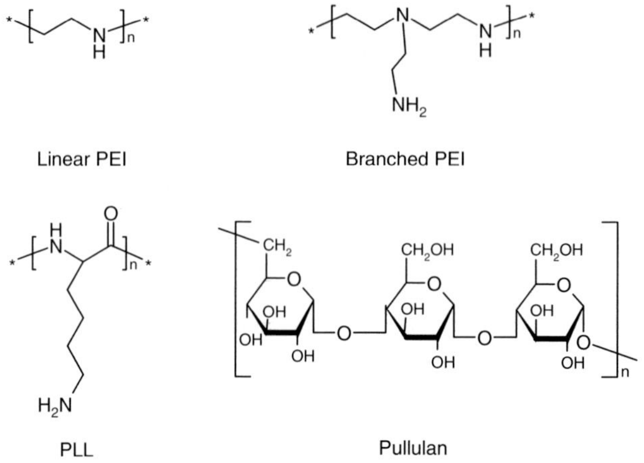

Fig. 3 Chemical structure of various cationic polymers

complexes coated with unmodified KALA formed much larger aggregates with more than 1,000 nm diameter. This result suggests that the presence of PEG segment on the surface of the polyelectrolyte complexes may prevent the interparticular aggregation (Kim et al. 2001).

Poly(L-lysine) (PLL) has been widely used as a nonviral gene carrier (Fig. 3). PLL having a molecular weight of less than 3,000 could not form stable complexes with DNA, suggesting that the number of primary amines in the PLL backbone is important for the complex formation (Kwoh et al. 1999). Although PLLs with high molecular weight have some properties suitable for a gene carrier, the PLL/DNA complexes showed a relatively high cytotoxicity (Choi et al. 1998) and a tendency to aggregate and precipitate depending on the ionic strength (Liu et al. 2001).

PLL–PEG block copolymer, synthesized by ring-opening polymerization of *N*-carboxyanhydride derivatized lysine monomers, could form complexes more than 100 nm in size. These complexes showed a significant reduction in surface ζ-potential and cell cytotoxicity. Polyion complex micelles using PLL–PEG block copolymer, in which PLL segments and plasmid DNA formed a hydrophobic core by charge neutralization and PEG plays a role as a surrounding hydrophilic shell layer, showed a prolonged in vivo circulation time after systemic administration (Harada et al. 2001).

Modified *collagen* has been extensively studied for gene transfer applications. Atelocollagen, collagen obtained after removal of telopeptides, has been investigated for a wide range of drug and gene delivery systems (Sano et al. 2003). Intramuscular administration of the atelocollagen led to long-term detection of the plasmid DNA in the peripheral blood (up to 40 days, compared with 6 h–14 days with naked pDNA injection) and significant serum HST-1 levels up to 60 days. Wang et al. compared the transfection levels achieved by native collagen and methylated collagen that were complexed with pRELuc (luciferase encoding plasmid), both in vitro and in vivo (Wang et al. 2004). Methylated collagen/DNA particles were more condensed and exhibited a higher charge density at pH 7.4 than did native collagen/DNA particles at pH 3. Once at neutral pH, native collagen/DNA particles aggregated significantly and became destabilized.

Gelatin, the denatured form of collagen, has also been demonstrated as a viable gene delivery vehicle. Cationic gelatin was used to facilitate delivery of small interference RNA (siRNA) to silence the TGF-h receptor gene. Plasmid DNA encoding this siRNA vector was complexed with cationic gelatin and introduced into the ureter for delivery into the kidney. This resulted in a reduction in collagen content and fibrotic tissue in the kidney interstitium for up to 10 days after injection (Dang and Leong 2006).

One of the novel gene carriers is *pullulan* (Fig. 3), which is a polysaccharide with an inherent affinity for the liver (Akiyoshi et al. 1993; Hosseinkhani et al. 2002). Cationized pullulan was prepared by chemical introduction of spermine (Jo et al. 2006). Intravenous injection of spermine-pullulan-DNA complexes to mice prolonged the survival time of the mice and suppressed the tumor growth in the liver, compared with free plasmid DNA.

Lipoplexes, complexes between *cationic lipids* and DNA, are believed to facilitate their interaction with the negatively charged cell membranes and improve the transfection efficiency. In particular, lipoplexes containing a fusogenic lipid (IDOPE) are able to fuse with the endosomal membrane. Various lipids, e.g., dioleoylphosphatidylethanolamine (DOPE), cholesterol, and N-[1-(2,3-dimyristyloxy) propyl]-N,N-dimethyl-N-(2-hydroxyethyl) ammonium bromide, are capable of facilitating the endosomal release of DNA by destabilizing the endosome membrane (El Ouahabi et al. 1997). Among the cationic lipids commercially available on the market are Lipofectine® (an equimolar mixture of N-[1-(2,3-dioleyloxy) propyl]-N,N,N-trimethylammonium chloride – DOTMA and DOPE), Oligofectamine®, Megafectine®, and DOTAP®.

2.2 Particulate Nanocarriers Made from Organic or Polymeric Materials

Nanotechnology focuses on formulating therapeutic agents in biocompatible nanocomposites such as nanoparticles, nanocapsules, micellar systems, and conjugates (Panyam and Labhasetwar 2003). Owing to their subcellular size and their biodegradable properties, nanoparticles can penetrate into tissues through fine capillaries and be efficiently taken up by the cells. By modulating polymer characteristics or surface properties, one can control the release of therapeutic agent or deliver it to distant target sites.

2.2.1 Nanoparticles

A number of polymers, either natural or synthetic, have been used in formulating biodegradable nanoparticles. Among the natural polymers, chitosan, albumin, alginate, collagen, and gelatin have been extensively applied, while polylactide–polyglycolide copolymers, polyacrylates, and polycaprolactones can be considered the most famous synthetic polymers (Panyam and Labhasetwar 2003).

Chitosan nanoparticles were widely used for the delivery of macromolecules such as vaccines and proteins across the oral and nasal mucosa (Illum 2007). For example, ovalbumin was adsorbed on the surface of chitosan particles to enhance their uptake by the M-cells of the Peyer's patches. Additional coating of particles with sodium alginate was able to prevent the burst release of the loaded antigen and improve the nanoparticle stability in gastrointestinal fluid. However, alginate-modified chitosan nanoparticles encountered many problems, including flocculation at low alginate to particle ratio, and desorption of ovalbumin, which led to significant decrease in the loading capacity of the coated particles (Borges et al. 2005). In another recent study, alginate-modified trimethyl chitosan nanoparticles were prepared for protein delivery. The results showed that nanoparticles with a lower degree of quaternarization showed an increase in particle size, a decrease in ζ-potential and a slower drug release profile, whereas for the alginate-modified particles a smaller size and lower ζ-potential were observed (Chen et al. 2007).

One other way to enhance the interaction of chitosan nanoparticles with M-cells is their coating with carbohydrates having an affinity for the mannose receptors in the epithelial cells, such as glucomannan. Coating was evidenced by the conversion of the ζ-potential to negative values. These particles were efficiently loaded with the inmunomodulatory protein complex P1 for the peroral administration (Cuna et al. 2006).

Poly(lactic-co-glycolic acid) (PLGA) is one of the most common biodegradable polymers used for drug delivery purposes. However, owing to its extremely slow rate of biodegradation, PLGA is considered unsuitable for pulmonary drug delivery, especially in cases where frequent dosing is required. A new class of rapidly degrading polymers based on amine-modified branched polyesters, such as diethyl-aminopropylamine poly(vinyl alcohol)-*graft*-poly(lactic-*co*-glycolic acid) (DEAPA-PVAL-*g*-PLGA), has been identified as suitable candidates for pulmonary drug delivery systems. These cationic biodegradable nanoparticles elicited a significantly lower pulmonary inflammatory response compared with polystyrene particles of similar hydrodynamic diameter (Dailey et al. 2006). Another novel approach was the synthesis (by ring-opening polymerization) of poly(lactide)–tocopheryl poly(ethylene glycol) succinate (PLA–TPGS) of desired hydrophobic–hydrophilic balance for the delivery of anticancer drugs (Zhang and Feng 2006). PLA–TPGS nanoparticles encapsulating paclitaxel have shown great advantages in cellular uptake over the traditional PLGA nanoparticles (Zhang and Feng 2006).

In the context of DNA or RNA delivery, the major limitation in the application of these nanoparticles is primarily their negative charge, which limits the interaction with the negatively charged DNA, in addition to the poor transport characteristics of the DNA-encapsulated PLGA nanoparticles through the cell membrane. PLGA nanoparticles with cationic surface modification can overcome these disadvantages and hence readily bind and condense DNA. Several polycations were used to accomplish this cationic surface modification, including PEI (Kim et al. 2005; Messai et al. 2003; Trimaille et al. 2003), chitosan (Ravi Kumar et al. 2004), cetyltrimethylamonium bromide (Singh et al. 2003), poly(2-dimethyl-amino)ethyl methacrylate (pDMAEMA) (Munier et al. 2005), and didodecyl dimethyl ammonium bromide (Kwon et al. 2001). Many of these systems proved to be efficient nonviral carriers to the lung epithelium (Bivas-Benita et al. 2004). Munier et al. (2005) described a versatile approach for the elaboration of cationic poly(D,L-lactide), PLA, nanoparticles, based on the use of preformed particles and subsequent adsorption of a model polycation. Among three different cationic polymers (chitosan, PEI, pDMAEMA) used, PLA–pDMAEMA cNP was the most promising system for in vitro DNA delivery.

A novel cationic modification of PLGA was recently developed based on poly[vinyl-3-(dialkylamino)alkylcarbamate-co-vinyl acetate-co-vinyl alcohol]-graft-poly(D,L-lactide-co-glycolide) (Unger et al. 2007). The branched polyesters are biodegradable and positively charged depending on the type and degree of amine substitution. Accordingly, they have been successfully used for DNA delivery (Oster et al. 2004), transmucosal insulin delivery (Simon et al. 2005), and pulmonary drug delivery (Dailey et al. 2003). In addition, PVA-graft-PLGA nanoparticles have been successfully used for the local delivery of paclitaxel for the treatment of restenosis (Westedt et al. 2007).

The potential of trimethyl chitosan derivatives as cationic surface modifiers for nanoparticles was investigated by many research groups. Poly ε-caprolactone nanoparticles coated with trimethyl chitosan proved to bind efficiently DNA and enhance its internalization by cos-1 cells (Haas et al. 2005). Besides, other in vivo uptake studies indicated the transport of FITC-ovalbumin-associated TMC nanoparticles as well as influenza subunit antigen-TMC nanoparticles across the nasal mucosa (Amidi et al. 2006, 2007). Nevertheless, highly quaternarized derivatives were found to be significantly cytotoxic (Haas et al. 2005).

The nanoparticulate carriers of gelatin have been used for efficient intracellular delivery of the encapsulated payload. Thiolated gelatin nanoparticles can result in rapid release of their content in a medium containing high glutathione concentration. The thiol content of gelatin is suggested to form disulfide bonds within the polymer structure, thus strengthening the tertiary and quaternary protein structure and stabilize the nanoparticles during systemic circulation. However, inside the cell, where the glutathione concentration is 1,000-fold higher, these disulfide bonds are broken and the biopolymer unfolds releasing its content. In a recent study, it was observed that surface modification of these gelatin/thiolated gelatin nanoparticles with hydrophilic polymers such as PEG affords longer circulation time in vivo (Kaul and Amiji 2005). PEGylated gelatin nanoparticles are characterized by a slight decrease in the negative charge on the surface, which could be attributed to the modification of the surface amino groups of the surface of gelatin nanoparticles (Kommareddy and Amiji 2007).

2.2.2 Polymeric Micelles

New triblock copolymers consisting of poly(ε-caprolactone)-b-chitooligosaccharide-b-poly(ethylene glycol) were synthesized and evaluated for the delivery of doxorubicin.

The copolymers can self-assemble to form polymeric micelles of 90 nm diameter, at the critical micelle concentration (CMC) of 1.0 μM. Doxorubicin-loaded micelles cross-linked with genipin were able to reduce the burst release of doxorubicin and sustain its release (Chung et al. 2006).

In another study, a water-soluble lipopolymer (WSLP) was synthesized by conjugating cholesteryl chloroformate to a low molecular weight PEI (1.8 kDa) (Han et al. 2001). The balance between hydrophobic cholesterol group and hydrophilic PEI in WSLP enables the polymer to form a micellar structure in aqueous milieu (CMC = 500 μg mL⁻¹). WSLP interacts with DNA to form stable colloidal particles (70 nm in diameter). The PEI moiety of WSLP confers a buffering effect (proton sponge effect), which could facilitate endosomal escape of the WSLP/DNA complex. In a similar way, the hydrophobic cholesterol moiety of WSLP would give a chance to form small and stable complexes, resulting in enhanced cellular uptake and transfection efficiency. WSLP showed higher transfection efficiency and much lower cytotoxicity than 25-kDa PEI, suggesting that WSLP has the advantages from both PEI and cholesterol (Han et al. 2001).

2.2.3 Dendrimers

Dendrimers are three-dimensional, highly branched monodisperse macromolecules, which are obtained by an iterative sequence of reaction steps producing a precise, unique branching structure, with controlled surface functionality, that add to their potential as new scaffolds for drug delivery (Gupta et al. 2006). Dendrimers have reported to release the loaded drug, either hydrophilic or hydrophobic, in a controlled manner (Baars et al. 2000) and have been recently used successfully in gene delivery (El-Sayed et al. 2003; Jevprasesphant et al. 2004), and as magnetic resonance imaging agents (Wiener et al. 1994) and solubilizing agents (Milhem et al. 2000). In an attempt to increase the loading capacity, and the sustaining and targeting properties, modified dendrimers such as pegylated dendrimers, glycodendrimers, polyester dendrimers, and dendrimer grafts have been synthesized (Bhadra et al. 2003, 2005; Ihre et al. 2002; Padilla De Jesus et al. 2002).

Cationic dendrimers and partial dendrimers (dendrons) can condense DNA and enhance its internalization (Dufes et al. 2005). Important attributes of dendrons include the molecular weight, charge, hydrophobicity, flexibility, and geometry of the molecule (Sakthivel et al. 1998; Shah et al. 2000). In a study that explored polyamidoamine (PAMAM) dendrimers as a gene delivery system, the modulation of complexes through the presence of other compounds during the complex formation was described (Godbey et al. 1999). The addition of the cationic transfection polymer DEAE-dextran to the dendrimer-DNA complexes appears to have an additive or possibly synergistic effect on the transfection efficiency observed with PAMAM. However, doubling of cell death was also noted (Lampela et al. 2004). On the other hand, the presence of anionic oligomers such as dextran sulphate or oligonucleotides together with plasmid DNA during the complex formation has been demonstrated to significantly influence the morphology of complexes formed

(Vincent et al. 2003). Complexes formed from phosphorous-containing dendrimers and activated PAMAM dendrimer (Superfect) were significantly less dense and appeared to be less prone to aggregation when the anionic oligomers were used (Maksimenko et al. 2003).

Another particular example is the quaternarization of the PPI dendrimers, which led to an improved DNA binding, complex stability, and in vivo safety (Schatzlein et al. 2005).

Although most dendritic polymers currently used are based on symmetric structure with multiple branches emanating from a central core, work by Florence and colleagues demonstrates that this is not obligatory: water-soluble amphiphilic dendritic polylysine or polyornithine peptides with a hydrophobic root (3_a-amino myristic acid) are asymmetric and have a relatively lower charge density than do PAMAM dendrimers but can still achieve transfection (Choi et al. 1999, 2000). Another polylysine "branch" structure was synthesized as an asymmetric methoxy(ethylene glycol)-*block*-PLL dendrimer or a symmetric "barbell shaped" PLL dendrimer-*block*-methoxy(ethylene glycol)-*block*-PLL dendrimer triblock copolymer which efficiently condenses DNA (Bosman et al. 1999).

Polylysine-based dendrons (having 16 free surface amino groups attached to 7 lysine groups) were used to condense a fluorescent protein vector, pRedN-1 DNA. Increased lipophilicity and molar charge ratios are prerequisites to obtain compact and reproducible dendriplexes. These dendriplexes were encapsulated in PLGA nanoparticles with a higher encapsulation efficiency compared with DNA alone (Ribeiro et al. 2005). Wiwattanapatapee et al. investigated the influence of charge on the mechanism of dendrimer transport across the gastrointestinal membrane (Malik et al. 2000). Using the everted rat intestinal sac method, it was found that for cationic PMAM the tissue uptake was higher than serosal transport whereas anionic PMAM dendrimers were assumed to be transported through third-phase endocytosis (Wiwattanapatapee et al. 2000).

2.2.4 Liposomes

Liposomes have been demonstrated to be an efficient drug and gene carrier system (Oku et al. 2001; Oussoren and Storm 2001). Surface modification of liposomes achieved by inserting PEG, dextran, galactose, or mannose to the surface was found to increase liposome stability and to introduce some specificity in binding to particular cell types (Hashida et al. 2001; Shigeru et al. 2000; Trubetskoy et al. 1995).

A novel nonviral gene transfer system was developed by Oku et al. (2001) by modifying liposomes with cetylated PEI. This polycation liposome (PCL) showed remarkable transfection efficiency in eight malignant and two normal cell lines tested as well as in cos-1 cells in comparison with conventional cationic liposome. In vivo testing of the effectiveness of gene transfer using PCL indicated that GFP and Luciferase genes were effectively expressed in mice. The effect of the molecular weight of PEI on PCL-mediated gene transfer showed that PEIs with a molecular weight of 600 and 1,800 Da were quite effective but PEI of 25,000 was far less effective (Oku et al. 2001).

Recently, pH-sensitive liposomes of 30 nm diameter containing cationic/anionic lipid combinations were synthesized for cancer therapy (Kawakami et al. 2007). Liposomes were composed of Fe-porphyrin (as a superoxide dismutase mimic, L-α-phosphatidylcholine (DMPC), dimethylditetradecyl-ammonium bromide, sodispdum oleate (OA$_{Na}$), and Tween-80. Results suggested that the liposome was highly cytotoxic toward the cancer cells, compared with cisplatin, the control drug (Kawakami et al. 2007). In another study, stable large unilamellar vesicles were prepared either from the ionizable anionic lipid, cholesteryl hemisuccinate, and the permanently cationic lipid, *N,N*-dioleoyl-*N,N*-dimethylammonium chloride, or from the acidic lipid dioleoylphosphatidic acid and the ionizable cationic lipid 3α-[*N*-(*N′,N′*-dimethylaminoethane)-carbamoyl] (Hafez et al. 2000). The critical pH of fusion was dependent on the cationic to anionic lipid ratio. This pH-dependent fusion occurs preferably when the surface charge is zero (Hafez et al. 2000). Further specificity of these liposomes was accomplished by association with enzymes such as elastase (Meers 2001). The presence of soluble proteases in high concentrations near the exterior of a desired target cell activates liposomes near the target to fuse with the cell surface before other extracellular factors can interfere.

2.3 Particulate Nanocarriers Made from Inorganic Materials

2.3.1 Silica-Based Particulate Matter

SiO_2 surfaces exhibit some favourable properties for the biological application: good biocompatibility, negative charge, and ease of covalent binding to the silica surface. There are several studies demonstrating that silica-based nanoparticles are of benign nature regarding their cytotoxicity (Kneuer et al. 2000a; Luo et al. 2004; Ravi Kumar et al. 2004; Zhu et al. 2004). The negative charge, as well as the ability to modify the surface by covalently attaching specific entities, is due to the existence of 4–6 Si–OH groups per nm^2 (Spierings et al. 1995; Zhuravlev 1987). Based on silane chemistry and the self-assembly of those molecules a wide variety of modifications is possible (Ulman 1990; Wang et al. 2006).

The inheriting negative charge of these particles is important for the stabilization of nanoparticulate suspensions since the particles easily aggregate because of the high surface area. In general, the stability of suspension is assumed if the ζ-potential of the particulate materials is $|\phi| > 30$ mV unless no other stabilizing mechanism is present (Wang et al. 2006).

A major reason for using positively charged particles is the loading of these carriers with genetic material (DNA, RNA, and oligonucleotides) bearing negative charges. Complexation of genetic material to the surface of such particles showed that the plasmid DNA is protected from enzymatic degradation and allows for in vitro transfection (Kneuer et al. 2000a, b; Peng et al. 2006). Recently, it was shown that organic-modified particles (Roy et al. 2003) were efficient for in vitro and in vivo nonviral gene delivery (Bharali et al. 2005; Roy et al. 2005). Silica beads were surface-modified to selectively bind to various cancers (He et al. 2003, 2004; Ow et al. 2005; Santra et al. 2001), bacterial cells (Zhao et al. 2004), and PTK2 cells (Qhobosheane et al. 2001). Furthermore, silica particles with sizes between 40 and 70 nm were found to penetrate into the nucleus, which might explain their transfection potential (Chen and Von Mikecz 2005).

In terms of toxicity, amorphous silica materials are considered as relatively safe (Barbé et al. 2004; Brunner et al. 2006). It has therefore been used as food additive for years and as negative control in cytotoxicity experiments (Brunner et al. 2006). Also, recent studies have shown that silica-overcoated magnetic particles are of benign nature and that a prolonged presence in the tissue is possible without toxic effect (Kim et al. 2006a). A bioactive effect is known as well (Hench and Wilson 1986; Klein et al. 1995; Li et al. 1994). From implantate research it is known that silica materials can be excreted in the urine through the kidneys or is actively phagocytosed (Lai et al. 1998, 2000, 2005). On the other side, SiO_2 nanoparticles were found to trigger subnuclear pathology (Chen and Von Mikecz 2005). In contrast, crystalline silica taken up via the airways induces pulmonary inflammation, oxidative stress, and mediator release from lung target cells (Chang et al. 2007; Nel et al. 2006).

Silica particles that show fluorescent and magnetic properties are promising multifunctional entities to investigate biological systems (Levy et al. 2002; Lu et al. 2002, 2004).

2.3.2 Magnetic Nanoparticles

Promising multifunctional nanocarriers or nanoparticles are magnetic systems. They can be used for diagnosis, localisation, drug delivery, and therapeutic applications (Dobson 2006; Ito et al. 2005; Jain et al. 2005). A magnetic core material can be used for magnetic resonance imaging, as a hyperthermia mediating component, or to be deposited by means of a field gradient at the target site (Arruebo et al. 2007). The surface of the particle, which may be modified by using different materials such as polymers, lipids, and silica (SiO_2), allows tuning the environmental interaction. Negative charge, as found on silicon-coated magnetic particles reduce aggregate formation because of electrostatic repulsion (Arruebo et al. 2007). These negative charges are typically related with the oxidic nature of the materials.

The modification of the carriers' surfaces can be based on hydrophobic forces facilitating adsorption onto the surface or insertion of hydrophobic parts in already existing hydrophobic cavities (Torchilin 1998, 2001; Yuan et al. 1995). Typically, chemical binding is the preferred route of modification (Klibanov et al. 2003; Torchilin et al. 2003). Therefore, originally charged surface due to the presence of free amino groups or carboxylic groups are intensively used.

3 Influence of the Charge on Biodistribution, Cellular Uptake, and Transport

For future application several crucial needs have to be fulfilled: appropriate circulation time without being excreted, direction to specific targets or biological compartments, and release of the active substances loaded on or in the carrier.

Systemic circulation is strongly influenced by the polarity of the particle. Most frequently used are PEG coatings (Milton Harris and Chess 2003) but negatively charged particulate carriers are known to show longer circulation times as well (McNeil 2005). Regarding targeting, two strategies are used: passive targeting and active targeting based on ligands. The passive targeting is, concerning cancer, based on the so-called enhanced permeability and retention effect. This effect is long since known to be dependent on the surface charge of the carrier systems (Dellian et al. 2000; Juliano and Stamp 1975).

Overall, the surface charge is assumed to play an important role for uptake into the cell and its further destiny after internalization. For cationic amphiphiles it was shown that they may induce a highly curved membrane and therefore were invaginated (Elouahabi and Ruysschaert 2005).

For polymeric complexes, the charge, tuned by the ratio of polymer to load, has been found to influence the transfection efficiency (Cherng et al. 1996; Dalluge et al. 2002; Kneuer et al. 2006). In addition, different accumulation behaviors have been observed related to the surface charge and its amount and are exploited for drug delivery (Blau et al. 2000; Borowik et al. 2005; Chiu et al. 2001, 2002). In this context, the unspecific cellular adhesion can be controlled by the density of PEG molecules on the surface (Satomi et al. 2007). Therefore, the transfer to colloidal surfaces allows controlling to a certain amount the adhesive potential of the nanoparticulate carriers.

Hence, there are studies showing that cellular uptake and association are possible for particles with any surface charge for in vitro models of the gastrointestinal tract (Baczynska et al. 2001). Similar results were found for negatively charged particles that were taken up via Peyer's Patches of mice (Shakweh et al. 2005). According to Patel, radioactively labeled small negatively charged liposomes are most efficient in targeting rat regional lymph nodes after subcutaneous administration (Patel et al. 1984); that is, negatively charged SUV were more effective than positively charged and both were more efficient than neutral carriers (Patel et al. 1984). But these data were as well contradicting earlier data received (Osborne et al. 1979). Obviously, different aspects are strongly influencing and masking the charge effect so that no clear statement is possible as long as the experimental procedures and the carriers are not chosen very carefully.

Let us take chitosan as an example, which is known to be a penetration enhancer in acidic environment toward monostratified and pluristratified epithelia both endowed with and lacking tight junctions (Dodane et al. 1999). The uptake of chitosan nanocarriers seems to be related to the size and the surface charge: the higher the ζ-potential, the stronger is the affinity between the nanoparticles and the negatively charged cell membranes and mucus respectively (Huang et al. 2004). At the same time the binding capacity, as well as the stability of such a suspension, is increased (Nafee et al. 2007; Pang et al. 2002).

The contact time of the carrier systems with the membrane increases uptake probability (Fahmy et al. 2005). This holds for charge-related effects (El-Shabouri 2002; Hariharan et al. 2006) as well as for targeting moieties (Hong et al. 2007).

Cellular uptake is definitely not the only obstacle when trying to deliver therapeutic substances. The uptake is typically accomplished via endocytotic pathways (macro- or pinocytose) (Fahmy et al. 2005; Felberbaum-Corti et al. 2003; Steimer et al. 2005; Watson et al. 2005) and therefore the carrier needs to escape the endo- and lysosome respectively (Panyam and Labhasetwar 2003).

The destabilization of the endosomes can be achieved applying cationic lipids or cationic polyelectrolytes. Relying on cationic lipids such as dioleoyl phospatidyleth-anolamine (IDOPE), which is known to be fusogenic, leads to incorporation into the membrane (Felgner et al. 1994). Another possible mechanism is the so-called sponge effect, which may be responsible for the necessary release of the carrier from the lysosome to the cytosol (Fig. 4) (Blau et al. 2000; Boussif et al. 1995; Mislick and Baldeschwieler 1996; Pang et al. 2002; Panyam and Labhasetwar 2003). However, this effect based on the buffering of macromolecules at low pH is still under debate (Dubruel et al. 2004; Funhoff et al. 2004). A further membrane-destabilizing effect is described for polymers exhibiting anionic carboxyl groups (Yessine and Leroux 2004). A strong pH-influenced conformational change of these polyelectrolytes is responsible for the membrane leakage (Yessine et al. 2003). The hydrophobicity of the polymers influences additionally the hemolytic activity but is not associated with

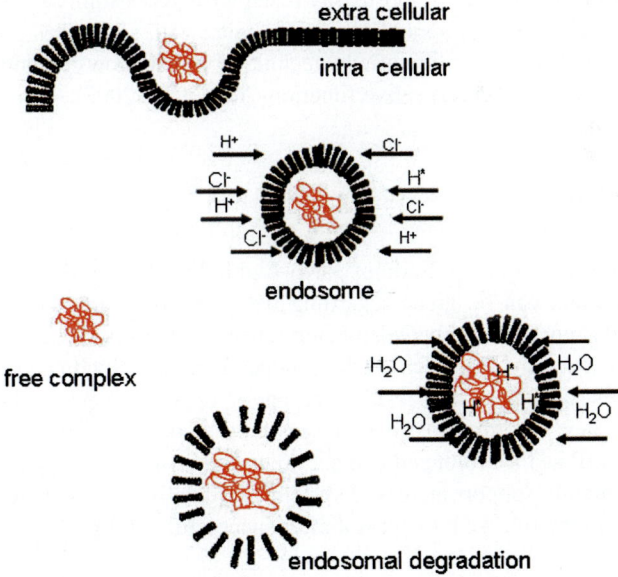

Fig. 4 Proton sponge effect. This sketch reflects the assumed mechanism of the destabilization of the endosomal membrane by polyplexes or nanoparticles with attached polyelectrolytes (repre-sented by the red polymer-like structure). After invagination of the object, the endosome is acidi-fied by an influx of protons and chloride. The buffering capacity of the complex leads to an increased ionic strength, followed by an increased uptake of water. Swelling of the endosome is the consequence, which results in destabilization of the endosome and release to the cytosol. Image modified but according to Nanochemistry homepage (Preece 2007)

membrane disintegration. Moreover, the hydrophobicity results in elevated macrophage cytotoxicity (Yessine et al. 2003). Hence, adjusting the hydrophilic–hydrophobic balance of synthetic polymers allows to tune the disruption of membranes in a pH-dependent manner (Yessine and Leroux 2004).

There are several studies that suggest that the type of charge plays a crucial role for uptake and toxicity. Quaternary ammonium groups seem to be benign compared to primary amino groups (Chanana et al. 2005; Chung et al. 2007), and from dendrimes it is known that amine groups are more cytotoxic than carboxylic groups (Malik et al. 2000).

Additionally, the interaction pattern depends on the cell type investigated. Stem cells show a lesser uptake of mesoporous silica particles when compared with 3T3-L1 cells (Chung et al. 2007), or A549 show higher uptake of modified PLGA particles when compared with Caco-2 cells in confluent monolayers (Weiss et al. 2007).

Negative charge in comparison with membrane-penetrating peptides is investigated but not evaluated with respect to combined effect of charge and surface-located peptides (Koch et al. 2005).

An interesting strategy for future applications besides using smart and responsive carriers might be the combination of steric hindrance, e.g., with PEG with an underlying charged layer that is not fully screened. In combination with flexible nanocarriers that change their architecture in dependence on the environment or other stimuli (Torchilin 2006). Let us say a stepwise accessibility of functions. In addition, the more classical idea of the magic bullet with all entities present at the time or the so-called particle lithography technique which allows to modify specific parts of the surface with a respective function (Yake et al. 2007).

4   Conclusion

Finally, looking at the present findings partly highlighted here, it is clear that to date no clear statement can be given regarding the effect of charge. There are clear behavioral differences found between neutral, negatively or positively charged carrier systems. Generally, the specific effects depend on the individual target and carrier system. Nevertheless, the results are partly opposing each other regarding uptake efficiency. Overall, positive charges may be used for transport via complexation as well as for prolonged contact times and unspecific targeting. Negative charges will mainly support increased systemic circulation. Both charges can play a crucial role in the release into cytosol after successful uptake.

References

Akiyoshi, K., Deguchi, S., Moriguchi, N., Yamaguchi, S. and Sunamoto, J. 1993. Self-aggregates of hydrophobized polysaccharides in water. Formation and characteristics of nanoparticles. Macromolecules 26:3062–3068.

Amidi, M., Romeijn, S. G., Borchard, G., Junginger, H. E., Hennink, W. E. and Jiskoot, W. 2006. Preparation and characterization of protein-loaded N-trimethyl chitosan nanoparticles as nasal delivery system. Journal of Controlled Release 111:107–116.

Amidi, M., Romeijn, S. G., Verhoef, J. C., Junginger, H. E., Bungener, L., Huckriede, A., Crommelin, D. J. A. and Jiskoot, W. 2007. N-Trimethyl chitosan (TMC) nanoparticles loaded with influenza subunit antigen for intranasal vaccination: Biological properties and immunogenicity in a mouse model. Vaccine 25:144–153.

Arruebo, M., Fernández-Pacheco, R., Ibarra, M. R. and Santamaría, J. 2007. Magnetic nanoparticles for drug delivery. Nano Today 2:22–32.

Baars, M. W. P. L., Kleppinger, R., Koch, M. H. J., Yeu, S.-L. and Meijer, E. W. 2000. The localization of guests in water-soluble oligoethyleneoxy-modified poly(propylene imine) dendrimers. Angewandte Chemie International Edition 39:1285–1288.

Baczynska, D., Widerak, K., Ugorski, M. and Langner, M. 2001. Surface charge and the association of liposomes with colon carcinoma cells. Zeitschrift fur Naturforschung C: A Journal of Biosciences 56:872–877.

Barbé, C., Bartlett, J., Kong, L., Finnie, K., Lin, H. Q., Larkin, M., Calleja, S., Bush, A. and Calleja, G. 2004. Silica particles: A novel drug-delivery system. Advanced Materials 16:1959–1966.

Bhadra, D., Bhadra, S., Jain, S. and Jain, N. K. 2003. A PEGylated dendritic nanoparticulate carrier of fluorouracil. International Journal of Pharmaceutics 257:111–124.

Bhadra, D., Yadav, A. K., Bhadra, S. and Jain, N. K. 2005. Glycodendrimeric nanoparticulate carriers of primaquine phosphate for liver targeting. International Journal of Pharmaceutics 295:221–233.

Bharali, D. J., Klejbor, I., Stachowiak, E. K., Dutta, P., Roy, I., Kaur, N., Bergey, E. J., Prasad, P. N. and Stachowiak, M. K. 2005. Organically modified silica nanoparticles: A nonviral vector for in vivo gene delivery and expression in the brain. Proceedings of the National Academy of Sciences of the United States of America 102:11539–11544.

Bivas-Benita, M., Romeijn, S., Junginger, H. E. and Borchard, G. 2004. PLGA-PEI nanoparticles for gene delivery to pulmonary epithelium. European Journal of Pharmaceutics and Biopharmaceutics 58:1–6.

Blau, S., Jubeh, T. T., Haupt, S. M. and Rubinstein, A. 2000. Drug targeting by surface cationization. Critical Reviews in Therapeutic Drug Carrier Systems 17:425–465.

Borges, O., Borchard, G., Verhoef, J. C., de Sousa, A. and Junginger, H. E. 2005. Preparation of coated nanoparticles for a new mucosal vaccine delivery system. International Journal of Pharmaceutics 299:155–166.

Borm, P. J. A., Robbins, D., Haubold, S., Kuhlbusch, T., Fissan, H., Donaldson, K., Schins, R., Stone, V., Kreyling, W., Lademann, J., Krutmann, J., Warheit, D. B. and Oberdorster, E. 2006. The potential risks of nanomaterials: A review carried out for ECETOC. Particle and Fibre Toxicology 3:

Borowik, T., Widerak, K., Ugorski, M. and Langner, M. 2005. Combined effect of surface electrostatic charge and poly(ethyl glycol) on the association of liposomes with colon carcinoma cells. Journal of Liposome Research 15:199–213.

Bosman, A. W., Janssen, H. M. and Meijer, E. W. 1999. About dendrimers: Structure, physical properties, and applications. Chemical Reviews 99:1665–1688.

Boussif, O., Lezoualc'h, F., Zanta, M. A., Mergny, M. D., Scherman, D., Demeneix, B. and Behr, J. P. 1995. A versatile vector for gene and oligonucleotide transfer into cells in culture and in vivo: Polyethylenimine. Proceedings of the National Academy of Sciences of the United States of America 92:7297–7301.

Brewster, L. P., Brey, E. M. and Greisler, H. P. 2006. Cardiovascular gene delivery: The good road is awaiting. Advanced Drug Delivery Reviews—Gene Delivery for Tissue Engineering 58:604–629.

Brunner, T. J., Wick, P., Manser, P., Spohn, P., Grass, R. N., Limbach, L. K., Bruinink, A. and Stark, W. J. 2006. In vitro cytotoxicity of oxide nanoparticles: Comparison to asbestos, silica, and the effect of particle solubility. Environmental Science and Technology 40:4374–4381.

Chanana, M., Gliozzi, A., Diaspro, A., Chodnevskaja, I., Huewel, S., Moskalenko, V., Ulrichs, K., Galla, H. J. and Krol, S. 2005. Interaction of polyelectrolytes and their composites with living cells. Nano Letters 5:2605–2612.

Chang, J. S., Chang, K. L. B., Hwang, D. F. and Kong, Z. L. 2007. In vitro cytotoxicity of silica nanoparticles at high concentrations strongly depends on the metabolic activity type of the cell line. Environmental Science and Technology 41:2064–2068.

Chen, F., Zhang, Z. R. and Huang, Y. 2007. Evaluation and modification of N-trimethyl chitosan chloride nanoparticles as protein carriers. International Journal of Pharmaceutics 336:166–173.

Chen, M. and Von Mikecz, A. 2005. Formation of nucleoplasmic protein aggregates impairs nuclear function in response to SiO_2 nanoparticles. Experimental Cell Research 305:51–62.

Cherng, J. Y., Van De Wetering, P., Talsma, H., Crommelin, D. J. A. and Hennink, W. E. 1996. Effect of size and serum proteins on transfection efficiency of poly((2-dimethylamino)ethyl methacrylate)-plasmid nanoparticles. Pharmaceutical Research 13:1038–1042.

Chiu, G. N. C., Bally, M. B. and Mayer, L. D. 2001. Selective protein interactions with phosphatidylserine containing liposomes alter the steric stabilization properties of poly(ethylene glycol). Biochimica et Biophysica Acta – Biomembranes 1510:56–69.

Chiu, G. N. C., Bally, M. B. and Mayer, L. D. 2002. Effects of phosphatidylserine on membrane incorporation and surface protection properties of exchangeable poly(ethylene glycol)-conjugated lipids. Biochimica et Biophysica Acta – Biomembranes 1560:37–50.

Choi, J. S., Lee, E. J., Choi, Y. H., Jeong, Y. J. and Park, J. S. 1999. Poly(ethylene glycol)-block-poly(L-lysine) dendrimer: Novel linear polymer/dendrimer block copolymer forming a spherical water-soluble polyionic complex with DNA. Bioconjugational Chemistry 10:62–65.

Choi, J. S., Joo, D. K., Kim, C. H., Kim, K. and Park, J. S. 2000. Synthesis of a barbell-like triblock copolymer, poly(L-lysine) dendrimer-block-poly(ethylene glycol)-block-poly(L-lysine) dendrimer, and its self-assembly with plasmid DNA. Journal of American Chemical Society 122:474–480.

Choi, Y. H., Liu, F., Kim, J.-S., Choi, Y. K., Jong Sang, P. and Kim, S. W. 1998. Polyethylene glycol-grafted poly-L-lysine as polymeric gene carrier. Journal of Controlled Release 54:39–48.

Chung, T. H., Wu, S. H., Yao, M., Lu, C. W., Lin, Y. S., Hung, Y., Mou, C. Y., Chen, Y. C. and Huang, D. M. 2007. The effect of surface charge on the uptake and biological function of mesoporous silica nanoparticles in 3T3-L1 cells and human mesenchymal stem cells. Biomaterials 28:2959–2966.

Chung, T.-W., Liu, D.-Z., Hsieh, J.-H., Fan, X.-C., Yang, J.-D. and Chen, J.-H. 2006. Characterizing poly(ε-caprolactone)-b-chitooligosaccharide-b-poly(ethylene glycol) (PCP) copolymer micelles for doxorubicin (DOX) delivery: Effects of crosslinked of amine groups. Journal of Nanoscience and Nanotechnology 6:2902–2911.

Cuna, M., Alonso-Sande, M., Remunan-Lopez, C., Pivel, J. P., Alonso-Lebrero, J. L. and Alonso, M. J. 2006. Development of phosphorylated glucomannan-coated chitosan nanoparticles as nanocarriers for protein delivery. Journal of Nanoscience and Nanotechnology 6:2887–2895.

Dailey, L. A., Schmehl, T., Gessler, T., Wittmar, M., Grimminger, F., Seeger, W. and Kissel, T. 2003. Nebulization of biodegradable nanoparticles: Impact of nebulizer technology and nanoparticle characteristics on aerosol features. Journal of Controlled Release 86:131–144.

Dailey, L. A., Jekel, N., Fink, L., Gessler, T., Schmehl, T., Wittmar, M., Kissel, T. and Seeger, W. 2006. Investigation of the proinflammatory potential of biodegradable nanoparticle drug delivery systems in the lung. Toxicology and Applied Pharmacology 215:100–108.

Dalluge, R., Haberland, A., Zaitsev, S., Schneider, M., Zastrow, H., Sukhorukov, G. and Bottger, M. 2002. Characterization of structure and mechanism of transfection-active peptide-DNA complexes. Biochimica et Biophysica Acta (BBA) – Gene Structure and Expression 1576:45–52.

Dang, J. M. and Leong, K. W. 2006. Natural polymers for gene delivery and tissue engineering. Advanced Drug Delivery Reviews Gene Delivery for Tissue Engineering 58:487–499.

Dellian, M., Yuan, F., Trubetskoy, V. S., Torchilin, V. P. and Jain, R. K. 2000. Vascular permeability in a human tumour xenograft: Molecular charge dependence. British Journal of Cancer 82:1513–1518.

Dobson, J. 2006. Magnetic nanoparticles for drug delivery. Drug Development Research 67:55–60.

Dodane, V., Amin Khan, M. and Merwin, J. R. 1999. Effect of chitosan on epithelial permeability and structure. International Journal of Pharmaceutics 182:21–32.

Dubruel, P., Christiaens, B., Rosseneu, M., Vandekerckhove, J., Grooten, J., Goossens, V. and Schacht, E. 2004. Buffering properties of cationic polymethacrylates are not the only key to successful gene delivery. Biomacromolecules 5:379–388.

Dufes, C., Uchegbu, I. F. and Schatzlein, A. G. 2005. Dendrimers in gene delivery. Advanced Drug Delivery Reviews 57:2177–2202.

Duguid, J. G., Li, C., Shi, M., Logan, M. J., Alila, H., Rolland, A., Tomlinson, E., Sparrow, J. T. and Smith, L. C. 1998. A physicochemical approach for predicting the effectiveness of peptide-based gene delivery systems for use in plasmid-based gene therapy. Biophysical Journal 74:2802–2814.

El-Sayed, M., Rhodes, C. A., Ginski, M. and Ghandehari, H. 2003. Transport mechanism(s) of poly(amidoamine) dendrimers across Caco-2 cell monolayers. International Journal of Pharmaceutics 265:151–157.

El-Shabouri, M. H. 2002. Positively charged nanoparticles for improving the oral bioavailability of cyclosporin-A. International Journal of Pharmaceutics 249:101–108.

El Ouahabi, A., Thiry, M., Pector, V., Fuks, R., Ruysschaert, J. M. and Vandenbranden, M. 1997. The role of endosome destabilizing activity in the gene transfer process mediated by cationic lipids. FEBS Letters 414:187–192.

Elouahabi, A. and Ruysschaert, J. M. 2005. Formation and intracellular trafficking of lipoplexes and polyplexes. Molecular Therapy 11:336–347.

Erbacher, P., Zou, S., Bettinger, T., Steffan, A. M. and Remy, J. S. 1998. Chitosan-based vector/DNA complexes for gene delivery: Biophysical characteristics and transfection ability. Pharmaceutical Research 15:1332–1339.

Fahmy, T. M., Fong, P. M., Goyal, A. and Saltzman, W. M. 2005. Targeted for drug delivery. Materials Today 8:18–26.

Felberbaum-Corti, M., Van Der Goot, F. G. and Gruenberg, J. 2003. Sliding doors: Clathrin-coated pits or caveolae? Nature Cell Biology 5:382–384.

Felgner, J. H., Kumar, R., Sridhar, C. N., Wheeler, C. J., Tsai, Y. J., Border, R., Ramsey, P., Martin, M. and Felgner, P. L. 1994. Enhanced gene delivery and mechanism studies with a novel series of cationic lipid formulations. Journal of Biological Chemistry 269:2550–2561.

Felgner, P. L. 1999. Progress in gene delivery research and development. In non-viral vectors for gene therapy, L. Huang, M. C. Hung, and E. Wagner (eds.). pp. 26–38. San Diego: Academic Press.

Felgner, P. L., Gadek, T. R., Holm, M., Roman, R., Chan, H. W., Wenz, M., Northrop, J. P., Ringold, G. M. and Danielsen, M. 1987. Lipofection: A highly efficient, lipid-mediated DNA-transfection procedure. Proceedings of the National Academy of Sciences of the United States of America 84:7413–7417.

Fenart, L., Casanova, A., Dehouck, B., Duhem, C., Slupek, S., Cecchelli, R. and Betbeder, D. 1999. Evaluation of effect of charge and lipid coating on ability of 60-nm nanoparticles to cross an in vitro model of the blood-brain barrier. Journal of Pharmacology and Experimental Therapeutics 291:1017–1022.

Ferrari, M. 2005. Cancer nanotechnology: Opportunities and challenges. Nature Reviews Cancer 5:161–171.

Funhoff, A. M., van Nostrum, C. F., Koning, G. A., Schuurmans-Nieuwenbroek, N. M. E., Crommelin, D. J. A. and Hennink, W. E. 2004. Endosomal escape of polymeric gene delivery complexes is not always enhanced by polymers buffering at low pH. Biomacromolecules 5:32–39.

Gao, S., Chen, J., Xu, X., Ding, Z., Yang, Y.-H., Hua, Z. and Zhang, J. 2003. Galactosylated low molecular weight chitosan as DNA carrier for hepatocyte-targeting. International Journal of Pharmaceutics 255:57–68.

Gao, S., Chen, J., Dong, L., Ding, Z., Yang, Y.-H. and Zhang, J. 2005. Targeting delivery of oligonucleotide and plasmid DNA to hepatocyte via galactosylated chitosan vector. European Journal of Pharmaceutics and Biopharmaceutics 60:327–334.

Garnett, M. C. 1999. Gene-delivery systems using cationic polymers. Critical Reviews in Therapeutic Drug Carrier Systems 16:147–207.

Godbey, W. T., Wu, K. K., Hirasaki, G. J. and Mikos, A. G. 1999. Improved packing of poly(ethylenimine)/DNA complexes increases transfection efficiency. Gene Therapy 6:1380–1388.

Goodman, C. M., McCusker, C. D., Yilmaz, T. and Rotello, V. M. 2004. Toxicity of gold nanoparticles functionalized with cationic and anionic side chains. Bioconjugate Chemistry 15:897–900.

Gupta, U., Agashe, H. B., Asthana, A. and Jain, N. K. 2006. A review of in vitro–in vivo investigations on dendrimers: The novel nanoscopic drug carriers. Nanomedicine: Nanotechnology, Biology and Medicine 2:66–73.

Haas, J., Ravi Kumar, M., Borchard, G., Bakowsky, U. and Lehr, C.-M. 2005. Preparation and characterization of chitosan and trimethyl-chitosan-modified poly-(ε-caprolactone) nanoparticles as DNA carriers. AAPS PharmSciTech 6:E22–E30.

Hafez, I. M., Ansell, S. and Cullis, P. R. 2000. Tunable pH-sensitive liposomes composed of mixtures of cationic and anionic lipids. Biophysical Journal 79:1438–1446.

Han, S.-o., Mahato, R. I. and Kim, S. W. 2001. Water-soluble lipopolymer for gene delivery. Bioconjugate Chemistry 12:337–345.

Harada, A., Togawa, H. and Kataoka, K. 2001. Physicochemical properties and nuclease resistance of antisense-oligodeoxynucleotides entrapped in the core of polyion complex micelles composed of poly(ethylene glycol)–poly(L-lysine) block copolymers. European Journal of Pharmaceutical Sciences 13:35–42.

Hardman, R. 2006. A toxicologic review of quantum dots: Toxicity depends on physicochemical and environmental factors. Environmental Health Perspectives 114:165–172.

Hariharan, S., Bhardwaj, V., Bala, I., Sitterberg, J., Bakowsky, U. and Ravi Kumar, M. N. V. 2006. Design of estradiol loaded PLGA nanoparticulate formulations: A potential oral delivery system for hormone therapy. Pharmaceutical Research 23:184–195.

Hashida, M., Nishikawa, M., Yamashita, F. and Takakura, Y. 2001. Cell-specific delivery of genes with glycosylated carriers. Advanced Drug Delivery Reviews 52:187–196.

He, X., Duan, J., Wang, K., Tan, W., Lin, X. and He, C. 2004. A novel fluorescent label based on organic dye-doped silica nanoparticles for HepG liver cancer cell recognition. Journal of Nanoscience and Nanotechnology 4:585–589.

He, X. X., Wang, K., Tan, W., Liu, B., Lin, X., He, C., Li, D., Huang, S. and Li, J. 2003. Bioconjugated nanoparticles for DNA protection from cleavage. Journal of the American Chemical Society 125:7168–7169.

Hench, L. L. and Wilson, J. 1986. Biocompatibility of silicates for medical use. In Silicon Biochemistry, D. Evered and M. O'Connor (eds.). pp. 231–246. Chichester: Wiley.

Hong, S., Leroueil, P. R., Majoros, I. J., Orr, B. G., Baker Jr, J. R. and Banaszak Holl, M. M. 2007. The binding avidity of a nanoparticle-based multivalent targeted drug delivery platform. Chemistry and Biology 14:107–115.

Hosseinkhani, H., Aoyama, T., Ogawa, O. and Tabata, Y. 2002. Liver targeting of plasmid DNA by pullulan conjugation based on metal coordination. Journal of Controlled Release 83:287–302.

Huang, L. and Viroonchatapan, E. 1999. Introduction. In Non-Viral Vectors for Gene Therapy, L. Huang, M. C. Hung, and E. Wagner (eds.). pp. 3–22. San Diego: Academic Press.

Huang, M., Khor, E. and Lim, L.-Y. 2004. Uptake and cytotoxicity of chitosan molecules and nanparticles: Effects of molecular weight and degree of deacetylation. Pharmaceutical Research 21:344–353.

Hwa Kim, S., Hoon Jeong, J., Chul Cho, K., Wan Kim, S. and Gwan Park, T. 2005. Target-specific gene silencing by siRNA plasmid DNA complexed with folate-modified poly(ethylenimine). Journal of Controlled Release 104:223–232.

Illum, L. 2007. Nanoparticulate systems for nasal delivery of drugs: A real improvement over simple systems? Journal of Pharmaceutical Sciences 96:473–483.

Issa, M. M., Koping-Hoggard, M. and Artursson, P. 2005. Chitosan and the mucosal delivery of biotechnology drugs. Drug Discovery Today: Technologies 2:1–6.

Ito, A., Shinkai, M., Honda, H. and Kobayashi, T. 2005. Medical application of functionalized magnetic nanoparticles. Journal of Bioscience and Bioengineering 100:1–11.

Jain, T. K., Morales, M. A., Sahoo, S. K., Leslie-Pelecky, D. L. and Labhasetwar, V. 2005. Iron oxide nanoparticles for sustained delivery of anticancer agents. Molecular Pharmaceutics 2:194–205.

Janes, K., Calvo, P. and Alonso, M. 2001. Polysaccharide colloidal particles as delivery systems for macromolecules. Advanced Drug Delivery 47:83–97.

Jevprasesphant, R., Penny, J., Attwood, D. and D'Emanuele, A. 2004. Transport of dendrimer nanocarriers through epithelial cells via the transcellular route. Journal of Controlled Release 97:259–267.

Jo, J.-i., Yamamoto, M., Matsumoto, K., Nakamura, T. and Tabata, Y. 2006. Liver targeting of plasmid DNA with a cationized pullulan for tumor suppression. Journal of Nanoscience and Nanotechnology 6:2853–2859.

Juliano, R. L. and Stamp, D. 1975. The effect of particle size and charge on the clearance rates of liposomes and liposome encapsulated drugs. Biochemical and Biophysical Research Communications 63:651–658.

Jung, T., Kamm, W., Breitenbach, A., Hungerer, K.-D., Hundt, E. and Kissel, T. 2001. Tetanus toxoid loaded nanoparticles from sulfobutylated poly(vinyl alcohol)-*graft*-poly(lactide-*co*-glycolide): Evaluation of antibody response after oral and nasal application in mice. Pharmaceutical Research 18:352–360.

Kaneda, Y. and Tabata, Y. 2006. Non-viral vectors for cancer therapy. Cancer Science 97:348–354.

Kaul, G. and Amiji, M. 2005. Tumor-targeted gene delivery using poly(ethylene glycol)-modified gelatin nanoparticles: In vitro and in vivo studies. Pharmaceutical Research 22:951–961.

Kawakami, H., Hiraka, K., Tamai, M., Horiuchi, A., Ogata, A., Hatsugai, T., Yamaguchi, A., Oyaizu, K. and Yuasa, M. 2007. pH-Sensitive liposome retaining Fe-porphyrin as SOD mimic for novel anticancer drug delivery system. Polymers for Advanced Technologies 18:82–87.

Kean, T., Roth, S. and Thanou, M. 2005. Trimethylated chitosans as non-viral gene delivery vectors: Cytotoxicity and transfection efficiency. Journal of Controlled Release 103:643–653.

Khalil, I. A., Kogure, K., Futaki, S., Hama, S., Akita, H., Ueno, M., Kishida, H., Kudoh, M., Mishina, Y., Kataoka, K., Yamada, M. and Harashima, H. 2007. Octaarginine-modified multifunctional envelope-type nanoparticles for gene delivery. Gene Therapy 14:682–689.

Kim, I.-S., Lee, S.-K., Park, Y.-M., Lee, Y.-B., Shin, S.-C., Lee, K. C. and Oh, I.-J. 2005. Physicochemical characterization of poly(L-lactic acid) and poly(D,L-lactide-*co*-glycolide) nanoparticles with polyethylenimine as gene delivery carrier. International Journal of Pharmaceutics 298:255–262.

Kim, J. S., Yoon, T. J., Yu, K. N., Kim, B. G., Park, S. J., Kim, H. W., Lee, K. H., Park, S. B., Lee, J. K. and Cho, M. H. 2006a. Toxicity and tissue distribution of magnetic nanoparticles in mice. Toxicological Sciences 89:338–347.

Kim, T. H., Nah, J. W., Cho, M. H., Park, T. G. and Cho, N. S. 2006b. Receptor-mediated gene delivery into antigen presenting cells using mannosylated chitosan/DNA nanoparticles. Journal of Nanoscience and Nanotechnology 6:2796–2803.

Kim, T. H., Park, I. K., Nah, J. W., Choi, Y. J. and Cho, C. S. 2004. Galactosylated chitosan/DNA nanoparticles prepared using water-soluble chitosan as a gene carrier. Biomaterials 25:3783–3792.

Klein, C. P. A. T., Li, P., De Blieck-Hogervorst, J. M. A. and De Groot, K. 1995. Effect of sintering temperature on silica gels and their bone bonding ability. Biomaterials 16:715–719.

Klibanov, A. L., Torchilin, V. P. and Zalipsky, S. 2003. Long-circulating sterically protected liposomes. In Liposomes, V. P. Torchilin and V. Weissig (eds.). 231 pp. Oxford, UK: Oxford University Press.

Kneuer, C., Sameti, M., Bakowsky, U., Schiestel, T., Schirra, H., Schmidt, H. and Lehr, C. M. 2000a. A nonviral DNA delivery system based on surface modified silica-nanoparticles can efficiently transfect cells in vitro. Bioconjugate Chemistry 11:926–932.

Kneuer, C., Sameti, M., Haltner, E. G., Schiestel, T., Schirra, H., Schmidt, H. and Lehr, C. M. 2000b. Silica nanoparticles modified with aminosilanes as carriers for plasmid DNA. International Journal of Pharmaceutics 196:257–261.

Kneuer, G., Ehrhardt, C., Bakowsky, H., Kumar, M. N. V. R., Oberle, V., Lehr, C. M., Hoekstra, D. and Bakowsky, U. 2006. The influence of physicochemical parameters on the efficacy of non-viral DNA transfection complexes: A comparative study. Journal of Nanoscience and Nanotechnology 6:2776–2782.

Koch, A. M., Reynolds, F., Merkle, H. P., Weissleder, R. and Josephson, L. 2005. Transport of surface-modified nanoparticles through cell monolayers. ChemBioChem 6:337–345.

Kommareddy, S. and Amiji, M. 2007. Poly(ethylene glycol)-modified thiolated gelatin nanoparticles for glutathione-responsive intracellular DNA delivery. Nanomedicine: Nanotechnology, Biology and Medicine 3:32–42.

Kreuter, J. 2004. Influence of the surface properties on nanoparticle-mediated transport of drugs to the brain. Journal of Nanoscience and Nanotechnology 4:484–488.

Kunath, K., von Harpe, A., Fischer, D. and Kissel, T. 2003. Galactose-PEI-DNA complexes for targeted gene delivery: Degree of substitution affects complex size and transfection efficiency. Journal of Controlled Release 88:159–172.

Kwoh, D. Y., Coffin, C. C., Lollo, C. P., Jovenal, J., Banaszczyk, M. G., Mullen, P., Phillips, A., Amini, A., Fabrycki, J., Bartholomew, R. M., Brostoff, S. W. and Carlo, D. J. 1999. Stabilization of poly-L-lysine/DNA polyplexes for in vivo gene delivery to the liver. Biochimica et Biophysica Acta (BBA) – Gene Structure and Expression 1444:171–190.

Kwon, H. Y., Lee, J. Y., Choi, S. W., Jang, Y. and Kim, J. H. 2001. Preparation of PLGA nanoparticles containing estrogen by emulsification-diffusion method. Colloids and Surfaces A: Physicochemical and Engineering Aspects 182:123–130.

Lai, W., Ducheyne, P. and Garino, J. 1998. Removal pathway of silicon released from bioactive glass granules in vivo. In Bioceramics, R. Z. LeGeros and J. P. LeGeros (eds.). pp. 383–386. New York: World Scientific.

Lai, W., Ducheyne, P., Garino, J. and Flaitz, C. M. 2000. Removal pathway of bioactive glass resorption products from the body. Materials Research Society Symposium – Proceedings 599:261–266.

Lai, W., Garino, J., Flaitz, C. and Ducheyne, P. 2005. Excretion of resorption products from bioactive glass implanted in rabbit muscle. Journal of Biomedical Materials Research Part A 75:398–407.

Lampela, P., Soininen, P., Urtti, A., Mannisto, P. T. and Raasmaja, A. 2004. Synergism in gene delivery by small PEIs and three different nonviral vectors. International Journal of Pharmaceutics 270:175–184.

Lehr, C. M., Bouwstra, J. A., Tukker, J. J. and Junginger, H. E. 1990. Intestinal transit of bioadhesive microspheres in an in situ loop in the rat – A comparative study with copolymers and blends based on poly(acrylic acid). Journal of Controlled Release 13:51–62.

Levy, L., Sahoo, Y., Kim, K. S., Bergey, E. J. and Prasad, P. N. 2002. Nanochemistry: Synthesis and characterization of multifunctional nanoclinics for biological applications. Chemistry of Materials 14:3715–3721.

Li, P., Ohtsuki, C., Kokubo, T., Nakanishi, K., Soga, N. and De Groot, K. 1994. The role of hydrated silica, titania, and alumina in inducing apatite on implants. Journal of Biomedical Materials Research 28:7–15.

Li, S. and Huang, L. 2000. Nonviral gene therapy: Promises and challenges. Gene Therapy 7:31–34.

Liu, W. G., Yao, K. D. and Liu, Q. G. 2001. Formation of a DNA/N-dodecylated chitosan complex and salt-induced gene delivery. Journal of Applied Polymer Sciences 82:3391–3395.

Lockman, P. R., Koziara, J. M., Mumper, R. J. and Allen, D. 2004. Nanoparticle surface charges alter blood-brain barrier integrity and permeability. Journal of Drug Targeting 12:635–641.

Lu, H., Yi, G., Zhao, S., Chen, D., Guo, L. H. and Cheng, J. 2004. Synthesis and characterization of multi-functional nanoparticles possessing magnetic, up-conversion fluorescence and bio-affinity properties. Journal of Materials Chemistry 14:1336–1341.

Lu, Y., Yin, Y., Mayers, B. T. and Xia, Y. 2002. Modifying the surface properties of superparamagnetic iron oxide nanoparticles through a sol-gel approach. Nano Letters 2:183–186.

Luo, D., Han, E., Belcheva, N. and Saltzman, W. M. 2004. A self-assembled, modular DNA delivery system mediated by silica nanoparticles. Journal of Controlled Release 95:333–341.

Lv, H., Zhang, S., Wang, B., Cui, S. and Yan, J. 2006. Toxicity of cationic lipids and cationic polymers in gene delivery. Journal of Controlled Release 114:100–109.

Maksimenko, A., Mandrouguine, V., Gottikh, M. B., Bertrand, J.-R., Majoral, J.-P. and Malvy, C. 2003. Optimisation of dendrimer-mediated gene transfer by anionic oligomers. Journal of Gene Medicine 5:61–71.

Malik, N., Wiwattanapatapee, R., Klopsch, R., Lorenz, K., Frey, H., Weener, J. W., Meijer, E. W., Paulus, W. and Duncan, R. 2000. Dendrimers: Relationship between structure and biocompatibility in vitro, and preliminary studies on the biodistribution of ^{125}I-labelled polyamidoamine dendrimers in vivo. Journal of Controlled Release 65:133–148.

Mansouri, S., Lavigne, P., Corsi, K., Benderdour, M., Beaumont, E. and Fernandes, J. C. 2004. Chitosan-DNA nanoparticles as non-viral vectors in gene therapy: Strategies to improve transfection efficacy. European Journal of Pharmaceutics and Biopharmaceutics 57:1–8.

Mao, S., Shuai, X., Unger, F., Wittmar, M., Xie, X. and Kissel, T. 2005. Synthesis, characterization and cytotoxicity of poly(ethylene glycol)-*graft*-trimethyl chitosan block copolymers. Biomaterials 26:6343–6356.

Meers, P. 2001. Enzyme-activated targeting of liposomes. Advanced Drug Delivery Reviews 53:265–272.

Messai, I., Munier, S., Ataman-Onal, Y., Verrier, B. and Delair, T. 2003. Elaboration of poly(ethyleneimine) coated poly(D,L-lactic acid) particles. Effect of ionic strength on the surface properties and DNA binding capabilities. Colloids and Surfaces B: Biointerfaces 32:293–305.

Milhem, O. M., Myles, C., McKeown, N. B., Attwood, D. and D'Emanuele, A. 2000. Polyamidoamine Starburst® dendrimers as solubility enhancers. International Journal of Pharmaceutics 197:239–241.

Milton Harris, J. and Chess, R. B. 2003. Effect of pegylation on pharmaceuticals. Nature Reviews Drug Discovery 2:214–221.

Mislick, K. A. and Baldeschwieler, J. D. 1996. Evidence for the role of proteoglycans in cation-mediated gene transfer. Proceedings of the National Academy of Sciences of the United States of America 93:12349–12354.

Munier, S., Messai, I., Delair, T., Verrier, B. and Ataman-Onal, Y. 2005. Cationic PLA nanoparticles for DNA delivery: Comparison of three surface polycations for DNA binding, protection and transfection properties. Colloids and Surfaces B: Biointerfaces 43:163–173.

Nafee, N., Taetz, S., Schneider, M., Schaefer, U. F. and Lehr, C. M. in press. Chitosan-coated PLGA nanoparticles for DNA/RNA delivery: Effect of the formulation parameters on complexation and transfection of antisense oligonucleotides. Nanomedicine: Nanotechnology, Biology, and Medicine.

Nel, A., Xia, T., Madler, L. and Li, N. 2006. Toxic potential of materials at the nanolevel. Science 311:622–627; DOI: 10.1126/science.1114397.

Nemmar, A., Vanbilloen, H., Hoylaerts, M. F., Hoet, P. H. M., Verbruggen, A. and Nemery, B. 2001. Passage of intratracheally instilled ultrafine particles from the lung into the systemic circulation in hamster. American Journal of Respiratory and Critical Care Medicine 164:1665–1668.

Ohya, Y., Cai, R., Nishizawa, H., Hara, K. and Ouchi, T. 1999. Preparation of PEG-grafted chitosan nano-particle for peptide drug carrier. Proceedings of the International Symposium on Controlled Release of Bioactive Materials 26:655–656.

Oku, N., Yamazaki, Y., Matsuura, M., Sugiyama, M., Hasegawa, M. and Nango, M. 2001. A novel non-viral gene transfer system, polycation liposomes. Advanced Drug Delivery Reviews 52:209–218.

Osborne, M. P., Richardson, V. J., Jeyasingh, K. and Ryman, B. E. 1979. Radionuclide-labelled liposomes. A new lymph node imaging agent. International Journal of Nuclear Medicine and Biology 6:75–83.

Oster, C., Wittmar, M., Unger, F., Barbu-Tudoran, L., Schaper, A. and Kissel, T. 2004. Design of amine-modified graft polyesters for effective gene delivery using DNA-loaded nanoparticles. Pharmaceutical Research 21:927–931.

Oussoren, C. and Storm, G. 2001. Liposomes to target the lymphatics by subcutaneous administration. Advanced Drug Delivery Reviews 50:143–156.

Ow, H., Larson, D. R., Srivastava, M., Baird, B. A., Webb, W. W. and Wiesnert, U. 2005. Bright and stable core-shell fluorescent silica nanoparticles. Nano Letters 5:113–117.

Pang, S. W., Park, H. Y., Jang, Y. S., Kim, W. S. and Kim, J. H. 2002. Effects of charge density and particle size of poly(styrene/(dimethylamino)ethyl methacrylate) nanoparticle for gene delivery in 293 cells. Colloids and Surfaces B: Biointerfaces 26:213–222.

Panyam, J. and Labhasetwar, V. 2003. Biodegradable nanoparticles for drug and gene delivery to cells and tissue. Advanced Drug Delivery Reviews 55:329–347.

Park, T. G., Jeong, J. H. and Kim, S. W. 2006. Current status of polymeric gene delivery systems. Advanced Drug Delivery Reviews Gene Delivery for Tissue Engineering 58:467–486.

Patel, H. M., Boodle, K. M. and Vaughan-Jones, R. 1984. Assessment of the potential uses of liposomes for lymphoscintigraphy and lymphatic drug delivery. Failure of 99m-technetium marker to represent intact liposomes in lymph nodes. Biochimica et Biophysica Acta – General Subjects 801:76–86.

Peng, J., He, X., Wang, K., Tan, W., Li, H., Xing, X. and Wang, Y. 2006. An antisense oligonucleotide carrier based on amino silica nanoparticles for antisense inhibition of cancer cells. Nanomedicine: Nanotechnology, Biology, and Medicine 2:113–120.

Petersen, H., Fechner, P. M., Martin, A. L., Kunath, K., Stolnik, S., Roberts, C. J., Fischer, D., Davies, M. C. and Kissel, T. 2002. Polyethylenimine-graft-poly(ethylene glycol) copolymers: Influence of copolymer block structure on DNA complexation and biological activities as gene delivery system. Bioconjugate Chemistry 13:845–854.

Preece, J. A. 2007. Proton sponge. Nanoscale Chemistry Research Group at University of Birmingham. http://www.nanochem.bham.ac.uk/gene_delivery/protonsponge.htm

Qhobosheane, M., Santra, S., Zhang, P. and Tan, W. 2001. Biochemically functionalized silica nanoparticles. Analyst 126:1274–1278.

Ravi Kumar, M. N. V., Sameti, M., Mohapatra, S. S., Kong, X., Lockey, R. F., Bakowsky, U., Lindenblatt, G., Schmidt, H. and Lehr, C. M. 2004. Cationic silica nanoparticles as gene carriers: Synthesis, characterization and transfection efficiency in vitro and in vivo. Journal of Nanoscience and Nanotechnology 4:876–881.

Ribeiro, S., Hussain, N. and Florence, A. T. 2005. Release of DNA from dendriplexes encapsulated in PLGA nanoparticles. International Journal of Pharmaceutics (Selected contributions from the 5th European Workshop on Particulate Systems) 298:354–360.

Roy, I., Ohulchanskyy, T. Y., Bharali, D. J., Pudavar, H. E., Mistretta, R. A., Kaur, N. and Prasad, P. N. 2005. Optical tracking of organically modified silica nanoparticles as DNA carriers: A nonviral, nanomedicine approach for gene delivery. Proceedings of the National Academy of Sciences of the United States of America 102:279–284.

Roy, I., Ohulchanskyy, T. Y., Pudavar, H. E., Bergey, E. J., Oseroff, A. R., Morgan, J., Dougherty, T. J. and Prasad, P. N. 2003. Ceramic-based nanoparticles entrapping water-insoluble photosensitizing anticancer drugs: A novel drug-carrier system for photodynamic therapy. Journal of the American Chemical Society 125:7860–7865.

Sakthivel, T., Istvan, T. and Alexander, T. F. 1998. Synthesis and physicochemical properties of lipophilic polyamide dendrimers. Pharmaceutical Research 15:776–782.

Sakurai, F., Inoue, R., Nishino, Y., Okuda, A., Matsumoto, O., Taga, T., Yamashita, F., Takakura, Y. and Hashida, M. 2000. Effect of DNA/liposome mixing ratio on the physicochemical characteristics, cellular uptake and intracellular trafficking of plasmid DNA/cationic liposome complexes and subsequent gene expression. Journal of Controlled Release 66:255–269.

Sano, A., Maeda, M., Nagahara, S., Ochiya, T., Honma, K., Itoh, H., Miyata, T. and Fujioka, K. 2003. Atelocollagen for protein and gene delivery. Advanced Drug Delivery Reviews 55:1651–1677.

Santra, S., Zhang, P., Wang, K., Tapec, R. and Tan, W. 2001. Conjugation of biomolecules with luminophore-doped silica nanoparticles for photostable biomarkers. Analytical Chemistry 73:4988–4993.

Schatzlein, A. G., Zinselmeyer, B. H., Elouzi, A., Dufes, C., Chim, Y. T. A., Roberts, C. J., Davies, M. C., Munro, A., Gray, A. I. and Uchegbu, I. F. 2005. Preferential liver gene expression with polypropylenimine dendrimers. Journal of Controlled Release 101:247–258.

Shah, D. S., Sakthivel, T., Toth, I., Florence, A. T. and Wilderspin, A. F. 2000. DNA transfection and transfected cell viability using amphipathic asymmetric dendrimers. International Journal of Pharmaceutics 208:41–48.

Shakweh, M., Besnard, M., Nicolas, V. and Fattal, E. 2005. Poly (lactide-*co*-glycolide) particles of different physicochemical properties and their uptake by Peyer's patches in mice. European Journal of Pharmaceutics and Biopharmaceutics 61:1–13.

Shi, C., Zhu, Y., Ran, X., Wang, M., Su, Y. and Cheng, T. 2006. Therapeutic potential of chitosan and its derivatives in regenerative medicine. Journal of Surgical Research 133:185–192.

Shigeru, K., Shintaro, F., Makiya, N., Fumiyoshi, Y. and Mitsuru, H. 2000. In vivo gene delivery to the liver using novel galactosylated cationic liposomes. Pharmaceutical Research 17:306–313.

Simon, M., Wittmar, M., Kissel, T. and Linn, T. 2005. Insulin containing nanocomplexes formed by self-assembly from biodegradable amine-modified poly(vinyl alcohol)-*graft*-poly(L-lactide): Bioavailability and nasal tolerability in rats. Pharmaceutical Research 22:1879–1886.

Singh, M., Ugozzoli, M., Briones, M., Kazzaz, J., Soenawan, E. and O'Hagan, D. T. 2003. The effect of CTAB concentration in cationic PLG microparticles on DNA adsorption and in vivo performance. Pharmaceutical Research 20:247–251.

Spierings, G. A. C. M., Haisma, J. and Michielsen, T. M. 1995. Surface-related phenomena in the direct bonding of silicon and fused-silica wafer pairs. Philips Journal of Research 49:47–63.

Steimer, A., Haltner, E. and Lehr, C.-M. 2005. Cell culture models of the respiratory tract relevant to pulmonary drug delivery. Journal of Aerosol Medicine: Deposition, Clearance, and Effects in the Lung 18:137–182.

Thanou, M., Florea, B. I., Geldof, M., Junginger, H. E. and Borchard, G. 2002. Quaternized chitosan oligomers as novel gene delivery vectors in epithelial cell lines. Biomaterials 23:153–159.

Torchilin, V. P. 1998. Polymer-coated long-circulating microparticulate pharmaceuticals. Journal of Microencapsulation 15:1–19.

Torchilin, V. P. 2001. Structure and design of polymeric surfactant-based drug delivery systems. Journal of Controlled Release 73:137–172.

Torchilin, V. P. 2006. Multifunctional nanocarriers. Advanced Drug Delivery Reviews 58:1532–1555.

Torchilin, V. P., Weissig, V., Martin, F. J. and Heath, T. D. 2003. Surface modifications of liposomes. In Liposomes, V. P. Torchilin and V. Weissig (eds.). pp. 193–230. Oxford, UK: Oxford University Press.

Trimaille, T., Pichot, C. and Delair, T. 2003. Surface functionalization of poly(D,L-lactic acid) nanoparticles with poly(ethylenimine) and plasmid DNA by the layer-by-layer approach. Colloids and Surfaces A: Physicochemical and Engineering Aspects 221:39–48.

Trubetskoy, V. S., Cannillo, J. A., Milshtein, A., Wolf, G. L. and Torchilin, V. P. 1995. Controlled delivery of Gd-containing liposomes to lymph nodes: Surface modification may enhance MRI contrast properties. Magnetic Resonance Imaging 13:31–37.

Uchegbu, I. F., Schätzlein, A. G., Tetley, L., Gray, A. I., Sludden, J., Siddique, S. and Mosha, E. 1998. Polymeric chitosan-based vesicles for drug delivery. Journal of Pharmacy and Pharmacology 50:453–458.

Ulman, A. 1990. Self-assembled monolayers of alkyltrichlorosilanes: Building blocks for future organic materials. Advanced Materials 2:573–582.

Unger, F., Wittmar, M. and Kissel, T. 2007. Branched polyesters based on poly[vinyl-3-(dialkylamino)alkylcarbamate-*co*-vinyl acetate-*co*-vinyl alcohol]-*graft*-poly(D,L-lactide-*co*-glycolide): Effects of polymer structure on cytotoxicity. Biomaterials 28:1610–1619.

Vila, A., Sanchez, A., Tobio, M., Calvo, P. and Alonso, M. J. 2002. Design of biodegradable particles for protein delivery. Journal of Controlled Release 78:15–24.

Vila, A., Gill, H., McCallion, O. and Alonso, M. J. 2004. Transport of PLA–PEG particles across the nasal mucosa: Effect of particle size and PEG coating density. Journal of Controlled Release 98:231–244.

Vincent, L., Varet, J., Pille, J.-Y., Bompais, H., Opolon, P., Maksimenko, A., Malvy, C., Mirshahi, M., Lu, H., Vannier, J.-P., Soria, C. and Li, H. 2003. Efficacy of dendrimer-mediated angiostatin and TIMP-2 gene delivery on inhibition of tumor growth and angiogenesis: In vitro and in vivo studies. International Journal of Cancer 105:419–429.

Wall, D. A., Wilson, G. and Hubbard, A. L. 1980. The galactose-specific recognition system of mammalian liver: The route of ligand internalization in rat hepatocytes. Cell 21:79–93.

Wang, J., Lee, I. L., Lim, W. S., Chia, S. M., Yu, H., Leong, K. W. and Mao, H.-Q. 2004. Evaluation of collagen and methylated collagen as gene carriers. International Journal of Pharmaceutics 279:115–126.

Wang, L., Wang, K., Santra, S., Zhao, X., Hilliard, L. R., Smith, J. E., Wu, Y. and Tan, W. 2006. Watching silica nanoparticles glow in the biological world. Analytical Chemistry 78:646–654.

Watson, P., Jones, A. T. and Stephens, D. J. 2005. Intracellular trafficking pathways and drug delivery: Fluorescence imaging of living and fixed cells. Advanced Drug Delivery Reviews 57:43–61.

Weiss, B., Schneider, M., Muys, L., Taetz, S., Neumann, D., Schaefer, U. F. and Lehr, C. M. in press. Coupling of biotin-(poly(ethylene glycol))amine to poly(D,L-lactide-co-glycolide) nano-particles for versatile surface modification. Bioconjugate Chemistry.

Westedt, U., Kalinowski, M., Wittmar, M., Merdan, T., Unger, F., Fuchs, J., Schaller, S., Bakowsky, U. and Kissel, T. 2007. Poly(vinyl alcohol)-graft-poly(lactide-co-glycolide) nanoparticles for local delivery of paclitaxel for restenosis treatment. Journal of Controlled Release 119:41–51.

Wiener, E., Brechbiel, M., Brothers, W. H., Magin, R. L., Gansow, O. A., Tomalia, D. A. and Lauterbur, P. C. 1994. Dendrimer-based metal chelates: A new class of magnetic resonance imaging contrast agents. 31:1–8.

Wiwattanapatapee, R., Carreño-Gómez, B., Malik, N. and Duncan, R. 2000. Anionic PAMAM dendrimers rapidly cross adult rat intestine in vitro: A potential oral delivery system? Pharmaceutical Research 17:991–998.

Wu, G. Y. and Wu, C. H. 1987. Receptor-mediated in vitro gene transformation by a soluble DNA carrier system. Journal of Biological Chemistry 262:4429–4432.

Yessine, M. A., Lafleur, M., Meier, C., Petereit, H. U. and Leroux, J. C. 2003. Characterization of the membrane-destabilizing properties of different pH-sensitive methacrylic acid copoly-mers. Biochimica et Biophysica Acta – Biomembranes 1613:28–38.

Yessine, M. A. and Leroux, J. C. 2004. Membrane-destabilizing polyanions: Interaction with lipid bilayers and endosomal escape of biomacromolecules. Advanced Drug Delivery Reviews 56:999–1021.

Yuan, F., Dellian, M., Fukumura, D., Leunig, M., Berk, D. A., Torchilin, V. P. and Jain, R. K. 1995. Vascular permeability in a human tumor xenograft: Molecular size dependence and cut-off size. Cancer Research 55:3752–3756.

Zhang, X. and Godbey, W. T. 2006. Viral vectors for gene delivery in tissue engineering. Advanced Drug Delivery Reviews Gene Delivery for Tissue Engineering 58:515–534.

Zhang, Z. and Feng, S.-S. 2006. The drug encapsulation efficiency, in vitro drug release, cellular uptake and cytotoxicity of paclitaxel-loaded poly(lactide)-tocopheryl polyethylene glycol suc-cinate nanoparticles. Biomaterials 27:4025–4033.

Zhao, X., Hilliard, L. R., Mechery, S. J., Wang, Y., Bagwe, R. P., Jin, S. and Tan, W. 2004. A rapid bioassay for single bacterial cell quantitation using bioconjugated nanoparticles. Proceedings of the National Academy of Sciences of the United States of America 101:15027–15032.

Zhu, S. G., Xiang, J. J., Li, X. L., Shen, S. R., Lu, H. B., Zhou, J., Xiong, W., Zhang, B. C., Nie, X. M., Zhou, M., Tang, K. and Li, G. Y. 2004. Poly(L-lysine)-modified silica nanoparticles for the delivery of antisense oligonucleotides. Biotechnology and Applied Biochemistry 39:179–187.

Zhuravlev, L. T. 1987. Concentration of hydroxyl-groups on the surface of amorphous silicas. Langmuir 3:316–318.

Funtionalization of Pharmaceutical Nanocarriers for Mitochondria-Targeted Drug and DNA Delivery

Volkmar Weissig, Sarathi Boddapati, Gerard D'Souza, and Richard W. Horobin

1 Mitochondria and Human Diseases

Every fifteen minutes, a child is born with a mitochondrial disease or will develop one by the age of five (Cohen, 2006). Mitochondrial diseases are characterized by a bewildering array of signs and symptoms (Naviaux, 2004). For example, one single-point mutation in mitochondrial DNA has been reported to contribute to over nine different disorders such as diabetes, congestive heart failure, chronic progressive external ophthalmoplegia (CPEO), schizophrenia, and kidney malfunction [reviewed in Naviaux (2004)]. Although mitochondrial involvement in the pathogenesis of human diseases had already been discussed as early as in 1962 (Luft et al., 1962), the causative link between mitochondrial defects and human diseases was identified for the first time only 26 years later. In 1988, Wallace et al. reported the association of a mitochondrial DNA mutation with Leber's hereditary optic neuropathy, and Holt et al. (1988) identified mitochondrial DNA deletions in patients suffering from myopathies. Since then, the number of human diseases that have been recognized to be caused by mitochondrial malfunctions has exploded. So far, 347 mitochondrial disorders have been identified (Naviaux, 2004). The majority of them display either neurodegenerative or neuromuscular symptoms. Mitochondrial medicine is currently one of the fastest growing areas in biomedical research (Naviaux, 2004) that has also given rise to new sub-disciplines such as mitochondrial pharmacology (Szewczyk & Wojtczak, 2002) and mitochondrial pharmaceutics (Weissig et al., 2004). The identification of new molecular mitochondrial drug targets in combination with the development of methods for selectively delivering biologically active molecules to the site of mitochondria will potentially launch new therapeutic approaches for the treatment of mitochondria-related diseases, based on the selective protection, or repair or eradication of cells.

2 Mitochondria as Cell Organelles

Although a single animal mitochondrion inside a cell was depicted on the front cover of the March 5, 1999, issue of *Science*, the widely accepted perception of individual mitochondria moving around the cytosol has to be thoroughly revised. It is now known that mitochondria form a complex tubular network, which is constantly changing its shape by undergoing fission and fusion (Chan, 2006; Chen & Chan, 2005; Polyakov et al., 2003). Moreover, the mitochondrial network is tightly associated and seems to interact with cytosolic microtubules. Using two-color 4Pi microscopy in combination with quantum dots, Medda et al. (2006) were able to provide insight into the intimate spatial intermingling of mitochondria with the microtubular network in mammalian cells (Medda et al., 2006). Figure 1 shows how the microtubular meshwork (in green) is strongly intertwined with the red colored mitochondrial tubular network. Images of higher resolution (not shown here) even depict features like ring-shaped mitochondria surrounding microtubules (Medda et al., 2006). Such a strong association between the mitochondrial and the microtubular network obviously has to be considered when designing nano drug carriers for mitochondrial-targeted drug and DNA delivery.

The mitochondrial mass in a single cell depends on cellular energy demand. Cells in metabolically active organs such as the liver, the brain, and cardiac and skeletal muscle tissues may have up to 100 times the mitochondrial mass of cells of somatic tissues with a lower demand for energy. Mitochondria are composed of two membranes, which together create two separate compartments, the internal matrix and the narrow intermembrane space. The outer membrane is permeable to molecules smaller than 5 kDa, whereas the inner membrane is highly impermeable and characterized by an unusually high content of membrane proteins as well as a unique lipid composition. The mitochondrial inner membrane proteins comprise components of the respiratory chain plus a large number of transport proteins. The

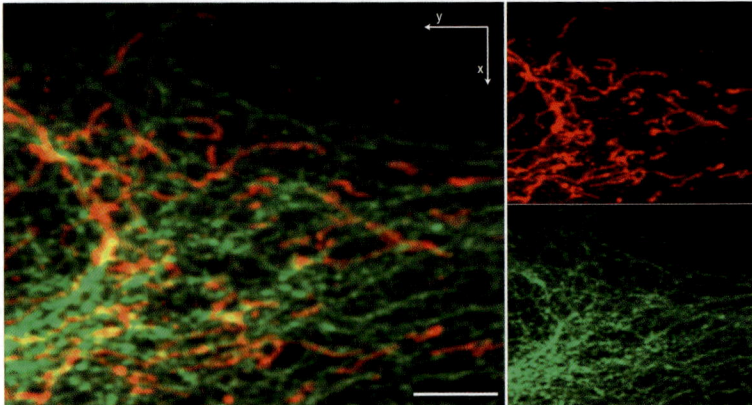

Fig. 1 Quantum dot labeling of the mitochondrial and the microtubular network within a mammalian cell. Reprinted in slightly modified form from Medda et al. (2006). Two-color 4Pi microscopy with quantum dot labeling: mitochondria (α-subunit of the F_1F_0-ATP-synthase, QDot 655, red) and microtubules (α-Tubulin, QDot 605, green). The figure shows the axial average projection of the whole 4Pi data stack; the red and green structures reveal the mitochondria and the microtubules, respectively. The *left panel* is the overlay of both images shown in the *right panel*. Scale bar, 5 µm. The image was graciously provided by Rebecca Medda and Joerg Bewersdorf; copyright permission was granted by the *Journal of Structural Biology* (Elsevier)

impermeability of the inner membrane is a prerequisite for the establishment of an imbalance in the distribution of protons between the mitochondrial matrix and the cytosol, which in turn is the driving force for the synthesis of ATP. Mitochondria are unique among animal cell organelles as they contain their own genome (mtDNA) and associated expression system.

While as the "power house" of the cell the mitochondrion is essential for the energy metabolism, as the "death switch" this organelle is central to the regulation and execution of apoptosis. Apoptosis plays a central role in tissue homeostasis and it is generally considered that inhibition of apoptosis may contribute to cell transformation (Costantini et al., 2000; Green & Reed, 1998; Gulbins et al., 2003; Waterhouse et al., 2001). Significant knowledge has been accumulated about how the apoptotic machinery is controlled. As each new regulatory mechanism had been identified, dysfunction of that mechanism has been linked to one or another type of cancer (Kaufmann & Gores, 2000).

Further, mitochondria are critically involved in the modulation of intracellular calcium concentration; the mitochondrial respiratory chain is the major source of damaging reactive oxygen species; and mitochondria play a crucial role in numerous catabolic and anabolic pathways.

3   Cytosolic Barriers to Mitochondrial Drug and DNA Delivery

Because of the presence of mitochondria in almost all cells, the view that once inside the cell, any drug or DNA molecule will eventually interact with components of the mitochondrial membrane and – depending on its physico-chemical properties – perhaps diffuse into the matrix, appears plausible. Without a doubt, statistical collision may indeed lead to an interaction of drug molecules with mitochondria. Yet, while barriers to drug delivery between the site of administration and the target tissue are well-defined and recognized, intracellular barriers preventing an even distribution of drug molecules throughout the cell, which the drug has to overcome in order to reach its subcellular target site, are less well-characterized.

Various factors slow, or even prevent, free diffusion of solutes in the cytosol. First is the fluid-phase viscosity of the cytoplasm; second are collisional interactions due to macromolecular crowding; and third is binding to intracellular components (Lukacs et al., 2000; Seksek et al., 1997). The impact on cytoplasmic solute diffusion of the combination of these three factors can be measured, and is expressed as the translational diffusion coefficient. Verkman and colleagues (Lukacs et al., 2000) have used spot photobleaching to measure the translational diffusion of fluorescein-labeled double-stranded DNA fragments of different sizes after microinjection into the cytoplasm of HeLa cells. While the ratio between the relative diffusion coefficients in water (D_w) and in cytoplasm (D_{cyto}) was about unity for small oligonucleotides, D_{cyto}/D_w progressively decreased to 0.19, 0.067, and 0.032 for DNA fragments of 100, 250, and 500 bp, respectively. The diffusion rate was further dramatically reduced as DNA size increased beyond 1 kb (660 kDa) and was immeasurably slow for DNAs of 3 kb or greater. The same group has shown in more recent work (Dauty & Verkman, 2005) that the size-dependent reduction in the mobility of large macromolecules or DNA fragments was abolished after disruption of actin microfilaments, thereby identifying this component of the cytoskeleton as the principal structure that limits passive diffusion through the cytoplasm.

In contrast, Vaughan and Dean (2006) did not see changes in intracellular plasmid trafficking following either stabilization or disassembly of the actin cytoskeleton. Instead, using microinjection and electroporation approaches in the presence of drugs that alter the dynamics and organization of the cytoskeleton, they showed that microtubules are involved in plasmid trafficking to the nucleus (Vaughan & Dean, 2006). In addition, by co-injecting inhibitory antibodies, these authors found that dynein likely facilitates this movement. These results were confirmed using an in vitro spin-down assay that demonstrated that plasmids bind to microtubules through adaptor proteins provided by cytoplasmic extracts. Since Vermerk's studies were focused on measuring rates of diffusion on small time scales (seconds to minutes), while Vaughan and Dean were assessing the movement of DNA over periods of hours, it is concluded that disruption of the actin network may play a significant role in local movement over very short times, but this role may be lessened over time, as microtubules play the predominant role in DNA movement (Vaughan & Dean, 2006).

Detailed insight into the mechanism of interaction between DNA and microtubules was recently provided by Mesika et al. (2005). These authors studied the intracellular transport of plasmid DNA microinjected into HeLa cell cytoplasm, alone or as a complex with intact or NLS-deleted NFκB p50. They found that association of NLS-carrying p50 with DNA facilitated not only nuclear entry of the DNA but also its migration through the cytoplasm towards the nucleus. Moreover, they also found that facilitated transport of p50-DNA complexes in the cytoplasm proceeded along microtubules in a dynein-dependent manner and was mediated by the heterodimeric nuclear transport receptor that recognizes the p50-born NLS (Mesika et al., 2005). Also, from work concerning protein transport in neurons, it has been recognized that the formation of multiprotein complexes is responsible for transport processes along the microtubule network (Hanz et al., 2003; Perlson et al., 2005). Likewise, it is well-known that a variety of viruses, i.e., "natural nanoparticulates", utilize microtubules to reach the nucleus [reviewed in (Vaughan & Dean, 2006)].

With particular respect to mitochondrial DNA delivery, and considering both the role the microtubular network plays for intracellular DNA transport and the intimate interaction between the microtubular and the mitochondrial network, one can speculate that all DNA in whatever form repeatedly passes near to mitochondrial membranes. Subsequently, one can hypothesize that linking the DNA with molecules that specifically accumulate at or in mitochondria will divert the traffic flow from the nucleus towards mitochondria.

In addition to the components of the cytoskeleton (actin filaments and microtubular network), the cytoplasm is filled with a large number of cell organelles with unique intralumenal pH values, resident molecules, electronic potentials, lipid-bilayer compositions and membrane-bound proteins. Obviously, such organelle-specific properties may also have a significant impact on the intracellular drug and DNA trafficking. Based on studying the intracellular localization for a wide variety of dyes and fluorescent probes, Horobin and colleagues have developed a QSAR modeling approach for predicting the cellular uptake, the intracellular distribution, and the site of intracellular accumulation for fluorescent probes and dyes (Horobin, 2001, 2002; Trapp & Horobin, 2005). Horobin's work demonstrates that the amphiphilic character, the size of the aromatic system, the electric charge, the overall size, the presence of planar aromatic moieties, the lipophilicity, and the acid–base properties can all influence the intracellular disposition of these molecules (as discussed in more detail later).

In overall conclusion it becomes apparent that, following cell entry, a random or statistical interaction of drug molecules or pDNA with mitochondria is not to be expected. The effective interaction of all kind of pharmaceuticals with or uptake by mitochondria will take place only when the drug either meets Horobin's QSAR criteria for mitochondrial localization or when the drug is being transported to mitochondria by an appropriate mitochondria-targeted delivery system.

4 Mitochondriotropics and Mitochondria-Targeted Nano Drug Carriers

The physico-chemical parameters required for mitochondrial targeting of low-molecular weight compounds (drugs, dyes, and other xenobiotics) can be numerically defined by utilizing available QSAR models. These specify the physico-chemical features of such agents and, by predicting possible localization in additional cellular structures, also indicate the degree of selectivity of the mitochondriotropic behavior.

Most recently it was found that the mitochondrial targeting of nearly 80% of a nonselected sample of over 100 compounds known to be localized, or active, in mitochondria could be predicted using two sets of QSAR models, one specifying accumulation of lipophilic cations and the other of lipophilic weak acids (Horobin et al., 2007). The author's approach allowed for uptake into additional intracellular sites, such as endoplasmic reticulum and generalized biomembranes. Table 1 summarizes numerically expressed physico-chemical criteria for entry into the cell and for mitochondrial uptake as well as for localization to other cell organelles.

However, this approach fails to predict the observed mitochondrial activity of certain compounds. Some of these may access the organelles without selective accumulation, and criteria for this are $8 > \log P > 0$. Nevertheless, as some mitochondriotropic compounds still remain outside the criteria, an alternative predictive procedure has been explored, namely a first-principles physico-chemical approach based on the Fick-Nernst-Planck equations (Trapp & Horobin, 2005). However, the details of this are beyond the scope of this chapter.

Table 1 Quantitative physico-chemical molecular parameters ("QSAR decision rules") allowing the prediction of the intracellular distribution of low-molecular compounds

Event	Physico-chemical criteria
Entry into the cell	$8 > AI$ or $log\,P > 0$; $CBN < 40$
Selective mitochondrial uptake of cations	$5 > log\,P_{cation} > 0$
Selective mitochondrial uptake of cations according to "First-principles model" (see text)	$5 > log\,P_{cation} > -2$
Selective mitochondrial uptake of acids	$pK_a = 7\pm3$; $5 > log\,P_{less\,ionized\,species} > 0$
Selective cation uptake in ER	$6 > AI > 3.5$; $6 > log\,P_{cation} > 0$
Non-selective uptake in membranes	$8 > log\,P > 5$

AI amphipathic index; *CBN* conjugated bond number; *ER* endoplasmic reticulum; *Log P* log of the octanol water partition coefficient [for details see references in Horobin et al. (2007)]

More important here is the fact that the QSAR decision rules outlined above have been validated only for low-molecular weight compounds. The critical question for this chapter is whether the summary criteria (Table 1) can be applied to mitochondrial localization of nano particulate systems. Since obviously the size of the nanoparticles should markedly influence organelle targeting, the answer to the critical question has to be negative. But unfortunately, numerical criteria for selective targeting of nanoparticles to mitochondria are unavailable, as these will evidently require either a better theory or more extensive experimentation (or both).

However, it has been demonstrated experimentally that either the self-assembly of mitochondriotropic molecules or the surface modification of nanoparticles with mitochondriotropics creates nano particulate systems with cationic surfaces, which exhibit a remarkable mitochondrial targeting specifity when applied to intact living cells. Figure 2 shows a schematic overview of nano particulate systems composed of or modified with mitochondriotropic molecules or other mitochondria-specific ligands.

Nanoparticulate systems displaying cationic head groups on their surfaces include bolasomes prepared from dequalinium, with weakly hydrophilic cations, and its cyclohexyl-dequalinium derivative (Weissig et al., 1998a, 2001b), liposomes with surface-linked lipophilic triphenylphosphonium cations (Boddapati et al., 2005) and liposomes modified with lipophilic cationic cetylated polyethylenimine (Takeuchi et al., 2004). The use of bolasomes and triphenylphosphonium-bearing liposomes for mitochondria-targeted drug and DNA delivery will be discussed in more detail later.

In contrast, the most widely used cationic lipids (Kostarelos & Miller, 2005) used for nuclear-targeted DNA delivery (lipoplexes) contain strongly hydrophilic head groups, and these do not target mitochondria. Reassuringly these examples are congruent with the criteria for small molecules given earlier; at least if using the Trapp & Horobin (2005) modification.

When modifying the surface of nano particulate systems with mitochondriotropic ligands, the amount of ligand per particle, i.e., the ligand's surface density, should be important. Although appropriate systematic quantitative studies are not yet available, Callahan & Kopecek (2006) have linked a single mitochondriotropic triphenylphosphonium group to hydroxypropylmethacrylamide polymers. With

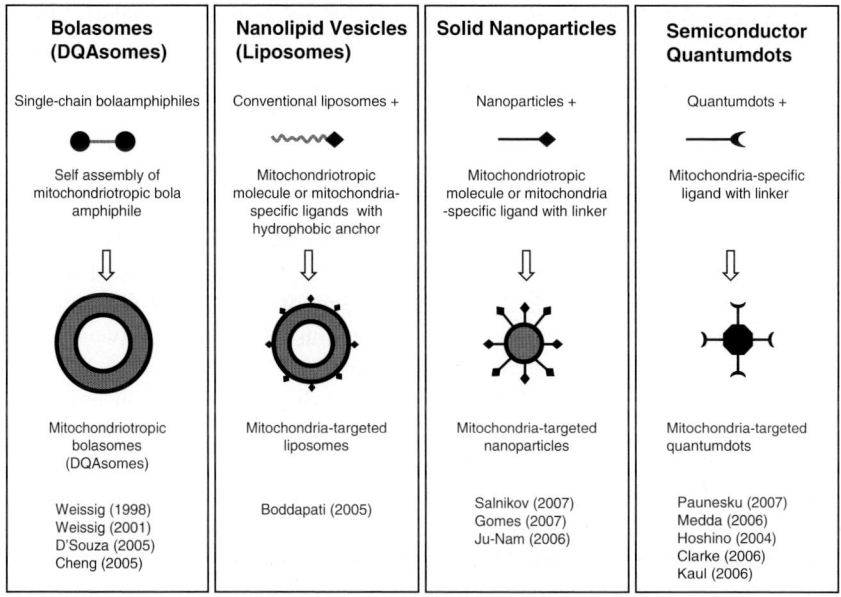

Bolasomes (DQAsomes)	Nanolipid Vesicles (Liposomes)	Solid Nanoparticles	Semiconductor Quantumdots
Single-chain bolaamphiphiles	Conventional liposomes +	Nanoparticles +	Quantumdots +
Self assembly of mitochondriotropic bola amphiphile	Mitochondriotropic molecule or mitochondria-specific ligands with hydrophobic anchor	Mitochondriotropic molecule or mitochondria -specific ligand with linker	Mitochondria-specific ligand with linker
Mitochondriotropic bolasomes (DQAsomes)	Mitochondria-targeted liposomes	Mitochondria-targeted nanoparticles	Mitochondria-targeted quantumdots
Weissig (1998) Weissig (2001) D'Souza (2005) Cheng (2005)	Boddapati (2005)	Salnikov (2007) Gomes (2007) Ju-Nam (2006)	Paunesku (2007) Medda (2006) Hoshino (2004) Clarke (2006) Kaul (2006)

Fig. 2 Mitochondria-specific nano particulate systems. Reprinted from Weissig et al. (2007), with permission

isolated mitochondria, there was evidence of mitochondrial localization, but only with the smallest (<5 kDa) species. With intact cells, even the smallest nanoparticles were restricted to the lysosomal compartment after extended as well as short exposure periods.

Figure 2 shows schematically that, in addition to bolasomes and liposomes, solid nanoparticles and quantum dots have also been rendered mitochondria-specific via surface attachment of appropriate ligands. The exploration of such systems for mitochondria targeted drug and DNA delivery, however, is in a very early stage and will therefore not be further discussed in this chapter.

5 Bolasome-Mediated Mitochondria-Specific DNA Delivery

The direct delivery of nucleic acids and/or their derivatives (pDNA, PNA, oligonucleotides, siRNA) into mitochondria constitutes one of several approaches towards a long-term therapy of mitochondrial DNA diseases (D'Souza & Weissig, 2004). However, while considerable effort has been invested in the optimization of nonviral nuclear-targeted DNA delivery, less work has been undertaken so far to achieve successful delivery of nucleic acids into mitochondria within living mammalian cells (D'Souza et al., 2007). Until the end of the 1990s, not one single mitochondria-specific vesicular or particular transfection vector had been described and the development of the first such system was actually based on an accidental discovery.

In the mid 1990s, while screening mitochondriotropic drugs potentially able to interfere with the mitochondrial DNA metabolism (Rowe et al., 2001) in *P. falciparum*, Weissig et al. found that dequalinium chloride, a bola-amphiphilic drug, tends to self-associate into nanovesicular structures when sonicated as an aqueous suspension. At the time of their discovery, these vesicles were called DQAsomes, i.e., dequalinium-based liposome-like vesicles; pronounced dequasomes (Weissig et al., 1998a).

Symmetric amphiphilic molecules, in which two hydrophilic residues are linked by a hydrophobic segment, are generally known as "bola-lipids" based on their resemblance to an old South American hunting weapon. Well-characterized bola-amphiphiles are archaebacterial lipids, which usually consist of two glycerol backbones connected by two hydrophobic chains. The self-assembly behavior of such bipolar archaeal lipids has been extensively studied and it has been shown that they can self-associate into mechanically very stable monolayer membranes (De Rosa et al., 1986; Gambacorta et al., 1995).

A structural difference between dequalinium and archaeal lipids, however, lies in the number of bridging hydrophobic chains between the polar head groups. In contrast to the common arachaeal lipids, dequalinium has only one alkyl chain that connects the two cationic hydrophilic head groups. Therefore, theoretically two different conformations within a self-assembled layer structure are imaginable. While the stretched conformation would give rise to the formation of a monolayer, assuming the horseshoe conformation would result in the formation of a bilayer. While analyzing the self-assembly behavior of dequalinium salts employing Monte Carlo Computer Simulations (Weissig et al., 1998b) it was found that both conformations, i.e., bola and horse shoe, could theoretically co-exist, although the balance between them appeared to be temperature dependent (Mögel & Wahab, unpublished).

Just like cationic liposomes used for transgene delivery into the nucleus, DQAsomes form complexes with DNA (called *DQAplexes*) following mixing of the positively charged vesicles with the negatively charged polynucleotide acid. Figure 3 shows an electron micrograph of DQAsomes before (*bottom panel*) and after adding pDNA (*top panel*).

Although the detailed structure of DQAplexes is yet to be determined, the appearance of the DQAplex shown in Fig. 3 strongly resembles so-called "spaghetti and meatball" structures observed by Sternberg et al for complexes formed by DC-Chol/DOTAP liposomes and pDNA (Sternberg et al., 1994).

The observation that DQAsomes were able to efficiently bind DNA (Weissig et al., 1998a) and protect the DNA from nuclease digestion as well as mediate its cellular uptake (Lasch et al., 1999) led to the first particulate carrier-based strategy proposed for direct mitochondrial transfection (Weissig & Torchilin, 2000, 2001a,b).

This approach requires the transport of a DNA-mitochondrial leader sequence (MLS) peptide conjugate to mitochondria using DQAsomes, the liberation of this conjugate from the cationic vector upon contact with the mitochondrial outer membrane followed by DNA uptake via the mitochondrial

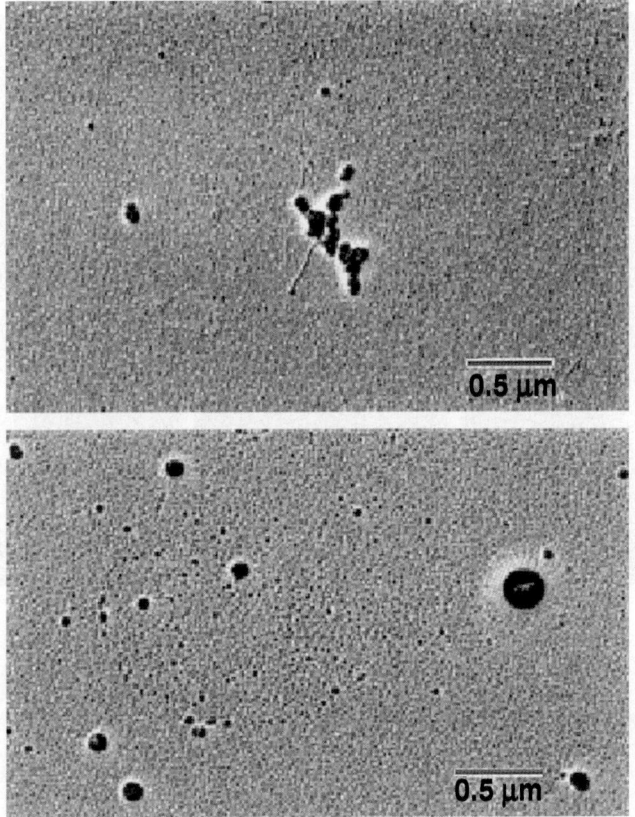

Fig. 3 Formation of complexes of DQAsomes and pDNA ("DQAplexes"). Electron photomicrograph of a DQAsome–DNA complex (rotary shadowed): *Bottom panel*: DQAsomes before adding plasmid DNA; *Top*: DQAsomes in the presence of DNA at a ratio DQA/pDNA = 12, w/w (image taken by G. Erdos, Gainesville, FL)

protein import machinery. DQAsome–DNA complexes ("DQAplexes") were shown to release their DNA cargo only upon contact with mitochondrial membranes (Weissig et al., 2000, 2001a). Further, utilizing a newly developed protocol for selectively staining free pDNA inside the cytosol, it was demonstrated that DQAsomes, upon their endosomal escape, selectively deliver pDNA to and release the pDNA exclusively at the site of mitochondria in living mammalian cells (D'Souza et al., 2003). Finally, utilizing confocal fluorescence microscopy and using DNA–MLS peptide conjugates it was shown that DQAsomes not only deliver oligonucleotides but are also able to deliver plasmid-sized DNA into mitochondria within living mammalian cells (D'Souza et al., 2005).

Figure 4 shows confocal fluorescence micrographs of cells exposed to circular (*top row*) and linearized pDNA conjugate (*bottom row*) complexed with DQAsomes.

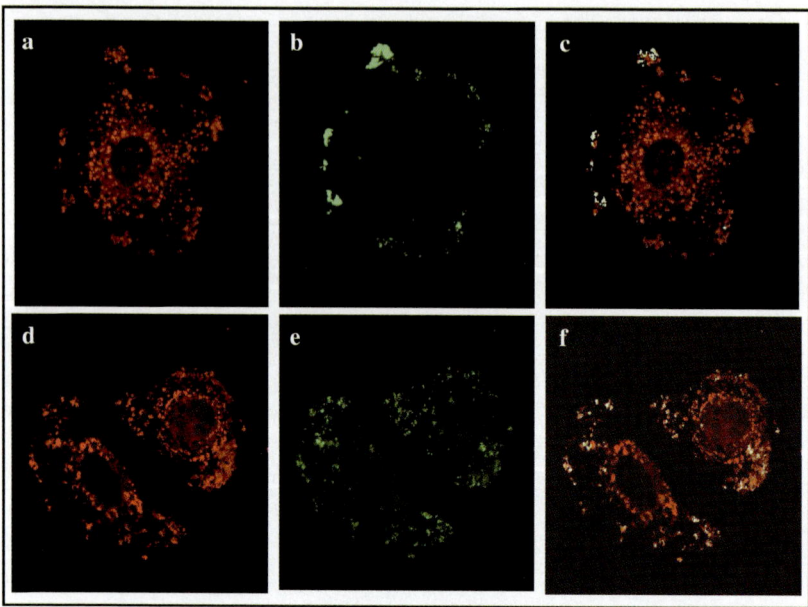

Fig. 4 DQAsome-mediated delivery of pDNA to mitochondria within living mammalian cells: Confocal fluorescence micrographs of BT20 cells stained with Mitotracker Red CMXRos (red) after exposure for 10 h to fluorescein labeled MLS-pDNA conjugate (green) complexed with cyclohexyl-DQAsomes; (**a–c**): circular pDNA conjugate, (**d–f**): linearized pDNA conjugate. (**a,d**) green channel, (**b,e**): red channel, (**c,f**): overlay of red and green channels with white indicating co-localization of red and green fluorescence. Reprinted with permission from D'Souza et al. (2005)

In the case of the linearized conjugate higher levels of co-localization (depicted in white) than with the same amount of nonlinearized conjugate can be seen, suggesting the possibility that a linear conjugate might be better suited to delivery and import into the mitochondria.

It should be stressed at this point that the use of physico-chemical methods is so far still the only way to demonstrate the import of transgene DNA into the mitochondrial matrix in living mammalian cells. Up to this date (June 2007) no functional expression of any transgene DNA inside mammalian mitochondria within living cells has been demonstrated. Moreover, the complete lack of a mitochondria-specific reporter plasmid designed for mitochondrial expression severely hampers all efforts towards the development of effective mitochondrial expression vectors. While any new nonviral transfection system (i.e., cationic lipids, polymers, and others) aimed at the nuclear-cytosolic expression of proteins can be systematically tested and subsequently improved by utilizing anyone of many commercially available reporter gene systems, such a methodical approach to develop mitochondrial transfection systems is currently impossible.

6   Mitochondria-Specific Liposomes

Phospholipid vesicles (liposomes) were developed during the 1970s and 1980s and subsequently brought to the clinic in the middle of the 1990s. The enormous potential of liposomes as a colloidal drug and DNA delivery system for biomedical applications was discussed in two prescient papers published in 1976 in the *New England Journal of Medicine* by Gregory Gregoriadis (Gregoriadis, 1976a,b); note that this was three decades before the term *nano* started replacing the term *colloidal* in the pharmaceutical literature. Widely discussed problems and issues linked to the biomedical applications of solid nanoparticles such as biocompatibility, biodegradability, biodistribution, toxicity, surface modification, drug encapsulation, drug retention, and release have been addressed and largely solved in the field of liposome technology (Gregoriadis, 1988, 1993, 2007; Janoff & NetLibrary Inc., 1999; Lasic, 1993; Lasic & Martin, 1995; Philippot & Schuber, 1995; Torchilin & Weissig, 2003; Woodle & Storm, 1998).

The preparation of liposomes in the presence of hydrophilic molecules, which have been artificially hydrophobized via linkage to fatty acid or phospholipid derivatives, results in the "anchoring" of the hydrophilic moiety to the liposomal surface (Niedermann et al., 1991; Torchilin et al., 2003; Weissig & Gregoriadis, 1993; Weissig et al., 1986, 1989). This established procedure was used to render liposomes mitochondria-specific, as schematically shown in Fig. 5.

First, triphenylphosphonium cations were hydrophobized by reacting triphenylphosphine with stearylbromide. The resulting stearyl-triphenylphosphonium bromide (STPP) was then added to the solution of phospholipids in chloroform used to prepare SUV liposomes. STPP liposomes, i.e., liposomes with surface-linked

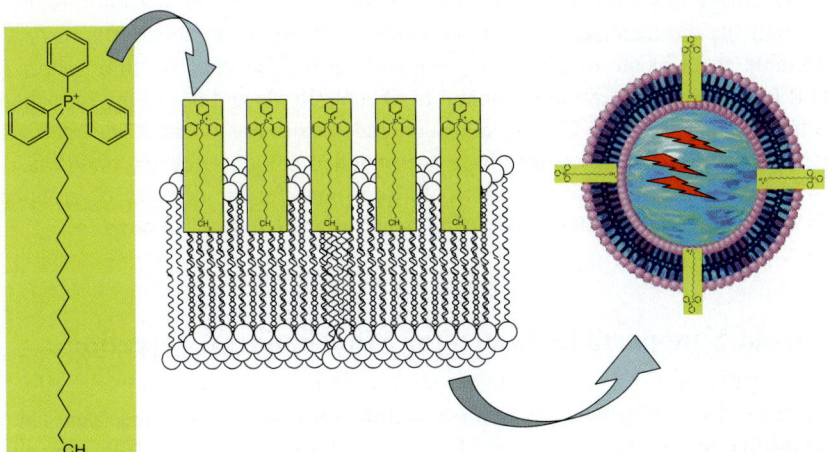

Fig. 5 Schematic depiction of the surface modification of liposomes with hydrophobized triphenylphosphonium cations. Stearyl-triphenylphosphonium bromide (STPP) was synthesized from triphenylphosphine and stearylbromide. SUV liposomes were then prepared in the presence of STPP. Triphenylphosphonium cations anchored in the inner monolayer are not shown

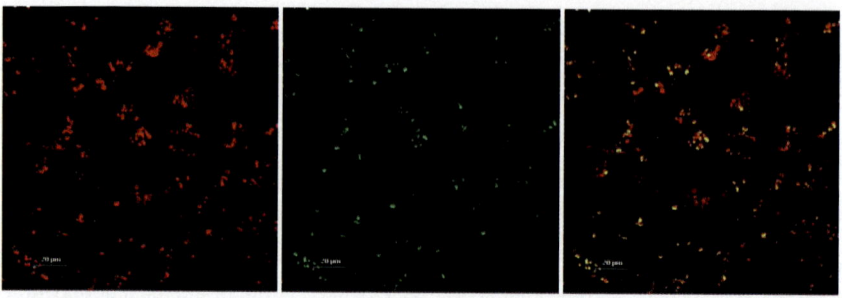

Fig. 6 Intracellular distribution of mitochondria-specific liposomes. Confocal fluorescence microscopic image of human breast cancer cells (MDA-MB231) following their incubation with NBD-PE labeled mitochondriotropic (20% STPP) liposomes. *Left panel*: Red channel, mitochondria stained with Mitotracker Red CMXROS; *Middle panel*: Green channel, NBD-PE; *Right panel*: Overlaid images, areas of co-localization of NBD-PE and mitochondria appear in yellow, bar = 20 μm. Reprinted with permission from Weissig et al. (2006)

triphenylphosphonium cations, were isolated using a Sepharose CL-4B column and characterized by ^{31}P-NMR, size distribution analysis, and zeta potential measurements [for details see Boddapati et al. (2005)]. Figure 6 shows a confocal fluorescence image of human breast cancer cells incubated with NBD-PE labeled STPP liposomes.

The *left panel* shows mitochondria stained in red and the *middle panel* shows NBD-PE in green. The *right panel* in Fig. 6 was obtained by overlaying the red and the green images, with mixed red and green color pixels being depicted in yellow. Strikingly, all of the green NBD fluorescence from the *middle panel* appears in the overlaid image in yellow, indicating that the amount of NBD-PE used in this study quantitatively co-localizes with mitochondria. There is no green fluorescence detectable at all in the overlaid image. Based on the fact that the fluorophore in STPP liposomes was covalently linked to phospholipids and not to the mitochondriotropic entity, i.e., to STPP, it can be concluded from Fig. 6 that at least partially intact phospholipid vesicles must have accumulated at the site of mitochondria.

The exploration of mitochondriotropic liposomes for the delivery of proapoptotic drugs known to act at the site of mitochondria is currently in progress.

7   Gold Nanoparticles Suited for Mitochondrial Targeting

Nanotechnology is increasingly applied to different areas of medicine. New nanodiagnostic tools, including quantum dots and gold nanoparticles (AuNPs), promise increased sensitivity, multiplexing capabilities, and reduced costs for many diagnostic applications (Azzazy et al., 2006). Imaging agents such as fluorescent dye-doped silica nanoparticles, quantum dots, and AuNPs have overcome many of the limitations of conventional contrast agents such as poor photostability, low quantum yield, and insufficient in vitro and in vivo stability (Sharma et al., 2006). Such particulates are now being developed for absorbance and emission in the near infrared region, which is expected to allow for real-time and deep tissue imaging via optical routes (Sharma et al., 2006). A particular promising application of nanotechnology lies in the area of anticancer therapy. In this regard, AuNPs can mediate hyperthermia induction and kill tumor cells upon laser irradiation, thereby functioning as a "thermal scalpel" (Everts, 2007).

Attempts to introduce nanotechnology into mitochondrial medicine have been very limited so far. Though considerable progress has been made in identifying molecular components of the mitochondrial machinery responsible for life and cell death, large gaps remain in our knowledge regarding their exact quantities, variations and locations, the timescales of their interactions, their precise internal movements, as well as their specific affinities. The recent exciting advances in nanotechnology hold the promise of helping to overcome the limitations of existing technologies for probing and controlling intracellular processes (Paunesku et al., 2003). Nanotechnology is bound to provide the needed tools for accessing, probing, and manipulating mitochondria and their sub-organellar components under physiological conditions.

A very first step towards merging nanotechnology with mitochondrial medicine was recently undertaken by coating the surface of AuNPs with mitochondriotropic triphenylphosphonium cations (Fig. 7). Ju-Nam et al. (2006) incubated potassium tetrachloroaurate with triphenyl-phosphonioalkylthiosulfate, a member of a whole group of newly synthesized zwitterions, and added dropwise an aqueous solution of sodium borohydride.

After vigorous stirring for 24 h, the authors were able to isolate 5–10-nm sized AuNPs with surface-attached triphenylphosphonium residues. Unfortunately, at the time of writing this chapter (June 2007) data regarding the intracellular distribution of such triphenylphosphonium cation bearing AuNPs were not yet available. It will be particularly interesting to compare plain 3-nm AuNPs as described by Parfenov

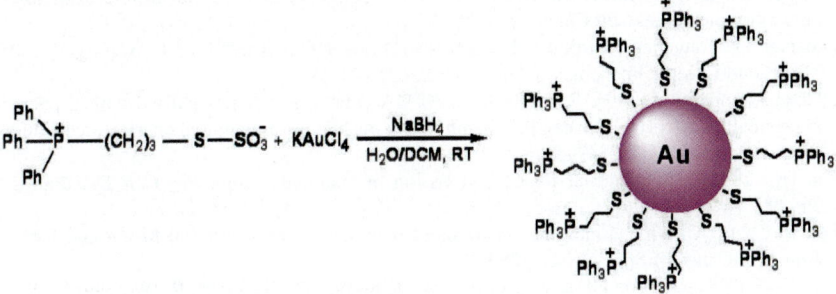

Fig. 7 Gold nanoparticles bearing surface-linked mitochondriotropic ligands. Reaction scheme for the surface modification of gold nanoparticles with triphenylphosphonium cations. The image was graciously provided by Neil Bricklebank and reprinted in slightly modified form from Ju-Nam et al. (2006); copyright permission was granted by *Organic and Biomolecular Chemistry* (The Royal Society of Chemistry)

et al. (2006) and triphenylphosphonium-capped AuNPs of the same size in terms of their mitochondrial accumulation inside living mammalian cells. Corresponding experiments are ongoing in the author's laboratory.

8 Summary

In recent years, the incursion of nanotechnology into mitochondrial research has begun. Despite the limited amount of published data, the potential of nanotechnological tools to either probe or to manipulate mitochondrial functions has been demonstrated. DQAsomes, liposomes, and solid nanoparticles can be targeted to mitochondria. Nanotechnology has also generated mitochondria-specific vectors capable of selectively delivering biologically active molecules to mitochondria within living mammalian cells, thereby enhancing the subcellular bioavailability of drugs. Although DQAsomes and modified liposomes have been established as the above-mentioned mitochondria-specific pDNA and drug vectors, solid nanoparticles also have the potential to be utilized as vector systems. Given the current pace of development, nanotechnology will have an integral role in the future of mitochondrial research, including mitochondrial biology, medicine, pharmacology, and pharmaceutics.

Acknowledgements We acknowledge the Muscular Dystrophy Association (Tucson, AZ) and the Mitochondrial Disease Foundation (Pittsburgh, PA) for their financial support (VW). We thank Prof V. P. Torchilin for his support and many helpful discussions (VW). Finally, we acknowledge Prof W. Martin, Division of Neuroscicence & Biomedical Systems, IBLS, The University of Glasgow, Scotland, for provision of facilities to one of us (RWH).

References

Azzazy HM, Mansour MM, Kazmierczak SC. 2006. Nanodiagnostics: a new frontier for clinical laboratory medicine. *Clin Chem* 52: 1238–46.

Boddapati SV, Tongcharoensirikul P, Hanson RN, D'Souza GG, Torchilin VP, Weissig V. 2005. Mitochondriotropic liposomes. *J Liposome Res* 15: 49–58.

Callahan J, Kopecek J. 2006. Semitelechelic HPMA copolymers functionalized with triphenyl-phosphonium as drug carriers for membrane transduction and mitochondrial localization. *Biomacromolecules* 7: 2347–56.

Chan DC. 2006. Mitochondrial fusion and fission in mammals. *Annu Rev Cell Dev Biol* 22: 79–99.

Chen H, Chan DC. 2005. Emerging functions of mammalian mitochondrial fusion and fission. *Hum Mol Genet* 14 Spec No. 2: R283–9.

Cheng SM, Pabba, S., Torchilin, V.P., Fowle, W., Kimpfler, A., Schubert, R., Weissig, V. 2005. Towards mitochondria-specific delivery of apoptosis-inducing agents: DQAsomal incorporated paclitaxel. *J Drug Deliv Sci Technol* 15: 81–6.

Clarke SJ, Hollmann CA, Zhang Z, Suffern D, Bradforth SE, et al. 2006. Photophysics of dopamine-modified quantum dots and effects on biological systems. *Nat Mater* 5: 409–17.

Cohen BH. 2006. Incidence and prevalence rates of mitochondrial diseases. *UMDF Mitochondrial News* 11: 5–16.

Costantini P, Jacotot, E., Decaudin, D., Kroemer, G. 2000. Mitochondrion as a novel target of anticancer chemotherapy. *J Natl Canc Inst* 92: 1042–53.

Dauty E, Verkman AS. 2005. Actin cytoskeleton as the principal determinant of size-dependent DNA mobility in cytoplasm: a new barrier for non-viral gene delivery. *J Biol Chem* 280: 7823–8.

De Rosa M, Gambacorta A, Gliozi A. 1986. Structure, biosynthesis, and physicochemical properties of archaebacterial lipds. *Microbiol Rev* 50: 70–80.

D'Souza GG, Boddapati SV, Weissig V. 2005. Mitochondrial leader sequence–plasmid DNA conjugates delivered into mammalian cells by DQAsomes co-localize with mitochondria. *Mitochondrion* 5: 352–8.

D'Souza GG, Boddapati SV, Weissig V. 2007. Gene therapy of the other genome: the challenges of treating mitochondrial DNA defects. *Pharm Res* 24: 228–38.

D'Souza GG, Rammohan R, Cheng SM, Torchilin VP, Weissig V. 2003. DQAsome-mediated delivery of plasmid DNA toward mitochondria in living cells. *J Contr Rel* 92: 189–97.

D'Souza GGM, Weissig, V. 2004. Approaches to mitochondrial gene therapy. *Curr Gene Ther* 4: 317–28.

Everts M. 2007. Thermal scalpel to target cancer. *Expert Rev Med Devices* 4: 131–6.

Gambacorta A, Gliozi A, De Rosa M. 1995. Archaeal lipids and their biotechnological applications. *World J Microbiol Biotechnol* 11: 115–31.

Gomes AJ, Faustino AS, Lunardi CN, Lunardi LO, Machado AE. 2007. Evaluation of nanoparticles loaded with benzopsoralen in rat peritoneal exudate cells. *Int J Pharm* 332: 153–60.

Green DR, Reed JC. 1998. Mitochondria and apoptosis. *Science* 281: 1309–12.

Gregoriadis G. 1976a. The carrier potential of liposomes in biology and medicine (first of two parts). *N Engl J Med* 295: 704–10.

Gregoriadis G. 1976b. The carrier potential of liposomes in biology and medicine (second of two parts). *N Engl J Med* 295: 765–70.

Gregoriadis G. 1988. *Liposomes as drug carriers*. Chichester, UK: Wiley.

Gregoriadis G. 1993. *Liposome technology*. Boca Raton, FL.: CRC Press. 3 v. pp.

Gregoriadis G. 2007. *Liposome technology*. New York: Informa Healthcare.

Gulbins E, Dreschers S, Bock J. 2003. Role of mitochondria in apoptosis. *Exp Physiol* 88: 85–90.

Hanz S, Perlson E, Willis D, Zheng JQ, Massarwa R, et al. 2003. Axoplasmic importins enable retrograde injury signaling in lesioned nerve. *Neuron* 40: 1095–104.

Holt IJ, Harding AE, Morgan-Hughes JA. 1988. Deletions of muscle mitochondrial DNA in patients with mitochondrial myopathies. *Nature* 331: 717–19.

Horobin RW. 2001. Uptake, distribution and accumulation of dyes and fluorescent probes within living cells: A structure–activity modelling approach. *Adv Colour Sci Technol* 4: 101–7.

Horobin RW. 2002. Biological staining: mechanisms and theory. *Biotech Histochem* 77: 3–13.

Horobin RW, Trapp, S., Weissig, V. 2007. Mitochondriotropics: A review of their mode of action, and their application for drug and DNA delivery to mammalian mitochondria. *J Cont Rel* 121: 125–136.

Hoshino A, Fujioka K, Oku T, Nakamura S, Suga M, et al. 2004. Quantum dots targeted to the assigned organelle in living cells. *Microbiol Immunol* 48: 985–94.

Janoff AS, NetLibrary Inc. 1999. *Liposomes rational design*. New York: M. Dekker. xxxi, 451 pp.

Ju-Nam Y, Bricklebank N, Allen DW, Gardiner PH, Light ME, Hursthouse MB. 2006. Phosphonioalkylthiosulfate zwitterions – new masked thiol ligands for the formation of cationic functionalised gold nanoparticles. *Org Biomol Chem* 4: 4345–51.

Kaufmann SH, Gores GJ. 2000. Apoptosis in cancer: cause and cure. *Bioessays* 22: 1007–17.

Kaul Z, Yaguchi T, Kaul SC, Wadhwa R. 2006. Quantum dot-based protein imaging and functional significance of two mitochondrial chaperones in cellular senescence and carcinogenesis. *Ann N Y Acad Sci* 1067: 469–73.

Kostarelos K, Miller AD. 2005. Synthetic, self-assembly ABCD nanoparticles; a structural paradigm for viable synthetic non-viral vectors. *Chem Soc Rev* 34: 970–94.

Lasch J, Meye A, Taubert H, Koelsch R, Mansa-ard J, Weissig V. 1999. Dequalinium vesicles form stable complexes with plasmid DNA which are protected from DNase attack. *Biol Chem* 380: 647–52.

Lasic DD. 1993. *Liposomes: From physics to applications.* Amsterdam, The Netherlands: Elsevier.

Lasic DD, Martin FJ. 1995. *Stealth liposomes.* Boca Raton: CRC Press. 289 pp.

Luft R, Ikkos D, Palmieri G, Ernster L, Afzelius B. 1962. A case of severe hypermetabolism of nonthyroid origin with a defect in the maintenance of mitochondrial respiratory control: a correlated clinical, biochemical, and morphological study. *J Clin Invest* 41: 1776–804.

Lukacs GL, Haggie P, Seksek O, Lechardeur D, Freedman N, Verkman AS. 2000. Size-dependent DNA mobility in cytoplasm and nucleus. *J Biol Chem* 275: 1625–9.

Medda R, Jakobs S, Hell SW, Bewersdorf J. 2006. 4Pi microscopy of quantum dot-labeled cellular structures. *J Struct Biol* 156: 517–23.

Mesika A, Kiss V, Brumfeld V, Ghosh G, Reich Z. 2005. Enhanced intracellular mobility and nuclear accumulation of DNA plasmids associated with a karyophilic protein. *Hum Gene Ther* 16: 200–8.

Naviaux RK. 2004. Developing a systematic approach to the diagnosis and classification of mitochondrial disease. *Mitochondrion* 4: 351–61.

Niedermann G, Weissig V, Sternberg B, Lasch J. 1991. Carboxyacyl derivatives of cardiolipin as four-tailed hydrophobic anchors for the covalent coupling of hydrophilic proteins to liposomes. *Biochim Biophys Acta* 1070: 401–8.

Parfenov AS, Salnikov V, Lederer WJ, Lukyanenko V. 2006. Aqueous diffusion pathways as a part of the ventricular cell ultrastructure. *Biophys J* 90: 1107–19.

Paunesku T, Rajh, T., Wiederrecht, G., Maser, J., Vogt, S., Stojicevic, N., Protic, M., Lai, B., Oryhon, J., Thurnauer, M., Woloschak, G.E. 2003. Biology of TiO2-oligonucleotide nanocomposites. *Nat Mater* 2: 343–6.

Paunesku T, Vogt S, Lai B, Maser J, Stojicevic N, et al. 2007. Intracellular distribution of TiO(2)-DNA oligonucleotide nanoconjugates directed to nucleolus and mitochondria indicates sequence specificity. *Nano Lett.*

Perlson E, Hanz S, Ben-Yaakov K, Segal-Ruder Y, Seger R, Fainzilber M. 2005. Vimentin-dependent spatial translocation of an activated MAP kinase in injured nerve. *Neuron* 45: 715–26.

Philippot JR, Schuber F. 1995. *Liposomes as tools in basic research and industry.* Boca Raton: CRC Press. 277 pp.

Polyakov VY, Soukhomlinova MY, Fais D. 2003. Fusion, fragmentation, and fission of mitochondria. *Biochemistry (Mosc)* 68: 838–49.

Rowe TC, Weissig V, Lawrence JW. 2001. Mitochondrial DNA metabolism targeting drugs. *Adv Drug Deliv Rev* 49: 175–87.

Salnikov V, Lukyanenko YO, Frederick CA, Lederer WJ, Lukyanenko V. 2007. Probing the outer mitochondrial membrane in cardiac mitochondria with nanoparticles. *Biophys J* 92: 1058–71.

Seksek O, Biwersi J, Verkman AS. 1997. Translational diffusion of macromolecule-sized solutes in cytoplasm and nucleus. *J Cell Biol* 138: 131–42.

Sharma P, Brown S, Walter G, Santra S, Moudgil B. 2006. Nanoparticles for bioimaging. *Adv Colloid Interface Sci* 123–126: 471–85.

Sternberg B, Sorgi FL, Huang L. 1994. New structures in complex formation between DNA and cationic liposomes visualized by freeze-fracture electron microscopy. *FEBS Lett* 356: 361–6.

Szewczyk A, Wojtczak L. 2002. Mitochondria as a pharmacological target. *Pharmacol Rev* 54: 101–27.

Takeuchi Y, Ichikawa K, Yonezawa S, Kurohane K, Koishi T, et al. 2004. Intracellular target for photosensitization in cancer antiangiogenic photodynamic therapy mediated by polycation liposome. *J Contr Rel* 97: 231–40.

Torchilin VP, Weissig V. 2003. *Liposomes: a practical approach.* Oxford: Oxford University Press. 396 pp.

Torchilin VP, Weissig V, Martin FJ, Heath TD, New RRC. 2003. Surface modification of liposomes. In *Liposomes – a practical approach,* ed. VP Torchilin, V Weissig, pp. 193–230. Oxford: Oxford University press.

Trapp S, Horobin RW. 2005. A predictive model for the selective accumulation of chemicals in tumor cells. *Eur Biophys J* 34: 959–66.

Vaughan EE, Dean DA. 2006. Intracellular trafficking of plasmids during transfection is mediated by microtubules. *Mol Ther* 13: 422–8.

Wallace DC, Singh G, Lott MT, Hodge JA, Schurr TG, et al. 1988. Mitochondrial DNA mutation associated with Leber's hereditary optic neuropathy. *Science* 242: 1427–30.

Waterhouse NJ, Goldstein JC, Kluck RM, Newmeyer DD, Green DR. 2001. The (Holey) study of mitochondria in apoptosis. *Methods Cell Biol* 66: 365–91.

Weissig V, Gregoriadis, G. 1993. Coupling of aminogroup-bearing ligands to liposomes. In *Liposome technology*, ed. G Gregoriadis, pp. 231–48. Boca Raton: CRC Press.

Weissig V, Torchilin VP. 2000. Mitochondriotropic cationic vesicles: a strategy towards mitochondrial gene therapy. *Curr Pharm Biotechnol* 1: 325–46.

Weissig V, Torchilin VP. 2001a. Cationic bolasomes with delocalized charge centers as mitochondria-specific DNA delivery systems. *Adv Drug Deliv Rev* 49: 127–49.

Weissig V, Torchilin VP. 2001b. Towards mitochondrial gene therapy: DQAsomes as a strategy. *J Drug Target* 9: 1–13.

Weissig V, Lasch J, Klibanov AL, Torchilin VP. 1986. A new hydrophobic anchor for the attachment of proteins to liposomal membranes. *FEBS Lett* 202: 86–90.

Weissig V, Lasch J, Gregoriadis G. 1989. Covalent coupling of sugars to liposomes. *Biochim Biophys Acta* 1003: 54–7.

Weissig V, Lasch J, Erdos G, Meyer HW, Rowe TC, Hughes J. 1998a. DQAsomes: a novel potential drug and gene delivery system made from Dequalinium. *Pharm Res* 15: 334–7.

Weissig V, Mogel HJ, Wahab M, Lasch J. 1998b. Computer simulations of DQAsomes. *Proceed Intl Symp Control Rel Bioact Mater* 25: 312.

Weissig V, Lizano C, Torchilin VP. 2000. Selective DNA release from DQAsome/DNA complexes at mitochondria-like membranes. *Drug Deliv* 7: 1–5.

Weissig V, D'Souza GG, Torchilin VP. 2001a. DQAsome/DNA complexes release DNA upon contact with isolated mouse liver mitochondria. *J Contr Rel* 75: 401–8.

Weissig V, Lizano C, Ganellin CR, Torchilin VP. 2001b. DNA binding cationic bolasomes with delocalized charge center: A structure–activity relationship study. *S.T.P. Pharma Sciences* 11: 91–6.

Weissig V, Cheng SM, D'Souza GG. 2004. Mitochondrial pharmaceutics. *Mitochondrion* 3: 229–44.

Weissig V, Boddapati SV, Cheng SM, D'Souza GG. 2006. Liposomes and liposome-like vesicles for drug and DNA delivery to mitochondria. *J Liposome Res* 16: 249–64.

Weissig V, Boddapati SV, Jabre L, D'Souza GGM. 2007. Mitochondria-specific nanotechnology. *Nanomedicine* 2: 275–85.

Woodle MC, Storm G. 1998. *Long circulating liposomes: old drugs, new therapeutics*. Austin: Landes Bioscience. 301 pp.

Multifunctional Magnetic Nanosystems for Tumor Imaging, Targeted Delivery, and Thermal Medicine

Dattatri Nagesha, Harikrishna Devalapally, Srinivas Sridhar, and Mansoor Amiji

1 Introduction

Cancer remains one of the leading causes of death in most regions of the world, including the United States (Beardsley 1994). According to the American Cancer Society (ACS 2007), 1.5 million new cases of cancer are expected to be diagnosed in the year 2007 with approximately 560,000 projected deaths. The current treatment options are not sufficient to deal with this influx. Therefore, there is a need for a paradigm shift in the approach to cancer prevention, diagnosis, and therapy. One approach that has shown significant promise is the field of nanotechnology (Brigger et al. 2002; Davis 1997).

Nanotechnology is the science of materials, in the size range of approximately 1–100 nm in diameter, that have unique physical, chemical, and biological properties. A variety of biological and medical processes occur in the nanometer length scales and nanotechnology offers a unique approach to probe and control these processes (Sridhar et al. 2005).

Nanoparticles can be made from organic molecules, such as biodegradable and nondegradable polymers as in polymeric nanoparticles and phospholipids as in liposomes. They can also be synthesized from inorganic materials such as metals and alloys as well as semiconductors, as in iron oxide, gold, and silver nanoparticles and quantum dots, respectively. One property that is common to all of the nanoparticles, irrespective of their chemical composition, is their ability to form multifunctional nanosystems that can be used for diagnosis, imaging, and therapeutic applications. These multifunctional nanoparticle-based approaches are, therefore, expected to make significant impacts in the field of cancer nanomedicine.

This chapter will discuss magnetic nanoparticle-based systems that have been used to improve diagnostics through better tumor imaging, for enhanced drug delivery, and in magnetothermal therapy.

1.1 Tumor-Targeted Molecular Imaging

Systemic administration of chemotherapeutic drugs and imaging agents often leads to undesirable side effects. In chemotherapy, this limits the total amount of drugs

V. Torchilin (ed.), *Multifunctional Pharmaceutical Nanocarriers*,
© Springer Science+Business Media, LLC 2008

that can be safely administered to patients, and in most cases, results in suboptimum dosage for effective treatment. Therefore for successful cancer treatment, both in diagnosis and therapy, it is imperative to have a vehicle for tumor site-specific delivery. Additionally, various clinical imaging modalities, such as magnetic resonance imaging (MRI), computed tomography, ultrasound, and others can be very useful in assessing the impact of therapy.

To address this issue, an integrated discipline that has emerged recently is "molecular imaging." This methodology has set the stage for an evolutionary leap in diagnostic imaging and therapy (Allport and Weissleder 2001). Molecular imaging is not a substitute for the traditional process of image formation and interpretation, but is intended to improve diagnostic accuracy and sensitivity by providing an in vivo analog of immunocytochemistry or in situ hybridization. Moreover, imaging will become crucial for in vivo characterization of the complex behaviors of disease in time and space that will tell us: where it is, how big it is, how fast it is developing, how many molecular processes are contributing simultaneously, what to treat it with, how it is responding to therapy, and how it is changing.

Because molecules themselves obviously are too small to be imaged directly with noninvasive techniques, specific and sensitive site-targeted contrast agents or imaging agents are typically employed as beacons to depict epitopes of interest. And, unlike traditional blood pool contrast agents, a site-targeted agent is intended to enhance a selected biomarker that otherwise might be impossible to distinguish from surrounding normal tissue. Molecular imaging actually has been a clinical reality for some time with the use of targeted radionuclides. Somatostatin receptor imaging for detection of neuroendocrine tumors is but one clinical example, as is fluorodeoxyglucose imaging by positron emission tomography for characterization of diverse disease states. Targeted in vivo detection of cellular apoptosis with the use of technetium-labeled annexin is now in clinical trials on the basis of binding to membrane phosphatidyl serine epitopes that are exposed during apoptosis (Blankenberg and Strauss 2001).

1.2   Nanosystems for Tumor-Specific Drug Delivery

Nanosystems or nanoparticles offer unique approaches in cancer diagnosis as imaging agents and in therapy as drug delivery vehicles. One fundamental property of nanoparticles that is exploited the most is large surface area to volume ratio which gives nanoparticles the ability to carry large payloads. A variety of different platforms are available which include metallic, semiconducting, polymeric nanoparticles, liposomes, and micellar systems (Sridhar et al. 2005). Nanoparticles or nanocarriers in general have at least a tripartite system, consisting of a core constituent material, a therapeutic and/or imaging payload, and biological surface modifiers, which enhance the biodistribution, circulation half-life, and tumor targeting of the nanoparticle formulation. Figure 1 shows schematic illustrations of typical multifunctional nanoparticle systems useful in cancer imaging and therapy. A major therapeutic

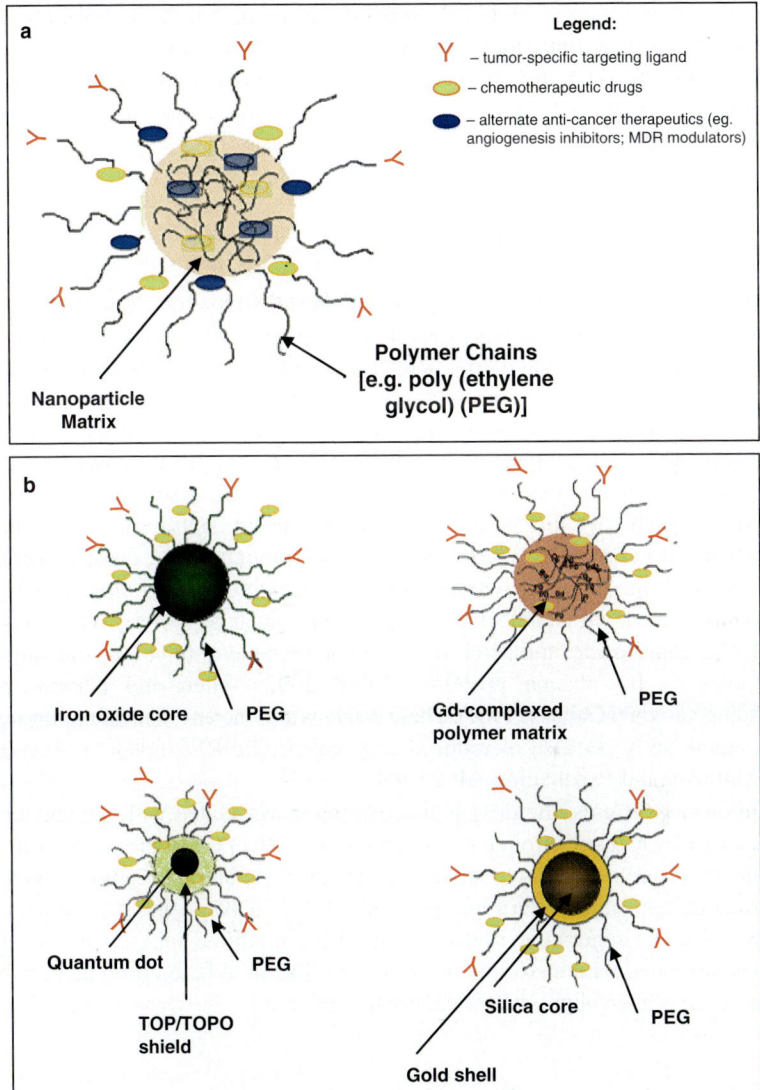

Fig. 1 Multifunctional nanoparticle systems for tumor-targeted delivery. (**a**) Simple multifunctional nanoparticles are formulated from a lipid or polymer in which chemotherapeutic drugs and/or alternate anticancer therapeutics are coencapsulated. (**b**) Complex multifunctional nanoparticles include iron oxide nanoparticles, gadolinium nanoparticles, gold nanoshells, and quantum dots. Surface modification allows for covalent attachment of tumor targeting ligands and simultaneous encapsulation of anticancer drugs

advantage gained by the use of nanocarriers over conventional formulations is the specific delivery of large amounts of therapeutic or imaging agents per targeting biorecognition event. Site specificity is achieved either through covalent binding of antibodies that are tumor specific (Duncan 2003; Nashat et al. 1998) or by mechanisms based on size and physicochemical properties of the nanocarriers (Decuzzi et al. 2008). Nanocarrier formulations are designed to reduce the clearance time, increase circulation half-life, and provide protection of active agents from enzymatic or environmental degradation. Two such systems that have been extensively studied are polymeric nanoparticles and liposome-based systems.

1.2.1 Liposomal Drug Delivery Systems

Liposomes are spherical vesicles composed of phospholipids and cholesterol bilayers within which drug molecules can be loaded either in the hydrophobic or the hydrophilic region. Liposomal drug delivery systems have been studied extensively to increase the solubility and therapeutic index of chemotherapeutic agents (Klibanov et al. 1991). However, several problems have occurred with these conventional vehicles, including nonspecific uptake by the reticuloendothelial system (RES) within a few minutes to few hours, rapid clearance, opsonization, and instability all of which have limited the use of these conventional liposomes in clinical applications (Ito et al. 2004; Park 2002). Novel formulations of these compounds have been developed recently with the aim of overcoming some of these drawbacks. Liposomes coated with some flexible polymers, such as poly(ethylene glycol) (PEG), G_{M1} ganglioside, and cerebroside sulfate are able to inhibit opsonization of the liposomes by plasma proteins (Allen 2002; Allen and Chonn 1987; Papahadjopoulos and Gabizon 1987). These coatings also increase half-life of liposomal drugs significantly because of reduced and nonspecific RES uptake (Allen et al. 1989; Gabizon and Papahadjopoulos 1988).

In liposome systems, the drug is inactive while associated with the carrier and escape either by leakage through the membranes of intact liposomes or by diffusion from degraded or destabilized liposomes is not possible. The physicochemical properties of liposomes determine their rate of drug leakage and natural degradation (Kirby and Gregoriadis 1980). Doxil®, PEG-modified long-circulating liposome encapsulated with doxorubicin, is the first liposome-based drug nanodelivery system that has been approved for cancer treatment.

1.2.2 Polymeric Nanoparticles

Another nanoparticle-based system used extensively in drug delivery and imaging are polymeric nanoparticles (Duncan 2003). Depending on the chemical composition, polymeric nanoparticles are differentiated as those derived from synthetic polymers (e.g., poly(epsilon-caprolactone), polyacrylamide, poly(methyl methacrylate)) or natural polymers (e.g., gelatin, chitosan, albumin). Based on their fate in vivo, they are further classified as biodegradable (e.g., poly(D,L-lactide-*co*-glycolides), polycaprolactone, polylactides) and nonbiodegradable (e.g., polyurethane). Depending on the type, site, and duration of application, different types of polymeric nanoparticles are employed.

Polymeric nanoparticles are generally coated with a layer of nonionic surfactants such as poloxamers and poloxamines to reduce opsonization and interparticulate attractive van der Waals forces (Amiji and Park 1992; Lasic et al. 1991; Moghimi and Hunter 2000). Additionally, advances in surface engineering principles have enabled the production of polymeric nanoparticles with enhanced circulation half-life. Many of the reported surface engineering strategies employ PEG or poly(ethylene oxide) (PEO) chains for surface modifications (Gref et al. 2000; Otsuka et al. 2003; Tan et al. 1993). Depending upon the particle size, surface charge, molecular weight, and type of the polymer, the half-lives vary from 2 to 24 h in mice and rats and can be as high as 45 h in human beings (Moghimi et al. 2001). Drug loading in these polymeric nanoparticles can be accomplished by absorption, adsorption, or encapsulation. Once the polymeric nanoparticles reach the target tissue, the drug may be released by desorption, diffusion through the polymer matrix or polymer wall, or nanoparticle erosion (Gupta et al. 1987; Kreuter 1994; Lockman et al. 2002).

1.2.3 Magnetic Nanoparticles

Inorganic nanoparticles are a new class of nanomaterials that is increasingly finding applications in nanomedicine. This is in part due to rapid advances in synthesis and characterization of novel metallic nanoparticles with excellent control on size, surface charge, optical and magnetic properties. These inorganic nanoparticles have been suitably modified to impart multifunctionality, improved biocompatibility, and used for various medically relevant applications. The last decade or so has seen numerous applications for these inorganic nanoparticles and three such developments that have revolutionized this field are the fluorescent semiconducting nanoparticles or quantum dots for imaging, gold-coated silica particles for thermal therapy of tumors, and dextran-coated superparamagnetic iron oxide nanoparticles system as MRI contrast enhancement agent.

Magnetic nanoparticles are unique in the sense that they are multifunctional by themselves and do not need any special treatment to render them with this property. Magnetic nanoparticles in addition to being an excellent contrast-enhancing agent in MRI, they can also be used as an energy delivery system. Similarly, using external magnetic fields, targeted delivery to desired sites is possible and its accumulation can be monitored through MRI.

The key applications of magnetic nanoparticles in diagnosis and therapy are the following.

Magnetic Nanosystems for Molecular Imaging

MRI applications have steadily increased over the last decade as a tool to image soft tissues. In the presence of strong external magnetic fields, protons in water molecule undergo excitation and their relaxations to ground state are measured as T_1 and T_2 time. The Quality of the image is enhanced through the use of contrast-enhancing agents and they work by affecting the T_1 or T_2 relaxation time. Iron oxide-based magnetic nanoparticles, with a size generally between 3 and 10 nm, have also been developed as T_2 contrast agents for MR imaging. These multifunctional magnetic nanoparticles are modified with ligands to make them site specific and used to image tumor cells. Investigations of ligand-mediated MR contrast agents for tumor diagnosis in small animals revealed that it is possible to achieve a high concentration of the magnetic label at the target.

Magnetic Nanosystems for Targeted Delivery

Magnetic nanosystems for targeted delivery approach involves systemic administration of a therapeutic agent bound or encapsulated in a magnetic carrier, which can then be directed and preferentially localized in the tumor tissue, upon application of an external localized magnetic field. Magnetic responsive drug carriers usually include magnetic materials in the form of magnetite, iron, nickel, cobalt, etc. These drug carriers include magnetic liposomes, microspheres, nanospheres, and colloidal iron oxide solution (magnetic ferrofluids) (Hafeli 2004; Lubbe et al. 1996a,b). A special carbohydrate which can reversibly bind drugs was coated with magnetic ferrofluid (particle size 100 nm) and explored for targeting tumor tissues by means of properly arranged external magnets (Lubbe et al. 1996a,b). Magnetic drug carriers are under active preclinical investigation for various chemotherapeutic agents such as mitoxantrone, etoposide, and paclitaxel (Alexiou et al. 2000). These magnetic drug carrier systems can be used to deliver targeted high dose of radiation to the tumor cells, by sparing the normal cells (Hafeli 2001; Jain et al. 2005). Due to high pressure exerted by convective flow in the tumor blood vessels, a large targeting magnetic force is required for effective targeting via systemic administration (Hafeli 2004). Thus, research has been directed toward preparing targeted carriers with high magnetic moment or developing magnets that can provide higher magnetic field gradients to externally direct these magnetic drug carriers to the tumor site.

Magnetic Nanosystems for Thermal Therapy

Hyperthermia treatment of tumor and metastasis has been demonstrated to be effective, either alone, or in combination with classical chemo- or radiotherapies (Engin 1996; Hildebrandt et al. 2002; van der Zee 2002). Despite some interesting preliminary results in preclinical tumor animal models (Gordon et al. 1979), hyperthermia using superparamagnetic iron oxide particles has, however, been overlooked for almost a decade. Since the early 1990s, a lot of progress has been made, especially in the field of superparamagnetic materials. The hyperthermia is induced by applying an alternating magnetic field of suitable frequency that produces heat dissipation through the oscillation of the internal magnetic moment of the superparamagnetic particles. The interest with this therapy lies in the specific delivery of heat to target cells causing selective damage to their cytoplasm structures. This approach is similar to laser-absorbing gold nanoparticles where the laser has been successfully applied to obtain specific targeting and killing of cancer cells. Unlike other hyperthermia systems, normal adjacent healthy cells are not affected by the oscillating magnetic fields.

In the following sections, we have discussed the properties and preparation of magnetic nanosystems, and their applications in diagnosis and therapy.

2 Magnetic Nanosystems for Biomedical Applications

The arrangement of magnetic dipoles within a material determines its behavior in the presence of external magnetic fields and hence its magnetic properties. Based on organization of these dipoles, magnetic materials are classified as diamagnetic, ferromagnetic, ferrimagnetic, antiferromagnetic, and paramagnetic (O'Handley 2000). Arrangement of magnetic dipoles in these systems is illustrated in Fig. 2. Diamagnetic materials possess very weak-induced dipoles in the presence of external fields and none in their absence. Para-, ferro-, ferri-, and antiferromagnetic materials all have an inherent magnetic dipole or domains in them and what differentiates them from each other is their behavior in the presence of external magnetic fields and the value of the net magnetization moment. Paramagnetic materials are made up of weak magnetic domains that are arranged in a random fashion and can be aligned in the direction of the external magnetic field. Ferromagnets have very strong magnetic moments and long range order of magnetic dipoles observed both in the presence and absence of the external magnetic field. Ferrimagnets and antiferrimagnets both have adjacent magnetic domains that are antiparallel to each other in the absence of an external field. However, in the case of antiferromagnets, these adjacent domains cancel out each other and in the case of ferrimagnetic material; there is a net magnetization moment.

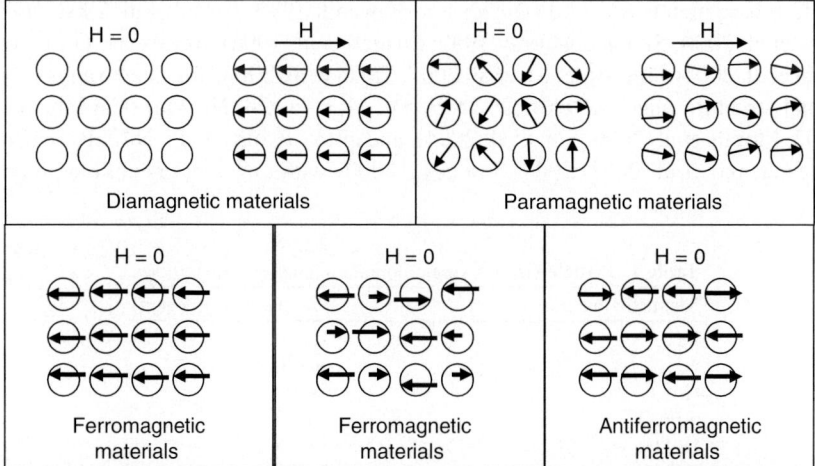

Fig. 2 Arrangement of magnetic dipoles for different kinds of magnetic materials in the presence and absence of magnetic field (*H*)

2.1   Superparamagnetism in Magnetic Nanoparticles

As discussed above, magnetic materials are made up of domains within which magnetic dipoles are arranged and in a typical bulk material, multiple such domains exist. As the particle size starts getting smaller and smaller, in the 10–20 nm size range, only a single domain is seen. In this regime, the energy required to change the orientation of these magnetic dipoles is comparable to the thermal energy. This leads to the phenomenon of superparamagnetism. These materials respond to an applied external magnetic field but do not have any residual magnetization upon removal of the magnetic field. Most of the magnetic materials used in biomedical applications are superparamagnetic in nature. Another important parameter in superparamagnetic behavior is the blocking temperature. This is defined as the temperature at which the energy required to change the magnetic spin states or domain is comparable to thermal energy $k_B T$, where k_B is the Boltzmann constant and T is the absolute temperature.

   In bulk material, a large number of domains exist with magnetic dipoles arising from each of these domains and orientation of these domains determining the overall magnetic properties of the material. However, as particle size decreases below a critical size, energy required to form a multidomain structure is greater than the energy to maintain a single-domain state. As shown in Table 1, the size limit is dependent on the nature of the material and can range from a few nanometers to hundreds of nanometers (Chung et al. 2004).

2.2   Types of Magnetic Nanosystems

A variety of magnetic nanoparticles have been synthesized of different composition such as bare metals (Fe, Co) (Dinega and Bawendi 1999; Farrell et al. 2003, 2005; Park et al. 2000; Sun and Murray 1999; Suslick et al. 1996), oxides (Fe_2O_3, Fe_3O_4) (Jain et al. 2005; Jung 1995; Lu et al. 2007; Tartaj et al. 2003; Teng and Yang 2004), spinel-structure magnetic particles (MFe_2O_4, M = Fe, Co, Ni, Mn) (Caruntu et al. 2002; Morais et al. 2006; Sun et al. 2004), and alloys ($CoPt_3$, FePt, NdFeB, $SmCo_5$) (Shevchenko et al. 2002, 2003; Sun et al. 2000) with different crystalline phases,

Table 1   Particle size for single-domain magnetic nanoparticles

Magnetic material	Particle diameter (nm)
Fe	15–50
Ni	30–70
Co	5–30
γ-Fe_2O_3	20–60
Fe_3O_4	15–55
FePt	5–50
$CoFe_2O_4$	10–90

controlled size, and morphology. Nanoparticles are produced that are monodisperse, with tunable size, extremely stable, and functionalizable for various applications. However, for biomedical applications, the choice of magnetic materials that can be used is very limited. There is severe cytotoxicity associated with some of these elements or there is no data on their fate in a biological system. Additionally, some of these magnetic materials such as cobalt and nickel are susceptible to easy oxidation, thereby making them highly unsuitable for biomedical applications. Therefore, for in vivo applications, magnetic nanoparticles that are noncytotoxic, do not have any immunogenic response, and are eventually resorbed by the bodies are used. Magnetic materials that meet this criteria are the iron oxide-based nanoparticles and to some extent gadolinium and manganese-based nanosystems.

2.3   Synthesis of Magnetic Nanosystems

In the last decade or so, there has been tremendous development in magnetic nanoparticle synthesis and characterization. The synthetic procedure can be simple, carried out at room temperature in aqueous medium, or be very elaborate involving high temperature and pressure in hydrophobic solvents using organometallic compounds. Metals (Fe, Co), metal oxides (Fe_3O_4, Fe_2O_3), ferrites ($CoFe_2O_4$, Fe_3O_4), alloys (FePt, $SmCo_5$), core–shell systems ($Au@Fe_3O_4$, $SiO_2@Fe_3O_4$), or in the form of composites have been synthesized with excellent size control and narrow size distribution. Discussion of all the synthetic routes for these magnetic nanoparticles is beyond the scope of this work. A few representative synthetic routes for the synthesis of biomedically relevant magnetic nanoparticles will be discussed here.

2.3.1   Coprecipitation Method

This is the simplest method to synthesize metal oxide and ferrite nanoparticles in aqueous medium (Bee et al. 1995; Jeong et al. 2007; Lu et al. 2007; Massart et al. 1995). In a typical experiment, metal salts are first dissolved in water under inert atmosphere. Nanoparticles are then precipitated by the addition of base either at room temperature or slightly elevated temperature. Stabilizer or surfactant molecules could be introduced in the solution along with the metal salts to help control particle size. The size and morphology of the nanoparticle synthesized is determined by ratio of metal salts added, nature of salt, and rate of mixing, temperature, pH, and ionic strength of the aqueous media. For example, 10-nm Fe_3O_4 nanoparticles are synthesized by the coprecipitation of $FeCl_2$ and $FeCl_3$ salts by sodium hydroxide at 70 °C. A representative transmission electron microscopy image is shown in Fig. 3. Similarly, $CoFe_2O_4$ is synthesized by the coprecipitation of $FeCl_2$, $FeCl_3$, and $CoCl_2$ by base. This method of nanoparticle synthesis is highly reproducible and can be used for large-scale synthesis. However, challenges in this method involve difficulty in controlling the particle size and maintaining narrow size distribution. Nanoparticles

Fig. 3 Transmission electron microscopy image of iron oxide nanoparticles synthesized by the coprecipitation method

synthesized by this method are generally subjected to surface modification or chemical modification for biomedical applications (Chen et al. 2003; Koh et al. 2006; Kohler et al. 2005; Mikhaylova et al. 2004; Pankhurst et al. 2003).

2.3.2   Thermal Decomposition Method

This is a nonaqueous method of synthesis involving, as the name suggests, decomposition of metal salt or precursors at high temperature in high-boiling organic solvent (Hyeon et al. 2001; Song and Zhang 2004; Sun and Zeng 2002; Sun et al. 2004). Using this method, a better control on size distribution and particle size distribution is possible. This method has been used to synthesize Fe_2O_3 (Hyeon et al. 2001), $CoFe_2O_4$ (Sun et al. 2004), and Fe nanoparticles (Farrell et al. 2003). Organometallic precursors include $Fe(CO)_5$, $Fe(acac)_3$, $Mn(CO)_5$, and salts such as $FeCl_2$, $MnSO_4$, and $CoCl_2$. High-boiling solvents used in this reaction include octyl ether (Park et al. 2004; Teng and Yang 2004) and phenyl ether (Sun and Zeng 2002). The stabilizer molecules used in this reaction are usually long chain fatty acid type molecules (Lu et al. 2007) with at least one functional group. This functional group, either a carboxylic acid or an amine group, usually bound to the nanoparticle surface is not available for conjugation. Fatty acids such as oleic acid, oleylamine, steric acid, and lauric acid have been used as stabilizer molecules (Hyeon et al. 2001; Li et al. 2006).

The most salient feature of magnetic nanoparticles synthesized by this method is excellent control on size, morphology, and size distribution (Caruntu et al. 2004; Hyeon et al. 2001; Jun et al. 2005; Park et al. 2004). This is achieved by regulating the ratio of precursor salts, stabilizer molecules, solvent and reaction time, and temperature. Figure 4 is an illustration of the experimental setup for iron oxide nanoparticle synthesis by this method. In a typical synthesis, octyl ether, oleic acid, lauric acid, and $(CH_3)_3NO$ are mixed and heated to $100\,°C$. $Fe(CO)_5$ is then introduced and the resulting solution heated to reflux for 1 h. After cooling to room temperature, iron oxide nanoparticles are precipitated using ethanol. Figure 5 is the TEM of 13-nm Fe_2O_3 NPs synthesized by this method.

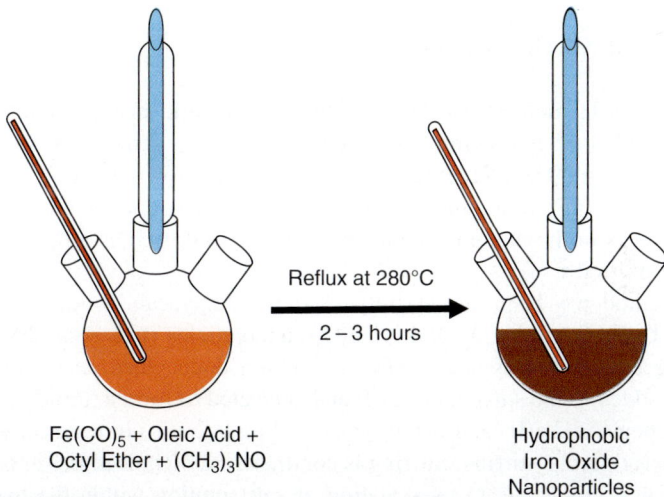

Reflux at 280°C

2 – 3 hours

$Fe(CO)_5$ + Oleic Acid +
Octyl Ether + $(CH_3)_3NO$

Hydrophobic
Iron Oxide
Nanoparticles

Fig. 4 Synthesis of iron oxide nanoparticles by thermal decomposition method

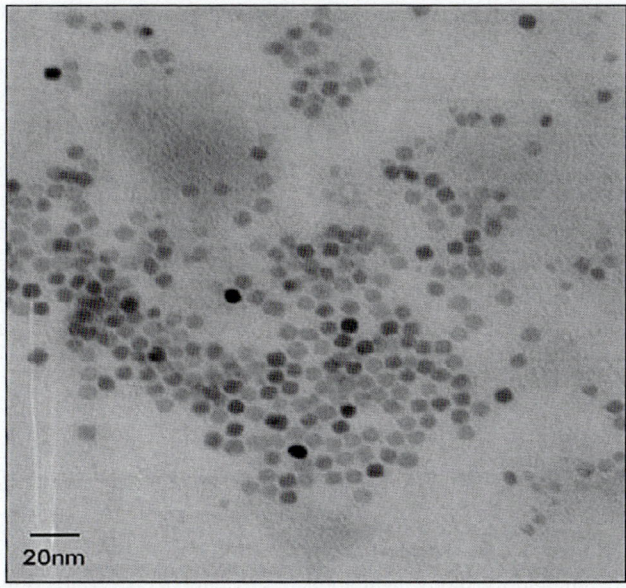

20nm

Fig. 5 Transmission electron microscopy image of hydrophobic 13-nm iron oxide nanoparticles stabilized with oleic acid

Nanoparticles synthesized by this method are soluble in a variety of hydrophobic or nonpolar solvents. However, for use in biomedical applications, these nanoparticles have to be subjected to surface modification techniques to convert them to hydrophilic nanoparticles. The methods used for this procedure will be discussed in subsequent sections.

2.3.3   Reverse Micelle Method

Another versatile method used to synthesize magnetic nanoparticles is using reverse micelles or microemulsion (Lee et al. 2005; Tartaj and Serna 2002; Vestal and Zhang 2003). Reverse micelle is a water-in-oil system formed by the mixing of surfactant in organic solvents to form a water pocket. The size of this water pocket is in the nanometer range (1–50 nm) and used as a reactor to form small nanoparticles.

This method has been successfully used for large-scale synthesis of Fe_3O_4, $MnFe_2O_4$, $CoFe_2O_4$, $ZnFe_2O_4$, and $NiFe_2O_4$ nanoparticles (Lee et al. 2005; Vestal and Zhang 2003). In a typical synthesis, sodium dodecylbenzenesulfonate (surfactant) is added to the solvent (xylene) and sonicated to form a turbid suspension. To this suspension, Fe(II) and Fe(III) salts (1:2 molar ratio) in water or ethanol is added under vigorous stirring. Stirring is continued for 12 h resulting in the formation of a clear liquid due to intercalation of salt solution within the hydrophilic pocket of reverse micelle. Under the inert atmosphere of argon, the solution is slowly heated to 90 °C followed by addition of the reducing agent, aqueous hydrazine solution. Instantaneous color change to brown-black indicates the formation of iron oxide nanoparticles and the solution is further refluxed for 5 h to complete the nanoparticle formation. The nanoparticles can be precipitated with ethanol, collected by centrifugation, and redispersed in organic solvents. To synthesize MFe_2O_4 (M = Co, Mn, Zn) nanoparticles, the corresponding MCl_2 salt is added instead of $FeCl_2$ in the above synthesis procedure.

The size of the water pocket is determined by ω, ratio of surfactant to water (Ngo and Pileni 2000). Larger the ω, larger the water pocket size and hence the size of nanoparticles. In the above method, ω was varied from 3.6 to 8.1, resulting in magnetite nanoparticle in the size range of 3–9 nm. NPs synthesized by this method are usually only soluble in organic solvent and to make them more biocompatible, they have to undergo ligand exchange.

2.4   *Functionalization for Targeted Delivery and Imaging*

As-synthesized magnetic nanoparticles do not have the right surface composition and are usually not suitable for biomedical applications (Neuberger et al. 2005; Pankhurst et al. 2003; Tartaj et al. 2003; Wang et al. 2005; Zhang and Zhang 2005). To facilitate conjugation of biomolecules such as antibodies, proteins, and targeting ligands, surface modification or functionalization needs to be carried out. In some instances, the magnetic nanoparticles are synthesized in hydrophobic solvents and for use in biomedical applications, it is necessary to first convert these nanoparticle suspensions into polar solvents such as water or alcohol. Surface modification or functionalization can be generally carried out through ligand exchange or through chemical treatment.

2.4.1 Ligand Exchange

As discussed earlier, nanoparticles are synthesized in the presence of surfactants or stabilizer molecules. Some of the properties associated with NPs such as stability, size distribution, solubility, and surface charges are directly influenced by the presence of these stabilizer molecules. In the ligand exchange method, as the name suggests, one ligand is exchanged with another that will help in either functionalization or to change the polarity. For example, fatty acids such as oleic acid and lauric acid are used in the thermal decomposition method to synthesize stable and monodisperse magnetic nanoparticles. The stabilizer molecule binds through the carboxylic acid group on the nanoparticle surface. The other end of fatty acid, the alkyl group, cannot be used for conjugation to other molecules. Therefore, ligand exchange is necessary to introduce functionality to the nanoparticles (DePalma et al. 2007). It is important to note that the properties of the nanoparticles with respect to particle size, magnetic susceptibility, and size distribution are not affected by successful ligand exchange.

A very simple yet highly effective ligand exchange reaction to convert nanoparticles from hydrophobic solvent to hydrophilic solvent is through reaction with 2,3-dimercaptosuccinic acid (DMSA) (Bertorelle et al. 2006; Jun et al. 2005; Lee et al. 2007). The two carboxylic acid residues on DMSA bind to nanoparticle surface to form a protective coating. Some of the sulfhydryl groups form intermolecular disulfide bonds to provide additional stability and the free thiol groups on DMSA can be used to conjugate various biomolecules.

2.4.2 Chemical Modification

In this method, instead of a simple ligand exchange, surfactant-coated NPs are reacted with a chemical compound that reacts with the nanoparticles and results in complete transfer of ligands. The details of some of these methods are given below.

Magnetic nanoparticles synthesized by the thermal decomposition method using fatty acid surfactants such as oleic acid are readily soluble in nonpolar solvents. Successful dispersion of these nanoparticles into polar solvents such as alcohol has been achieved through ozonolysis (Lee and Harris 2006). Oleic acid, $CH_3(CH_2)_7C H=CH(CH_2)_7COOH$, has two distinct features, a double bond and an end carboxylic group that is bound to nanoparticle surface. During ozonolysis, the double bond is cleaved and converted to a carboxylic acid end group (azelaic acid) on the nanoparticle surface making them soluble in polar solvents such as alcohols. This resulting stable ferrous alcohol can be used in various biomedical applications.

A chemical treatment method used to introduce functionality to nanoparticle surface is through the silanization process (Mikhaylova et al. 2004; Zhang and Zhang 2005; Zhang et al. 2002). This involves reacting nanoparticles with an organosilane compound, also known as silane coupling agent. The general formula of these organosilanes is $R_xSiY_{(4-x)}$, wherein R = organic functional group (amine, thiol, carbonyl, etc.) and Y = reactive functional group (ethoxy, methoxy, etc.). In the reaction, the functional group Y hydrolyzes to form siloxane bond with the nanoparticle surface exposing the functional group R on the outside. The nanoparticle–silane construct can now be covalently conjugated with various proteins (Mikhaylova et al. 2004), biomolecules (Koh et al. 2006), etc. This surface modification method has been used to conjugate bovine serum albumin (BSA) (Koh et al. 2006), horseradish peroxidase (HRP) (Ma et al. 2003), and folic acid (Zhang and Zhang 2005; Zhang et al. 2004).

2.4.3   Core–Shell System

One method that has been used to improve stability and introduce functionality to magnetic nanoparticles is through thin inorganic coatings. Magnetic nanoparticles form the core and different types of inorganic materials form a shell around it to form core–shell morphology. Depending on the material and application, the coating material could be inorganic, polymeric, or metallic in nature.

Silica-Coated Magnetic Nanoparticles

Silica is a biocompatible material that has been used to coat the surface of magnetic nanoparticles (Ohmori and Matijevic 1993). This method of having a magnetic nanoparticle in the center with a silica coating is known as core–shell system. The geometry of this core–shell system can be varied to incorporate a few nanoparticles within a shell or coat individual nanoparticles with silica shell and, the thickness of the shell coated can also be precisely controlled. In addition to providing stability to the nanoparticle system, silanol groups on the surface of silica can be easily derivatized to introduce various functional groups. Presence of these functional groups makes them amicable to conjugation to various biomolecules for applications (Koh et al. 2006; Kohler et al. 2005; Zhang et al. 2002).

The most common method used to introduce silica coatings is through the sol-gel process even though other techniques such as reverse micelle and aerosol pyrolysis method exist. Iron oxide nanoparticles synthesized by coprecipitation method can be directly coated with silica by mixing the nanoparticles in water with 2-proponol, 30% aqueous ammonia, and silanization agent tetraethoxyorthosilane (TEOS) (Lu et al. 2002). Formation of bare silica nanoparticles can be prevented and control on thickness of silica shell can be regulated by the ratio of TEOS to water. In addition, during the shell formation, organic fluorescent dyes can be mixed with the silanization agent and incorporated within the shell to introduce both magnetic and fluorescent properties in the system.

The other experimental method that is used for silica coating on nanoparticles is through the reverse micelle method. This method can be used for both simultaneous, in situ, formation of silica coating on nanoparticle as it is being synthesized (Tartaj and Serna 2003) or form silica coating on premade nanoparticles (Vestal and Zhang 2003; Yi et al. 2005). This method has been successfully used to coat α-Fe, MFe_2O_4 (M = Fe, Co, Mn, Ni) (Vestal and Zhang 2003). Using this method, quantum dots and magnetite have been coentrapped within the silica shell to form a core–shell system that is both magnetic and fluorescent (Yi et al. 2005).

Metal Coating on Magnetic Nanoparticles

A method of surface coating that is gaining popularity is to put a thin coating of metals on magnetic nanoparticles (Ban et al. 2005). Precious metal, especially gold, is used as it provides stability and introduces functionality such that the well-established gold-thiol chemistry can be used for various biomedical applications (Ban et al. 2005; Chen et al. 2003; Mizukoshi et al. 2005; Sonvico et al. 2005; Storhoff et al. 1998).

However, direct coating of gold on surface of magnetic nanoparticles by chemical methods is difficult due to mismatch in their crystal structure. With some clever chemistry, there have been some successes in this front using iterative hydroxylamine seeding (Liu et al. 1998; Lyon et al. 2004) and reverse micelle method. In the reverse micelle method, using a mixture of cetyltetramethylammonium bromide (CTAB), 1-butanol, and octane, $FeSO_4$ was reduced by sodium borohydride ($NaBH_4$), followed by addition of $HAuCl_4$ which resulted in the formation of Fe_3O_4 NPs coated with gold. In the iteration method, citrate-stabilized aqueous iron oxide nanoparticles were mixed with hydroxylamine and chloroauric acid ($HAuCl_4$) solution. Hydroxylamine, a mild reducing agent, reduces gold on the surface of iron oxide surface. Iterative addition of hydroxylamine and $HAuCl_4$ solution resulted in the formation of gold on nanoparticle surface to form core–shell morphology. The thickness of shell formed was controlled by the number of iterations carried out and controlled between 10 and 60 nm.

Polymer-Magnetic Nanoparticle Composites

Stability of magnetic nanoparticles can also be improved by forming thin organic polymer coatings (Shen et al. 1999; Sousa et al. 2001). This is achieved either through natural polymers such as proteins (Butterworth et al. 1996) and sugars (Neuberger et al. 2005) or through synthetic, biocompatible polymers such as polylactic acid, poly(alkylcyanoacrylates), polycaprolactone, and poly(ethylene imine) (Thunemann Andreas et al. 2006). Using the layer-by-layer technique, these polymers can be coated on magnetic nanoparticles to introduce suitable surface charge (positive or negative), stability, functionality, and biocompatibility (Caruso 2001).

Iron Oxide Nanoparticles with Organic Polymer Coatings

Using the well-established synthetic organic polymer chemistry, magnetic nano-particles have been encapsulated within polymer matrices to form robust core–shell systems. Atomic transfer radical polymerization (ATRP) has been used successfully to form monodisperse polymeric systems and using this technique, magnetic nanoparticles have been incorporated within these systems during their synthesis (Li et al. 2004; Vestal and Zhang 2002; Wang et al. 2003). Nanoparticles are mixed with an initiator of ATRP (2-bromo-2-methyl propionic acid, 3-chloro propionic acid), solvent (xylene), monomer solution (styrene, divinyl benzene, methyl methacrylate) and allowed to react to form the core–shell system. The function group, usually carboxylic acid, of the initiator reacts with the nanoparticle and anchors on to its surface. It then converts it to a reactive species known as the macroinitiators of ATRP. Subsequent addition of monomer leads to polymeriza-tion through these activated sites resulting in the core–shell system. This method of polymer coating is very robust and permanently traps the magnetic nanoparti-cles within the matrix making them extremely stable under a variety of chemical and physiological conditions.

Magnetomicelles

Amphiphilic block copolymers are known to self-assemble, based on the type of polymer and its concentration, to form uniform nanostructures or micelles (Ai et al. 2005; Kim et al. 2005). When magnetic nanoparticles are dispersed in the medium during self-assembly process, they can get entrapped within these micellar structures to form "magnetomicelles," as depicted in Fig. 6. γ-Fe$_2$O$_3$ have been coassembled with amphiphilic poly(styrene$_{250}$-*block*-acrylic acid$_{13}$) (PS$_{250}$-*b*-PAA$_{13}$) and block copolypeptide – poly(N_e-(ACS 2007)acetyl-L-lysine)$_{100}$-*b*-poly(L-aspartic acid, sodium salt) (poly(EG$_2$-lys)$_{100}$-*b*-poly(asp)$_{30}$) (Berret et al. 2006; Euliss et al. 2003;

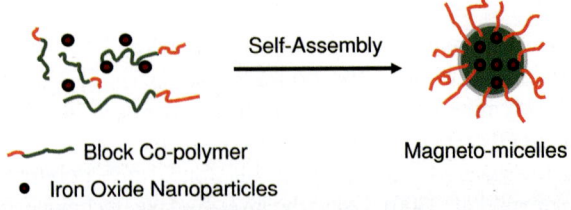

Self-Assembly

⎯ Block Co-polymer
● Iron Oxide Nanoparticles

Magneto-micelles

Fig. 6 Synthesis of magnetomicelle systems

Kim et al. 2005). These composite structures can be made water-soluble and stable, and offer functional surface to conjugate biomolecules. The number of encapsulated nanoparticles within the micelle can be regulated by altering the ratio of nanoparticles to polymer. Magnetic nanoparticles synthesized in hydrophobic solvents can be readily made water-soluble by this method without the need for any surface modification techniques on the nanoparticles.

2.5 Examples of Applications for Magnetic Nanosystems

Biomedical applications of magnetic nanoparticles can be classified as in vivo or in vitro and based on application, as therapeutic or diagnostic in nature. Therapeutic application of magnetic nanoparticles include drug delivery (Kohler et al. 2005; Lanza et al. 2002; Neuberger et al. 2005; Zhang et al. 2004) and in hyperthermia (Jordan et al. 2001; Moroz et al. 2002; Sonvico et al. 2005). They are also used in diagnostics as in vivo MRI contrast agent (Reimer et al. 1992; Schellenberger et al. 2002; Seneterre et al. 1991). In vitro applications (Kohler et al. 2005) of magnetic nanoparticles include the isolation of cells and biomolecules as a diagnostics tool (Haukanes and Kvam 1993; Moeser et al. 2004; Stoeva et al. 2005; Stoltenburg et al. 2005).

External magnets can be applied to create magnetic field gradients to guide magnetic nanoparticles for localized delivery and is the basis of magnetic drug targeting (Alexiou et al. 2000; Lubbe et al. 1999; Neuberger et al. 2005). In the presence of alternating magnetic fields, magnetic nanoparticles can be selectively heated and this is used in magnetic hyperthermia as a new modality in cancer therapy. The above applications are elaborated in the following sections.

3 Magnetic Nanosystems for Molecular Imaging

MRI is increasingly being used as a noninvasive diagnostic tool to image soft tissue. In this method, a radiofrequency (RF) pulse is applied which effects the magnetization of protons in water molecules. The change in this magnetization and its subsequent return to the ground state is measured in terms of the T_1 and T_2 relaxation times. T_1 corresponds to the relaxation along the z-direction and results in the loss of energy in the form of heat. This is also known as longitudinal relaxation or spin–lattice relaxation. T_2, also known as transverse relaxation or spin–spin relaxation, is along the x–y plane and is a consequence of dephasing.

Contrast agents are routinely employed in MRI to obtain better images. Presence of these contrast agents affect the T_1 and T_2 relaxation time of water molecules and hence improve quality of image. Most of these contrast agents are magnetic in nature and, depending on the material, can lead to either enhancement or decrease in contrast.

Contrast agents are classified as either positive or negative agents. Positive contrast agents are usually Gadolinium-based systems resulting in increase of T_1 relaxation time and cause the image to brighten in the MRI. On the other hand, iron oxide-based systems result in darkening of image in MRI and increase in T_2 relaxation. Depending on the type of tissue, area of imaging, and degree of contrast needed, different agents are used.

3.1 Paramagnetic Systems

Gadolinium with seven unpaired electrons is an excellent paramagnetic material with very high magnetic moment. Toxicity associated with Gd^{3+} ions prevents direct administration and instead are introduced in the form of chelates such as diethyltriaminepentaacetic acid (DTPA) and 1,4,7,10-tetraazacyclododecane-tetraacetic acid (DOTA). The first approved intravenous MRI contrast agent was Gd-DTPA (gadolinium (III) – diethyltriaminepentaacetic acid) (Flacke et al. 2001; Lanza et al. 2002) and since then several chelated versions of gadolinium-based contrast agents have been approved by the FDA for clinical use in central nervous system and whole body imaging. Another paramagnetic contrast agent that is available is Teslascan® and is composed of a manganese-chelate – mangafodipir or Mn^{2+}-DPDP (manganese dipyridoxyl diphophate) to specifically image lesions of the liver. Targeting of specific regions in the body or tissue, using the chelates, is not possible with these contrast agents. Additionally, Gd^{3+}-based systems undergo rapid clearance from the body and this limits long-term observation using these contrast agents. Newer generation of Gd^{3+} nanoparticle-based systems is being investigated to address this issue (Flacke et al. 2001; Oyewumi and Mumper 2004; Reynolds et al. 2000).

3.2 Superparamagnetic Systems

The other type of MRI contrast agents is iron oxide nanoparticle-based superparamagnetic system. Iron oxide nanoparticles when introduced as contrast agent cause enhancement of T_2 relaxation resulting in darkening of area being imaged. For use as MRI contrast agent, superparamagnetic iron oxide nanoparticles in the size range of 5–10nm is usually coated with a protective layer of dextran (Josephson et al. 2001; Schellenberger et al. 2002). This increases the hydrodynamic diameter to about 50nm. These iron oxide-based nanoparticles have been used to image the spleen, liver, gastrointestinal tract, and lymph nodes (Lee et al. 1991). Different generations of these dextran-coated iron oxide nanoparticles are available and these include SPION (superparamagnetic iron oxide nanoparticles), MION (monocrystalline superparamagnetic iron oxide nanoparticles), USPIO (ultrasmall superparamagnetic iron oxide nanoparticles), and CLIO (crosslinked iron oxide nanoparticles).

Imaging of a specific area in the body through localization is possible using these magnetic nanoparticles. They can readily be conjugated with antibodies, oligonucleotides, and small molecules that target specific receptor sites on cell surface. CLIO modified with Tat peptide was used to target lymphocytes (Wunderbaldinger et al. 2002) and streptavidin-coated iron oxide nanoparticles modified with anti-HER2/neu antibody for imaging the HER2/neu overexpressing breast cancer cell lines (Artemov et al. 2003).

4   Magnetic Nanosystems for Targeted Delivery

One of the biggest hurdles in chemotherapy is nonspecific targeting of drug molecules. Intravenous administration of drugs leads to systemic distribution through out the body resulting in undesirable side effects and as a consequence only a suboptimal dosage of drugs reaches the desired target site. New concepts are being developed in chemotherapy for drug delivery to tumor sites using nanotechnology. Some of these methodologies include using polymeric nanoparticles, liposomes, and micelles as drug carriers.

One method to improve localized delivery of drugs is through the use of magnetic carriers and magnetic targeting (Goodwin et al. 1999; Jain et al. 2005; Widder et al. 1979). Drug molecules can be attached or embedded in a matrix along with the magnetic nanoparticles to form magnetic carrier systems. Magnetic field gradient can then be generated in the area of interest, using external magnets, to attract these magnetic carriers. The advantages of this approach are that systemic distribution of drug can be substantially reduced resulting in less side effect. Added benefit includes better efficacy from the chemotherapeutic drug as a result of localization (Lubbe et al. 1996a,b).

Chemotherapeutic drug delivery using magnetic carriers was first reported in 1980 by Widder et al. (1980) wherein, doxorubicin and magnetite were encapsulated with albumin. Since then, numerous magnetic carriers–drug composites have been designed, developed, and tested. There are a few basic criteria for effective drug delivery through the use of magnetic targeting delivery (Lubbe et al. 1999). The size of magnetic particle, its magnetic susceptibility, and magnetic strength of external magnet used for concentrating is very important. If the particle is small and has a small magnetic susceptibility, a much larger and stronger external magnet will be required to attract these particles toward the magnet. Additionally, the magnetic field strength or gradient drops inversely to square of the distance between the particle and magnet. Hence, the magnetic field strength is felt more by the particle closer to the magnet and starts decreasing as it gets away from the magnet.

Penetration depth of magnetic field, using external magnet, is limited by the size of the magnet and is usually in the order of 5–10 cm using standard bar magnets. Permanent magnets in the form of disks and rectangles and field strength in the range of 0.1–0.8 T are usually employed to achieve targeting

(Dandamudi and Campbell 2007). The penetration depth for magnetic flux density is a few centimeters. Focusing magnetic flux density to organs and tissue deep within the body, beyond 10 cm, is extremely difficult. Beyond this distance the magnetic field strength is not sufficient enough to be able to attract the magnetic carrier to the target site. This limits the use of magnetic drug targeting to subcutaneous tumor and cancer grown on the skin. This problem can be addressed by either using larger particles with higher magnetic susceptibility or through use of stronger magnets.

Large particles are attracted more efficiently toward magnetic field when compared to smaller particles. However, larger particles tend to undergo embolization which in principle could prevent it from reaching the target site. It has been reported that particles in the size range of 100 nm can be administered without any embolization.

In addition to physical demands from the magnetic carriers, there are other physiological conditions in the subject that need to be considered for magnetic drug targeting to work. This includes the size and capacity of the subject, blood volume, flow, and velocity at the target site. The magnetic carrier must be able to overcome these forces to be attracted toward the external magnet. For drug delivery applications, these magnetic carriers should be able to carry a large payload of drug molecules to the tumor site aided by the magnetic field and subsequently release the drug at the targeted site. To achieve this, surface chemistry of magnetic particles needs to be selectively tuned to help facilitate reversible binding of drug molecules.

Recently, Furlani and Furlani (2007) have developed mathematical models to predict magnetic targeting using magnetic particles and external magnets to aid in the design of new magnetic carriers in drug delivery applications.

5   Magnetic Nanosystems for Thermal Therapy

Living tissues when heated to a temperature range of 42–46 °C will result in disruption of normal cellular activity and this phenomenon is known as hyperthermia. Heating to temperatures above 46 °C is known as thermal ablation and results in complete necrosis of the cells. The concept of hyperthermia as a therapeutic modality in treatment of cancer has been extensively explored (Engin 1996). A variety of frequencies from the electromagnetic spectrum including ultrasound, infrared, radiofrequency, and microwave radiation has been used to deliver energy for hyperthermia. Using near infrared (NIR) light, therapeutic dosage of heat was selectively delivered to human breast carcinoma cells incubated with gold-coated silica nanoparticles (Hirsch et al. 2003). Penetration depth of NIR for effective ablation was in the range of 1–6 mm and can be further increased with longer exposure time and higher laser power. Another method to deliver localized energy is through the use of catheters and probes placed interstitially in cancerous tissues followed by using RF ablation. Each of these radiation sources has met with success and limitations.

The challenge in thermal therapy or hyperthermia, however, is to deliver energy only to cancerous cells and not to adjacent healthy cells. An approach that has shown significant potential for localized thermal energy delivery is the use of alternating magnetic fields and magnetic nanoparticles.

Magnetic hyperthermia in conjugation with radiation or chemotherapy of a variety of tumors has been extensively investigated in the past 30 years. This can be associated with understanding the biology of hyperthermia, development of better diagnostic tools to measure infinitesimal fluctuations in temperature, and advancements in the synthesis of magnetic nanoparticles with excellent magnetic properties. The first reported use of iron oxide nanoparticles in hyperthermia was reported by Gilchrist et al. in 1960 (Gilchrist 1960; Gilchrist et al. 1965) and since then there has been tremendous progress in this field.

For magnetic hyperthermia to work, first the magnetic carriers (in the form of nanoparticles) are introduced throughout the tumor site followed by application of magnetic field of required strength and frequency to induce selective heating through these carriers (Ivkov et al. 2005). Since the tumor site is embedded with these nanoparticles, heat is dissipated from nanoparticles only to cancerous cells. If sufficient quantities of nanoparticles are present in the tumor site to maintain a temperature of $>42\,^{\circ}C$ for over 30 min, tumor cells can be destroyed. The greatest advantage of this method of thermal therapy is that only cells with magnetic nanoparticles are killed and not adjacent healthy cells.

Albeit this method has been successfully demonstrated in animal models, there is no widespread therapeutic use of this method in human subjects. There are several challenges to successful implementation of magnetic hyperthermia in humans. The first challenge is to selectively localize these magnetic carriers in tumor cells alone and not in healthy cells. The other challenge is to be able to deliver sufficient concentration of these magnetic carriers so as to generate enough heat to kill the tumor cells. Additionally, the magnetic field strength to generate the process of hyperthermia should be safe in humans. The biggest challenge, however, is not in generating the heat but rather sustaining this heat generated by hyperthermia. Heat generated in the tissue has to overcome the natural tissue cooling process – blood flow and tissue perfusion. Both of these vary significantly within various tissues in the body and between individuals.

Some of these challenges have been addressed for implementation of magnetic hyperthermia in humans. Magnetic carriers have been surface-modified with biomolecules that specifically target and recognize overexpressed receptors on tumor cell surface (DeNardo et al. 2005). This method of active targeting has been used to selectively distinguish and localize in tumor cells. Passive targeting to tumor cells using magnetic carriers has also been achieved through some clever designing. Magnetic carriers have been encapsulated within polymeric, liposomes, or micellar structures to form a composite that can be localized within the tumor. This method exploits the leaky vasculature of the tumor to selectively target cancerous tissues. Another method to introduce the magnetic carriers is to directly inject in the tumor area. This method will guarantee the highest concentration of magnetic carriers compared to any other method.

The optimum magnetic field strength that is tolerated by humans has been established to be in the range of $f = 0.05$–$1.2\,MHz$ and $H = 0$–$15\,kA\ m^{-1}$. The overall magnetic field strength $H \times f$ should not exceed $4.85 \times 10^8\ Am^{-1}\ s^{-1}$ (Atkinson et al. 1984).

As mentioned earlier, the most studied magnetic material for biomedical use is iron oxide in the form of either magnetite (Fe_3O_4) or maghemite (γ-Fe_2O_3). These nanomaterials have been extensively investigated for cytotoxicity, biocompatibility, and biological fate in body.

6 Conclusions

Organic and inorganic nanoparticle systems offer significant opportunity in diagnosis, imaging, and therapy of cancer. In this chapter, we have discussed the role of magnetic nanoparticle systems in molecular imaging, targeted delivery, and thermal therapy. Magnetic nanoparticles based on paramagnetic or ferromagnetic materials and engineered for specific biorecognition of cancer targets offer tremendous opportunities in this field. With better understanding of disease biology combined with opportunities for development of targeted imaging and therapeutic systems, magnetic nanosystems will continue to play an important role in cancer nanomedicine.

References

ACS. 2007. http://www.cancer.org/downloads/STT/CAFF2007PWSecured.pdf

Ai, H., Flask, C., Weinberg, B., Shuai, X., Pagel, M. D., Farrell, D., Duerk, J., and Gao, J. 2005. Magnetite-loaded polymeric micelles as ultrasensitive magnetic-resonance probes, Adv. Mater. (Weinheim, Germany) 17:1949–1952.

Alexiou, C., Arnold, W., Klein, R. J., Parak, F. G., Hulin, P., Bergemann, C., Erhardt, W., Wagenpfeil, S., and Lubbe, A. S. 2000. Locoregional cancer treatment with magnetic drug targeting, Cancer Res. 60:6641–6648.

Allen, T. M. 2002. Ligand-targeted therapeutics in anticancer therapy, Nat. Rev. Cancer 2:750–763.

Allen, T. M. and Chonn, A. 1987. Large unilamellar liposomes with low uptake into the reticuloendothelial system, FEBS Lett. 223:42–46.

Allen, T. M., Hansen, C., and Rutledge, J. 1989. Liposomes with prolonged circulation times: factors affecting uptake by reticuloendothelial and other tissues, Biochim. Biophys. Acta 981:27–35.

Allport, J. R. and Weissleder, R. 2001. In vivo imaging of gene and cell therapies, Exp. Hematol. 29:1237–1246.

Amiji, M. and Park, K. 1992. Prevention of protein adsorption and platelet adhesion on surfaces by PEO/PPO/PEO triblock copolymers, Biomaterials 13:682–692.

Artemov, D., Mori, N., Okollie, B., and Bhujwalla, Z. M. 2003. MR molecular imaging of the Her-2/neu receptor in breast cancer cells using targeted iron oxide nanoparticles, Magn. Reson. Med. 49:403–408.

Atkinson, W. J., Brezovich, I. A., and Chakraborty, D. P. 1984. Usable frequencies in hyperthermia with thermal seeds, IEEE Trans. Biomed. Eng. 31:70–75.

Ban, Z., Barnakov, Y. A., Li, F., Golub, V. O., and O'Connor, C. J. 2005. The synthesis of core–shell iron@gold nanoparticles and their characterization, J. Mater. Chem. 15:4660–4662.

Beardsley, T. 1994. A war not won. Sci. Am. 270:130–138.

Bee, A., Massart, R., and Neveu, S. 1995. Synthesis of very fine maghemite particles, J. Magn. Magn. Mater. 149:6–9.

Berret, J.-F., Schonbeck, N., Gazeau, F., El Kharrat, D., Sandre, O., Vacher, A., and Airiau, M. 2006. Controlled clustering of superparamagnetic nanoparticles using block copolymers: design of new contrast agents for magnetic resonance imaging, J. Am. Chem. Soc. 128:1755–1761.

Bertorelle, F., Wilhelm, C., Roger, J., Gazeau, F., Menager, C., and Cabuil, V. 2006. Fluorescence-modified superparamagnetic nanoparticles: intracellular uptake and use in cellular imaging, Langmuir 22:5385–5391.

Blankenberg, F. G. and Strauss, H. W. 2001. Noninvasive strategies to image cardiovascular apoptosis, Cardiol. Clin. 19:165–172.

Brigger, I., Dubernet, C., and Couvreur, P. 2002. Nanoparticles in cancer therapy and diagnosis, Adv. Drug Deliv. Rev. 54:631–651.

Butterworth, M. D., Bell, S. A., Armes, S. P., and Simpson, A. W. 1996. Synthesis and characterization of polypyrrole–magnetite–silica particles, J. Colloid Interface Sci. 183:91–99.

Caruntu, D., Caruntu, G., Chen, Y., O'Connor, C. J., Goloverda, G., and Kolesnichenko, V. L. 2004. Synthesis of variable-sized nanocrystals of Fe_3O_4 with high surface reactivity, Chem. Mater. 16:5527–5534.

Caruntu, D., Remond, Y., Chou, N. H., Jun, M. J., Caruntu, G., He, J., Goloverda, G., O'Connor, C., and Kolesnichenko, V. 2002. Reactivity of 3d transition metal cations in diethylene glycol solutions. Synthesis of transition metal ferrites with the structure of discrete nanoparticles complexed with long-chain carboxylate anions, Inorg. Chem. 41:6137–6146.

Caruso, F. 2001. Nanoengineering of particle surfaces, Adv. Mater. 13:11–22.

Chen, M., Yamamuro, S., Farrell, D., and Majetich, S. A. 2003. Gold-coated iron nanoparticles for biomedical applications, J. Appl. Phys. 93:7551–7553.

Chung, S. H., Hoffmann, A., Bader, S. D., Liu, C., Kay, B., Makowski, L., and Chen, L. 2004. Biological sensors based on Brownian relaxation of magnetic nanoparticles, Appl. Phys. Lett. 85:2971–2973.

Dandamudi, S. and Campbell, R. B. 2007. Development and characterization of magnetic cationic liposomes for targeting tumor microvasculature, Biochim. Biophys. Acta 1768:427–438.

Davis, S. S. 1997. Biomedical applications of nanotechnology – implications for drug targeting and gene therapy, Trends Biotechnol. 15:217–224.

Decuzzi, P., Lee, S., Bhushan, B., and Ferrari, M. 2005. A theoretical model for the margination of particles within blood vessels. Ann. Biomed. Eng. 33:179–190.

DeNardo, S. J., DeNardo, G. L., Miers, L. A., Natarajan, A., Foreman, A. R., Gruettner, C., Adamson, G. N., and Ivkov, R. 2005. Development of tumor targeting bioprobes (^{111}In-chimeric L6 monoclonal antibody nanoparticles) for alternating magnetic field cancer therapy, Clin. Cancer Res. 11:7087–7092.

DePalma, R., Peeters, S., VanBael, M. J., VandenRul, H., Bonroy, K., Laureyn, W., Mullens, J., Borghs, G., and Maes, G. 2007. Silane ligand exchange to make hydrophobic superparamagnetic nanoparticles water-dispersible, Chem. Mater. 19:1821–1831.

Dinega, D. P. and Bawendi, M. G. 1999. A solution-phase chemical approach to a new crystal structure of cobalt, Angew. Chem. Int. Ed. 38:1788–1791.

Duncan, R. 2003. The dawning era of polymer therapeutics, Nat. Rev. Drug Discov. 2:347–360.

Engin, K. 1996. Biological rationale and clinical experience with hyperthermia, Control Clin. Trials 17:316–342.

Euliss, L. E., Grancharov, S. G., O'Brien, S., Deming, T. J., Stucky, G. D., Murray, C. B., and Held, G. A. 2003. Cooperative assembly of magnetic nanoparticles and block copolypeptides in aqueous media, Nano Lett. 3:1489–1493.

Farrell, D., Cheng, Y., McCallum, R. W., Sachan, M., and Majetich, S. A. 2005. Magnetic interactions of iron nanoparticles in arrays and dilute dispersions, J. Phys. Chem. B 109: 13409–13419.

Farrell, D., Majetich, S. A., and Wilcoxon, J. P. 2003. Preparation and characterization of mono-disperse Fe nanoparticles, J. Phys. Chem. B 107:11022–11030.

Flacke, S., Fischer, S., Scott, M. J., Fuhrhop, R. J., Allen, J. S., McLean, M., Winter, P., Sicard, G. A., Gaffney, P. J., Wickline, S. A., and Lanza, G. M. 2001. Novel MRI contrast agent for molecular imaging of fibrin: implications for detecting vulnerable plaques, Circulation 104:1280–1285.

Furlani, E. J. and Furlani, E. P. 2007. A model for predicting magnetic targeting of multifunctional particles in the microvasculature, J. Magn. Magn. Mater. 312:187–193.

Gabizon, A. and Papahadjopoulos, D. 1988. Liposome formulations with prolonged circulation time in blood and enhanced uptake by tumors, Proc. Natl Acad. Sci. USA 85:6949–6953.

Gilchrist, R. K. 1960. Potential treatment of cancer by electromagnetic heating, Surg. Gynecol. Obstet. 110:499–500.

Gilchrist, R. K., Shorey, W. D., Hanselman, R. C., Depeyster, F. A., Yang, J., and Medal, R. 1965. Effects of electromagnetic heating on internal viscera: a preliminary to the treatment of human tumors, Ann. Surg. 161:890–896.

Goodwin, S., Peterson, C., Hoh, C., and Bittner, C. 1999. Targeting and retention of magnetic targeted carriers (MTCs) enhancing intra-arterial chemotherapy, J. Magn. Magn. Mater. 194:132–139.

Gordon, R. T., Hines, J. R., and Gordon, D. 1979. Intracellular hyperthermia. A biophysical approach to cancer treatment via intracellular temperature and biophysical alterations, Med. Hypotheses 5:915–936.

Gref, R., Luck, M., Quellec, P., Marchand, M., Dellacherie, E., Harnisch, S., Blunk, T., and Muller, R. H. 2000. 'Stealth' corona–core nanoparticles surface modified by polyethylene glycol (PEG): influences of the corona (PEG chain length and surface density) and of the core composition on phagocytic uptake and plasma protein adsorption, Colloids Surf. B Biointerfaces 18:301–313.

Gupta, P. K., Hung, C. T., and Perrier, D. G. 1987. Quantitation of the release of doxorubicin from colloidal dosage forms using dynamic dialysis, J. Pharm. Sci. 76:141–145.

Hafeli, U. 2001. Radioactive magnetic microspheres. In: Microspheres, Microcapsules & Liposomes (Arshady, R. ed.), Citus Books, London, pp. 559–584.

Hafeli, U. O. 2004. Magnetically modulated therapeutic systems, Int. J. Pharm. 277:19–24.

Haukanes, B. I. and Kvam, C. 1993. Application of magnetic beads in bioassays, Biotechnology 11:60–63.

Hildebrandt, B., Wust, P., Ahlers, O., Dieing, A., Sreenivasa, G., Kerner, T., Felix, R., and Riess, H. 2002. The cellular and molecular basis of hyperthermia, Crit. Rev. Oncol. Hematol. 43:33–56.

Hirsch, L. R., Stafford, R. J., Bankson, J. A., Sershen, S. R., Rivera, B., Price, R. E., Hazle, J. D., Halas, N. J., and West, J. L. 2003. Nanoshell-mediated near-infrared thermal therapy of tumors under magnetic resonance guidance, Proc. Natl Acad. Sci. USA 100:13549–13554.

Hyeon, T., Lee, S. S., Park, J., Chung, Y., and Na, H. B. 2001. Synthesis of highly crystalline and monodisperse maghemite nanocrystallites without a size-selection process, J. Am. Chem. Soc. 123:12798–12801.

Ito, I., Ji, L., Tanaka, F., Saito, Y., Gopalan, B., Branch, C. D., Xu, K., Atkinson, E. N., Bekele, B. N., Stephens, L. C., Minna, J. D., Roth, J. A., and Ramesh, R. 2004. Liposomal vector mediated delivery of the 3p FUS1 gene demonstrates potent antitumor activity against human lung cancer in vivo, Cancer Gene Ther. 11:733–739.

Ivkov, R., DeNardo, S. J., Daum, W., Foreman, A. R., Goldstein, R. C., Nemkov, V. S., and DeNardo, G. L. 2005. Application of high amplitude alternating magnetic fields for heat induction of nanoparticles localized in cancer, Clin. Cancer Res. 11:7093–7103.

Jain, T. K., Morales, M. A., Sahoo, S. K., Leslie-Pelecky, D. L., and Labhasetwar, V. 2005. Iron oxide nanoparticles for sustained delivery of anticancer agents, Mol. Pharm. 2:194–205.

Jeong, U., Teng, X., Wang, Y., Yang, H., and Xia, Y. 2007. Superparamagnetic colloids: controlled synthesis and niche applications, Adv. Mater. 19:33–60.

Jordan, A., Scholz, R., Maier-Hauff, K., Johannsen, M. Wust, P., Nadobny, J., Schirra, H., Schmidt, H., Deger, S., Loening, S., Lanksch, W., and Felix, R. 2001. Presentation of a new magnetic field therapy system for the treatment of human solid tumors with magnetic fluid hyperthermia, J. Magn. Magn. Mater. 225:118–126.

Josephson, L., Perez, J. M., and Weissleder, R. 2001. Magnetic nanosensors for the detection of oligonucleotide sequences, Angew. Chem. Int. Ed. 40:3204–3206.

Jun, Y. W., Huh, Y. M., Choi, J. S., Lee, J. H., Song, H. T., Kim, S., Yoon, S., Kim, K. S., Shin, J. S., Suh, J. S., and Cheon, J. 2005. Nanoscale size effect of magnetic nanocrystals and their utilization for cancer diagnosis via magnetic resonance imaging, J. Am. Chem. Soc. 127:5732–5733.

Jung, C. 1995. Surface properties of superparamagnetic iron oxide MR contrast agents: ferumoxides, ferumoxtran, ferumoxsil, Magn. Reson. Imaging 13:675–691.

Kim, B. S., Qiu, J. M., Wang, J. P., and Taton, T. A. 2005. Magnetomicelles: composite nanostructures from magnetic nanoparticles and cross-linked amphiphilic block copolymers, Nano Lett. 5:1987–1991.

Kirby, C. and Gregoriadis, G. 1980. The effect of the cholesterol content of small unilamellar liposomes on the fate of their lipid components in vitro, Life Sci. 27:2223–2230.

Klibanov, A. L., Maruyama, K., Beckerleg, A. M., Torchilin, V. P., and Huang, L. 1991. Activity of amphipathic poly(ethylene glycol) 5000 to prolong the circulation time of liposomes depends on the liposome size and is unfavorable for immunoliposome binding to target, Biochim. Biophys. Acta 1062:142–148.

Koh, I., Wang, X., Varughese, B., Isaacs, L., Ehrman, S. H., and English, D. S. 2006. Magnetic iron oxide nanoparticles for biorecognition: evaluation of surface coverage and activity, J. Phys. Chem. B 110:1553–1558.

Kohler, N., Sun, C., Wang, J., and Zhang, M. 2005. Methotrexate-modified superparamagnetic nanoparticles and their intracellular uptake into human cancer cells, Langmuir 21:8858–8864.

Kreuter, J. 1994. Nanoparticles, Dekker, New York, pp. 165–190.

Lanza, G. M., Yu, X., Winter, P. M., Abendschein, D. R., Karukstis, K. K., Scott, M. J., Chinen, L. K., Fuhrhop, R. W., Scherrer, D. E., and Wickline, S. A. 2002. Targeted antiproliferative drug delivery to vascular smooth muscle cells with a magnetic resonance imaging nanoparticle contrast agent: implications for rational therapy of restenosis, Circulation 106:2842–2847.

Lasic, D. D., Martin, F. J., Gabizon, A., Huang, S. K., and Papahadjopoulos, D. 1991. Sterically stabilized liposomes: a hypothesis on the molecular origin of the extended circulation times, Biochim. Biophys. Acta 1070:187–192.

Lee, A. S., Weissleder, R., Brady, T. J., and Wittenberg, J. 1991. Lymph nodes: microstructural anatomy at MR imaging, Radiology 178:519–522.

Lee, J.-H., Huh, Y.-M., Jun, Y.-W., Seo, J.-W., Jang, J.-T., Song, H.-T., Kim, S., Cho, E.-J., Yoon, H.-G., Suh, J.-S., and Cheon, J. 2007. Artificially engineered magnetic nanoparticles for ultrasensitive molecular imaging, Nat. Med. 13:95–99.

Lee, S. Y. and Harris, M. T. 2006. Surface modification of magnetic nanoparticles capped by oleic acids: characterization and colloidal stability in polar solvents, J. Colloid Interface Sci. 293:401–408.

Lee, Y., Lee, J., Bae, C. J., Park, J.-G., Noh, H.-J., Park, J.-H., and Hyeon, T. 2005. Large-scale synthesis of uniform and crystalline magnetite nanoparticles using reverse micelles as nanoreactors under reflux conditions, Adv. Funct. Mater. 15:2036.

Li, G., Fan, J., Jiang, R., and Gao, Y. 2004. Cross-linking the linear polymeric chains in the ATRP synthesis of iron oxide/polystyrene core/shell nanoparticles, Chem. Mater. 16:1835–1837.

Li, Y., Afzaal, M., and O'Brien, P. 2006. The synthesis of amine-capped magnetic (Fe, Mn, Co, Ni) oxide nanocrystals and their surface modification for aqueous dispersibility, J. Mater. Chem. 16:2175–2180.

Liu, Q., Xu, Z., Finch, J. A., and Egerton, R. 1998. A novel two-step silica-coating process for engineering magnetic nanocomposites, Chem. Mater. 10:3936–3940.

Lockman, P. R., Mumper, R. J., Khan, M. A., and Allen, D. D. 2002. Nanoparticle technology for drug delivery across the blood–brain barrier, Drug Dev. Ind. Pharm. 28:1–13.

Lu, A.-H., Salabas, E. L., and Schüth, F.. 2007. Magnetic nanoparticles: synthesis, protection, functionalization, and application, Angew. Chem. Int. Ed. 46:1222–1244.

Lu, Y., Yin, Y., Mayers, B. T., and Xia, Y. 2002. Modifying the surface properties of superparamagnetic iron oxide nanoparticles through a sol-gel approach, Nano Lett. 2:183–186.

Lubbe, A. S., Bergemann, C., Brock, J., and McClure, D. G. 1999. Physiological aspects in magnetic drug-targeting, J. Magn. Magn. Mater. 194:149–155.

Lubbe, A. S., Bergemann, C., Huhnt, W., Fricke, T., Riess, H., Brock, J. W., and Huhn, D. 1996a. Preclinical experiences with magnetic drug targeting: tolerance and efficacy, Cancer Res. 56:4694–4701.

Lubbe, A. S., Bergemann, C., Riess, H., Schriever, F., Reichardt, P., Possinger, K., Matthias, M., Dorken, B., Herrmann, F., Gurtler, R., Hohenberger, P., Haas, N., Sohr, R., Sander, B., Lemke, A. J., Ohlendorf, D., Huhnt, W., and Huhn, D. 1996b. Clinical experiences with magnetic drug targeting: a phase I study with 4 -epidoxorubicin in 14 patients with advanced solid tumors, Cancer Res. 56:4686–4693.

Lyon, J. L., Fleming, D. A., Stone, M. B., Schiffer, P., and Williams, M. E. 2004. Synthesis of Fe oxide core/Au shell nanoparticles by iterative hydroxylamine seeding, Nano Lett. 4:719–723.

Ma, M., Zhang, Y., Yu, W., Shen, H. Y., Zhang, H. Q., and Gu, N. 2003. Preparation and characterization of magnetite nanoparticles coated by amino silane, Colloids Surf. A: Physicochem. Eng. Asp. 212:219–226.

Massart, R., Dubois, E., Cabuil, V., and Hasmonay, E. 1995. Preparation and properties of monodisperse magnetic fluids, J. Magn. Magn. Mater. 149:1–5.

Mikhaylova, M., Kim, D. K., Berry, C. C., Zagorodni, A., Toprak, M., Curtis, A. S. G., and Muhammed, M. 2004. BSA immobilization on amine-functionalized superparamagnetic iron oxide nanoparticles, Chem. Mater. 16:2344–2354.

Mizukoshi, Y., Seino, S., Okitsu, K., Kinoshita, T., Otome, Y., Nakagawa, T., and Yamamoto, T. A. 2005. Sonochemical preparation of composite nanoparticles of $Au/g-Fe_2O_3$ and magnetic separation of glutathione, Ultrason. Sonochem. 12:191–195.

Moeser, G. D., Roach, K. A., Green, W. H., Hatton, T. A., and Laibinis, P. E. 2004. High-gradient magnetic separation of coated magnetic nanoparticles, AIChE J. 50:2835–2848.

Moghimi, S. M. and Hunter, A. C. 2000. Poloxamers and poloxamines in nanoparticle engineering and experimental medicine, Trends Biotechnol. 18:412–420.

Moghimi, S. M., Hunter, A. C., and Murray, J. C. 2001. Long-circulating and target-specific nanoparticles: theory to practice, Pharmacol. Rev. 53:283–318.

Morais, P. C., Santos, R. L., Pimenta, A. C. M., Azevedo, R. B., and Lima, E. C. D. 2006. Preparation and characterization of ultra-stable biocompatible magnetic fluids using citratecoated cobalt ferrite nanoparticles, Thin Solid Films 515:266–270.

Moroz, P., Jones, S. K., and Gray, B. N. 2002. Magnetically mediated hyperthermia: current status and future directions, Int. J. Hyperthermia 18:267–284.

Nashat, A. H., Moronne, M., and Ferrari, M. 1998. Detection of functional groups and antibodies on microfabricated surfaces by confocal microscopy, Biotechnol. Bioeng. 60:137–146.

Neuberger, T., Schopf, B., Hofmann, H., Hofmann, M., and von Rechenberg, B. 2005. Superparamagnetic nanoparticles for biomedical applications: possibilities and limitations of a new drug delivery system, J. Magn. Magn. Mater. 293:483–496.

Ngo, A. T. and Pileni, M. P. P. 2000. Nanoparticles of cobalt ferrite: influence of the applied field on the organization of the nanocrystals on a substrate and on their magnetic properties, Adv. Mater. 12:276–279.

O'Handley, R. C. 2000, Modern Magnetic Materials: Principles and Applications, Wiley, New York.

Ohmori, M. and Matijevic, E. 1993. Preparation and properties of uniform coated inorganic colloidal particles: 8. Silica on iron, J. Colloid Interface Sci. 160:288–292.

Otsuka, H., Nagasaki, Y., and Kataoka, K. 2003. PEGylated nanoparticles for biological and pharmaceutical applications, Adv. Drug Deliv. Rev. 55:403–419.

Oyewumi, M. O. and Mumper, R. J. 2004. Comparison of cell uptake, biodistribution and tumor retention of folate-coated and PEG-coated gadolinium nanoparticles in tumor-bearing mice, J. Control. Release 24:613–626.

Pankhurst, Q. A., Connolly, J., Jones, S. K., and Dobson, J. 2003. Applications of magnetic nanoparticles in biomedicine, J. Phys. D: Appl. Phys. 36:R167–R181.

Papahadjopoulos, D. and Gabizon, A. 1987. Targeting of liposomes to tumor cells in vivo, Ann. NY Acad. Sci. 507:64–74.

Park, J., An, K., Hwang, Y., Park, J.-G., Noh, H.-J., Kim, J.-Y., Park, J.-H., Hwang, N.-M., and Hyeon, T. 2004. Ultra-large-scale syntheses of monodisperse nanocrystals, Nat. Mater. 3:891–895.

Park, J. W. 2002. Liposome-based drug delivery in breast cancer treatment, Breast Cancer Res. 4:95–99.

Park, S. J., Kim, S., Lee, S., Khim, Z. G., Char, K., and Hyeon, T. 2000. Synthesis and magnetic studies of uniform iron nanorods and nanospheres, J. Am. Chem. Soc. 122:8581–8582.

Reimer, P., Weissleder, R., Wittenberg, J., and Brady, T. J. 1992. Receptor-directed contrast agents for MR imaging: preclinical evaluation with affinity assays, Radiology 182:565–569.

Reynolds, C. H., Annan, N., Beshah, K., Huber, J. H., Shaber, S. H., Lenkinski, R. E., and Wortman, J. A. 2000. Gadolinium-loaded nanoparticles: new contrast agents for magnetic resonance imaging, J. Am. Chem. Soc. 122:8940–8945.

Schellenberger, E. A., Bogdanov, A., Jr., Hogemann, D., Tait, J., Weissleder, R., and Josephson, L. 2002. Annexin V-CLIO: a nanoparticle for detecting apoptosis by MRI, Mol. Imaging 1:102–107.

Seneterre, E., Weissleder, R., Jaramillo, D., Reimer, P., Lee, A. S., Brady, T. J., and Wittenberg, J. 1991. Bone marrow: ultrasmall superparamagnetic iron oxide for MR imaging, Radiology 179:529–533.

Shen, L., Laibinis, P. E., and Hatton, T. A. 1999. Bilayer surfactant stabilized magnetic fluids: synthesis and interactions at interfaces, Langmuir 15:447–453.

Shevchenko, E. V., Talapin, D. V., Rogach, A. L., Kornowski, A., Haase, M., and Weller, H. 2002. Colloidal synthesis and self-assembly of $CoPt_3$ nanocrystals, J. Am. Chem. Soc. 124:11480–11485.

Shevchenko, E. V., Talapin, D. V., Schnablegger, H., Kornowski, A., Festin, O., Svedlindh, P., Haase, M., and Weller, H. 2003. Study of nucleation and growth in the organometallic synthesis of magnetic alloy nanocrystals: the role of nucleation rate in size control of $CoPt_3$ nanocrystals, J. Am. Chem. Soc. 125:9090–9101.

Song, Q. and Zhang, Z. J. 2004. Shape control and associated magnetic properties of spinel cobalt ferrite nanocrystals, J. Am. Chem. Soc. 126:6164–6168.

Sonvico, F., Dubernet, C., Colombo, P., and Couvreur, P. 2005. Metallic colloid nanotechnology, applications in diagnosis and therapeutics, Curr. Pharm. Des. 11:2095–2105.

Sousa, M. H., Tourinho, F. A., Depeyrot, J., da Silva, G. J., and Lara, M. C. F. L. 2001. New electric double-layered magnetic fluids based on copper, nickel, and zinc ferrite nanostructures, J. Phys. Chem. B 105:1168–1175.

Sridhar, S., Amiji, M., Shenoy, D., Nagesha, D., Weissig, V., and Fu, W. 2005. Nanomedicine: a new paradigm in diagnosis and therapy, Proc. SPIE Int. Soc. Opt. Eng. 6008, 600816.

Stoeva, S. I., Huo, F., Lee, J.-S., and Mirkin, C. A. 2005. Three-layer composite magnetic nanoparticle probes for DNA, J. Am. Chem. Soc. 127:15362–15363.

Stoltenburg, R., Reinemann, C., and Strehlitz, B. 2005. FluMag-SELEX as an advantageous method for DNA aptamer selection, Anal. Bioanal. Chem. 383:83–91.

Storhoff, J. J., Elghanian, R., Mucic, R. C., Mirkin, C. A., and Letsinger, R. L. 1998. One-pot colorimetric differentiation of polynucleotides with single base imperfections using gold nanoparticle probes, J. Am. Chem. Soc. 120:1959–1964.

Sun, S. and Murray, C. B. 1999. Synthesis of monodisperse cobalt nanocrystals and their assembly into magnetic superlattices (invited), J. Appl. Phys. 85:4325–4330.

Sun, S., Murray, C. B., Weller, D., Folks, L., and Moser, A. 2000. Monodisperse FePt nanoparticles and ferromagnetic FePt nanocrystal superlattices, Science 287:1989–1992.

Sun, S. and Zeng, H. 2002. Size-controlled synthesis of magnetite nanoparticles, J. Am. Chem. Soc. 124:8204–8205.

Sun, S., Zeng, H., Robinson, D. B., Raoux, S., Rice, P. M., Wang, S. X., and Li, G. 2004. Monodisperse MFe2O4 (M=Fe, Co, Mn) nanoparticles, J. Am. Chem. Soc. 126:273–279.

Suslick, K. S., Fang, M., and Hyeon, T. 1996. Sonochemical synthesis of iron colloids, J. Am. Chem. Soc. 118:11960–11961.

Tan, J. S., Butterfield, D. E., Voycheck, C. L., Caldwell, K. D., and Li, J. T. 1993. Surface modification of nanoparticles by PEO/PPO block copolymers to minimize interactions with blood components and prolong blood circulation in rats, Biomaterials 14:823–833.

Tartaj, P., Morales, M. D. P., Veintemillas-Verdaguer, S., Gonzalez-Carreno, T., and Serna, C. J. 2003. The preparation of magnetic nanoparticles for applications in biomedicine, J. Phys. D: Appl. Phys. 36:182–197.

Tartaj, P. and Serna, C. J. 2002. Microemulsion-assisted synthesis of tunable superparamagnetic composites, Chem. Mater. 14:4396–4402.

Tartaj, P. and Serna, C. J. 2003. Synthesis of monodisperse superparamagnetic Fe/silica nanospherical composites, J. Am. Chem. Soc. 125:15754–15755.

Teng, X. and Yang, H. 2004. Effects of surfactants and synthetic conditions on the sizes and self-assembly of monodisperse iron oxide nanoparticles, J. Mater. Chem. 14:774–779.

Thunemann Andreas, F., Schutt, D., Kaufner, L., Pison, U., and Mohwald, H. 2006. Maghemite nanoparticles protectively coated with poly(ethylene imine) and poly(ethylene oxide)-block-poly(glutamic acid), Langmuir 22:2351–2357.

Vestal, C. R. and Zhang, Z. J. 2002. Atom transfer radical polymerization synthesis and magnetic characterization of $MnFe_2O_4$/polystyrene core/shell nanoparticles, J. Am. Chem. Soc. 124:14312–14313.

Vestal, C. R. and Zhang, Z. J. 2003. Synthesis and magnetic characterization of Mn and Co spinel ferrite–silica nanoparticles with tunable magnetic core, Nano Lett. 3:1739–1743.

Wang, G.-P., Song, E.-Q., Xie, H. Y., Zhang, Z. L., Tian, Z. Q., Zuo, C., Pang, D. W., Wu, D. C., and Shi, Y. B. 2005. Biofunctionalization of fluorescent-magnetic-bifunctional nanospheres and their applications, Chem. Commun. 34:4276–4278.

Wang, Y., Teng, X., Wang, J. S., and Yang, H. 2003. Solvent-free atom transfer radical polymerization in the synthesis of Fe_2O_3@polystyrene core–shell nanoparticles, Nano Lett. 3:789–793.

Widder, K., Flouret, G., and Senyei, A. 1979. Magnetic microspheres: synthesis of a novel parenteral drug carrier, J. Pharm. Sci. 68:79–82.

Widder, K. J., Senyei, A. E., and Ranney, D. F. 1980. In vitro release of biologically active adriamycin by magnetically responsive albumin microspheres, Cancer Res. 40:3512–3517.

Wunderbaldinger, P., Josephson, L., and Weissleder, R. 2002. Tat peptide directs enhanced clearance and hepatic permeability of magnetic nanoparticles, Bioconjug. Chem. 13:264–268.

Yi, D. K., Selvan, S. T., Lee, S. S., Papaefthymiou, G. C., Kundaliya, D., and Ying, J. Y. 2005. Silica-coated nanocomposites of magnetic nanoparticles and quantum dots, J. Am. Chem. Soc. 127:4990–4991.

van der Zee, J. 2002. Heating the patient: a promising approach? Ann. Oncol. 13:1173–1184.

Zhang, Y., Kohler, N., and Zhang, M. 2002. Surface modification of superparamagnetic magnetite nanoparticles and their intracellular uptake, Biomaterials 23:1553–1561.

Zhang, Y., Sun, C., Kohler, N., and Zhang, M. 2004. Self-assembled coatings on individual monodisperse magnetite nanoparticles for efficient intracellular uptake, Biomed. Microdevices 6:33–40.

Zhang, Y. and Zhang, J. 2005. Surface modification of monodisperse magnetite nanoparticles for improved intracellular uptake to breast cancer cells, J. Colloid Interface Sci. 283:352–357.

Nanosystems for Multimodality In vivo Imaging

Jinzi Zheng, David A. Jaffray, and Christine Allen

1 Introduction

The inherent differences and the complementary nature of existing imaging systems have prompted the quest for multimodality imaging platforms that allow for integration of images acquired at different scales (i.e., whole organism, organ, suborgan, cell, and subcellular) and at various stages of disease treatment (i.e., diagnostic, preoperative, intraoperative, and follow-up images). The successful development of a contrast agent platform that is able to provide persistent and colocalized signal enhancements across multiple imaging systems has the potential to seamlessly bridge wide ranges of spatial, temporal, and sensitivity scales and to be employed throughout a variety of clinical scenarios.

CT, MR, SPECT, and PET are whole body imaging systems with differing sensitivities that range from millimolar to picomolar and with a wide range of spatial resolutions ranging from sub-micrometer to millimeter (Fischman and Alpert, 2002; Krause, 1999) (Fig. 1). Conversely, optical imaging techniques are primarily utilized for ex vivo applications and have the ability to detect a single probe molecule and resolve its localization at the subcellular level. The various imaging systems also differ in their modes of operation, possible requirement for radiation exposure, cost, and predisposition for quantitation (Table 1). As a result, in routine clinical practice, specific imaging modalities are employed at different stages of medical diagnosis and treatment (Fig. 2). For example, combinations of anatomical and functional imaging techniques such as CT, MR, PET, SPECT, and US are often utilized for disease detection, staging, and treatment planning. X-ray, MR, US, and optical imaging systems are also found in the treatment room to guide the delivery of various interventions such as surgery, radiotherapy, and radiofrequency ablation with the aim of improving the accuracy of these procedures (Hashizume, 2007; Hsu, 2005; Jaffray, 2002; Lindner, 2006; McVeigh, 2006; Rygh, 2006; Uematsu, 1999).

Hardware-based integration of different imaging systems (i.e., CT/PET (Townsend, 2004; Townsend and Beyer, 2002; Townsend, 2004), CT/SPECT (Bocher, 2000), MR/PET (Lucas, 2006; Wagenaar, 2006), X-ray/MR (Fahrig et al., 2001, 2003)) has addressed some of the needs by allowing simultaneous or near-simultaneous acquisition of two complementary image sets from the same patient. However, not all

V. Torchilin (ed.), *Multifunctional Pharmaceutical Nanocarriers*,
© Springer Science+Business Media, LLC 2008

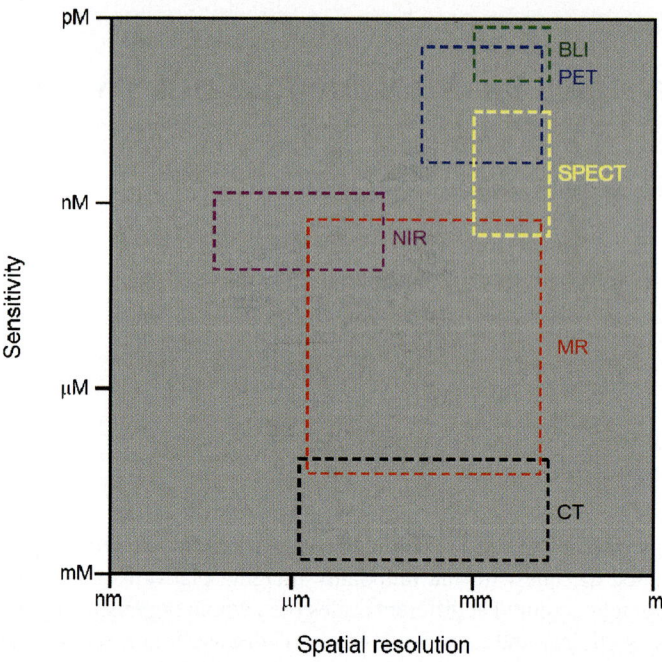

Fig. 1 Sensitivity and spatial resolution ranges of the different imaging modalities (figure adapted from Cassidy et al. (Cassidy and Radda, 2005)). Efforts by various groups are focused on extending the current sensitivity and spatial limits of each modality

Table 1 Summary of the strengths and limitations of different imaging modalities

Imaging modality	Strengths	Limitations
X-ray computed tomography (CT)	– Best bone structure visualization	– Radiation exposure
	– High temporal resolution	
Magnetic resonance (MR)	– Best soft tissue visualization	– Compatibility issues
	– No radiation dose	– Trade-off between temporal, spatial resolution and sensitivity
		– High cost
Radionuclide imaging (PET, SPECT)	– Whole body, high sensitivity functional imaging	– Radiation exposure
	– Multilabeling capability	– Lack of anatomic information but can be combined with CT or MR
Ultrasound (US)	– Real time imaging	– Image quality is operator dependent
	– No radiation dose	– Limited penetration depth (especially in the presence of bony structures)
	– Low cost	– Limited imageable area
Optical Imaging	– Single probe detection	– Limited penetration depth
	– No radiation dose	– Semiquantitative
	– Subcellular resolution	
	– Multilabeling capability	
	– Low cost	

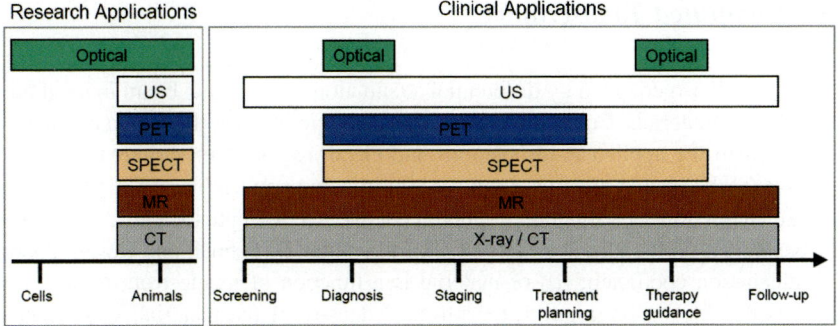

Fig. 2 Research and clinical applications that rely on the various imaging modalities

useful combinations of imaging modalities have been hardware integrated. In addition, many clinical scenarios would benefit from the registration of images that are acquired at different times from distinct modalities. The development of a multimodal contrast agent that enhances common structures in the different modalities has the potential to impact conventional practices by easing the alignment and common display of complementary image sets. As a matter of fact, single modality contrast agents are already required as part of many routine imaging sessions to increase visibility (Exadaktylos, 2005; Saeed, 2005) or to generate an imageable signal (i.e., radionuclide tracers).

Contrast-enhanced multimodality imaging may be performed using simultaneous or sequential administration of multiple modality-specific imaging probes. However, this is not a suitable solution when different probes have their own distinct pharmacokinetics and biodistribution profiles in vivo. Consequently, they cannot provide common enhancement features across different modalities. In addition, existing imaging agents are generally small molecular weight molecules and have rapid clearance profiles (Mutzel and Speck, 1980; Tweedle, 1997). Therefore, they do not allow for long (i.e., high-resolution scans) or repeated imaging sessions (i.e., longitudinal studies) to be performed following a single administration. For these reasons, nano-sized systems, such as nanoparticles (Josephson, 2002; Kircher, 2003; Schellenberger, 2004; Sosnovik, 2005), nanocrystals (Huh, 2005), liposomes (Zheng, 2006, 2007; Zielhuis, 2006) and dendrimers (Koyama, 2007), have been employed as delivery strategies for multiple imaging probes to provide colocalized and persistent contrast enhancement across imaging systems.

2   Imaging Modalities and Modality-Specific Imaging Probes

CT, MR, SPECT, PET, US, and optical imaging provide complementary information on structure and/or function. This is due to the fact that their signal generation processes differ, making each more suitable for a specific application (Table 1). Before discussing the novel developments in multimodal contrast agents, there is value in a brief primer on how different imaging probes modify the signal generated by these imaging modalities.

2.1   Computed Tomography

Contrast in CT is generated by differential attenuation of the X-ray beam in neighboring tissues or materials. Compton (σ_c) and photoelectric (σ_{pe}) are the two main interactions between the applied X-ray photons and electrons found in various tissues and materials (at the photon energies employed for imaging applications). The probability of those interactions is a function of photon energy (E), with photoelectric interaction also dependent on the atomic number (Z) of the material (Schnall and Rosen, 2006). The attenuation coefficient (μ) of material is a function of its electron density (ρ_e), which is roughly proportional to the physical density of the material, with relative contribution from Compton and photoelectric cross sections (Hsieh, 2003) (1).

$$\mu(E) = \rho_e[\sigma_c(E) + \sigma_{pe}(E, Z)]. \tag{1}$$

At the low photon energies (~70 keV) used for CT imaging, photoelectric absorption is the dominant photon–electron interaction, with the cross section per atom being proportional to Z^4 (Evans, 1955). Contrast enhancement is therefore achieved by increasing the difference in attenuation of two neighboring structures by the addition of high atomic number material within the volume. As a result, elements with a high Z, such as iodine ($Z = 53$) and barium ($Z = 56$), are attractive CT contrast agents and have been adopted in the clinic. Although iodine-based agents have well-characterized toxicity and relatively good safety profiles, they still put a small number of patients (<1%) at risk of allergic reactions and nephrotoxicity (Namasivayam, 2006). Recent attempts have been made to use other elements that have even higher Z (i.e., gadolinium (Bonvento, 2006) ($Z = 64$), gold (Hainfeld, 2006) ($Z = 79$), and bismuth (Rabin, 2006) ($Z = 83$)). However, the toxicity and safety of these agents need to be further investigated.

It is important to note that the signal in CT imaging is usually referred to as Hounsfield units (HU), after Sir Godfrey Hounsfield, the developer of the first CT scanner, where HU is defined as $\mathrm{HU} = \dfrac{\mu - \mu_{water}}{\mu_{water}} \times 1{,}000$ (Hsieh, 2003). The attenuation value in HU for water is therefore by definition zero. CT attenuation values range between −1,000 HU for vacuum, through zero for water and as high as +2,000 HU for hard bone. The noise level in CT images acquired in a typical clinical setting is approximately ±10 HU. Therefore, a relative difference of 30–50 HU is desirable to significantly discriminate between neighboring structures.

2.2   Magnetic Resonance Imaging

There are multiple mechanisms responsible for inducing contrast in MR imaging. It is critical to select appropriate imaging parameters to fully exploit the presence of inherent contrasts as well as any applied contrast agents. There are four main parameters that determine signal intensity and contrast in MR: spin density, relaxivity, magnetic susceptibility, as well as diffusion and perfusion (Weishaupt et al., 2003). Contrast agent visualization is optimized when the imaging technique selected suppresses the contrast generated from unchanged tissue parameters, and accentuates the signal that is altered by the presence of the contrast agent (Froehlich, 2006).

2.2.1 Spin Density

The magnetic moment of the proton is the fundamental signaling species in MR. Spin density is the number of protons per volume found in a specific tissue. The tissue proton density is dominated by the number of water protons rather than the protons associated with organic compounds that are present in the tissue. Because an exogenous agent cannot significantly alter the tissue proton density, this parameter is not readily altered by the introduction of a contrast agent.

2.2.2 Relaxivity

The relaxation of the protons from their excited state can be measured. There are two proton relaxation parameters in MR. T_1 is the longitudinal or spin-lattice relaxation time. It is the amount of time required for the magnetized proton nuclei to return to their equilibrium state in the longitudinal direction of the main magnetic field after a radiofrequency (RF) excitation pulse. T_2 is the transverse or spin–spin relaxation time. During the T_2 relaxation process, the energy deposited by the RF pulse results in loss of spin phase coherence in the transverse plane as well as spin dephasing (Koenig and Brown, 1994). Most MR contrast agents alter these two relaxivity parameters, usually affecting one relaxation time more than the other. Positive, or T_1 agents (i.e., gadolinium), reduce the longitudinal relaxation time and increase the signal intensity on a T_1-weighted MR image. While negative, or T_2 agents (i.e., iron), decrease the signal intensity on a T_2-weighted image through reduction of the transverse relaxation time. The presence of unpaired electrons in a paramagnetic ion is needed to cause a change in the T_1 and T_2 relaxation rates of the surrounding water protons. In addition, the relaxation rate is proportional to the spin quantum number of an atom. For example, gadolinium (Gd^{3+}) with a spin quantum number of 7/2 is a better relaxation agent than manganese (Mn^{3+}), which has a spin quantum number of 5/2.

2.2.3 Magnetic Susceptibility

Magnetic susceptibility is the degree of magnetization induced in the domain of a substance subjected to a magnetic field. There are four different classes of magnetically susceptible compounds: diamagnetic, paramagnetic, superparamagnetic, and ferromagnetic (Schnall and Rosen, 2006; Weishaupt et al., 2003). Diamagnetic substances weaken the magnetic field because they have a small negative magnetic susceptibility. Paramagnetic compounds have a net positive magnetic susceptibility which causes the magnetic field to strengthen. However, a high paramagnetic agent concentration is required for magnetic susceptibility to affect signal intensity. In the case of gadolinium, a paramagnetic ion, it is the presence of its unpaired electrons that affects tissue contrast by changing the relaxation properties of the surrounding water protons, not its paramagnetic susceptibility. Superparamagnetic materials acquire magnetic moments that are several orders of magnitude greater than those acquired by paramagnetic species in the same external magnetic field. Unlike paramagnetic ions, for which spin alignment is directly proportional to the strength of the external magnetic field, superparamagnetic ions align all available spins even at low field strengths (Sandler, 1989). Therefore, no further magnetization is achieved with greater increase in the strength of the external magnetic field. Iron oxide and magnetite particles are superparamagnetic contrast agents. The large magnetic moments associated with these superparamagnetic particles cause local field inhomogeneities, which produce rapid proton dephasing and correspondingly fast shortening of the T_2 relaxation time. As a result, superparamagnetic agents may be detected at concentrations as low as μmol kg^{-1}, while mmol kg^{-1} concentrations are needed for paramagnetic agents (Gimi, 2005). Ferromagnetic materials such as iron, nickel, and cobalt have a large positive magnetic susceptibility. They are generally not employed as MR contrast agents because they tend to distort the linear magnetic field gradients causing susceptibility artifacts (i.e., local signal loss and spatial distortion) (Sandler, 1989).

2.2.4   Diffusion and Perfusion

The signal intensity in MR is maximal when all of the transverse spins are in phase. Movement or diffusion of water protons can lead to spin dephasing on the transverse plane and produce loss of MR signal (Schnall and Rosen, 2006; Weishaupt et al., 2003). Consequently, perfusion of blood in tissues causes spin dephasing, resulting in a signal decrease and contributes to contrast in MR. Contrast agents are not generally used to directly manipulate the water proton diffusion process. However, the presence of MR contrast agents such as gadolinium has been shown to affect the calculation of the apparent diffusion coefficients from diffusion weighted images (Firat, 2006).

2.3   Single Photon Emission Computed Tomography

Gamma rays emitted by the process of radioisotope decay can be counted by a camera-like detector. In SPECT imaging, such a detector is rotated around the subject and detects gamma photons that are emitted by a radioisotope. The distribution of the radioisotope is then estimated through a reconstruction algorithm. The attachment of radioisotopes to molecules that target specific cell populations results in the production of radiotracers. It is undesirable to use radiotracers that emit electron or alpha particles, since this will increase the radiation dose given to patients. The effective half-life (T_{eff}) is a crucial factor when choosing a SPECT agent. T_{eff} of a radiopharmaceutical in a biological system is a function of both its physical half-life (T_p) and its biological half-life (T_b). Their relationship is described by (2) (Sandler, 1989).

$$T_{eff} = \frac{T_p - T_b}{T_p + T_b} \qquad (2)$$

The effective half-life of a SPECT tracer should match the time window required to image the desired physiological process (i.e., uptake of the tracer by the desired target structure). Table 2 lists the physical characteristics of the most commonly used SPECT radioisotopes (Sandler, 1989; Tsui, 1996). Advanced SPECT systems allow for count scoring of photons emitted at different energy windows. This is very valuable for applications requiring simultaneous imaging of multiple radiotracers.

2.4 Positron Emission Tomography

Some radioisotopes decay by emission of positrons $(\beta+)$ (Cassidy and Radda, 2005). An annihilation event occurs when the emitted positron encounters an electron $(\beta-)$ within the subject. This produces two high-energy photons of $511\,keV$ that travel in opposite directions (~180° apart) (Levin, 2005, Phelps, 2003). A positron emission event is recorded only when the opposite crystals in the PET detector ring detect two photons within a short time window (~10ns) (Phelps, 2003). This requirement results in improved scatter rejection and noise reduction, as well as better spatial resolution in human imaging applications, when compared to SPECT. However, some PET radioisotopes, such as ^{15}O and ^{11}C, have very limited physical half-lives (Table 3) and this limits their use in imaging applications involving slower physiological processes.

Table 2 List of the physical characteristics of the common SPECT radioisotopes

Radionuclide	Half-life (h)	Energy of primary photons (keV)
Gallium-67 (^{67}Ga)	78	93 (40%), 184 (24%), 300 (22%)
Indium-111 (^{111}In)	68	173 (89%), 247 (94%)
Iodine-123 (^{123}I)	13.1	159 (83%)
Iodine-131 (^{131}I)	8.08	364 (82%)
Technetium-99 m (^{99m}Tc)	6.03	140 (90%)
Thallium-201 (^{201}Tl)	73	69–81 (98%)

Table 3 Half-lives of common radionuclides used for PET imaging

Radionuclide	Half-life (min)	Half-life (h)
Oxygen-15 (^{15}O)	2.0	0.03
Nitrogen-13 (^{13}N)	10.0	0.17
Carbon-11 (^{11}C)	20.4	0.34
Fluorine-18 (^{18}F)	110	1.83
Bromine-76 (^{76}Br)	972	16.2
Iodine-124 (^{124}I)	6,048	101

2.5   Ultrasound

Contrast in ultrasound is generated by the difference in acoustic impedance of two neighboring tissues (McCulloch, 2000). The majority of currently available US contrast agents comprise gas-filled microbubbles (i.e., $<7\,\mu m$ in size for capillary access) (Calliada, 1998; Kaul, 2004). The ultrasonic properties of the microbubble agents are dependent on their size, as well as the composition of the bubble shell and gas. The outer shell of the microbubbles is most often composed of phospholipid, palmitic acid, albumin, or polymer, while the gas contained within the shell may be air or a heavy gas. The elasticity of the shell determines whether the microbubble breaks (stiffer shell) or resonates (more elastic shell) to produce backscatter when exposed to ultrasonic waves. Air-filled microbubbles are highly soluble and are strong reflectors; however, their shrinkage following air diffusion makes them less reflective over time. Heavy gas-filled microbubbles are less soluble, but they tend to be more stable (Calliada, 1998; Kaul 2004). Unlike conventional MR, CT, and nuclear medicine contrast agents, micron-sized conventional US agents do not provide interstitial signal enhancement because they do not cross the vascular endothelium. This feature makes them good blood pool agents. However, the vascular enhancement that they provide is only transient due to bubble destruction (Calliada 1998; McCulloch 2000).

2.6   Optical Imaging

In vivo optical imaging involves the detection of light photons transmitted through tissues. Two processes affect the propagation of light through tissues: absorption and scattering (Alfano et al., 1997; Schnall and Rosen, 2006). Absorption occurs when the photon energy is lost because of its frequency match to an energy transition state within a tissue. Scattering occurs when the photon path is deflected by particles in a tissue (Alfano et al., 1997; Schnall and Rosen, 2006). Both absorption and scattering affect the sensitivity of optical imaging and are wavelength dependent (Gillies, 2002). The optimal tissue imaging wavelength window is in the near-infrared (IR) region (700–1000 nm) (Cassidy and Radda, 2005; Gillies, 2002). The two main in vivo optical imaging techniques are near-IR (NIR) fluorescence and bioluminescence.

2.6.1   Fluorescence Imaging

Most in vivo fluorescence imaging is performed at the NIR wavelength region because most biological tissues exhibit low inherent scattering and minimal absorption at these wavelengths (Anderson and Parrish, 1981). The two types of exogenous contrast agents used in fluorescence imaging are organic fluorophores and inorganic fluorescent semiconductor nanocrystals, also known as quantum dots (Frangioni, 2003).

The most common organic fluorophores are polymethines. A major class of the polymethines is the heptamethine cyanines, which include the following subclasses: benzoxazole, benzothiazole, indolyl, 2-quinoline, and 4-quinoline (Frangioni, 2003). The range of peak excitation for compounds in this class is 760–800 nm, and the range for peak emission is 790–830 nm. The indocyanines (i.e., indocyanine green, IRDye78) are the most widely used compounds from this class (Frangioni, 2003; Sima and Kanofsky, 2000). However, these conventional organic fluorophores have many limitations, including difficulty to control their excitation and emission wavelengths, low quantum yield (QY) in aqueous environments, and high susceptibility to photobleaching (Frangioni, 2003).

Inorganic fluorescent semiconductor nanocrystals, quantum dots (QDs), typically comprise an inorganic core and a metal shell, allowing for fine-tuning of their emission wavelength. QDs have very high quantum yield in organic solvents, and are extremely resistant to photobleaching (Gao, 2005; Gao et al., 2007; Smith et al., 2004). The surface of the QDs may be further modified with an appropriate organic hydrophilic coating to increase their biocompatibility for in vivo imaging. These properties make QDs desirable for a wide range of optical imaging applications. A major difference between organic fluorophores and QDs is their size and molecular weight. NIR organic fluorophores are typically less than 1,200 Da, while QDs have molecular weights greater than 100 kDa and hydrodynamic diameters ranging between 3–20 nm (Frangioni, 2003). This size difference is significant enough to alter the pharmacokinetics, biodistribution, and clearance profiles of these two classes of agents when administered on their own. Therefore, when choosing an optical agent for a nanoparticulate system, size and molecular weight concerns must be taken into account. The incorporation of a QD may impact the in vivo distribution of the overall nanoplatform, while the use of organic fluorophores may not achieve optimal imaging properties.

Another important issue to be considered when selecting the appropriate optical agent for in vivo imaging is its toxicity profile. Organic fluorophores, such as heptamethine indocyanines, without charged groups, are reported to be quite toxic owing to their intracellular accumulation (Frangioni, 2003). However, disulfonated indocyanine green (ICG) has been approved by the FDA and used routinely in humans for over 40 years and has an excellent safety profile. At this point, there is limited information on the toxicity of other heptamethine indocyanines, yet studies in mice have shown that the LD50 of these agents increases with an increase in the degree of sulfonation (Frangioni, 2003). The in vivo toxicity of QDs also remains to be fully elucidated. QDs typically comprise metals (i.e., cadmium, arsenic, and lead), which, in their elemental form, are known to be toxic. It is suggested that in the nanocrystalline form, since the metals are present as a less reactive salt, the QDs may be relatively inert (Frangioni, 2003). However, further investigation is required to fully understand the acute and chronic toxicity profiles of QDs in vivo.

2.6.2 Bioluminescence Imaging

Bioluminescence imaging detects the visible light produced by the cellular expression of the enzyme luciferase. The gene responsible for the production of luciferase is naturally expressed in select bacteria, insects, and jellyfish (WT, 1999). In the research setting, the same gene is tagged onto one or more genes of interest and then incorporated into the DNA of a desired cell population. The expression of the luciferase gene is then used as a surrogate marker for the expression of the gene(s) of interest. However, the appropriate substrate must be available for the enzyme luciferase to react with and emit bioluminescence. For example, in order for the firefly luciferase to produce biolumi-nescence, the substrate D-luciferin, oxygen, and adenosine triphosphate (ATP) must all be present (Levin, 2005). In general, bioluminescence reactions have a very low quantum yield and the photons generated are subject to scatter and absorption while propagating through tissue. Therefore, relatively long exposure times are necessary in order to collect sufficiently high signal to noise ratios. Mice are the animal model of choice for bioluminescence studies due to the limited tissue thickness that biolumi-nescence photons need to travel through in order to reach the detector (Levin, 2005).

3 Nanosystems for Multimodality In Vivo Imaging

There are four key criteria that must be fulfilled in the design and development of multimodal contrast agent nanosystems. Firstly, the contrast agent must provide and maintain detectable signal levels for a time that is sufficient to allow for image acquisition in the multiple imaging modalities (i.e., including the time required to move the subject from one scanner to another when the imaging systems are not physically integrated). Secondly, if multiple imaging probes are present in the same nanosystem, one must not interfere with the signal generation or contrast enhance-ment ability of the other. Thirdly, the ratios between the different signaling moieties employed in the contrast agent system must remain constant throughout the imaging time window. As a result, the successful multimodality nanosystem will generate colocalized and sustained signal enhancements in all imaging modalities, thus allowing for spatial and temporal integration of the complementary information that is obtained from the different imaging systems.

Although there are numerous combinations of modalities that are possible, the main developments have been carried forward in the areas of MR/optical, SPECT/optical, PET/optical, CT/MR, and SPECT/MR imaging.

3.1 Nanosystems for Combined MR and Optical Imaging

To date, MR and fluorescence is the multimodality imaging combination that has been explored most extensively. A compelling justification for this combination is that it enables contrast-enhanced preoperative MR imaging for disease localization, identification, and surgical procedure planning, and then optical-based intraoperative target delineation. The principal strategy (Fig. 3) used by various research groups to engineer combined MR and fluorescent nanosystems involves the conjugation of an optical dye (i.e., Cy5.5) to an MR agent (i.e., iron oxide nanoparticle, iron oxide nanocrystal, gadolinium labeled dendrimer). In some cases, an additional step has been taken to conjugate a ligand or moiety (i.e., peptide, antibody) to the surface of the nanosystem in order to actively target specific cell populations in vivo. This strategy allows for colocalized signal changes to be generated in both MR and optical imaging modalities. In addition, the final size of these macromolecular nanosystems, which are usually greater than 10 nm, significantly decreases their clearance through the kidneys, and consequently increases their overall circulation lifetime in vivo. This enables imaging of the same animal with two distinct modalities following a single administration of contrast-generating media, and in some cases even allows follow-up scans for up to 24 h (Sosnovik, 2005; Trehin, 2006). The physical and imaging characteristics of the various MR/optical systems are summarized in Table 4.

The two MR contrast-enhancing agents employed are iron oxide (T_2-shortening agent, causes signal decrease) and gadolinium (T_1-shortening agent, causes signal increase). It is worth pointing out that there is a higher number (Table 4, > 10-fold) of MR agents per particle in the iron oxide systems in comparison to the existing gadolinium chelate systems. In fact, iron oxide preparations, such as the superparamagnetic iron oxide (SPIO), can result in high R2 relaxivity values of up to 200 mM^{-1} s^{-1} (Strable, 2001), which is between two- and threefold higher than the

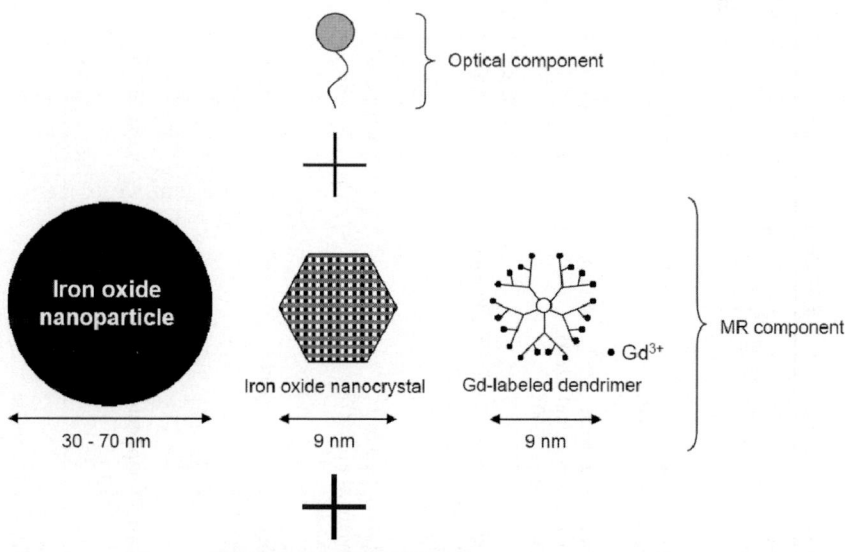

Fig. 3 Three possible components of an MR/optical dual-modality imaging nanosystem

Table 4 Physical and imaging properties of the various nanosystems employed in vivo for dual MR and optical imaging

Particle type (reference)	Particle size	Relaxivity (0.47T)	Optical probe/particle	MR agent/particle	Peptide/particle	Suggested applications
Iron oxide nanoparticle (Josephson et al., 2002)	62–68 nm	R1 = 27.8–29.9 mM^{-1} s^{-1}; R2 = 91.2–92.5 mM^{-1} s^{-1}	1.19–1.79 (Cy5.5)	2064[a] (iron)	11.9–15.5 (R4[b])	Imaging of macrophage activity
Iron oxide nanoparticle (Kircher et al., 2003)	32 nm	Not reported	~1 (Cy5.5)	2064[a] (iron)	–	Preoperative MR imaging and intraoperative optical imaging-based disease delineation
Iron oxide nanoparticle (Schellenberger et al., 2004; Sosnovik et al., 2005)	50 nm	R1 = 19 mM^{-1} s^{-1}; R2 = 48 mM^{-1} s^{-1}	1.8 (Cy5.5)	2064[a] (iron)	3.5 (Annexin V)	
Iron oxide nanoparticle (Moore et al., 2004; Medarova et al., 2006)	35.8 nm	R1 = 26.43 mM^{-1} s^{-1}; R2 = 53.44 mM^{-1} s^{-1}	1.8–5 (Cy5.5)	2064[a] (iron)	7.8–14 (EPPT[c])	Detection of primary and meta-static uMUC–1 + tumors
Generation 6 dendrimer (Koyama et al., 2007)	~9 nm	Not reported	2 (Cy5.5)	172–191 (gadolinium)	–	Imaging and image-guided biopsy of sentinel lymph node
Iron oxide nanocrystal (Huh et al., 2005)	9 nm (alone); 28 nm (with Herceptin)	R2 = 2.5x > CLIO at 1 μmol L^{-1} of iron (1.5 T)	0 (used fluorescein-labeled antihu-man IgG to stain the nanoconjugate)	Not reported (iron)	Not reported (Herceptin)	Cancer diagnosis
Low-density lipoprotein (Li et al., 2004)	26 nm	R1 = 26.7 M^{-1} s^{-1}; R2 = 25.5 M^{-1} s^{-1}	50 (dicarbocyanine)	50 (gadolinium)	–	In vivo cell tracking

[a]Assumed value
[b]R4 = RRRRGC peptide
[c]EPPT = YCAREPPTRTFAYWG peptide

R1 relaxivities of gadolinium chelates (5–$20\,mM^{-1}\,s^{-1}$) (Koenig and Brown, 1994). In addition, the intrinsic T_2 of tissue is usually shorter than its T_1 (Gimi, 2005). Therefore, even if a gadolinium-based contrast agent were to cause the same absolute T_1 shortening as an iron oxide agent does in T_2 shortening, the resulting percentage change in tissue T_2 would still be greater than the percentage change in tissue T_1. As a result, T_2-weighted MR can detect µmol to nmol concentrations of iron oxide particles (Gimi, 2005), while mmol to µmol levels of gadolinium chelates are required for T_1-weighted MR imaging. Yet, T_1 agents have the advantage of not significantly affecting the bulk tissue magnetic susceptibility. They are also more commonly employed in the clinical setting because the majority of the faster and higher resolution imaging protocols employ T_1-weighted sequences (i.e., dynamic contrast-enhanced MR) (Caravan, 1999). When selecting the appropriate MR contrast agent for a multimodal nanosystem one must understand the trade-off between sensitivity and temporal as well as spatial resolution.

Most of the research groups listed in Table 4 selected Cy5.5 as their optical imaging probe because it allows imaging in the NIR wavelength range. More recently, quantum dots with combined high relaxivity and high quantum yield luminescence have been explored for multimodality MR and optical imaging of cells in culture (Wang, 2007). Appropriate surface modifications may be needed for their widespread adoption in in vivo imaging applications.

3.2 Nanosystem for Combined CT and MR Imaging

Image-guided procedures are being increasingly integrated into routine clinical practice. CT and MR imaging are two core modalities employed in image guidance. An excellent example of this is in the context of radiation therapy, where CT and MR images are acquired and registered for the purpose of treatment planning. X-ray imaging-based guidance techniques (i.e., cone-beam CT) are then adopted at the time of radiation treatment delivery to ensure that patients are accurately positioned and that the target volume has not moved or changed from the planning stage. A common CT and MR contrast agent could facilitate registration of the two planning image sets by providing high contrast to common structures, since tissues and bones are visualized very differently in these two modalities. Furthermore, if the dual-modality agent is able to remain stable in vivo for a prolonged time, the patient could undergo planning in CT and MR, as well as multiple fractions of cone-beam CT guided radiotherapy with a single administration of contrast agent.

Zheng et al. reported on the successful development of a liposome-based system (Fig. 4) that co-encapsulates iohexol, an iodine-based CT agent, and gadoteridol, a gadolinium-based MR agent. The liposomal system allowed for imaging of colocalized contrast enhancement in both modalities for up to 7 days postadministration in rabbits (Fig. 5). (Zheng, 2006, 2007). The physicochemical characteristics and imaging properties of this dual-modality system are summarized in Table 5. A limitation of this liposome encapsulation strategy is the lower iodine-loading levels, which

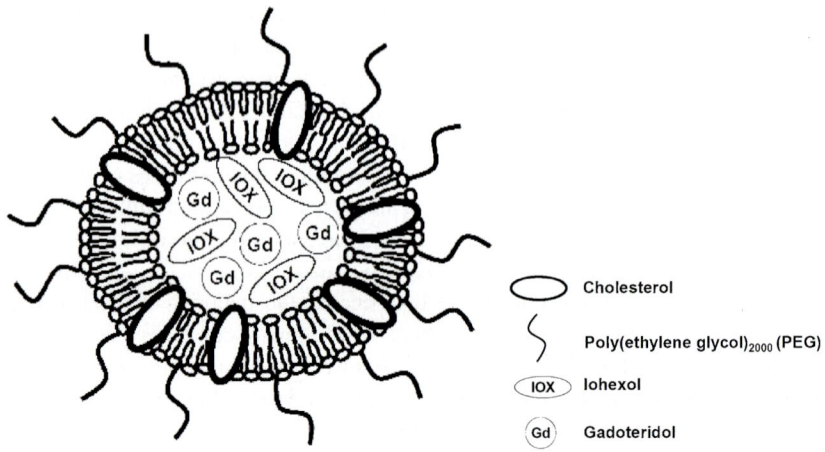

Cholesterol

Poly(ethylene glycol)$_{2000}$ (PEG)

IOX Iohexol

Gd Gadoteridol

Fig. 4 Dual-modality liposomes encapsulating iohexol and gadoteridol (adapted from Zheng et al. (Zheng, 2007))

are about tenfold below the concentration of iodine found in the commercially available formulation (i.e., Omnipaque®). However, the macromolecular nature of the liposomes ensures vascular containment following intravenous administration, and consequently decreases the volume of distribution of the agent by a factor of three (equivalent to the ratio of extravascular space to intravascular space), in comparison to the volume of distribution of the agent when administered as Omnipaque®. The longer vascular circulation lifetime of the liposomal agent enables longer imaging times, which results in improved contrast and spatial resolution even at lower contrast agent concentrations.

Furthermore, for tumor-imaging applications, the single lamellar liposomes, with their optimal size (i.e., $<200\,\text{nm}$), are able to take advantage of the enhanced permeability and retention (EPR) effect (Hobbs, 1998; Maeda, 2000; Torchilin, 2000) and preferentially accumulate in solid tumors following systemic intravenous injection. As a result, high local levels of iodine and gadolinium can be detected at the tumor site through the liposome-based delivery strategy (Zheng, 2007).

Although the liposomes can load similar gadoteridol levels as the commercially available formulations (i.e., Prohance®), the encapsulated gadolinium atoms experience decreased relaxivity compared to their un-encapsulated counterparts (Zheng, 2006). Two interaction processes occur between the unpaired gadolinium electrons and the water protons: inner-sphere relaxation, which refers to the interaction with bound water molecules, and outer-sphere relaxation, which results from the diffusion of nearby water protons (Froehlich, 2006). The presence of a lipid membrane, depending on its permeability, can affect the diffusion of water across this boundary, and as a consequence limit outer-sphere relaxation. Fossheim et al. (Fossheim, 1999) expressed the relaxivity of a liposome encapsulating a paramagnetic species as $r_1 = \dfrac{1}{C_{\text{eff}}\left(\tau + T_{i1}\right)}$, where C_{eff} is the effective gadolinium concentration encapsu-

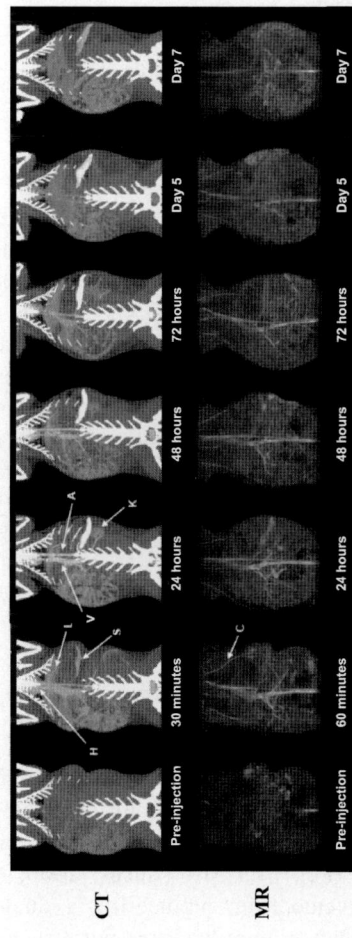

Fig. 5 Colocalized signal enhancement in CT and MR and the extended imaging window achieved with liposomal coencapsulation of iohexol and gadoteridol. Abbreviations: heart (H), aorta (A), vena cava (V), carotid artery (C), kidney (K), liver (L), and spleen (S). Reproduced from Zheng et al. (Zheng, 2007)

Table 5 Physical and imaging properties of the combined CT and MR imaging nanoplatform

Particle type (reference)	Particle size	Relaxivity (1.5 T)	CT agent/ particle	MR agent/ particle	Suggested applications
Unilamellar liposomes (Zheng et al., 2006, 2007)	70–85 nm	$R1 = 1.23 \pm 0.02\,\text{mM}^{-1}\,\text{s}^{-1}$	700,000[a] (iodine)	130,000[a] (gadolinium)	Image-guided therapy and in vivo noninvasive assessment of liposome distribution
		$R2 = 1.46 \pm 0.02\,\text{mM}^{-1}\,\text{s}^{-1}$			

[a] Values calculated using information provided by Zheng et al. (2006, 2007)

lated in the liposomes, τ is the water exchange time between the interior and exterior of the liposome, and T_{1i} is the intrinsic T_1 relaxation time of the encapsulated paramagnetic species. The entrapped gadolinium atoms will have similar relaxation properties as in solution only when $\tau = T_{1i}$, that is when the liposome membrane is permeable enough to allow for very rapid water exchange between the interior and exterior aqueous compartments. However, an increase in the permeability of the lipid membrane will reduce the retention of entrapped molecules and decrease the overall stability of the system. A viable strategy to improve the relaxation properties of the gadolinium atoms in the nanosystem is to attach gadolinium chelates such as diethylene triamine pentaacetic acid (DTPA) to the outer surface of the lipid bilayer. In this way, greater amounts of iodine could be loaded into the liposome core and a less permeable lipid membrane could be employed to increase the retention of the cargo.

3.3 Nanocarrier for Combined SPECT and MR Imaging

The SPECT and MR combination is an attempt to bring together high sensitivity and exquisite anatomical imaging. Although a lot of effort has gone into the hardware integration of radionuclide and MR imaging systems for both clinical and preclinical applications (i.e., PET/MR (Lucas, 2006; Wagenaar, 2006)), there has been little advancement in engineering a dual-modality PET or SPECT/MR imaging agent. One attempt has been put forward by Zielhuis et al. (Zielhuis, 2006) that relies on liposomes (Fig. 6, Table 6) to carry gadolinium acetylacetonates (GdAcAc) for MR imaging and radionuclides (holmium[166] or technetium[99m]) for SPECT imaging. The system demonstrated adequate relaxivity properties in vitro and good radiochemical stability in human serum. However, in vivo evaluation of this system has not been reported to this point and therefore the retention of GdAcAc in the bilayer following administration remains to be confirmed.

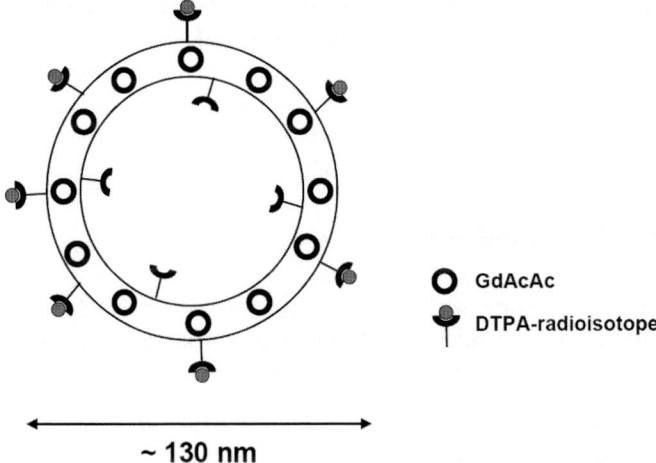

Fig. 6 Drawing (not to scale) of the liposome formulation loaded with MR and SPECT agents. Adapted from Zielhuis et al. (Zielhuis, 2006)

Table 6 Physical and imaging properties of the combined SPECT and MR imaging nanoplatform

Particle type (reference)	Particle size	Relaxivity	Radionuclide/ particle	MR agent/ particle	Suggested applications
Unilamellar liposomes (Zielhuis et al., 2006)	130 nm	Varies with paramagnetic species and formulation	70[a] (holmium[166])	140–280[a] (gadolinium)	Merger of high sensitivity SPECT images with anatomical MR images
			20[a] (technicium[99m])		

[a] Values calculated using information provided by Zielhuis et al. (2006)

3.4 Agents for Combined Radionuclide (PET or SPECT) and Optical Imaging

The main rationale behind early developments of a dual SPECT or PET and fluorescence imaging agent is the desire to use a clinically established radionuclide imaging method to validate the emerging NIR fluorescence imaging technique, since the sensitivities of fluorescence and radionuclide imaging methods are comparable (Houston, 2005; Pandey, 2005; Zhang, 2005). In addition, a radioisotope tagged photodynamic therapy (PDT) agent has the potential to allow image-based monitoring of the in vivo distribution of the drug (Pandey, 2005). Another difference between the dual radionuclide and optical imaging agents developed to date and other dual-modality imaging systems is that they do not rely on colloidal carriers and have much lower molecular weights. This trend may have resulted from the fact that a very low agent payload is needed due to the high sensitivity of both radionuclide and optical imaging systems. The lower molecular weight of these agents allows for rapid vascular elimination which is needed to maximize target (i.e., tumor) to background (i.e., blood) signal ratios. Table 7 summarizes the physical characteristics of the various dual radionuclide and fluorescence agents employed for in vivo imaging.

4   Trends and Opportunities

The importance of imaging has been recognized not only in the clinical setting for diagnosis and treatment guidance, but also in a vast array of research areas. Its noninvasive nature allows for visualization and characterization of anatomy and physiology without affecting the biological systems that are under investigation. In addition, contrast-generating or -enhancing molecules can be tagged to therapeutic agents with little or no perturbation of their activity or biodistribution. In this way, longitudinal noninvasive monitoring of the in vivo behavior of therapeutic agents may be achieved in the same patient, providing information on the pathway of drug delivery vehicles as well as composition–performance relationships. Furthermore, with the increased adoption of imaging systems for guiding interventions such as surgery and radiotherapy, the ability to image drug administration is advantageous for the optimization of combination therapy regimens.

Advances in the development of multimodal imaging agents may facilitate the integration of complementary information that is obtained from the same patient using different combinations of imaging modalities. This chapter described currently available nanosystems that support imaging in at least two distinct modalities. A few of these systems have been designed to actively target specific cell populations (i.e., Annexin V to target apoptotic cells (Schellenberger, 2004; Sosnovik, 2005) and Herceptin to target HER-2 overexpressing tumor cells (Huh, 2005)). As this field matures, the knowledge and experience gained in the area of molecular targeting (i. e., molecularly targeted therapeutics) may be applied to the multimodal imaging nanosystems to increase their specificity and sensitivity for molecular imaging applications. In addition to imaging the location of specific cell populations in vivo, targeted nanosystems may also be employed as a tracking device for monitoring the time-dependent distribution of cells of interest within the body. This is especially useful for designing cell-based therapies and stem cell-based regenerative medicine.

Another interesting direction for future generations of multimodality imaging agents is the design of nanosystems that are either fully or in part sensitive to their environment (i.e., temperature, pH, oxygen levels) or responsive to a specific trigger (i.e., the presence or absence of a signaling molecule). Many of these advances are being investigated for therapeutic or single-modality imaging nanosystems (Kale and Torchilin, 2007; Kim, 2007; Ponce. 2006), and they may be rapidly translated to support multimodality imaging nanosystems.

The multimodal nanosystems described in this chapter have shed a new light on the older and oversimplified view that contrast agents are simply small molecules that transiently perturb the signal generated by a given imaging system. They have shown

Table 7 Physicochemical and imaging properties of the combined radionuclide and optical imaging systems

Tracer-Probe system (reference)	Particle size	Radioisotope/particle	Optical probe/particle	Peptide/particle	Suggested applications
^{111}In-DOTA-Gly-Ser-Gly-Lys-(ε-cypate)-Ahx-NH₂ (Zhang et al., 2005)	<6,000 Da	1 (indium[111])	1 (cypate)	N/A	Improved registration of SPECT and optical images for disease diagnosis
^{111}In-DTPA-Lys(IRDye800)-c(KRGDf) (Houston et al., 2005; Li et al., 2006)	Not reported	~40 (indium[111])	1 (IRDye800)	1 (Lys-c(KRGDf))	Validation and clinical translation of NIR imaging techniques with nuclear imaging
[111]In –HPPT[a] (Pandey et al., 2005)	Not reported	1 (iodine[124])	1 (pyropheo-phorbide analogue)	N/A	PET-based validation of NIR imaging and monitoring of PDT agent biodistribution

[a] HPPH = 3-(1'-*m*-hexyloxyethyl)-3-devinylpyropheophorbide-*a*

that novel contrast agents that provide relevant signal generation and enhancement have now become a more essential component in the imaging world, along with hardware devices employed for detection and software systems used for analysis. There remain huge opportunities of great variety and richness in the contrast agent development space: a fantastic voyage to be continued!

References

Alfano, R.R., Demos, S.G., and Gayen, S.K., 1997. Advances in optical imaging of biomedical media. Ann N Y Acad Sci, **820**: 248–270; discussion 271.

Anderson, R.R. and Parrish, J.A., 1981. The optics of human skin. J Invest Dermatol, 13–19.

Bocher, M., et al., 2000. Gamma camera-mounted anatomical X-ray tomography: technology, system characteristics and first images. Eur J Nucl Med, 619–627.

Bonvento, M.J., et al., 2006. CT angiography with gadolinium-based contrast media. Acad Radiol, **13**: 979–985.

Calliada, F., et al., 1998. Ultrasound contrast agents: basic principles. Eur J Radiol, **27 Suppl 2**: S157–S160.

Caravan, P., et al., 1999. Gadolinium(III) chelates as MRI contrast agents: structure, dynamics, and applications. Chem Rev, **99**: 2293–2352.

Cassidy, P.J. and Radda, G.K., 2005. Molecular imaging perspectives. J R Soc Interface, **2**: 133–144.

Evans, E., 1955. The Atomic Nucleus. New York: McGraw-Hill.

Exadaktylos, A.K., et al., 2005. The role of contrast-enhanced spiral CT imaging versus chest X-rays in surgical therapeutic concepts and thoracic aortic injury: a 29-year Swiss retrospective analysis of aortic surgery. Cardiovasc J S Afr, 162–165

Fahrig, R., et al., 2003. First use of a truly-hybrid X-ray/MR imaging system for guidance of brain biopsy. Acta Neurochir (Wien), 995–997; discussion 997.

Fahrig, R., et al., 2001. Truly hybrid interventional MR/X-ray system: investigation of in vivo applications. Acad Radiol, 1200–1207.

Fahrig, R., et al., 2001. A truly hybrid interventional MR/X-ray system: feasibility demonstration. J Magn Reson Imaging, 294–300.

Firat, A.K., et al., 2006. The effect of intravenous gadolinium-DTPA on diffusion-weighted imaging. Neuroradiology, **48**: 465–470.

Fischman, A.J., Alpert, N.M., and Rubin, R.H., 2002. Pharmacokinetic imaging: a noninvasive method for determining drug distribution and action. Clin Pharmacokinet, **41**: 581–602.

Fossheim, S.L., et al., 1999. Paramagnetic liposomes as MRI contrast agents: influence of liposomal physicochemical properties on the in vitro relaxivity. Magn Reson Imaging, **17**: 83–89.

Frangioni, J.V., 2003. In vivo near-infrared fluorescence imaging. Curr Opin Chem Biol, **7**: 626–634.

Froehlich, J., 2006. MR contrast agents, in How Does MRI Work? An Introduction to the Physics and Function of Magnetic Resonance Imaging, D. Weishaupt, V. Kochli, and B. Marincek, Eds. Berlin Heidelberg New York: Springer, pp. 103–118.

Gao, X., et al., 2005. In vivo molecular and cellular imaging with quantum dots. Curr Opin Biotechnol, **16**: 63–72.

Gao, X., Chung, L.W., and Nie, S., 2007. Quantum dots for in vivo molecular and cellular imaging. Methods Mol Biol, **374**: 135–146.

Gillies, R.J., 2002. In vivo molecular imaging. J Cell Biochem Suppl, **39**: 231–238.

Gimi, B., et al., 2005. Molecular imaging of cancer: applications of magnetic resonance methods. Proc IEEE, **93**: 784–799.

Hainfeld, J.F., et al., 2006. Gold nanoparticles: a new X-ray contrast agent. Br J Radiol,**79**: 248–253.

Hashizume, M., 2007. MRI-guided laparoscopic and robotic surgery for malignancies. Int J Clin Oncol, **12**: 94–98.

Hobbs, S.K., et al., 1998. Regulation of transport pathways in tumor vessels: role of tumor type and microenvironment. Proc Natl Acad Sci U S A, 4607–4612.

Houston, J.P., et al., 2005. Quality analysis of in vivo near-infrared fluorescence and conventional gamma images acquired using a dual-labeled tumor-targeting probe. J Biomed Opt, **10**: 054010.

Hsieh, J., 2003. Computed Tomography: Principles, Design, Artifacts, and Recent Advances. Bellingham, Washington: SPIE – The International Society for Optical Engineering.

Hsu, C.P., et al., 2005. Liver tumor gross margin identification and ablation monitoring during liver radiofrequency treatment. J Vasc Interv Radiol, **16**: 1473–1478.

Huh, Y.M., et al., 2005. In vivo magnetic resonance detection of cancer by using multifunctional magnetic nanocrystals. J Am Chem Soc, **127**: 12387–12391.

Jaffray, D.A., et al., 2002. Flat-panel cone-beam computed tomography for image-guided radiation therapy. Int J Radiat Oncol Biol Phys, 1337–1349.

Josephson, L., et al., 2002. Near-infrared fluorescent nanoparticles as combined MR/optical imaging probes. Bioconjug Chem, **13**: 554–560.

Kale, A.A. and Torchilin, V.P., 2007. Design, synthesis, and characterization of pH-sensitive PEG-PE conjugates for stimuli-sensitive pharmaceutical nanocarriers: the effect of substitutes at the hydrazone linkage on the ph stability of PEG-PE conjugates. Bioconjug Chem, **18**: 363–370.

Kaul, S., 2004. Microbubbles and ultrasound: a bird's eye view. Trans Am Clin Climatol Assoc, **115**: 137–148.

Kim, K.Y., 2007. Nanotechnology platforms and physiological challenges for cancer therapeutics. Nanomedicine, **3**: 103–110.

Kircher, M.F., et al., 2003. A multimodal nanoparticle for preoperative magnetic resonance imaging and intraoperative optical brain tumor delineation. Cancer Res, **63**: 8122–8125.

Koenig, S.H. and Brown, R.D., 1994. Relaxometry and MRI, in NMR in Physiology and Biomedicine, R.J. Gillies, Ed. San Diego, CA: Academic, p. 70.

Koyama, Y., et al., 2007. A dendrimer-based nanosized contrast agent dual-labeled for magnetic resonance and optical fluorescence imaging to localize the sentinel lymph node in mice. J Magn Reson Imaging, **25**: 866–871.

Krause, W., 1999. Delivery of diagnostic agents in computed tomography. Adv Drug Deliv Rev, 159–173.

Levin, C.S., 2005. Primer on molecular imaging technology. Eur J Nucl Med Mol Imaging, **32 Suppl 2**: S325–S345.

Li, C., et al., 2006. Dual optical and nuclear imaging in human melanoma xenografts using a single targeted imaging probe. Nucl Med Biol, **33**: 349–358.

Li, H., et al., 2004. MR and fluorescent imaging of low-density lipoprotein receptors. Acad Radiol, **11**: 1251–1259.

Lindner, D., et al., 2006. Application of intraoperative 3D ultrasound during navigated tumor resection. Minim Invasive Neurosurg, **49**: 197–202.

Lucas, A.J., et al., 2006. Development of a combined microPET-MR system. Technol Cancer Res Treat, **5**: 337–341.

Maeda, H., et al., 2000. Tumor vascular permeability and the EPR effect in macromolecular therapeutics: a review. J Control Release, 271–284.

McCulloch, M., et al., 2000. Ultrasound contrast physics: a series on contrast echocardiography, article 3. J Am Soc Echocardiogr, **13**: 959–967.

McVeigh, E.R., et al., 2006. Real-time interactive MRI-guided cardiac surgery: aortic valve replacement using a direct apical approach. Magn Reson Med, **56**: 958–964.

Medarova, Z., et al., 2006. In vivo imaging of tumor response to therapy using a dual-modality imaging strategy. Int J Cancer, **118**: 2796–2802.

Moore, A., et al., 2004. In vivo targeting of underglycosylated MUC-1 tumor antigen using a multimodal imaging probe. Cancer Res, **64**: 1821–1827.

Mutzel, W. and U. Speck, 1980. Pharmacokinetics and biotransformation of iohexol in the rat and the dog. Acta Radiol Suppl, 87–92.

Namasivayam, S., et al., 2006. Adverse reactions to intravenous iodinated contrast media: a primer for radiologists. Emerg Radiol, **12**: 210–215.

Pandey, S.K., et al., 2005. Multimodality agents for tumor imaging (PET, fluorescence) and photodynamic therapy. A possible "see and treat" approach. J Med Chem, **48**: 6286–6295.

Phelps, M., 2003. PET: Molecular Imaging and Its Biological Applications. Berlin Heidelberg New York: Springer.

Ponce, A.M., et al., 2006. Hyperthermia mediated liposomal drug delivery. Int J Hyperthermia, **22**: 205–213.

Rabin, O., et al., 2006. An X-ray computed tomography imaging agent based on long-circulating bismuth sulphide nanoparticles. Nat Mater, **5**: 118–122.

Rygh, O.M., et al., 2006. Intraoperative navigated 3-dimensional ultrasound angiography in tumor surgery. Surg Neurol, **66**: 581–592; discussion 592.

Saeed, M., et al., 2005. MRI in guiding and assessing intramyocardial therapy, Eur Radiol, 851–863.

Sandler, M., et al., 1989. Correlative Imaging – Nuclear Medicine, Magnetic Resonance, Computed Tomography and Ultrasound. Baltimore, MD: Williams & Wilkins.

Schellenberger, E.A., et al., 2004. Magneto/optical annexin V, a multimodal protein. Bioconjug Chem, **15**: 1062–1067.

Schnall, M. and Rosen, M. 2006. Primer on imaging technologies for cancer. J Clin Oncol, **24**: 3225–3233.

Sima, P.D. and J.R. Kanofsky, 2000.Cyanine dyes as protectors of K562 cells from photosensitized cell damage. Photochem Photobiol, **71**: 413–421.

Smith, A.M., Gao, X., and Nie, S., 2004. Quantum dot nanocrystals for in vivo molecular and cellular imaging. Photochem Photobiol, **80**: 377–385.

Sosnovik, D.E., et al., 2005. Magnetic resonance imaging of cardiomyocyte apoptosis with a novel magneto-optical nanoparticle. Magn Reson Med, **54**: 718–724.

Strable, E., et al., 2001. Synthesis and characterization of soluble iron oxide-dendrimer composites. Chem Mater, **13**: 2201–2209.

Torchilin, V.P., 2000. Drug targeting. Eur J Pharm Sci, S81–S91.

Townsend, D.W. and Beyer, T., 2002. A combined PET/CT scanner: the path to true image fusion. Br J Radiol, S24–S30.

Townsend, D.W., et al., 2004. PET/CT today and tomorrow. J Nucl Med, S4–S14.

Trehin, R., et al., 2006. Fluorescent nanoparticle uptake for brain tumor visualization. Neoplasia, **8**: 302–311.

Tsui, M., 1996. Physics of SPECT. Radiographics, **16**: 173–183.

Tweedle, M.F., et al., 1997. A noninvasive method for monitoring renal status at bedside. Invest Radiol, 802–805.

Uematsu, M., et al., 1999. Daily positioning accuracy of frameless stereotactic radiation therapy with a fusion of computed tomography and linear accelerator (focal) unit: evaluation of z-axis with a z-marker. Radiother Oncol, 337–339.

Wagenaar, D.J., et al., 2006. Rationale for the combination of nuclear medicine with magnetic resonance for pre-clinical imaging. Technol Cancer Res Treat, **5**: 343–350.

Wang, S., et al., 2007. Core/shell quantum dots with high relaxivity and photoluminescence for multimodality imaging. J Am Chem Soc, **129**: 3848–3856.

Weishaupt, D., Kochli, V., and Marincek, B. 2003. How Does MRI Work? An Introduction to the Physics and Function of Magnetic Resonance Imaging. Berlin Heidelberg New York: Springer.

WT, M., 1999. Fluorescent and Luminescent Probes for Biological Activities. London: Academic Press.

Zhang, Z., et al., 2005. Monomolecular multimodal fluorescence-radioisotope imaging agents. Bioconjug Chem, **16**: 1232–1239.

Zheng, J., et al., 2006. Multimodal contrast agent for combined computed tomography and magnetic resonance imaging applications. Invest Radiol, **41**: 339–348.

Zheng, J., et al., 2007. 2007. Longitudinal vascular imaging using a novel nano-encapsulated CT and MR contrast agent. Proc SPIE, **6511**(Medical Imaging 2007).

Zheng, J., et al., 2007. In vivo performance of a liposomal vascular contrast agent for CT and MR-based image guidance applications. Pharm Res, **24**(6): 1193–1201.

Zielhuis, S.W., et al., 2006. Lanthanide-loaded liposomes for multimodality imaging and therapy. Cancer Biother Radiopharm, **21**: 520–527.

Long-Circulating Liposomes with Attached Diagnostic Moieties: Application for Gamma and MR Imaging

William Phillips, Beth Goins, and Ande Bao

1 Long-Circulating Liposomes

Liposomes, spontaneously forming lipid nanoparticles, have long been recognized to have great potential as carriers for diagnostic and therapeutic agents since their initial discovery more than 40 years ago (Gregoriadis and Ryman 1971; Bangham 1993). In the last 30 years, remarkable progress in liposome technology has been accomplished in the following areas (1) methods of large-scale manufacture, (2) methods of stably encapsulating sufficient quantities of therapeutic agents within the liposomes, (3) methods of producing homogeneously sized liposomes, (4) effective methods of noninvasively tracking the distribution of liposomes in the body, and (5) development of methodology to prolong the circulation time of liposomes in the blood (Allen et al. 1991; Phillips et al. 1999; Torchilin and Papisov 1994).

Early liposome formulations were generally cleared rapidly from the blood by macrophages in the liver and spleen, typically in less than a few hours. Eventually, small stable liposome formulations with longer circulation times were developed for clinical use. The development of long-circulating liposomes has significant advantages in terms of drug delivery. When administered intravenously, long-circulating liposomes remain in the blood circulation longer, allowing a greater percentage of the liposomes to come in contact with and become trapped in the intended targeted tissue of the body. The initial long-circulating liposome formulations were effective for accumulation in certain diseases by passive targeting that takes advantage of the enhanced permeability inherent in certain disease processes. Initial clinically successful applications were for liposomes with diameters of around 100 nm and that contained the antifungal agent, amphotericin B, and the anticancer agent, doxorubicin. Prolonged circulation of formulations of liposomes composed of phospholipids with two 18-carbon fatty acid chains, distearoyl phosphatidylcholine (DSPC), and cholesterol can be achieved without any type of surface modification. These nonsurface-modified liposomes are required to be of a small size (generally less than 100 nm) to be effective.

Another approach to prolonging the circulation of larger liposomes (up to 250 nm in size) is to coat the surface of the liposomes with hydrophilic molecules such as polyethylene glycol (PEG). These liposomes generally contain a PEG-lipid

V. Torchilin (ed.), *Multifunctional Pharmaceutical Nanocarriers*,
© Springer Science+Business Media, LLC 2008

at a concentration of 5–10 mol% and a PEG chain length of 2,000–5,000 molecular weight. When PEG is placed on the surface of liposomes, it significantly prolongs the circulation of the liposomes. This prolonged circulation has been thought to be due to the steric hindrance that the PEG molecules provide to the liposomes (Papahadjopoulos et al. 1991; Torchilin and Trubetskoy 1995), so that free immunological access to the liposomes is blocked and the attachment of complement to the liposomes is reduced, resulting in a slower removal of these liposomes by macrophages located in the liver and spleen. However, the effectiveness of PEG on prolonged liposomal circulation was less remarkable at low lipid doses, which is the typical case in the use of liposomes for diagnostic imaging (Gabizon et al. 1989; Dams et al. 2000; Laverman et al. 2001; Ishida et al. 2005, 2006a). Although PEG modification appears quite effective at prolonging the circulation time of liposomes when given in sufficient lipid doses (4–400 μmol kg^{-1}), this prolonged circulation effect is not observed at lower lipid doses. When administered at very low lipid dose (0.3–1 μmol kg^{-1}), the rapid clearance from the circulation was associated with a rapid uptake in the liver and spleen macrophages. It has also been observed that, when PEG liposomes are reinjected within a 4-week period, there was a decrease in the circulation time, that was not observed after 6 weeks (Dams et al. 2000; Ishida et al. 2005, 2006b). This behavior of PEG-coated liposomes is particularly important for diagnostic imaging applications, where it is anticipated that relatively low doses of liposomes will be administered. As a solution to this problem with PEG, a new liposome surface modification has been described by Romberg et al. (2007a,b). With this method, the liposome surface was coated with poly(hydroxyethyl-L-asparagine). With this new surface modification, circulation times in the blood were similar to those of PEG-coated liposomes at doses of 25 μmol kg^{-1} (Romberg et al. 2007b). The poly(hydroxyethyl-L-asparagine) liposomes had significantly prolonged circulation times even when low doses and repeated injections were adopted (Romberg et al. 2007b). Although these poly(hydroxyethyl-L-asparagine) liposomes had improved characteristics, they still appear to activate the complement (Romberg et al. 2007b). It has also been recently shown that PEG liposomes stimulate the production of anti-PEG antibodies (Sroda et al. 2005; Ishida et al. 2006a) and the presence of the anti-PEG IgM antibody enhanced the clearance of the PEG liposomes on second injection (Ishida et al. 2006b).

It is likely that further improvements in methods to prolong the circulation time of liposomes will be forthcoming as our understanding of the mechanisms related to liposome circulation increases. For instance, complement activation by liposomes is becoming better understood (Moghimi and Szebeni 2003; Szebeni et al. 2003, 2006; Moghimi et al. 2006). In a recent article, Moghimi, Szebeni, and colleagues have shown that methylation of the oxygen moiety in the phospholipid-methoxy(polyethylene glycol) prevents complement activation and anaphylatoxin production when PEGylated liposomes with this modified PEG-phospholipid are administered (Moghimi et al. 2006). It is likely that decreasing the complement activation will result in a longer circulation time. It is also apparent that this work with liposomes will have wide application to research with other nanoparticles. It is also important to note that much of this research was conducted using imaging in which the liposomes were labeled with a minute quantity of radioactive tracer and noninvasive imaging as well as the counting of radioactive blood samples greatly helped in the determination of the effect of low doses and repeated injections of the circulation time of intravenously injected liposomes (Dams et al. 2000; Laverman et al. 2001).

2 Specifically Targeted Liposomes

When liposomes with attached diagnostic moieties are used as imaging agents, the formulation of the liposomes is generally customized to the particular diagnostic application. Liposomes with specific surface characteristics have been shown to have an improved ability to target specific tissues or disease processes. One approach is to simply change the charge of the lipid used in the liposome formulation. For example, negatively-charged liposomes containing 5% dimyristoyl phosphatidylglycerol appear to accumulate more effectively in infectious processes than comparable liposome formulations containing a neutral charge, whereas small liposomes with a neutral charge were found to be more useful for tumor drug delivery due to their longer circulation persistence (Goins et al. 1994).

Another approach for achieving specific tissue targeting is to add a targeting ligand to the surface of long-circulating liposomes, allowing these liposomes to specifically bind to a particular location or disease process in the body through a molecular interaction. This ligand can be coembedded in the liposomes or attached to the distal end of the PEG molecules (Bendas et al. 1999). Generally, this targeting ligand has been a monoclonal antibody which binds to a disease-specific biomolecule, although other ligands can also be used. The approach of attaching the ligand to the distal end of the PEG-lipid appears to be the most promising with the majority of published articles using this approach for the specific targeting of liposomes. The PEG molecules on the surface of the liposomes attenuate the rapid uptake by macrophages, while the specific targeting ligand can significantly increase the uptake of the liposomes in the targeted tissue in either absolute quantity or speed of target accumulation. These specifically targeted liposomes can therefore be considered to be multifunctional as they not only carry a drug that is encapsulated in the liposomes, but also have PEG molecules to increase the circulation time of the liposomes as well as a specific ligand to target the liposomes to a target of interest in the body.

3 Potential Uses and Advantages of Imaging Technology as Applied to Liposomes and Other Nanoparticles

The attachment of contrast agents to nanoparticles for the purpose of noninvasive imaging can provide useful information for either clinical diagnostic imaging or research purposes. For diagnostic imaging purposes, the distribution of a targeted liposome containing a diagnostic agent can indicate that the person has a particular disease process. The ability to noninvasively determine the location of the targeted tissue in the body can also provide information regarding the distribution of a liposomal drug in the body for drug dose calculations. This dosimetry can be for either chemotherapeutic agents or radionuclide therapy (Bao et al. 2005; Viglianti et al. 2006). For research and drug development purposes, imaging can be used to determine the in vivo pharmacokinetics and organ distribution of drug-containing liposomes, thus making it possible to carry out multiple assessments in the same animal across different points in time (Gabizon et al. 2003). Scintigraphic gamma photon imaging can provide an image of the entire body in a quantitative manner which permits the determination of total body biodistribution over time, so that the portion of an injected agent that actually arrives at its targeted destination can be determined. In animal models, high-resolution imaging with special pinhole collimators can also be used under specific situations to determine the distribution of a radiolabeled liposome formulation within a pathological process such as a tumor (Bao et al. 2006). The excellent resolution and contrast properties of contrast-enhanced magnetic resonance (MR) imaging can also clearly show the accumulation of liposomes containing an MR contrast agent as well as the release of drugs in a specific local region with a high resolution which can be less than 0.5 mm (Viglianti et al. 2006).

4   Gamma Photon Imaging

Gamma photon imaging, also known as *scintigraphic imaging*, is a noninvasive imaging technique commonly used in the practice of nuclear medicine. Radiolabeled compounds (called radiopharmaceuticals or radiotracers) are administered intravenously to patients for diagnostic or in some cases for therapeutic purposes. The in vivo distribution of these radionuclides can provide important physiological information about tissue function.

There are two types of gamma photon imaging, which detect either single photons or dual photons produced following positron emission. The latter imaging technique is known as *positron emission tomography* (PET). Single photon imaging can be performed with a wide variety of agents which are primarily radionuclides with an excess of neutrons and decay with β^- mode or radionuclides in metastable nuclear state, for example technetium-99 m (^{99m}Tc). These nuclides usually decay and emit a single gamma photon.

For diagnostic applications using single photon imaging, the tracer is labeled with a radionuclide emitting photons with energies ranging from 100 to 300 keV (Table 1). These photon energies are high enough to allow detection from outside the body by a gamma camera. The location of the emitted photons is determined by detection of gamma photons by the camera. In a gamma camera, a high-energy photon emitted from an administered radionuclide interacts with a crystal detector resulting in the emission of lower energy secondary photons, known as *scintillations*.

Table 1 Physical characteristics of some commonly used single photon imaging agents used for labeling liposomes

Radionuclide	Half-life	Photons (keV) (abundance (%))
Gallium-67	3.3 days	93 (38)
		185 (21)
		300 (17)
Technetium-99 m	6.01 h	141 (89)
Indium-111	2.8 days	171 (91)
		245 (94)
Iodine-123	13.2 h	159 (83)

Table 2 Physical characteristics of some commonly used PET agents for the potential application of labeling liposomes

Radionuclide	Half-life	Photons (keV)
Fluorine-18	109 min	511
Copper-64	12.7 h	511
Copper-62	3.5 h	511
Iodine-124	4.2 days	511

These scintillations are detected and amplified by either the traditional photomultiplier tubes or more recently avalanche photodiode detectors (Catana et al. 2006). With single photon-emitting agents, the most ideal energies suitable for single photon emission computed tomography (SPECT) imaging range from approximately 120 to 200 keV. These energies are high enough to penetrate through a human body while low enough to be easily collimated by lead and detected.

PET gamma photon-emitting agents decay by the emission of a positron (positively charged electron) that travels a short distance of millimeters where it collides with a shell electron. This collision results in the emission of two annihilation photons that travel approximately 180° apart. This 180° emission can be used to localize the position of the positron decay along a line, so that multiple emission lines can be iteratively localized to a particular position in the body. The most commonly used radionuclides for PET imaging are short lived such as oxygen-15 (2 min), carbon-11 (20 min), and fluorine-18 (2 h) (Table 2). Radionuclides with a half-life longer than 6 h, such as copper-64 (^{64}Cu) and iodine-124 (^{124}I), may be particularly promising for tracking the distribution of long-circulating nanoparticles such as liposomes.

5 Methods of Labeling Liposomes with Gamma Photon-Emitting Agents

The most commonly used gamma photon-emitting radionuclides that have been used to label liposomes for scintigraphic imaging studies have been technetium-99 m (^{99m}Tc), indium-111 (^{111}In), and gallium-67 (^{67}Ga) and several detailed reviews have been recently written describing the methodology of these labeling methods (Laverman et al. 2003; Goins and Phillips 2007; Laverman et al. 2007) (Table 1). Ideally, liposomes should be labeled after their preparation and just prior to the experiments. This is because the radionuclides used in clinical and research studies are fairly short lived (on the order of hours to days) (Table 1). This type of post-manufacture labeling is known as "after loading" or "remote-labeling" in which the preformed liposomes are labeled just prior to the start of the experiment.

It is important that the radiolabeled liposomal formulations have a high stability from a radiolabeling point of view. Release of radiolabel from the liposomes during or after injection will lead to inaccurate biodistribution results. The commonly used radionuclides for scintigraphic imaging are all widely available (Table 1) from local clinical radiopharmacies. However, generally ^{99m}Tc is preferred over the other diagnostic imaging radionuclides due to its optimal imaging characteristics including an ideal photon energy. It is generally available in most nuclear medicine departments and relatively inexpensive because it can be eluted daily from a commercially available molybdenum-99 (^{99}Mo)/^{99m}Tc generator. Because ^{67}Ga, ^{111}In, and ^{123}I are cyclotron products, they are more expensive and not always available in every nuclear medicine department.

A relatively simple method of labeling liposomes is to incubate the premanufactured liposomes with a lipophilic radiolabel, resulting in association of the label within the lipid bilayer. This approach appears to yield very unstable radiolabeled liposome preparations (Love et al. 1989) and is therefore not preferable. In contrast, the following two "after loading" approaches have proven to yield radiolabeled liposomes with high efficiency and good radiochemical stability. With these approaches, the radionuclide is either (1) trapped in the aqueous phase after the manufacturing of the liposomes or (2) coordinated to a lipid-conjugated chelator incorporated in the lipid bilayer of preformed liposomes.

One "after loading" method that has been developed for labeling liposomes uses radionuclide chelators attached to the liposomal surface (Hnatowich et al. 1981; Laverman et al. 1998; Torchilin et al. 2003). A method that has been found to be stable for the labeling of liposomes with ^{111}In uses phosphatidylethanolamine (PE) that is conjugated to the metal chelator, diethylenetriamine pentacetic acid (DTPA). A detailed description of the preparation and application of radiolabeled DTPA-PE liposomes has been published (Torchilin et al. 2003).

DTPA-PE is not effective, however, for labeling liposomes with ^{99m}Tc (Tilcock et al. 1994). For this purpose, a new chelation method based on the technetium chelator, N-hydroxysuccinimidyl hydrazino nicotinate hydrochloride (HYNIC), was developed by Laverman et al. (1998). With this method, the HYNIC ligand was conjugated to the free amino group of distearoyl phosphatidylethanolamine (DSPE) and subsequently incorporated in the lipid bilayer during the liposome preparation. Just prior to the study, the liposomes are incubated with ^{99m}Tc for use in imaging. This HYNIC-labeling methodology has been used to investigate the potential use of ^{99m}Tc liposomes for detection of infection. It has also been used to investigate the relatively short circulation times that occur when very low doses of lipids are used as well as the greatly shortened circulation time of PEG-coated liposomes when they are reinjected at frequent intervals (Dams et al. 2000; Laverman et al. 2001, 2002).

In the second "after loading" method, a radionuclide is coordinated with a lipophilic chelator and then mixed with an aliquot of liposomes encapsulating a second molecule. Once the lipophilic chelator carries the radionuclide across the lipid bilayer, the second molecule interacts with the radionuclide complex causing the radionuclide to become trapped within the interior of the liposome. Several methods have been developed that use this second approach in which the radiolabel is attached to a lipophilic chelator which carries it through the lipid bilayer of the liposome where the radiolabel becomes trapped. This approach to labeling preformed liposomes has the advantage of not leaving the label on the surface where it might interfere with specific targeting molecules or where it has the oppor- tunity to interact with natural metal-binding molecules in the body.

This second approach has been used with ^{111}In, ^{99m}Tc, and ^{67}Ga (Mauk and Gamble 1979; Hwang et al. 1982; Gabizon et al. 1989; Phillips et al. 1992; Corvo et al. 1999; Bao et al. 2003a). One ideal use of this method with the radionuclide, ^{111}In, has been widely used for liposome imaging studies (Harrington et al. 2001; Metselaar et al. 2003; Gaspar et al. 2007). In this method, a clinically available ^{111}In complex, ^{111}In-oxine, is used for liposome labeling. The ^{111}In-oxine is incubated with premanufactured liposomes that encapsulate DTPA. The ^{111}In-oxine migrates into the aqueous interior of a liposome and is trapped by transchelation onto DTPA. The ^{111}In-labeled liposomes are stable and can be used for long-term tracking of liposomes because of the long half-life of ^{111}In (68 h).

One method for labeling liposomes using the lipophilic ^{99m}Tc complex, ^{99m}Tc- hexamethylpropyleneamine oxime (^{99m}Tc-HMPAO), has been successfully used by many investigators (Phillips et al. 1992). With this method, lipophilic ^{99m}Tc- HMPAO enters the liposome where it interacts with glutathione and converts to the hydrophilic form, and thus is trapped in the liposome.

A similar method for radiolabeling liposomes uses ^{99m}Tc-N,N-bis(2-mercaptoethyl)- N',N'-diethylethylenediamine (^{99m}Tc-BMEDA) (Bao et al. 2003a) and this method has been shown to have good in vitro and in vivo stability with a variety of preformed liposome formulations. In addition, an advantage of the BMEDA-labeling method is that it can also be used for labeling liposomes with therapeutic rhenium radionuclides (Bao et al. 2003b). A second advantage of the BMEDA method is that it can be used to label commercially available liposome formulations that use the pH gradient method for loading drugs into the liposomes (Bao et al. 2004). This feature is useful for drug delivery studies in animals and it may eventually be approved for use in clinical studies. BMEDA can currently be purchased in non-GMP form for research studies from a commercial vendor (ABX, GMBH, Germany).

A new method has also been described to label preformed liposomes with radio- halogenated agents such as iodine-123 (^{123}I) and iodine-124 (^{124}I) (Mougin-Degraef et al. 2006). With this method, a Bolton–Hunter reagent, in the form of an activated ester, crosses the liposome membrane and reacts with encapsulated arginine. Imaging of ^{123}I may have some advantages as a single photon emitter with a longer half-life of 13.2 h. This isotope is readily available from the local radiopharmacy in many locations although it is fairly expensive as it must be produced by a cyclotron. This method could also be used to labeled liposomes with [124]I which is a long-lived positron emitter with a half-life of 4.2 days. The labeling of liposomes with this long-lived PET radionuclide may have useful applications in research. For clinical studies, one consideration in using this agent will be the possible need to reduce the dose to the patient due to radiation safety concerns because of its long 4-day half-life. This required reduction in dose may impact image quality due to the lower photon flux. However, concerns about the higher dose levels may be counterbalanced by the ability to track the liposomes for a prolonged period.

A new method of labeling liposomes with the PET agent, fluorine-18 (^{18}F), has been recently described (Marik et al. 2007). With this method, ^{18}F is incorporated into the dipalmitoylglycerol lipid molecule by nucleophilic substitution of the p-toluenesulfonyl moiety. This procedure had a decay corrected yield of 43%. Following this procedure, long-circulating liposomes were then rapidly manufactured and used for PET imaging. Although this procedure in which the liposome is labeled as part of the manufacturing process is not as ideal as procedures for labeling preformed liposomes, it produces very nice quality images. In this particular study, a new dedicated small animal PET camera with much higher resolution than previously available was used to acquire a very high-quality image, thus demonstrating the great future potential for resolution improvements in scintigraphic imaging that will further delineate the anatomical location of labeled liposomes and other nanoparticles (Marik et al. 2007). A problem with the use of ^{18}F is that it has a short half-life of 109 min, so that it would only be useful for tracking liposomes for a short time period of no more than 8 h; however, this short time period of imaging may still prove useful for certain applications.

6 Specific Uses of Gamma Photon Imaging for Liposome Drug Development

Gamma photon imaging of liposomes has been successfully used in the last 15 years for studies of drug delivery and as potential diagnostic imaging agents. Several comprehensive reviews have been written describing in detail these liposome-labeling methods and uses (Phillips and Goins 2002; Goins and Phillips 2003; Laverman et al. 2003; Goins 2008). In the following section, some specific examples of the use of radiolabeled liposomes will be given with emphasis on research carried out in the last 3 years.

6.1 Noninvasive Determination of Drug Concentrations

The tracking of radiolabeled liposomes using gamma photon imaging can be used as a method to noninvasively and quantitatively determine how much of a drug is taken up by the targeted site. A validation of this approach has been recently shown by Kleiter et al. (2006). In this research, liposomes were labeled with ^{99m}Tc using the HMPAO method previously discussed (Phillips et al. 1992). The purpose of the study was to determine if noninvasive imaging could be used to predict the amount of doxorubicin that accumulated in a rat fibrosarcoma tumor following the administration of long-circulating PEG liposomes containing doxorubicin (Doxil™) without having to biopsy the tumor. Scintigraphic images were obtained at 18 h following injection of the ^{99m}Tc-labeled liposomes. Following imaging at 18 h, tumors were removed for radioactivity counting and for measurement of the doxorubicin concentration in the tumors. The results of this study demonstrated that there was a significant positive correlation of the intratumoral radiolabeled tracer and the amount of doxorubicin in the tumor. This study was also interesting in that these studies were also performed on rats that had received local hyperthermia directed at the tumor. The images clearly demonstrated the increased uptake in the hyperthermia treated tumor as compared with the tumors were not treated with hyperthermia. The hyperthermia increased the uptake of the radiolabel associated with the liposomes by approximately fourfold and it increased the uptake of measured doxorubicin concentrations in the tumor by a slightly less increment that ranged from 2.6- to 3-fold (Kleiter et al. 2006).

6.2   Studies of Liposomes Containing Anticancer Agents

Biodistribution studies in rats of the clinical formulation of Doxil™ labeled using the previously described ^{99m}Tc-BMEDA method have been performed by the authors (Bao et al. 2004). This study demonstrated that ^{99m}Tc-Doxil™ remained associated in vivo and that significant ^{99m}Tc activity was visualized in the heart and blood at 48 h. In this first report, only planar images were acquired using a gamma camera. Since that time, our group has been using a specialized small animal imaging camera that can simultaneously obtain SPECT and X-ray computed tomography (CT) images. The excellent qualities of these SPECT/CT images are shown in Fig. 1. These images demonstrate the tumor uptake of ^{99m}Tc-Doxil™ following intravenous injection in a nude rat xenograft model of human head and neck cancer. The uptake in the tumor is clearly visualized in the SPECT images and it can be correlated with the CT anatomic images. When using high-resolution pinhole collimators, this type of SPECT imaging can provide information about the intratumoral distribution of the radiolabeled Doxil™. Figure 2 is a volume-rendered image in which the SPECT volume-rendered images are superimposed on the volume-rendered CT images that have been windowed to display only the bone. This type of imaging provides a three-dimensional view of the tumor uptake in relationship to the whole body bony anatomy. Serial images such as these can be obtained over time in the same animal for up to three half-lives of the radionuclide used to label the liposome.

Liposomes labeled by the ^{99m}Tc-BMEDA method have been used to investigate the distribution of liposomes directly injected into solid tumors (Bao et al. 2006). Following intratumoral administration of the ^{99m}Tc-labeled liposomes, images were

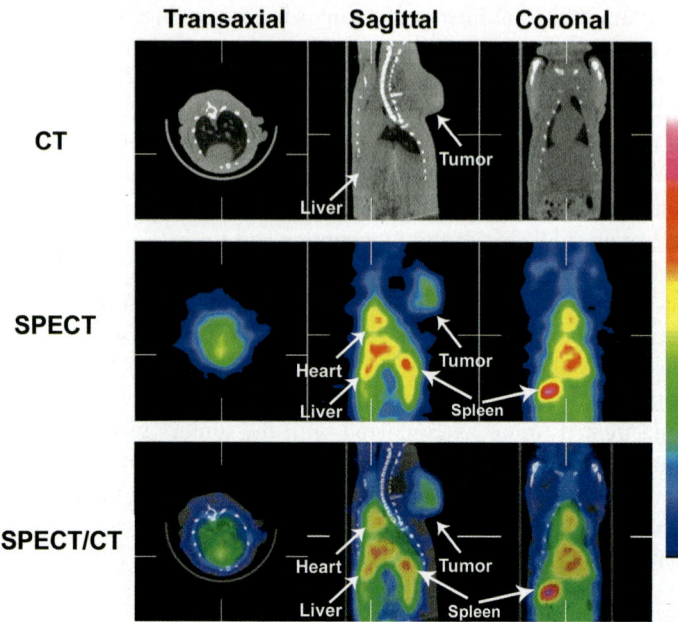

Fig. 1 Images obtained with multimodality SPECT/CT camera of ⁹⁹ᵐTc-labeled liposomes encapsulating doxorubicin (Doxil™) in a nude rat with a 1.5-cm head and neck xenograft tumor at 24-h postinjection. The *top panel* shows the CT images, the *middle panel* the SPECT images, and the *bottom panel* the fused SPECT/CT images. The three types of images are displayed as transaxial slice images on the *left*, sagittal slices in the *middle*, and coronal slices on the *right*. The uptake in the tumor of the labeled Doxil™ can be clearly visualized on the SPECT images. The fusion of the CT images with the SPECT images demonstrates the location of the ⁹⁹ᵐTc-Doxil™ in relationship to other body organs such as the liver, spleen, and heart

acquired with a SPECT/CT camera. Approximately 50% of the injected activity was retained in the tumor immediately following injection and 40% of the initial dose still remained in the tumor at 20 h. The high-resolution pinhole imaging made it possible to track the intratumoral distribution of the liposomes in tumors of 1.5-cm diameter. Using this methodology, it should be possible to develop better methods of achieving a homogeneous spread to liposomes throughout the tumor for a more effective local treatment of the tumor.

6.3   Specifically Targeted Anticancer Imaging

Imaging has also been used to study the efficacy of specifically targeted liposomes. Several studies have used imaging to show the promise of using specific targeted surface modification of liposomes for the treatment of cancer (Elbayoumi and Torchilin 2006; Erdogan et al. 2006b). In these studies, liposomes were formed to

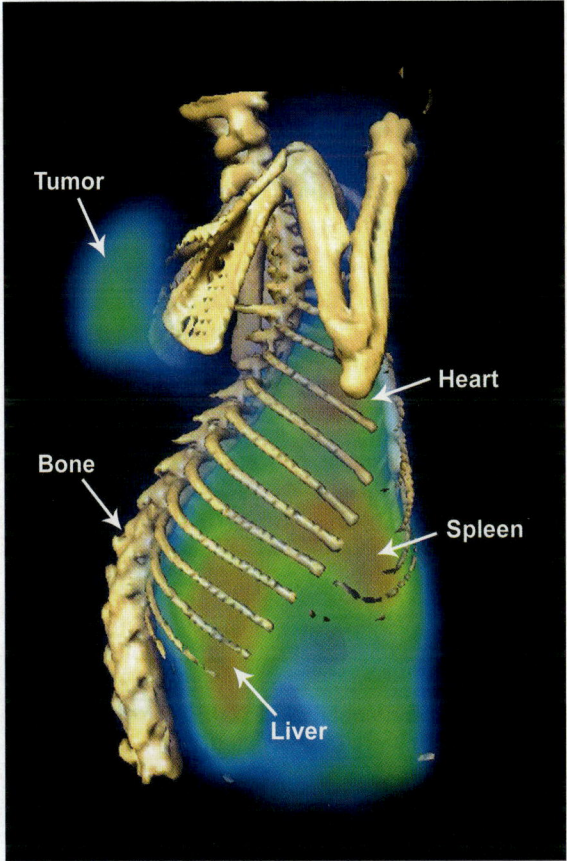

Fig. 2 A volume-rendered image of the SPECT images in Fig. 1 superimposed on the volume-rendered CT images that were windowed for bone imaging. The uptake in the body and the tumor are clearly visualized in relationship to the bony anatomy. These images can be easily dynamically rotated in movie format to provide researchers with clear 3D visualization of the distribution of the labeled liposomes such as the ^{99m}Tc-Doxil™ shown here

contain both a polychelating amphiphilic polymer (PAP) and a specific targeting antibody on their surface. The specific polychelating polymer was composed of hydrophilic blocks carrying multiple side chains of the metal chelating agent, DTPA. For specific targeting, a monoclonal antibody against nucleosomes that are expressed on the surface of a wide variety of cancer cells was also attached to these liposomes. Tumor-to-muscle ratio for the liposomes with the specific antinucleosome antibodies was 13.9 at 24 h vs. lower ratios determined for the nonspecifically targeted liposomes that were either modified with nonspecific polyclonal antibodies (4.3) or liposomes with an unmodified surface (3.0) in rats that had Lewis lung carcinomas implanted on their thighs. This comparison with both types of control liposomes provides convincing evidence of the specificity of this antibody for targeting the tumor. This same antibody was also shown to have a significantly increased tumor uptake in rats that had been implanted with colon carcinoma cells, demonstrating that antinucleosome antibodies may enhance the targeting of liposomes to a wide variety of cancer types.

6.4 Liposomes for Treating Arthritis

Liposome imaging has been used in the preclinical development of liposomes containing prednisolone for the therapy of an animal model of rheumatoid arthritis (Metselaar et al. 2003). Long-circulating PEG-coated liposomes containing prednisolone as well as DTPA were labeled with ^{111}In-oxine using the method previously described. These liposomes were found to accumulate in the inflamed joints by imaging. The increased uptake was very obvious on the images and the gamma photon labeling of the liposomes was used to determine the clearance from the blood and the biodistribution in the tissues. The inflamed hind paws had sevenfold increased activity in comparison with the hind paws of normal rats. A single dose of liposome-encapsulated prednisolone resulted in complete remission of the inflammatory response for 1 week.

A similar use of imaging has also been recently reported for liposomes that carry superoxide dismutase on their surface (Gaspar et al. 2007). These liposomes were used to treat experimental arthritis. The use of liposomes as a carrier for superoxide dismutase greatly improved the pharmacokinetic behavior of superoxide dismutase, allowing this powerful antioxidant to reach the site of inflammation in the joint. A comparison of the therapeutic efficiency was made between liposomes that had no superoxide dismutase on their surface and liposomes in which the superoxide dismutase was carried on their surface. Imaging using the ^{111}In-oxine/DTPA method demonstrated that, even though both liposome formulations had similar accumulations in the inflamed joints, the liposomes with the superoxide dismutase on their surface were significantly more effective for the treatment of arthritis (Gaspar et al. 2007). The investigators have named this type of liposome, an *enzymosome*.

6.5 Bone Marrow-Targeted Liposomes

Another recent paper has reported that liposomes with a special surface modification have very high uptake in bone marrow (Sou et al. 2007). With this special surface modification, rabbits injected with these liposomes, also referred to as *vesicles*, had a very high uptake in the bone marrow as revealed by imaging with the ^{99m}Tc-HMPAO method. The special modification to the liposome surface was the addition of a negative charge to the liposome surface by adding a nonphospholipid anionic amphiphile, L-glutamic acid, *N*-(3-carboxy-1-oxopropyl)-, 1,5-dihexadecyl ester (abbreviated SA for succinic acid) to the liposome composition. The addition of a small amount of PEG to this liposome surface also appeared to modestly enhance the already high uptake of the liposomes in the marrow. Imaging was used to determine the most ideal concentration of PEG to have on the surface of these SA liposomes to maximized bone marrow uptake. Based on imaging studies, the uptake in the marrow was very rapid with more than 60% of the infused dose taken up by the marrow at 6-h postintravenous administration (Sou et al. 2007). Although the precise mechanism by which these liposomes accumulate in marrow is unknown, electron micrograph studies as well as fluorescently labeled liposome studies have demonstrated that the liposomes in the marrow are located in marrow macrophages. It is believed that the special surface modification of the liposomes results in specific targeting of the scavenger receptors on the surface of the bone marrow macrophages. The specific targeting of therapeutic agents to bone marrow using these SA liposomes may be a promising approach for the delivery of drugs to the bone marrow (Fig. 3).

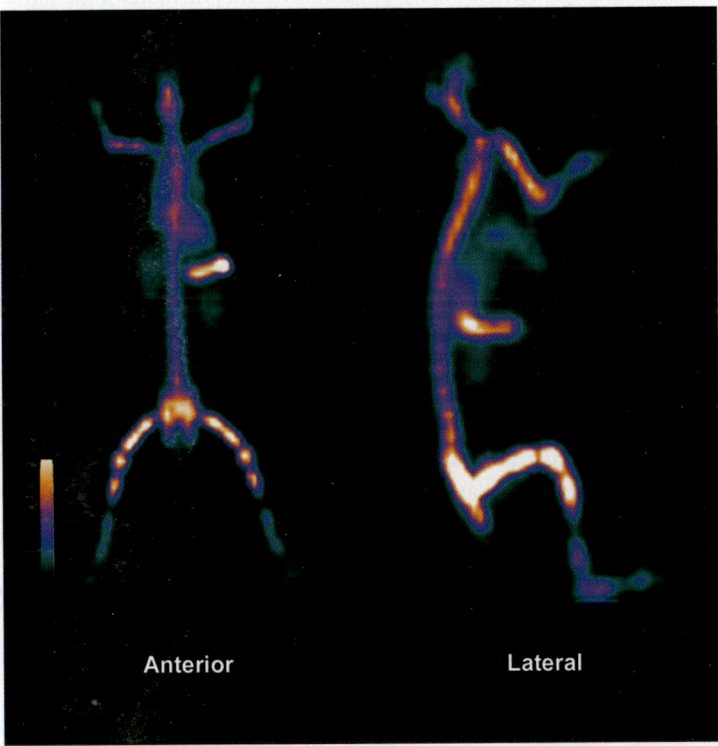

Fig. 3 Gamma photon images of a rabbit acquired 6 h after intravenous injection with bone marrow-targeted liposomes labeled with ^{99m}Tc

6.6   *Intracavitary Liposomes*

Liposomes with high retention in body cavities and in the lymph nodes that drain from these cavities have been developed for drug delivery applications using imaging as a tool (Phillips et al. 2002; Medina et al. 2004a,b; Zavaleta et al. 2007a,b). These liposomes contain biotin on their surface and when these biotin-coated liposomes are injected into a body cavity such as the peritoneum or the pleural space within 2 h after the injection of avidin into the same cavity, the liposomes will have high retention in this cavity. The mechanism by which this retention occurs is thought to be due to the aggregation of the liposomes by the high affinity that the multivalent avidin has for the biotin liposomes causing the liposomes to become aggregated. Liposome imaging using the ^{99m}Tc-HMPAO methodology has been used to determine the best timing for the injection of the avidin.

When this methodology is used, the majority of the liposomes administered in the body cavity are retained in that cavity for a prolonged time, whereas the same biotin-liposome formulation without the avidin is rapidly cleared from the cavity over a 4–6 h period during which time the liposomes return to the blood from the lymphatics that drain the cavity. This approach has great potential for the intracavitary retention of drugs encapsulated in liposomes. One potential use for these liposomes is in the treatment of ovarian cancer in which the cancer cells are generally disseminated into the peritoneal cavity and lymph nodes that drain this cavity at the time of diagnosis. Another use could be in the treatment of lung cancer in which the cancer frequently drains into the mediastinal nodes. Imaging studies using the avidin/biotin procedure have shown that a large liposome dose can be targeted to the mediastinal nodes following administration of biotin liposomes along with avidin into the pleural cavity.

7   MR Imaging

Compared with gamma photon imaging, MR imaging has its distinct advantages. These advantages include (1) higher spatial resolution providing anatomic information with excellent soft tissue contrast and (2) greater temporal resolution, making it practical to observe a rapid dynamic procedure. Pulse sequences are currently available to achieve temporal resolutions required for liposome studies.

MR molecular imaging utilizes the signal changes caused by the contrast agent. These signal changes mainly result from the changes to spin–lattice relaxation time (T1), or spin–spin relaxation time (T2), or a combination of both effects. Good reviews concerning MR imaging with contrast agents can be found in the following references (Gupta and Weissleder 1996; Torchilin 2000; Caruthers et al. 2006; Mulder et al. 2006). These reviews also include liposome-based MR contrast agents and their current applications.

Although the strengths of MR imaging are many, it also has several disadvantages relative to gamma photon imaging. One disadvantage is that MR contrast imaging is less directly a quantitative procedure. However, it can be approximately quantitative since there is a linear relationship between relaxation changes ($\Delta(1/Ti)$, $i = 1, 2$) and local contrast agent concentration, which can be expressed as

$$\Delta\left(\frac{1}{Ti}\right) = Ri[M], i = 1, 2.$$

Here, Ri is called T1 or T2 relaxivity of a contrast agent with the unit of $(\text{mM s})^{-1}$ and $[M]$ is the local concentration of contrast agent in unit of mM. Ri is a measure of the power of an MR contrast agent. A second disadvantage of MR imaging is its lower sensitivity for detecting contrast agents. This means that the physical concentration of contrast agent required to generate enough image contrast is approximately 1,000-fold greater than that required for gamma photon imaging. This lower sensitivity makes it difficult to achieve effective contrast by simply attaching a single contrast molecule to a single targeting ligand, such as an antibody as is frequently performed in gamma photon imaging. This requirement for high amounts of material to arrive at a particular target does, however, make the use of nanoparticles, such as liposomes, promising carriers for achieving targeted MR contrast. This is because a liposome particle can carry a large amount of contrast molecules in a small volume of space or on the surface of the particle.

MR contrast agents can be classified into two types, paramagnetic agents and superparamagnetic agents, based on their mechanism of achieving image contrast. These contrast agents act to perturb the local magnetic uniformity thus decreasing T1 and T2 values of the tissue. Paramagnetic contrast agents use ions which have unpaired shell electrons. These elements include the ions of gadolinium (Gd), manganese (Mn), and iron (Fe). These unpaired electrons expand their MR relaxation perturbation to the local water molecules mainly through a water exchange mechanism. The direct administration of a relatively large quantity of these metallic MR contrast molecules into a living body is highly toxic. Fortunately, the safe use of these elements can be achieved after conjugating these ions with certain chelating compounds to form a stable complex, while the water binding and exchange space of the ion can be preserved to ensure a high relaxivity. The most common representative of this kind of MR contrast agents is gadolinium chelated with DTPA (Gd-DTPA). These kinds of contrast agents have stronger impact on T1 relaxivity than T2 relaxivity. Liposomes composed of Gd-DTPA-lipid have been shown to make excellent T1 MR contrast. Specific examples of these liposomes will be discussed subsequently.

Superparamagnetic agents use nanoscale micromagnets, generally iron oxide nanoparticles. Superparamagnetic agents have a stronger T2 relaxation effect than a T1 relaxation effect. Due to these facts, paramagnetic agents usually result in increased signal strength as the consequence of decreased T1 value, while iron oxide nanoparticle contrast agents produce a negative contrast effect. Unfortunately, iron oxide nanoparticles are also very toxic. To utilize this kind of contrast agent safely, the iron oxide nanoparticles need to be coated with safe chemical moieties to isolate iron oxide from direct interaction with in vivo biomolecules. For example, the FDA-approved MR contrast agent, Feridex™, is heavily dextran-coated iron oxide nanoparticles. Although the average particle diameter of Feridex™ particles is around 150 nm, the iron oxide core is only 20–30 nm in diameter (Shapiro et al. 2006). Another strategy of coating iron oxide nanoparticles is encapsulating them into liposomes, which will be further discussed later in this chapter.

8   Methods of Labeling Liposomes with MR Contrast Agents

Previous investigations of liposome-based MR contrast enhancement mainly include (1) liposomes with nitroxide spin labels (Keana and Pou 1985; Grant et al. 1987; Bacic et al. 1988), nitroxide spin labeling has very low relaxivity and weak contrast effect (Grant et al. 1987; Karlik et al. 1991); (2) liposomes serving as carriers of hyperpolarized xenon-129 gas (Grant et al. 1987; Karlik et al. 1991; Venkatesh et al. 2000, 2002); (3) liposomes encapsulating paramagnetic ions or their complexes; (4) membrane surface conjugation of chelated paramagnetic ions; and (5) liposome-coated iron oxide nanoparticles. In this section, we primarily review in detail liposomes carrying paramagnetic agents and iron oxide nanoparticles.

To enable the visualization of the liposome-based contrast effect by MR imaging, the incorporation of paramagnetic ions into liposomes with diameters of 100 nm or smaller has utilized the following two routes (1) the conjugation of paramagnetic ions to the liposome surface (Grant et al. 1987; Kabalka et al. 1987) and (2) the encapsulation of paramagnetic ions or their complexes into the inner space of the liposome (Devoisselle et al. 1988; Unger et al. 1988; Turski et al. 1998).

The first method can use a functional lipid which conjugates a chelating group, such as DTPA or its analogues, where each paramagnetic ion will bind. Coating the surface of the liposomes with Gd-DTPA has been shown to make excellent contrast. Researchers have further enhanced this method by creating a polymer of multiple DTPA chelates to achieve excellent contrast (Erdogan et al. 2006a).

The second approach includes the encapsulation of paramagnetic ions or their complexes into liposomes, so that a relatively large amount of paramagnetic agents can be trapped into liposomes. However, MR contrast from paramagnetic agents is largely a result of water exchange between the water molecule binding to paramagnetic ions and the surrounding bulk water. The encapsulation of paramagnetic ions or their complexes into the nanoscale area of the liposomal inner space may greatly decrease the water mobility and water exchange kinetics; thus resulting in greatly decreased relaxivity compared with the same concentration of free paramagnetic components (Barsky et al. 1992). This phenomenon may be utilized to advantage in controlled drug release studies to demonstrate in vivo when a liposome begins releasing an MR contrast agent from its interior. This increased MR contrast due to the release of MR contrast agent can be used as a model for drug release, triggered by heat, pH change, or other changes in the in vivo local microenvironment (Bacic et al. 1988; Fossheim et al. 2000; Lokling et al. 2001). Furthermore, this decreased relaxivity occurring when MR contrast agents are encapsulated entirely within liposomes may be partly or mostly recovered by choosing specific liposome formulations such as liposomes of smaller size in which there is more freedom of water migration and the confinement effect from the liposome bilayer is greatly decreased (Bacic et al. 1988; Tilcock et al. 1992; Strijkers et al. 2005). It is also possible to use liposomes containing unsaturated lipids which allow more water exchange through their membranes.

Iron oxide nanoparticles can also be trapped into liposomes (Chan et al. 1992; Pauser et al. 1997). However, to ensure the stability of this entrapment and a low in vivo toxicity, a delicately designed entrapment system may need to be used.

9 Specific Applications of Liposome-Based MR Imaging

9.1 Hepatic Metastases and Blood Perfusion Imaging

As discussed previously, liposomes with conventional formulations are mainly cleared through the liver and spleen after intravenous administration. From this behavior emerges the possibility for using MR imaging to enhance the diagnosis and detection of liver lesions/metastasis (Leander 1995). A study by Unger et al. (1988) in rat hepatic metastases has shown that paramagnetic liposomes had high normal liver concentration, while liver metastases had low accumulation. Another animal study has shown a similar behavior of liposome-based MR contrast enhancement characteristics (Niesman et al. 1990). However, the diagnostic significance and its applicability in human liver cancer still need to be proven.

By using long-circulating liposomes, liposome-based contrast enhancement can be used in blood pool or perfusion imaging. Previous ex vivo and in vivo animal studies have proven the potential of using paramagnetic liposomes in blood pool and perfusion imaging (Storrs et al. 1995; Weissig et al. 2000; Suga et al. 2001; Ghaghada et al. 2007).

9.2 Lymph Node Imaging

Clinical lymphoscintigraphy commonly uses gamma photon imaging after a local injection of ^{99m}Tc-sulfur colloid (Alex 2004; Krynyckyi et al. 2004). Liposomes may also be applied in lymphatic imaging by subcutaneous administration (Oussoren and Storm 2001). To enable liposomes to accumulate in lymph nodes, specifically designed strategies may also need to be used (Phillips et al. 2001, 2002). However, since MR imaging has high temporal resolution, it may be possible to image rapidly clearing liposome particles moving through the lymphatics. Previous animal studies have shown the feasibility of using paramagnetic liposomes to perform lymphoscintigraphy (Fig. 4) (Trubetskoy et al. 1995; Misselwitz and Sachse 1997; Fujimoto et al. 2000).

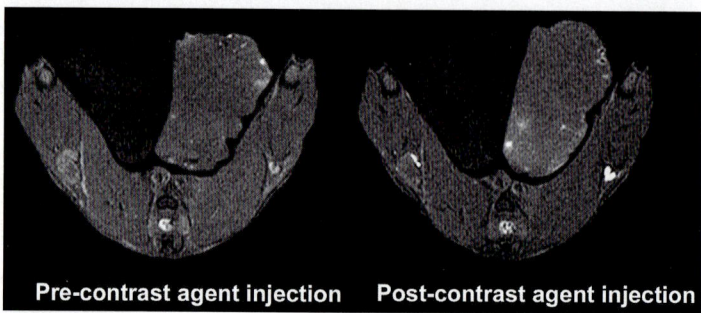

Fig. 4 Visualization of VX2 tumor in rabbit popliteal lymph node using MRI after subcutaneous injection of PEG-modified Gd-liposomes. The lymph node on the right side of the postcontrast image has greatly increased brightness after the administration of the PEG-modified Gd-liposomes compared with precontrast image (courtesy of V. P. Torchilin, Ph.D., Northeastern University)

9.3 Targeted MR Molecular Imaging

Compared with molecular imaging using radiolabeled liposomes and gamma photon imaging, MR molecular imaging with targeted paramagnetic/superparamagnetic liposomes may be more challenging due to its lower sensitivity. However, recent studies have shown that MR molecular imaging with liposome systems is practical via targeting to certain biomolecules (Mulder et al. 2004, 2005, 2006, 2007; Strijkers et al. 2005).

Mulder et al. (2004) introduced a multifunctional liposome system. With this system, lipid-DTPA was utilized to chelate Gd for MR contrast, rhodamine B lipid was added into the liposome formulation for fluorescence imaging, while Mal-PEG$_{2000}$-lipid was incorporated into the liposome as the site for conjugation of specific targeting molecules. Using this system, E-selectin antibody-conjugated paramagnetic liposomes were studied with in vitro relaxivity and HUVEC cell-specific binding after TNF-α treatment. With 7-T magnetic field strength, paramagnetic liposomes had significantly higher T1 relaxivity. In another in vitro study, annexin-5 targeted paramagnetic liposomes were found to accumulate in apoptotic cells, resulting in greatly increased relaxation rates of these cells (van Tilborg et al. 2006).

Recently, $\alpha_v\beta_3$-targeted paramagnetic RGD peptide liposomes were reported for use in MR molecular imaging of tumor angiogenesis (Mulder et al. 2007). It was shown that the targeted liposomes accumulated in the tumor blood vessels. MR imaging has also shown a higher level of contrast enhancement in tumor. The association of this increased MR contrast enhancement to tumor angiogenesis was proven via fluorescence imaging and blood vessel histochemistry.

9.4   Drug Delivery Monitoring

Another application of MR imaging is the trafficking of liposomal drug delivery during organ distribution and microdistribution studies. By labeling liposomes carrying drugs with paramagnetic agents, MR imaging has been successfully used in the trafficking of the in vivo distribution of liposomal chemotherapeutic agents and liposomal gene delivery using various liposomal drug administration methods in animals (Rubesova et al. 2002; Leclercq et al. 2003; Pirollo et al. 2006).

The rationale of using paramagnetic agents to track drug delivery lies in the good correlation between drug and paramagnetic agent concentrations. A related study was recently reported with liposomes carrying both Mn^{2+} ion and doxorubicin (Viglianti et al. 2006). A positive correlation between doxorubicin and relaxation change in tumor xenografts was observed. In another study, the simultaneous release of both encapsulated fludarabine monophosphate and Gd-DTPA from liposomes was observed with ^1H MR imaging and ^{19}F MR spectroscopy (Port et al. 2006).

9.5   Thermo/pH-Sensitive Liposomes

Due to the lower relaxivity of encapsulated paramagnetic agents in liposomes, the sudden release of the paramagnetic agent from the liposome interior can result in a remarkable change in MR contrast. This behavior can be used in (1) hyperthermia-triggered local drug release; in this situation, thermosensitive liposomes were designed and used for the local release of chemotherapeutic agents from liposomes to initiate a therapeutic effect and/or combined with thermal therapy (Abdel-Wahab et al. 2004; Frich et al. 2004; McDannold et al. 2004; Viglianti et al. 2004; Bos et al. 2005; Ponce et al. 2007) and (2) controlled local drug release due to the differences in local pH value. Accordingly, pH-sensitive liposomes could respond to lower pH environment in tumors for assessing acidity/pH range in a tumor or for the enhancement of tumor drug delivery (Lokling et al. 2001).

Another application of thermosensitive liposomes is in thermometry (Lindner et al. 2005). In hyperthermia tumor therapy, the coverage of heat of sufficient duration to kill tumor cells is necessary. Thermosensitive liposomes provide a tool to evaluate intratumoral temperature distribution for evaluation of therapy outcome.

10   Future Developments in Imaging Technology Applicable to Liposome Research

Recently, several clinical imaging camera systems have been introduced that simultaneously perform two different types of imaging that are then fused into a single image. This permits the formation of one image that combines the strengths of each type of imaging. For instance, clinical systems combining PET imaging and CT imaging have become widely available. This combined system uses PET for physiologic imaging with traditional CT imaging to provide very high anatomic detail. Imaging systems that combine SPECT imaging and CT have also been recently introduced. Currently under development is a combined PET and MRI system for imaging (Catana et al. 2006). Prior to the recent development of avalanche photon detectors, the very high magnetic field interfered with traditional gamma photon detectors that rely on photomultiplier tubes. These combined imaging systems will permit the imaging of liposomes and other nanoparticles that have been dual labeled with both radionuclides and MR contrast agents. This type of dual PET/MRI imaging would offer the possibility of quantitatively assessing the whole body distribution of the radiolabeled liposomes using PET as well as higher resolution anatomic imaging provided by MR imaging to determine the location of the liposomes in the body. Simultaneous MR imaging will also be able to monitor the release of drugs from the liposomes by the increase in contrast that occurs when agents such as Gd-DTPA are released from the liposome. These types of studies offer the potential to better understand drug distribution and release over time by noninvasive imaging.

11 Summary

Continuing rapid progress in both liposome technology and imaging technology will continue to provide new synergistic applications for the development of liposomes and other nanoparticle systems as effective diagnostic and therapeutic agents.

Acknowledgments The authors would like to thank Anuradha Soundararajan and Jonathan Sumner for their kind assistance in the preparation of the images used in this chapter.

References

Abdel-Wahab, O. I., Grubbs, E., Viglianti, B. L., Cheng, T. Y., Ueno, T., Ko, S., Rabbani, Z., Curtis, S., Pruitt, S. K., Dewhirst, M. W. and Tyler, D. S. 2004. The role of hyperthermia in regional alkylating agent chemotherapy. Clin Cancer Res 10:5919–5929.

Alex, J. C. 2004. The application of sentinel node radiolocalization to solid tumors of the head and neck: a 10-year experience. Laryngoscope 114:2–19.

Allen, T. M., Hansen, C., Martin, F., Redemann, C. and Yau-Young, A. 1991. Liposomes containing synthetic lipid derivative of poly(ethylene) show prolonged circulation half-lives *in vivo*. Biochim Biophys Acta 1066:29–36.

Bacic, G., Niesman, M. R., Bennett, H. F., Magin, R. L. and Swartz, H. M. 1988. Modulation of water proton relaxation rates by liposomes containing paramagnetic materials. Magn Reson Med 6:445–458.

Bangham, A. D. 1993. Liposomes: the Babraham connection. Chem Phys Lipids 64:275–285.

Bao, A., Goins, B., Klipper, R., Negrete, G., Mahindaratne, M. and Phillips, W. T. 2003a. A novel liposome radiolabeling method using 99 mTc-"SNS/S" complexes: in vitro and in vivo evaluation. J Pharm Sci 92:1893–1904.

Bao, A., Goins, B., Klipper, R., Negrete, G. and Phillips, W. T. 2003b. 186Re-liposome labeling using 186Re-SNS/S complexes: in vitro stability, imaging, and biodistribution in rats. J Nucl Med 44:1992–1999.

Bao, A., Goins, B., Klipper, R., Negrete, G. and Phillips, W. T. 2004. Direct 99mTc labeling of pegylated liposomal doxorubicin (Doxil) for pharmacokinetic and non-invasive imaging studies. J Pharmacol Exp Ther 308:419–425.

Bao, A., Phillips, W. T., Goins, B., Zheng, X., Sabour, S., Natarajan, M., Ross Woolley, F., Zavaleta, C. and Otto, R. A. 2006. Potential use of drug carried-liposomes for cancer therapy via direct intratumoral injection. Int J Pharm 316:162–169.

Bao, A., Zhao, X., Phillips, W. T., Woolley, F. R., Otto, R. A., Goins, B. and Hevezi, J. M. 2005. Theoretical study of the influence of a heterogeneous activity distribution on intratumoral absorbed dose distribution. Med Phys 32:200–208.

Barsky, D., Putz, B., Schulten, K. and Magin, R. L. 1992. Theory of paramagnetic contrast agents in liposome systems. Magn Reson Med 24:1–13.

Bendas, G., Krause, A., Bakowsky, U., Vogel, J. and Rothe, U. 1999. Targetability of novel immunoliposomes prepared by a new antibody conjugation technique. Int J Pharm 181:79–93.

Bos, C., Lepetit-Coiffe, M., Quesson, B. and Moonen, C. T. 2005. Simultaneous monitoring of temperature and T1: methods and preliminary results of application to drug delivery using thermosensitive liposomes. Magn Reson Med 54:1020–1024.

Caruthers, S. D., Winter, P. M., Wickline, S. A. and Lanza, G. M. 2006. Targeted magnetic resonance imaging contrast agents. Meth Mol Med 124:387–400.

Catana, C., Wu, Y., Judenhofer, M. S., Qi, J., Pichler, B. J. and Cherry, S. R. 2006. Simultaneous acquisition of multislice PET and MR images: initial results with a MR-compatible PET scanner. J Nucl Med 47:1968–1976.

Chan, T. W., Eley, C., Liberti, P., So, A. and Kressel, H. Y. 1992. Magnetic resonance imaging of abscesses using lipid-coated iron oxide particles. Invest Radiol 27:443–449.

Corvo, M. L., Boerman, O. C., Oyen, W. J., Van Bloois, L., Cruz, M. E., Crommelin, D. J. and Storm, G. 1999. Intravenous administration of superoxide dismutase entrapped in long circulating liposomes. II. In vivo fate in a rat model of adjuvant arthritis. Biochim Biophys Acta 1419:325–334.

Dams, E. T., Laverman, P., Oyen, W. J., Storm, G., Scherphof, G. L., van Der Meer, J. W., Corstens, F. H. and Boerman, O. C. 2000. Accelerated blood clearance and altered biodistribution of repeated injections of sterically stabilized liposomes. J Pharmacol Exp Ther 292:1071–1079.

Devoisselle, J. M., Vion-Dury, J., Galons, J. P., Confort-Gouny, S., Coustaut, D., Canioni, P. and Cozzone, P. J. 1988. Entrapment of gadolinium-DTPA in liposomes. Characterization of vesicles by P-31 NMR spectroscopy. Invest Radiol 23:719–724.

Elbayoumi, T. A. and Torchilin, V. P. 2006. Enhanced accumulation of long-circulating liposomes modified with the nucleosome-specific monoclonal antibody 2C5 in various tumours in mice: gamma-imaging studies. Eur J Nucl Med Mol Imaging 33:1196–1205.

Erdogan, S., Roby, A., Sawant, R., Hurley, J. and Torchilin, V. P. 2006a. Gadolinium-loaded polychelating polymer-containing cancer cell-specific immunoliposomes. J Liposome Res 16:45–55.

Erdogan, S., Roby, A. and Torchilin, V. P. 2006b. Enhanced tumor visualization by gamma-scintigraphy with 111In-labeled polychelating-polymer-containing immunoliposomes. Mol Pharmacol 3:525–530.

Fossheim, S. L., Il'yasov, K. A., Hennig, J. and Bjornerud, A. 2000. Thermosensitive paramagnetic liposomes for temperature control during MR imaging-guided hyperthermia: in vitro feasibility studies. Acad Radiol 7:1107–1115.

Frich, L., Bjornerud, A., Fossheim, S., Tillung, T. and Gladhaug, I. 2004. Experimental application of thermosensitive paramagnetic liposomes for monitoring magnetic resonance imaging guided thermal ablation. Magn Reson Med 52:1302–1309.

Fujimoto, Y., Okuhata, Y., Tyngi, S., Namba, Y. and Oku, N. 2000. Magnetic resonance lymphography of profundus lymph nodes with liposomal gadolinium-diethylenetriamine pentaacetic acid. Biol Pharm Bull 23:97–100.

Gabizon, A., Huberty, J., Straubinger, R. M., Price, D. C. and Papahadjpulos, D. 1989. An improved method for in vivo tracing and imaging of liposomes using a gallium-67-deferoxamine complex. J Liposome Res 1:123–135.

Gabizon, A., Shmeeda, H., Barenholz, Y. 2003. Pharmacokinetics of pegylated liposomal Doxirubicin: Review of animal and human studies. Clinical Pharmacokinetics 42:418–436.

Gaspar, M. M., Boerman, O. C., Laverman, P., Corvo, M. L., Storm, G. and Cruz, M. E. 2007. Enzymosomes with surface-exposed superoxide dismutase: in vivo behaviour and therapeutic activity in a model of adjuvant arthritis. J Control Release 117:186–195.

Ghaghada, K. B., Bockhorst, K. H., Mukundan, S., Jr., Annapragada, A. V. and Narayana, P. A. 2007. High-resolution vascular imaging of the rat spine using liposomal blood pool MR agent. Am J Neuroradiol 28:48–53.

Goins, B., Klipper, R., Rudolph, A. S. and Phillips, W. T. 1994. Use of technetium-99 m-liposomes in tumor imaging. J Nucl Med 35:1491–1498.

Goins, B. A. 2008. Radiolabeled lipid nanoparticles for cancer diagnosis and treatment. Handbook of Particulate Drug Delivery. Ed. M. N. V. R. Kumar, American Scientific Publishers, Stevenson Ranch, CA (in press).

Goins, B. A. and Phillips, W. T. 2003. Radiolabelled liposomes for imaging and biodistribution studies. Liposomes: A Practical Approach. Eds. V. P. Torchilin and V. Weissig. Oxford University Press, Oxford, pp. 319–336.

Goins, B. A. and Phillips, W. T. 2007. Methods for tracking radiolabeled after injection in the body. Liposome Technology, vol. 3. Ed. G. Gregoriadis. Informa Healthcare, New York, pp. 191–209.

Grant, C. W., Barber, K. R., Florio, E. and Karlik, S. 1987. A phospholipid spin label used as a liposome-associated MRI contrast agent. Magn Reson Med 5:371–376.

Gregoriadis, G. and Ryman, B. E. 1971. Liposomes as carriers of enzymes or drugs: a new approach to the treatment of storage diseases. Biochem J 124:58P.

Gupta, H. and Weissleder, R. 1996. Targeted contrast agents in MR imaging. Magn Reson Imaging Clin N Am 4:171–184.

Harrington, K. J., Mohammadtaghi, S., Uster, P. S., Glass, D., Peters, A. M., Vile, R. G. and Stewart, J. S. 2001. Effective targeting of solild tumors in patients with locally advanced cancers by radiolabeled pegylated liposomes. Clin Cancer Res 7:243–254.

Hnatowich, D. J., Friedman, B., Clancy, B. and Novak, M. 1981. Labeling of preformed liposomes with Ga-67 and Tc-99 m by chelation. J Nucl Med 22:810–814.

Hwang, K. J., Merriam, J. E., Beaumier, P. L. and Luk, K. F. 1982. Encapsulation, with high efficiency, of radioactive metal ions in liposomes. Biochim Biophys Acta 716:101–109.

Ishida, T., Atobe, K., Wang, X. and Kiwada, H. 2006a. Accelerated blood clearance of PEGylated liposomes upon repeated injections: effect of doxorubicin-encapsulation and high-dose first injection. J Control Release 115:251–258.

Ishida, T., Harada, M., Wang, X. Y., Ichihara, M., Irimura, K. and Kiwada, H. 2005. Accelerated blood clearance of PEGylated liposomes following preceding liposome injection: effects of lipid dose and PEG surface-density and chain length of the first-dose liposomes. J Control Release 105:305–317.

Ishida, T., Ichihara, M., Wang, X., Yamamoto, K., Kimura, J., Majima, E. and Kiwada, H. 2006b. Injection of PEGylated liposomes in rats elicits PEG-specific IgM, which is responsible for rapid elimination of a second dose of PEGylated liposomes. J Control Release 112:15–25.

Kabalka, G., Buonocore, E., Hubner, K., Moss, T., Norley, N. and Huang, L. 1987. Gadolinium-labeled liposomes: targeted MR contrast agents for the liver and spleen. Radiology 163:255–258.

Karlik, S., Florio, E. and Grant, C. W. 1991. Comparative evaluation of two membrane-based liposomal MRI contrast agents. Magn Reson Med 19:56–66.

Keana, J. F. and Pou, S. 1985. Nitroxide-doped liposomes containing entrapped oxidant: an approach to the "reduction problem" of nitroxides as MRI contrast agents. Physiol Chem Phys Med NMR 17:235–240.

Kleiter, M. M., Yu, D., Mohammadian, L. A., Niehaus, N., Spasojevic, I., Sanders, L., Viglianti, B. L., Yarmolenko, P. S., Hauck, M., Petry, N. A., Wong, T. Z., Dewhirst, M. W. and Thrall,

D. E. 2006. A tracer dose of technetium-99 m-labeled liposomes can estimate the effect of hyperthermia on intratumoral doxil extravasation. Clin Cancer Res 12:6800–6807.

Krynyckyi, B. R., Kim, C. K., Goyenechea, M. R., Chan, P. T., Zhang, Z. Y. and Machac, J. 2004. Clinical breast lymphoscintigraphy: optimal techniques for performing studies, image atlas, and analysis of images. Radiographics 24:121–145 (discussion 139–145).

Laverman, P., Boerman, O. C., Oyen, W. J. G., Corstens, F. H. M. and Storm, G. 2001. In vivo applications of PEG liposomes: unexpected observations. Crit Rev Ther Drug Carrier Syst 18:551–566.

Laverman, P., Boerman, O. C. and Storm, G. 2003. Radiolabeling of liposomes for scintigraphic imaging. Meth Enzymol 373:234–248.

Laverman, P., Boerman, O. C., Storm, G. and Oyen, W. J. 2002. (99 m)Tc-labelled Stealth liposomal doxorubicin (Caelyx) in glioblastomas and metastatic brain tumours. Br J Cancer 86:659–661.

Laverman, P., Dams, E. T. M., Oyen, W. J. G., Storm, G., Koenders, E. B., Prevost, R., Corstens, F. H. M. and Storm, G. 1998. A novel method to label liposomes with Tc-99 m via the hydrazino nicotinyl derivative. J Nucl Med 39:7P.

Laverman, P., Storm, G., Phillips, W. T., Bao, A. and Goins, B. 2007. Radiolabeling of liposome for scintigraphic imaging. Liposome Technology, vol. 3. Ed. G. Gregoriadis. Informa Healthcare, New York, pp. 169–185.

Leander, P. 1995. Liver-specific contrast media for MRI and CT experimental studies. Acta Radiol Suppl 396:1–36.

Leclercq, F., Cohen-Ohana, M., Mignet, N., Sbarbati, A., Herscovici, J., Scherman, D. and Byk, G. 2003. Design, synthesis, and evaluation of gadolinium cationic lipids as tools for biodistribution studies of gene delivery complexes. Bioconjug Chem 14:112–119.

Lindner, L. H., Reinl, H. M., Schlemmer, M., Stahl, R. and Peller, M. 2005. Paramagnetic thermosensitive liposomes for MR-thermometry. Int J Hyperther 21:575–588.

Lokling, K. E., Fossheim, S. L., Skurtveit, R., Bjornerud, A. and Klaveness, J. 2001. pH-sensitive paramagnetic liposomes as MRI contrast agents: in vitro feasibility studies. Magn Reson Imaging 19:731–738.

Love, W. G., Amos, N., Williams, B. D. and Kellaway, I. W. 1989. Effect of liposome surface charge on the stability of technetium (^{99m}Tc) radiolabelled liposomes. Ann Rheum Dis 6:105–113.

Marik, J., Tartis, M. S., Zhang, H., Fung, J. Y., Kheirolomoom, A., Sutcliffe, J. L. and Ferrara, K. W. 2007. Long-circulating liposomes radiolabeled with [18F]fluorodipalmitin ([18F]FDP). Nucl Med Biol 34:165–171.

Mauk, M. R. and Gamble, R. C. 1979. Preparation of lipid vesicles containing high levels of entrapped radioactive cations. Anal Biochem 94:302–307.

McDannold, N., Fossheim, S. L., Rasmussen, H., Martin, H., Vykhodtseva, N. and Hynynen, K. 2004. Heat-activated liposomal MR contrast agent: initial in vivo results in rabbit liver and kidney. Radiology 230:743–752.

Medina, L. A., Calixto, S. M., Klipper, R., Phillips, W. T. and Goins, B. 2004a. Avidin/biotin-liposome system injected in the pleural space for drug delivery to mediastinal lymph nodes. J Pharm Sci 93:2595–2608.

Medina, L. A., Klipper, R., Phillips, W. T. and Goins, B. 2004b. Pharmacokinetics and biodistribution of [111In]-avidin and [99mTc]-biotin-liposomes injected in the pleural space for the targeting of mediastinal nodes. Nucl Med Biol 31:41–51.

Metselaar, J. M., Wauben, M. H., Wagenaar-Hilbers, J. P., Boerman, O. C. and Storm, G. 2003. Complete remission of experimental arthritis by joint targeting of glucocorticoids with long-circulating liposomes. Arthritis Rheum 48:2059–2066.

Misselwitz, B. and Sachse, A. 1997. Interstitial MR lymphography using Gd-carrying liposomes. Acta Radiol Suppl 412:51–55.

Moghimi, S. M., Hamad, I., Andresen, T. L., Jorgensen, K. and Szebeni, J. 2006. Methylation of the phosphate oxygen moiety of phospholipid-methoxy(polyethylene glycol) conjugate prevents PEGylated liposome-mediated complement activation and anaphylatoxin production. FASEB J 20:2591–2593.

Moghimi, S. M. and Szebeni, J. 2003. Stealth liposomes and long circulating nanoparticles: critical issues in pharmacokinetics, opsonization and protein-binding properties. Prog Lipid Res
42:463–478.

Mougin-Degraef, M., Jestin, E., Bruel, D., Remaud-Le Saec, P., Morandeau, L., Faivre-Chauvet,
A. and Barbet, J. 2006. High-activity radio-iodine labeling of conventional and stealth liposomes. J Liposome Res 16:91–102.

Mulder, W. J., van der Schaft, D. W., Hautvast, P. A., Strijkers, G. J., Koning, G. A., Storm, G.,
Mayo, K. H., Griffioen, A. W. and Nicolay, K. 2007. Early in vivo assessment of angiostatic
therapy efficacy by molecular MRI. FASEB J 21:378–383.

Mulder, W. J., Strijkers, G. J., Griffioen, A. W., van Bloois, L., Molema, G., Storm, G., Koning,
G. A. and Nicolay, K. 2004. A liposomal system for contrast-enhanced magnetic resonance
imaging of molecular targets. Bioconjug Chem 15:799–806.

Mulder, W. J., Strijkers, G. J., Habets, J. W., Bleeker, E. J., van der Schaft, D. W., Storm, G.,
Koning, G. A., Griffioen, A. W. and Nicolay, K. 2005. MR molecular imaging and fluorescence microscopy for identification of activated tumor endothelium using a bimodal lipidic
nanoparticle. FASEB J 19:2008–2010.

Mulder, W. J., Strijkers, G. J., van Tilborg, G. A., Griffioen, A. W. and Nicolay, K. 2006. Lipid-
based nanoparticles for contrast-enhanced MRI and molecular imaging. NMR Biomed
19:142–164.

Niesman, M. R., Bacic, G. G., Wright, S. M., Swartz, H. J. and Magin, R. L. 1990. Liposome
encapsulated MnCl2 as a liver specific contrast agent for magnetic resonance imaging. Invest
Radiol 25:545–551.

Oussoren, C., Storm, G. 2001. Liposomes to target the lymphatics by subcutaneous administration.
Advanced Drug Delivery Reviews. 50:143–156.

Papahadjopoulos, D., Allen, T. M., Gabizon, A., Mayhew, E., Matthay, K., Huang, S. K., Lee, K.
D., Woodle, M. C., Lasic, D. D., Redemann, C. et al. 1991. Sterically stabilized liposomes:
improvements in pharmacokinetics and antitumor therapeutic efficacy. Proc Natl Acad Sci
USA 88:11460–11464.

Pauser, S., Reszka, R., Wagner, S., Wolf, K. J., Buhr, H. J. and Berger, G. 1997. Liposome-encapsulated superparamagnetic iron oxide particles as markers in an MRI-guided search for tumor-
specific drug carriers. Anticancer Drug Des 12:125–135.

Phillips, W. T., Klipper, R. W., Awasthi, V. D., Rudolph, A. S., Cliff, R., Kwasiborski, V., Goin,
B. A. 1999. Polyethylene glycol-modified liposome-encapsulated hemoglobin: a long circulating red cell substitute. 288:655–670.

Phillips, W. T. and Goins, B. 2002. Assessment of liposome delivery using scintigraphic imaging.
J Liposome Res 12:71–80.

Phillips, W. T., Klipper, R. and Goins, B. 2001. Use of (99m)Tc-labeled liposomes encapsulating
blue dye for identification of the sentinel lymph node. J Nucl Med 42:446–451.

Phillips, W. T., Medina, L. A., Klipper, R. and Goins, B. 2002. A novel approach for the increased
delivery of pharmaceutical agents to peritoneum and associated lymph nodes. J Pharmacol Exp
Ther 303:11–16.

Phillips, W. T., Rudolph, A. S., Goins, B., Timmons, J. H., Klipper, R. and Blumhardt, R. 1992.
A simple method for producing a technetium-99m-labeled liposome which is stable in vivo.
Nucl Med Biol 19:539–547.

Pirollo, K. F., Dagata, J., Wang, P., Freedman, M., Vladar, A., Fricke, S., Ileva, L., Zhou, Q. and
Chang, E. H. 2006. A tumor-targeted nanodelivery system to improve early MRI detection of
cancer. Mol Imaging 5:41–52.

Ponce, A. M., Viglianti, B. L., Yu, D., Yarmolenko, P. S., Michelich, C. R., Woo, J., Bally, M. B.
and Dewhirst, M. W. 2007. Magnetic resonance imaging of temperature-sensitive liposome
release: drug dose painting and antitumor effects. J Natl Cancer Inst 99:53–63.

Port, R. E., Schuster, C., Port, C. R. and Bachert, P. 2006. Simultaneous sustained release of
fludarabine monophosphate and Gd-DTPA from an interstitial liposome depot in rats: potential
for indirect monitoring of drug release by magnetic resonance imaging. Cancer Chemother
Pharmacol 58:607–617.

Romberg, B., Metselaar, J. M., Baranyi, L., Snel, C. J., Bunger, R., Hennink, W. E., Szebeni, J. and Storm, G. 2007a. Poly(amino acid)s: promising enzymatically degradable stealth coatings for liposomes. Int J Pharm 331:186–189.

Romberg, B., Oussoren, C., Snel, C. J., Carstens, M. G., Hennink, W. E. and Storm, G. 2007b. Pharmacokinetics of poly(hydroxyethyl-L-asparagine)-coated liposomes is superior over that of PEG-coated liposomes at low lipid dose and upon repeated administration. Biochim Biophys Acta 1768:737–743.

Rubesova, E., Berger, F., Wendland, M. F., Hong, K., Stevens, K. J., Gooding, C. A. and Lang, P. 2002. Gd-labeled liposomes for monitoring liposome-encapsulated chemotherapy: quantification of regional uptake in tumor and effect on drug delivery. Acad Radiol 9 Suppl 2: S525–S527.

Shapiro, E. M., Sharer, K., Skrtic, S. and Koretsky, A. P. 2006. In vivo detection of single cells by MRI. Magn Reson Med 55:242–249.

Sou, K., Goins, B., Takeoka, S., Tsuchida, E. and Phillips, W. T. 2007. Selective uptake of surface-modified phospholipid vesicles by bone marrow macrophages in vivo. Biomaterials 28:2655–2666.

Sroda, K., Rydlewski, J., Langner, M., Kozubek, A., Grzybek, M. and Sikorski, A. F. 2005. Repeated injections of PEG-PE liposomes generate anti-PEG antibodies. Cell Mol Biol Lett 10:37–47.

Storrs, R. W., Tropper, F. D., Li, H. Y., Song, C. K., Sipkins, D. A., Kuniyoshi, J. K., Bednarski, M. D., Strauss, H. W. and Li, K. C. 1995. Paramagnetic polymerized liposomes as new recirculating MR contrast agents. J Magn Reson Imaging 5:719–724.

Strijkers, G. J., Mulder, W. J., van Heeswijk, R. B., Frederik, P. M., Bomans, P., Magusin, P. C. and Nicolay, K. 2005. Relaxivity of liposomal paramagnetic MRI contrast agents. MAGMA 18:186–192.

Suga, K., Mikawa, M., Ogasawara, N., Okazaki, H. and Matsunaga, N. 2001. Potential of Gd-DTPA-mannan liposome particles as a pulmonary perfusion MRI contrast agent: an initial animal study. Invest Radiol 36:136–145.

Szebeni, J., Baranyi, L., Savay, S., Bodo, M., Milosevits, J., Alving, C. R. and Bunger, R. 2006. Complement activation-related cardiac anaphylaxis in pigs: role of C5a anaphylatoxin and adenosine in liposome-induced abnormalities in ECG and heart function. Am J Physiol Heart Circ Physiol 290:H1050–H1058.

Szebeni, J., Baranyi, L., Savay, S., Milosevits, J., Bodo, M., Bunger, R. and Alving, C. R. 2003. The interaction of liposomes with the complement system: in vitro and in vivo assays. Meth Enzymol 373:136–154.

Tilcock, C., Ahkong, Q. F. and Fisher, D. 1994. ^{99m}Tc-labeling of lipid vesicles containing the lipophilic chelator PE-DTTA: effect of tin-to-chelate ratio, chelate content and surface polymer on labeling efficiency and biodistribution behavior. Nucl Med Biol 21:89–96.

Tilcock, C., Ahkong, Q. F., Koenig, S. H., Brown, R. D., III, Davis, M. and Kabalka, G. 1992. The design of liposomal paramagnetic MR agents: effect of vesicle size upon the relaxivity of surface-incorporated lipophilic chelates. Magn Reson Med 27:44–51.

Torchilin, V. and Trubetskoy, V. 1995. Which polymers can make nanoparticulate drug carriers long-circulating? Adv Drug Deliv Rev 16:141–155.

Torchilin, V. P. 2000. Polymeric contrast agents for medical imaging. Curr Pharm Biotechnol 1:183–215.

Torchilin, V. P. and Papisov, M. I. 1994. Why do polyethylene glycol-coated liposomes circulate so long? J Liposome Res 4:725–739.

Torchilin, V. P., Weissig, V., Martin, F. J. and Heath, T. D. 2003. Surface modification of liposomes. Liposomes: A Practical Approach. Eds. V. P. Torchilin and V. Weissig. Oxford University Press, Oxford, p. 193.

Trubetskoy, V. S., Cannillo, J. A., Milshtein, A., Wolf, G. L. and Torchilin, V. P. 1995. Controlled delivery of Gd-containing liposomes to lymph nodes: surface modification may enhance MRI contrast properties. Magn Reson Imaging 13:31–37.

Turski, P. A., Cordes, D., Mock, B., Wendt, G., Sorenson, J. A., Fitzurka-Quigley, M., Strother, C. and Dempsey, R. 1998. Basic concepts of functional magnetic resonance imaging and arteriovenous malformations. Magn Reson Imaging Clin N Am 6:801–810.

Unger, E., Needleman, P., Cullis, P. and Tilcock, C. 1988. Gadolinium-DTPA liposomes as a potential MRI contrast agent. Work in progress. Invest Radiol 23:928–932.

van Tilborg, G. A., Mulder, W. J., Deckers, N., Storm, G., Reutelingsperger, C. P., Strijkers, G. J. and Nicolay, K. 2006. Annexin A5-functionalized bimodal lipid-based contrast agents for the detection of apoptosis. Bioconjug Chem 17:741–749.

Venkatesh, A. K., Zhao, L., Balamore, D., Jolesz, F. A. and Albert, M. S. 2000. Evaluation of carrier agents for hyperpolarized xenon MRI. NMR Biomed 13:245–252.

Venkatesh, A. K., Zhao, L., Balamore, D., Jolesz, F. A. and Albert, M. S. 2002. Hyperpolarized 129Xe MRI using gas-filled liposomes. Acad Radiol 9 Suppl 1:S270–S274.

Viglianti, B. L., Abraham, S. A., Michelich, C. R., Yarmolenko, P. S., MacFall, J. R., Bally, M. B. and Dewhirst, M. W. 2004. In vivo monitoring of tissue pharmacokinetics of liposome/drug using MRI: illustration of targeted delivery. Magn Reson Med 51:1153–1162.

Viglianti, B. L., Ponce, A. M., Michelich, C. R., Yu, D., Abraham, S. A., Sanders, L., Yarmolenko, P. S., Schroeder, T., MacFall, J. R., Barboriak, D. P., Colvin, O. M., Bally, M. B. and Dewhirst, M. W. 2006. Chemodosimetry of in vivo tumor liposomal drug concentration using MRI. Magn Reson Med 56:1011–1018.

Weissig, V. V., Babich, J. and Torchilin, V. V. 2000. Long-circulating gadolinium-loaded liposomes: potential use for magnetic resonance imaging of the blood pool. Colloids Surf B Biointerfaces 18:293–299.

Zavaleta, C. L., Phillips, W. T., Bradley, Y. C., McManus, L. M., Jerabek, P. A. and Goins, B. A. 2007a. Characterization of an intraperitoneal ovarian cancer xenograft model in nude rats using noninvasive microPET imaging. Int J Gynecol Cancer 17:407–417.

Zavaleta, C. L., Phillips, W. T., Soundararajan, A. and Goins, B. A. 2007b. Use of avidin/biotin-liposome system for enhanced peritoneal drug delivery in an ovarian cancer model. Int J Pharm 337:316–328.

Index

Printed in the United States of America